American Mathematical Society
Colloquium Publications
Volume 30

Length and Area

Tibor Radó

American Mathematical Society
Providence, Rhode Island

2000 *Mathematics Subject Classification.* Primary 28-02.

Library of Congress Cataloging-in-Publication Data

Radó, Tibor, 1895–1965.
Length and area.
p. cm. — (American Mathematical Society Colloquium publications, ISSN 0065-9258 ; v. 30)
New York, American mathematical society, 1948.
Includes bibliography.
ISBN 978-0-8218-4621-6 (alk. paper)
1. Topology. I. Title. II. Colloquium publications (American Mathematical Society) ; v. 30.

QA1 .A5225 vol. 30
513.83 48003259

∞ The paper used in this book is acid-free and falls within the guidelines established to ensure permanence and durability.
Visit the AMS home page at `http://www.ams.org/`

10 9 8 7 6 5 4 3 2 13 12 11 10 09 08

PREFACE

The terms *curve, surface, length, area* have been used by many generations of mathematicians. And yet, no general agreement has been arrived at concerning the selection of precise formal definitions for these terms, as a basis of a general comprehensive theory. The theory of length and area which is presented in this book is based upon fundamental ideas of Lebesgue and Geöcze. As compared with other equally plausible and relevant approaches proposed during the past fifty years, this theory is distinguished by a high degree of completeness, achieved through the sustained efforts of many mathematicians. However, the number of difficult and fascinating problems is still large, and application of the results to other fields, especially to Calculus of Variations, is hardly begun. Hence the writer hopes that he was justified in emphasizing methods and results that seemed most relevant for further researches.

This book is an amplification of a set of four Colloquium lectures which the writer had the honor to deliver at the annual meeting of the American Mathematical Society in Chicago, November 1945. War-time conditions created delays and complications that were resolved in the most sympathetic manner by the officers of the Society. Several colleagues rendered invaluable help by reading and preparing for print the manuscript of this book. The Institute for Advanced Study in Princeton and the Ohio State University cooperated in providing leisure and material assistance. The writer wishes to express his appreciation of the cooperation he received from all these sources.

TIBOR RADÓ

COLUMBUS, OHIO,
March 1946

TABLE OF CONTENTS

PART I. BACKGROUND MATERIAL

CHAPTER I.1. INTRODUCTION

I.1.1. The concepts of *curve, surface, length, area* are of general occurrence in mathematics and its applications, and indeed in everyday life. As indicated in the preface, the theory studied in this book represents only one of several plausible approaches to the difficult and fascinating problems that arise in connection with these fundamental concepts. The purpose of this introductory chapter is to give a brief survey of the field that we propose to cover and to acquaint the reader with the general plan of exposition. In this preliminary survey, the terms curve, surface, length, area will be used in the same vague sense as in Calculus, for example. Precise definitions will be formulated and studied in subsequent chapters.

I.1.2. Given a curve C in terms of parametric equations $x = x(u)$, $y = y(u)$, $z = z(u)$, $a \leqq u \leqq b$, its length L is determined by the formula

$$L = \int_a^b \left[\left(\frac{dx}{du}\right)^2 + \left(\frac{dy}{du}\right)^2 + \left(\frac{dz}{du}\right)^2\right]^{1/2} du. \tag{1}$$

Similarly, if a surface S is given by parametric equations $x = x(u, v)$, $y = y(u, v)$, $z = z(u, v)$, $a \leqq u \leqq b$, $c \leqq v \leqq d$, then its area A is determined by the formula

$$A = \int_a^b\int_c^d \left[\left(\frac{\partial(y, z)}{\partial(u, v)}\right)^2 + \left(\frac{\partial(z, x)}{\partial(u, v)}\right)^2 + \left(\frac{\partial(x, y)}{\partial(u, v)}\right)^2\right]^{1/2} du\, dv. \tag{2}$$

These formulas are familiar to every mathematician. They are established, or rather made plausible in Calculus under very restrictive assumptions. As these assumptions are gradually relaxed, the validity, and even the meaning, of the formulas (1) and (2) becomes less and less clear. Thus there arises a fundamental problem that we shall term, for convenient reference, *the representation problem: determine the precise range of validity of the formulas* (1) *and* (2).

The representation problem is merely one of several fundamental issues in the theory studied in this book. We selected it as the starting point of this preliminary survey because even a cursory discussion of this problem reveals the variety and complexity of the conceptual and technical difficulties that we shall have to deal with. It will be convenient to write the formulas (1), (2) in a more concise form. As regards (1), we introduce the vector function $\mathfrak{x}(u)$ with components $x(u)$, $y(u)$, $z(u)$, and we denote by $\mathfrak{x}'(u)$ the vector function with components $x'(u)$, $y'(u)$, $z'(u)$. The integrand in (1) is then equal to the length $|\mathfrak{x}'(u)|$ of $\mathfrak{x}'(u)$. Let us also observe that L may be thought of as a functional whose argument is a vector function $\mathfrak{x}(u)$. Accordingly, we write $L(\mathfrak{x})$ for L. The formula (1) appears then in the form

$$L(\mathfrak{x}) = \int_a^b |\mathfrak{x}'(u)|\, du. \tag{1a}$$

Similarly, to condense formula (2), we introduce the vector function $\mathfrak{x}(u, v) = [x(u, v), y(u, v), z(u, v)]$, and we denote by $\mathfrak{x}_u$, $\mathfrak{x}_v$ the vector functions $[x_u, y_u, z_u]$, $[x_v, y_v, z_v]$ respectively, where the subscripts refer to partial differentiation. The integrand in (2) is then equal to the length $|\mathfrak{x}_u \times \mathfrak{x}_v|$ of the vector product $\mathfrak{x}_u \times \mathfrak{x}_v$. It is customary to put $W = |\mathfrak{x}_u \times \mathfrak{x}_v|$. In analogy with (1a), we can now write (2) in the form

$$A(\mathfrak{x}) = \int_a^b\int_c^d W\, du\, dv. \tag{2a}$$

In (1a), the length L appears as a functional depending upon a representation of the curve C, rather than a functional depending upon the curve C itself. A similar remark applies to (2a). Thus there arises the question of invariance under changes of the representation. This point will be studied carefully later on. For the moment, the use of the symbols $L(\mathfrak{x})$, $A(\mathfrak{x})$ instead of the symbols $L(C)$, $A(S)$ permits us to postpone a discussion of the concepts of curve and surface, a discussion that will constitute one of the major issues of the theory (see part II).

I.1.3. A striking feature of the formulas (1) and (2) is their obvious formal analogy. In fact, (1) and (2) are special cases of a general formula concerned with n-dimensional varieties in N-dimensional space (see Eisenhart [1]). The formulas (1) and (2) represent merely the special cases $n = 1$ and $n = 2$, the surrounding space being Euclidean three-space. Thus the question is justified whether it would not be more appropriate to undertake the construction of a uniform theory for general values of n, N, instead of concerning ourselves merely with the two simplest special cases. The construction of a uniform general theory is indeed an ultimate objective of great interest. However, even the case $n = 2$ presents difficulties that seem to indicate that the case $n > 2$ is perhaps beyond the reach of our present resources in Analysis and in Topology.

Let us recall that the terms curve, length, and so on, are also used to refer to geometrical objects in the complex domain, giving rise to theories of great interest and importance. However, *we shall restrict ourselves to the real domain throughout.*

I.1.4. Returning to the formula (1a), it would seem that a first step should consist of defining the length $L(\mathfrak{x})$. And yet, it is general practice in several mathematical fields, especially in Differential Geometry in its higher phases, to consider the formula (1a) as the definition of $L(\mathfrak{x})$. In the mathematical fields just referred to, $\mathfrak{x}(u)$ is assumed to possess all the properties of smoothness that may be needed. However, if these assumptions are relaxed, the formula (1a) tends to become doubtful. In the first place, the integrand of (1) may be discontinuous, and in fact it is apparent that even in comparatively simple cases the concept of the definite integral, as understood in mathematical disciplines of a classical type, soon becomes inadequate. One of the reasons advanced by Lebesgue [2] for introducing the concept of his integral was precisely the inadequacy of the classical concept of the definite integral for the purpose of evaluating the length of a curve. From this point on, we shall use the term *integral* in the sense of *Lebesgue integral.* A degree of familiarity with this integral will be assumed on the part of the reader (for excellent presentations, see for example McShane [7], Saks [6]). For the moment, the following two remarks concerning the Lebesgue integral are relevant.

(i) As regards scope, that is, range of applicability, the Riemann integral is definitely inadequate, while the Lebesgue integral gives complete coverage as far as the theory studied in this book is concerned.

(ii) For the purpose of integration in the Lebesgue sense, the integrand need not be defined on the whole range of integration. Indeed, the values of the integrand may be arbitrarily changed on a set of measure zero without affecting the value of its Lebesgue integral. As a consequence, the Lebesgue process of integration applies even though the integrand is undefined on a subset of measure zero of the range of integration.

Now let $f(x)$ denote the familiar Cantor ternary function in the interval $I: 0 \leqq x \leqq 1$. Then $f(x)$ is continuous and nondecreasing in I, and $f(0) = 0$, $f(1) = 1$. Furthermore, there exists in I a sequence of nonoverlapping intervals $I_1, \cdots, I_n, \cdots$, such that $f(x)$ is constant in each I_n and the sum of the lengths of the intervals I_n is equal to 1. Now let us define the vector function $\mathfrak{x}(u)$ by the formula $\mathfrak{x}(u) = [u, f(u), 0]$, $0 \leqq u \leqq 1$. The corresponding curve C is then a simply covered arc which can be also represented in the nonparametric form $y = f(x)$, $0 \leqq x \leqq 1$. The length L of C may be readily determined by using inscribed polygons, and one finds that $L = 2$. On the other hand, $f'(u)$ exists and is equal to zero at every point u that is interior to one of the intervals I_n. Thus $|\,\mathfrak{x}'(u)\,| = 1$ with the exception of a set of measure zero in I, and hence the integral of $|\,\mathfrak{x}'(u)\,|$ over I is equal to 1. Hence the formula (1a) fails in this relatively simple case. In other words, the formula (1a) cannot be used as a definition of the arc length outside of the range usually considered in mathematical disciplines of a classical type. Similar remarks apply to the formula (2a).

I.1.5. Proceeding to the discussion of the definition of $L(\mathfrak{x})$, we assume from now on that $\mathfrak{x}(u)$ is merely continuous in the interval $a \leqq u \leqq b$. Even though the use of the symbol $L(\mathfrak{x})$ relieves us, temporarily, of the obligation to discuss the geometrical aspects of the situation, a few remarks concerning the geometrical interpretation are in order. While u describes the interval $a \leqq u \leqq b$, the corresponding point P, whose coordinates coincide with the components $x(u)$, $y(u)$, $z(u)$ of the continuous vector function $\mathfrak{x}(u)$, describes a certain point set $\sigma(\mathfrak{x})$, but we do not exclude the case where P crosses the same point of $\sigma(\mathfrak{x})$ several times. For example, if we first choose $\mathfrak{x}(u) = (\cos u, \sin u, 0)$, $0 \leqq u \leqq 2\pi$, then $\sigma(\mathfrak{x})$ coincides with the perimeter K of the unit circle in the xy-plane. By inspection, or from the formula (1a), it is clear that $L(\mathfrak{x}) = 2\pi$. Let us choose next $\mathfrak{x}(u) = (\cos u, \sin u, 0)$, $0 \leqq u \leqq 4\pi$. Then $\sigma(\mathfrak{x})$ coincides with K as before, but this time clearly $L(\mathfrak{x}) = 4\pi$, both by formula (1a) and by inspection, since K is now described twice. Thus we shall think of $L(\mathfrak{x})$ not as representing the length of a point set but rather as representing *the length of a trip* over a point set. Note that mere inspection of the point set $\sigma(\mathfrak{x})$ cannot by itself reveal the value of $L(\mathfrak{x})$, as the preceding simple examples show. While this conception of $L(\mathfrak{x})$ may seem most natural, especially from the point of view of applications in Analysis, it is important to observe that the alternative point of view, where length is considered as a quantity attached to a point set itself rather than to a continuous trip over a point set, is of equal interest and importance. In deciding to accept the point of view that $L(\mathfrak{x})$ should stand for the length of a trip rather than the length of a point set, we made one of several major decisions that we shall have to face yet. Similar remarks will apply in connection with our study of surface area.

I.1.6. For convenience, let us define a *quasi-linear* vector function $q(u)$ as follows: $q(u)$ is quasi-linear in the interval $I : a \leqq u \leqq b$ if there exists a subdivision of I, by points $a = u_0 < u_1 < \cdots < u_{k-1} < u_k < \cdots < u_n = b$, such that the components of $q(u)$ are continuous in I and are linear in each interval of the indicated subdivision. As regards the geometrical picture, $q(u)$ may be thought of as determining, in xyz-space, a polygon which may intersect itself and for which some, or even all, of the sides may reduce to single points. The quantity $l(q) = \sum | q(u_k) - q(u_{k-1})|$, $k = 1, \cdots, n$, may be thought of as the length, in the elementary sense, of the corresponding polygon. Now let $\mathfrak{x}(u)$ be a continuous vector function in the same interval I. If $q(u_k) = \mathfrak{x}(u_k)$, $k = 0, 1, \cdots, n$, then we shall say that $q(u)$ is inscribed in $\mathfrak{x}(u)$, the geometrical interpretation being obvious. We define now

$$(1) \qquad L(\mathfrak{x}) = \text{l.u.b. } l(q),$$

where the least upper bound is taken with respect to all inscribed quasi-linear vector functions $q(u)$. This definition originates in the intuitive observation that the length of an inscribed polygon is a lower bound for the length of the curve. Thus (1) may be thought of as a definition of $L(\mathfrak{x})$ *from below*. Elementary examples show that $L(\mathfrak{x})$ may be infinite. An equivalent definition is represented by the formula

$$(2) \qquad L(\mathfrak{x}) = \lim l(q_n),$$

where $q_n(u)$ is any sequence of inscribed quasi-linear vector functions such that $q_n(u) \to \mathfrak{x}(u)$ uniformly in I. While the preceding alternative definitions of

$L(\mathfrak{x})$ are generally known and accepted, an equally important equivalent definition in terms of upper bounds is perhaps less familiar. Let $\mathfrak{x}_n(u)$ be any sequence of continuous vector functions in I such that $\mathfrak{x}_n(u) \to \mathfrak{x}(u)$ uniformly in I. As a ready consequence of the definition of $L(\mathfrak{x})$, there follows the inequality

$$(3) \qquad L(\mathfrak{x}) \leqq \liminf L(\mathfrak{x}_n),$$

which expresses the fundamental fact that $L(\mathfrak{x})$ is a *lower semi-continuous functional.* Let then $q_n(u)$ be any sequence of (*not necessarily inscribed*) quasi-linear vector functions in I such that $q_n(u) \to \mathfrak{x}(u)$ uniformly in I. Then we have, in view of the lower semi-continuity property, the inequality $L(\mathfrak{x}) \leqq \liminf l(q_n)$. In other words, each sequence $q_n(u)$, with the properties just described, yields an upper bound $\liminf l(q_n)$ for $L(\mathfrak{x})$. An easy argument leads to the formula

$$(4) \qquad L(\mathfrak{x}) = \text{gr.l.b.} \liminf l(q_n),$$

where the greatest lower bound is taken with respect to all (not necessarily inscribed) quasi-linear sequences $q_n(u)$ such that $q_n(u) \to \mathfrak{x}(u)$ uniformly in I. The formula (4) yields an equivalent definition *from above* for $L(\mathfrak{x})$.

I.1.7. It is immediate that $L(\mathfrak{x})$ is invariant under changes of the Cartesian coordinate system xyz. Since we did not define yet the concept of a curve, the assertion that $L(\mathfrak{x})$ has the same value for all representations of a curve is meaningless as yet, but this point will be carefully discussed later on. For the moment, let us state only the important and obvious fact that $L(\mathfrak{x})$ is a *convex functional.* That is, if $\mathfrak{x}_1(u)$, $\mathfrak{x}_2(u)$ are continuous vector functions in the interval $I : a \leqq u \leqq b$, then $L(\mathfrak{x}_1) + L(\mathfrak{x}_2) \geqq L(\mathfrak{x}_1 + \mathfrak{x}_2)$. This is the so-called *inequality of Steiner.* This inequality implies the following geometrical statement. Let C_1, C_2 be simple arcs, and let T designate any biunique and continuous correspondence between the points of C_1 and C_2. If P_1, P_2 are corresponding points of C_1, C_2, then the locus of the mid-point of the segment with end points P_1, P_2 is a third curve C, and the inequality of Steiner implies that the length of C is less than or equal to the arithmetic mean of the lengths of C_1, C_2.

I.1.8. Let $\mathfrak{x}(u)$ denote a continuous vector function in the interval $I : a \leqq u \leqq b$. We have then the following fundamental theorems (see chapter III.3 for a detailed discussion of the topics considered in the present section).

(i) $L(\mathfrak{x}) < \infty$ if and only if the components of $\mathfrak{x}(u)$ are of bounded variation.

(ii) If $L(\mathfrak{x}) < \infty$, then $\mathfrak{x}'(u)$ exists in I except possibly for a set of measure zero, and $|\mathfrak{x}'(u)|$ is summable in I. Furthermore, we have the inequality

$$L(\mathfrak{x}) \geqq \int_a^b |\mathfrak{x}'(u)| \, du,$$

where the sign of equality holds if and only if the components of $\mathfrak{x}(u)$ are absolutely continuous in I.

Of course, there are many further important results concerning $L(\mathfrak{x})$ (see part

III), but the theorems (i) and (ii) suffice to indicate the nature of the representation problem (see I.1.2). It is now apparent that the solution of this problem depends upon several concepts of fundamental importance throughout Analysis (Lebesgue integral, bounded variation, absolute continuity). Furthermore, it is apparent that there will arise various interesting geometrical questions in connection with the concept of a curve. Turning presently to surface area, we naturally expect an analogous set of theorems, depending upon concepts of comparable importance. However, a little reflection will make it clear that mere formal analogy, such as that obtaining between the formulas (1) and (2) in I.1.2, may turn out to be superficial. Let us consider, for example, the theory of harmonic functions $h(x_1, x_2, \cdots, x_n)$ of n variables. Such a function is, by definition, a solution of the differential equation $\partial^2 h/\partial x_1^2 + \cdots + \partial^2 h/\partial x_n^2 = 0$. For $n = 1$, h is simply a linear function $ax_1 + b$, and thus the theory of harmonic functions of a single variable is a triviality. On the other hand, the theory of harmonic functions of two variables represents one of the great accomplishments in mathematics. Further instances of superficial analogies abound in the theory of Differential Equations, in Calculus of Variations, and in many other fields. We shall discuss presently certain phenomena that seem to indicate that the analogy between arc length and surface area is of an equally superficial character, a situation that did not discourage, however, the persistent efforts of many mathematicians to develop analogous theories of length and area. The present status of the theory supports the view that far-reaching analogies do exist. But the analogies lie deep, while the discrepancies are conspicuous, as we now shall see.

I.1.9. The preceding remark may suggest that we propose to discuss presently so-called *pathological cases*. On the contrary, our main objective at this time is to call attention to the fact that the concept of surface area involves complications even if we consider only relatively simple situations. Thus the phenomena to be discussed presently may be of interest not merely to the specialist in the field but perhaps also to the general mathematical community. Furthermore, it is clear that a general theory of surface area will have to rely, to a substantial extent, upon preliminary observations carried out in the elementary range. For these reasons, the following somewhat detailed discussion of elementary phenomena is perhaps justified.

I.1.10. We shall first consider an interesting example due to H. A. Schwarz [1]. Let S denote the cylindrical surface given by the formulas $x^2 + y^2 = 1, 0 \leqq z \leqq 1$. Then the area $A(S)$ of S is equal to 2π. Now cut S along a generator and spread S upon a plane. The result is a rectangle R whose sides have the lengths 1 and 2π respectively. Subdivide the sides of R into m and n equal parts respectively, and subdivide R, by lines parallel to the sides through the points of division, into mn congruent rectangles r. Subdivide each of these rectangles r into four triangles by drawing both diagonals. Bend R so as to obtain S, and use the vertices of the $4mn$ triangles as the vertices of an inscribed polyhedron with $4mn$ (*rectilinear*) triangular faces. Let A_{mn} denote the area of this inscribed polyhedron. An elementary calculation yields the formula

$$A_{mn} = 2n \sin \frac{\pi}{2n} + \left[\frac{1}{4} + \frac{4m^2}{n^4} \left(n \sin \frac{\pi}{2n} \right)^4 \right]^{1/2} \cdot 2n \sin \frac{\pi}{n}.$$

Inspection yields the following remarks.

(i) If we choose $m = n^3$, then $A_{mn} \to \infty$ for $n \to \infty$. If we choose $m = n$, then $A_{mn} \to 2\pi = A(S)$ for $n \to \infty$. As a matter of fact, if k is any number such that $2\pi \leqq k < \infty$, then we can make A_{mn} approach k by properly coordinating m and n. It follows that surface area cannot be defined as the limit of the areas of inscribed polyhedra. This definition would be logically inconsistent, since the limit in question does not exist, as the Schwarz example shows. Neither can surface area be defined as the least upper bound of the areas of inscribed polyhedra. Such a definition would be logically consistent, but it would be unacceptable because it would fail to agree with generally accepted formulas for surface area in elementary cases. Thus the definitions of arc length, represented by the formulas (1) and (2) in I.1.6, do not admit of direct analogues in the theory of surface area.

(ii) If m and n converge to ∞ in any manner, then clearly A_{mn} never approaches a value less than 2π, since $A_{mn} \geqq 2n \sin (\pi/2n) + n \sin (\pi/n)$. Thus $A(S) = 2\pi$ is the smallest limit that A_{mn} may approach. This may be construed as an indication, even though very faint, that the definition of arc length represented by formula (4) in I.1.6 may admit of an analogue in the theory of surface area.

I.1.11. It is perhaps difficult to understand how mathematicians could ever have been surprised by an example as simple as that devised by H. A. Schwarz. In any case, an overwhelming number of definitions have been subsequently proposed for surface area, and the end is apparently not yet in sight. Now since any definition of surface area must agree with the generally accepted values in elementary cases, it is reasonable to assume that each new definition is based upon observations in the range of those simple cases where one may speak of a generally accepted value of the area. A brief review of some such cases may therefore be in order at this time.

Case 1. Let $\mathfrak{x}(u, v) = [x(u, v), y(u, v), z(u, v)]$ be a continuous vector function in the unit square $Q: 0 \leqq u \leqq 1, 0 \leqq v \leqq 1$, and let σ denote the set of all those points in xyz-space that correspond to the points (u, v) of Q by means of the equations $x = x(u, v)$, $y = y(u, v)$, $z = z(u, v)$. Suppose further that these equations determine a 1-1 correspondence between the points of Q and the points of σ. Then we may say that σ is a simply covered topological 2-cell (cf. part II).

Case 1a. The set σ is some simple figure like a square, a rectangle, a circular disc, a spherical cap, and so forth. Briefly, let σ be any simple figure whose area is given by some standard formula. Then the functional $A(\mathfrak{x})$ must agree with the value furnished by the standard formula. Note that in such cases it is entirely unnecessary to refer, for the calculation of $A(\mathfrak{x})$, to the formula (2a) in I.1.2, and in fact that formula may be altogether meaningless even though the generally accepted value of $A(\mathfrak{x})$ is beyond doubt. The following example is instructive from this point of view. Let $f(x)$ and $g(x)$ be two continuous functions in the interval $0 \leqq x \leqq 1$, such that $0 < f(x) < g(x)$ for $0 \leqq x \leqq 1$, and let σ denote the figure in the xy-plane that is bounded by the graphs $y = f(x)$, $y = g(x)$, and by the lines $x = 0$, $x = 1$. By a standard formula in Calculus, the area of σ is then given by the integral

$$\int_0^1 [g(x) - f(x)]\, dx.$$

In particular, if we choose $g(x) = f(x) + 1$, then the area of σ is equal to 1. Now let us consider the vector function $\mathfrak{x}(u, v) = [u, f(u) + v, 0]$ in the unit square $Q: 0 \leqq u \leqq 1, 0 \leqq v \leqq 1$. The equations $x = u$, $y = f(u) + v$, $z = 0$ determine then a 1-1 continuous correspondence between Q and σ, and hence we must have $A(\mathfrak{x}) = 1$. On the other hand, if we choose $f(x)$ as a continuous function without a derivative, then the partial derivative y_u does not exist anywhere in Q. Let us now introduce new parameters α, β by means of the formulas $u = \alpha + \beta$, $v = \alpha - \beta$, and let us also introduce a new Cartesian coordinate system in a general position relative to the system xyz. There results an example where $A(\mathfrak{x}) = 1$ by a standard formula in Calculus, while the integral formula for surface area (see I.1.2) is altogether meaningless because none of the six first partial derivatives involved exists anywhere.

Case 1b. The components $x(u, v)$, $y(u, v)$, $z(u, v)$ of $\mathfrak{x}(u, v)$ have continuous first partial derivatives in some square that contains the unit square Q in its interior, and the quantity W of the formula (2a) in I.1.2 is positive in Q. Under these conditions, $A(\mathfrak{x})$ is given by the formula (2a) in I.1.2, by general agreement of all mathematicians. As a matter of fact, in this case many mathematicians consider that formula as the definition of surface area for this particular case.

Case 1c. Let us now assume merely that the components of $\mathfrak{x}(u, v)$ have continuous first partial derivatives in the interior of the unit square Q and that $W > 0$ in the interior of Q. In other words, we make no assumptions concerning the existence or behavior of the first partial derivatives on the perimeter of Q. From the Calculus point of view, the integral in formula (2a) in I.1.2 is then to be treated as an improper integral, and thus that formula should be written in the form

$$A(\mathfrak{x}) = \lim_{n\to\infty} \iint_{Q_n} W\, du\, dv, \tag{1}$$

where Q_n is the sequence

$$Q_n: 1/n \leqq u \leqq 1 - 1/n,\ 1/n \leqq v \leqq 1 - 1/n.$$

If the improper integral (in the Riemann sense) of W over Q fails to exist, then probably most mathematicians would agree that $A(\mathfrak{x}) = \infty$, as indicated by (1). If the improper integral exists, then the situation may be doubtful. In any case, textbooks dealing with this matter from the classical point of view fail, as a rule, to state assumptions with sufficient clarity for the purpose of a clear-cut decision concerning this point. So case 1c should be classified, for the moment, as a *doubtful case.*

Case 1d. Let the set σ (see above) coincide with a bounded, simply-connected Jordan region $\mathfrak{R}$ in the xy-plane, and let $\mathfrak{R}^0$ denote the set of the interior points of $\mathfrak{R}$. Now we may choose the boundary curve of $\mathfrak{R}$ as an *Osgood curve* (a simple closed continuous curve of positive two-dimensional measure, see Osgood [1]). Then the measure $|\mathfrak{R}^0|$ of $\mathfrak{R}^0$ is less than the measure $|\mathfrak{R}|$ of $\mathfrak{R}$. Hence, a decision must be reached whether $A(\mathfrak{x})$ should be equal to $|\mathfrak{R}^0|$ or else to $|\mathfrak{R}|$. Of course, we may say that in such cases we should rather speak of the area of $\mathfrak{R}$ and of the area of $\mathfrak{R}^0$, instead of insisting upon a definite choice. However, it is clear that once a general definition of surface area has been adopted, $A(\mathfrak{x})$ will be equal to either $|\mathfrak{R}|$ or else to $|\mathfrak{R}^0|$. Thus adoption of a general definition of surface area implies decisions in cases that are doubtful as regards a priori general agreements.

Incidentally, if a decision in favor of formula (1) above has been reached in the preceding doubtful case 1c, then there follows for the present doubtful case the decision $A(\mathfrak{x}) = |\mathfrak{R}^0|$. Indeed, by a fundamental theorem on conformal mapping (see Carathéodory [1]), we can choose $\mathfrak{x}(u, v) = [x(u, v), y(u, v), 0]$ in such a manner that the equations $x = x(u, v)$, $y = y(u, v)$ define a 1-1 continuous mapping from the unit square Q onto $\mathfrak{R}$ which is conformal in the interior of Q. The assumptions made in case 1c are then satisfied, and the formula (1) yields immediately the relation $A(\mathfrak{x}) = |\mathfrak{R}^0|$. Now a decision in case 1c in favor of the formula (1) seems to be quite natural, but the preceding remark shows that more is implied by such a decision than one may first expect.

Case 2. Using σ in the same sense as in case 1, let us drop now the assumption that the correspondence between Q and σ is biunique. The geometrical picture becomes then quite vague, and we mention only one special case that is relevant for our immediate purposes. Let us suppose that $\mathfrak{x}(u, v)$ is continuous and quasi-linear in Q. That is, there exists a subdivision of Q into rectilinear triangles $t_1, \cdots, t_n$ without common interior points, such that the components of $\mathfrak{x}(u, v)$ are linear functions of u, v in each one of the triangles $t_1, \cdots, t_n$. By means of the equations $x = x(u, v)$, $y = y(u, v)$, $z = z(u, v)$ there corresponds then to each triangle t_k a rectilinear triangle Δ_k in xyz-space, where Δ_k may reduce to a segment or even to a single point. Then the literature contains overwhelming evidence in favor of the agreement $A(\mathfrak{x}) = |\Delta_1| + \cdots + |\Delta_n|$, where $|\Delta_k|$ denotes the area of Δ_k in the elementary sense if Δ_k is a nondegenerate triangle, and $|\Delta_k| = 0$ if Δ_k reduces to a single point or to a segment. Incidentally, the quantity W of formula (2a) in I.1.2 is now integrable in the Lebesgue sense (and even in the Riemann sense if certain obvious agreements are made), and one finds readily that the integral formula leads also to the relation $A(\mathfrak{x}) = |\Delta_1| + \cdots + |\Delta_n|$. Of course, the set σ need not be simply covered, and taken by itself cannot be used to gain information concerning $A(\mathfrak{x})$.

I.1.12. The preceding brief and incomplete review of what may be termed the elementary range yields some relevant information concerning the difficulties that should be anticipated in a general theory of surface area. For conciseness, let us introduce the following terminology. The unit square $0 \leqq u \leqq 1$, $0 \leqq v \leqq 1$ will be denoted by Q. The notation $\mathfrak{x}(u, v)$ will refer to a vector function, defined in Q, with components $x(u, v)$, $y(u, v)$, $z(u, v)$ that are continuous in Q. Given $\mathfrak{x}(u, v)$, the formulas $x = x(u, v)$, $y = y(u, v)$, $z = z(u, v)$, $(u, v) \in Q$, define a continuous mapping from Q onto a certain subset $\sigma(\mathfrak{x})$ of Euclidean three-space. This mapping will be denoted by $T(u, v, \mathfrak{x})$. Then every point (x, y, z) is the image of a certain number of distinct points (u, v) in Q. The number of these points (u, v) will be denoted by $N(x, y, z, \mathfrak{x})$. Then $N(x, y, z, \mathfrak{x}) = 0$ if $(x, y, z) \notin \sigma(\mathfrak{x})$. If $(x, y, z) \in \sigma(\mathfrak{x})$, then $N(x, y, z, \mathfrak{x})$ may be infinite. If the mapping $T(u, v, \mathfrak{x})$ is topological, then $N(x, y, z, \mathfrak{x}) = 1$ on $\sigma(\mathfrak{x})$.

Let K denote the class of all continuous vector functions $\mathfrak{x}(u, v)$ in Q, and let K_t denote the class of those continuous vector functions $\mathfrak{x}(u, v)$ in Q for which the associated mapping $T(u, v, \mathfrak{x})$ is *topological*. Any general definition of surface area should then yield a functional $A(\mathfrak{x})$ with the following properties.

(i) $A(\mathfrak{x})$ is defined and non-negative for every $\mathfrak{x} \in K_t$ (it is understood that $A(\mathfrak{x})$ may be infinite). Of course, $A(\mathfrak{x})$ may be defined also for $\mathfrak{x} \notin K_t$.

(ii) $A(\mathfrak{x})$ agrees with the generally accepted formulas for surface area in the cases 1a and 1b described in I.1.11.

The properties (i), (ii) represent minimum requirements that every general definition of surface area may be expected to meet. Let us note that the assumption $\mathfrak{x} \in K_t$ implies that the set $\sigma(\mathfrak{x})$ is a simply covered surface, where the use of the term surface is not subject to the doubts that may arise if one only assumes that $\mathfrak{x} \in K$. We proceed to consider certain difficulties that arise for every choice of the functional $A(\mathfrak{x})$ subject to the conditions (i), (ii).

I.1.13. In the theory of arc length we had the fundamental theorem: if $L(\mathfrak{x}) < \infty$, then the first derivatives of the components of $\mathfrak{x}$ exist in the interval of definition, with the possible exception of a set of measure zero. Furthermore, the definite integral occurring in the formula for arc length exists in the Lebesgue sense (see I.1.8). The analogous statement for $A(\mathfrak{x})$ is however generally false. Indeed, in our discussion of case 1a in I.1.11 we exhibited an example where $A(\mathfrak{x}) = 1$ (by virtue of condition (ii) in I.1.12), and yet none of the six first partial derivatives of the components of $\mathfrak{x}(u, v)$ exists anywhere in Q. Thus we cannot even begin to consider the integral occurring in the formula for surface area.

This example indicates that a surface may possess more unfavorable parametric representations than those that a curve can possibly have. However, the nonexistence of derivatives may have more fundamental reasons. Returning to the case of arc length, it is intuitive that the nonexistence, locally, of derivatives is generally a consequence of oscillations that will appreciably increase the arc length. Conversely, the assumption that $L(\mathfrak{x}) < \infty$ will have a dampening influence upon the oscillation of $\mathfrak{x}(u)$, and as a matter of fact the proof of the theorem (ii) in I.1.8, is based upon this intuitive idea, as far as the existence of derivatives is concerned. Now take a circular disc S and modify it as follows: cut out from S a small circular disc and replace it by a circular cone. Let S^* be the resulting surface. Clearly, if the base circle of the cone is sufficiently small, we may choose the height of the cone very large, and yet keep the area of S^* as close to the area of S as we please. Of course, we can replace any number of small discs in S by very long and very thin cones without appreciably increasing the area of S. These remarks indicate a fundamental discrepancy between the dampening effect of the assumptions $L < \infty$ and $A < \infty$.

The long thin cones of the preceding remark point to another significant discrepancy. A short curve may be enclosed in a small sphere. But inspection of a long and narrow rectangle reveals that no similar statement holds for surface area.

Finally, let us consider the following situation. Let U denote the unit cube $0 \leqq x \leqq 1$, $0 \leqq y \leqq 1$, $0 \leqq z \leqq 1$, and let $\epsilon > 0$ be small. If a curve passes within ϵ of every point of U, then its length must obviously be large. On the other hand, by properly folding a very narrow and very long rectangle, it is easy to obtain an example of a surface that passes within ϵ of every point of U and has an area less than ϵ.

The experienced reader will readily visualize the paradoxical phenomena that result from the preceding remarks by applying the method of the condensation of singularities. He will also realize that the preceding examples, in spite of their trivial character, are related to some of the most fundamental difficulties that arise in various fields of mathematics. Clearly, any general theory of surface area will have to contend with difficulties of this type.

To illustrate a somewhat different type of difficulties, let us consider, in the xy-plane, a (rectilinear) triangle Δ_1 with vertices at $(0, 0)$, $(\epsilon, 0)$, $(0, 1)$, and a second triangle Δ_2 with vertices at $(0, 0)$, $(1, 0)$, $(0, \epsilon)$, where $\epsilon > 0$ is small. Let T denote the affine transformation that carries the vertices of Δ_1 into the vertices of Δ_2 in the indicated order. Let P_1, P_2 be corresponding points of Δ_1, Δ_2 under T, and let P be the mid-point of the segment that joins P_1 and P_2. Then the locus of P is a triangle Δ with vertices at $(0, 0)$, $((1 + \epsilon)/2, 0)$, $(0, (1 + \epsilon)/2)$. For the areas of these triangles we have the formulas

$$| \Delta | = \frac{(1 + \epsilon)^2}{8}, \quad \Delta_1 = \Delta_2 = \frac{\epsilon}{2},$$

and hence the inequality $| \Delta | \leqq (| \Delta_1 | + | \Delta_2 |)/2$ fails to hold if ϵ is sufficiently small. Comparison with I.1.7 shows that the inequality of Steiner does not admit of an unqualified generalization to surface area. The preceding remark is due to L. Fejér. Now the inequality of Steiner is a most efficient tool in the theory of arc length, and the lack of a corresponding tool in the theory of surface area is the source of endless difficulties. Further instances could be exhibited to illustrate the point that the technical difficulties to be anticipated in surface area theory are far greater than those encountered in the theory of arc length.

I.1.14. In view of the examples discussed in I.1.13 and I.1.10, it is apparent that a program of developing analogous theories of arc length and surface area is not a routine undertaking by any means. A second set of difficulties arises from the fact that equally plausible definitions of surface area may, and indeed generally must, conflict beyond the elementary range. We proceed to explain this significant point.

One of the most obvious observations in the elementary range may be stated in the form of the *projection principle.* Let $\sigma(\mathfrak{x})_p$ denote the orthogonal projection of the set $\sigma(\mathfrak{x})$ (see I.1.12) upon the plane p in xyz-space, and let $|\sigma(\mathfrak{x})_p|$ denote the (two-dimensional) measure of $\sigma(\mathfrak{x})_p$. Then, in simple cases, $A(\mathfrak{x}) \geqq |\sigma(\mathfrak{x})_p|$, for every choice of the plane p. Several definitions of surface area are based upon this projection principle. Let $A(\mathfrak{x}, \text{proj})$ denote *any* definition of this type. Let us now consider the doubtful case 1d in I.1.11. In that case, if we choose p as the xy-plane, we have $\sigma(\mathfrak{x})_p = \mathfrak{R}$, and hence there follows the inequality

$$(1) \qquad A(\mathfrak{x}, \text{proj}) \geqq |\mathfrak{R}|.$$

A second, though less obvious, observation in the elementary range yields the *lower semi-continuity principle* in the following form: if $\mathfrak{x}_n(u, v)$, $n = 0, 1, 2, \cdots$, is a sequence of continuous vector functions in the unit square Q, such that $\mathfrak{x}_n \to \mathfrak{x}_0$ uniformly in Q, then $A(\mathfrak{x}_0) \leqq \liminf A(\mathfrak{x}_n)$. This principle is clearly suggested by the analogous property of arc length (see I.1.6). Its validity in the elementary range does not seem to follow by altogether elementary methods, but is rigorously established at present (in particular, this book contains ample proof of the principle). Furthermore, the property of lower semi-continuity is shared by important double integrals occurring in Calculus of Variations. Also, the lower semi-continuity principle represents one of the few clear-cut analogies between arc length and surface area. In fact, several important definitions of surface area are based upon this principle. Let $A(\mathfrak{x}, \text{lsc})$ represent *any* definition of this type. Let us consider again the doubtful case 1d in I.1.11. By approximating to $\mathfrak{R}$ in terms of simply-connected polygonal regions comprized in $\mathfrak{R}^0$, the lower semi-continuity property yields readily the inequality (cf. the argument in V.2.68)

$$(2) \qquad A(\mathfrak{x}, \text{lsc}) \leqq |\mathfrak{R}^0|.$$

If we choose the boundary curve of $\mathfrak{R}$ as an Osgood curve of positive two-dimensional measure, then it follows from (1) and (2) that $A(\mathfrak{x}, \text{lsc}) < A(\mathfrak{x}, \text{proj})$. *Thus the projection principle and the lower semi-continuity principle conflict beyond the elementary range.*

We already mentioned that the example of Schwarz was the starting point of a large and still growing number of definitions for surface area. It may be natural to assume that the choice of a definition for surface area is a matter of taste. The preceding remarks show that the choice of the definition represents a major decision. Another significant conflict of this type should be mentioned here. We considered, at least in this preliminary survey, surface area as a functional $A(\mathfrak{x})$ depending upon a vector function. A theory based upon this conception may be called an analytic theory, in contradistinction to certain measure-theoretical theories. In theories of this latter type, one starts with a measure-function $M(E)$ defined, for example, for all subsets of Euclidean three-space. If $\mathfrak{x} \in K_t$ (see I.1.12), then the theory requires that $A(\mathfrak{x}) = M[\sigma(\mathfrak{x})]$. Let us consider again the doubtful case 1d in I.1.11. According to every measure-theoretical theory known to the writer, we have then $M(\mathfrak{R}) = | \mathfrak{R} |$, and hence also $A(\mathfrak{x}) = | \mathfrak{R} |$. Thus a measure-theoretical theory selects $| \mathfrak{R} |$ as the area, and hence, in view of (2), *any analytic theory based upon the lower semi-continuity principle will present an inevitable conflict with every measure-theoretical theory, and vice versa.*

I.1.15. *The theory of surface area, initiated by Lebesgue and Geöcze and presented in this book, may now be characterized as an analytic theory based upon the lower semi-continuity principle.* The area-functional $A(\mathfrak{x})$ will be required to possess the following properties, in addition to the properties (i) and (ii) stated in I.1.12.

(iii) $A(\mathfrak{x})$ is defined for every continuous vector function $\mathfrak{x} \in K$ (see I.1.12), and is lower semi-continuous. That is, if $\mathfrak{x}_n(u, v) \rightarrow \mathfrak{x}_0(u, v)$ uniformly in Q, and $\mathfrak{x}_n \in K$, $n = 0, 1, 2, \cdots$, then $A(\mathfrak{x}_0) \leqq \liminf A(\mathfrak{x}_n)$.

(iv) If $q(u, v)$ is a quasi-linear vector function, then $A(q)$ agrees with the value described in I.1.11, case 2.

It is important to note that the properties stated so far do not univocally determine the functional $A(\mathfrak{x})$. As a matter of fact, there are known at present at least four such functionals, all based upon the lower semi-continuity principle, that are clearly revelant for the general theory. The functional proposed by Lebesgue is distinguished by the fact that it is the largest of all functionals $A(\mathfrak{x})$ satisfying the conditions (i)-(iv) stated in I.1.12, I.1.15. We noted before that the lower semi-continuity principle conflicts with the projection principle (see I.1.14). One of the many fundamental discoveries of Geöcze consists of a *modified projection principle* that is compatible with the lower semi-continuity principle and in combination with the latter leads to further concepts of great importance in the development of the theory. The study of the modified projection principle of Geöcze constitutes one of the major topics in this book. It is entirely possible that an analogous modification of the measure-theoretical concepts of surface area (see I.1.14) may lead yet to a unified theory.

I.1.16. CONTINUATION. According to (iii) in I.1.15, our area-functionals $A(\mathfrak{x})$ will be defined for all continuous vector functions $\mathfrak{x} \in K$ (see I.1.12). Clearly, we shall need therefore a concept of surface of extreme generality (observe that the mapping $T(u, v, \mathfrak{x})$ of I.1.12 is not required to be biunique). This leads to topological problems of a high order of interest and difficulty (see part II), and also to phenomena that must be constantly kept in mind by a worker in this field. In the way of illustration, let us return to the trivial remark, made in I.1.13, that for every $\epsilon > 0$ we can readily construct a surface whose area is less than ϵ and which passes within ϵ of every point of the unit cube. In view of the general setting of our theory, we can carry out a passage to the limit, obtaining the Geöcze example of a cube-filling surface of zero area (see V.2.71). Next, using the notations of I.1.12, let us consider two vector functions $\mathfrak{x}_1 \in K$, $\mathfrak{x}_2 \in K$, such that $N(x, y, z, \mathfrak{x}_1) \geqq N(x, y, z, \mathfrak{x}_2)$ for every point (x, y, z). One would then expect that the area-functional $A(\mathfrak{x})$ should satisfy the inequality $A(\mathfrak{x}_1) \geqq A(\mathfrak{x}_2)$. And yet, as a ready consequence of the properties (iii), (iv) stated in I.1.15, there follows the existence of simple cases, analogous to the cube-filling surface of zero area, where $A(\mathfrak{x}_1) < A(\mathfrak{x}_2)$ (see V.2.71). In all these instances, the examples are obtained by a passage to the limit from perfectly obvious elementary situations, similar to those discussed in I.1.13, and perhaps the initial elementary situation is more remarkable than the final product. Still, there will be little room in our theory for intuition based on lack of observations. Incidentally, the example of a cube-filling surface of zero area was the starting point of profound topological investigations of Geöcze which led to results of great importance in the subsequent development of the theory. One is reminded of the square-filling curve of Peano which was the starting point of the present theory of Peano spaces. In a general way, the theory presented in this book contains many instances where an apparent paradox turns out to be the source of essentially new insight, the most elementary case being, perhaps, the example of Schwarz (see I.1.10).

I.1.17. The analytic theory of surface area, presented in this book, has been initiated by Lebesgue [2] in his famous memoir, *Intégrale, Longueur, Aire,* which appeared in 1902. In a series of profound investigations, published between 1908 and 1917, Geöcze undertook a first systematic study of surface area, based upon the lower semi-continuity principle proposed by Lebesgue and upon a modified projection principle (see I.1.15) discovered by Geöcze himself. In the light of these initial fundamental contributions of Lebesgue and Geöcze, it was clear that the development of an adequate theory of surface area, along the lines just indicated, will depend upon the solution of a number of problems of a high order of difficulty in Analysis and in Topology. Furthermore, most of the inauspicious phenomena described in I.1.9-I.1.16 were known to and fully appreciated by Geöcze. In view of the discouraging conceptual and technical difficulties revealed by the work of Lebesgue and Geöcze, it seemed doubtful whether the theory would attract the general interest necessary for continuous development. And yet, the progress was steady, due in particular to the sustained effort of workers primarily interested in Analysis, a curious situation in view of the fact, clearly indicated by the work of Geöcze, that some of the most essential difficulties of the theory are of a topological character. The explanation of this somewhat surprising course of events seems to be rather obvious. The *nonparametric case,* that is the case of surfaces represented by an equation of the form $z = f(x, y)$ where $f(x, y)$ is single-valued and continuous, was a natural starting point on account of its relative simplicity. In the first place, the geometrical picture of the surface is quite clear, the topological difficulties disappear, and as a consequence the theory of this case is essentially an application of the theory of functions of real variables. From the point of view of Analysis, a striking simplification arises due to the fact that the inequality of Steiner admits of an adequate extension in the nonparametric case (cf. I.1.7, I.1.13). As a consequence, the nonparametric case attracted the attention of many Analysts whose joint efforts resulted in a beautiful theory of a high degree of completeness. Furthermore, this special theory showed far-reaching analogies with the theory of the arc length, thus serving as a stimulus for a renewed attack upon the parametric case which seemed to present so many discouraging aspects. As regards developments in the *parametric case*, many fundamental contributions were made by Analysts primarily interested in Calculus of Variations. Clearly, the double integral occurring in the formula for surface area (see I.1.2) may be thought of as a very simple special case of the integrals that occur in double integral problems in Calculus of Variations, a circumstance which explains the interest of specialists in this field in surface area theory. Even though the results achieved by the use of analytic methods are of great beauty and importance, the fundamental role of Topology seems to be clearly established, both by the pioneer work of Geöcze and by the latest developments. By a most fortunate coincidence, many of the topological problems relevant for surface area theory arose, independently, in that important branch of Topology which is known at present as Analytic Topology (see G. T. Whyburn [3]). Thus the student of surface area theory can, and perhaps must, draw upon the combined resources of Analysis and Topology in attacking the problems that still await solution.

I.1.18. In concluding this preliminary survey, let us observe that we discussed so far only the difficulties to be anticipated. As regards the development of the theory, the reader interested in first obtaining a general picture may now turn to the Chapters II.5, III.4, IV.5, V.4 which contain information of a general character concerning results achieved, problems yet open, and bibliography. In particular, these chapters should be consulted for detailed references concerning the origin of the various results included in the main text where references are given only for the purpose of technical information. In view of the extensive use of both Analysis and Topology, an entirely self-contained presentation could not be achieved. For this reason, background material that could not be adequately discussed has been assembled in chapters I.2 and I.3. A reader who feels that his background in either Analysis or in Topology may be inadequate has many excellent texts at his disposal. Amongst these, Saks [6] and G. T. Whyburn [3] are admirably suited to ease the task of the reader.

CHAPTER I.2. BACKGROUND IN TOPOLOGY

I.2.1. The purpose of this chapter is to list definitions and theorems in Topology which will be needed in the sequel but could not be discussed adequately in this book. Proofs will be omitted. In most cases, Whyburn [3] may be used as general reference, and in such cases no specific reference will be given. Many theorems in this chapter may be used as exercises by a reader whose previous training did not include an adequate amount of Topology.

I.2.2. A *topological space* S consists of an aggregate K of elements, to be termed the *points* of S, and of an aggregate Ω of subsets of K, to be termed the *open sets* of S, such that the following conditions hold. (i) The empty set, to be denoted by 0, and K itself belong to Ω. (ii) If the sets E_1, E_2 belong to Ω, then their common part E_1E_2 also belongs to Ω. As a consequence, if $E_1, E_2, \cdots, E_n$ is any finite system of sets in Ω, then their common part $E_1E_2 \cdots E_n$ also belongs to Ω. (iii) If F is any family of sets that belong to Ω, then their sum also belongs to Ω. (iv) If p_1, p_2 are distinct elements of K, then there exists a pair of sets E_1, E_2 in Ω, such that p_1 is contained in E_1, p_2 is contained in E_2, and $E_1E_2 = 0$.

If it is desired to refer explicitly to the aggregates K and Ω, one may use the symbol $S(K, \Omega)$ to refer to the topological space determined by K and Ω. Two topological spaces $S(K_1, \Omega_1)$, $S(K_2, \Omega_2)$ will be considered different unless $K_1 = K_2$ and $\Omega_1 = \Omega_2$. If K and Ω are clearly identified by the context, then we may write S instead of $S(K, \Omega)$.

The term topological space is being used in several texts in a more general sense, and this fact should be carefully noted by the reader in using such texts. The concept of a *topological space*, defined above, is due to Hausdorff [1], and seems to be best suited for our purposes.

I.2.3. We shall list presently various terms and notations that will be used in the sequel. Let K be any aggregate of elements to be termed points. A subset E of K is termed *nondegenerate* if it contains more than one point. If the subsets E_1, E_2 have no common points, then E_1 and E_2 are said to be *disjoint*. The empty subset is denoted by 0 (zero). A subset E which contains at most a denumerable infinity of distinct points is termed *countable*. Thus the empty set 0 is countable, and every finite subset E is countable. The symbol $\in$ is used in the sense of "element of". Thus if p is a point and E is a subset of K, then $p \in E$ means that p lies in E. The negation of $\in$ is denoted by $\notin$. The inclusion sign $\subset$ means "subset of". If F is a family of subsets of K, then the sum $\sum E$, $E \in F$, denotes the set of all those points p for which the relation $p \in E$ holds for at least one set $E \in F$, and the product $\prod E$, $E \in F$, denotes the set of all those points p for which the relation $p \in E$ holds for every set $E \in F$. If the family F is countable, and is comprised of a (finite or infinite) sequence of subsets $E_1, \cdots, E_n, \cdots$, then the notations $\sum E_n$, $\prod E_n$, $n =$

1, 2, $\cdots$, may be used. If E_1 , E_2 are subsets of K, then $E_1 - E_2$ is the set of those points p which satisfy the simultaneous relations $p \in E_1$, $p \notin E_2$. In particular, the difference $K - E$ is termed the *complement* of the subset E (with respect to K). The various concepts just defined give rise to an algebra of sets which will be freely used in the sequel. For a detailed discussion, see Kuratowski [3], Newman [1].

I.2.4. CONTINUATION. Let p_1 , p_2 , $\cdots$, p_n , $\cdots$ be a finite or infinite sequence of points of K which are not necessarily distinct. Then a point p of K may occur in the sequence either not at all or else a finite or infinite number of times. Let $N(p)$ be the number of occurrences of p in the sequence. Thus $N(p)$ may be equal to zero, or to a finite positive integer, or to ∞. If p_1^* , p_2^* , $\cdots$, p_n^* , $\cdots$ is a second sequence, and $N^*(p)$ denotes the number of occurrences of p in this sequence, then we shall say that the sequence p_n^* is a *rearrangement* of the sequence p_n provided that $N(p) = N^*(p)$ for every point p of K, and we shall write $[p_1 , \cdots , p_n , \cdots] = [p_1^* , \cdots , p_n^* , \cdots]$, or more concisely $[p_n] = [p_n^*]$. Thus this equation does *not* mean that $p_n = p_n^*$ for every n, but merely that the sequences p_n , p_n^* are rearrangements of each other. The symbol $[p_n]$ by itself will be used to express the fact that the particular arrangement of the sequence p_n is immaterial in a certain situation. If $\overset{1}{p}_n$, $\overset{2}{p}_n$, $\overset{3}{p}_n$ are three sequences, and if the corresponding functions $N^1(p)$, $N^2(p)$, $N^3(p)$ satisfy the relation $N^3(p) \equiv N^1(p) + N^2(p)$, then we shall write $[\overset{3}{p}_n] = [\overset{1}{p}_n] + [\overset{2}{p}_n]$. Thus this formula means that the sequence $\overset{3}{p}_n$ is obtained by dovetailing the sequences $\overset{1}{p}_n$ and $\overset{2}{p}_n$, and it does *not* mean that $\overset{3}{p}_n = \overset{1}{p}_n + \overset{2}{p}_n$ for every n.

I.2.5. CONTINUATION. Let there be given a second aggregate K^*. Let us use p, p^* as generic notations for points of K, K^* respectively. A *single-valued transformation*, or *mapping*, $p^* = T(p)$ from K *into* K^* arises if for every point $p \in K$ there is assigned a unique image point $T(p) = p^* \in K^*$. Distinct points p_1 , p_2 of K may have the same image in K^* under T. It is not assumed that every point $p^* \in K^*$ is the image of some point $p \in K$. If E is a subset of K, then $T(E)$ denotes the image of E under T. In particular, $T(K)$ denotes the image of K itself. Generally $T(K)$ is a proper subset of K^*, but if $T(K) = K^*$, then we shall say that T is a transformation from K *onto* (and not merely *into*) K^*. In the sequel, a formula $T(K) = K^*$ will mean that T is a single-valued mapping from K *onto* K^*, and the formula $T(K) \subset K^*$ will mean that T is a single-valued mapping from K *into* K^*. The term single-valued will be usually omitted.

If E^* is a subset of K^*, then $T^{-1}(E^*)$ denotes the set of those points $p \in K$ for which $T(p) \in E^*$. The set $T^{-1}(E^*)$ is termed the *inverse set* of E^* (under T in K). In particular, if E^* reduces to a single point p^*, then $T^{-1}(p^*)$ is the inverse set of the point p^*. If $T(K) \subset K^*$, then $T^{-1}(p^*)$ may be empty for certain points p^*. The following formulas and facts will be useful in the sequel (T denotes a mapping from K into K^*).

(1) If F is any family of subsets E of K, then $T(\sum E) = \sum T(E)$, $T(\prod E) \subset \prod T(E)$, $E \in F$.

(2) If E_1, E_2 are subsets of K, and $E_1 \subset E_2$, then $T(E_1) \subset T(E_2)$.

(3) If E_1, E_2 are any two subsets of K, then $T(E_1) - T(E_2) \subset T(E_1 - E_2)$.

(4) If E_1^*, E_2^* are subsets of K^*, and $E_1^* \subset E_2^*$, then $T^{-1}(E_1^*) \subset T^{-1}(E_2^*)$.

(5) If E_1^*, E_2^* are any two subsets of K^*, then $T^{-1}(E_1^*) - T^{-1}(E_2^*) \subset T^{-1}(E_1^* - E_2^*)$.

(6) If E_1^* is a subset of K^*, and $E_2^* = K^* - E_1^*$ is the complement of E_1^*, then $T^{-1}(E_2^*)$ is the complement of $T^{-1}(E_1^*)$ in K.

(7) If E is any subset of K, then $E \subset T^{-1}T(E)$.

(8) If E^* is any subset of K^*, then $TT^{-1}(E^*) = E^*$.

(9) $T^{-1}(0) = 0$, $T(0) = 0$.

(10) $T^{-1}(K^*) = K$, $T^{-1}T(K) = K$, but the relation $TT^{-1}(K^*) = K^*$ holds if and only if T maps K onto K^*.

(11) If $T(K) = K^*$, then $TT^{-1}(E^*) = E^*$ for every subset E^* of K^*.

(12) If $T(K) = K^*$, and $0 \neq E^* \subset K^*$, then $T^{-1}(E^*) \neq 0$.

(13) If F^* is any family of subsets E^* of K^*, then $T^{-1}(\sum E^*) = \sum T^{-1}(E^*)$, $T^{-1}(\prod E^*) = \prod T^{-1}(E^*)$, $E^* \in F^*$.

For further interesting formulas, see Kuratowski [3].

I.2.6. Continuation. A set $E \subset K$ is termed an inverse set (under T) if and only if there exists a set $E^* \subset K^*$ such that $E = T^{-1}(E^*)$. Assuming that $T(K) = K^*$, we have then the following statements.

(1) A set $E \subset K$ is an inverse set if and only if $E = T^{-1}T(E)$.

(2) If E_1, E_2 are subsets of K, then $T(E_1E_2) \subset T(E_1)T(E_2)$. But if at least one of E_1, E_2 is an inverse set, then $T(E_1E_2) = T(E_1)T(E_2)$.

I.2.7. Continuation. Suppose that $T(K) = K^*$. If $T(p_1) \neq T(p_2)$ whenever p_1, p_2 are distinct points of K, then T is termed *biunique*. Clearly, T is biunique if and only if $T^{-1}(p^*)$ consists of a single point for every choice of the point $p^* \in K^*$. If T is biunique, then $T^{-1}(p^*)$ is a single-valued mapping from K^* onto K which is termed the *inverse mapping* to T.

I.2.8. Let K, K_1, K_2 be three aggregates, and let $T_1(K) = K_1$, $T_2(K) = K_2$, $T_{12}(K_1) = K_2$ be single-valued mappings. If $T_2(p) = T_{12}T_1(p)$ for every point $p \in K$, then we shall say that T_2 is the product of T_1 and T_{12}, and we shall write $T_2 = T_{12}T_1$. *Note that products of mappings should be read from the right to the left.* A relation $T_2 = T_{12}T_1$ is referred to as a *factorization* of T_2.

I.2.9. A topological space $S = S(K, \Omega)$ may be thought of as obtained in two steps. First, the aggregate K is assigned, and then a class Ω of subsets of K is chosen, subject to the conditions stated in I.2.2. The second step (selection of the class Ω) will be termed *topologization* of K. Before the selection of Ω, the aggregate K may be thought of as an untopologized space. Let us emphasize again that the class Ω is required to satisfy all the conditions (i)-(iv) of I.2.2. This remark is very relevant, since the term topologization is used in several texts in a less exacting sense.

Given an untopologized space, it may be that one of several possible topologizations appears to be the most convenient. For example, let $S = S(K, \Omega)$ be a topological space, and let K^* be a subset of S. Let Ω^* be the class of subsets of

K^* defined as follows: a subset E^* of K^* belongs to Ω^* if and only if there exists a subset G of S such that $G \in \Omega$ and $E^* = GK^*$. The class Ω^* is readily seen to satisfy the conditions (i)-(iv) of I.2.2 (relative to K^*). There arises the *topological subspace* $S^* = S(K^*, \Omega^*)$ of S. In this particular manner, each subset of S gives rise to a topological space which is termed a *subspace* of S, it being understood that the topologization of the subset is selected in the manner just described.

I.2.10. CONTINUATION. As a second important example, we consider *metric spaces*. A metric space $\mathfrak{M} = \mathfrak{M}(K, \rho)$ consists of an aggregate K of elements termed points, and of a function $\rho(p, q)$ defined for all pairs of points p, q of K, such that the following conditions hold. (i) $\rho(p, q) = \rho(q, p)$. (ii) $0 \leqq \rho(p, q) < \infty$. (iii). $\rho(p, q) = 0$ if and only if $p = q$. (iv) $\rho(p_1, p_3) \leqq \rho(p_1, p_2) + \rho(p_2, p_3)$ (*triangle inequality*).

The *spherical neighborhood* $U(p, r)$ of radius r of a point p (where $0 < r < \infty$) is defined as the set of those points x that satisfy the inequality $\rho(p, x) < r$. A point p_0 of a set $E \subset \mathfrak{M}$ is called an *interior point* of E if there exists some spherical neighborhood $U(p_0, r)$ such that $U(p_0, r) \subset E$. Let us define a class Ω of subsets of $\mathfrak{M}$ as follows: (a) The empty set belongs to Ω. (b) A non-empty set $E \subset \mathfrak{M}$ belongs to Ω if and only if every point of E is an interior point of E. The fact that Ω satisfies the conditions (i)-(iv) of I.2.2 is readily verified, and thus we obtain a topological space $S(K, \Omega)$. In this manner, every metric space $\mathfrak{M}(K, \rho)$ gives rise to a topological space that we may denote also by $\mathfrak{M}(K, \rho)$. In fact, we shall use the term metric space to refer to a topological space that has been derived from a metric space in the manner just described.

If E is a subset of a metric space $\mathfrak{M}$, then the set E jointly with the distance function ρ of $\mathfrak{M}$ clearly constitutes a *metric subspace* of $\mathfrak{M}$.

Given two metric spaces $\mathfrak{M}(K, \rho_1)$, $\mathfrak{M}(K, \rho_2)$ which differ only in the choice of the distance functions, we shall consider them identical if and only if the topological spaces are identical which are derived from them in the manner described above. If $U^1(p, r)$, $U^2(p, r)$ are generic notations for spherical neighborhoods in the metric spaces $\mathfrak{M}(K, \rho_1)$, $\mathfrak{M}(K, \rho_2)$ respectively, then the following condition is necessary and sufficient for these metric spaces to be identical: every $U^1(p_0, r)$ contains some $U^2(p_0, r^*)$ and vice versa.

I.2.11. Given a subset E of a metric space $\mathfrak{M} = \mathfrak{M}(K, \rho)$, the *diameter* $d(E)$ of E is defined by the formula $d(E) = \text{l.u.b. } \rho(p, q)$, where the least upper bound is taken with respect to all pairs of points p, q in E. For the empty set 0 we agree to put $d(0) = 0$.

If p is a point and E is a set in $\mathfrak{M}$, then their distance $\rho(p, E)$ is defined by the formula $\rho(p, E) = \text{gr.l.b. } \rho(p, x)$, where the greatest lower bound is taken with respect to all points $x \in E$. If E_1, E_2 are two subsets of $\mathfrak{M}$, then their distance $\rho(E_1, E_2)$ is defined by the formula $\rho(E_1, E_2) = \text{gr.l.b. } \rho(p_1, p_2)$, where the greatest lower bound is taken with respect to all pairs of points $p_1 \in E_1$, $p_2 \in E_2$. The quantity $d(E)$ may be infinite.

In the sequel, several metric spaces will be considered simultaneously in certain situations. We shall use then ρ to denote distance in any one of the spaces concerned, since it will be clear from the context which one of the spaces is involved. A similar remark applies to the diameter d.

I.2.12. Let $p_1, \cdots, p_n, \cdots$ be an infinite sequence of (not necessarily distinct) points in a topological space S. Then p_n is said to converge to a point x, in symbols $p_n \to x$, if and only if for every open set $G \subset S$ that contains x the relation $p_n \in G$ holds for all but a finite number of subscripts n. The following statements hold.

(1) If $p_n \to x'$ and $p_n \to x''$, then $x' = x''$.

(2) If $p_n = x$ for every n, then $p_n \to x$.

(3) If $p_n \to x$, then every infinite subsequence also converges to x.

(4) If $p_n \to x$, $q_n \to x$, then the sequence $p_1, q_1, p_2, q_2, \cdots$ also converges to x.

(5) If $p_n \to x$ and q is any point of S, then the sequence $q, p_1, p_2, \cdots$ also converges to x.

(6) If $p_n \to x$, and the sequence q_n is a rearrangement of the sequence p_n, then $q_n \to x$.

I.2.13. In a topological space S, let $E_1, \cdots, E_n, \cdots$ be an infinite sequence of (not necessarily distinct) sets. Then lim sup E_n is defined as the set of those points x that satisfy the following condition: if G is any open set that contains x, then the relation $GE_n \neq 0$ holds for infinitely many values of n. Furthermore, lim inf E_n is defined as the set of those points x that satisfy the following conditions: if G is any open set that contains x, then the relation $GE_n \neq 0$ holds for all but a finite number of subscripts n. The sequence E_n is said to converge to the set E (*which may be empty*), if and only if lim sup E_n = lim inf $E_n = E$. We shall write then $E_n \to E$ or $E = \lim E_n$. The sets lim sup E_n, lim inf E_n may be empty.

In the special case when each E_n reduces to a single point, the preceding definition of convergence of sequences of sets presents a slight conflict with the previously introduced concept of the convergence of a sequence of points. The reader should analyze this situation, in order to avoid misunderstandings. A convenient device consists of using a symbol like (p) to denote the set that consists of the single point p, if there is danger of ambiguity.

I.2.14. A subset E of a topological space S is termed *closed* if and only if its complement $S - E$ is open. The following statements hold.

(1) The empty set is closed, and S itself is closed.

(2) The sum of a finite number of closed sets is closed.

(3) Every finite set is closed.

(4) If F is any family of closed sets, then the product $\prod E$, $E \in F$, is closed.

Now let E be a subset of S. The *closure* $c(E)$ of E is then defined by the formula $c(E) = \prod A$, $E \subset A$, A closed. The following statements hold.

(5) $c(E)$ is closed.

(6) E is closed if and only if $E = c(E)$.

(7) If p is a point and E is a set in S, then $p \in c(E)$ if and only if the following condition holds: if G is any open set that contains p, then $GE \neq 0$.

(8) Always $c[c(E)] = c(E)$.

(9) If $E_1, \cdots, E_n$ is any finite system of sets in S, then $c(\sum E_j) = \sum c(E_j)$, $j = 1, \cdots, n$.

(10) If (p) denotes the set consisting of the single point p, then $c[(p)] = (p)$.

(11) If F is any family of sets in S, then $c(\prod E) \subset \prod c(E)$, $E \in F$, and $\sum c(E) \subset c(\sum E)$, $E \in F$.

I.2.15. If E is a set in a topological space S, then the *frontier* $\text{fr}(E)$ of E is defined by the formula $\text{fr}(E) = c(E)c(S - E)$. The *interior* of E, to be denoted by E^0, is then defined by the formula $E^0 = E - \text{fr}(E)$. The following statements hold:

(1) $\text{fr}(0) = 0$, $\text{fr}(S) = 0$.

(2) Always $\text{fr}(E) = \text{fr}(S - E)$.

(3) If E is open, then $E\,\text{fr}(E) = 0$.

(4) Always $E + \text{fr}\,(E) = c(E)$.

I.2.16. A topological space S is termed *normal* if and only if the following condition holds: given any two disjoint closed sets E_1, E_2 in S, there exist two open sets G_1, G_2 in S such that $E_1 \subset G_1$, $E_2 \subset G_2$, $G_1G_2 = 0$.

I.2.17. Let E be a subset of a topological space S. By an *open covering* of E we mean a family F of open sets in S such that for every point $x \in E$ there exists some set $G \in F$ such that $x \in G$. If the family F is finite, then the open covering is termed finite. If F^* is a subfamily of F which also yields an open covering of E, then this latter covering is termed a sub-covering of the covering given by F.

A subset E of S is termed *compact* if and only if every open covering of E contains a finite sub-covering. Accordingly, S itself is termed compact if and only if this condition holds for $E = S$.

I.2.18. A topological space S is said to satisfy the *first countability axiom* if and only if for every point $x \in S$ there exists a sequence $G_1, \cdots, G_n, \cdots$ of open sets containing x, such that if G is any open set that contains x, then $G_n \subset G$ for some n.

I.2.19. Given a topological space S, a family F of open sets in S is called an *open base* for S if and only if every open set $G \subset S$ is a sum of sets $E \in F$. If the family F is countable, then we speak of a *countable open base*.

A topological space S is said to satisfy the *second countability axiom* if and only if S possesses a countable open base. Clearly, if S satisfies the second countability axiom, then it also satisfies the first countability axiom.

Suppose that S satisfies the second countability axiom, and let Φ be any family of disjoint (nonempty) open sets in S. Then Φ is countable.

I.2.20. A topological space S is termed *separable* if and only if it contains some countable set E such that $c(E) = S$. If S satisfies the second countability axiom, then S is separable.

I.2.21. Let E be a subset of a topological space S. Suppose that there exist two sets A, B in S such that $E = A + B$, $A \neq 0$, $B \neq 0$, $Ac(B) + Bc(A) = 0$. We say then that the sets A, B constitute a *separation* of E, and we shall write $E = A \mid B$ to express this fact.

A subset E of S is termed *disconnected* if and only if there exists some separation $E = A \mid B$. Otherwise E is termed *connected.* If E is open and connected, then E is termed a *domain.* The following statements hold.

(1) The empty set is connected. A set consisting of a single point is connected.

(2) S itself is connected if and only if it cannot be represented as the sum of two nonempty, disjoint open sets.

I.2.22. A subset E of a topological space S is termed a *continuum* if and only if E is nonempty, compact, and connected. *Thus a set consisting of a single point is a continuum.* In several texts on Topology, a continuum is required to contain more than one point, but for our purposes it is convenient to permit a continuum to reduce to a single point.

I.2.23. Let E be a subset of a topological space S. A (nonempty) subset G of E is termed a *component* of E if and only if the following conditions hold: (i) G is connected. (ii) G is not a proper subset of any connected subset of E.

If G_1, G_2 are components of E, then either $G_1 = G_2$ or else $G_1G_2 = 0$. The set E is the sum of its components.

A set E is termed *totally disconnected* if and only if every component of E reduces to a single point. A set consisting of a single point is thus both connected and totally disconnected.

I.2.24. A continuum G in a topological space S is termed *multicoherent* if and only if there exist two continua G_1, G_2 such that $G_1 + G_2 = G$ and G_1G_2 is disconnected. Otherwise G is termed *unicoherent.*

I.2.25. A topological space S is termed *locally connected* if and only if for every open set $G \subset S$ it is true that the components of G are open.

I.2.26. Let S be a topological space and let S^* be a subspace of S, in the sense of I.2.9. If E is a subset of S^*, then a statement like "E is closed" is ambiguous, since it may mean either that E is closed relative to the topology of S or else relative to the topology of S^*. A similar ambiguity arises in connection with a number of other statements. To clarify this issue, let us adopt temporarily terms like S-closed, S^*-closed, S-connected, S^*-connected, and so forth, to mean closed (or connected) relative to S and to S^* respectively. Also, if E is a subset of S^*, let us use temporarily the symbols $c(E)$, $c^*(E)$ to denote the closure of E relative to S and S^* respectively. We have then the following statements.

(1) If $E \subset S^*$, then $c^*(E) = S^*c(E)$.

(2) A set $E \subset S^*$ is S^*-open if and only if there exists an S-open set G such that $E = S^*G$. Similarly, a set $E \subset S^*$ is S^*-closed if and only if there exists an S-closed set F such that $E = S^*F$.

(3) If E, A, B are subsets of S^*, such that A, B constitute an S^*-separation of E, then they also constitute an S-separation of E, and the converse is also true (cf. I.2.21).

(4) A set $E \subset S^*$ is S^*-connected if and only if it is S-connected.

(5) If $G \subset E \subset S^*$, then G is an S^*-component of E if and only if it is an S-component of E.

(6) A set $E \subset S^*$ is S^*-compact if and only if E is S-compact.

I.2.27. CONTINUATION. If $E = A \mid B$ is a separation of E relative to S, then it is also a separation of E relative to any subspace S^* that contains E, by I.2.26,(3). Hence it is unnecessary to specify the space relative to which the statement "$E = A \mid B$ is a separation of E" is made. In view of I.2.26,(4),(5),(6), a similar remark applies to the following statements: "E is connected", "G is a component of E", "E is compact". As regards the symbol $c(E)$ and the statements "E is closed", "E is open", it will be understood in the sequel that the space S is identified by the context, or else qualifications like "closed relative to the subspace S^*" will be made.

I.2.28. Let G be a subset of a topological space S. Then G is termed *locally connected* if and only if it is locally connected if considered as a space by itself (see I.2.9, I.2.25). We have then the following statements:

(1) G is locally connected if and only if for every set $H \subset G$ that is open relative to G it is true that the components of H are open relative to G.

(2) G is locally connected if and only if the following condition holds for every point $p \in G$: if H is any open set containing p and C is the component of GH that contains p, then there exists an open set V such that $p \in GV \subset C$.

(3) If $G \subset S^* \subset S$, then G considered as a subset of the space S^* is locally connected if and only if G is locally connected considered as a subset of the space S. In other words, in a statement "G is locally connected", it is unnecessary to specify the containing space.

I.2.29. Given two topological spaces S, S^*, and a single-valued transformation $T(S) \subset S^*$, we shall say that T is *continuous* at a point $p_0 \in S$ if and only if for every open set $G^* \subset S^*$ that contains the point $T(p_0)$, there exists an open set $G \subset S$ such that $p_0 \in G$ and $T(G) \subset G^*$. If T is continuous at every point of S, then T is termed continuous on S.

A transformation $T(S) = S^*$ is termed *topological* if and only if the following conditions hold. (i) T is biunique and hence $T^{-1}(S^*) = S$ is also a biunique and single-valued transformation. (ii) T is continuous on S and T^{-1} is continuous on S^*. A topological transformation is also referred to as a *homeomorphism.* Two topological spaces S, S^* are termed *homeomorphic* if and only if there exists some homeomorphism $T(S) = S^*$. Homeomorphism is an equivalence relation, that is, a binary relation (between topological spaces) which is reflexive, symmetric, and transitive.

I.2.30. Given a metric space $\mathfrak{M}$, we indicated in I.2.10 a standard process to derive from $\mathfrak{M}$ a topological space. There arises the *metrization problem*: given a topological space $S = S(K, \Omega)$ (see I.2.2), under what conditions is it possible to define a distance function $\rho(p, q)$ in S such that $\rho(p, q)$ satisfies the conditions (i)-(iv) of I.2.10 and the corresponding metric space $\mathfrak{M}(K, \rho)$ yields precisely the given topological space S, in the sense of I.2.10. If this is possible, then S is termed *metrizable.* The following fundamental theorems hold (for a very concise treatment, see Lefschetz [1]).

(1) Let S be a topological space that satisfies the second countability axiom. Then S is metrizable if and only if it is normal.

(2) Let S be a compact topological space. Then S is metrizable if and only if it satisfies the second countability axiom.

I.2.31. Topological spaces of a special character include the following.

(1) The *linear interval* $I : a \leqq x \leqq b$, with the distance function $\rho(x_1, x_2) = | x_1 - x_2 |$. A topological space homeomorphic with the linear interval is termed a *simple arc* or a *topological* 1-*cell,* or simply a 1-*cell.* Let γ be a simple arc and let $T(I) = \gamma$ be a homeomorphism. The points $T(a)$, $T(b)$ can be shown to be independent of the particular choice of the homeomorphism T. These points are termed the *end points* of the simple arc γ. If the end points of γ are p and q, then each of the two arrangements p, q and q, p determines an orientation of γ. An oriented simple arc arises if one of the two orientations is assigned.

(2) The *two-dimensional interval* $R : a \leqq x \leqq b$, $c \leqq y \leqq d$, with the ordinary Euclidean distance. A topological space homeomorphic with R is termed a *topological* 2-*cell* or simply a 2-*cell.* Let S be a 2-cell in this sense, and let $T(R) = S$ be a homeomorphism. Then the image, under T, of the perimeter of R is a subset of S which may be shown to be independent of the choice of the homeomorphism T. This subset of S is termed the *boundary* of S. The complement of the boundary is the *interior* of S.

(3) The *circle* $x^2 + y^2 = 1$, with the ordinary Euclidean distance. A topological space homeomorphic with this circle is termed a *topological* 1-*sphere* or simply a 1-*sphere,* or *simple closed curve.* The term oriented 1-sphere is self-explanatory.

(4) The *sphere* $x^2 + y^2 + z^2 = 1$, with the ordinary Euclidean distance. A topological space S homeomorphic with this sphere is termed a *topological* 2-*sphere* or simply a 2-*sphere.*

(5) *Euclidean space of n dimensions.* The points of this space are the ordered sets of n real numbers $(x_1, \cdots, x_n)$, distance being defined as in Euclidean analytic geometry.

I.2.32. Continuation. The following statements hold.

(1) Let E be a nondegenerate subset of a simple arc S. Then E is a continuum if and only if it is a simple arc.

(2) Let E be a nondegenerate proper subset of a simple closed curve S. Then E is a continuum if and only if it is a simple arc.

(3) Let C_1, C_2 be disjoint continua on a simple closed curve S. Then the set $S - (C_1 + C_2)$ is disconnected.

(4) Let S be either a 2-cell or a 2-sphere, and let E be a totally disconnected closed subset of S. Then $S - E$ is connected.

I.2.33. A topological space S is termed a *Peano space* if and only if it is compact, connected, locally connected, and satisfies the second countability axiom. If S is a Peano space, then S is metrizable by I.2.30. Hence we shall assume in the sequel, in dealing with a Peano space S, that a distance function $\rho(p, q)$ is assigned (cf. I.2.30). The following statements hold.

(1) A metric space $\mathfrak{M}$ is a Peano space if and only if it is compact, connected, and locally connected.

(2) A topological space S is a Peano space if and only if it is a continuous image of the unit interval $I : 0 \leqq x \leqq 1$. More explicitly, S is a Peano space if and only if there exists a continuous mapping $T(I) = S$. A similar statement holds if I is replaced by the unit square $0 \leqq x \leqq 1$, $0 \leqq y \leqq 1$ of the Euclidean plane.

(3) Let S be a Peano space. Then a distance function $\rho(p, q)$ can be so chosen that every spherical neighborhood is connected (see I.2.10).

(4) Let S be a Peano space, with a distance function $\rho(p, q)$. For every $\epsilon > 0$ there exists a $\delta = \delta(\epsilon) > 0$ such that the following holds. If p_1, p_2 are any two points of S such that $\rho(p_1, p_2) < \delta$, then there exists a continuum $C \subset S$ such that $p_1 \in C$, $p_2 \in C$, $d(C) < \epsilon$.

I.2.34. Let x be a point of a topological space S. If $S - x$ is disconnected, then x is termed a *cut point* of S, otherwise x is termed a *non-cut point*. A topological space is termed *cyclic* if it has no cut points.

I.2.35. It will be a matter of great importance in the sequel to recognize whether or not a given topological space S is a simple arc, a simple closed curve, or a 2-sphere. The following *characterization theorems* will be used:

(1) If S is a nondegenerate Peano space, then S is a simple arc if and only if it has not more than two non-cut points.

(2) If S is a nondegenerate Peano space, then S is a simple closed curve if and only if for every choice of two distinct points x, y in S, the set $S - (x + y)$ is disconnected.

(3) If S is a nondegenerate Peano space, then S is a (topological) 2-sphere if and only if the following conditions hold simultaneously: (i) S is cyclic (see I.2.34). (ii) if C is any continuum in S such that $S - C$ is connected, then C is unicoherent (see I.2.24).

Let us note that there are several topological characterizations of the 2-sphere. The one given above is especially convenient for our purposes. This particular characterization theorem is due to Kuratowski [1].

I.2.36. Let S be a metric space. Concerning convergent sequences of points in S we have the following statements (cf. I.2.12).

(1) The relation $p_n \to x$ is equivalent to the relation $\rho(p_n, x) \to 0$.

(2) Let E be a closed subset of S, and let p_n be a convergent sequence of points in E. Then the point $\lim p_n$ is also in E.

(3) Let E be a subset of S such that the following condition holds: whenever $p_n \to x$ and $p_n \in E$, $n = 1, 2, \cdots$, then $x \in E$. Then E is closed.

(4) If S is compact, then every (infinite) sequence of points p_n contains a convergent subsequence.

I.2.37. Let S be a topological space. The following statements hold concerning (infinite) sequences of subsets E_n of S (cf. I.2.13).

(1) The sets lim inf E_n, lim sup E_n are both closed (possibly empty). Hence, if lim E_n exists, then it is also closed (possibly empty).

(2) Suppose that S is compact, and let E_n be a sequence of nonempty subsets of S. Then the set $E = \limsup E_n$ is nonempty, and if G is any open set containing E, then $E_n \subset G$ for n sufficiently large.

(3) Suppose that S is compact, and let E_n be a sequence of connected subsets of S such that lim inf $E_n \neq 0$. Then lim sup E_n is connected.

(4) Suppose that S satisfies the second countability axiom. Then every (infinite) sequence of subsets E_n of S contains a convergent subsequence (where the limit set may be empty). However, if S is also compact and $E_n \neq 0$ for every n, then there exists a convergent subsequence with a *nonempty* limit set. If, in addition, each E_n is connected, then this limit set is a continuum.

I.2.38. Let S be a topological space. Then the concepts *compact* and *closed* are related by the following statements.

(1) If E is a compact subset of S, then E is closed.

(2) If S is compact, and E is a closed subset of S, then E is compact.

(3) If S is compact, then it is also normal.

I.2.39. The following statements, where S denotes a topological space, are concerned with the concept of connectedness.

(1) Let $E = A \mid B$ be a separation of the set $E \subset S$. If E is closed, then A, B are both closed.

(2) If G_1, G_2 are nonempty, disjoint open sets in S, then $G_1 + G_2 = G_1 \mid G_2$.

(3) If $E \neq 0$ is a subset and x is a point of S, such that $E + x$ is connected, then $x \in c(E)$.

(4) If E is a connected subset of S, and G is any set such that $E \subset G \subset c(E)$, then G is connected.

(5) Let F be a family of nonempty, disjoint open sets in S, and put $G = \sum E$, $E \in F$. If A is a connected subset of G, then A is a subset of some set $E \in F$. If, in addition, each set $E \in F$ is connected, then the components of G are precisely the sets $E \in F$.

(6) If E is a closed subset of S, then each component of E is closed.

(7) Let $E = A \mid B$ be a separation of a subset E of S. If G is a connected subset of E, then G is either a subset of A or a subset of B.

(8) Let G_0 be a connected subset of S, and let F be a family of connected subsets of S such that $EG_0 \neq 0$ for each $E \in F$. Then the set $G_0 + \sum E$, $E \in F$, is connected.

(9) Let F be a family of connected subsets of S, and put $G = \sum E$, $E \in F$. If C is a component of G, then C is a sum of sets of the family F.

(10) Suppose that S is compact. Let $G_1, G_2, \cdots, G_n, \cdots$ be a sequence of continua in S, such that $G_1 \supset G_2 \supset \cdots \supset G_n \supset \cdots$. Then the product $E = \prod G_n$ is a continuum (in particular, $E \neq 0$).

I.2.40. Let S be a topological space. The following statements are concerned with the *frontier* of a set (see I.2.15).

(1) Let G be a connected subset of S, and let E be a subset of S such that $EG \neq 0$, $(S - E)G \neq 0$. Then $G \operatorname{fr}(E) \neq 0$.

(2) Let E be a nonempty proper subset of a connected set $G \subset S$. Then $G \operatorname{fr}(E) \neq 0$.

(3) If S is connected, and E is a nonempty proper subset of S, then $\operatorname{fr}(E) \neq 0$.

(4) Let G be a connected open set in S such that $\operatorname{fr}(G)$ consists of a single point p. Then G is a component of $S - p$.

I.2.41. The following statements are concerned with the concept of *local connectedness.*

(1) Let S be a locally connected topological space that satisfies the second countability axiom. If G is any open set in S, then the components of G constitute a countable aggregate.

(2) Let S be a metric space. Suppose that for every $\epsilon > 0$ there exists a finite system of connected sets $G_1, \cdots, G_n$, such that $G_1 + \cdots + G_n = S$ and $d(G_i) < \epsilon$, $i = 1, \cdots, n$. Then S is locally connected.

(3) Let E be a subset of a metric space S. Suppose that for every $\epsilon > 0$ there exists a finite system of connected sets $G_1, \cdots, G_n$, such that $E = G_1 + \cdots + G_n$, and $d(G_i) < \epsilon$, $i = 1, \cdots, n$. Then E is locally connected.

(4) If E is a compact, locally connected subset of a metric space S, then for every $\epsilon > 0$ there exists a finite system of connected sets $G_1, \cdots, G_n$ with the properties described under (3).

(5) Let $E_1, \cdots, E_m$ be a finite system of compact, locally connected sets in a metric space S. Then $E_1 + \cdots + E_m$ is compact and locally connected.

(6) Let S be a locally connected metric space, and let x be any point of S. Then there exists a sequence of connected open sets $G_1 \supset G_2 \supset \cdots \supset G_n \supset \cdots$, such that $d(G_n) \to 0$ and $x \in G_n$ for every n.

(7) Let S be a locally connected topological space that satisfies the second countability axiom. Then S possesses a countable open base comprised of connected open sets.

(8) Let S be a Peano space, x a point of S, and G a component of $S - x$. Then $\operatorname{fr}(G) = x$.

(9) Let S be a Peano space. If E is a nonempty, closed proper subset of S, and G is a component of $S - E$, then $\operatorname{fr}(G) \subset E$.

(10) Let $S_1, \cdots, S_n$ be a finite system of Peano subspaces of a Peano space S, such that $S^* = S_1 + \cdots + S_n$ is connected. Then S^* is a Peano subspace of S.

(11) Let $T(S) \subset S^*$ be a mapping where S is a Peano space and S^* is a metric space (note that S is also a metric space). Suppose that for every $\epsilon > 0$ there exists an $\eta = \eta(\epsilon) > 0$, such that $d[T(E)] < \epsilon$ whenever E is a connected subset of S and $d(E) < \eta$. Then T is continuous on S.

(12) Let G be a connected open subset of a Peano space S, and let x, y be any two distinct points of G. Then there exists in G a simple arc γ with end points x, y.

I.2.42. The following statements are concerned with *metric spaces.*

(1) In a metric space S, let $F_0, F_1, \cdots, F_n, \cdots$ be closed sets such that $F_0F_n \neq 0$ for every n, and $d(F_n) \to 0$. Then $F_0 + F_1 + \cdots + F_n + \cdots$ is closed.

(2) Let S be a metric space. Then S satisfies the first countability axiom. If, in addition, S is compact, then it satisfies the second countability axiom.

(3) Let S be a metric space. Then the distance function $\rho(p, q)$ is continuous in the following sense: if $p_n \to p_0$, $q_n \to q_0$, then $\rho(p_n, q_n) \to \rho(p_0, q_0)$.

(4) If E is a nondegenerate connected subset of a metric space S, then E is uncountable.

(5) If S is a compact metric space, then the distance function $\rho(p, q)$ is bounded.

(6) If E is a subset of a metric space S, then $d[c(E)] = d(E)$.

(7) If E_1, E_2 are subsets of a metric space S, and $E_1E_2 \neq 0$, then $d(E_1 + E_2) \leqq d(E_1) + d(E_2)$.

(8) In a compact metric space S, let E_n be a sequence of nonempty sets. Then

$$d(\liminf E_n) \leqq \liminf d(E_n), \qquad d(\limsup E_n) \geqq \limsup d(E_n).$$

If the sequence E_n is convergent, then $d(\lim E_n) = \lim d(E_n)$.

(9) Let $T(S) \subset S^*$ be a mapping, where S, S^* are metric spaces and S is compact. Then T is continuous on S if and only if for every $\epsilon > 0$ there exists an $\eta = \eta(\epsilon) > 0$ such that $d(E) < \eta$ implies $d[T(E)] < \epsilon$, where E is a generic notation for a subset of S.

I.2.43. The following statements are concerned with single-valued transformations. Let us recall that *transformation* and *mapping* are equivalent terms, and that single-valuedness is always assumed unless the contrary is explicitly stated.

(1) Let $T(S) \subset S^*$ be a mapping, where S, S^* are topological spaces. Then T is continuous on S if and only if for every open set $G^* \subset S^*$ the inverse set $T^{-1}(G^*)$ is an open set of S. In this statement, open sets may be replaced throughout by closed sets.

(2) Let $T(S) \subset S^*$ be a continuous mapping from a topological space S into a topological space S^*. If p_n is a sequence of points in S that converges to a point x, then $T(p_n) \to T(x)$. If E_n is a sequence of subsets of S, then we have $T(\liminf E_n) \subset \liminf T(E_n)$, $T(\limsup E_n) \subset \limsup T(E_n)$.

(3) Let $T(S) = S^*$ be a topological mapping from a topological space S onto a topological space S^*. Then a set $E \subset S$ is open if and only if $T(E)$ is open.

(4) Let $T(S) \subset S^*$ be a continuous mapping from a topological space S into a topological space S^*. Let E be a subset of S, and let $E^* = T(E)$ be the image of E. If E, E^* are topologized as subspaces of S, S^* respectively, then the mapping $T(E) = E^*$ is continuous on E.

(5) Let $T(S) \subset S^*$ be a continuous mapping, where S, S^* are topological

spaces. If E is a compact subset of S, then $T(E)$ is a compact subset of S^*. Hence, if $T(S) = S^*$ and S is compact, then S^* is also compact.

(6) Let $T(S) = S^*$ be a continuous mapping, where S, S^* are topological spaces. Suppose that S is compact. If E is a closed set in S, then $T(E)$ is also closed. If E^* is a subset of S^*, then E^* is closed if and only if $T^{-1}(E^*)$ is closed. Similarly, a subset G^* of S^* is open if and only if $T^{-1}(G^*)$ is open.

(7) Let $T(S) \subset S^*$ be a continuous mapping, where S, S^* are topological spaces. If G is a connected subset of S, then $T(G)$ is connected. If G is a continuum, then $T(G)$ is also a continuum.

(8) Let $T(S) = S^*$ be a continuous mapping, where S, S^* are topological spaces. Suppose that S is compact. If E_n is a convergent sequence of subsets of S, then the sequence $T(E_n)$ is also convergent, and $\lim T(E_n) = T(\lim E_n)$.

(9) Let $T(S) = S^*$ be a continuous mapping, where S, S^* are topological spaces. If S is compact and satisfies the second countability axiom, then the same holds for S^*.

(10) Let $T(S) = S^*$ be a continuous mapping, where S, S^* are topological spaces. If S is separable, then S^* is also separable. If S is locally connected, then S^* is also locally connected. If S is a Peano space, then S^* is also a Peano space (and hence, in particular, S^* is metrizable by I.2.33).

I.2.44. The following statements are concerned with sequences of continuous mappings. As a general reference, Hahn [2] may be used. Even though the following statements hold under more general conditions, we assume that all spaces involved are Peano spaces, since this will be the case in the situations relevant for our purposes. We shall use $\mathcal{P}$ as a generic notation for a Peano space. As noted in I.2.33, we can assume that a distance function ρ is given on $\mathcal{P}$ (see I.2.11 concerning the use of the symbols ρ, d). In the following statements, S^* denotes a fixed metric space.

(1) If $T(\mathcal{P}) \subset S^*$ is a continuous mapping, then T is *uniformly continuous* on $\mathcal{P}$ in the following sense. Given $\epsilon > 0$, there exists a $\delta = \delta(\epsilon) > 0$, such that $d[T(E)] < \epsilon$ whenever the set $E \subset \mathcal{P}$ satisfies the condition $d(E) < \delta$. In particular, if the points p_1, p_2 of $\mathcal{P}$ satisfy the condition $\rho(p_1, p_2) < \delta$, then $\rho[T(p_1), T(p_2)] < \epsilon$.

(2) Let $T_n(\mathcal{P}) \subset S^*$ be a sequence of continuous mappings. Then the sequence is termed *equicontinuous* on $\mathcal{P}$ if for every $\epsilon > 0$ there exists a $\delta = \delta(\epsilon) > 0$, such that $d[T_n(E)] < \epsilon$ whenever the set $E \subset \mathcal{P}$ satisfies the condition $d(E) < \delta$.

(3) Let $T'(\mathcal{P}) \subset S^*$, $T''(\mathcal{P}) \subset S^*$ be two (not necessarily continuous) mappings. We put $\rho(T', T'') = \text{l.u.b. } \rho[T'(x), T''(x)]$, where the least upper bound is taken with respect to all points $x \in \mathcal{P}$. Thus $\rho(T', T'')$ may be infinite. If $T_n(\mathcal{P}) \subset S^*$, $n = 0, 1, 2, \cdots$, is a sequence of mappings such that $\rho(T_n, T_0) \to 0$, then we shall say that T_n converges to T_0 *uniformly* on $\mathcal{P}$.

(4) Let $T_n(\mathcal{P}) \subset S^*$, $n = 0, 1, 2, \cdots$, be a sequence of mappings such that $T_n \to T_0$ uniformly on $\mathcal{P}$. If each T_n, $n = 1, 2, \cdots$, is continuous, then T_0 is also continuous, and the sequence is equicontinuous on $\mathcal{P}$.

(5) Let $T_n(\mathcal{P}) \subset S^*$ be an equicontinuous sequence of mappings. Then the sequence contains a uniformly convergent subsequence.

I.2.45. Let S be a topological space, and let us denote by R^1 the number line $-\infty < x < \infty$ (one-dimensional Euclidean space, with the distance function $\rho(x_1, x_2) = |x_1 - x_2|$). Let p be a generic notation for a point of S, and let $f(p)$ denote a single-valued, real-valued function on S. Then the formula $x = f(p)$, $p \in S$, defines a mapping $T(S) \subset R^1$. The function $f(p)$ is termed continuous on S if and only if the associated mapping $T(S) \subset R^1$ is continuous on S. The following statements hold.

(1) If $f(p)$ is continuous on S, then for every real number c the sets where $f(p) > c$, $f(p) < c$ respectively are open.

(2) If S is compact and $f(p)$ is continuous on S, then $f(p)$ takes on a finite maximum and a finite minimum on S. In particular, $|f(p)|$ is bounded on S.

(3) If S is connected and $f(p)$ is continuous on S, and if $f(p)$ takes on the values a and $b \neq a$ on S, then it takes on every value between a and b. Hence, if it is known that the values taken on by $f(p)$ on S form a countable aggregate, then it follows that $f(p)$ is constant on S.

(4) If E is a subset of a topological space S, then E itself is a topological space, in the sense of I.2.9. Hence the preceding statements can be extended to functions $f(p)$ defined only on E (cf. I.2.26, I.2.27).

(5) By means of the associated mapping $T : x = f(p)$, the concepts and theorems of I.2.44 can be extended, in an obvious manner, to real-valued functions $f(p)$.

I.2.46. We shall consider presently *Borel sets* in a metric space S. The following terminology will be used. A set $E \subset S$ is termed an F_σ if it is the sum of a countable aggregate of closed sets, and it is termed a G_δ if it is the product of a countable aggregate of open sets. A sequence of points p_n in S is termed a *Cauchy sequence* if and only if for every $\epsilon > 0$ there exists an integer $j = j(\epsilon)$ such that $\rho(p_n, p_m) < \epsilon$ if m and n both exceed $j(\epsilon)$. The metric space S is termed *complete* if and only if every Cauchy sequence in S is convergent. In particular, Euclidean n-space is a complete metric space.

Let K be a generic notation for an aggregate of subsets of S, with the following properties. (i) Every closed set belongs to K. (ii) If a set E belongs to K, then its complement $S - E$ also belongs to K. (iii) If $E = \sum E_n$, $n = 1, 2, \cdots$, and if each E_n belongs to K, then E also belongs to K. (iv) As a consequence of (ii) and (iii), if $G = \prod E_n$, $n = 1, 2, \cdots$, and every E_n belongs to K, then G belongs to K. (v) As a consequence of (i) and (ii), every open set belongs to K.

Clearly, if K_0 denotes the product of all the aggregates K, then K_0 itself is an aggregate K, and in fact K_0 is the smallest aggregate K. The sets $E \subset K_0$ are termed the *Borel sets* of S, and K_0 itself is termed the *Borel class* in S. As a general reference for the theory of Borel sets, one may use Hahn [2], Sierpinski [1], or Kuratowski [3]. The following facts will be needed.

(1) Every F_σ and every G_δ is a Borel set.

(2) Given S as a complete, separable metric space, let $\mathfrak{A}$ be an aggregate of subsets of S with the following properties. (i) $\mathfrak{A}$ contains all closed sets. (ii) If a set E belongs to $\mathfrak{A}$, then its complement $S - E$ also belongs to $\mathfrak{A}$, and hence all open sets belong to $\mathfrak{A}$. (iii) If E_n is a sequence of sets, such that $E_iE_j = 0$ for $i \neq j$ and each E_n belongs to $\mathfrak{A}$, then $\sum E_n$, $n = 1, 2, \cdots$, also belongs to $\mathfrak{A}$. Under these conditions, it is true that every Borel set belongs to $\mathfrak{A}$.

(3) Let S^* be a subspace of S, and let K_0^* be the Borel class relative to S^*. Then a set $E^* \subset S^*$ belongs to K_0^* if and only if there exists a set $E \in K_0$ such that $E^* = S^*E$.

(4) Let $T(S) = S^*$ be a topological mapping, where S, S^* are metric spaces. If E is a Borel set in S, then $T(E)$ is a Borel set in S^*.

(5) Let $T(S) \subset R^n$ be a continuous mapping, where S is a complete and separable metric space and R^n is the Euclidean n-space. If E is a Borel set in S, such that its image $T(E)$ is a bounded subset of R^n, then $T(E)$ is measurable in the Lebesgue sense (however, $T(E)$ need *not* be a Borel set). In particular, the theorem holds if S itself is a Euclidean space.

I.2.47. The closing sections of this chapter are devoted to statements concerned with the topology of the Euclidean plane and of the 2-sphere. Newman [1] may be used as general reference. The term plane will mean the Euclidean plane, and an individual point will be given in the form (u, v), where u, v are thought of as Cartesian coordinates. Alternatively, we shall speak also of the uv-plane, meaning Euclidean plane. A subset E of the plane may be thought of as a subspace of the plane (see I.2.9). We shall also be concerned with mappings T given by formulas $T : x = x(u, v)$, $y = y(u, v)$, $(u, v) \in E$, where the image point (x, y) lies in a Euclidean xy-plane (which may or may not coincide with the uv-plane).

A *domain* in the uv-plane is a connected open set (note this agreement carefully, since the term domain is used in various senses in different texts). If the frontier of a bounded domain $\mathfrak{D}$ consists of a finite number of simple closed curves $C_1, \cdots, C_n$, such that $C_iC_j = 0$ for $i \neq j$, then the set $\mathfrak{D} + C_1 + \cdots + C_n$ is termed a bounded, finitely connected *Jordan region.* The letter $\mathfrak{R}$ will be used as a standard notation for such a region. The set $C_1 + \cdots + C_n$ is termed the *boundary* B of $\mathfrak{R}$, and $\mathfrak{R} - B$ is termed the *interior* of $\mathfrak{R}$ and will be denoted by $\mathfrak{R}^0$. If $n = 1$, then $\mathfrak{R}$ is said to be *simply connected.* A simply connected Jordan region is a 2-cell (see I.2.31). If each boundary curve of a Jordan region $\mathfrak{R}$ is a simple closed polygon, then $\mathfrak{R}$ is termed a *polygonal region.*

I.2.48. Continuation. Let $\mathfrak{D}$ be a domain in the uv-plane. Then there exists a sequence of finitely connected polygonal regions $\mathfrak{R}_n$ that fill up $\mathfrak{D}$ in the following sense. (i) $\mathfrak{R}_n \subset \mathfrak{R}_{n+1}^0$, $n = 1, 2, \cdots$. (ii) $\sum \mathfrak{R}_n = \mathfrak{D}$.

I.2.49. Let $\mathfrak{D}$ be a domain in the uv-plane, and let $T : x = x(u, v)$, $y = y(u, v)$, $(u, v) \in \mathfrak{D}$, be a single-valued continuous mapping from $\mathfrak{D}$ into the xy-plane (that is, $T(\mathfrak{D})$ lies in the xy-plane, cf. I.2.5). The following statements hold.

(1) If T is biunique in $\mathfrak{D}$, then T is a homeomorphism.

(2) If T is a homeomorphism, then $T(\mathfrak{D})$ is again a domain.

(3) Similar statements hold if the term domain is replaced by bounded, finitely connected Jordan region.

(4) Let $T(\mathfrak{D}) = \mathfrak{D}^*$ be a homeomorphism, where $\mathfrak{D}$, $\mathfrak{D}^*$ are domains in the uv- and xy-planes respectively. Let $C \subset \mathfrak{D}$ be a simple closed curve oriented in the counterclockwise sense. Then $T(C)$ is a simple closed curve in $\mathfrak{D}^*$, and the orientation assigned on C induces on $T(C)$, by means of T, an orientation in an obvious manner. This induced orientation is either clockwise for every choice of $C \subset \mathfrak{D}$, or else it is counterclockwise for every choice of $C \subset \mathfrak{D}$. In the first case, T is termed *sense-reversing,* in the second case it is termed *sense-preserving.*

I.2.50. Let $\mathfrak{R}$ be a bounded, finitely connected Jordan region in the uv-plane, and let n be the number of its boundary curves. Then there exists a homeomorphism $T : x = x(u, v),\ y = y(u, v),\ (u, v) \in \mathfrak{R}$, such that $T(\mathfrak{R})$ coincides with an arbitrarily assigned bounded, finitely connected Jordan region with n boundary curves in the xy-plane.

I.2.51. Let $\mathfrak{D}$ be a domain in the uv-plane. Then $\mathfrak{D}$ is termed *simply connected* if and only if the following condition holds: if C is any simple closed curve in $\mathfrak{D}$, then the interior of C is a subset of $\mathfrak{D}$. The following statements hold.

(1) $\mathfrak{D}$ is simply connected if and only if it is homeomorphic with the open unit disc $u^2 + v^2 < 1$.

(2) If $\mathfrak{D}$ is simply connected, then the regions $\mathfrak{R}_n$ occurring in I.2.48 can be chosen as simply connected Jordan regions.

I.2.52. Let $\tau(C) = C^*$ be a homeomorphism, where C, C^* are simple closed curves in the uv- and xy-planes respectively. Let $\mathfrak{R}$, $\mathfrak{R}^*$ be the (bounded) Jordan regions bounded by C, C^* respectively. Then there exists a homeomorphism $T(\mathfrak{R}) = \mathfrak{R}^*$ such that T agrees with τ on C (that is, $\tau(p) = T(p)$ for every point p on C). The following facts follow readily from the preceding statement.

(1) Let $\mathfrak{R}$ be a bounded, simply-connected Jordan region in the uv-plane, bounded by a simple closed curve C. Let p, q be two distinct points on C, and let γ be a simple arc with end points p, q that lies in $\mathfrak{R}^0$ except for its end points. Let $\mathfrak{R}^*$, C^*, p^*, q^*, γ^* be similarly related in the xy-plane. Given then a homeomorphism $\tau(C) = C^*$ such that $\tau(p) = p^*$, $\tau(q) = q^*$, there exists a homeomorphism $T(\mathfrak{R}) = \mathfrak{R}^*$ that agrees on C with τ and maps γ onto γ^*.

(2) Let S be a topological 2-sphere, and let C be a simple closed curve on S. Then C is the common boundary of two 2-cells A, B on S. Let S^* be a second 2-sphere, and let C^*, A^*, B^* have analogous meanings relative to S^*. Finally, let $\tau(A) = A^*$ be a homeomorphism. Then there exists a homeomorphism $T(S) = S^*$ that agrees with τ on A.

I.2.53. Let $\mathfrak{R}$ be a simply-connected, bounded Jordan region in the uv-plane. A *curvilinear triangulation* $\mathfrak{I}$ of $\mathfrak{R}$ consists of a finite system of curvilinear triangles $t_1, \cdots, t_n$, subject to the following conditions. (i) $\mathfrak{R} = t_1 + \cdots + t_n$. (ii) If $i \neq j$, then $t_i t_j$ is either empty, or is a common vertex of t_i and t_j, or else it is a common side of t_i and t_j. Then there exists a homeomorphism $T(\mathfrak{R}) = \mathfrak{R}^*$ such that the following conditions hold. (a) $\mathfrak{R}^*$ is a bounded, convex polygonal region in the xy-plane. (b) $T(t_j)$ is a *rectilinear* triangle, $j = 1, \cdots, n$ (see Huskey [3]).

I.2.54. Let $R : u_1 \leqq u \leqq u_2$, $v_1 \leqq v \leqq v_2$, $R^* : x_1 \leqq x \leqq x_2$, $y_1 \leqq y \leqq y_2$, be rectangles in the uv- and xy-planes respectively, and let $\tau(R) = R^*$ be a homeomorphism. Given $\epsilon > 0$, there exists then a homeomorphism $T_\epsilon(R) = R^*$ with the following properties. (i) $\rho(\tau, T_\epsilon) < \epsilon$ (cf. I.2.44). (ii) T_ϵ is *quasi-linear* in the following sense: there exists a rectilinear triangulation $\mathfrak{T}$ of R such that T_ϵ is an affine transformation on each one of the triangles of $\mathfrak{T}$ (see Franklin-Wiener [1]; the discussion given there reveals that R and R^* may be replaced by convex polygonal regions).

CHAPTER I.3. BACKGROUND IN ANALYSIS

I.3.1. In this chapter, certain concepts and theorems in Analysis will be briefly reviewed. Proofs will be omitted. Saks [6] may be used as general reference. We shall first review certain aspects of so-called Lebesgue theory in Euclidean n-space. For definiteness, we take $n = 2$. General familiarity with Lebesgue theory is presupposed on the part of the reader (McShane [7] gives an excellent introduction to this theory).

I.3.2. In the Euclidean xy-plane, let $\mathfrak{D}$ be a domain. Let R be a generic notation for an *oriented rectangle*, that is, a rectangle given in the form R : $a \leqq x \leqq b$, $c \leqq y \leqq d$. The interior of R will be denoted by R^0; thus R^0 : $a < x < b$, $c < y < d$. We have then the statement: there exists a sequence of rectangles R_n, such that $\mathfrak{D} = \sum R_n$, $n = 1, 2, \cdots$, and $R_j^0 R_k^0 = 0$ for $j \neq k$.

I.3.3. In the Euclidean xy-plane, let E be a bounded measurable set, and let F be any family of oriented squares s (that is, squares with sides parallel to the coordinate axes) that cover E in the following sense: given any point $(x_0, y_0) \in E$ and any $\delta > 0$, there exists a square $s \in F$ such that $(x_0, y_0) \in s$ and $| s | < \delta$ (generally, if G is a measurable set, then $| G |$ denotes its measure). Then the family F contains a (finite or infinite) sequence of squares $s_1, \cdots, s_n, \cdots$, such that $s_j s_k = 0$ for $j \neq k$ and $| E - \sum s_n | = 0$. This is a very special form of the *Vitali covering theorem.* Under the conditions just stated, the squares of the family F are said to cover E in the Vitali sense.

I.3.4. Given a bounded set E in the xy-plane, let Q be an oriented square (sides parallel to the coordinate axes) that contains E. For each positive integer n, let us subdivide Q into 4^n congruent oriented squares. Let N denote the family of all those oriented squares that occur in these subdivisions. Then N is termed a *network of oriented squares.* Clearly, the squares of N cover E in the Vitali sense. Due to the special character of N, we have the following statement: let F be any subfamily of N, such that the squares of F cover E. Then F contains a (finite or infinite) sequence of squares $s_1, \cdots, s_n, \cdots$ such that $s_j^0 s_k^0 = 0$ for $j \neq k$ and $E \subset \sum s_n$ (compare this statement carefully with that in I.3.3).

I.3.5. Given a set E in the xy-plane, the *characteristic function* $c(x, y)$ of E is defined as follows: $c(x, y) = 0$ if $(x, y) \notin E$, and $c(x, y) = 1$ if $(x, y) \in E$. Then E is measurable if and only if $c(x, y)$ is measurable. If E is measurable and $f(x, y)$ is a (real-valued) measurable function in an oriented square Q containing E, then

$$(1) \qquad \iint_E f(x, y)\, dx\, dy = \iint_Q f(x, y) c(x, y)\, dx\, dy,$$

provided that $f(x, y)c(x, y)$ is summable on Q. If this is the case, then $f(x, y)$ is summable on E. In particular, we have

$$| E | = \iint_Q c(x, y)\, dx\, dy.$$

In the applications, the following situation arises frequently. A function $g(x, y)$ is defined and measurable on a measurable subset E of an oriented square Q. We extend then the definition of $g(x, y)$ by setting $g(x, y) = 0$ on $Q - E$. Then the extended function $g(x, y)$ is measurable on Q, and the integral of $g(x, y)$ on E, if it exists, satisfies the formula (1). In the special case when $| Q - E | = 0$, it follows that

$$\iint_E g(x, y)\, dx\, dy = \iint_Q g(x, y)\, dx\, dy.$$

Thus we can speak of the integral of $g(x, y)$ on Q, even though $g(x, y)$ is defined only on E. Similar remarks apply if Q is replaced by a bounded measurable set.

I.3.6. Let E be a bounded set in the xy-plane. In many cases a statement involving the points of E will be found to hold for all points of E with the exception of a subset of measure zero. The statement will then be said to hold a.e. (*almost everywhere*) on E. Caution must be exercised in deciding whether or not a set is of measure zero, the following two remarks being especially relevant for our purposes.

(1) Let C be a simple arc in the xy-plane (see I.2.31). According to an example found by Osgood [1], the (two-dimensional) measure of C may be positive. The same remark applies to simple closed curves.

(2) Let E be a subset of a rectangle $R : a \leqq x \leqq b,\ c \leqq y \leqq d$, with the following property: for a.e. y_0 in the interval $c \leqq y \leqq d$, the intersection of E with the segment $a \leqq x \leqq b,\ y = y_0$ is of linear measure zero. If E is known to be measurable, then this property is sufficient (and also necessary) for E to be of measure zero. However, as noted by Sierpinski, there exist nonmeasurable sets E that possess the above property and have positive exterior measure.

I.3.7. Let E be a bounded measurable set in the xy-plane. Then there exists a Borel set E^* such that $E \subset E^*$ and $| E | = | E^* |$. More precisely, E^* may be chosen as a G_δ. Similarly, E contains a set F^* which is an F_σ and for which $| E - F^* | = 0$ (see I.2.46). Furthermore, for every $\epsilon > 0$ there exists an open set G such that $E \subset G$ and $| G | < | E | + \epsilon$. Furthermore, if $| E | > 0$, then E contains some nonmeasurable set.

If E is an unbounded set in the xy-plane, then E is said to be of measure zero if and only if $| KE | = 0$ for every choice of a circular disc K in the xy-plane.

I.3.8. Let $f(x, y)$ be a measurable function on a Borel measurable set E. Then there exist two Borel measurable functions $f_1(x, y)$, $f_2(x, y)$, such that $f_1(x, y) \leqq f(x, y) \leqq f_2(x, y)$ on E and the sign of equality holds a.e. on E (a function $g(x, y)$ is Borel measurable on E if for every real number λ the subset of E where $g(x, y) \geqq \lambda$ is a Borel set).

I.3.9. Borel measurable functions $g(x, y)$ (see I.3.8) will play an important role in the sequel. A sample theorem (stated for functions of a single variable) is as follows. Let $x(t)$ be a (real-valued) continuous function on an interval $I : a \leqq t \leqq b$, and let $N(x)$ denote, for given x, the number of distinct solutions of the equation $x = x(t)$ in I. Then $N(x)$ is a Borel measurable function of x. For many similar theorems, see Hahn [2].

I.3.10. The following statements are useful in testing a function for summability.

(1) Let $\mathfrak{D}$ be a bounded domain in the xy-plane, and let $R_1, \cdots, R_n, \cdots$ be a sequence of oriented rectangles such that $\sum R_n = \mathfrak{D}$, $R_i^0 R_j^0 = 0$ for $i \neq j$. If a measurable function $f(x, y)$, defined in $\mathfrak{D}$, is summable on each R_n and if the series

$$\iint_{R_1} | f | \, dx \, dy + \cdots + \iint_{R_n} | f | \, dx \, dy + \cdots$$

is convergent, then f is summable on $\mathfrak{D}$ and

$$\iint_{\mathfrak{D}} f \, dx \, dy = \sum_n \iint_{R_n} f \, dx \, dy$$

(2) On a bounded measurable set E in the xy-plane, let $f_n(x, y)$, $n = 0, 1, 2, \cdots$, be a sequence of non-negative measurable functions with the following properties: $f_0 \leqq \liminf f_n$ a.e. on E, each f_n is summable on E for $n > 0$, and

$$\liminf \iint_E f_n \, dx \, dy < +\infty .$$

Then f_0 is summable on E and

$$\iint_E f_0 \, dx \, dy \leqq \liminf \iint_E f_n \, dx \, dy.$$

This is the theorem (or lemma) of Fatou.

(3) A very special but very useful form of the lemma of Fatou is as follows. Let $f_n(x, y)$, $n = 0, 1, 2, \cdots$, be a sequence of non-negative, measurable functions on the bounded measurable set E, with the following properties: $f_n \to f_0$ a.e. on E, each f_n is summable on E for $n > 0$, and there exists a constant M such that

$$\iint_E f_n \, dx \, dy \leqq M \qquad \text{for } n > 0.$$

Then f_0 is summable on E, and

$$\iint_E f_0 \, dx \, dy \leqq M .$$

(4) Let $f(x, y)$ be a non-negative summable function on an oriented rectangle R, and let $D(R)$ be a subdivision of R. Suppose that there exists a number λ, $0 < \lambda < \infty$, such that

$$\lambda \geqq \sum \frac{\left[\iint_r f\,dx\,dy\right]^2}{|r|}, \qquad r \in D(R),$$

for every subdivision $D(R)$. Then f^2 is summable on R, and

$$\iint_R f^2\,dx\,dy \leqq \lambda.$$

(5) Let $f(x, y)$ be a summable function on an oriented rectangle $R : a \leqq x \leqq b$, $c \leqq y \leqq d$. Then for a.e. y in the interval $c \leqq y \leqq d$, $f(x, y)$ is a summable function, as a function of x, on the interval $a \leqq x \leqq b$, and the integral

$$\int_a^b f(x, y)\,dx$$

is a summable function of y in the interval $c \leqq y \leqq d$. A similar statement holds with x and y exchanged, and we have the formulas

$$(1) \qquad \int_c^d \left[\int_a^b f(x, y)\,dx\right] dy = \int_a^b \left[\int_c^d f(x, y)\,dy\right] dx = \iint_R f(x, y)\,dx\,dy.$$

However, if one only knows that the iterated integrals in (1) exist, then it does not generally follow that $f(x, y)$ is summable on R. On the other hand, if $f(x, y)$ is non-negative and if one of the iterated integrals in (1) exists, then $f(x, y)$ is summable and (1) holds. Similar theorems hold for multiple integrals in n-space (theorem of Fubini-Tonelli). In all these statements, it is assumed that the functions involved are measurable.

(6) Let $f(x, y)$ be a measurable function on a bounded measurable set in the xy-plane, and let p be a positive number. Then $f(x, y)$ is said to belong to the Lebesgue class L^p on E if and only if $|f|^p$ is summable on E. The Lebesgue classes L^p, L^q are termed associated Lebesgue classes if $(1/p) + (1/q) = 1$, $p > 0$, $q > 0$. We have then the following statements. If f and g belong to associated Lebesgue classes L^p, L^q on E, then fg is summable on E, and (Hölder inequality)

$$\left|\iint_E fg\,dx\,dy\right| \leqq \left[\iint_E |f|^p\,dx\,dy\right]^{1/p} \left[\iint_E |g|^q\,dx\,dy\right]^{1/q}.$$

Let now $f_1(x, y), \cdots, f_n(x, y)$ be non-negative functions of the Lebesgue class L^p

on E, where $p > 1$. Then $f_1(x, y) + \cdots + f_n(x, y)$ is also of the Lebesgue class L^p, and (Minkowski inequality)

$$\left[\iint_E (f_1 + \cdots + f_n)^p \, dx\, dy\right]^{1/p} \leqq \left[\iint_E f_1^p \, dx\, dy\right]^{1/p} + \cdots + \left[\iint_E f_n^p \, dx\, dy\right]^{1/p}.$$

For $0 < p < 1$, the same relation holds except that the sign of inequality is reversed. For $p = 1/2$ there follows the useful inequality

$$\left\{\left[\iint_E g_1 \, dx\, dy\right]^2 + \cdots + \left[\iint_E g_n \, dx\, dy\right]^2\right\}^{1/2} \leqq \iint_E (g_1^2 + \cdots + g_n^2)^{1/2} \, dx\, dy,$$

where $g_1(x, y), \cdots, g_n(x, y)$ are summable functions (of arbitrary sign) on E. Many similar inequalities are discussed in Hardy-Littlewood-Pólya [1]. Analogous inequalities hold if the integrals are replaced by finite sums. Furthermore, the Lebesgue integral may be replaced by integrals relative to a general non-negative measure.

(7) If $f(x, y)$ is non-negative and summable on E, then $f < +\infty$ a.e. on E.

I.3.11. Let E be a bounded measurable set in the xy-plane, and let $f_n(x, y)$, $n = 0, 1, 2, \cdots$, be a sequence of measurable functions on E such that $f_n \to f_0$ a.e. on E. The following statements are concerned with conditions for the term-wise integrability of such a sequence.

(1) A fundamental concept in this connection is due to Vitali. Let G be any family (not necessarily countable) of summable functions $g(x, y)$ on E. Then the family G is said to possess the property (V) (the Vitali property) on the set E if and only if for every $\epsilon > 0$ there exists a $\delta = \delta(\epsilon) > 0$ such that

$$\iint_e |g| \, dx\, dy < \epsilon \qquad \text{if} \qquad |e| < \delta,$$

for every function $g(x, y)$ of the family G (where e is a generic notation for a measurable subset of E).

(2) Let $f_n(x, y)$, $n = 0, 1, 2, \cdots$, be a sequence of summable functions on E such that (i) $f_n \to f_0$ a.e. on E, and (ii) the sequence possesses the property (V) (that is, the family consisting of the functions $f_n(x, y)$ possesses the property (V)). Then, for every measurable set $S \subset E$,

$$\iint_S f_n \, dx\, dy \to \iint_S f_0 \, dx\, dy.$$

(3) Let $f_n(x, y)$, $n = 0, 1, 2, \cdots$, be a sequence of summable functions on E such that $f_n \to f_0$ a.e. on E. Then the relation

$$\iint_E |f_n| \, dx\, dy \to \iint_E |f_0| \, dx\, dy$$

holds if and only if the sequence possesses the property (V).

(4) Let $f_n(x, y)$, $n = 0, 1, 2, \cdots$, be a sequence of summable functions on E such that $f_n \to f_0$ a.e. on E. Then the relations

$$\iint_E |f_n - f_0| \, dx \, dy \to 0, \qquad \iint_E |f_n| \, dx \, dy \to \iint_E |f_0| \, dx \, dy$$

are equivalent (that is, either both hold or else both fail to hold).

(5) Let $f_n(x, y)$, $n = 1, 2, \cdots$, be a sequence of non-negative summable functions on E, such that the series $\sum \iint f_n \, dx \, dy$, where the integrals are taken over E, is convergent. Then the series $\sum f_n(x, y)$ converges a.e. on E, its sum is summable on E, and

$$\iint_E (\sum f_n) \, dx \, dy = \sum \iint_E f_n \, dx \, dy.$$

(6) Let $f_n(x, y)$, $n = 0, 1, 2, \cdots$, be a sequence of non-negative summable functions on E such that $f_0(x, y)$ is summable on E and $f_0 = \sum f_n$, $n > 0$, a.e. on E. Then each f_n is summable on E, and

$$\iint_E f_0 \, dx \, dy = \sum_{n=1}^{\infty} \iint_E f_n \, dx \, dy.$$

(7) Let $f_n(x, y)$, $n = 1, 2, \cdots$, be a sequence of summable functions on E, and let $F(x, y)$ be a summable function on E, such that $f_1 \leqq f_2 \leqq \cdots$ and $f_n \to F$ a.e. on E. Then $\iint f_n \, dx \, dy \to \iint F \, dx \, dy$, where the integrals are extended over E.

(8) The preceding statement is a very special case of the following classical theorem of Lebesgue. Let $f_n(x, y)$, $n = 0, 1, 2, \cdots$, be summable functions on E such that $f_n \to f_0$ a.e. on E. Suppose there exist two summable functions $\phi(x, y)$, $\Phi(x, y)$ on E such that $\phi \leqq f_n \leqq \Phi$ a.e. on E, $n = 1, 2, \cdots$. Then

$$\iint_E f_n \, dx \, dy \to \iint_E f_0 \, dx \, dy.$$

I.3.12. Let $f(x, y)$ be a non-negative, summable function on a bounded measurable set E. If $\iint f(x, y) \, dx \, dy = 0$, where the integration is extended over E, then clearly $f = 0$ a.e. on E. Now let $f_n(x, y)$, $n = 1, 2, \cdots$, be a sequence of non-negative summable functions on E such that $\iint f_n \, dx \, dy \to 0$, where the integrals are taken over E. Then it does *not* follow that $f_n \to 0$ a.e. on E. However, it follows that there exists an infinite sequence of positive integers $n_1 < n_2 < \cdots < n_k < \cdots$, such that $f_{n_k} \to 0$ a.e. on E for $k \to \infty$. A second useful inference may be stated in terms of the concept of *convergence in measure.* A sequence f_n of measurable functions on E is said to converge in measure to zero on E if and only if on denoting by $E_n(\epsilon)$ the subset of E where $|f_n|$ is greater than or equal to a given $\epsilon > 0$, it is true that $|E_n(\epsilon)| \to 0$ for $n \to \infty$, for every choice of $\epsilon > 0$. Then, if a sequence f_n satisfies on E the conditions

$$f_n \geqq 0 \text{ a.e. on } E, \qquad \iint_E f_n \, dx \, dy \to 0,$$

it follows that f_n converges in measure to zero on E.

I.3.13. Let $f(x, y)$ be summable in a bounded domain $\mathfrak{D}$, and let E be a generic notation for a measurable subset of $\mathfrak{D}$. Then f is also summable on every $E \subset \mathfrak{D}$, and for every $\epsilon > 0$ there exists a $\delta = \delta(\epsilon) > 0$ such that

$$\iint_E |f| \, dx \, dy < \epsilon \qquad \text{if} \qquad |E| < \delta.$$

That is, the indefinite Lebesgue integral of f is an absolutely continuous set function. Let now (x_0, y_0) be an interior point of $\mathfrak{D}$, and let s_n be any sequence of squares in $\mathfrak{D}$ such that $(x_0, y_0) \in s_n$ and $|s_n| \to 0$. Then there exists a subset e, of measure zero, of $\mathfrak{D}$ such that

$$\frac{1}{|s_n|} \iint_{s_n} f \, dx \, dy \to f(x_0, y_0) \qquad \text{if} \qquad (x_0, y_0) \notin e.$$

For functions of a single variable x, the corresponding theorem implies the following result. Let $g(x)$ be a summable function in an interval $I : a \leqq x \leqq b$. Then

$$\frac{d}{dx} \int_a^x g(\xi) \, d\xi = g(x) \text{ a.e. in } I.$$

I.3.14. Let $f(x, y)$ be continuous in a domain $\mathfrak{D}$, and let (x_0, y_0) be a point of $\mathfrak{D}$. Then $f(x, y)$ is termed *totally differentiable* at the point (x_0, y_0) if and only if there exist two numbers a and b such that on setting

$$\phi(x, y, x_0, y_0) = \frac{|f(x, y) - f(x_0, y_0) - a(x - x_0) - b(y - y_0)|}{[(x - x_0)^2 + (y - y_0)^2]^{1/2}}, \tag{1}$$

$$(x, y) \neq (x_0, y_0),$$

the relation $\phi(x, y, x_0, y_0) \to 0$ holds for $(x, y) \to (x_0, y_0)$. Clearly, if this condition holds, then the partial derivatives f_x, f_y exist at (x_0, y_0), and $a = f_x(x_0, y_0)$, $b = f_y(x_0, y_0)$. However, if $f_x(x_0, y_0)$, $f_y(x_0, y_0)$ exist, it does *not* follow generally that $f(x, y)$ is totally differentiable at (x_0, y_0). The following facts are relevant for our purposes.

(1) If $f(x, y)$ is continuous in $\mathfrak{D}$ and if the partial derivatives f_x, f_y exist a.e. in $\mathfrak{D}$, then f_x, f_y are measurable in $\mathfrak{D}$.

(2) Suppose that $f(x, y)$ is *Lipschitzian* in $\mathfrak{D}$. That is, suppose that there exists a finite constant M such that the following condition holds: if (x_1, y_1), (x_2, y_2) are any two points in $\mathfrak{D}$ such that the straight segment that joins them is contained in $\mathfrak{D}$, then

$$| f(x_2, y_2) - f(x_1, y_1) | \leqq M[(x_2 - x_1)^2 + (y_2 - y_1)^2]^{1/2}.$$

If this condition holds, then f_x, f_y exist a.e. in $\mathfrak{D}$ and $f(x, y)$ is totally differentiable almost everywhere in $\mathfrak{D}$ (see Rademacher [2]).

(3) The preceding fundamental result of Rademacher was the starting point of far-reaching studies. We mention only two special theorems. Let $f(x, y)$ be continuous in $\mathfrak{D}$. Let us define, for (x, y) in $\mathfrak{D}$,

$$L(x, y) = \limsup \frac{| f(x + h, y + k) - f(x, y) |}{(h^2 + k^2)^{1/2}}, \quad 0 < (h^2 + k^2)^{1/2} \to 0.$$

As a special case of a theorem of Stepanoff (see, also for further related results, Burkill and Haslam-Jones [3, 4]), the following statement holds: $f(x, y)$ is totally differentiable a.e. in $\mathfrak{D}$ if and only if $L(x, y) < \infty$ a.e. in $\mathfrak{D}$. The methods used in proving results of this type yield readily a number of miscellaneous statements. One such statement, useful in the sequel, is the following.

(4) Let $f_j(x, y)$, $j = 1, \cdots, N$, be continuous in $\mathfrak{D}$, and let (x_0, y_0) be a point of $\mathfrak{D}$ where f_{jx}, f_{jy} both exist, $j = 1, \cdots, N$. Let us say that the system of functions $f_j(x, y)$ satisfies the condition (C) at (x_0, y_0) if there exists a sequence of squares s_n, with sides parallel to the coordinate axes, with the following properties. (i) $s_n \subset \mathfrak{D}$, $n = 1, 2, \cdots$. (ii) $(x_0, y_0) \in s_n^0$, $n = 1, 2, \cdots$ (iii) $| s_n | \to 0$. (iv) On denoting by m_{jn} the maximum, on the perimeter of s_n, of the function

$$\frac{| f_j(x, y) - f_j(x_0, y_0) - f_{jx}(x_0, y_0)(x - x_0) - f_{jy}(x_0, y_0)(y - y_0) |}{[(x - x_0)^2 + (y - y_0)^2]^{1/2}},$$

the relation $m_{jn} \to 0$ holds for $n \to \infty$, $j = 1, \cdots, N$. We have then the theorem: if f_{jx}, f_{jy}, $j = 1, \cdots, N$, exist a.e. in $\mathfrak{D}$, then the system $f_j(x, y)$ satisfies the condition (C) a.e. in $\mathfrak{D}$ (for an explicit proof, see T. Radó [12]).

I.3.15. The following statements are concerned with the Riemann-Stieltjes integral of functions of a single real variable. Hobson [1] may be used as a general reference.

(1) Let $f(x)$, $g(x)$ be continuous in the interval $I : a \leqq x \leqq b$. If one at least of these two functions is of bounded variation in I, then both of the integrals

$$\int_a^b f\, dg, \qquad \int_a^b g\, df$$

exist and are related by the formula

$$\int_a^b f\, dg + \int_a^b g\, df = f(b)g(b) - f(a)g(a).$$

(2) If $f(x)$ is continuous and $g(x)$ is absolutely continuous in $a \leqq x \leqq b$, then we have the formula

$$\int_a^b f\,dg = \int_a^b f(x)g'(x)\,dx,$$

where the integral on the right is taken in the Lebesgue sense.

I.3.16. The concepts of Lebesgue measure and of Lebesgue integral, originally developed for Euclidean spaces, may be extended to abstract spaces. An excellent presentation may be found in Saks [6]; we restrict ourselves to a few especially relevant topics, beginning with completely additive functions of Borel sets.

Let S be a metric space (see I.2.10), and let $\Phi(B)$ denote a real-valued, finite-valued function that is defined for every Borel set $B \subset S$. For the empty set 0, we agree that $\Phi(0) = 0$. Then $\Phi(B)$ is termed a *completely additive function of Borel sets* in S (cf. I.2.46) if and only if the following condition holds: if $B_1, \cdots, B_n, \cdots$ is any sequence of disjoint Borel sets in S, then $\Phi(\sum B_n) = \sum \Phi(B_n)$.

Let $\Phi(B)$, $\mu(B)$ be two completely additive functions of Borel sets in S, and suppose that $\mu(B) \geqq 0$ for every Borel set $B \subset S$ (briefly, μ is non-negative). Then $\Phi(B)$ is termed AC relative to μ (*absolutely continuous relative to* μ) if and only if $\Phi(B) = 0$ for every Borel set $B \subset S$ such that $\mu(B) = 0$, and $\Phi(B)$ is termed *singular with respect to* μ if and only if there exists a Borel set $B_0 \subset S$ such that $\mu(B_0) = 0$ and $\Phi(B) = \Phi(BB_0)$ for every Borel set $B \subset S$. In the special case when S is a bounded Borel set in a Euclidean space, it will be understood that μ is chosen as the Lebesgue measure, unless a different μ is explicitly specified. We have then the following statements.

(1) If $\Phi(B)$ is a completely additive function of Borel sets in S, then there exists a pair of completely additive non-negative functions $\Phi_1(B)$, $\Phi_2(B)$ of Borel sets in S, such that $\Phi(B) = \Phi_1(B) - \Phi_2(B)$ for every Borel set $B \subset S$ (*Jordan decomposition*).

(2) A completely additive function $\Phi(B)$ of Borel sets in S is AC relative to μ if and only if for every $\epsilon > 0$ there exists a $\delta = \delta(\epsilon) > 0$ such that $| \Phi(B) | < \epsilon$ whenever the Borel set B satisfies the inequality $\mu(B) < \delta$.

(3) Let $\mu(B)$ be given as above (that is, μ is completely additive and non-negative). Given then a Borel set $B \subset S$ and an $\epsilon > 0$, there exists a closed set $F \subset B$ and an open set $G \supset B$ such that $\mu(B) < \mu(F) + \epsilon$ and $\mu(G) < \mu(B) + \epsilon$.

(4) Given $\mu(B)$ as under (3), let $\Phi(B)$ be a completely additive, non-negative function of Borel sets in S. Then there exists a Borel set B_0 such that $\mu(B_0) = 0$ and $\Phi(B_0) \geqq \Phi(B)$ whenever the Borel set B satisfies the condition $\mu(B) = 0$.

(5) Given $\mu(B)$ as under (3), let $\Phi(B)$ be a completely additive function of Borel sets in S. Then $\Phi(B)$ possesses a representation of the form $\Phi(B) = \Phi_a(B) + \Phi_s(B)$, where $\Phi_a(B)$, $\Phi_s(B)$ are completely additive functions of Borel sets in S, and Φ_a is AC and Φ_s is singular with respect to μ (*Lebesgue decomposition*). Furthermore, this representation is unique, and Φ is AC if and only if $\Phi_s(B) = 0$ for every Borel set $B \subset S$.

I.3.17. Let $\mu(B)$ be a completely additive, non-negative function of Borel sets in a metric space S. A real-valued, finite-valued function $f(p)$ of the variable point $p \in S$ is termed *Borel measurable* if and only if for every real number a the set where $f \geqq a$ is a Borel set. Using μ as a measure, a μ-integral may be defined, in complete analogy with the Lebesgue integral in Euclidean space, for Borel measurable functions $f(p)$. This integral will be denoted by $\int_B f\, d\mu$, where B is a Borel set in S. A Borel measurable function $f(p)$ will be termed μ-summable over the Borel set $B \subset S$ if its μ-integral over B exists and is finite, in contradistinction with Saks [6], where infinite μ-integrals are also admitted under certain conditions. The restriction to Borel measurable functions is merely a matter of convenience, and is adequate for our purposes. The μ-integral gives rise to a theory entirely analogous to the theory of the Lebesgue integral in Euclidean space. We note only a few additional facts.

(1) Suppose that $f(p)$ is Borel measurable and μ-summable on S. Given $\epsilon > 0$, there exists then a finite system of disjoint Borel sets $B_1, \cdots, B_n$ and constants $c_1, \cdots, c_n$ such that $B_1 + \cdots + B_n = S$ and

$$\left| \int_S f\, d\mu - [c_1\mu(B_1) + \cdots + c_n\mu(B_n)] \right| < \epsilon.$$

(2) Given $f(p)$ as under (1), $f(p)$ is also μ-summable on every Borel set $B \subset S$, and on setting

$$\Phi(B) = \int_B f\, d\mu,$$

$\Phi(B)$ is completely additive on Borel sets and AC relative to μ.

(3) Let $\Phi(B)$ be a completely additive function of Borel sets in S, such that $\Phi(B)$ is AC relative to μ. Then there exists a function $g(p)$ which is Borel measurable and μ-summable on S, such that, for every Borel set $B \subset S$,

$$\Phi(B) = \int_B g\, d\mu.$$

If Φ is non-negative, then g may be chosen as non-negative.

(4) Given two completely additive, non-negative functions $\mu(B)$, $\mu^*(B)$ of Borel sets in S, suppose that μ^* is AC relative to μ. As noted under (3), there exists then a Borel measurable, μ-summable function $g(p)$ such that

$$\mu^*(B) = \int_B g\, d\mu$$

for every Borel set $B \subset S$. We have the theorem: if $f(p)$ is Borel measurable on S, then

$$\int_B f\, d\mu^* = \int_B fg\, d\mu,$$

for every Borel set $B \subset S$, as soon as one of the two integrals involved exists.

(5) Given a completely additive, non-negative function $\mu(B)$ of Borel sets in S, let $f_n(p)$, $n = 0, 1, 2, \cdots$, be a sequence of Borel measurable, finite-valued functions on a Borel set B_0, such that $f_n \to f_0$ on B_0. Then for every $\epsilon > 0$ there exists a Borel set $Q \subset B_0$ such that $\mu(B_0 - Q) < \epsilon$ and the convergence of f_n to f_0 is uniform on Q (*Egoroff's theorem*).

I.3.18. Let S be a metric space, and let $\Gamma(E)$ be a finite-valued, non-negative function defined for every subset E of S (not only for Borel sets). Then $\Gamma(E)$ will be called a *Carathéodory outer measure* if and only if the following conditions hold. (C_1) $\Gamma(0) = 0$, and $\Gamma(E_1) \leqq \Gamma(E_2)$ whenever $E_1 \subset E_2$. (C_2) $\Gamma(\sum E_n) \leqq \sum \Gamma(E_n)$ for every sequence $E_1, \cdots, E_n, \cdots$ of subsets of S. (C_3) $\Gamma(E_1 + E_2) = \Gamma(E_1) + \Gamma(E_2)$ whenever the distance $\rho(E_1, E_2)$ of the sets E_1, E_2 is positive (see I.2.11). As far as our needs are concerned, the principal theorem is as follows: if $\Gamma(E)$ is a Carathéodory outer measure in S, then Γ is completely additive on Borel sets in S. In other words, if we consider Γ only on Borel sets and if we write $\Gamma(B)$ to express this point of view, then $\Gamma(B)$ is a completely additive, non-negative function of Borel sets in S.

I.3.19. While we shall be concerned all the time with problems stated for the real domain, we shall use complex numbers in an auxiliary role on many occasions. In dealing with situations in a Euclidean uv-plane, it will be a device of great convenience to introduce the complex variable $w = u + iv$, and in a similar manner the complex variable $z = x + iy$ if the Cartesian coordinates are denoted by x, y. Furthermore, we shall have occasion to use, in studying transformations in the plane, certain simple facts from the theory of functions of a complex variable (cf. II.4.21). To avoid unwelcome complications, we shall also use the following fundamental theorem on conformal mapping. Let $\mathfrak{R}$ be a (bounded) simply-connected Jordan region (cf. I.2.47) in the $w = u + iv$ plane, and let $\mathfrak{R}_*$ be a region of the same type in the $z = x + iy$ plane. Then there exists a function $w = f(z)$, $z \in \mathfrak{R}_*$, with the following properties. (i) $f(z)$ is continuous in $\mathfrak{R}_*$ and analytic in $\mathfrak{R}_*^0$. (ii) The formula $w = f(z)$, $z \in \mathfrak{R}_*$, defines a homeomorphism from $\mathfrak{R}_*$ onto $\mathfrak{R}$. A beautiful presentation of this subject may be found in Carathéodory [1]. A more involved theorem, concerned with the conformal mapping of polyhedra, will be used in V.2.33. For polyhedra free of self-intersections this theorem is one of the classical results of Schwarz, and a proof may be found in Carathéodory [1]. For polyhedra with self-intersections, the case needed in V.2.33, the result follows either by appropriate modifications of the method used in the classical case, or else by an appeal to the general concept of an abstract Riemann surface (see Weyl [1]). A further proof, perhaps more direct, may be obtained by induction upon the number n of the faces of the polyhedron. Indeed, for $n = 1$ the theorem needed reduces to the theorem on the conformal mapping of simply-connected Jordan regions, stated above. The two *welding processes*, discussed in T. Radó [1], lend themselves readily to carry out the passage from n to $n + 1$.

PART II. CURVES AND SURFACES

CHAPTER II.1. CONTINUOUS TRANSFORMATIONS OF PEANO SPACES.

II.1.1. We shall be concerned with single-valued continuous transformations of the form $T(\mathcal{P}) = \mathcal{P}^*$, where $\mathcal{P}$ and $\mathcal{P}^*$ are Peano spaces (see I.2.33). T is *not* assumed to be biunique; but if it happens to be biunique, then it is topological, on account of the compactness of the spaces involved (see I.2.29). Many of the theorems discussed below remain valid for more general spaces, but for our purposes only Peano spaces are relevant.

If x^* is any point of $\mathcal{P}^*$, then $T^{-1}(x^*)$ is clearly a closed and hence compact set. We shall use the symbol $I(T)$ to denote the class of the sets of the form $T^{-1}(x^*)$, where $x^* \in \mathcal{P}^*$. Thus $E \in I(T)$ means that (1) $E \subset \mathcal{P}$, and (2) there exists a point $x^* \in \mathcal{P}^*$ such that $E = T^{-1}(x^*)$. Since T maps $\mathcal{P}$ onto $\mathcal{P}^*$ (cf. I.2.5), $T^{-1}(x^*) \neq 0$ for every point $x^* \in \mathcal{P}^*$. If every set $E \in I(T)$ is connected (and hence a continuum), then T will be termed *monotone*. If every set $E \in I(T)$ is totally disconnected, then T will be termed *light*. Clearly, if T is both monotone and light, then T is topological, the converse being even more obvious.

II.1.2. Suppose that $T(\mathcal{P}) = \mathcal{P}^*$ is monotone (see II.1.1). If E^* is a connected set in $\mathcal{P}^*$, then $T^{-1}(E^*)$ is connected.

PROOF. In contradiction with the assertion, assume that there exists a separation (cf. I.2.21)

$$(1) \qquad T^{-1}(E^*) = A \mid B.$$

We assert that (cf. I.2.14)

$$(2) \qquad T(A)c[T(B)] = 0, \qquad T(B)c[T(A)] = 0.$$

Let us deny the first one of these relations. Then we should have a point $a_0^* \in T(A)$ and a sequence of points $b_n^* \in T(B)$ such that $b_n^* \to a_0^*$. We have then points a_0, b_n such that

$$(3) \qquad a_0 \in A, \qquad b_n \in B, \qquad T(a_0) = a_0^*, \qquad T(b_n) = b_n^*.$$

Since $\mathcal{P}$ is compact, we can assume without loss of generality that the sequence b_n is convergent, say $b_n \to x_0$ (cf. I.2.36). We have then, since T is continuous, the relations

$$(4) \qquad x_0 \in c(B), \qquad T(x_0) = a_0^* \in T(A) \subset E^*.$$

(1), (3), (4) imply that $x_0 + a_0 \in T^{-1}(a_0^*) \subset A + B$. Since T is monotone and hence $T^{-1}(a_0^*)$ is connected, it follows that $T^{-1}(a_0^*) \subset A$, and consequently $x_0 \in$

A. Hence, in view of (4), $x_0 \in Ac(B)$, in contradiction with (1). Thus the first part of (2) holds, and the second part of (2) is verified in a similar manner. Clearly

$$(5) \qquad T(A) \neq 0, \qquad T(B) \neq 0, \qquad T(A) + T(B) = E^*.$$

From (2), (5) there follows the separation $E^* = T(A) \mid T(B)$, in contradiction with the assumption that E^* is connected.

II.1.3. Suppose that $T(\mathcal{P}) = \mathcal{P}^*$ is monotone (see II.1.1). Then a set $E^* \subset \mathcal{P}^*$ is connected if and only if $T^{-1}(E^*)$ is connected.

PROOF. The necessity follows from II.1.2; the sufficiency follows from I.2.43 since $E^* = TT^{-1}(E^*)$.

II.1.4. Let $T_1(\mathcal{P}_1) = \mathcal{P}_2$, $T_2(\mathcal{P}_2) = \mathcal{P}_3$ be monotone transformations, where $\mathcal{P}_1$, $\mathcal{P}_2$, $\mathcal{P}_3$ are Peano spaces. Then $T = T_2T_1$ is monotone (cf. I.2.8).

PROOF. If x_3 is any point of $\mathcal{P}_3$, then we have $T^{-1}(x_3) = T_1^{-1}T_2^{-1}(x_3)$. Since T_2 is monotone, $T_2^{-1}(x_3)$ is a connected subset of $\mathcal{P}_2$ by definition, and hence $T_1^{-1}T_2^{-1}(x_3)$ is connected by II.1.2.

II.1.5. Let $T(\mathcal{P}) = \mathcal{P}^*$ be monotone (see II.1.1), and let F^* be a subset of $\mathcal{P}^*$. Then the components of $T^{-1}(F^*)$ coincide with the inverse sets of the components of F^* (cf. I.2.23).

PROOF. (i) Let C^* be a component of F^*. Then $T^{-1}(C^*)$ is a connected subset of $T^{-1}(F^*)$ (see II.1.2), and hence $T^{-1}(C^*)$ is a subset of a component Γ of $T^{-1}(F^*)$. Then $C^* \subset T(\Gamma) \subset F^*$. But $T(\Gamma)$ is connected by I.2.43 and C^* is a component of F^*. Hence $C^* = T(\Gamma)$, and thus $\Gamma \subset T^{-1}(C^*)$. Since $T^{-1}(C^*) \subset \Gamma$, it follows that $T^{-1}(C^*) = \Gamma$.

(ii) Let Γ be a component of $T^{-1}(F^*)$. Put $\Gamma^* = T(\Gamma)$. By I.2.43, Γ^* is then a connected subset of F^*, and hence Γ^* is a subset of a component C^* of F^*. Then $\Gamma \subset T^{-1}(C^*) \subset T^{-1}(F^*)$. But $T^{-1}(C^*)$ is a component of $T^{-1}(F^*)$ by (i) above, and hence $\Gamma = T^{-1}(C^*)$.

II.1.6. Let $T(\mathcal{P}) = \mathcal{P}^*$ be monotone (see II.1.1), and let G^* be a continuum in $\mathcal{P}^*$. If $T^{-1}(G^*)$ is unicoherent, then G^* is also unicoherent (cf. I.2.24).

PROOF. Let G_1^*, G_2^* be any two continua such that $G^* = G_1^* + G_2^*$. Then $T^{-1}(G^*) = T^{-1}(G_1^*) + T^{-1}(G_2^*)$. Since $T^{-1}(G^*)$ is unicoherent by assumption, and $T^{-1}(G_1^*)$, $T^{-1}(G_2^*)$ are continua by II.1.2, it follows that $T^{-1}(G_1^*)T^{-1}(G_2^*) = T^{-1}(G_1^*G_2^*)$ is connected. Hence $G_1^*G_2^*$ is connected by I.2.43.

II.1.7. Let $T(\mathcal{P}) = \mathcal{P}^*$ be monotone (see II.1.1), and suppose that $\mathcal{P}$ is unicoherent. Then $\mathcal{P}^*$ is also unicoherent Indeed, $T^{-1}(\mathcal{P}^*) = \mathcal{P}$, and hence the assertion follows directly from II.1.6.

II.1.8. Given $T(\mathcal{P}) = \mathcal{P}^*$ as in II.1.1, T is light if and only if for every $\epsilon > 0$ there exists a $\delta(\epsilon) > 0$, such that the following statement holds: if C is a continuum in $\mathcal{P}$ and $d[T(C)] < \delta(\epsilon)$, then $d(C) < \epsilon$ (cf. I.2.11).

PROOF. *Sufficiency.* Suppose for every $\epsilon > 0$ there exists a $\delta(\epsilon) > 0$ with the property described above. Let x^* be any point of $\mathcal{P}^*$, and let C be a component of $T^{-1}(x^*)$. Then $T(C) = x^*$, and hence $d[T(C)] = 0 < \delta(1/n)$ for every positive integer n. Consequently $d(C) < 1/n$ for every positive integer n. Thus C reduces to a single point. In other words, $T^{-1}(x^*)$ is totally disconnected.

Necessity. Suppose that T is light. Assume that for a certain $\epsilon > 0$ there exists no $\delta(\epsilon) > 0$ with the desired property. Then for every positive integer n there exists a continuum $C_n \subset \mathcal{P}$ such that $d[T(C_n)] < 1/n$ and $d(C_n) \geqq \epsilon$. By I.2.37, there exists an infinite sequence of positive integers $k_1 < k_2 < \cdots < k_n < \cdots$, such that the sequence C_{k_n} is convergent, say $C_{k_n} \to C_0$. By I.2.37, C_0 is a (nonempty) continuum, and $d(C_{k_n}) \to d(C_0)$. Thus $d(C_0) \geqq \epsilon$. By I.2.43, I.2.42, it follows that $d[T(C_{k_n})] \to d[T(C_0)]$. Hence $d[T(C_0)] = 0$. In other words, $T(C_0)$ reduces to a single point x_0^* of $\mathcal{P}^*$. We have then $C_0 \subset T^{-1}(x_0^*)$, $d(C_0) > 0$, in contradiction with the assumption that T is light.

II.1.9. Let P, $\mathfrak{P}$, $\mathcal{P}$ be Peano spaces. Let $T^n(P) = \mathfrak{P}$ be a sequence of continuous transformations, and let $L(\mathfrak{P}) = \mathcal{P}$ be a light transformation. If the sequence LT^n is equicontinuous on P, then the sequence T^n is also equicontinuous on P (cf. I.2.44).

PROOF. Give $\epsilon > 0$. Since L is light, there exists a $\sigma > 0$ such that

$$(1) \qquad d(\mathfrak{C}) < \epsilon \qquad \text{if } d[L(\mathfrak{C})] < \sigma,$$

where $\mathfrak{C}$ is a generic notation for a continuum in $\mathfrak{P}$ (see II.1.8). Since the sequence LT^n is equicontinuous on P, there exists $\eta > 0$ such that

$$(2) \qquad \rho[LT^n(q_1), LT^n(q_2)] < \sigma \qquad \text{if } q_1, q_2 \in P,\ \rho(q_1, q_2) < \eta.$$

As a consequence, we have

$$(3) \qquad d[LT^n(C)] < \sigma \qquad \text{if } d(C) < \eta,$$

where C is a generic notation for a continuum in P. Since P is a Peano space, we have $\delta > 0$ such that the conditions $p_1, p_2 \in P$, $\rho(p_1, p_2) < \delta$ imply the existence of a continuum C such that $p_1 + p_2 \subset C \subset P$, $d(C) < \eta$ (see I.2.33). Now take any two points

$$(4) \qquad x_1, x_2 \in P, \qquad \rho(x_1, x_2) < \delta.$$

By the choice of δ, we have a continuum C such that

$$(5) \qquad x_1 + x_2 \subset C \subset P, \qquad d(C) < \eta.$$

In view of (3), we obtain from (5)

$$(6) \qquad \rho[T^n(x_1), T^n(x_2)] \leqq d[T^n(C)],$$

$$(7) \qquad d[LT^n(C)] < \sigma.$$

Now $T^n(C)$ is a continuum in $\mathfrak{P}$, and hence (1) and (7) yield

$$(8) \qquad d[T^n(C)] < \epsilon.$$

(4), (6), and (8) yield

$$\rho[T^n(x_1), T^n(x_2)] < \epsilon \qquad \text{if } x_1, x_2 \in P,\ \rho(x_1, x_2) < \delta.$$

Since δ is independent of x_1, x_2, n, the equicontinuity of the sequence T^n is established.

II.1.10. Let $\mathcal{P}$ be a Peano space. A class K of subsets of $\mathcal{P}$ will be termed a *disjoint closed covering* of $\mathcal{P}$ if (i) every set $E \in K$ is closed and nonempty, (ii) every point $x \in \mathcal{P}$ is contained in exactly one set $E \in K$. We shall then denote by $E(x)$ the unique set $E \in K$ that contains the point $x \in \mathcal{P}$.

The disjoint closed covering K will be termed an *upper semicontinuous collection of closed sets*, to be abbreviated to u.s.c.c. of closed sets, if the relation $x_n \to x_0$ implies the relation lim sup $E(x_n) \subset E(x_0)$ (cf. I.2.13). An u.s.c.c. of continua is then an u.s.c.c. of closed sets such that every set of the collection is a continuum (which may reduce to a single point).

Let us note explicitly that an u.s.c.c. of closed sets (or of continua) is also a disjoint closed covering of $\mathcal{P}$, while the converse is generally false. The definition given here is more restrictive than the one generally used in the literature, but it is quite adequate for our purposes.

II.1.11. Let K be an u.s.c.c. of closed sets in the Peano space $\mathcal{P}$ (see II.1.10). Let K_C denote the class of all components of all the sets $E \in K$. That is, a set C belongs to K_C if and only if there exists a set $E \in K$ such that C is a component of E. By I.2.39, K_C is then a disjoint closed covering of $\mathcal{P}$, and every set $C \in K_C$ is a continuum. We assert that K_C is an u.s.c.c. of continua.

PROOF. Let x_n be a sequence of points in $\mathcal{P}$ converging to a point x_0. Let $C(x_n)$, $n = 0, 1, 2, \cdots$, denote the (unique) set of K_C that contains the point x_n, and let $E(x_n)$, $n = 0, 1, 2, \cdots$, denote the (unique) set of K that contains x_n. Then for each $n = 0, 1, 2, \cdots$, $C(x_n)$ is a component of $E(x_n)$. We have to show that

$$\text{(1)} \qquad \limsup C(x_n) \subset C(x_0).$$

Now since K is an u.s.c.c., and $C(x_n) \subset E(x_n)$, we have lim sup $C(x_n) \subset$ lim sup $E(x_n) \subset E(x_0)$. Since $x_n \in C(x_n)$ and $x_n \to x_0$, the set lim inf $C(x_n)$ is nonempty, and hence lim sup $C(x_n)$ is a continuum (see I.2.37). Thus we have $x_0 \in$ lim sup $C(x_n) \subset E(x_0)$. In other words, lim sup $C(x_n)$ is a connected subset of $E(x_0)$ that contains x_0. Since $C(x_0)$ is the component of $E(x_0)$ that contains x_0, the inclusion (1) follows.

II.1.12. Let K be a disjoint closed covering of the Peano space $\mathcal{P}$ (see II.1.10). We propose to prove that the following three properties are equivalent.

Property (a). K is an u.s.c.c. of closed sets.

Property (b). If F is any closed set in $\mathcal{P}$, then the set $\sum E$, where the summation is extended over all sets $E \in K$ such that $EF \neq 0$, is closed.

Property (c). If G is any open set in $\mathcal{P}$, then the set $\sum E$, where the summation is extended over all sets $E \in K$ such that $E \subset G$, is open.

PROOF. (i) Property (a) implies property (b). Indeed, let us put

$$\text{(1)} \qquad E' = \sum E, \qquad E \in K, EF \neq 0.$$

Let x_0 be any point of the closure $c(E')$ of E'. Then we have in E' a sequence of points x_n' such that $x_n' \to x_0$. Using the notation $E(x)$ in the sense of II.1.10, we have then $E(x_n') \subset E'$ and hence $E(x_n')F \neq 0$ by (1). Thus for each n we have a point $y_n' \in E(x_n')F$. Then $E(x_n') = E(y_n')$ (cf. II.1.10). Since $\mathcal{P}$ is compact, we can assume without loss of generality that the sequence y_n' is convergent, say $y_n' \to y_0$. Since F is closed, we have $y_0 \in F$. Summing up, we have the following relations

$$(2) \qquad x_n' \to x_0, \qquad y_n' \to y_0 \in F, \qquad y_n' \in E(x_n'), \qquad E(x_n') = E(y_n').$$

These relations imply that

$$(3) \qquad x_0 + y_0 \subset \liminf E(x_n') \subset \limsup E(x_n') = \limsup E(y_n').$$

Since K is u.s.c.c. by assumption, (2) implies that

$$(4) \qquad \limsup E(x_n') \subset E(y_0).$$

(1) and (2) imply that

$$(5) \qquad E(y_0) \subset E'.$$

(3), (4), (5), imply that $x_0 \in E'$. Since x_0 was any point of $c(E')$, it follows that E' is closed.

(ii) Property (b) implies property (c). Indeed, let us put

$$G' = \sum E, \qquad E \in K, E \subset G.$$

The set $\mathcal{P} - G = F$ being closed, the set

$$F' = \sum E, \qquad E \in K, E(\mathcal{P} - G) \neq 0,$$

is also closed, since we assumed that K possesses the property (b). Clearly $G' = \mathcal{P} - F'$, and hence G' is open.

(iii) Property (c) implies property (a). Indeed, let x_n be a convergent sequence of points in $\mathcal{P}$ with limit x_0. Using the notation $E(x)$ in the sense of II.1.10, we have to show that $\limsup E(x_n) \subset E(x_0)$. If this inclusion is denied, then there should exist a point y_0 such that

$$(6) \qquad y_0 \in \limsup E(x_n), \qquad E(x_0) \subset \mathcal{P} - y_0.$$

Since $E(x_0)$ is closed, we have two open sets G_1, G_2 such that

$$(7) \qquad E(x_0) \subset G_1, \qquad y_0 \in G_2, G_1G_2 = 0.$$

Since we assumed property (c), the set

$$(8) \qquad G_1' = \sum E, \qquad E \in K, E \subset G_1,$$

is open, and in view of (7) clearly $x_0 \in G_1'$. Since $x_n \to x_0$, it follows that $x_n \in G_1'$ for n large, and hence by (8), (6)

$$(9) \qquad E(x_n) \subset G_1 \subset \mathcal{P} - G_2 \qquad \text{for } n \text{ large}.$$

(9) implies, by I.2.13, that $G_2 \limsup E(x_n) = 0$, in contradiction with (6), (7).

In view of (i), (ii), (iii), it is clear that any two of the properties (a), (b), (c) are equivalent.

II.1.13. Given a Peano space $\mathcal{P}$, we have the following examples involving an u.s.c.c. of closed sets.

(i) Let $T(\mathcal{P}) = \mathcal{P}^*$ be a continuous transformation. Then the class $I(T)$ (see II.1.1) is an u.s.c.c. of closed sets. Indeed, $I(T)$ is clearly a disjoint closed covering of $\mathcal{P}$. Let x_n be a sequence of points in $\mathcal{P}$ with limit x_0. Using the notation $E(x)$ in the sense of II.1.10, we have by I.2.43

$$(1) \qquad T[\limsup E(x_n)] \subset \limsup T[E(x_n)].$$

But $T[E(x_n)] = T(x_n)$, $T[E(x_0)] = T(x_0)$. Furthermore, $T(x_n) \to T(x_0)$ since $x_n \to x_0$. Hence $\limsup T[E(x_n)] = T(x_0)$, and thus (1) implies $T[\limsup E(x_n)] = T(x_0)$. Consequently $\limsup E(x_n) \subset T^{-1}T(x_0) = E(x_0)$.

(ii) Let $T(\mathcal{P}) = \mathcal{P}^*$ be a continuous transformation, and let $I_C(T)$ denote the class of all the components of all the sets $E \in I(T)$. Then $I_C(T)$ is an u.s.c.c. of continua. This is an immediate consequence of (i) and II.1.11. Let us note that obviously T is monotone if and only if $I(T) \equiv I_C(T)$, and T is light if and only if every set $E \in I_C(T)$ reduces to a single point.

(iii) Let K be any u.s.c.c. of closed sets in $\mathcal{P}$, and let G be an open set such that $G = \sum E$, $E \in K$, $E \subset G$. In other words the open set G is a sum of sets $E \in K$. Let K_G denote the class of those subsets E of $\mathcal{P}$ that satisfy one of the following two conditions. (α) $E \in K$ and $E \subset \mathcal{P} - G$. ($\beta$) E is a single point of G. Obviously, K_G is again an u.s.c.c. of closed sets.

II.1.14. Let K be a disjoint closed covering in the Peano space $\mathcal{P}$. We consider an (untopologized) space $\mathcal{P}_K^*$ whose points are the sets $E \in K$. Thus every set $E \in K$ is a subset of $\mathcal{P}$ and a point of $\mathcal{P}_K^*$. For clarity, we shall use the symbol $[E]^*$ to signify that the set $E \in K$ is interpreted as a point of the space $\mathcal{P}_K^*$. Using the notation $E(x)$ in the sense of II.1.10, we define a transformation $T_K(\mathcal{P}) = \mathcal{P}_K^*$ by the formula

$$(1) \qquad T_K(x) = [E(x)]^*.$$

Clearly, for every set $E \in K$

$$(2) \qquad T_K^{-1}([E]^*) = E, \qquad T_K(E) = [E]^*.$$

Thus K determines, in the sense just explained, an (untopologized) space $\mathcal{P}_K^*$ and a single-valued transformation $T_K(\mathcal{P}) = \mathcal{P}_K^*$. The following question is relevant for our purposes: under what further conditions upon K is it possible to topologize $\mathcal{P}_K^*$ (cf. I.2.9) in such a manner that the transformation $T_K(\mathcal{P}) = \mathcal{P}_K^*$ becomes continuous? The following statements will be proved presently.

(i) A topologization with the required property exists if and only if K is an u.s.c.c. of closed sets.

(ii) There exists at most one topologization with the required property.

(iii) If such a topologization exists, then the convergent sequences of the (topologized) space $\mathcal{P}_K^*$ may be characterized as follows: $[E_n]^* \to [E_0]^*$ if and only if $\limsup E_n \subset E_0$.

Statement (ii) is obvious. Indeed, if T_K is continuous, then a subset G^* of $\mathcal{P}_K^*$ is open if and only if $T_K^{-1}(G^*)$ is open (see I.2.43). Now by (2), $T_K^{-1}(G^*) = \sum E$, $E \in K$, $[E]^* \in G^*$. Since $G^* = T_K T_K^{-1}(G^*)$, it follows that G^* is open if and only if it is of the form $T_K(G)$, where G is an open subset of $\mathcal{P}$ that is a sum of sets $E \in K$. Thus the class of open sets of $\mathcal{P}_K^*$, and hence the topologization of $\mathcal{P}_K^*$, is univocally determined by K if T_K is to be continuous. To verify statement (iii), suppose first that $[E_n]^* \to [E_0]^*$. Let x_0 be any point of $\limsup E_n$. Then by I.2.43, I.2.13, $T_K(x_0) \in T_K(\limsup E_n) \subset \limsup T_K(E_n) = \limsup [E_n]^* = [E_0]^*$. Hence $x_0 \in T_K^{-1}([E_0]^*) = E_0$. Since x_0 was an arbitrary point of $\limsup E_n$, it follows that $\limsup E_n \subset E_0$. Suppose conversely that

$$\limsup E_n \subset E_0 . \tag{3}$$

Let G^* be any open set in $\mathcal{P}_K^*$ containing the point $[E_0]^*$. Since T_K is continuous by assumption, the set $G = T_K^{-1}(G^*)$ is open, and $E_0 = T_K^{-1}([E_0]^*) \subset T_K^{-1}(G^*) = G$. Hence (3) implies that (cf. I.2.37)

$$E_n \subset T_K^{-1}(G^*) \qquad \text{for } n \text{ large.}$$

Consequently

$$T_K(E_n) = [E_n]^* \in T_K T_K^{-1}(G^*) = G^* \qquad \text{for } n \text{ large.} \tag{4}$$

Since G^* was any open subset of $\mathcal{P}_K^*$ containing $[E_0]^*$, (4) shows that $[E_n]^* \to [E_0]^*$.

As regards statement (i), let us note that by (2) we have $I(T_K) \equiv K$ (cf. II.1.1), and hence the necessity part of statement (i) is a direct consequence of II.1.13(i). To prove the sufficiency part, let us assume that K is an u.s.c.c. of closed sets in $\mathcal{P}$ (see II.1.10). Let Ω^* be the class of those subsets G^* of $\mathcal{P}_K^*$ for which $T_K^{-1}(G^*)$ is open (in $\mathcal{P}$). We proceed to show that Ω^* satisfies (relative to $\mathcal{P}_K^*$) the postulates for the class of open subsets of a topological space (see I.2.2).

(a) Clearly, the empty set and $\mathcal{P}_K^*$ itself belong to Ω^*.

(b) Let $G_1^*, \cdots, G_n^*$ be a finite system of sets belonging to Ω^*. Then $T_K^{-1}(G_1^*), \cdots, T_K^{-1}(G_n^*)$ are open sets by assumption, and hence $T_K^{-1}(G_1^* \cdots G_n^*) = T_K^{-1}(G_1^*) \cdots T_K^{-1}(G_n^*)$ is also open. Thus $G_1^* \cdots G_n^* \in \Omega^*$.

(c) Let $\mathfrak{F}^*$ be any family of sets belonging to Ω^*. Then $T_K^{-1}(G^*)$ is open if $G^* \in \mathfrak{F}^*$, and hence on setting $F^* = \sum G^*$, $G^* \in \mathfrak{F}^*$, it follows that $T_K^{-1}(F^*) = \sum T_K^{-1}(G^*)$, $G^* \in \mathfrak{F}^*$, is open also. Hence $F^* \in \Omega^*$.

(d) Let $[E_1]^*$, $[E_2]^*$ be any two distinct points of $\mathcal{P}_K^*$. Then $E_1 \neq E_2$ by (2), and hence E_1, E_2 are disjoint closed subsets of $\mathcal{P}$, both of which belong to K. Hence (see I.2.38) we have two open sets G_1, G_2, such that $E_1 \subset G_1$, $E_2 \subset G_2$, $G_1G_2 = 0$. Let us put

$$G_1' = \sum E, \qquad E \in K, E \subset G_1 ,$$

$$G_2' = \sum E, \qquad E \in K, E \subset G_2 .$$

Then we have by (2)

$$T_K(G_1') = \sum T_K(E) = \sum [E]^*, \qquad E \in K, E \subset G_1 ,$$

and consequently

$$T_K^{-1}T_K(G_1') = \sum T_K^{-1}([E]^*) = \sum E, \qquad E \in K, E \subset G_1 .$$

Thus $T_K^{-1}T_K(G_1') = G_1'$. Now G_1' is open by II.1.12, and hence $T_K(G_1') \in \Omega^*$. Similarly $T_K(G_2') \in \Omega^*$. Obviously $E_1 \subset G_1'$, and hence $[E_1]^* = T_K(E_1) \in T_K(G_1')$. Similarly $[E_2]^* \in T_K(G_2')$. As noted above, $T_K^{-1}T_K(G_1') = G_1'$, and similarly $T_K^{-1}T_K(G_2') = G_2'$. Hence $T_K^{-1}(G_1'G_2') = T_K^{-1}(G_1')T_K^{-1}(G_2') = G_1'G_2' \subset G_1G_2 = 0$, and thus $G_1'G_2' = 0$. Summing up: for any two distinct points $[E_1]^*$, $[E_2]^*$ of $\mathcal{P}_K^*$ there exist two sets G_1^*, G_2^* in $\mathcal{P}_K^*$ such that $[E_1]^* \in G_1^*$, $[E_2]^* \in G_2^*$, $G_1^*G_2^* = 0$, $G_1^* \in \Omega^*$, $G_2^* \in \Omega^*$. Indeed, the sets $G_1^* = T_K(G_1')$, $G_2^* = T_K(G_2')$ satisfy all these conditions as we have just shown.

Thus if we use the class Ω^* as the class of open sets, then $\mathcal{P}_K^*$ becomes a topological space (see I.2.2). Since $T_K^{-1}(G^*)$ is open if $G^* \in \Omega^*$, by the definition of the class Ω^*, the transformation T_K is obviously continuous with respect to this topologization of $\mathcal{P}_K^*$ (cf. I.2.29). Let us note again that $I(T_K) \equiv K$ as a consequence of (1) (cf. II.1.1).

II.1.15. Theorem. *Let $\mathcal{P}$ be a Peano space, and let K be an u.s.c.c. of closed sets in $\mathcal{P}$* (see II.1.10). *Then there exists a continuous transformation $T(\mathcal{P}) = \mathcal{P}^*$ such that $I(T) \equiv K$* (cf. II.1.1). *Conversely, if $T(\mathcal{P}) = \mathcal{P}^*$ is any continuous transformation, then $I(T)$ is an u.s.c.c. of closed sets in $\mathcal{P}$.*

Indeed, the first assertion follows from II.1.14 and the second assertion is merely a repetition of II.1.13(i).

II.1.16. To avoid unnecessary complications in the sequel, we derive a simple lemma for spaces more general than Peano spaces. Let $T_1(\Sigma) = \Sigma_1^*$, $T_2(\Sigma) = \Sigma_2^*$ be continuous transformations from a compact topological space Σ onto the (necessarily compact) topological spaces Σ_1^*, Σ_2^* (see I.2.43). Let us use the symbols $I(T_1)$, $I(T_2)$ in the sense explained in II.1.1, and let us suppose that $I(T_1) \equiv I(T_2)$. Then for each point $x_1^* \in \Sigma_1^*$, the set $T_2T_1^{-1}(x_1^*)$ is a single point of Σ_2^*. Indeed, $T_1^{-1}(x_1^*)$ belongs to $I(T_1)$ and hence, by assumption, also to $I(T_2)$, and the assertion follows. Thus the formula

$$(1) \qquad x_2^* = H_1(x_1^*) = T_2T_1^{-1}(x_1^*), \qquad x_1^* \in \Sigma_1^* ,$$

defines a single-valued transformation. Similarly, the formula

$$(2) \qquad y_1^* = H_2(y_2^*) = T_1T_2^{-1}(y_2^*), \qquad y_2^* \in \Sigma_2^* ,$$

defines a single-valued transformation. We assert that H_1 is a topological transformation from Σ_1^* onto Σ_2^*, H_2 is a topological transformation from Σ_2^* onto Σ_1^*, and $H_2 = H_1^{-1}$ (note that these statements show that Σ_1^*, Σ_2^* are homeomorphic).

Proof. The relation $x_2^* = H_1(x_1^*)$ yields $T_2^{-1}(x_2^*) = T_2^{-1}T_2T_1^{-1}(x_1^*) \supset T_1^{-1}(x_1^*)$

and hence $H_2(x_2^*) = T_1T_2^{-1}(x_2^*) \supset T_1T_1^{-1}(x_1^*) = x_1^*$. Since $H_2(x_2^*)$ is a single point, it follows that $H_2(x_2^*) = x_1^*$. As x_1^* was an arbitrary point of Σ_1^*, and the situation is symmetric with respect to the subscripts 1 and 2, the following statements are immediate. (i) $H_1(\Sigma_1^*) = \Sigma_2^*$, and similarly $H_2(\Sigma_2^*) = \Sigma_1^*$. (ii) H_1, H_2 are both biunique. (iii) $H_2 = H_1^{-1}, H_1 = H_2^{-1}$. We proceed to verify the continuity of H_1. Let G_2^* be any open set in Σ_2^*. Then the set $G = T_2^{-1}(G_2^*)$ is open, and clearly $G = \sum E,\ E \in I(T_2),\ E \subset G$. Since $I(T_1) \equiv I(T_2)$, we have therefore $G = \sum E,\ E \in I(T_1),\ E \subset G$. Since clearly $E \in I(T_1)$ implies that $T_1^{-1}T_1(E) = E$, it follows that $G = T_1^{-1}T_1(G)$. As G is open, it follows that $T_1(G)$ is also open. But $T_1(G) = T_1T_2^{-1}(G_2^*) = H_2(G^*) = H_1^{-1}(G_2^*)$. Thus $H_1^{-1}(G_2^*)$ is open for every open set $G_2^* \subset \Sigma_2^*$, and hence H_1 is continuous (see I.2.43). The continuity of H_2 is verified in a similar manner.

II.1.17. THEOREM. *Given* $T(\mathcal{P}) = \mathcal{P}^*$ *as in* II.1.1, *there exists a factorization*

$$T = LM, \qquad M(\mathcal{P}) = \mathfrak{M}, \qquad L(\mathfrak{M}) = \mathcal{P}^*,$$

where M is monotone and L is light (see II.1.1).

PROOF. Let us consider the classes $I(T)$ and $I_C(T)$ (see II.1.1, II.1.13(ii)). Then $I(T)$ is an u.s.c.c. of closed sets and $I_C(T)$ is an u.s.c.c. of continua by II.1.13. By II.1.15 there exists a continuous transformation $T_1(\mathcal{P}) = \mathcal{P}_1$ such that $I(T_1) \equiv I_C(T)$. Then T_1 is clearly monotone (see II.1.1). Let now x_1 be any point of $\mathcal{P}_1$. Then the set $T_1^{-1}(x_1)$ is a component of a certain set $E \in I(T)$, and hence $TT_1^{-1}(E)$ reduces to a single point of $\mathcal{P}^*$. Thus the formula

$$x^* = TT_1^{-1}(x_1), \qquad x_1 \in \mathcal{P}_1,$$

defines a single-valued transformation. On setting $T_2 = TT_1^{-1}$, we obtain

$$(1) \qquad T_2T_1(x) = TT_1^{-1}T_1(x) \supset T(x), \qquad x \in \mathcal{P}.$$

Since T_1 and T_2 are both single-valued, it follows from (1) that

$$(2) \qquad T(x) = T_2T_1(x), \qquad x \in \mathcal{P}.$$

Since $T_1(\mathcal{P}) = \mathcal{P}_1$ and $T(\mathcal{P}) = \mathcal{P}^*$, (2) implies that $T_2(\mathcal{P}_1) = \mathcal{P}^*$. We assert that T_2 is continuous. Indeed, let G^* be any open set in $\mathcal{P}^*$. Then $T_2^{-1}(G^*) = T_1T^{-1}(G^*)$. Clearly

$$(3) \qquad T^{-1}(G^*) = \sum E, \qquad E \in I(T), E \subset T^{-1}(G^*).$$

Since each set $E \in I(T)$ is a sum of sets $C \in I_C(T)$ and $I_C(T) \equiv I(T_1)$, (3) implies that

$$(4) \qquad T^{-1}(G^*) = \sum C, \qquad C \in I(T_1), C \subset T^{-1}(G^*).$$

Since clearly $T_1^{-1}T_1(C) = C$ for every set $C \in I(T_1)$, (4) yields

$$(5) \qquad T_1^{-1}T_2^{-1}(G^*) = T_1^{-1}T_1T^{-1}(G^*) = T^{-1}(G^*).$$

But $T^{-1}(G^*)$ is open, since T is continuous and G^* is open. Thus (5) shows that

$T_1^{-1}T_2^{-1}(G^*)$ is open, and hence it follows that $T_2^{-1}(G^*)$ is open for every open set $G^* \subset \mathcal{P}^*$. The continuity of T_2 follows by I.2.43. We assert finally that T_2 is light. Indeed, let $C_1 \subset \mathcal{P}_1$ be a continuum such that $T_2(C_1) = TT_1^{-1}(C_1)$ is a single point x_0^* of $\mathcal{P}^*$. Then $T_1^{-1}(C_1) \subset T^{-1}(x_0^*)$. Now T_1 is monotone, and hence $T_1^{-1}(C_1)$ is a continuum by II.1.2. Also, $T^{-1}(x_0^*) \in I(T)$. Hence $T_1^{-1}(C_1)$ is a subset of a component C_0 of $T_1^{-1}(x_0^*)$. Thus $T_1^{-1}(C_1) \subset C_0 \in I_c(T) \equiv I(T_1)$, and hence $T_1(C_0)$ reduces to a single point y_1 of $\mathcal{P}_1$. Since $C_1 = T_1T_1^{-1}(C_1) \subset T_1(C_0) = y_1$, it follows that C_1 also reduces to the same single point y_1. In other words, if T_2 is constant on a continuum $C_1 \subset \mathcal{P}_1$, then C_1 must be a single point. Thus T_2 is light. On setting $\mathfrak{M} = \mathcal{P}_1$, $M = T_1$, $L = T_2$, we obtain therefore, in view of (2), a factorization of the desired character.

II.1.18. Given $T(\mathcal{P}) = \mathcal{P}^*$ as in II.1.1, a factorization

$$T = LM, \qquad M(\mathcal{P}) = \mathfrak{M}, \qquad L(\mathfrak{M}) = \mathcal{P}^*,$$

where M is monotone and L is light, is termed a *monotone-light factorization* of T. The space $\mathfrak{M}$, which is a Peano space by I.2.43, will be termed *the middle-space*. According to II.1.17, T admits of at least one monotone-light factorization. We propose to determine all such factorizations of T.

II.1.19. Given $T(\mathcal{P}) = \mathcal{P}^*$ as in II.1.1, let $T = LM$, $M(\mathcal{P}) = \mathfrak{M}$, $L(\mathfrak{M}) = \mathcal{P}^*$ be any monotone-light factorization of T. Then $I(M) \equiv I_c(T)$ (cf. II.1.1, II.1.13(ii)).

Proof. We establish first two auxiliary statements.

(i) Let X_0 be a continuum that belongs to $I(M)$. Then there exists a continuum $X_0' \in I_c(T)$ such that $X_0 \subset X_0'$. Indeed, by assumption we have a point $\mathfrak{x}_0 \in \mathfrak{M}$ such that $X_0 = M^{-1}(\mathfrak{x}_0)$. Put $L(\mathfrak{x}_0) = x_0^*$. Then $T(X_0) = LM(X_0) = L(\mathfrak{x}_0) = x_0^*$, and hence $X_0 \subset T^{-1}(x_0^*)$. Since X_0 is connected, it is contained in a component X_0' of $T^{-1}(x_0^*)$. Since $T^{-1}(x_0^*) \in I(T)$, we have $X_0' \in I_c(T)$. Thus $X_0 \subset X_0' \in I_c(T)$.

(ii) Let Y_0' be a continuum that belongs to $I_c(T)$. Then there exists a continuum $Y_0 \in I(M)$ such that $Y_0' \subset Y_0$. Indeed, by assumption Y_0' is a component of a set of the form $T^{-1}(y_0^*)$, $y_0^* \in \mathcal{P}^*$. Then $LM(Y_0') = T(Y_0') \subset TT^{-1}(y_0^*) = y_0^*$. Hence $M(Y_0') \subset L^{-1}(y_0^*)$. Since $M(Y_0')$ is connected (cf. I.2.43) and $L^{-1}(y_0^*)$ is totally disconnected, it follows that $M(Y_0') = \mathfrak{y}_0$, a single point of $\mathfrak{M}$. Setting $M^{-1}(\mathfrak{y}_0) = Y_0$, it follows that $Y_0' \subset Y_0 \in I(M)$.

Now let $X_0 \in I(M)$. By (i) we have a continuum X_0' such that $X_0 \subset X_0' \in I_c(T)$. By (ii) we have a continuum X_0'' such that $X_0' \subset X_0'' \in I(M)$. Thus $X_0 \subset X_0' \subset X_0''$, $X_0 \in I(M)$, $X_0'' \in I(M)$. Since $X_0X_0'' = X_0 \neq 0$, it follows that $X_0 = X_0''$ (cf. II.1.10). Hence $X_0 = X_0' \in I_c(T)$. Thus $I(M) \subset I_c(T)$.

Next let $Y_0' \in I_c(T)$. By (ii) we have a continuum Y_0 such that $Y_0' \subset Y_0 \in I(M)$. By (i) we have a continuum Y_0'' such that $Y_0 \subset Y_0'' \in I_c(T)$. Thus $Y_0' \subset Y_0 \subset Y_0''$, $Y_0' \in I_c(T)$, $Y_0'' \in I_c(T)$. It follows that $Y_0' = Y_0''$ (cf. II.1.10), and hence $Y_0' = Y_0 \in I(M)$. Thus we obtain the complementary inclusion $I_c(T) \subset I(M)$, and the proof is complete.

II.1.20. Given $T(\mathcal{P}) = \mathcal{P}^*$ as in II.1.1, let

$$(1) \qquad T = L_1 M_1 , \qquad M_1(\mathcal{P}) = \mathfrak{M}_1 , \qquad L_1(\mathfrak{M}_1) = \mathcal{P}^*,$$

$$(2) \qquad T = L_2 M_2 , \qquad M_2(\mathcal{P}) = \mathfrak{M}_2 , \qquad L_2(\mathfrak{M}_2) = \mathcal{P}^*,$$

be any two monotone-light factorizations of T. Then the middle-spaces $\mathfrak{M}_1$, $\mathfrak{M}_2$ are homeomorphic. Furthermore, there exists a (unique) topological transformation $h(\mathfrak{M}_1) = \mathfrak{M}_2$ that satisfies the relations

$$(3) \qquad L_1(\mathfrak{z}_1) = L_2 h(\mathfrak{z}_1), \qquad \mathfrak{z}_1 \in \mathfrak{M}_1 ,$$

$$(4) \qquad M_2(x) = h M_1(x), \qquad x \in \mathcal{P}.$$

PROOF. By II.1.19 we have $I(M_1) \equiv I_C(T) \equiv I(M_2)$. By II.1.16 (applied to the transformations M_1, M_2), the formula

$$(5) \qquad \mathfrak{z}_2 = h(\mathfrak{z}_1) = M_2 M_1^{-1}(\mathfrak{z}_1), \qquad \mathfrak{z}_1 \in \mathfrak{M}_1 ,$$

defines a topological transformation from $\mathfrak{M}_1$ onto $\mathfrak{M}_2$. Thus the spaces $\mathfrak{M}_1$, $\mathfrak{M}_2$ are homeomorphic. Now let x be any point of $\mathcal{P}$. Put $M_1(x) = \mathfrak{z}_1$. Then $h(\mathfrak{z}_1) = M_2 M_1^{-1} M_1(x) \supset M_2(x)$. Since $h(\mathfrak{z}_1)$ is a single point, it follows that $M_2(x) = h(\mathfrak{z}_1) = h M_1(x)$, and (4) is proved. To prove (3), note that (5), (2), (1), yield

$$L_2 h(\mathfrak{z}_1) = L_2 M_2 M_1^{-1}(\mathfrak{z}_1) = T M_1^{-1}(\mathfrak{z}_1) = L_1 M_1 M_1^{-1}(\mathfrak{z}_1) = L_1(\mathfrak{z}_1).$$

Thus (3) is verified. As regards the uniqueness of h, it is clear that h is univocally determined by the condition (4) alone.

II.1.21. Given $T(\mathcal{P}) = \mathcal{P}^*$ as in II.1.1, and a monotone-light factorization $T = L_1 M_1$, $M_1(\mathcal{P}) = \mathfrak{M}_1$, $L_1(\mathfrak{M}_1) = \mathcal{P}^*$, let $\mathfrak{M}_2$ be any (Peano) space homeomorphic with $\mathfrak{M}_1$. Then there exists a monotone-light factorization of T with the middle-space $\mathfrak{M}_2$. Indeed, if $h(\mathfrak{M}_1) = \mathfrak{M}_2$ is any topological transformation, then clearly $T = L_2 M_2$, where $L_2 = L_1 h^{-1}$, $M_2 = h M_1$ is a factorization with the desired properties. In view of II.1.20, we obtain in this manner all possible monotone-light factorizations of T, provided one such factorization is known. On the other hand, II.1.17 yields an initial monotone-light factorization explicitly.

II.1.22. Let $M_n(\mathcal{P}) = \mathfrak{M}$, $n = 1, 2, \cdots$, be an infinite sequence of continuous monotone transformations from the Peano space $\mathcal{P}$ onto the Peano space $\mathfrak{M}$. Suppose that the sequence M_n is uniformly convergent on $\mathcal{P}$, and let $\mathfrak{z} = T(x)$, $x \in \mathcal{P}$, be the limit transformation. Then T is a continuous transformation from $\mathcal{P}$ onto $\mathfrak{M}$ (cf. I.2.44). We assert that T is monotone.

PROOF. Deny the assertion. Then there should exist a point $\mathfrak{z}_0 \in \mathfrak{M}$ such that the set $E_0 = T^{-1}(\mathfrak{z}_0)$ is disconnected. We have then a separation $E_0 = E_0' \mid E_0''$ (see I.2.21). Since E_0 is closed, E_0' and E_0'' are also closed (see I.2.39). By I.2.38, we have therefore two open sets G', G'' such that

$$(1) \qquad E_0' \subset G', \qquad E_0'' \subset G'', \qquad G'G'' = 0.$$

Since $\mathcal{P}$ is connected, the set $\mathcal{P} - (G' + G'')$ is a nonempty set which is clearly closed. On $\mathcal{P} - (G' + G'')$, the function $\rho[T(x), \mathfrak{z}_0]$ is continuous and different from zero; hence we have (cf. I.2.45) a positive number δ such that

$$\rho[T(x), \mathfrak{z}_0] \geqq 2\delta > 0 \qquad \text{for } x \in \mathcal{P} - (G' + G''). \tag{2}$$

Since $M_n \to T$ uniformly on $\mathcal{P}$, we shall have $\rho[T(x), M_n(x)] < \delta$ on $\mathcal{P}$ for n exceeding a certain $\nu(\delta)$. In view of (2) it follows that

$$\rho[M_n(x), \mathfrak{z}_0] > \delta > 0 \qquad \text{for } x \in \mathcal{P} - (G' + G''), n > \nu(\delta). \tag{3}$$

Now let $\mathfrak{A}(\mathfrak{z}_0, \delta)$ denote the set of those points $\mathfrak{z} \in \mathfrak{M}$ that satisfy the inequality $\rho(\mathfrak{z}, \mathfrak{z}_0) < \delta$. Then (3) yields

$$M_n^{-1}[\mathfrak{A}(\mathfrak{z}_0, \delta)] \subset G' + G'' \qquad \text{for } n > \nu(\delta). \tag{4}$$

The set $\mathfrak{A}(\mathfrak{z}_0, \delta)$ is open and contains $\mathfrak{z}_0$. Let $\mathfrak{C}_\delta$ be the component of $\mathfrak{A}(\mathfrak{z}_0, \delta)$ that contains $\mathfrak{z}_0$. Then $\mathfrak{C}_\delta$ is a connected open set (see I.2.25), and we have therefore an $\eta > 0$ such that

$$\rho(\mathfrak{z}, \mathfrak{z}_0) < \eta \text{ implies } \mathfrak{z} \in \mathfrak{C}_\delta. \tag{5}$$

Since $M_n \to T$ uniformly on $\mathcal{P}$, we have $\rho[T(x), M_n(x)] < \eta$ for n exceeding a certain $\nu(\eta)$. As $T(E_0' + E_0'') = \mathfrak{z}_0$, it follows that (cf. (5))

$$M_n(E_0' + E_0'') \subset \mathfrak{C}_\delta \qquad \text{for } n > \nu(\eta). \tag{6}$$

(4) and (6) yield (note that $\mathfrak{C}_\delta \subset \mathfrak{A}(\mathfrak{z}_0, \delta)$)

$$E_0' + E_0'' \subset M_n^{-1}(\mathfrak{C}_\delta) \subset G' + G'' \qquad \text{for } n > \nu(\delta) + \nu(\eta). \tag{7}$$

Now $M_n^{-1}(\mathfrak{C}_\delta)$ is connected (see II.1.2). According to (1) and (7), we have

$$G'M_n^{-1}(\mathfrak{C}_\delta) \neq 0, \qquad G''M_n^{-1}(\mathfrak{C}_\delta) \neq 0, \qquad M_n^{-1}(\mathfrak{C}_\delta) \subset G' + G'', \tag{8}$$

for n sufficiently large. Since G', G'' are disjoint open sets and $M_n^{-1}(\mathfrak{C}_\delta)$ is connected, (8) is impossible.

II.1.23. Given $T(\mathcal{P}) = \mathcal{P}^*$ as in II.1.1, let

$$T = LM, \qquad M(\mathcal{P}) = \mathfrak{M}, \qquad L(\mathfrak{M}) = \mathcal{P}^* \tag{1}$$

be a monotone-light factorization of T. Let $Z(\mathfrak{M}) = \mathfrak{M}^*$ be any continuous transformation from $\mathfrak{M}$ into $\mathfrak{M}$ (that is, $\mathfrak{M}^* \subset \mathfrak{M}$). Then LZM is clearly a continuous transformation from $\mathcal{P}$ into $\mathcal{P}^*$. A continuous transformation obtained in this manner will be termed a *partial transformation* associated with T. We assert that any partial transformation LZM can be derived, in the manner just described, from any fixed monotone-light factorization

$$T = L_0M_0, \qquad M_0(\mathcal{P}) = \mathfrak{M}_0, \qquad L_0(\mathfrak{M}_0) = \mathcal{P}^*. \tag{2}$$

Indeed, by II.1.20 there exists a topological transformation $h(\mathfrak{M}) = \mathfrak{M}_0$ such that $L = L_0h$, $M_0 = hM$. Then $hZh^{-1} = Z_0$ is clearly a continuous transformation from $\mathfrak{M}_0$ into $\mathfrak{M}_0$, and $L_0Z_0M_0 = Lh^{-1}(hZh^{-1})hM = LZM$.

II.1.24. We shall be concerned in the sequel with a pair of continuous transformations $T_1(\mathcal{P}_1) = \mathcal{P}^*$, $T_2(\mathcal{P}_2) = \mathcal{P}^*$ such that (i) $\mathcal{P}_1$, $\mathcal{P}_2$, $\mathcal{P}^*$ are Peano spaces, and (ii) $\mathcal{P}_1$ and $\mathcal{P}_2$ are homeomorphic. For example, $\mathcal{P}_1$, $\mathcal{P}_2$ may be both simple arcs, or both simple closed curves, or both 2-cells, or both 2-spheres, and so forth. In fact, the four special cases just mentioned will be of especial interest for our purposes.

II.1.25. Given T_1, T_2 as in II.1.24, we shall say that T_1 and T_2 are *F-equivalent (equivalent in the Fréchet sense)*, if and only if for every $\epsilon > 0$ there exists a topological transformation $H_\epsilon(\mathcal{P}_1) = \mathcal{P}_2$ such that $\rho[T_1(x_1), T_2H_\epsilon(x_1)] < \epsilon$ for every point $x_1 \in \mathcal{P}_1$. If this condition is satisfied, then we shall write $T_1 \sim T_2(F)$. Clearly, the relations $T_1 \sim T_2(F)$ and $T_2 \sim T_1(F)$ mutually imply each other.

II.1.26. Given T_1, T_2 as in II.1.24, we shall say that T_1, T_2 are *topologically similar*, in symbols $T_1 \sim T_2(ts)$, if and only if there exists a topological transformation $H(\mathcal{P}_1) = \mathcal{P}_2$ such that $T_1(x_1) = T_2H(x_1)$ for every point $x_1 \in \mathcal{P}_1$. Clearly the relations $T_1 \sim T_2(ts)$ and $T_2 \sim T_1(ts)$ mutually imply each other. It is also obvious that the relation $T_1 \sim T_2(ts)$ implies the relation $T_1 \sim T_2(F)$. Simple examples show that the converse is generally false.

II.1.27. Given T_1, T_2 as in II.1.24, suppose that $T_1 \sim T_2(F)$. Let

$$(1) \qquad T_1 = L_1M_1, \qquad M_1(\mathcal{P}_1) = \mathfrak{M}_1, \qquad L_1(\mathfrak{M}_1) = \mathcal{P}^*,$$

$$(2) \qquad T_2 = L_2M_2, \qquad M_2(\mathcal{P}_2) = \mathfrak{M}_2, \qquad L_2(\mathfrak{M}_2) = \mathcal{P}^*$$

be arbitrary monotone-light factorizations of T_1, T_2 respectively. Then there exists a sequence of topological transformations $H_n(\mathcal{P}_1) = \mathcal{P}_2$, such that the following conditions hold.

(i) $\lambda_n = \max \rho[T_1(x_1), T_2H_n(x_1)] \to 0$ for $n \to \infty$, where the maximum is taken for x_1 varying on $\mathcal{P}_1$. Note that the use of the maximum is justified since $\mathcal{P}_1$ is compact (see I.2.45).

(ii) The sequence M_2H_n is uniformly convergent on $\mathcal{P}_1$.

PROOF. Let ϵ_n be a sequence of positive numbers converging to zero. Since $T_1 \sim T_2(F)$, we have for every n a topological transformation $\bar{H}_n(\mathcal{P}_1) = \mathcal{P}_2$, such that on setting

$$\bar{\lambda}_n = \max \rho[T_1(x_1), T_2\bar{H}_n(x_1)], \qquad x_1 \in \mathcal{P}_1,$$

the inequality $\bar{\lambda}_n < \epsilon_n$ holds. Thus the sequence $T_2\bar{H}_n$ is uniformly convergent on $\mathcal{P}_1$ (the limit transformation being T_1). Hence the sequence $T_2\bar{H}_n$ is equicontinuous on $\mathcal{P}_1$ (see I.2.44). Since $T_2\bar{H}_n = L_2M_2\bar{H}_n$ by (2), and since $M_2\bar{H}_n$ is clearly monotone, it follows by II.1.9 that the sequence $M_2\bar{H}_n$ is also equicontinuous. By I.2.44 the sequence $M_2\bar{H}_n$ contains therefore a uniformly convergent infinite subsequence. If $k_1 < k_2 < \cdots < k_n < \cdots$ are the subscripts occurring (in this order) in such a subsequence, and if we put $\bar{H}_{k_n} = H_n$, then the sequence H_n satisfies the conditions (i), (ii).

II.1.28. THEOREM. *Given* T_1, T_2 *as in* II.1.24, *suppose that* $T_1 \sim T_2(F)$. *Let*

$$(1) \qquad T_1 = L_1M_1, \qquad M_1(\mathcal{P}_1) = \mathfrak{M}_1, \qquad L_1(\mathfrak{M}_1) = \mathcal{P}^*,$$

(2) $$T_2 = L_2M_2\,, \qquad M_2(\mathcal{P}_2) = \mathfrak{M}_2\,, \qquad L_2(\mathfrak{M}_2) = \mathcal{P}^*$$

be arbitrary monotone-light factorizations of T_1, T_2 *respectively. Then the middle-spaces* $\mathfrak{M}_1$, $\mathfrak{M}_2$ *are homeomorphic, and there exists a topological transformation* $h(\mathfrak{M}_1) = \mathfrak{M}_2$ *with the following properties.*

(i) $L_1(\mathfrak{z}_1) = L_2h(\mathfrak{z}_1)$ *for every point* $\mathfrak{z}_1 \in \mathfrak{M}_1$.

(ii) *If* Z_1 *is any continuous transformation from* $\mathfrak{M}_1$ *into* $\mathfrak{M}_1$ *(that is,* $Z_1(\mathfrak{M}_1) \subset \mathfrak{M}_1$*), then* $L_1Z_1M_1 \sim L_2hZ_1h^{-1}M_2(F)$.

Proof. By II.1.27 there exists a sequence of topological transformations $H_n(\mathcal{P}_1) = \mathcal{P}_2$ with the properties II.1.27(i), II.1.27(ii). If $\overline{M}_1$ denotes the limit transformation of the (uniformly convergent) sequence M_2H_n, then by II.1.22 (applied to the sequence M_2H_n) it follows that $\overline{M}_1$ is a continuous monotone transformation from $\mathcal{P}_1$ onto $\mathfrak{M}_2$. Since M_2H_n converges uniformly to $\overline{M}_1$, it follows that $L_2M_2H_n$ converges uniformly to $L_2\overline{M}_1$. On the other hand, $L_2M_2H_n = T_2H_n$ converges uniformly to T_1. Thus $L_2\overline{M}_1 = T_1$. In other words, we have for T_1 the monotone-light factorization

(3) $$T_1 = L_2\overline{M}_1\,, \qquad \overline{M}_1(\mathcal{P}_1) = \mathfrak{M}_2\,, \qquad L_2(\mathfrak{M}_2) = \mathcal{P}^*.$$

Comparison of (1) and (3) yields, by II.1.20, the existence of a topological transformation h with the properties

(4) $$h(\mathfrak{M}_1) = \mathfrak{M}_2\,, \qquad L_1 = L_2h, \qquad \overline{M}_1 = hM_1\,.$$

Thus $\mathfrak{M}_1$, $\mathfrak{M}_2$ are homeomorphic, and h satisfies the condition (i). Now let Z_1 be any continuous transformation from $\mathfrak{M}_1$ into $\mathfrak{M}_1$. Let us put

(5) $$X_2 = L_2hZ_1h^{-1}, \qquad T_1^* = L_1Z_1M_1\,, \qquad T_2^* = L_2hZ_1h^{-1}M_2\,.$$

Then (4) and (5) yield $T_2^*H_n = X_2M_2H_n$. Since M_2H_n converges uniformly to $\overline{M}_1$ and X_2 is uniformly continuous on $\mathfrak{M}_2$, it follows that $T_2^*H_n$ converges uniformly to $X_2\overline{M}_1 = L_2hZ_1h^{-1}\overline{M}_1 = L_1Z_1M_1 = T_1^*$ (see (4), (5)). Hence, given $\epsilon > 0$, we have an integer $n(\epsilon)$ such that

$$\rho[T_1^*(x_1), T_2^*H_{n(\epsilon)}(x_1)] < \epsilon \qquad \text{for } x_1 \in \mathcal{P}_1\,.$$

Hence $T_1^* \sim T_2^*(F)$ by II.1.26. In view of (5) it follows that h satisfies condition (ii) also.

II.1.29. Continuation. Let us add a few remarks concerning the results obtained in II.1.28.

(a) Simple examples show that the property II.1.28(ii) is *not* a consequence of the property II.1.28(i). In other words, there may exist a topological transformation $h(\mathfrak{M}_1) = \mathfrak{M}_2$ that possesses the property II.1.28(i) but fails to possess the property II.1.28(ii).

(b) Comparison of the formulas II.1.28(2) and II.1.28(3) yields the following statement. If $T_1 \sim T_2(F)$, then there exist simultaneous monotone-light factorizations of the form

$$(1) \qquad T_1 = LM_1, \quad M_1(\mathcal{P}_1) = \mathfrak{M}, \quad L(\mathfrak{M}) = \mathcal{P}^*,$$

$$(2) \qquad T_2 = LM_2, \quad M_2(\mathcal{P}_2) = \mathfrak{M}, \quad L(\mathfrak{M}) = \mathcal{P}^*.$$

That is, there exist simultaneous monotone-light factorizations for T_1, T_2 with the same middle-space and the same light factor.

(c) Using the terminology of II.1.23, we obtain from II.1.28(ii) the following statement. If $T_1 \sim T_2(F)$, then for every partial transformation T_1^*, associated with T_1, there exists a partial transformation T_2^*, associated with T_2, such that $T_1^* \sim T_2^*(F)$.

(d) Suppose that $T_1 \sim T_2(F)$. Then by (b) above we have simultaneous factorizations of the form (1), (2), with the same middle-space $\mathfrak{M}$ and the same light factor L. Then the identity transformation on $\mathfrak{M}$, if denoted by h, clearly satisfies the condition II.1.28(i) relative to the factorizations II.1.29(1), II.1.29(2), and there may be a temptation to expect that the identity also satisfies, relative to these same factorizations, the condition II.1.28(ii). In other words, one may expect that for every continuous transformation Z from $\mathfrak{M}$ into $\mathfrak{M}$ the relation $LZM_1 \sim LZM_2(F)$ will hold. Simple examples show that this is generally not the case. The theorem of II.1.28, for special simultaneous factorizations of the form II.1.29(1), II.1.29(2), merely implies the existence of some topological transformation $h(\mathfrak{M}) = \mathfrak{M}$ such that $LZM_1 \sim LhZh^{-1}M_2(F)$, for every continuous transformation $Z(\mathfrak{M}) \subset \mathfrak{M}$ (provided that $T_1 \sim T_2(F)$). Generally, h cannot be chosen as the identity.

II.1.30. Given T_1, T_2 as in II.1.24, we shall say that T_1 and T_2 are *K-equivalent*, in symbols $T_1 \sim T_2(K)$, if and only if T_1, T_2 admit of simultaneous monotone-light factorizations with the same middle-space and the same light factor. The following remarks will be useful.

(a) Suppose that $T_1 \sim T_2(K)$. Let

$$(1) \qquad T_1 = L_1M_1, \quad M_1(\mathcal{P}_1) = \mathfrak{M}_1, \quad L_1(\mathfrak{M}_1) = \mathcal{P}^*,$$

$$(2) \qquad T_2 = L_2M_2, \quad M_2(\mathcal{P}_2) = \mathfrak{M}_2, \quad L_2(\mathfrak{M}_2) = \mathcal{P}^*$$

be arbitrary monotone-light factorizations of T_1, T_2 respectively. Then there exists a topological transformation $h(\mathfrak{M}_1) = \mathfrak{M}_2$, such that $L_1 = L_2h$.

Proof. By assumption, we have monotone-light factorizations of the form

$$(3) \qquad T_1 = LM_1^*, \quad M_1^*(\mathcal{P}_1) = \mathfrak{M}^*, \quad L(\mathfrak{M}^*) = \mathcal{P}^*,$$

$$(4) \qquad T_2 = LM_2^*, \quad M_2^*(\mathcal{P}_2) = \mathfrak{M}^*, \quad L(\mathfrak{M}^*) = \mathcal{P}^*.$$

By II.1.20, applied to the factorizations (1), (3) and (2), (4), respectively, there follows the existence of topological transformations h_1, h_2 such that

$$h_1(\mathfrak{M}_1) = \mathfrak{M}^*, \quad h_2(\mathfrak{M}_2) = \mathfrak{M}^*, \quad L_1 = Lh_1,$$
$$L_2 = Lh_2, \quad M_1^* = h_1M_1, \quad M_2^* = h_2M_2.$$

Clearly $h = h_2^{-1}h_1$ satisfies our requirements.

(b) Suppose, conversely, that T_1, T_2 (given as in II.1.24) admit of factorizations of the form (1), (2), such that there exists a topological transformation $h(\mathfrak{M}_1) = \mathfrak{M}_2$ satisfying the condition $L_1 = L_2 h$. Then $T_1 \sim T_2(K)$.

PROOF. On setting $\overline{M}_1 = hM_1$, we have $T_1 = L_1 M_1 = L_2 h M_1 = L_2 \overline{M}_1$. Since $\overline{M}_1$ is clearly monotone, we obtain thus for T_1 the monotone-light factorization

$$(5) \qquad T_1 = L_2 \overline{M}_1, \qquad \overline{M}_1(\mathcal{P}_1) = \mathfrak{M}_2, \qquad L_2(\mathfrak{M}_2) = \mathcal{P}^*.$$

Comparison of (2) and (5) yields $T_1 \sim T_2(K)$.

II.1.31. Let us consider the mutual implications of the relations $T_1 \sim T_2(ts)$, $T_1 \sim T_2(F)$, $T_1 \sim T_2(K)$, defined in II.1.26, II.1.25, II.1.30. Clearly, $T_1 \sim T_2(ts)$ implies $T_1 \sim T_2(F)$, the converse being generally false. By II.1.29(b), $T_1 \sim T_2(F)$ implies $T_1 \sim T_2(K)$, but *the converse is generally false*, as shown by the following simple examples.

(a) Let us choose $\mathcal{P}_1$ as the unit disc $u_1^2 + v_1^2 \leqq 1$, $\mathcal{P}_2$ as the unit square $0 \leqq u_2 \leqq 1$, $0 \leqq v_2 \leqq 1$, and $\mathcal{P}^*$ as the unit segment $0 \leqq t \leqq 1$. Let us define the continuous transformations $T_1(\mathcal{P}_1) = \mathcal{P}^*$, $T_2(\mathcal{P}_2) = \mathcal{P}^*$ by the formulas

$$T_1 : t = (u_1^2 + v_1^2)^{1/2}, \qquad (u_1, v_1) \in \mathcal{P}_1,$$

$$T_2 : t = u_2, \qquad (u_2, v_2) \in \mathcal{P}_2.$$

Clearly, T_1 and T_2 are monotone. If L denotes the identity transformation on $\mathcal{P}^*$, we have therefore the monotone-light factorizations $T_1 = LT_1$, $T_2 = LT_2$ with the same middle-space $\mathcal{P}^*$ and the same light factor L. Thus $T_1 \sim T_2(K)$. On the other hand, the relation $T_1 \sim T_2(F)$ does not hold. Indeed, consider *any* topological transformation $H(\mathcal{P}_1) = \mathcal{P}_2$. Then H carries the boundary of $\mathcal{P}_1$ into the boundary of $\mathcal{P}_2$, and hence we have on the boundary of $\mathcal{P}_1$ a point (u_1^0, v_1^0) that is carried by H into the point $u_2 = 0$, $v_2 = 0$. Then $\rho[T_1(u_1^0, v_1^0), T_2 H(u_1^0, v_1^0)] = 1$. Thus for $\epsilon < 1$ there exists no topological transformation $H_\epsilon(\mathcal{P}_1) = \mathcal{P}_2$ such that $\rho[T_1(u_1, v_1), T_2 H_\epsilon(u_1, v_1)] < \epsilon$ for every point $(u_1, v_1) \in \mathcal{P}_1$.

(b) Let us choose $\mathcal{P}_1$ as the unit disc $u_1^2 + v_1^2 \leqq 1$, $\mathcal{P}_2$ as the unit disc $u_2^2 + v_2^2 \leqq 1$, and $\mathcal{P}^*$ as the unit sphere $x^2 + y^2 + z^2 = 1$. It is easy to see that there exists a continuous monotone transformation $T_1(\mathcal{P}_1) = \mathcal{P}^*$ such that the boundary of $\mathcal{P}_1$ is carried by T_1 into the north pole $(0, 0, 1)$. Similarly, there exists a continuous monotone transformation $T_2(\mathcal{P}_2) = \mathcal{P}^*$ that carries the boundary of $\mathcal{P}_2$ into the south pole $(0, 0, -1)$. If we denote by L the identity transformation on $\mathcal{P}^*$, we have the monotone-light factorizations $T_1 = LT_1$, $T_2 = LT_2$ with the same middle-space $\mathcal{P}^*$ and the same light factor L. Thus $T_1 \sim T_2(K)$. On the other hand, the relation $T_1 \sim T_2(F)$ does not hold. Indeed, if $H(\mathcal{P}_1) = \mathcal{P}_2$ is *any* topological transformation, then H carries the boundary of $\mathcal{P}_1$ into the boundary of $\mathcal{P}_2$, and hence $\rho[T_1(u_1, v_1), T_2 H(u_1, v_1)] = 2$ for every point (u_1, v_1) on the boundary of $\mathcal{P}_1$. Hence for $\epsilon < 2$ there exists no topological transformation $H_\epsilon(\mathcal{P}_1) = \mathcal{P}_2$ such that $\rho[T_1(u_1, v_1), T_2 H_\epsilon(u_1, v_1)] < \epsilon$ for every point $(u_1, v_1) \in \mathcal{P}_1$.

We proceed now to exhibit several important special instances where the relation $T_1 \sim T_2(K)$ does imply the relation $T_1 \sim T_2(F)$. A detailed study of monotone transformations of simple arcs, simple closed curves, 2-cells and 2-spheres seems to be necessary for this purpose. This study will be carried out only to the extent needed in the sequel.

II.1.32. We shall be concerned with continuous monotone transformations of the form $M(\mathcal{P}) = \mathfrak{M}$, where $\mathcal{P}$, $\mathfrak{M}$ will be Peano spaces of some very simple type. The 2-sphere will play an important role, and it will be convenient to agree on certain notations for this case. S will denote the unit sphere $x^2 + y^2 + z^2 = 1$ in Euclidean three-space. E will denote the *equator*, that is the great circle in the xy-plane. The *north pole* $(0, 0, 1)$ and the *south pole* $(0, 0, -1)$ will be denoted by ν and σ respectively. The set of points of S satisfying $z \geqq 0$ and $z > 0$ will be called the closed and open *northern hemispheres* and will be denoted by S_ν, S_ν^0 respectively. The closed and open *southern hemispheres* S_σ, S_σ^0 are defined similarly in terms of the inequalities $z \leqq 0$, $z < 0$ respectively. We shall use a second unit sphere $\bar{S}$ in a Euclidean three-space $\bar{x}\bar{y}\bar{z}$. The notations $\bar{\nu}$, $\bar{\sigma}$, $\bar{S}_\nu$, $\bar{S}_\nu^0$, $\bar{E}$, and so forth, will have analogous meaning relative to $\bar{S}$.

II.1.33. Given $M(\mathcal{P}) = \mathfrak{M}$ as in II.1.32, suppose that $\mathcal{P}$ is a simple arc (see I.2.31). Then $\mathfrak{M}$ is either a single point or a simple arc.

PROOF. Suppose that $\mathfrak{M}$ is not a single point. Then $\mathfrak{M}$ has infinitely many points. Let a, b be the end points of the simple arc $\mathcal{P}$, and let $\mathfrak{x}_0$ be any point of $\mathfrak{M}$ distinct from $M(a)$ and $M(b)$. Then $M^{-1}(\mathfrak{x}_0)$ is a continuum in $\mathcal{P}$ that does not contain either a or b (see II.1.2). Thus $\mathcal{P} - M^{-1}(\mathfrak{x}_0)$ is disconnected, and hence $\mathfrak{M} - \mathfrak{x}_0$ is also disconnected, since $\mathcal{P} - M^{-1}(\mathfrak{x}_0) = M^{-1}(\mathfrak{M} - \mathfrak{x}_0)$ (cf. II.1.3). Thus $\mathfrak{M}$ has at most two non-cut points, namely $M(a)$ and $M(b)$. Hence $\mathfrak{M}$ is a simple arc (see I.2.35).

II.1.34. Given $M(\mathcal{P}) = \mathfrak{M}$ as in II.1.32, suppose that $\mathcal{P}$ is a simple closed curve (see I.2.31). Then $\mathfrak{M}$ is either a single point or a simple closed curve.

PROOF. Suppose $\mathfrak{M}$ is not a single point. Let $\mathfrak{x}_1$, $\mathfrak{x}_2$ be any two distinct points of $\mathfrak{M}$. By II.1.2 the sets $M^{-1}(\mathfrak{x}_1)$, $M^{-1}(\mathfrak{x}_2)$ are then disjoint continua on the simple closed curve $\mathcal{P}$. By I.2.32 it follows that $\mathcal{P} - (M^{-1}(\mathfrak{x}_1) + M^{-1}(\mathfrak{x}_2))$ is disconnected. Since $\mathcal{P} - (M^{-1}(\mathfrak{x}_1) + M^{-1}(\mathfrak{x}_2)) = M^{-1}(\mathfrak{M} - (\mathfrak{x}_1 + \mathfrak{x}_2))$, it follows by II.1.3 that $M - (\mathfrak{x}_1 + \mathfrak{x}_2)$ is disconnected. Hence $\mathfrak{M}$ is a simple closed curve by I.2.35.

II.1.35. Given $M(\mathcal{P}) = \mathfrak{M}$ as in II.1.32, suppose that $\mathcal{P}$ is a 2-sphere. Simple examples show that $\mathfrak{M}$ need not be a single point or a 2-sphere. On the other hand, if we assume that $\mathfrak{M}$ is cyclic (see I.2.34), then it follows that $\mathfrak{M}$ is either a single point or a 2-sphere.

PROOF. Suppose that $\mathfrak{M}$ is not a single point. Let $\mathfrak{C}$ be any continuum in $\mathfrak{M}$ such that $\mathfrak{M} - \mathfrak{C}$ is connected. Then $M^{-1}(\mathfrak{C})$ is a continuum in $\mathcal{P}$ and $\mathcal{P} - M^{-1}(\mathfrak{C}) = M^{-1}(\mathfrak{M} - \mathfrak{C})$ is connected (see II.1.2). Since $\mathcal{P}$ is a 2-sphere, it follows that $M^{-1}(\mathfrak{C})$ is unicoherent (see I.2.35). Hence $\mathfrak{C}$ itself is unicoherent by II.1.6. Thus every non-separating continuum of the cyclic Peano space $\mathfrak{M}$ is unicoherent. By I.2.35 it follows that $\mathfrak{M}$ is a 2-sphere.

II.1.36. Given $T(\mathcal{P}) = \mathcal{P}^*$ as in II.1.1, let G be a Peano subspace of $\mathcal{P}$, and let us put $T(G) = G^*$. For clarity, let us use the symbol T_G to refer to T thought of as operating from G only. We shall say that T is monotone on G if and only if the transformation $T_G(G) = G^*$ is monotone in the sense of II.1.1 Clearly, T is monotone on G if and only if $GT^{-1}(x^*)$ is connected (possibly empty) for every point $x^* \in \mathcal{P}^*$.

II.1.37. Given $M(\mathcal{P}) = \mathfrak{M}$ as in II.1.32, suppose that $\mathcal{P}$ is a 2-cell and $\mathfrak{M}$ is cyclic (see I.2.34). Then M is monotone on the boundary of $\mathcal{P}$ (see I.2.31, II.1.36).

Proof. *Case* 1. M is constant on the boundary of $\mathcal{P}$. The assertion is then obvious.

Case 2. M is not constant on the boundary of $\mathcal{P}$. Let us assume first that $\mathcal{P}$ coincides with the closed southern hemisphere S_σ on the unit sphere S (see II.1.32). In view of II.1.36, we have to show that $EM^{-1}(\mathfrak{x}_0)$ is connected (possibly empty) for every point $\mathfrak{x}_0 \in \mathfrak{M}$ (E is the equator). If $EM^{-1}(\mathfrak{x}_0) = 0$, the assertion is obvious. So we can assume that

$$EM^{-1}(\mathfrak{x}_0) = e \neq 0. \tag{1}$$

Let us put $M^{-1}(\mathfrak{x}_0) = C_\sigma$. Then C_σ is a continuum on the southern hemisphere S_σ (cf. II.1.2). Let C_ν denote the image of C_σ under the reflection on the plane of the equator. Then C_ν is a continuum on the northern hemisphere S_ν, and clearly

$$C_\nu C_\sigma = e \neq 0, \tag{2}$$

$$S - (C_\nu + C_\sigma) = (S_\nu - C_\nu) + (S_\sigma - C_\sigma), \tag{3}$$

$$(S_\nu - C_\nu)(S_\sigma - C_\sigma) = E - e \neq 0. \tag{4}$$

In particular, the relation $E - e \neq 0$ follows from the assumption that M is not constant on E and hence E is not a subset of $M^{-1}(\mathfrak{x}_0) = C_\sigma$. Now since $\mathfrak{M}$ is cyclic, $\mathfrak{M} - \mathfrak{x}_0$ is connected, and hence $S_\sigma - C_\sigma = M^{-1}(\mathfrak{M} - \mathfrak{x}_0)$ is also connected (see II.1.2). As $S_\nu - C_\nu$ is derived from $S_\sigma - C_\sigma$ by reflection upon the xy-plane, $S_\nu - C_\nu$ is also connected. By (3), (4), it follows that $S - (C_\nu + C_\sigma)$ is connected. Since $C_\nu + C_\sigma$ is a continuum by (2), it follows that $C_\nu + C_\sigma$ is unicoherent (see I.2.35). Hence $C_\nu C_\sigma = e$ is connected. Since clearly $EM^{-1}(\mathfrak{x}_0) = EC_\sigma = C_\nu C_\sigma$, the connectedness of $EM^{-1}(\mathfrak{x}_0)$ is established.

The case when $\mathcal{P}$ is a general 2-cell may now be treated as follows. By I.2.31 we have a topological transformation $H(\mathcal{P}) = S_\sigma$. Then $MH^{-1}(S_\sigma) = \mathfrak{M}$ is clearly a continuous monotone transformation from S_σ onto $\mathfrak{M}$, and hence MH^{-1} is monotone on E by the preceding discussion. Since H carries the boundary of $\mathcal{P}$ topologically into E, it follows immediately that M is monotone on the boundary of $\mathcal{P}$.

II.1.38. Given $M(\mathcal{P}) = \mathfrak{M}$ as in II.1.32, suppose that $\mathcal{P}$ is a 2-cell, $\mathfrak{M}$ is cyclic, and M is constant on the boundary of $\mathcal{P}$. Then $\mathfrak{M}$ is either a single point or a 2-sphere.

PROOF. By an argument similar to that used in II.1.37, we can assume without loss of generality that $\mathcal{P}$ coincides with the southern hemisphere S_σ on the unit sphere S (see II.1.32). By assumption, M carries the equator E into a single point $\mathfrak{x}_0 \in \mathfrak{M}$. Let us define the transformation $\overline{M}(S) = \mathfrak{M}$ as follows: $\overline{M}(x) = M(x)$ if $x \in S_\sigma$, and $\overline{M}(x) = \mathfrak{x}_0$ if $x \in S_\nu^0$. Clearly $\overline{M}$ is continuous and monotone on S, and hence $\mathfrak{M}$ is either a single point or a 2-sphere by II.1.35.

II.1.39. Given $M(\mathcal{P}) = \mathfrak{M}$ as in II.1.32, suppose that $\mathcal{P}$ is a 2-cell, $\mathfrak{M}$ is cyclic, and M is not constant on the boundary of $\mathcal{P}$. Then $\mathfrak{M}$ is a 2-cell.

PROOF. By an argument similar to that used in II.1.37, we can assume without loss of generality that $\mathcal{P}$ coincides with the southern hemisphere S_σ on the unit sphere S (see II.1.32). The transformation $M(S_\sigma) = \mathfrak{M}$ gives rise to an u.s.c.c. $I(M)$ of continua (see II.1.13(ii), II.1.13(i)) in S_σ. Let K_1 be the collection of the continua of $I(M)$ plus the individual points of S_ν^0. Clearly, K_1 is an u.s.c.c. of continua in S (cf. II.1.10). Let us note that no continuum $\gamma \in K_1$ separates S. Indeed, if γ is a single point, then the assertion is obvious. If γ is not a single point, then $\gamma \in I(M)$, and hence $\gamma = M^{-1}(\mathfrak{x}_0)$, where $\mathfrak{x}_0$ is some point of $\mathfrak{M}$. Now $\mathfrak{M}$ is cyclic by assumption, and hence $S_\sigma - \gamma = M^{-1}(\mathfrak{M} - \mathfrak{x}_0)$ is connected (see II.1.2). Since $S - \gamma = (S_\sigma - \gamma) + S_\nu^0$ (note that $\gamma \subset S_\sigma$), the connectedness of $S - \gamma$ will be established if we can show that $(S_\sigma - \gamma)c(S_\nu^0) = (S_\sigma - \gamma)S_\nu = (E - \gamma)S_\nu = E - \gamma \neq 0$. Now the assumption $E - \gamma = 0$ implies that $E \subset \gamma$ and hence $M(E) \subset M(\gamma) = \mathfrak{x}_0$, in contradiction with the assumption that M is not constant on E.

Thus K_1 is an u.s.c.c. of continua in S, such that no continuum of K_1 separates S. Let us add that clearly no continuum of K_1 coincides with S. By II.1.15, there exists a continuous transformation M_1 from S onto some Peano space $\mathfrak{M}_1$, such that $I(M_1) \equiv K_1$. Since K_1 is an u.s.c.c. of continua, M_1 is clearly monotone. We assert that $\mathfrak{M}_1$ is cyclic. Indeed, if $\mathfrak{x}_1$ is any point of $\mathfrak{M}_1$, then $M_1^{-1}(\mathfrak{x}_1) = \gamma_1$ is a continuum of $I(M_1) \equiv K_1$, and hence $S - \gamma_1$ is connected (see above). Consequently $\mathfrak{M}_1 - \mathfrak{x}_1 = M_1(S - \gamma_1)$ is also connected. Furthermore, M_1 is not constant on S, since no continuum of $I(M_1) \equiv K_1$ coincides with S. Thus $\mathfrak{M}_1$ is cyclic and does not reduce to a single point. Since the transformation $M_1(S) = \mathfrak{M}_1$ is monotone, it follows that $\mathfrak{M}_1$ is a 2-sphere (see II.1.35).

We assert that M_1 is monotone on the equator E of S. Indeed, let $\mathfrak{z}_1$ be any point of $\mathfrak{M}_1$. If $M_1^{-1}(\mathfrak{z}_1)$ is a single point, then $EM_1^{-1}(\mathfrak{z}_1)$ is clearly connected. If $M_1^{-1}(\mathfrak{z}_1) = \gamma$ is not a single point, then $\gamma \in I(M)$ by the definition of M_1, and hence $EM_1^{-1}(\mathfrak{z}_1) = E\gamma$ is connected since M itself is monotone on E (see II.1.37, II.1.36). Thus M_1 is monotone on E by II.1.36. By II.1.34 it follows that $M_1(E)$ is either a single point or a simple closed curve. In the first case, M_1 should be constant on E, and hence E should be a subset of a continuum $\Gamma \in I(M_1) \equiv K_1$. By the definition of K_1 it follows that $\Gamma \in I(M)$ and hence M would be constant on E, contrary to our assumptions. Hence $M_1(E) = \mathfrak{C}_1$ is a simple closed curve on the 2-sphere $\mathfrak{M}_1$. Thus $\mathfrak{C}_1$ is the common boundary of two 2-cells $\mathfrak{A}_1$, $\mathfrak{B}_1$ on $\mathfrak{M}_1$. If the interiors of these 2-cells are denoted by $\mathfrak{A}_1^0$, $\mathfrak{B}_1^0$ respectively, then the components of $\mathfrak{M}_1 - \mathfrak{C}_1$ are precisely $\mathfrak{A}_1^0$, $\mathfrak{B}_1^0$.

Hence the components of $S - M_1^{-1}(\mathfrak{C}_1)$ are precisely $M_1^{-1}(\mathfrak{A}_1^0)$, $M_1^{-1}(\mathfrak{B}_1^0)$ by II.1.5. We assert that

$$(1) \qquad E \subset M_1^{-1}(\mathfrak{C}_1) \subset S_\sigma .$$

Indeed, the first inclusion follows from the fact that $M_1(E) = \mathfrak{C}_1$. Suppose now that there exists a point $x_0 \in S_\nu^0$ such that $x_0 \in M_1^{-1}(\mathfrak{C}_1)$. Since $\mathfrak{C}_1 = M_1(E)$, we should have then a point $y_0 \in E$ such that $M_1(y_0) = M_1(x_0)$. It follows that x_0 and y_0 are points of a continuum $\gamma_0 \in I(M_1) \equiv K_1$. Since γ_0 does not reduce to a single point, we should have $\gamma_0 \in I(M)$ and hence $\gamma_0 \subset S_\sigma$, in contradiction with the assumption that γ_0 contains the point $x_0 \in S_\nu^0$. Thus (1) is established. Clearly, (1) implies that S_ν^0 is a component of $S - M_1^{-1}(\mathfrak{C}_1)$. Hence, by a preceding remark, S_ν^0 coincides with either $M_1^{-1}(\mathfrak{A}_1^0)$ or $M_1^{-1}(\mathfrak{B}_1^0)$, say $S_\nu^0 = M_1^{-1}(\mathfrak{B}_1^0)$. Then $S_\sigma = S - S_\nu^0 = M_1^{-1}(\mathfrak{M}_1 - \mathfrak{B}_1^0) = M_1^{-1}(\mathfrak{A}_1)$. Consequently $M_1(S_\sigma) = \mathfrak{A}_1$. Now compare the mappings $M_1(S_\sigma) = \mathfrak{A}_1$, $M(S_\sigma) = \mathfrak{M}$. These two mappings determine, by the definition of M_1, the same u.s.c.c. of continua on S_σ. By II.1.16, there follows the existence of a topological transformation $H(\mathfrak{A}_1) = \mathfrak{M}$. Since $\mathfrak{A}_1$ is a 2-cell, it follows that $\mathfrak{M}$ is also a 2-cell, and the proof is complete.

II.1.40. CONTINUATION. The preceding argument yields a further result. Comparison of the mappings $M_1(S_\sigma) = \mathfrak{A}_1$, $M(S_\sigma) = \mathfrak{M}$ shows, in view of II.1.16, that the mapping $H = MM_1^{-1}$ is a topological mapping from $\mathfrak{A}_1$ onto $\mathfrak{M}$. Hence $H(\mathfrak{C}_1) = \mathfrak{C}$ is the boundary curve of $\mathfrak{M}$ (note that we have already proved that $\mathfrak{M}$ is a 2-cell). Now $M_1(E) = \mathfrak{C}_1$, and hence $M(E) = HM_1(E) = H(\mathfrak{C}_1) = \mathfrak{C}$. Thus M carries the boundary E of S_σ into the boundary $\mathfrak{C}$ of $\mathfrak{M}$.

II.1.41. Given $M(\mathcal{P}) = \mathfrak{M}$ as in II.1.32, suppose that $\mathcal{P}$ and $\mathfrak{M}$ are both 2-cells. Let C and $\mathfrak{C}$ denote the boundaries of $\mathcal{P}$ and $\mathfrak{M}$ respectively. Then $M(C) = \mathfrak{C}$, and M is monotone on C.

PROOF. As in II.1.39, we can assume without loss of generality that $\mathcal{P}$ coincides with the southern hemisphere S_σ on the unit sphere S. Then the boundary C of $\mathcal{P}$ coincides with the equator E (cf. II.1.32). We assert that M is not constant on E. Indeed, since $\mathfrak{M}$ is a 2-cell, $\mathfrak{M}$ is surely cyclic. The assumption that M is constant on E would therefore lead to the conclusion that $\mathfrak{M}$ is a 2-sphere or a single point (see II.1.38), while $\mathfrak{M}$ is a 2-cell by assumption. Thus M is not constant on E. Since $\mathfrak{M}$ is cyclic, the results derived in II.1.37 and II.1.40 are available. Hence $M(E) = \mathfrak{C}$ by II.1.40, and M is monotone on E by II.1.37.

II.1.42. Given $M(\mathcal{P}) = \mathfrak{M}$ as in II.1.32, suppose that $\mathcal{P}$ is a 2-cell and $\mathfrak{M}$ is cyclic. Then $\mathfrak{M}$ is either a single point, or a 2-cell, or a 2-sphere. The first case occurs if and only if M is constant on $\mathcal{P}$. The second case occurs if and only if M is not constant on the boundary of $\mathcal{P}$. The third case occurs if and only if M is constant on the boundary of $\mathcal{P}$ without being constant on $\mathcal{P}$. In the second case, M is monotone on the boundary of $\mathcal{P}$, and carries the boundary of $\mathcal{P}$ into the boundary of $\mathfrak{M}$.

This statement is merely a summary of the results in II.1.39, II.1.38, II.1.41.

II.1.43. Let $\mathcal{P}$ be a 2-sphere, and let K be an u.s.c.c. of continua on $\mathcal{P}$. By

II.1.15, there exists a continuous transformation $M(\mathcal{P}) = \mathfrak{M}$ from $\mathcal{P}$ onto some Peano space $\mathfrak{M}$, such that $I(M) \equiv K$. Since the elements of K are continua, M is clearly monotone. The following corollaries of preceding results will be useful.

(a) $\mathfrak{M}$ is cyclic if and only if $\mathcal{P} - \gamma$ is connected for every $\gamma \in K$. Indeed, if $\mathfrak{x}_0$ is any point of $\mathfrak{M}$, then $M^{-1}(\mathfrak{x}_0) \in I(M) \equiv K$ and $M^{-1}(\mathfrak{M} - \mathfrak{x}_0) = \mathcal{P} - M^{-1}(\mathfrak{x}_0)$. Hence, by II.1.3, $\mathfrak{x}_0$ is a cut point of $\mathfrak{M}$ if and only if $\mathcal{P} - M^{-1}(\mathfrak{x}_0)$ is disconnected. Thus if $\mathcal{P} - \gamma$ is connected, then $\mathfrak{M}$ is cyclic. The converse is equally immediate.

(b) Suppose that $\mathcal{P} - \gamma$ is connected and nonempty for every $\gamma \in K$. Then $\mathfrak{M}$ is a 2-sphere. Indeed, $\mathfrak{M}$ is cyclic and does not reduce to a single point (see (a)), and hence $\mathfrak{M}$ is a 2-sphere by II.1.35.

II.1.44. Let $\mathcal{P}$ be a 2-cell, and let K be an u.s.c.c. of continua on $\mathcal{P}$. By II.1.15, there exists a continuous transformation $M(\mathcal{P}) = \mathfrak{M}$ from $\mathcal{P}$ onto some Peano space $\mathfrak{M}$, such that $I(M) \equiv K$. Clearly M is monotone.

(a) $\mathfrak{M}$ is cyclic if and only if $\mathcal{P} - \gamma$ is connected for every $\gamma \in K$. The reasoning is the same as in II.1.43.

(b) If $\mathcal{P} - \gamma$ is connected for every $\gamma \in K$, then $\mathfrak{M}$ is either a single point, or a 2-cell, or a 2-sphere. If the boundary of $\mathcal{P}$ is not a subset of any $\gamma \in K$, then $\mathfrak{M}$ is a 2-cell. If the boundary of $\mathcal{P}$ is a subset of some $\gamma_0 \in K$, then $\mathfrak{M}$ is a 2-sphere if $\gamma_0 \not\equiv \mathcal{P}$ and $\mathfrak{M}$ is a single point if $\gamma_0 = \mathcal{P}$. In view of (a), these statements follow directly from II.1.42.

II.1.45. Let $\mathcal{P}_1$, $\mathcal{P}_2$ be 2-cells with boundaries C_1, C_2 and let $m_{12}(C_1) = C_2$ be a continuous monotone transformation. Then there exists a continuous monotone transformation $M_{12}(\mathcal{P}_1) = \mathcal{P}_2$ with the following properties.

(i) $M_{12}(x_1) = m_{12}(x_1)$ for $x_1 \in C_1$.

(ii) M_{12} maps the interior $\mathcal{P}_1^0$ of $\mathcal{P}_1$ topologically onto the interior $\mathcal{P}_2^0$ of $\mathcal{P}_2$.

Proof. The monotone transformation m_{12} determines on C_1 an u.s.c.c. of continua in the sense of II.1.13(i). Let us denote this collection by k_1. Let us define the collection K_1 as follows: K_1 consists of the continua of k_1 and of the individual points of $\mathcal{P}_1^0$. Clearly, K_1 is again an u.s.c.c. of continua in $\mathcal{P}_1$, and $\mathcal{P}_1 - \gamma_1$ is connected for every $\gamma_1 \in K_1$. Note also that C_1 is not a subset of any continuum of K_1, since $m_{12}(C_1)$ is not a single point. By II.1.44 there follows the existence of a continuous monotone transformation $M_{13}(\mathcal{P}_1) = \mathcal{P}_3$, such that $I(M_{13}) \equiv K_1$ and $\mathcal{P}_3$ is a 2-cell. Let C_3 denote the boundary of the 2-cell $\mathcal{P}_3$. Let us compare the transformations $m_{12}(C_1) = C_2$ and $M_{13}(C_1) = C_3$ (cf. II.1.42). These two transformations determine on C_1 the same u.s.c.c. of continua, namely k_1. By II.1.16 it follows that the formula $x_2 = h_{32}(x_3) = m_{12}M_{13}^{-1}(x_3)$, $x_3 \in C_3$, determines a topological transformation from C_3 onto C_2. By I.2.52, there follows the existence of a topological transformation $H_{32}(\mathcal{P}_3) = \mathcal{P}_2$ such that $H_{32} = h_{32}$ on C_3. Then $M_{12} = H_{32}M_{13}$ is a continuous monotone transformation from $\mathcal{P}_1$ onto $\mathcal{P}_2$. Let x_1 be any point of C_1. Then $M_{13}(x_1) = x_3$ is a point on C_3, and hence $M_{12}(x_1) = H_{32}M_{13}(x_1) = h_{32}M_{13}(x_1) = m_{12}M_{13}^{-1}M_{13}(x_1) \supset m_{12}(x_1)$. Since $M_{12}(x_1)$ is a single point, it follows that $M_{12}(x_1) = m_{12}(x_1)$ for $x_1 \in C_1$.

Clearly $I(M_{12}) \equiv I(M_{13}) \equiv K_1$ (note that H_{32} is topological). Since $\gamma \in K_1$,

$\gamma \mathcal{P}_1^0 \neq 0$ implies that γ is a single point, and since $M_{12}(C_1) = C_2$, it follows that M_{12} maps $\mathcal{P}_1^0$ biuniquely onto $\mathcal{P}_2^0$. Thus M_{12}^{-1} is single-valued on $\mathcal{P}_2^0$ and there remains to show only that M_{12}^{-1} is continuous on $\mathcal{P}_2^0$. Now let x_2^n be a convergent sequence of points in $\mathcal{P}_2^0$ with limit $x_2 \in \mathcal{P}_2^0$. Consider the points $x_1^n = M_{12}^{-1}(x_2^n)$, $x_1 = M_{12}^{-1}(x_2)$, and suppose that the relation $x_1^n \to x_1$ does not hold. Since $\mathcal{P}_1$ is compact, we can then assume without loss of generality that $x_1^n \to y_1 \neq x_1$. Then $y_1 \in \mathcal{P}_1$, and $M_{12}(x_1^n) \to M_{12}(y_1)$. On the other hand, $M_{12}(x_1^n) = x_2^n \to x_2$. Hence $M_{12}(y_1) = x_2 = M_{12}(x_1)$. Since M_{12} is biuinque on $\mathcal{P}_1^0$ and $x_1 \in \mathcal{P}_1^0$, it follows that y_1 must lie on C_1. But then $M_{12}(y_1) = x_2$ would lie on C_2, while $x_2 \in \mathcal{P}_2^0$ by assumption. This contradiction shows that M_{12}^{-1} is continuous on $\mathcal{P}_2^0$.

II.1.46. Using the terminology of II.1.32, let $M(S_\sigma) = \overline{S}_\sigma$ be a continuous monotone transformation from the southern hemisphere S_σ onto the southern hemisphere $\overline{S}_\sigma$. Then there exists a continuous monotone transformation $\tilde{M}(S) = \overline{S}$ with the following properties.

(i) $\tilde{M}(p) = M(p)$ for $p \in S_\sigma$.

(ii) $\tilde{M}$ maps S_ν^0 topologically onto $\overline{S}_\nu^0$

Proof. By II.1.42, M maps the equator E onto the equator $\overline{E}$, and the transformation $M(E) = \overline{E}$ is monotone (on E). By II.1.45, applied to the 2-cells S_ν, $\overline{S}_\nu$, there follows the existence of a continuous monotone transformation $M_\nu(S_\nu) = \overline{S}_\nu$, such that M_ν agrees with M on E and maps S_ν^0 topologically onto $\overline{S}_\nu^0$. If we define

$$\tilde{M}(p) = \begin{cases} M(p) & \text{for} \quad p \in S_\sigma, \\ M_\nu(p) & \text{for} \quad p \in S_\nu, \end{cases}$$

then clearly $\tilde{M}$ satisfies our requirements.

II.1.47. Using the terminology of II.1.32, *let $M(S) = \overline{S}$ be a continuous monotone transformation from the unit sphere S onto the unit sphere $\overline{S}$. Let $\overline{A}$ be a 2-cell on $\overline{S}$, and let $\overline{A}^0$ be the interior of $\overline{A}$. Then there exists a continuous monotone transformation $\tilde{M}(S) = \overline{S}$ with the following properties.*

(i) $\tilde{M}(x) = M(x)$ *for* $x \notin M^{-1}(\overline{A}^0)$.

(ii) $\tilde{M}$ *maps* $M^{-1}(\overline{A}^0)$ *topologically onto* $\overline{A}^0$.

Proof. Let us put $\overline{B} = \overline{S} - \overline{A}^0$. Then $\overline{B}$ is a 2-cell, and $M^{-1}(\overline{B}) = S - M^{-1}(\overline{A}^0)$. Let us consider the u.s.c.c. of continua $I(M)$ on S (cf. II.1.13). We define a new collection K_1 as follows: K_1 consists of the continua $\gamma \in I(M)$ that are subsets of $M^{-1}(\overline{B})$, and of the individual points of $M^{-1}(\overline{A}^0)$. Clearly K_1 is an u.s.c.c. of continua (cf. II.1.13(iii)), and no continuum of K_1 coincides with S. We assert that $S - \gamma$ is connected for every $\gamma \in K_1$. Indeed, this is obvious if γ is a single point. If γ is not a single point, then $\gamma \in I(M)$, and $S - \gamma$ is connected since $M(S) = \overline{S}$ is cyclic (cf. II.1.43). By II.1.15 there exists a continuous monotone transformation $M_1(S) = \Sigma_1$ such that $I(M_1) \equiv K_1$. By II.1.43 it follows that Σ_1 is a 2-sphere. Let us put $M_1[M^{-1}(\overline{B})] = B_1$, and let us compare the transformations M_1 and M on the set $M^{-1}(\overline{B})$. Thought of as operating from $M^{-1}(\overline{B})$ only, M and M_1 are related to each other as the transformations T_1, T_2 of II.1.16 (with T_1, T_2, Σ, Σ_1^*, Σ_2^* replaced by M, M_1, $M^{-1}(\bar{B})$, $\bar{B}$, B_1). Hence the transformation $H_1 = M_1M^{-1}$ is a single-valued topological transformation from $\bar{B}$ onto B_1. As a consequence, the set $B_1 = M_1[M^{-1}(\bar{B})]$ is also a 2-cell. Thus we have a topological transformation $H_1(\bar{B}) = B_1$. By I.2.52, H_1 can be extended to a topological transformation from $\bar{S}$ onto Σ_1. Using H_1 to refer also to the extended transformation, we can sum up our present information as follows.

(i) $H_1(\bar{S}) = \Sigma_1$, $H_1(\bar{B}) = B_1$, $H_1(\bar{A}^0) = \Sigma_1 - B_1$.

(ii) $H_1(\bar{x}) = M_1M^{-1}(\bar{x})$ for $\bar{x} \in \bar{B}$.

Now let us consider the transformation $\tilde{M}(S) = \bar{S}$ defined by $\tilde{M}(x) = H_1^{-1}M_1(x)$, $x \in S$. The relations (i) and (ii) yield

(i*) $\tilde{M}(S) = \bar{S}$, $\tilde{M}[M^{-1}(\bar{B})] = \bar{B}$, $\tilde{M}[M^{-1}(\bar{A}^0)] = \bar{A}^0$,

(ii*) $\tilde{M}(x) = MM_1^{-1}M_1(x) \supset M(x)$, and hence $\tilde{M}(x) = M(x)$ for $x \in M^{-1}(\bar{B})$.

Now let $\bar{x}_0$ be any point of $\bar{A}^0$. Then $\tilde{M}^{-1}(\bar{x}_0) = M_1^{-1}H_1(\bar{x}_0)$, where $H_1(\bar{x}_0) \in \Sigma_1 - B_1$. By the definition of M_1, it follows that $M_1^{-1}H_1(\bar{x}_0)$ is a single point in $M^{-1}(\bar{A}^0)$. Thus for $\bar{x}_0 \in \bar{A}^0$, the set $\tilde{M}^{-1}(\bar{x}_0)$ reduces to a single point in $M^{-1}(\bar{A}^0)$. In view of (i*) it follows that $\tilde{M}$ maps $M^{-1}(\bar{A}^0)$ biuniquely and continuously onto $\bar{A}^0$. The continuity of $\tilde{M}^{-1}$ on $\bar{A}^0$ follows now by a reasoning analogous to that used at the end of II.1.45.

II.1.48. For convenient reference, we list some obvious consequences of the properties (i), (ii) of the transformation $\tilde{M}$ of II.1.47.

(iii) If $\bar{x}$ is a point of $\bar{A}^0$, then $\tilde{M}^{-1}(\bar{x})$ is a single point in $M^{-1}(\bar{A}^0)$.

(iv) If $\bar{x} \notin \bar{A}^0$, then $\tilde{M}^{-1}(\bar{x}) = M^{-1}(\bar{x})$.

(v) If F is any set in S, then $(\bar{S} - \bar{A}^0)\tilde{M}(F) = (\bar{S} - \bar{A}^0)M(F)$.

(vi) If $\tilde{M}(x) \neq M(x)$, then $\rho[\tilde{M}(x), M(x)] \leqq d(\bar{A})$.

(vii) $d[\tilde{M}^{-1}(\bar{x})] \leqq d[M^{-1}(\bar{x})]$ for every point $\bar{x} \in \bar{S}$ (see (iii), (iv)).

(viii) If G is a continuum such that $G \subset S - M^{-1}(\bar{A}^0)$ and M is monotone on G, then $\tilde{M}$ is also monotone on G (see (iv) and II.1.36).

II.1.49. Let $H_n(\mathcal{P}) = \mathfrak{M}$ be a sequence of topological transformations from the Peano space $\mathcal{P}$ onto the Peano space $\mathfrak{M}$. Suppose that the sequence H_n converges on $\mathcal{P}$ uniformly to a transformation $M(\mathcal{P}) = \mathfrak{M}$. Then M is continuous and monotone, as a special case of II.1.22. Briefly, the uniform limit of homeomorphisms is monotone. The converse is true only in certain special cases, some of which will be discussed presently.

II.1.50. Let $M(\mathcal{P}) = \mathfrak{M}$ be a continuous monotone transformation, where $\mathcal{P}$ is a simple arc and M is not constant on $\mathcal{P}$. Given $\epsilon > 0$, there exists a homeomorphism $H(\mathcal{P}) = \mathfrak{M}$ such that $\rho[M(x), H(x)] < \epsilon$ for every $x \in \mathcal{P}$.

Proof. $\mathfrak{M}$ is a simple arc by II.1.33. This point being established, the rest of the proof is practically trivial, but we carry it out in some detail to illustrate a device that will be used frequently in the sequel. We make the proof in two steps.

Case (1). $\mathcal{P}$ coincides with the segment $0 \leqq u \leqq 1$, and $\mathfrak{M}$ coincides with the segment $0 \leqq v \leqq 1$. Then M can be represented by an equation $f(u) = v$, where $f(u)$ is a continuous function of u on the segment $0 \leqq u \leqq 1$. Clearly the monotone character of M implies that f is either nondecreasing or nonincreasing. Suppose that $f(u)$ is nondecreasing. Then clearly $f(0) = 0, f(1) = 1$. For each positive integer n, let us put

$$f_n(u) = \frac{f(u) + u/n}{1 + 1/n}.$$

Then $f_n(u)$ is continuous, strictly increasing, and $f_n(0) = 0, f_n(1) = 1$. Thus the transformation H_n defined by $v = f_n(u)$ is a topological transformation from the segment $0 \leqq u \leqq 1$ onto the segment $0 \leqq v \leqq 1$, and $\rho[M(u), H_n(u)] = | f(u) - f_n(u) | < 1/n$ for $0 \leqq u \leqq 1$. Hence H_n converges to M uniformly on the segment $0 \leqq u \leqq 1$, and thus H_n satisfies our requirements for n sufficiently large. The case when $f(u)$ is nonincreasing is treated similarly.

Case 2. $\mathcal{P}$ and $\mathfrak{M}$ are general simple arcs. Let σ_1, σ_2 denote the segments $0 \leqq u \leqq 1, 0 \leqq v \leqq 1$ respectively. We have then homeomorphisms $h_1(\sigma_1) = \mathcal{P}$, $h_2(\sigma_2) = \mathfrak{M}$. Clearly, $h_2^{-1}Mh_1$ is then a continuous monotone transformation from σ_1 onto σ_2. Hence, by the argument used in case (1), we have a sequence of homeomorphisms $H_n^*(\sigma_1) = \sigma_2$ such that H_n^* converges on σ_1 uniformly to $h_2^{-1}Mh_1$. Clearly, $H_n = h_2H_n^*h_1^{-1}$ is then a topological transformation from $\mathcal{P}$ onto $\mathfrak{M}$, and H_n converges on $\mathcal{P}$ uniformly to $h_2h_2^{-1}Mh_1h_1^{-1} = M$. Thus H_n satisfies our requirements for n sufficiently large.

II.1.51. Let $M(\mathcal{P}) = \mathfrak{M}$ be a continuous monotone transformation, where $\mathcal{P}$ is a simple closed curve and M is not constant on $\mathcal{P}$. Given $\epsilon > 0$, there exists a homeomorphism $H(\mathcal{P}) = \mathfrak{M}$ such that $\rho[M(x), H(x)] < \epsilon$ for every $x \in \mathcal{P}$.

Proof. $\mathfrak{M}$ is a simple closed curve by II.1.34. If $\mathfrak{x}$ is a point of $\mathfrak{M}$, then $M^{-1}(\mathfrak{x})$ is a continuum in $\mathcal{P}$. Since $\mathcal{P}$ is a simple closed curve, it follows that $M^{-1}(\mathfrak{x})$ is either a single point or a simple arc. Clearly, the second alternative can happen for at most a countably infinite number of points $\mathfrak{x} \in \mathfrak{M}$. We can therefore choose two distinct points $\mathfrak{x}_1$, $\mathfrak{x}_2$ in $\mathfrak{M}$, such that $M^{-1}(\mathfrak{x}_1)$ and $M^{-1}(\mathfrak{x}_2)$ reduce to single points x_1, x_2 of $\mathcal{P}$, where clearly $x_1 \neq x_2$. Let α^1, β^1 be the simple arcs determined in $\mathfrak{M}$ by the points $\mathfrak{x}_1$ and $\mathfrak{x}_2$ as end points. From II.1.2, II.1.5 it follows that $\alpha = M^{-1}(\alpha^1)$ and $\beta = M^{-1}(\beta^1)$ are simple arcs in $\mathcal{P}$ which have only their end points x_1, x_2 in common. Clearly, M is monotone on both α and β. By II.1.50 there follows the existence of homeomorphisms $H_\alpha(\alpha) = \alpha^1$, $H_\beta(\beta) = \beta^1$ such that $H_\alpha(x_1) = \mathfrak{x}_1$, $H_\alpha(x_2) = \mathfrak{x}_2$, $H_\beta(x_1) = \mathfrak{x}_1$, $H_\beta(x_2) = \mathfrak{x}_2$, $\rho[M(x), H_\alpha(x)] < \epsilon$ for $x \in \alpha$, $\rho[M(x), H_\beta(x)] < \epsilon$ for $x \in \beta$. If we define $H(x) = H_\alpha(x)$ for $x \in \alpha$, $H(x) = H_\beta(x)$ for $x \in \beta$, then clearly H has the required properties.

II.1.52. Let $M(\mathcal{P}) = \mathfrak{M}$ be a continuous monotone transformation, where $\mathcal{P}$ is a 2-sphere and M is not constant on $\mathcal{P}$. Then $\mathfrak{M}$ need not be homeomorphic with $\mathcal{P}$, and hence no results similar to those in II.1.50, II.1.51 can be expected unless further restrictions are introduced. Similar remarks apply to the case when $\mathcal{P}$ is a 2-cell. We proceed to discuss approximation theorems of the type indicated by these observations. In preparation, we introduce some further definitions. Using the terminology of II.1.32, let $\overline{A}$ be a 2-cell on $\overline{S}$, and let $\overline{A}^0$

be the interior of $\bar{A}$. The 2-cell $\bar{A}$ will be termed *admissible* if either $\bar{A}^0\bar{E} = 0$, or if $\bar{A}^0\bar{E} \neq 0$ and the boundary curve of $\bar{A}$ intersects the equator $\bar{E}$ in exactly two points. A finite system of 2-cells $\bar{A}_1, \cdots, \bar{A}_a$ on $\bar{S}$ will be termed admissible if $\bar{A}_i^0\bar{A}_j^0 = 0$ for $i \neq j$, and each $\bar{A}_i$ is admissible.

Let $\eta > 0$ be given. We assert the existence of three admissible systems $\bar{A}_1, \cdots, \bar{A}_a; \bar{B}_1, \cdots, \bar{B}_b; \bar{C}_1, \cdots, \bar{C}_c$ of 2-cells on $\bar{S}$ with the following properties. (1) Each of the 2-cells involved has a diameter less than η. (2) $\bar{S} = \bar{A}_1^0 + \cdots + \bar{C}_c^0$.

Proof. Using arcs of great circles on $\bar{S}$, we can clearly triangulate $\bar{S}$ in such a manner that each triangle has a diameter less than $\eta/2$, the equator $\bar{E}$ is a sum of sides of triangles, and the north pole $\bar{\nu}$ of $\bar{S}$ is a vertex in the triangulation. If $\bar{A}_1, \cdots, \bar{A}_a$ are the triangles of such a triangulation, then clearly $\bar{A}_1, \cdots, \bar{A}_a$ is an admissible system of 2-cells. In each $\bar{A}_i$, let us select an interior point $\bar{p}_i$, and on each side occurring in the triangulation, let us select a point, different from the end point, which will be referred to as the mid-point of that side. We subdivide each $\bar{A}_i$ into six curvilinear triangles by drawing simple arcs from $\bar{p}_i$ to the vertices of $\bar{A}_i$ and to the mid-points of the sides of $\bar{A}_i$. Let $\tilde{T}$ denote the triangulation of $\bar{S}$ consisting of $\bar{A}_1, \cdots, \bar{A}_a$, and let $\tilde{t}$ denote the triangulation consisting of the triangles into which $\bar{A}_1, \cdots, \bar{A}_a$ have been subdivided. Each vertex of $\tilde{T}$ is then a common vertex of a certain number of triangles of $\tilde{t}$ whose sum is a 2-cell. Let us denote by $\bar{B}_1, \cdots, \bar{B}_b$ the 2-cells associated in this manner with the vertices of $\tilde{T}$, and let us choose the notations in such a manner that the north pole $\bar{\nu}$ is interior to $\bar{B}_b$. Clearly $\bar{B}_1, \cdots, \bar{B}_b$ is an admissible system of 2-cells, and each $\bar{B}_j$ has a diameter less than η. Each side occurring in $\tilde{T}$ contains sides of exactly four triangles of $\tilde{t}$; the sum of these four triangles is a 2-cell. Let us denote by $\bar{C}_1, \cdots, \bar{C}_c$ the 2-cells associated in this manner with the various sides occurring in $\tilde{T}$. Then $\bar{C}_1, \cdots, \bar{C}_c$ is again an admissible system of 2-cells, and each $\bar{C}_k$ has a diameter less than η. Obviously, the systems $\bar{A}_1, \cdots, \bar{A}_a; \bar{B}_1, \cdots, \bar{B}_b; \bar{C}_1, \cdots, \bar{C}_c$ satisfy our requirements.

Remark. In view of the choice of $\bar{B}_b$, it is clear that on omitting $\bar{B}_b$ we are left with three admissible systems $\bar{A}_1, \cdots, \bar{A}_a; \bar{B}_1, \cdots, \bar{B}_{b-1}; \bar{C}_1, \cdots, \bar{C}_c$, such that $\bar{S} - \bar{\nu} = \bar{A}_1^0 + \cdots + \bar{A}_a^0 + \bar{B}_1^0 + \cdots + \bar{B}_{b-1}^0 + \bar{C}_1^0 + \cdots + \bar{C}_c^0$.

II.1.53. Using the notations of II.1.32, let $M(S) = \bar{S}$ be a continuous monotone transformation, such that $M(E) = \bar{E}$, and M is monotone on E (cf. II.1.36). Let $\bar{A}$ be an admissible 2-cell on $\bar{S}$ (see II.1.52). Then there exists a continuous monotone transformation $\bar{M}(S) = \bar{S}$ with the following properties.

(1) $\bar{M}(E) = \bar{E}$, and $\bar{M}$ is monotone on E.

(2) $d[\bar{M}^{-1}(\bar{x})] \leqq d[M^{-1}(\bar{x})]$ for every point $\bar{x} \in \bar{S}$.

(3) If $\bar{x} \in \bar{A}^0$, then $\bar{M}^{-1}(\bar{x})$ is a single point in $M^{-1}(\bar{A}^0)$.

(4) If $x \notin M^{-1}(\bar{A}^0)$, then $\bar{M}(x) = M(x)$.

(5) If $x \in M^{-1}(\bar{A}^0)$, then $\bar{M}(x) \in \bar{A}^0$.

Proof. Let $\tilde{M}(S) = \bar{S}$ be a continuous monotone transformation, related to M in the manner described in II.1.47. Then $\tilde{M}$ possesses the properties (i)-(viii) described in II.1.47, II.1.48. Noting that $\bar{A}$ is admissible, we distinguish between the following two cases.

Case I. $\bar{A}^0\bar{E} = 0$. Then the transformation $\bar{M} = \tilde{M}$ obviously satisfies our requirements.

Case II. $\bar{A}^0\bar{E} \neq 0$. Then the boundary curve of $\bar{A}$ intersects the equator $\bar{E}$ in exactly two points, since $\bar{A}$ is admissible. Let $\bar{p}$, $\bar{q}$ be these points of intersection, and let $\bar{\alpha}$, $\bar{\beta}$ be the two sub-arcs of $\bar{E}$ with end points $\bar{p}$, $\bar{q}$, where $\bar{\alpha} \subset \bar{A}$. Let us denote by $\bar{B}$ the 2-cell $\bar{S} - \bar{A}^0$. Then we have the relations

$$\bar{\alpha} \subset \bar{A}, \qquad \bar{\beta} \subset \bar{B}, \qquad \bar{\alpha}\bar{B} = \bar{\beta}\bar{A} = \bar{p} + \bar{q}.$$

We proceed to verify the following facts.

(a) $\tilde{M}$ is monotone on E.

(b) $\tilde{M}(E) = \bar{E}_1$ is a simple closed curve.

(c) $\bar{E}_1 = \bar{\beta} + \bar{\alpha}_1$, where $\bar{\alpha}_1$ is a simple arc with end points $\bar{p}$, $\bar{q}$ and $\bar{\alpha}_1\bar{B} = \bar{p} + \bar{q}$. In other words, $\bar{\alpha}_1$ is contained in $\bar{A}^0$ except for its end points $\bar{p}$, $\bar{q}$.

Proof of (a). Let $\bar{x}$ be any point in $\bar{S}$. If $\bar{x} \in \bar{A}^0$, then $\tilde{M}^{-1}(\bar{x})$ is a single point by II.1.48(iii), and hence $E\tilde{M}^{-1}(\bar{x})$ is either empty or else it is a single point. If $\bar{x} \notin \bar{A}^0$, then $E\tilde{M}^{-1}(\bar{x}) = EM^{-1}(\bar{x})$ by II.1.48(iv), and hence $E\tilde{M}^{-1}(\bar{x})$ is connected since M is monotone on E (cf. II.1.36). Thus $E\tilde{M}^{-1}(\bar{x})$ is connected (possibly empty) for every point $\bar{x} \in \bar{S}$, and hence $\tilde{M}$ is monotone on E.

Proof of (b). Note first that $\tilde{M}(E)\bar{B} = M(E)\bar{B} = \bar{E}\bar{B} = \bar{\beta}$ (cf. II.1.48(v)). Thus $\bar{\beta} \subset \tilde{M}(E)$, and hence $\tilde{M}(E)$ does not reduce to a single point. Since $\tilde{M}$ is monotone on E, and E is a simple closed curve, it follows that $\tilde{M}(E) = \bar{E}_1$ is also a simple closed curve.

Proof of (c). We have already proved that $\bar{E}_1$ is a simple closed curve that contains the simple arc $\bar{\beta}$. Hence $\bar{E}_1 = \bar{\beta} + \bar{\alpha}_1$, where $\bar{\alpha}_1$ is a simple arc with end points $\bar{p}$, $\bar{q}$, and $\bar{\alpha}_1\bar{\beta} = \bar{p} + \bar{q}$. There remains to show that $\bar{\alpha}_1\bar{B} = \bar{p} + \bar{q}$. Now we noted, while proving (b), that $\bar{E}_1\bar{B} = \tilde{M}(E)\bar{B} = \bar{\beta}$. Since $\bar{\alpha}_1 \subset \bar{E}_1$, it follows that $\bar{\alpha}_1\bar{B} = \bar{\alpha}_1\bar{E}_1\bar{B} = \bar{\alpha}_1\bar{\beta} = \bar{p} + \bar{q}$.

Now that (a), (b), (c) are verified, let us consider the 2-cell $\bar{A}$. The points $\bar{p}$, $\bar{q}$ lie on the boundary curve of $\bar{A}$, and are joined by two simple arcs $\bar{\alpha}$, $\bar{\alpha}_1$ that are contained in $\bar{A}^0$ except for their common end points $\bar{p}$, $\bar{q}$. There follows the existence of a topological transformation $\bar{h}(\bar{A}) = \bar{A}$ that maps $\bar{\alpha}_1$ onto $\bar{\alpha}$ and reduces to the identity on the boundary curve of $\bar{A}$ (see I.2.52). Then the transformation $\bar{H}(\bar{S}) = \bar{S}$, defined by $\bar{H}(\bar{x}) = \bar{x}$ for $\bar{x} \notin \bar{A}^0$, $\bar{H}(\bar{x}) = \bar{h}(\bar{x})$ for $\bar{x} \in \bar{A}^0$, is a topological transformation from $\bar{S}$ onto $\bar{S}$. Let us define now a transformation $\bar{M}(S) = \bar{S}$ by the formula $\bar{M}(x) = \bar{H}\tilde{M}(x)$, $x \in S$. In view of (a), (b), (c), the transformation $\bar{M}$ obviously satisfies our requirements.

II.1.54. Using the notations of II.1.32, let $M(S) = \bar{S}$ be a continuous monotone transformation, such that $M(E) = \bar{E}$ and M is monotone on E. Let $\bar{A}_1, \cdots, \bar{A}_a$ be an admissible system of 2-cells on $\bar{S}$ (cf. II.1.52). Then there exists a continuous monotone transformation $M_*(S) = \bar{S}$ with the following properties.

(1) $M_*(E) = \bar{E}$, and M_* is monotone on E.

(2) $d[M_*^{-1}(\bar{x})] \leqq d[M^{-1}(\bar{x})]$ for every point $\bar{x} \in \bar{S}$.
(3) If $\bar{x} \in \bar{A}_1^0 + \cdots + \bar{A}_a^0$, then $M_*^{-1}(\bar{x})$ is a single point.
(4) $\rho[M(x), M_*(x)] \leqq \max [d(\bar{A}_1), \cdots, d(\bar{A}_a)]$ for every point $x \in S$.

PROOF. Apply the result of II.1.53 to M, using the 2-cell $\bar{A}_1$, obtaining a transformation $\bar{M}_1$. Apply the same process to $\bar{M}_1$, using the 2-cell $\bar{A}_2$, obtaining a transformation $\bar{M}_2$, and so forth. There results a sequence of transformations $\bar{M}_1, \bar{M}_2, \cdots, \bar{M}_a$. We assert that the transformation $M_* = \bar{M}_a$ satisfies our requirements. Indeed, (1), (2), (3) are immediate consequences of II.1.53(1), II.1.53(2), II.1.53(3). As regards (4), observe that the sets $M^{-1}(\bar{A}_1^0), \cdots, M^{-1}(\bar{A}_a^0)$ are disjoint, and hence a point $x \in S$ lies in at most one of these sets. If x does not lie in the sum of these sets, then $M_*(x) = M(x)$ in view of II.1.53(4). If $x \in M^{-1}(\bar{A}_i^0)$ for a certain i, then clearly $M_*(x) = \bar{M}_i(x)$, and hence, in view of II.1.53(5), $M_*(x)$ and $M(x)$ both lie in $\bar{A}_i^0$. Thus (4) follows.

II.1.55. APPROXIMATION THEOREM. *Using the notations of* II.1.32, *let* $M(S) = \bar{S}$ *be a continuous monotone transformation, such that* $M(E) = \bar{E}$ *and* M *is monotone on* E. *Let* $\epsilon > 0$ *be given. Then there exists a topological transformation* $H(S) = \bar{S}$, *such that* $H(E) = \bar{E}$ *and* $\rho[M(x), H(x)] < \epsilon$ *for every point* $x \in S$.

PROOF. Put $\eta = \epsilon/3$. Choose three admissible systems $\bar{A}_1, \cdots, \bar{A}_a; \bar{B}_1, \cdots, \bar{B}_b; \bar{C}_1, \cdots, \bar{C}_c$ of 2-cells in $\bar{S}$ as described in II.1.52. Apply the result of II.1.54 to M and the system $\bar{A}_1, \cdots, \bar{A}_a$, obtaining a transformation M_{*1}. Apply the same process to M_{*1} and $\bar{B}_1, \cdots, \bar{B}_b$, obtaining a transformation M_{*2}. Apply the same process to M_{*2} and $\bar{C}_1, \cdots, \bar{C}_c$, obtaining a transformation M_{*3}. We assert that $H = M_{*3}$ satisfies our requirements. In view of II.1.54(1), II.1.54(4), clearly $H(E) = \bar{E}$ and $\rho[M(x), H(x)] < 3\eta = \epsilon$ for every $x \in S$. There remains to show that H is biunique. Now let $\bar{x}$ be any point of $\bar{S}$. Since $\bar{A}_1^0 + \cdots + \bar{C}_c^0 = \bar{S}$, it follows by II.1.54(3) that one at least of the sets $M_{*1}^{-1}(\bar{x})$, $M_{*2}^{-1}(\bar{x})$, $M_{*3}^{-1}(\bar{x})$ reduces to a single point. Since $d[M_{*3}^{-1}(\bar{x})] \leqq d[M_{*2}^{-1}(\bar{x})] \leqq d[M_{*1}^{-1}(\bar{x})]$ (cf. II.1.54(2)), it follows that $H^{-1}(\bar{x}) = M_{*3}^{-1}(\bar{x})$ reduces to a single point for every $\bar{x} \in \bar{S}$. Thus H is biunique. Since the spaces S, $\bar{S}$ are compact, it follows that H is a homeomorphism.

II.1.56. APPROXIMATION THEOREM. *Let* $M(\mathcal{P}) = \mathcal{P}^*$ *be a continuous monotone transformation, where* $\mathcal{P}$ *and* $\mathcal{P}^*$ *are both 2-cells. Let* $\epsilon > 0$ *be given. Then there exists a topological transformation* $H(\mathcal{P}) = \mathcal{P}^*$, *such that* $\rho[M(x), H(x)] < \epsilon$ *for every point* $x \in \mathcal{P}$.

PROOF. Using the terminology of II.1.32, we can assume without loss of generality that $\mathcal{P}$, $\mathcal{P}^*$ coincide with the hemispheres S_σ, $\bar{S}_\sigma$ respectively (cf. the argument in II.1.50, case (2)). Obviously, we can also assume that $0 < \epsilon < 1$. By II.1.46, we have a continuous monotone transformation $\tilde{M}(S) = \bar{S}$ such that $\tilde{M}(x) = M(x)$ for $x \in S_\sigma$. By II.1.41, $M(E) = \bar{E}$ and M is monotone on E. Since $\tilde{M} = M$ on E, it follows that $\tilde{M}(E) = \bar{E}$ and $\tilde{M}$ is monotone on E. By II.1.55, applied to $\tilde{M}$, there follows the existence of a topological transformation $H(S) = \bar{S}$, such that $H(E) = \bar{E}$ and $\rho[\tilde{M}(x), H(x)] < \epsilon$ for every point $x \in S$.

In particular, since $\tilde{M} = M$ on S_σ, it follows that $\rho[M(x), H(x)] < \epsilon$ for every point $x \in S_\sigma$. There remains to show that $H(S_\sigma) = \bar{S}_\sigma$. Now since H is topological and $H(E) = \bar{E}$, we have either $H(S_\sigma) = \bar{S}_\sigma$ or $H(S_\sigma) = \bar{S}_\nu$. The second alternative is clearly incompatible with the inequalities

$$\rho[M(x), H(x)] < \epsilon < 1 \qquad \text{for } x \in S_\sigma .$$

II.1.57. Approximation theorem. *Let $M(\mathcal{P}) = \mathcal{P}^*$ be a continuous monotone transformation, where $\mathcal{P}$ and $\mathcal{P}^*$ are both 2-spheres. Let $\epsilon > 0$ be given. Then there exists a topological transformation $H(\mathcal{P}) = \mathcal{P}^*$ such that $\rho[M(x), H(x)] < \epsilon$ for every point $x \in \mathcal{P}$.*

Proof. Without loss of generality, we can assume that $\mathcal{P}$, $\mathcal{P}^*$ coincide with the unit spheres S, $\bar{S}$ respectively (cf. II.1.32 and the argument in II.1.50, case (2)). The proof follows then from II.1.47 in the same manner as the approximation theorem of II.1.55 followed from II.1.53, except for substantial simplifications arising from the fact that no additional conditions concerning the equators E, $\bar{E}$ are involved at present.

II.1.58. Approximation theorem. *Let $M(\mathcal{P}) = \mathcal{P}^*$ be a continuous monotone transformation, where $\mathcal{P}$ and $\mathcal{P}^*$ are both 2-spheres, and let x_0^* be a point of $\mathcal{P}^*$. Let $\epsilon > 0$ be given. Then there exists a continuous monotone transformation $M_*(\mathcal{P}) = \mathcal{P}^*$ with the following properties.* (i) *$M_*(x) = M(x)$ for $x \in M^{-1}(x_0^*)$.* (ii) *M_* maps the set $\mathcal{P} - M^{-1}(x_0^*)$ topologically onto the set $\mathcal{P}^* - x_0^*$.* (iii) *$\rho[M(x), M_*(x)] < \epsilon$ for every $x \in S$.*

Proof. Without loss of generality, we can assume that $\mathcal{P}$, $\mathcal{P}^*$ coincide with the unit spheres S, $\bar{S}$ respectively and x_0^* coincides with the north pole $\bar{\nu}$ of $\bar{S}$ (cf. II.1.32). The proof is then entirely analogous to that outlined in II.1.57, except that we use now the systems $\bar{A}_1, \cdots, \bar{A}_a; \bar{B}_1, \cdots, \bar{B}_{b-1}; \bar{C}_1, \cdots, \bar{C}_c$ (see the remark at the end of II.1.52).

II.1.59. Approximation theorem. *Let $M(\mathcal{P}) = \mathcal{P}^*$ be a continuous monotone transformation, where $\mathcal{P}$ is a 2-cell and $\mathcal{P}^*$ is a 2-sphere; as noted in* II.1.42, *the boundary curve of $\mathcal{P}$ is then mapped onto a single point x_0^* of $\mathcal{P}^*$. Let $\epsilon > 0$ be given. Then there exists a continuous monotone transformation $M_*(\mathcal{P}) = \mathcal{P}^*$ with the following properties.* (i) *M_* maps the boundary curve of $\mathcal{P}$ onto the single point x_0^*.* (ii) *M_* maps the set $\mathcal{P} - M^{-1}(x_0^*)$ topologically onto the set $\mathcal{P}^* - x_0^*$.* (iii) *$\rho[M(x), M_*(x)] < \epsilon$ for every point $x \in \mathcal{P}$.*

Proof. We can assume without loss of generality that $\mathcal{P}$ coincides with the southern hemisphere S_σ on the unit sphere S (see II.1.32). Let us extend the transformation $M(S_\sigma) = \mathcal{P}^*$ by setting $M(x) = x_0^*$ for $x \in S_\nu^0$. Then the extended transformation $M(S) = \mathcal{P}^*$ is clearly continuous and monotone on S, and the present approximation theorem appears as an immediate consequence of II.1.58.

II.1.60. Given a pair of continuous transformations $T_1(\mathcal{P}_1) = \mathcal{P}^*$, $T_2(\mathcal{P}_2) =$

$\mathcal{P}^*$ as in II.1.24, let us suppose that $T_1 \sim T_2(K)$ (see II.1.30). In other words, we assume that T_1, T_2 admit of simultaneous monotone-light factorizations

$$T_1 = LM_1, \qquad M_1(\mathcal{P}_1) = \mathfrak{M}, \qquad L(\mathfrak{M}) = \mathcal{P}^*,$$

$$T_2 = LM_2, \qquad M_2(\mathcal{P}_2) = \mathfrak{M}, \qquad L(\mathfrak{M}) = \mathcal{P}^*,$$

with the same middle-space $\mathfrak{M}$ and the same light factor L. As we observed in II.1.31, it does *not* generally follow that $T_1 \sim T_2(F)$. We proceed to discuss some special cases where the relation $T_1 \sim T_2(K)$ implies the relation $T_1 \sim T_2(F)$.

II.1.61. *Using the notations and assumptions of* II.1.60, *suppose that* $\mathcal{P}_1$, $\mathcal{P}_2$, $\mathfrak{M}$ *are simple arcs. Then* $T_1 \sim T_2(F)$.

PROOF. Since L is continuous on $\mathfrak{M}$, and $\mathfrak{M}$ is compact, we have (see I.2.42) for given $\epsilon > 0$ an $\eta > 0$ such that

$$(1) \qquad \rho[L(\mathfrak{x}'), L(\mathfrak{x}'')] < \epsilon/2 \qquad \text{for } \rho(\mathfrak{x}', \mathfrak{x}'') < \eta, \ \mathfrak{x}' \in \mathfrak{M}, \ \mathfrak{x}'' \in \mathfrak{M}.$$

By II.1.50 we have topological transformations $H_1(\mathcal{P}_1) = \mathfrak{M}$, $H_2(\mathcal{P}_2) = \mathfrak{M}$ such that

$$(2) \qquad \rho[H_1(x_1), M_1(x_1)] < \eta \qquad \text{for } x_1 \in \mathcal{P}_1,$$

$$(3) \qquad \rho[H_2(x_2), M_2(x_2)] < \eta \qquad \text{for } x_2 \in \mathcal{P}_2.$$

Consider the transformation $H(\mathcal{P}_1) = \mathcal{P}_2$ defined by

$$(4) \qquad H(x_1) = H_2^{-1}H_1(x_1), \qquad x_1 \in \mathcal{P}_1.$$

Clearly H is topological, and $\rho[T_1(x_1), T_2H(x_1)] = \rho[LM_1(x_1), LM_2H(x_1)] \leqq \rho[LM_1(x_1), LH_1(x_1)] + \rho[LH_1(x_1), LM_2H(x_1)]$. Hence by (1), (2),

$$(5) \qquad \rho[T_1(x_1), T_2H(x_1)] < \epsilon/2 + \rho[LH_1(x_1), LM_2H(x_1)], \qquad x_1 \in \mathcal{P}_1.$$

Let us put $H(x_1) = x_2$. Then $H_2(x_2) = H_2H(x_1) = H_1(x_1)$ by (4), and hence by (3)

$$\rho[H_1(x_1), M_2H(x_1)] = \rho[H_2(x_2), M_2(x_2)] < \eta.$$

By (1) it follows that

$$(6) \qquad \rho[LH_1(x_1), LM_2H(x_1)] < \epsilon/2, \qquad x_1 \in \mathcal{P}_1.$$

(5), (6) yield $\rho[T_1(x_1), T_2H(x_1)] < \epsilon$ for $x_1 \in \mathcal{P}_1$. Since ϵ was arbitrary, the relation $T_1 \sim T_2(F)$ is established (see II.1.25).

II.1.62. *Using the notations and assumptions of* II.1.60, *suppose that* $\mathcal{P}_1$, $\mathcal{P}_2$, $\mathfrak{M}$ *are simple closed curves. Then* $T_1 \sim T_2(F)$.

The proof is the same as in II.1.61, except that II.1.51 is used instead of II.1.50.

II.1.63. *Using the notations and assumptions of* II.1.60, *suppose that* $\mathcal{P}_1$, $\mathcal{P}_2$, $\mathfrak{M}$ *are 2-spheres. Then* $T_1 \sim T_2(F)$.

The proof is the same as in II.1.61, except that II.1.57 (applied to the transformations M_1, M_2) is used instead of II.1.50.

II.1.64. *Using the notations and assumptions of* II.1.60, *suppose that* $\mathcal{P}_1$, $\mathcal{P}_2$, $\mathfrak{M}$ *are 2-cells. Then* $T_1 \sim T_2(F)$.

The proof is the same as in II.1.61, except that II.1.56 (applied to M_1, M_2) is used instead of II.1.50.

II.1.65. *Using the notations and assumptions of* II.1.60, *suppose that* $\mathcal{P}_1$, $\mathcal{P}_2$ *are 2-cells and* $\mathfrak{M}$ *is a 2-sphere. By* II.1.42, M_1 *maps the boundary curve of* $\mathcal{P}_1$ *onto a single point* $\mathfrak{x}_0^1$ *of* $\mathfrak{M}$, *and* M_2 *maps the boundary curve of* $\mathcal{P}_2$ *onto a single point* $\mathfrak{x}_0^2$ *of* $\mathfrak{M}$. *We assert that the relation* $T_1 \sim T_2(F)$ *holds if* $\mathfrak{x}_0^1 = \mathfrak{x}_0^2$.

PROOF. Let us put $\mathfrak{a}_0 = \mathfrak{x}_0^1 = \mathfrak{x}_0^2$. Give $\epsilon > 0$. Since L is uniformly continuous on $\mathfrak{M}$, we have an $\eta > 0$ such that

$$(1) \qquad \rho[L(\mathfrak{x}'), L(\mathfrak{x}'')] < \epsilon \qquad \text{if } \rho(\mathfrak{x}', \mathfrak{x}'') < \eta,\ \mathfrak{x}' \in \mathfrak{M},\ \mathfrak{x}'' \in \mathfrak{M}.$$

By II.1.59 (applied to M_1) we have a continuous monotone transformation $\tilde{M}_1(\mathcal{P}_1) = \mathfrak{M}$ with the following properties. (i) $\tilde{M}_1$ maps the boundary curve of $\mathcal{P}_1$ onto the single point $\mathfrak{a}_0$. (ii) $\tilde{M}_1$ maps the set $\mathcal{P}_1 - \tilde{M}_1^{-1}(\mathfrak{a}_0)$ topologically onto the set $\mathfrak{M} - \mathfrak{a}_0$. (iii) $\rho[M_1(x_1), \tilde{M}_1(x_1)] < \eta/3$ for $x_1 \in \mathcal{P}_1$. Similarly, we have a continuous monotone transformation $\tilde{M}_2(\mathcal{P}) = \mathfrak{M}$ with analogous properties relative to M_2. Now let $\mathfrak{C}$ be a simple closed curve on $\mathfrak{M}$ that does not pass through $\mathfrak{a}_0$. Then $\mathfrak{C}$ determines two 2-cells $\mathfrak{A}$, $\mathfrak{B}$ on $\mathfrak{M}$, where the notation is so chosen that $\mathfrak{a}_0 \in \mathfrak{A}$. Clearly we can choose $\mathfrak{C}$ so as to have $d(\mathfrak{A}) < \eta/3$. Let us put

$$(2) \qquad \tilde{M}_1^{-1}(\mathfrak{B}) = B_1, \qquad \tilde{M}_2^{-1}(\mathfrak{B}) = B_2.$$

Then B_1 is a 2-cell contained in the interior $\mathcal{P}_1^0$ of $\mathcal{P}_1$, and B_2 is a 2-cell contained in the interior $\mathcal{P}_2^0$ of $\mathcal{P}_2$. Let us consider the transformation $H(x_1) = \tilde{M}_2^{-1}\tilde{M}_1(x_1)$, $x_1 \in B_1$. Clearly H is a topological transformation from B_1 onto B_2, and hence H can be extended to a topological transformation $H(\mathcal{P}_1) = \mathcal{P}_2$ (see I.2.52). The extended transformation $H(\mathcal{P}_1) = \mathcal{P}_2$ possesses then the following properties.

$$(3) \qquad H(\mathcal{P}_1) = \mathcal{P}_2, \qquad H(B_1) = B_2, \qquad H(\mathcal{P}_1 - B_1) = \mathcal{P}_2 - B_2.$$

$$(4) \qquad H(x_1) = \tilde{M}_2^{-1}\tilde{M}_1(x_1) \qquad \text{for } x_1 \in B_1.$$

We assert that

$$(5) \qquad \rho[T_1(x_1), T_2H(x_1)] < \epsilon \qquad \text{for } x_1 \in \mathcal{P}_1.$$

Case (1). $x_1 \in B_1$. Let us put $H_1(x_1) = x_2$. Then, by (3), (4), $x_2 \in B_2$ and $x_2 = \tilde{M}_2^{-1}\tilde{M}_1(x_1)$. Since $\tilde{M}_1$, $\tilde{M}_2$ are topological in B_1, B_2 respectively, it follows that $\tilde{M}_2(x_2) = \tilde{M}_1(x_1)$ and hence

$$\rho[\tilde{M}_1(x_1), M_2H(x_1)] = \rho[\tilde{M}_2(x_2), M_2(x_2)] < \eta/3,$$

$$\rho[M_1(x_1), M_2H(x_1)] \leqq \rho[M_1(x_1), \tilde{M}_1(x_1)] + \rho[\tilde{M}_1(x_1), M_2H(x_1)] < \eta/3 + \eta/3 < \eta.$$

In view of (1) we obtain finally

$$\rho[T_1(x_1), T_2H(x_1)] = \rho[LM_1(x_1), LM_2H(x_1)] < \epsilon.$$

Case (2). $x_1 \in \mathcal{P}_1 - B_1$. Then $x_2 = H(x_1) \in \mathcal{P}_2 - B_2$ by (3), and hence $\tilde{M}_1(x_1) \in \mathfrak{A}^0$, $\tilde{M}_2H(x_1) \in \mathfrak{A}^0$ by (2). Consequently $\rho[\tilde{M}_1(x_1), \tilde{M}_2H(x_1)] < d(\mathfrak{A}) < \eta/3$. It follows that

$$\rho[M_1(x_1), M_2H(x_1)] \leqq \rho[M_1(x_1), \tilde{M}_1(x_1)] + \rho[\tilde{M}_1(x_1), \tilde{M}_2H(x_1)]$$

$$+ \rho[\tilde{M}_2H(x_1), M_2H(x_1)] < \eta/3 + \eta/3 + \eta/3 = \eta.$$

Hence, by (1), $\rho[T_1(x_1), T_2H(x_1)] = \rho[LM_1(x_1), LM_2H(x_1)] < \epsilon$.

Thus (5) is established. Since ϵ was arbitrary, the relation $T_1 \sim T_2(F)$ is proved.

II.1.66. Let $T(\mathcal{P}) = \mathcal{P}^*$ be a continuous transformation, where $\mathcal{P}$, $\mathcal{P}^*$ are Peano spaces. Let $\mathcal{P}_*$ be a Peano space homeomorphic with $\mathcal{P}$. If $H_*(\mathcal{P}_*) = \mathcal{P}$ is a homeomorphism, and if we put $T_* = TH_*$, then clearly $T_* \sim T(ts)$ and hence also $T_* \sim T(F)$ (cf. II.1.26). Now let us consider, more generally, a continuous monotone transformation $M_*(\mathcal{P}_*) = \mathcal{P}$, and let us put again $T_* = TM_*$. *We assert that we have* $T_* \sim T(F)$ *in the following cases.* (1) $\mathcal{P}$ *is a simple arc.* (2) $\mathcal{P}$ *is a simple closed curve.* (3) $\mathcal{P}$ *is a 2-sphere.* (4) $\mathcal{P}$ *is a 2-cell.*

Proof. Give $\epsilon > 0$. Since T is uniformly continuous on $\mathcal{P}$, we have an $\eta > 0$ such that

$$(1) \qquad \rho[T(x_1), T(x_2)] < \epsilon \qquad \text{for } \rho(x_1, x_2) < \eta,\ x_1 \in \mathcal{P},\ x_2 \in \mathcal{P}.$$

In all four cases, we have then a homeomorphism $H_*(\mathcal{P}_*) = \mathcal{P}$, such that

$$(2) \qquad \rho[M_*(x_*), H_*(x_*)] < \eta \qquad \text{for } x_* \in \mathcal{P}_*.$$

The existence of H_* follows from II.1.50, II.1.51, II.1.57, II.1.56 respectively. If x_* is any point of $\mathcal{P}_*$, then it follows that (see (1), (2))

$$\rho[T_*(x_*), TH_*(x_*)] = \rho[TM_*(x_*), TH_*(x_*)] < \epsilon.$$

Since ϵ was arbitrary, the relation $T_* \sim T(F)$ is proved.

II.1.67. Given $T_1(\mathcal{P}_1) = \mathcal{P}^*$, $T_2(\mathcal{P}_2) = \mathcal{P}^*$ as in II.1.24, we noted that the relation $T_1 \sim T_2(F)$ does not generally imply the relation $T_1 \sim T_2(ts)$ (see II.1.26). *However, if* $T_1 \sim T_2(F)$ *and* T_1, T_2 *are both light, then we assert that* $T_1 \sim T_2(ts)$. Indeed, in this case we can choose, in the theorem of II.1.28, M_1 and M_2 as the identity transformations on $\mathcal{P}_1$, $\mathcal{P}_2$ respectively, and thus there follows the existence of a homeomorphism $h(\mathcal{P}_1) = \mathcal{P}_2$ such that $T_1(x_1) = T_2h(x_1)$ for every point $x_1 \in \mathcal{P}_1$.

CHAPTER II.2. CYCLIC DECOMPOSITION

II.2.1. We shall study in this chapter the *structure of Peano spaces with respect to their cut points*, and we shall apply the results of this study to derive additivity properties of functions of continuous transformations. The topological results discussed in this chapter belong to the *cyclic element theory* of topological spaces. An excellent presentation of this theory, in its most general aspects, is given in a recent book of Whyburn [3]. For our purposes, only the case of Peano spaces is relevant, and we shall give a concise discussion of this important special case.

Let us recall a few definitions. Given a Peano space $\mathcal{P}$, a point $x \in \mathcal{P}$ is a *cut point* if $\mathcal{P} - x$ is disconnected. If $\mathcal{P} - x$ is connected, then x is a *non-cut point.* If E_1, E_2 are two subsets of $\mathcal{P}$, then a point x is said to cut between E_1 and E_2, or to separate E_1 and E_2, if E_1 and E_2 lie in different components of $\mathcal{P} - x$. Note that if x cuts between E_1 and E_2, then necessarily $E_1 \neq 0$, $E_2 \neq 0$, $E_1E_2 = 0$, $x \notin E_1 + E_2$. In particular, if p_1, p_2 are two points of $\mathcal{P}$, then a third point x cuts between p_1, p_2, or separates p_1 and p_2, if p_1 and p_2 lie in different components of $\mathcal{P} - x$.

A Peano space that has no cut points is termed *cyclic.* For example, a 2-sphere is a cyclic Peano space. A *dendrite* is a Peano space $\mathcal{P}$ such that for every pair p_1, p_2 of distinct points in $\mathcal{P}$, there exists a point x that cuts between p_1 and p_2. For example, a simple arc is a dendrite.

II.2.2. Let $\mathcal{P}$ be a Peano space. It will be convenient to use the term *semi-connected set* in the following sense. A subset E of $\mathcal{P}$ is *semi-connected* if and only if for every choice of a point x in $\mathcal{P} - E$, the set E is a subset of some component of $\mathcal{P} - x$. Clearly, every connected set is also semi-connected, while the converse is generally false. For example, if $\mathcal{P}$ is cyclic, then clearly every subset of $\mathcal{P}$ is semi-connected. The following statements are more or less obvious consequences of the definition of a semi-connected set.

(i) The empty set is semi-connected. $\mathcal{P}$ itself is semi-connected. Every connected set is semi-connected.

(ii) If E_0 is semi-connected, and Φ is any family of semi-connected subsets of $\mathcal{P}$ such that $EE_0 \neq 0$ for $E \in \Phi$, then the set $S = E_0 + \sum E$, $E \in \Phi$, is also semi-connected.

(iii) If Φ is any family of semi-connected subsets of $\mathcal{P}$, then the set $S = \prod E$, $E \in \Phi$, is also semi-connected.

(iv) If E is semi-connected, and if F is any set such that $E \subset F \subset c(E)$, where $c(E)$ is the closure of E, then F is also semi-connected.

(v) Let p, q be two distinct points of $\mathcal{P}$, and let E be a semi-connected set that contains p and q. If a point x cuts between p and q, then $x \in E$. Indeed, if we assume that $x \notin E$, then E lies in a certain component G_x of $\mathcal{P} - x$, and hence $p + q \in G_x$, in contradiction with the assumption that x cuts between p and q.

(vi) Given two distinct points p, q of $\mathcal{P}$, we shall denote by $K(p, q)$ the set of all those points of $\mathcal{P}$ that cut between p and q (of course $K(p, q)$ may be empty). Then $p + q + K(p, q)$ is semi-connected. Indeed, let x be any point such that $x \notin p + q + K(p, q)$. Then x does not cut between p and q, and hence p and q lie in the same component G_x of $\mathcal{P} - x$. By (i) and (v) it follows that every point $y \in K(p, q)$ lies in G_x. Hence $p + q + K(p, q) \subset G_x$.

II.2.3. Given a Peano space $\mathcal{P}$, two (not necessarily distinct) points p, q of $\mathcal{P}$ will be termed *conjugate* if and only if there exists no point $x \in \mathcal{P}$ that cuts between p and q. We shall write $p \bigcirc q$ to express the fact that p and q are conjugate points. The relation $p \bigcirc q$ is a binary relation that is clearly reflexive and symmetric. That is, $p \bigcirc p$ for every $p \in \mathcal{P}$, and $p \bigcirc q$ implies $q \bigcirc p$. On the other hand, this binary relation is *not transitive* generally. Indeed, let $\mathcal{P}$ consist of the two circular discs $(u + 1)^2 + v^2 \leqq 1$ and $(u - 1)^2 + v^2 \leqq 1$ in the Cartesian uv-plane. If p, q, r denote the points $(-2, 0)$, $(0, 0)$, $(1, 0)$, then clearly $p \bigcirc q$, $q \bigcirc r$, but $p \bigcirc r$ fails to hold, since q cuts between p and r.

We shall see a little later that the binary relation of conjugacy possesses a certain weaker transitivity property which will be useful in discussing the more elementary parts of cyclic element theory.

II.2.4. Using the terminology of II.2.3, let p, q be two points such that $p \bigcirc q$. Then $p + q$ is semi-connected. Indeed, the set $K(p, q)$ is now empty, and hence the assertion is a special case of II.2.2(vi).

Let $a_1, \cdots, a_n$ be a finite sequence of distinct points in $\mathcal{P}$. These points, taken in the given order, will be said to constitute a $\bigcirc$*-chain* if $a_i \bigcirc a_{i+1}$, $i = 1, \cdots, n - 1$. If, in addition, $a_n \bigcirc a_1$ then the $\bigcirc$-chain will be called a *simple closed* $\bigcirc$*-chain.* Otherwise, the $\bigcirc$-chain will be termed *open. Note that the points* $a_1, \cdots, a_n$ *are required to be distinct.*

If $a_1, \cdots, a_n$ is a $\bigcirc$-chain, then the set $a_1 + \cdots + a_n$ is semi-connected. Indeed, for $n = 2$ this is true by the remark made earlier in this section. Assuming that the assertion holds for $\bigcirc$-chains of not more than $n - 1$ points, $a_1 + \cdots + a_{n-1}$ and $a_{n-1} + a_n$ are semi-connected, and hence $a_1 + \cdots + a_n$ is also semi-connected by II.2.2(ii).

II.2.5. Let $a_1, \cdots, a_n$ be a simple closed $\bigcirc$-chain. Then any two of the points $a_1, \cdots, a_n$ are conjugate. Indeed, if the points $a_1, \cdots, a_n$ are subjected to a cyclic permutation, then clearly in their new arrangement they constitute again a simple closed $\bigcirc$-chain. Hence it is sufficient to show that $a_1 \bigcirc a_i$ for $1 < i < n$. Now suppose that there exists a point x that cuts between a_1 and a_i. Since $a_1, \cdots, a_i$ is a $\bigcirc$-chain and hence (see II.2.4) $a_1 + \cdots + a_i$ is a semi-connected set, it follows by II.2.2(v) that $x \in a_1 + \cdots + a_i$. Similarly, $a_i, \cdots, a_n, a_1$ is a $\bigcirc$-chain, and hence $x \in a_i + \cdots + a_n + a_1$. Thus $x \in a_1 + a_i$, which is a contradiction, since x cuts between a_1 and a_i by assumption and hence cannot coincide with either a_1 or a_i.

The result just proved may be restated as follows. If $a_1, \cdots, a_n$ are distinct points such that $a_1 \bigcirc a_2 \bigcirc \cdots \bigcirc a_n \bigcirc a_1$, then $a_i \bigcirc a_j$ for every choice of

$i, j = 1, 2, \cdots, n$. We shall express this fact by the statement that the binary relation $p \bigcirc q$ is *cyclicly transitive.*

II.2.6. Let there be given, in the Peano space $\mathcal{P}$, a finite system of distinct points $p_1, \cdots, p_n$ and an equal number of sets $S_1, \cdots, S_n$, such that the following conditions hold. (i) The sets $S_1, \cdots, S_n$ are semi-connected. (ii) $S_1S_2 = p_2$, $S_2S_3 = p_3$, $\cdots$, $S_{n-1}S_n = p_n$, $S_nS_1 = p_1$. (iii) $S_iS_j = 0$ if $1 < |i - j| < n - 1$. Under these conditions, we shall say that the points $p_1, \cdots, p_n$ and the sets $S_1, \cdots, S_n$ constitute, in the given arrangement, a *simple closed* $\bigcirc$*-polygon.* The points $p_1, \cdots, p_n$ are the *vertices* and the sets $S_1, \cdots, S_n$ are the *sides* of the $\bigcirc$-polygon. The $\bigcirc$-polygon will be denoted by the symbol $(p_1, \cdots, p_n, S_1, \cdots, S_n)$.

In particular, if the points $p_1, \cdots, p_n$ constitute a simple closed $\bigcirc$-chain, then they give rise to the simple closed $\bigcirc$-polygon $(p_1, \cdots, p_n, p_1 + p_2, p_2 + p_3, \cdots, p_{n-1} + p_n, p_n + p_1)$. Thus the statement in the next section is a generalization of II.2.5.

II.2.7. Any two vertices of a simple closed $\bigcirc$-polygon are conjugate. Indeed, let $(p_1, \cdots, p_n, S_1, \cdots, S_n)$ be such a polygon. In view of II.2.5 it is sufficient to show that the vertices $p_1, \cdots, p_n$ form, in this order, a simple closed $\bigcirc$-chain. Since the situation is clearly unaffected by a cyclic permutation of the subscripts, it is sufficient to verify that $p_1 \bigcirc p_2$. Suppose there exists a point x that cuts between p_1 and p_2. Since $p_1 + p_2 \subset S_1$ and S_1 is semi-connected, we should have then $x \in S_1$ (cf. II.2.2(v)). Since $p_1 + p_2 \subset S_2 + \cdots + S_n$ and $S_2 + \cdots + S_n$ is clearly semi-connected (cf. II.2.2(ii)), we should have also $x \in S_2 + \cdots + S_n$. But $(S_2 + \cdots + S_n)S_1 = p_1 + p_2$. Hence x should coincide with either p_1 or p_2, in contradiction with the assumption that x cuts between p_1 and p_2.

II.2.8. A subset E of the Peano space $\mathcal{P}$ will be termed $\bigcirc$*-coherent* if and only if the following conditions hold. (i) E is nondegenerate (see I.2.3). (ii) Any two points of E are conjugate. Clearly, if $p_1 \neq p_2$, then the set $p_1 + p_2$ is $\bigcirc$-coherent if and only if $p_1 \bigcirc p_2$. If $(p_1, \cdots, p_n, S_1, \cdots, S_n)$ is a simple closed $\bigcirc$-polygon, then the set $p_1 + \cdots + p_n$ is $\bigcirc$-coherent by II.2.7.

II.2.9. A subset E of the Peano space $\mathcal{P}$ will be termed $\bigcirc$*-complete* if and only if the following conditions hold. (i) E is nondegenerate. (ii) If a point x is conjugate to two distinct points of E, then $x \in E$.

II.2.10. A subset of the Peano space $\mathcal{P}$ which is both $\bigcirc$-coherent and $\bigcirc$-complete will be termed a *proper cyclic element* of $\mathcal{P}$. Thus a proper cyclic element is necessarily nondegenerate. If C is a proper cyclic element, then C is semi-connected. Indeed, if p_0 is a fixed point of C, then $p_0 \bigcirc p$ for every point $p \in C$, and since $C = \sum (p_0 + p)$, $p \in C$, it follows by II.2.4 and II.2.2(ii) that C is semi-connected.

II.2.11. If C_1, C_2 are proper cyclic elements and C_1C_2 is nondegenerate, then $C_1 = C_2$. Indeed, suppose that C_1C_2 contains two distinct points p, q. Let x_1 be any point of C_1. Then $p \bigcirc x_1 \bigcirc q$ since C_1 is $\bigcirc$-coherent, and hence $x_1 \in C_2$

since C_2 is $\bigcirc$-complete. Thus $C_1 \subset C_2$. The complementary inclusion $C_2 \subset C_1$ follows in a similar way.

II.2.12. If $p \bigcirc q$, where p, q are distinct points of the Peano space $\mathcal{P}$, then there exists exactly one proper cyclic element that contains both p and q.

PROOF. Let E be the set of all those points $x \in \mathcal{P}$ that satisfy the condition $p \bigcirc x \bigcirc q$. Then $p + q \subset E$, and hence E is nondegenerate. The set E is $\bigcirc$-coherent. Indeed, let x_1, x_2 be any two points of E. The relation $x_1 \bigcirc x_2$ is obvious if $x_1 = x_2$, or if x_1 coincides with either p or q, or if x_2 coincides with either p or q. Thus we can assume that p, x_1, q, x_2 are distinct points. By assumption $p \bigcirc x_1 \bigcirc q \bigcirc x_2 \bigcirc p$. By II.2.5 it follows that $x_1 \bigcirc x_2$. The set E is also $\bigcirc$-complete. Indeed, let y be a point that is conjugate to two distinct points x_1, x_2 of E. Then

$$(1) \qquad p \bigcirc x_1 \bigcirc y \bigcirc x_2 \bigcirc q \bigcirc p.$$

We have to show that $y \in E$. If y coincides with one of the points p, x_1, x_2, q, then the assertion is obvious. So we can assume that $y \neq p, x_1, x_2, q$. Since $p \neq q$ and $x_1 \neq x_2$, we can assume that $x_1 \neq q$ and $x_2 \neq p$. Thus we have the following possibilities to consider.

(i) The points p, x_1, y, x_2, q are distinct. Then (1) implies, by II.2.5, that $p \bigcirc y \bigcirc q$. Hence $y \in E$.

(ii) Exactly one of the relations $p = x_1$, $q = x_2$ holds, say $p = x_1$ and $q \neq x_2$. Then (1) yields $p \bigcirc y \bigcirc x_2 \bigcirc q \bigcirc p$. Since the points p, y, x_2, q are distinct, it follows by II.2.5 that $p \bigcirc y \bigcirc q$. Hence $y \in E$.

(iii) $p = x_1$ and $q = x_2$. Then (1) yields $p \bigcirc y \bigcirc q$, and hence $y \in E$.

Thus E is nondegenerate, $\bigcirc$-coherent, and $\bigcirc$-complete, and hence E is a proper cyclic element containing p and q. The uniqueness follows from II.2.11.

II.2.13. If p is a non-cut point of the Peano space $\mathcal{P}$, then there exists at most one proper cyclic element containing p.

PROOF. Suppose there exist two different proper cyclic elements C_1, C_2 containing p. Then $C_1C_2 = p$ by II.2.11. Since C_1 is nondegenerate, we have a point x_1 such that $x_1 \in C_1$, $x_1 \neq p$. Similarly, we have a point x_2 such that $x_2 \in C_2$, $x_2 \neq p$. Then necessarily $x_1 \neq x_2$. Now since p is not a cut point, the set $\mathcal{P} - p$ is connected and hence also semi-connected. Since x_1 and p lie in C_1, and hence $x_1 \bigcirc p$, the set $x_1 + p$ is also semi-connected (see II.2.4). Similarly the set $x_2 + p$ is semi-connected. We have therefore the simple closed $\bigcirc$-polygon $(x_1, p, x_2, x_1 + p, p + x_2, \mathcal{P} - p)$. By II.2.7 it follows that $x_1 \bigcirc x_2$. Thus $p \bigcirc x_1 \bigcirc x_2$. Since C_2 is $\bigcirc$-complete, it follows that $x_1 \in C_2$, in contradiction with the fact that $C_1C_2 = p$.

II.2.14. A nondegenerate Peano space $\mathcal{P}$ is cyclic if and only if it reduces to a single proper cyclic element. Indeed, if $\mathcal{P}$ is cyclic, then clearly any two points of $\mathcal{P}$ are conjugate. Thus $\mathcal{P}$ is $\bigcirc$-coherent, and obviously also $\bigcirc$-complete. Thus $\mathcal{P}$ itself is a proper cyclic element. Conversely, suppose that $\mathcal{P}$ itself is a proper cyclic element. Assume that $\mathcal{P}$ has a cut point y. Then we can choose two points x_1, x_2 that lie in different components of $\mathcal{P} - y$. Clearly, the relation $x_1 \bigcirc x_2$ does not hold, since y cuts between x_1 and x_2. This contradicts however the fact that $\mathcal{P}$ is $\bigcirc$-coherent. Thus $\mathcal{P}$ has no cut point.

Similarly, a Peano space $\mathcal{P}$ is a dendrite if and only if it possesses no proper cyclic element. Indeed, if $\mathcal{P}$ is a dendrite, then no two distinct points of $\mathcal{P}$ are conjugate (see II.2.1, II.2.3), and hence no nondegenerate subset of $\mathcal{P}$ is $\bigcirc$-coherent. Hence a dendrite possesses no proper cyclic element. Conversely, suppose $\mathcal{P}$ does not possess any proper cyclic element. If p_1, p_2 are any two distinct points of $\mathcal{P}$, then it follows from II.2.12 that the relation $p_1 \bigcirc p_2$ cannot hold. Hence there exists some point x that cuts between p_1 and p_2. Thus $\mathcal{P}$ is a dendrite.

II.2.15. Let E_1, E_2 be nonempty, disjoint, semi-connected sets in the Peano space $\mathcal{P}$. Then there exists at most one proper cyclic element that intersects both E_1 and E_2.

PROOF. *Case* (i). One at least of the sets E_1, E_2, say E_1, reduces to a single point p_1. Suppose there exist two different proper cyclic elements C', C'' that contain p_1 and intersect E_2. Then there exist points x_2', x_2'' such that $x_2' \in C'E_2$, $x_2'' \in C''E_2$. Since $C' \neq C''$, the points x_2', x_2'' are distinct by II.2.11. We have $p_1 \bigcirc x_2'$, $p_1 \bigcirc x_2''$, and hence we have the simple closed $\bigcirc$-polygon $(p_1, x_2', x_2'', p_1 + x_2', E_2, x_2'' + p_1)$. Thus $x_2' \bigcirc x_2''$ by II.2.7. The relations $p_1 \bigcirc x_2'' \bigcirc x_2'$ imply that $x_2'' \in C'$, and hence $p_1 + x_2'' \subset C'C''$. This contradicts the assumption that $C' \neq C''$ (see II.2.11).

Case (ii). *General case.* Suppose there exist two different proper cyclic elements C', C'' that intersect both E_1 and E_2. Then we have points x_1', x_1'', x_2', x_2'' such that $x_1' \in C'E_1$, $x_2' \in C'E_2$, $x_1'' \in C''E_1$, $x_2'' \in C''E_2$. If the points x_1', x_1'', x_2', x_2'' are distinct, then we have the simple closed $\bigcirc$-polygon $(x_1', x_2', x_2'', x_1'', x_1' + x_2', E_2, x_2'' + x_1'', E_1)$, and hence $x_1' \bigcirc x_1'' \bigcirc x_2'$, $x_1' \bigcirc x_2'' \bigcirc x_2'$ by II.2.7. Since C' is $\bigcirc$-complete, it follows that $x_1'' + x_2'' \subset C'C''$, in contradiction with the assumption that $C' \neq C''$ (cf. II.2.11). If the points x_1', x_2', x_1'', x_2'' are not distinct, then one at least of the relations $x_1' = x_1''$, $x_2' = x_2''$ must hold. Suppose that $x_1' = x_1'' = p_1$. Then $p_1 \notin E_2$, and $p_1 \in C'C''$, $C'E_2 \neq 0$, $C''E_2 \neq 0$. This is precisely case (i), and hence this situation is impossible.

II.2.16. A subset E of the Peano space $\mathcal{P}$ will be termed an *H-set* if and only if E is both $\bigcirc$-complete and semi-connected. *Thus an H-set is necessarily nondegenerate.*

II.2.17. If Φ is any family of H-sets in the Peano space $\mathcal{P}$, and if their product $G = \prod E$, $E \in \Phi$, is nondegenerate, then G is also an H-set.

PROOF. G is semi-connected by II.2.2(iii). Now let x be any point that is conjugate to two distinct points p_1, p_2 of G. If E is any set of Φ, then $p_1 + p_2 \subset G \subset E$, and hence $x \in E$, since E is $\bigcirc$-complete. Thus $x \in E$ for every $E \in \Phi$, and hence $x \in G$. Thus G is $\bigcirc$-complete.

II.2.18. In the Peano space $\mathcal{P}$, let $E_1, \cdots, E_n, \cdots$ be a (finite or infinite) sequence of H-sets such that $E_1 \subset \cdots \subset E_n \subset \cdots$. Then $G = E_1 + \cdots + E_n + \cdots$ is again an H-set.

PROOF. G is semi-connected by II.2.2(ii). Let now x be a point that is conjugate to two distinct points p', p'' of G. Then $p' + p'' \subset E_n$ for some n, and hence $x \in E_n \subset G$ since E_n is O-complete. Thus G is O-complete.

II.2.19. In the Peano space $\mathcal{P}$, let E_1, E_2 be H-sets such that E_1E_2 is a single point p_0. Then $E = E_1 + E_2$ is an H-set.

PROOF. E is semi-connected by II.2.2(ii). Let now x be a point that is conjugate to two distinct points a, b of E, and suppose that $x \notin E$. Then clearly $x \neq a, b, p_0$. Furthermore, $p_0 \neq a, b$. Indeed, if one of the points a, b coincides with p_0, then clearly a and b lie in one of the sets E_1, E_2, say in E_1. Since E_1 is O-complete, it would follow that $x \in E_1 \subset E$, while we assumed that $x \notin E$. This argument also shows that a and b cannot lie both in one of the sets E_1, E_2. Thus we see that the points p_0, a, x, b are distinct, and that we can assume that $a \in E_1$, $b \in E_2$. We have therefore the simple closed O-polygon (p_0, a, x, b, E_1, $a + x$, $x + b$, E_2). By II.2.7 it follows that $x \bigcirc p_0$. Thus $a \bigcirc x \bigcirc p_0$. Since $a \neq p_0$ and E_1 is O-complete, it follows that $x \in E_1 \subset E$, in contradiction with our assumption that $x \notin E$.

II.2.20. In the Peano space $\mathcal{P}$, let S be a nondegenerate set. Since $\mathcal{P}$ itself is then an H-set containing S, the class of all H-sets containing S is not empty. The product of all H-sets containing S is then an H-set by II.2.17. This product, to be denoted by $H(S)$, is clearly the smallest H-set that contains S.

II.2.21. If p, q are distinct points of the Peano space $\mathcal{P}$, then the set $H(p + q)$ (see II.2.20) will be termed the *cyclic chain joining p and q*, and will be denoted by $C(p, q)$. Clearly $C(p, q) = C(q, p)$.

II.2.22. The cyclic chain $C(p, q)$ is a proper cyclic element if and only if $p \bigcirc q$.

PROOF. Suppose that $p \bigcirc q$. By II.2.12 there exists a proper cyclic element C such that $p + q \subset C$. Since C is clearly an H-set (cf. II.2.10), it follows that $C(p, q) \subset C$. On the other hand, if x is any point of C, then $p \bigcirc x \bigcirc q$, and hence $x \in C(p, q)$, since $C(p, q)$ is an H-set and hence O-complete. Thus $C(p, q) = C$. Conversely, if $C(p, q)$ is a proper cyclic element, then $C(p, q)$ is O-coherent and hence $p \bigcirc q$.

II.2.23. If S is a nondegenerate subset of the Peano space $\mathcal{P}$, then $\Psi(S)$ will denote the class of all those proper cyclic elements of $\mathcal{P}$ that intersect S in at least two distinct points. Of course, the class $\Psi(S)$ may be empty.

II.2.24. In the Peano space $\mathcal{P}$, let S be a nondegenerate semi-connected set. If C_1, C_2 are two different proper cyclic elements of the class $\Psi(S)$ (see II.2.23), then C_1C_2 is either empty or reduces to a single point. In the second case, the point $p_0 = C_1C_2$ is a cut point, and $p_0 \in S$.

PROOF. In view of II.2.12, II.2.13 we have to verify only that $p_0 \in S$. Now suppose that $p_0 \notin S$. Then p_0 and S are two disjoint semi-connected sets, and the distinct proper cyclic elements C_1, C_2 intersect both of them. This is impossible by II.2.15.

II.2.25. In the Peano space $\mathcal{P}$, let S be a nondegenerate semi-connected set.

Let us put $E = S + \sum C$, $C \in \Psi(S)$ (cf. II.2.23). Let C^* be a proper cyclic element such that C^*E is nondegenerate. Then $C^* \in \Psi(S)$.

PROOF. By assumption, we have two distinct points p^*, q^* such that $p^* + q^* \subset C^*E$.

Case (i). $p^* + q^* \subset S$. Then $C^* \in \Psi(S)$ by definition.

Case (ii). Exactly one of the points p^*, q^* lies in S, say $p^* \in S$, $q^* \notin S$. Since $q^* \in E$, we have a proper cyclic element C such that $q^* \in C \in \Psi(S)$. Then q^* and S are disjoint semi-connected sets, and C, C^* intersect both of them. Hence $C = C^*$ by II.2.15, and thus $C^* \in \Psi(S)$.

Case (iii). $p^* \notin S$, $q^* \notin S$. Since $p^* \in E$, $q^* \in E$, we have proper cyclic elements C_1, C_2 such that $p^* \in C_1 \in \Psi(S)$, $q^* \in C_2 \in \Psi(S)$. Since $p^* + q^* \in C^*$, the points p^* and q^* are conjugate. Thus $p^* + q^*$ and S are disjoint semi-connected sets, and C_1, C_2 intersect both of them. By II.2.15 it follows that $C_1 = C_2$. Hence $p^* + q^* \subset C^*C_1$, and therefore $C^* = C_1$ by II.2.11. Thus $C^* \in \Psi(S)$.

II.2.26. Let S and E have the same meaning as in II.2.25. Then E is an H-set.

PROOF. Since $E \supset S$ and S is nondegenerate, E is nondegenerate. Since S is semi-connected, E is also semi-connected by II.2.2(ii), II.2.10. There remains to show that E is $\bigcirc$-complete. Let x be a point that is conjugate to two distinct points p, q of E. Suppose that $x \notin E$. Then we have the simple closed $\bigcirc$-polygon $(p, x, q, p + x, x + q, E)$. Hence $p \bigcirc q$ by II.2.7. By II.2.12 we have therefore a proper cyclic element C^* containing p and q. Since $p \bigcirc x \bigcirc q$, it follows that $x \in C^*$. Since $p + q \subset C^*E$, we have $C^* \in \Psi(S)$ by II.2.25. Hence $x \in C^* \subset E$, in contradiction with the assumption that $x \notin E$.

II.2.27. In the Peano space $\mathcal{P}$, let S be a nondegenerate semi-connected set. Then $H(S) = S + \sum C$, $C \in \Psi(S)$ (cf. II.2.23, II.2.20).

PROOF Let us put $E = S + \sum C$, $C \in \Psi(S)$. By II.2.26, E is an H-set containing S. Hence $H(S) \subset E$. Thus it is sufficient to show that

$$H(S) \supset S + \sum C, \qquad C \in \Psi(S). \tag{1}$$

Now if $C \in \Psi(S)$, then we have two distinct points p, q such that $p + q \subset CS \subset CH(S)$. Since C is $\bigcirc$-coherent and $H(S)$ is $\bigcirc$-complete, it follows that $C \subset H(S)$. Since C was any element of $\Psi(S)$, the inclusion (1) follows.

II.2.28. In the Peano space $\mathcal{P}$, let S be a nondegenerate semi-connected set, and let C be a proper cyclic element such that $C \subset H(S)$. Then $C \in \Psi(S)$ (cf. II.2.20, II.2.23). This is an immediate corollary of II.2.27, II.2.25.

II.2.29. In the Peano space $\mathcal{P}$, let S be a nondegenerate semi-connected set. Let C^* be a proper cyclic element such that C^*S reduces to a single point p_0. Then $C^*H(S) = p_0$. This is a direct consequence of II.2.27, II.2.25.

II.2.30. If p, q are distinct points of the Peano space $\mathcal{P}$, then $C(p, q) = H[p + q + K(p, q)]$ (cf. II.2.2(vi), II.2.21, II.2.20).

PROOF. By definition $C(p, q) = H(p + q)$, and hence clearly

$$C(p, q) \subset H[p + q + K(p, q)]. \tag{1}$$

Now let us observe that

$$(2) \qquad p + q \subset C(p, q).$$

Since $C(p, q)$ is semi-connected, (2) implies (cf. II.2.2(vi)) that

$$(3) \qquad p + q + K(p, q) \subset C(p, q).$$

Since $C(p, q)$ is an H-set, (3) implies that

$$(4) \qquad H[p + q + K(p, q)] \subset C(p, q).$$

(1) and (4) show that $C(p, q) = H[p + q + K(p, q)]$.

II.2.31. A subset E of the Peano space $\mathcal{P}$ will be termed an *A-set* if and only if E is a closed H-set. Explicitly, E is an A-set if and only if it is nondegenerate, semi-connected, $\bigcirc$-complete, and closed.

II.2.32. We have now at our disposal all of the fundamental concepts of cyclic element theory. Let us observe that we did not actually use so far the assumption that the space $\mathcal{P}$ is a Peano space. From this point on, we shall make full use of the properties of Peano spaces. In particular, the following fundamental theorem will play an important role: If G is a connected open set in a Peano space $\mathcal{P}$, and if p_1, p_2 are any two distinct points of G, then there exists a simple arc $\gamma \subset G$ with end points p_1, p_2 (cf. I.2.41).

A set E is said to be *arc-wise connected* if and only if for every choice of two distinct points p_1, p_2 in E there exists in E some simple arc with end points p_1, p_2. Thus every connected open subset of a Peano space $\mathcal{P}$ is arc-wise connected. We shall see in the sequel further important instances of arc-wise connected subsets of $\mathcal{P}$. Clearly, an arc-wise connected set is also connected.

II.2.33. Let A be an A-set in the Peano space $\mathcal{P}$. Then A is arc-wise connected (and hence A is a continuum). Furthermore, if γ is any simple arc in $\mathcal{P}$ whose end points p_1, p_2 lie in A, then $\gamma \subset A$.

Proof. Since $\mathcal{P}$ itself is arc-wise connected, it is clearly sufficient to prove the second statement. Let γ be a simple arc in $\mathcal{P}$ with end points p_1, p_2 such that $p_1 + p_2 \subset A$. If we deny the inclusion $\gamma \subset A$, then there should exist on γ a point x such that $x \notin A$. Since A is closed, we have on the simple arc xp_1 (the sub-arc of γ with end points x, p_1) a first point q_1 in the order from x to p_1 such that $q_1 \in A$. Let γ_1 be the sub-arc of γ with end points x, q_1. Similarly, the sub-arc xp_2 gives rise to an analogously defined sub-arc γ_2 with end points x, q_2. We have then the simple closed $\bigcirc$-polygon $(q_1, x, q_2, \gamma_1, \gamma_2, A)$, and hence $q_1 \bigcirc x \bigcirc q_2$ by II.2.7. Since A is $\bigcirc$-complete, it follows that $x \in A$, in contradiction with the assumption that $x \notin A$.

II.2.34. In the Peano space $\mathcal{P}$, let A be an A-set and G be a connected open set. Then AG is connected.

Proof. The assertion is obvious if AG is empty or reduces to a single point. So let us assume that AG is nondegenerate. Let p_1, p_2 be any two distinct points of AG. Since G itself is arc-wise connected (cf. II.2.32), we have some simple arc $\gamma \subset G$ with end points p_1, p_2. By II.2.33, $\gamma \subset A$ and hence $\gamma \subset AG$. Thus AG is arc-wise connected and hence also connected.

II.2.35. Let A be an A-set in the Peano space $\mathcal{P}$, such that $\mathcal{P} - A \neq 0$, and let S be a component of $\mathcal{P} - A$. Then the frontier $fr(S)$ of S is a single point $p \in A$ (cf. I.2.15).

PROOF. Since $\mathcal{P}$ is connected and $\mathcal{P} - S \neq 0$, the set $fr(S)$ is not empty (see I.2.40). Since $\mathcal{P}$ is locally connected, and S is a component of the open set $\mathcal{P} - A$, it follows that $fr(S) \subset A$ (cf. I.2.41). Let us suppose that, contrary to our assertion, $fr(S)$ does not reduce to a single point. Let then p_1, p_2 be two distinct points of $fr(S)$. Since $fr(S) \subset A$, we have then $p_1 + p_2 \subset Afr(S)$. Since $p_1 \neq p_2$, we have two connected open sets G_1, G_2, such that $p_1 \in G_1$, $p_2 \in G_2$, $G_1G_2 = 0$. Then $G_1 + S + G_2 = G$ is a connected open set (see I.2.25), and hence AG should be connected in view of II.2.34. But this is not the case, since $AG = AG_1 + AG_2$ is clearly disconnected (note that G_1, G_2 are nonempty, disjoint open sets).

II.2.36. Let A be an A-set in the Peano space $\mathcal{P}$, such that $\mathcal{P} - A \neq 0$. Then $\mathcal{P} - A$ has at most a finite number of components S such that $d(S)$ is greater than or equal to an assigned positive number δ.

PROOF. If we deny the assertion, then there should exist an infinite sequence $S_1, \cdots, S_n, \cdots$ of components of $\mathcal{P} - A$ such that $d(S_n) \geqq \delta > 0, n = 1, 2, \cdots$. For each n, we can choose then two points p_n, q_n in S_n, such that $\rho(p_n, q_n) > \delta - (1/n)$. Since $\mathcal{P}$ is compact, we can assume without loss of generality that the sequences p_n, q_n are convergent. If p, q are their respective limits, then clearly $p \neq q$. We have therefore two connected open sets P, Q such that $p \in P, q \in Q, PQ = 0$ (see I.2.33). For n sufficiently large, we shall have then $S_nP \neq 0, S_nQ \neq 0$. Now $P + S_n$ is again a connected set, and hence the inclusion $P + S_n \subset \mathcal{P} - A$ cannot hold (since S_n is a component of $\mathcal{P} - A$), unless $P \subset S_n$. Since the sets $S_1, \cdots, S_n, \cdots$ are disjoint, the inclusion $P \subset S_n$ can hold for at most one value of n. Thus $(P + S_n)A \neq 0$ for n sufficiently large. Since $S_nA = 0$, it follows that $PA \neq 0$. Similarly $QA \neq 0$. Now $G = P + S_n + Q$ is a connected open set for n large, and hence $AG = AP + AQ$ should be connected by II.2.34. This is however clearly not the case, since P and Q are disjoint open sets, and $AP \neq 0, AQ \neq 0$.

II.2.37. CONTINUATION. The preceding result implies that $\mathcal{P} - A$ has at most a countably infinite number of components. Suppose that $\mathcal{P} - A$ has infinitely many components. According to the preceding remark, these components can then be arranged into a sequence $S_1, \cdots, S_n, \cdots$, and the result of II.2.36 yields immediately the relation $d(S_n) \to 0$ for $n \to \infty$.

II.2.38. Let S be a component of $\mathcal{P} - A$, where A is an A-set in the Peano space $\mathcal{P}$. According to II.2.35, the frontier of S is then a single point $p \in A$. Let now G be a connected set that intersects both S and A. Then $p \in G$.

PROOF. Let us write

$$G = SG + (\mathcal{P} - S)G. \tag{1}$$

Since S is open (see I.2.25), and by assumption G intersects both S and A, we have then

$$(2) \qquad SG \neq 0, \qquad (\mathcal{P} - S)G \neq 0, \qquad SGc[(\mathcal{P} - S)G] = 0.$$

Since G is connected, we must have therefore $0 \neq (\mathcal{P} - S)Gc(SG) \subset (\mathcal{P} - S)Gc(S) = (\mathcal{P} - S)G(p + S) = Gp$. Hence $p \in G$.

II.2.39. A subset E of a topological space Σ is termed a *retract* of Σ if and only if there exists a continuous transformation $T(\Sigma) = E$ such that $T(p) = p$ for $p \in E$. Such a transformation T is then termed a *retraction* from Σ onto E. If the retraction $T(\Sigma) = E$ is *monotone* (see II.1.1), then T is called a *monotone retraction* and E a *monotone retract* of Σ.

II.2.40. Let A be an A-set of the Peano space $\mathcal{P}$. Then A is a monotone retract of $\mathcal{P}$.

Proof. If $A = \mathcal{P}$, then the identity transformation is clearly a monotone retraction from $\mathcal{P}$ onto A. So let us assume that $\mathcal{P} - A \neq 0$. If $x \in \mathcal{P} - A$, then let S_x denote the component of $\mathcal{P} - A$ that contains x. Let us put

$$\mu_A(x) = \begin{cases} x & \text{if} \quad x \in A, \\ fr(S_x) & \text{if} \quad x \in \mathcal{P} - A. \end{cases}$$

We assert that μ_A is a monotone retraction from $\mathcal{P}$ onto A.

Proof. If $x \in \mathcal{P} - A$, then $fr(S_x)$ is a single point of A by II.2.35. Thus μ_A is single-valued, and clearly $\mu_A(\mathcal{P}) = A$, $\mu_A(x) = x$ for $x \in A$. If we deny the continuity of μ_A, then in view of the compactness of $\mathcal{P}$ there should exist in $\mathcal{P}$ a point x_0 and a sequence x_n such that

$$(1) \qquad x_n \to x_0, \; \mu_A(x_n) \to y_0 \neq \mu_A(x_0).$$

Case (i). $x_0 \in \mathcal{P} - A$. Then $x_n \in S_{x_0}$ for n large, since S_{x_0} is open, and hence $S_{x_n} = S_{x_0}$. Thus $\mu_A(x_n) = fr(S_{x_n}) = fr(S_{x_0}) = \mu_A(x_0)$ for n large, in contradiction with (1).

Case (ii). $x_0 \in A$, and $x_n \in A$ for infinitely many values of n. Then $\rho[\mu_A(x_n), \mu_A(x_0)] = \rho(x_n, x_0)$ for infinitely many values of n, and hence lim inf $\rho[\mu_A(x_n), \mu_A(x_0)] = 0$, in contradiction with (1).

Case (iii). $x_0 \in A$, and $x_n \in \mathcal{P} - A$ for n exceeding a certain n_0. Then $\mu_A(x_n) = fr(S_{x_n})$ for $n > n_0$, and $\mu_A(x_0) = x_0$. From (1) it follows now that $x_0 \neq y_0$ and $\rho[\mu_A(x_n), x_n] > \rho(x_0, y_0)/2 > 0$ for n large. Hence *a fortiori* $d(S_{x_n}) > \rho(x_0, y_0)/2 > 0$ for n large. In view of II.2.36 it follows that a certain component S of $\mathcal{P} - A$ occurs an infinite number of times in the sequence $S_{x_1}, \cdots, S_{x_n}, \cdots$. If p denotes the unique frontier point of S, then we shall have therefore $x_n \in S$ and hence $\mu_A(x_n) = p$ for infinitely many values of n. Since $x_n \to x_0 \in A$, it follows in view of (1) that $x_0 = p = y_0$. As $x_0 \in A$ and hence $x_0 = \mu_A(x_0) \neq y_0$ by (1), we reached again a contradiction.

Thus μ_A is continuous. There remains to show that $\mu_A^{-1}(\bar{x})$ is connected for every point $\bar{x} \in A$. Now clearly $\mu_A^{-1}(\bar{x}) = \bar{x} + \sum S$, where the summation is extended over all those components S of $\mathcal{P} - A$ for which $fr(S) = \bar{x}$. If the summation is vacuous, then $\mu_A^{-1}(\bar{x}) = \bar{x}$ is connected. If the summation is non-vacuous, then $S + \bar{x} = c(S)$ for each S occurring in the summation. Thus we can write $\mu_A^{-1}(\bar{x}) = \sum c(S)$, where each $c(S)$ is connected (see I.2.39) and contains the point $\bar{x}$. Hence $\mu_A^{-1}(\bar{x})$ is connected by I.2.39. The fact that $\mu_A^{-1}(\bar{x})$ is a continuum follows now readily. Indeed, $\mu_A^{-1}(\bar{x})$ is closed and hence compact, since μ_A is continuous and $\mathcal{P}$ is compact.

II.2.41. If A is an A-set in the Peano space $\mathcal{P}$, then A is a Peano subspace of $\mathcal{P}$. Indeed, A is the image of $\mathcal{P}$ under the continuous transformation μ_A (see II.2.40, 1.2.43).

II.2.42. If A is an A-set and G is a connected set in a Peano space $\mathcal{P}$, then AG is connected (cf. the weaker statement in II.2.34).

PROOF. If either $A = \mathcal{P}$ or $AG = 0$, then the statement is obvious. So we can assume that

(1) $$\mathcal{P} - A \neq 0, \qquad AG \neq 0.$$

We shall use the monotone retraction $\mu_A(\mathcal{P}) = A$ defined in II.2.40. Since G is connected and μ_A is continuous, the set $\bar{G} = \mu_A(G)$ is connected (see I.2.43). Hence the connectedness of AG will be established if we can show that

(2) $$\mu_A(G) = AG.$$

Since $AG \subset A$ and $\mu_A(x) = x$ for $x \in A$, clearly

(3) $$AG = \mu_A(AG) \subset \mu_A(G).$$

Now let $\bar{x}_0$ be any point of $\mu_A(G)$. Then we have a point $x_0 \in G$ such that $\bar{x}_0 = \mu_A(x_0)$. Thus we have the relations

(4) $$\bar{x}_0 \in \mu_A(G) \subset A, \qquad \bar{x}_0 = \mu_A(x_0), \qquad x_0 \in G.$$

Case (i). $x_0 \in A$. Then $\bar{x}_0 = x_0$, and hence by (4) it follows that $\bar{x}_0 \in AG$.

Case (ii). $x_0 \in \mathcal{P} - A$. Then $x_0 \in S_{x_0}$, where S_{x_0} is a component of $\mathcal{P} - A$. By the definition of μ_A we have then

(5) $$\bar{x}_0 = fr(S_{x_0}).$$

In view of (1) we have by II.2.38

(6) $$fr(S_{x_0}) \in G.$$

(4), (5), (6) yield the inclusion $\bar{x}_0 \in AG$.

Thus in either case we have $\bar{x}_0 \in AG$. Since $\bar{x}_0$ was an arbitrary point of $\mu_A(G)$, it follows that $\mu_A(G) \subset AG$. In view of (3), the relation (2) follows.

II.2.43. The retraction μ_A, defined in II.2.40, is the *unique* monotone retraction from $\mathcal{P}$ onto A.

PROOF. Let $\mu(\mathcal{P}) = A$ be a monotone retraction from $\mathcal{P}$ onto A (if $\mathcal{P} = A$, then obviously μ reduces to the identity and the assertion is trivial). If $x \in A$, then $\mu(x) = x = \mu_A(x)$. Suppose now that $x \in \mathcal{P} - A$. Then the point $\bar{x} = \mu(x)$ lies in A, and clearly

$$(1) \qquad x + \bar{x} \in \mu^{-1}(\bar{x}), \qquad A\mu^{-1}(\bar{x}) = \bar{x}.$$

Let S_x be the component of $\mathcal{P} - A$ that contains x. Since μ is monotone, $\mu^{-1}(\bar{x})$ is connected, and the set $c(S_x) = S_x + fr(S_x)$ is also connected. Since both sets contain the point x, the set $S_x + fr(S_x) + \mu^{-1}(\bar{x})$ is connected. By II.2.42, it follows that $[S_x + fr(S_x) + \mu^{-1}(\bar{x})]A$ is also connected. Hence, in view of (1), $\bar{x} + fr(S_x)$ is connected (cf. II.2.35). Since $fr(S_x)$ is a single point, it follows finally that $\mu(x) = \bar{x} = fr(S_x) = \mu_A(x)$.

II.2.44. Let A be an A-set of the Peano space $\mathcal{P}$, such that $\mathcal{P} - A \neq 0$, and let S be a component of $\mathcal{P} - A$. Then $fr(S)$ is a single point $p_0 \in A$ by II.2.35. We assert that p_0 is a cut point, and S is also a component of $\mathcal{P} - p_0$.

PROOF. Suppose that p_0 is not a cut point. Then $\mathcal{P} - p_0$ is connected, and S is a nonempty proper subset of $\mathcal{P} - p_0$. Hence, by I.2.40, $(\mathcal{P} - p_0)\,fr(S) \neq 0$, while as a matter of fact $(\mathcal{P} - p_0)fr(S) = (\mathcal{P} - p_0)p_0 = 0$. Thus p_0 is a cut point. Now suppose S is not a component of $\mathcal{P} - p_0$. Then S would be a proper subset of some component S^* of $\mathcal{P} - p_0$, and by I.2.40 we should have $S^*fr(S) \neq 0$, while actually $S^*fr(S) \subset (\mathcal{P} - p_0)p_0 = 0$. Hence S is a component of $\mathcal{P} - p_0$.

II.2.45. Let H be an H-set of the Peano space $\mathcal{P}$, and let C be a proper cyclic element of $\mathcal{P}$. If CH is nondegenerate, then $C \subset H$. This follows directly from the fact that C is O-coherent and H is O-complete. In particular, if A is a A-set and CA is nondegenerate, then $C \subset A$.

II.2.46. In the Peano space $\mathcal{P}$, let Φ be a family of A-sets such that for any two different sets A', A'' of Φ the product $A'A''$ is either empty or reduces to a single point. Then for given $\epsilon > 0$ the family Φ contains at most a finite number of sets with a diameter exceeding ϵ.

PROOF. If the assertion is denied, then for some $\epsilon > 0$ there should exist an infinite sequence $A_1, \cdots, A_n, \cdots$ of distinct sets in Φ such that $d(A_n) > \epsilon$ for every n. For each n, there exists then a pair of points p_n, q_n in A_n such that $\rho(p_n, q_n) > \epsilon$. Since $\mathcal{P}$ is compact, we can assume without loss of generality that the sequences p_n, q_n are convergent, say $p_n \to p_0, q_n \to q_0$. Then clearly $p_0 \neq q_0$, and hence (see I.2.33) we have two connected open sets G', G'' such that $p_0 \in G'$, $q_0 \in G''$, $c(G')c(G'') = 0$. Since $p_n \to p_0$, $q_n \to q_0$, $p_n + q_n \subset A_n$, we have an n_0 such that $G'A_n \neq 0$, $G''A_n \neq 0$ for $n > n_0$. Let $n > n_0$, and consider the set $B_n = A_n + c(G') + c(G'')$. Then B_n is connected, and hence $B_nA_{n+1} = A_nA_{n+1} + A_{n+1}c(G') + A_{n+1}c(G'')$ should be connected by II.2.42. Now $A_{n+1}c(G')$, $A_{n+1}c(G'')$ are nonempty, disjoint, closed sets, and A_nA_{n+1} is either empty or reduces to a single point. Hence the set B_nA_{n+1} cannot be connected, in contradiction with II.2.42.

II.2.47. CONTINUATION. It follows now obviously that Φ is countable. If the family Φ is infinite, we can therefore arrange the sets of Φ into a (countable) sequence $A_1, \cdots, A_n, \cdots$, and clearly the result of II.2.46 implies that $d(A_n) \to 0$.

II.2.48. In the Peano space $\mathcal{P}$, let $A_1, \cdots, A_n, \cdots$ be a sequence of A-sets such that $A_1 \subset A_2 \subset \cdots \subset A_n \subset \cdots$, and $c(\sum A_n) = \mathcal{P}$. For each n, let δ_n denote the maximum diameter of the components of $\mathcal{P} - A_n$, where we put $\delta_n = 0$ in case $\mathcal{P} - A_n = 0$. Then $\delta_n \to 0$.

Proof. Let us first note that if $\mathcal{P} - A_n \neq 0$, then $\mathcal{P} - A_n$ has at most a countably infinite number of components. If the number of these components is infinite, then their diameters converge to zero by II.2.37, and hence the use of the maximum diameter of the components of $\mathcal{P} - A_n$ is justified.

Clearly $\delta_1 \geqq \delta_2 \geqq \cdots \geqq \delta_n \geqq \cdots$. Hence if the relation $\delta_n \to 0$ is denied, then we should have an $\epsilon > 0$ such that there exists, for each n, a component S_n of $\mathcal{P} - A_n$ that satisfies the inequality $d(S_n) > \epsilon$. For each n, we can choose then in S_n a pair of points p_n, q_n such that $\rho(p_n, q_n) > \epsilon$. Since $\mathcal{P}$ is compact, we have an infinite sequence of positive integers $k_1 < \cdots < k_n < \cdots$ such that the sequences p_{k_n}, q_{k_n} are convergent, say $p_{k_n} \to p_0, q_{k_n} \to q_0$. Clearly $p_0 \neq q_0$. By I.2.33 we have therefore two connected open sets G', G'' such that $p_0 \in G'$, $q_0 \in G''$, $G'G'' = 0$. Since $p_{k_n} \to p_0, q_{k_n} \to q_0, p_{k_n} + q_{k_n} \subset S_{k_n}$, we have an n_0 such that $S_{k_n}G' \neq 0$, $S_{k_n}G'' \neq 0$ for $n > n_0$. Since $c(\sum A_n) = \mathcal{P}$ and $A_1 \subset \cdots \subset A_n \subset \cdots$, we have an $\bar{n}_0$ such that $G'A_n \neq 0$, $G''A_n \neq 0$ for $n > \bar{n}_0$. Let now n be an integer that exceeds $n_0 + \bar{n}_0$. Then $S_{k_n} + G' + G''$ is connected, and hence by II.2.42 the set $(S_{k_n} + G' + G'')A_{k_n} = G'A_{k_n} + G''A_{k_n}$ should be connected. This is however clearly not the case, since $G'A_{k_n} \neq 0$, $G''A_{k_n} \neq 0$ and G', G'' are disjoint open sets.

II.2.49. Let x_0 be a point of the Peano space $\mathcal{P}$, and let E be the set of those points of $\mathcal{P}$ that are conjugate to x_0. Then E is closed (possibly degenerate).

Proof. Let $y \in \mathcal{P} - E$. Then y is not conjugate to x_0, and hence there exists a point z that cuts between y and x_0. In other words, x_0 and y are in different components S', S'' of $\mathcal{P} - z$. Then z also cuts between x_0 and any point u of S''. Thus no point of S'' is conjugate to x_0, and hence $S'' \subset \mathcal{P} - E$. Since $\mathcal{P}$ is a Peano space, S'' is open. Thus for every point $y \in \mathcal{P} - E$ there exists an open set containing y and contained in $\mathcal{P} - E$. It follows that $\mathcal{P} - E$ is open and hence E itself is closed.

II.2.50. If C is a proper cyclic element of the Peano space $\mathcal{P}$, then C is closed. Consequently, C is an A-set (cf. II.2.10, II.2.31).

Proof. Let p_1, p_2 be two distinct points of C, and let E_1, E_2 be the sets of those points that are conjugate to p_1, p_2 respectively. By II.2.12, C coincides with the set of those points x that satisfy the relations $p_1 \bigcirc x \bigcirc p_2$. In other words, $C = E_1E_2$. Since E_1, E_2 are closed by II.2.49, it follows that C is closed.

II.2.51. If C is a proper cyclic element of the Peano space $\mathcal{P}$, and if p is any point of $\mathcal{P}$, then $C - p$ is connected. Thus C is a cyclic subspace of $\mathcal{P}$ (cf. I.2.34).

Proof. If $p \in \mathcal{P} - C$, then $C - p = C$ is connected by II.2.50, II.2.33. So assume that $p \in C$. If x_0 is a fixed point of $C - p$ and x is any point of $C - p$, then $x \bigcirc x_0$, and hence x and x_0 lie in the same component S_0 of $\mathcal{P} - p$. Now S_0 is independent of x; indeed, S_0 is the (unique) component of $\mathcal{P} - p$ that contains

x_0. Thus $C - p = CS_0$. Since C is an A-set (see II.2.50) and S_0 is connected, it follows by II.2.42 that $C - p$ is connected.

II.2.52. A Peano space $\mathcal{P}$ has at most a countably infinite number of proper cyclic elements. Indeed, every proper cyclic element is an A-set by II.2.50, and two distinct proper cyclic elements cannot have a nondegenerate intersection by II.2.11. Hence the assertion follows directly from II.2.46. If $\mathcal{P}$ has any proper cyclic elements at all, then it follows from II.2.46 that they can be arranged into a (finite or countably infinite) sequence $C_1, \cdots, C_n, \cdots$, and $d(C_n) \to 0$ if this sequence is infinite.

II.2.53. Let C be a proper cyclic element of the Peano space $\mathcal{P}$. If a point $p \in C$ is a cut point of $\mathcal{P}$, then p is the frontier of some component of $\mathcal{P} - C$.

Proof. Since $C - p$ is connected by II.2.51 and $\mathcal{P} - p$ is disconnected by assumption, it follows that $\mathcal{P} - C \neq 0$. Let Φ be the class of the components of $\mathcal{P} - C$. By II.2.50 and II.2.35, for each $S \in \Phi$ the set $fr(S)$ consists of a single point of C. Suppose that $fr(S) \neq p$ for every $S \in \Phi$. Then $fr(S) \in C - p$ for every $S \in \Phi$ and consequently $(C - p)c(S) = (C - p)[S + fr(S)] \neq 0$ for every $S \in \Phi$. By II.2.51 it follows that $\mathcal{P} - p = (C - p) + \sum S$, $S \in \Phi$, is connected, in contradiction with the assumption that p is a cut point of $\mathcal{P}$.

II.2.54. Let C be a proper cyclic element of the Peano space $\mathcal{P}$. By II.2.50, II.2.37, the complement $\mathcal{P} - C$ of C has at most a countably infinite number of components. By II.2.53, II.2.50, II.2.35 it follows that C contains at most a countably infinite number of cut points of $\mathcal{P}$. Since C is a nondegenerate continuum, it follows that C contains a noncountable set of non-cut points of $\mathcal{P}$. Let us consider some implications of this fact.

(a) Let $p \in C$ be a non-cut point of $\mathcal{P}$, and let M_p denote the set of all those points of $\mathcal{P}$ that are conjugate to p. Since C is $\bigcirc$-coherent, clearly $C \subset M_p$. Let q be any point of M_p distinct from p. Then $q \bigcirc p$, and hence by II.2.12 there exists a proper cyclic element $\bar{C}$ that contains p and q. Then $p \in C\bar{C}$. Since p is a non-cut point, $\bar{C} = C$ by II.2.13. Thus $q \in C$. Since q was any point of M_p distinct from p, it follows that $C = M_p$.

(b) Conversely, let p be a non-cut point of $\mathcal{P}$, and let again M_p denote the set of all those points of $\mathcal{P}$ that are conjugate to p. The set M_p may reduce to the point p (for example, if $\mathcal{P}$ is a simple arc and p is one of the end points of this arc). Suppose that M_p does not reduce to p. Then M_p is a proper cyclic element that contains p. Indeed, $p \in M_p$ since $p \bigcirc p$. Let now q be any point of M_p distinct from p. Then $q \bigcirc p$, and hence by II.2.12 we have a proper cyclic element C that contains p and q. By (a) it follows that $M_p = C$.

(c) According to (a) and (b), a proper cyclic element may be characterized as a nondegenerate set M_p corresponding to a non-cut point p of $\mathcal{P}$. This fact yields an alternative definition of proper cyclic elements (in fact, this alternative definition represents the original definition of a proper cyclic element). Of course, the theory as a whole is unaffected by the choice between equivalent definitions of the fundamental concepts.

II.2.55. In the Peano space $\mathcal{P}$, let S be a closed nondegenerate, semi-connected set. Then $H(S)$ is closed (cf. II.2.20).

PROOF. We have $H(S) = S + \sum C, C \in \Psi(S)$, by II.2.27. Let x_0 be a point in $\mathcal{P} - H(S)$. Since S itself is closed, and $x_0 \in \mathcal{P} - S$, we have (cf. I.2.33) a connected open set G such that $x_0 \in G \subset \mathcal{P} - S$. Since S and G are semi-connected and disjoint, we have at most one proper cyclic element in $\Psi(S)$ that intersects G (see II.2.15). If such a proper cyclic element C_0 exists, then $G - H(S) = G - C_0$. If no such proper cyclic element exists, then $G - H(S) = G$. Since every proper cyclic element is closed (see II.2.50), it follows in either case that $G - H(S)$ is open. Thus $x_0 \in G - H(S) \subset \mathcal{P} - H(S)$, where $G - H(S)$ is open. Thus $\mathcal{P} - H(S)$ is open, and hence $H(S)$ itself is closed.

II.2.56. In the Peano space $\mathcal{P}$, let p, q be two distinct points, and let γ be any simple arc with end points p, q. Then $K(p, q) \subset \gamma$ (see II.2.2(vi)).

PROOF. Let $x \in \mathcal{P} - \gamma$. Since γ is connected, it follows that x does not cut between p and q (cf. II.2.2(v)). Hence $x \notin K(p, q)$.

II.2.57. If p and q are any two distinct points of the Peano space $\mathcal{P}$, then the set $p + q + K(p, q)$ is closed (cf. II.2.2(vi)).

PROOF. Let $x \notin p + q + K(p, q)$. Then x does not cut between p and q, and hence p and q lie in the same component S of $\mathcal{P} - x$. Thus we have a simple arc γ with end points p, q such that $\gamma \subset S$. By II.2.56 it follows that $x \in \mathcal{P} - \gamma \subset \mathcal{P} - [p + q + K(p, q)]$. Since $\mathcal{P} - \gamma$ is open, and x was any point of $\mathcal{P} - [p + q + K(p, q)]$, it follows that $p + q + K(p, q)$ is closed.

II.2.58. If p and q are distinct points of the Peano space $\mathcal{P}$, then the cyclic chain $C(p, q)$ is closed, and hence $C(p, q)$ is an A-set (see II.2.21).

PROOF. Since $C(p, q) = H[p + q + K(p, q)]$ by II.2.30, and $p + q + K(p, q)$ is closed by II.2.57 and semi-connected by II.2.2(vi), the assertion follows directly from II.2.55.

II.2.59. If p and q are distinct points of the Peano space $\mathcal{P}$, and γ is any simple arc with end points p, q, then $C(p, q) = H(\gamma)$ (cf. II.2.20).

PROOF. Since $C(p, q)$ is an A-set by II.2.58, we have $\gamma \subset C(p, q)$ by II.2.33. Thus $H(\gamma) \subset H[C(p, q)] = C(p, q)$, since $C(p, q)$ is itself an H-set. To obtain the complementary inclusion, note that $p + q \subset \gamma$, and hence (see II.2.21) $C(p, q) = H(p + q) \subset H(\gamma)$.

II.2.60. Let p and q be distinct points of the Peano space $\mathcal{P}$. We proceed to discuss the structure of the cyclic chain $C(p, q)$. Let γ be any simple arc with end points p, q. By II.2.59, II.2.27

$$C(p, q) = \gamma + \sum C, \qquad C \in \Psi(\gamma). \tag{1}$$

In other words, $C(p, q)$ consists of the simple arc γ and of all those proper cyclic elements C that have a nondegenerate intersection with γ. By II.2.30, II.2.27 we have the further formula

$$C(p, q) = p + q + K(p, q) + \sum C, \qquad C \in \Psi[p + q + K(p, q)]. \tag{2}$$

In other words, $C(p, q)$ consists of the set $p + q + K(p, q)$ and of all those proper cyclic elements C that have a nondegenerate intersection with $p + q + K(p, q)$. By II.2.56 we have the inclusion

(3) $$p + q + K(p, q) \subset \gamma.$$

We assert further that

(4) $$\Psi(\gamma) \equiv \Psi[p + q + K(p, q)].$$

Indeed, if $C \in \Psi(\gamma)$ then $C \subset C(p, q)$ by (1), and hence $C \in \Psi[p + q + K(p, q)]$ by II.2.30, II.2.28. Conversely, if $C \in \Psi[p + q + K(p, q)]$, then $C \in \Psi(\gamma)$ as an obvious consequence of (3).

II.2.61. CONTINUATION. *Case* (i). $K(p, q) = 0$. This happens if and only if $p \bigcirc q$, and by II.2.22 the cyclic chain $C(p, q)$ reduces then to the (unique) proper cyclic element that contains p and q.

Case (ii). $p + q + K(p, q) = \gamma$. We assert that in this case

(1) $$C(p, q) = \gamma.$$

Indeed, if (1) is denied, then we should have (see II.2.60(1)) at least one proper cyclic element C_0 that has a nondegenerate intersection with γ. Let then p_0, q_0 be two distinct points of $C_0\gamma$, where p_0, q_0 follow upon each other in the order from p to q on γ. Then $p_0 \bigcirc q_0$. Let x be any point between p_0 and q_0 on γ. By assumption $x \in K(p, q)$, and hence x should cut between p and q. On the other hand, let α denote the sub-arc of γ with end points p, p_0 if $p \neq p_0$ and let α denote the point p if $p = p_0$. Let β have a similar meaning with respect to q_0 and q. Then $\alpha + (p_0 + q_0) + \beta$ is clearly a semi-connected set that contains p and q and does not contain x, in contradiction with II.2.2(v).

Case (iii). *The general case* (*neither* (i) *nor* (ii) *holds*). Then the set $\gamma - [p + q + K(p, q)]$ is a nonempty, relatively open subset of γ. This set has at most a countably infinite number of components $\gamma_1^0, \cdots, \gamma_n^0, \cdots$, where each γ_n^0 is an open sub-arc of γ. The sequence γ_n^0 may reduce to a single term γ_1^0, but in this case $\gamma_1^0 \neq \gamma - (p + q)$, since otherwise we should have $K(p, q) = 0$. Clearly $\gamma_i^0\gamma_j^0 = 0$ for $i \neq j$. If p_n, q_n are the end points of γ_n^0 in the order from p to q on γ, then clearly $p_n + q_n \in p + q + K(p, q)$ (note that $p + q + K(p, q)$ is closed by II.2.57). We proceed to verify a series of simple statements concerning the sequence of simple arcs $\gamma_n = \gamma_n^0 + p_n + q_n$.

(a) $p_n \bigcirc q_n$ for every n. Indeed, for a point x to cut between p_n and q_n, the point x should lie on γ_n^0. But then $x \notin K(p, q)$, and hence p and q lie in the same component S of $\mathcal{P} - x$. Obviously, $S + (\gamma - \gamma_n^0)$ is then a connected set that contains p_n and q_n but does not contain x. Hence x does not cut between p_n and q_n.

(b) If x and y are distinct points of γ such that $x \bigcirc y$, then x and y lie on the same simple arc γ_n. Indeed, if this were not the case, then we would have a point $z \in K(p, q)$ between x and y on γ. Let α be the sub-arc of γ with end points x, y. Since $x \bigcirc y$, the points x and y lie in the same component S_z of $\mathcal{P} - z$. Clearly, on setting $\alpha^0 = \alpha - (x + y)$, the set $S_z + (\gamma - \alpha^0)$ is a connected set that contains p and q but does not contain z, in contradiction with the fact that z cuts between p and q.

(c) If C^* is a proper cyclic element contained in $C(p, q)$, then $C^*\gamma$ coincides with one of the simple arcs $\gamma_1, \cdots, \gamma_n, \cdots$. Indeed, by II.2.59 and II.2.25, it follows that $C^* \in \Psi(\gamma)$, and hence $C^*\gamma$ is nondegenerate. By II.2.42, the set $C^*\gamma$ is connected, and hence $C^*\gamma$ is a nondegenerate continuum by II.2.50. Since $C^*\gamma \subset \gamma$, it follows that $C^*\gamma$ is a simple sub-arc of γ (see I.2.32). Let us put $C^*\gamma = \gamma^*$, and let p^*, q^* be the end points of γ^*. Then $p^* \bigcirc q^*$, and hence by (b) above p^* and q^* lie on the same simple arc γ_n. Thus $\gamma^* \subset \gamma_n$. By (a) above there exists, in view of II.2.12, a unique proper cyclic element $\bar{C}$ such that $p_n + q_n \subset \bar{C}$. By II.2.58, II.2.33 it follows that $\gamma_n \subset \bar{C}$. Thus $\bar{C}C^*$ contains γ^*, and hence $\bar{C} = C^*$ by II.2.11. Thus $\gamma_n \subset \bar{C}\gamma = C^*\gamma = \gamma^*$. Since we also have $\gamma^* \subset \gamma_n$, it follows that $\gamma^* = C^*\gamma = \gamma_n$.

(d) For every k there exists a unique proper cyclic element C_k such that $C_k\gamma = \gamma_k$. Indeed, by (a) above the end points p_k and q_k of γ_k are conjugate, and hence by II.2.12 there exists a unique proper cyclic element C^* such that $p_k + q_k \subset C^*$. Then $C^* \subset C(p, q)$ by II.2.60(1), and hence $C^*\gamma$ coincides with one of the simple arcs γ_n, say γ_{n_0}. Since $p_k + q_k \subset C^*\gamma = \gamma_{n_0}$, it follows that $n_0 = k$.

(e) Now let $C_1, \cdots, C_n, \cdots$ be the proper cyclic elements that are determined by the condition $C_n\gamma = \gamma_n$ (see (d)). By (c) above, each $C^* \subset C(p, q)$ coincides then with a term of the sequence $C_1, \cdots, C_n, \cdots$, and by II.2.60(1), we have $C_n \subset C(p, q)$ for every n. Thus we have, in view of II.2.60(1), the fundamental formula $C(p, q) = \gamma + C_1 + \cdots + C_n + \cdots$. If $i \neq j$, then $C_i\gamma = \gamma_i \neq \gamma_j = C_j\gamma$, and hence $C_i \neq C_j$. By II.2.24 it follows that C_iC_j is either empty or else it reduces to a single point p_{ij} of γ. In the latter case, clearly $p_{ij} = C_iC_j = C_i\gamma C_j\gamma = \gamma_i\gamma_j$. Thus p_{ij} is a common end point of γ_i and γ_j, and hence $p_{ij} \in K(p, q)$ and $p_{ij} \neq p, q$. In particular it follows that there is at most one subscript n such that $p \in C_n$, with a similar statement for q. Let us also note that $C(p, q) - p$ is connected. Indeed, $C(p, q) - p = (\gamma - p) + (C_1 - p) + \cdots + (C_n - p) + \cdots$. Now $\gamma - p$ is clearly connected, and each $C_n - p$ is connected by II.2.51. Since $C_n\gamma$ is nondegenerate, $(C_n - p)(\gamma - p)$ is not empty, and the connectedness of $C(p, q) - p$ follows. Similarly $C(p, q) - q$ is connected.

II.2.62. CONTINUATION. In the general case II.2.61(iii), the cyclic chain $C(p, q)$ admits of subdivisions of a particular type that will be useful in the sequel. Let us consider the simple arc $\gamma_1 = p_1 + q_1 + \gamma_1^0$ (see II.2.61(iii)). Let us first suppose that $p \neq p_1, q \neq q_1$. Let α_1 be the sub-arc of γ with end points p and p_1, and let β_1 be the sub-arc of γ with end points q_1, q. Then

(1) $$\alpha_1 + \gamma_1 + \beta_1 = \gamma, \alpha_1(\gamma_1 + \beta_1) = p_1, (\alpha_1 + \gamma_1)\beta_1 = q_1.$$

We shall consider presently the cyclic chains $C(p, p_1)$, $C(p_1, q_1)$, $C(q_1, q)$. Since $p_1 + q_1 \subset C_1$, it is immediate that (see II.2.61(i))

(2) $$C(p_1, q_1) = C_1.$$

Now $C(p, p_1) = \alpha_1 + \sum C$, $C \in \Psi(\alpha_1)$ (cf. II.2.60(1)). Hence if $C \subset C(p, p_1)$, then we have also $C \subset C(p, q)$. Conversely, if $C \subset C(p, q)$ and $C\alpha_1$ is non-degenerate, then $C \subset C(p, p_1)$ (cf. II.2.60(1)). In view of II.2.61(e) it follows that

(3) $$C(p, p_1) = \alpha_1 + \sum C_n , \qquad C_n\gamma = \gamma_n \subset \alpha_1 .$$

Similarly we obtain

(4) $$C(q_1 , q) = \beta_1 + \sum C_n , \qquad C_n\gamma = \gamma_n \subset \beta_1 .$$

Of course, the summation occurring in (3), for example, may be vacuous, and then $C(p, p_1)$ reduces to α_1. The formulas (2), (3), (4) yield, in view of II.2.61(e), the relations

(5) $$C(p, q) = C(p, p_1) + C_1 + C(q_1 , q),$$

(6) $$C(p, p_1)C_1 = p_1 , \qquad C_1C(q_1 , q) = q_1 , \qquad C(p, p_1)C(q_1 , q) = 0$$

In the preceding discussion, we assumed that $p \neq p_1$, $q \neq q_1$. If $p = p_1$ for instance, then the cyclic chain $C(p, p_1)$ is absent, and otherwise the reasoning remains the same. The result just obtained will be generalized in the next section.

II.2.63. Using the assumptions and notations of II.2.61(iii), let $C^1, \cdots , C^n$ be any n terms of the sequence $C_1 , C_2 , \cdots$. Then there exists a finite system $\mathfrak{C}_1 , \cdots , \mathfrak{C}_N$ of cyclic chains, such that the following holds.

(a) $\mathfrak{C}_1 + \cdots + \mathfrak{C}_N = C(p, q)$.

(b) $\mathfrak{C}_i\mathfrak{C}_{i+1}$ is a single point of γ, $i = 1, \cdots , N - 1$, and otherwise $\mathfrak{C}_i\mathfrak{C}_j = 0$ for $i \neq j$.

(c) Each C^k, $k = 1, \cdots , n$, coincides with a term $\mathfrak{C}_{i_k}$ of the system $\mathfrak{C}_1 , \cdots , \mathfrak{C}_N$.

Proof. For $n = 1$, the assertion follows directly from II.2.62. Proceeding by induction, assume that the assertion has been already verified for $n - 1$. Let p^1, q^1 be the end points of C^1. If we apply the process used in II.2.62 with C_1 replaced by C^1, neither one of the resulting cyclic chains $C(p, p^1)$, $C(q^1, q)$ contains more than $n - 1$ terms of the system $C^2, \cdots , C^n$, and hence both of these chains can be subdivided into cyclic chains in the manner described. Clearly, there follows the existence of a subdivision of $C(p, q)$ itself with the desired properties.

II.2.64. Let H be an H-set in the Peano space $\mathcal{P}$, and let p, q be two distinct points of H. If γ is any simple arc with end points p, q, then $\gamma \subset H$.

Proof. Consider the cyclic chain $C(p, q)$. By II.2.58, II.2.33 we have $\gamma \subset C(p, q)$. By definition, $C(p, q) = H(p + q)$, and hence clearly $C(p, q) \subset H$ (see II.2.20, II.2.21). Thus $\gamma \subset H$.

II.2.65. If H is an H-set and G is a connected set in the Peano space $\mathcal{P}$, then GH is connected. In particular, H is connected, since $H = \mathcal{P}H$.

Proof. The assertion is obvious if GH is empty or reduces to a single point.

So let us assume that GH is nondegenerate. Let then p, x be any two distinct points of GH. By the argument used in II.2.64, the cyclic chain $C(p, x)$ is then a subset of H, and hence $p + x \subset GC(p, x) \subset GH$. Keeping p fixed and varying x in GH, we obtain

$$GH = \sum GC(p, x), \qquad x \in GH - p.$$

Since $C(p, x)$ is an A-set (see II.2.58), $GC(p, x)$ is connected by II.2.42, and the connectedness of GH follows.

II.2.66. If H is an H-set in the Peano space $\mathcal{P}$, and if p is a point in the closure $c(H)$ of H, then $H + p$ is also an H-set.

PROOF. $H + p$ is clearly nondegenerate and connected (cf. I.2.39, II.2.65). There remains to show that $H + p$ is $\bigcirc$-complete. Since H itself is $\bigcirc$-complete, it is clearly sufficient to establish the following fact: if q is any point of H different from p, and $p \bigcirc x \bigcirc q$, then $x \in H + p$. Suppose that $x \notin H + p$. Then we have the simple closed $\bigcirc$-polygon $[p, x, q, p + x, x + q, H + p]$. Hence $p \bigcirc q$ (see II.2.7). By II.2.12 it follows that p, x, q lie on a proper cyclic element C. Since $x \in C - (H + p)$, it follows by II.2.45 that $CH = q$, and hence $C(H + p) = p + q$. Thus $C(H + p)$ is disconnected. Since C is an A-set (see II.2.50) and $H + p$ is connected, this conclusion contradicts II.2.42.

II.2.67. If $c(H)$ is the closure of the H-set H in the Peano space $\mathcal{P}$, and if $H \subset E \subset c(H)$, then E is also an H-set.

PROOF. Clearly E is nondegenerate and connected (cf. I.2.39). There remains to show that E is $\bigcirc$-complete. So let p, q be any two distinct points of E and let $p \bigcirc x \bigcirc q$. Since $p \in E \subset c(H)$, the set $H + p$ is an H-set by II.2.66. Since $q \in E \subset c(H) \subset c(H + p)$, the set $H + p + q$ is also an H-set by II.2.66. Thus $H + p + q$ is $\bigcirc$-complete, and therefore $x \in H + p + q \subset E$.

II.2.68. If H is an H-set and G a connected set in the Peano space $\mathcal{P}$, and if $Gc(H)$ is nondegenerate, then GH is noncountable.

PROOF. By assumption, we have two distinct points p, q such that $p + q \in Gc(H)$. By II.2.67, $H + p + q$ is an H-set, and hence by II.2.65 the set $(H + p + q)G = HG + p + q$ is also connected, and hence noncountable (see I.2.42). Thus HG is clearly noncountable.

II.2.69. If H is an H-set in the Peano space $\mathcal{P}$, then $c(H) - H$ is totally disconnected.

PROOF. Suppose that $c(H) - H$ contains some nondegenerate connected set G. Let p, q be two distinct points of G. Then $p + q \subset G \subset c(H)$, and hence $(H + p + q)G = p + q$ should be connected by II.2.65, II.2.67. Since $p \neq q$, this is clearly a contradiction.

II.2.70. Let x be a point of the Peano space $\mathcal{P}$, and let S be a component of $\mathcal{P} - x$. Then $S + x$ is an A-set.

PROOF. We have $x = fr(S)$ by I.2.41, and thus $S + x = S + fr(S) = c(S)$ is closed. Also, $S + x = c(S)$ is connected by I.2.39. There remains to show that $S + x$ is $\bigcirc$-complete. Let p be any point in $\mathcal{P} - (S + x)$, and let q be any point of S. Since S is a component of $\mathcal{P} - x$, it is clear that p and q lie in different components of $\mathcal{P} - x$. Thus x cuts between p and q, and hence q is not conjugate to p. In other words, the only point of $S + x$ that is possibly conjugate to p is x. Since this holds for every point $p \in \mathcal{P} - (S + x)$, it follows that $S + x$ is O-complete.

II.2.71. Cyclic chain approximation theorem. *If $\mathcal{P}$ is a nondegenerate Peano space, then there exists in $\mathcal{P}$ a (finite or infinite) sequence of cyclic chains $\mathfrak{C}_1, \cdots, \mathfrak{C}_n, \cdots$ with the following properties.*

(i) $(\mathfrak{C}_1 + \cdots + \mathfrak{C}_n)\mathfrak{C}_{n+1}$ *is a single point for every n.*

(ii) *On setting* $H = \mathfrak{C}_1 + \cdots + \mathfrak{C}_n + \cdots$, *we have* $c(H) = \mathcal{P}$.

PROOF. Since $\mathcal{P}$ is separable, we have in $\mathcal{P}$ a sequence of points $p_1, \cdots, p_n, \cdots$, such that the closure of the set $p_1 + \cdots + p_n + \cdots$ coincides with $\mathcal{P}$. Let q_1 be any point distinct from p_1. We put $\mathfrak{C}_1 = C(p_1, q_1)$ (see II.2.21). If $\mathfrak{C}_1 = \mathcal{P}$, then the theorem is established. If $\mathcal{P} - \mathfrak{C}_1 \neq 0$, then $\mathcal{P} - \mathfrak{C}_1$ is a nonempty open set by II.2.58, and hence some of the points $p_2, \cdots, p_n, \cdots$ lie in $\mathcal{P} - \mathfrak{C}_1$. Let p_{k_2} be the first point of this sequence that lies in $\mathcal{P} - \mathfrak{C}_1$, and let S_1 be the component of $\mathcal{P} - \mathfrak{C}_1$ that contains p_{k_2}. By II.2.35, $fr(S_1)$ is then a single point q_2 of $\mathfrak{C}_1$, and by II.2.44 the set S_1 is a component of $\mathcal{P} - fr(S_1)$. By II.2.70 it follows that $S_1 + q_2$ is an A-set. Hence on setting $\mathfrak{C}_2 = C(p_{k_2}, q_2)$, we have $\mathfrak{C}_2 \subset S_1 + q_2$ (cf. II.2.31, II.2.21), and consequently $\mathfrak{C}_1\mathfrak{C}_2$ reduces to the single point q_2. If $\mathfrak{C}_1 + \mathfrak{C}_2 = \mathcal{P}$, then the theorem is established. Otherwise we continue the process. To see how this is done, let us assume that we have succeeded in exhibiting a finite sequence $\mathfrak{C}_1, \cdots, \mathfrak{C}_m$ of cyclic chains with the following properties.

(a) $(\mathfrak{C}_1 + \cdots + \mathfrak{C}_i)\mathfrak{C}_{i+1}$ reduces to a single point, $i = 1, \cdots, m - 1$.

(b) $p_1 + \cdots + p_m \subset \mathfrak{C}_1 + \cdots + \mathfrak{C}_m$.

If $\mathfrak{C}_1 + \cdots + \mathfrak{C}_m = \mathcal{P}$, then the theorem is proved. If $\mathfrak{C}_1 + \cdots + \mathfrak{C}_m \neq \mathcal{P}$, then let us put $A_m = \mathfrak{C}_1 + \cdots + \mathfrak{C}_m$. In view of condition (a) and II.2.31, II.2.19, it is clear that A_m is an A-set. Then $\mathcal{P} - A_m$ is a nonempty open set, and hence some of the points $p_{m+1}, p_{m+2}, \cdots$ lie in $\mathcal{P} - A_m$. Let $p_{k_{m+1}}$ be the first one of these points that lies in $\mathcal{P} - A_m$. Clearly $k_{m+1} \geqq m + 1$ in view of (b) above. Let S_m be the component of $\mathcal{P} - A_m$ that contains $p_{k_{m+1}}$. By an argument entirely similar to that used in deriving $\mathfrak{C}_2$ from $\mathfrak{C}_1, p_{k_2}, S_1$, we obtain a cyclic chain $\mathfrak{C}_{m+1}$, such that $(\mathfrak{C}_1 + \cdots + \mathfrak{C}_m)\mathfrak{C}_{m+1}$ reduces to a single point and $p_1 + \cdots + p_m + p_{m+1} \subset \mathfrak{C}_1 + \cdots + \mathfrak{C}_{m+1}$.

It is now clear that the process just described either comes to an end after a finite number N of steps (in which case $\mathfrak{C}_1 + \cdots + \mathfrak{C}_N = \mathcal{P}$ and the theorem is established), or else we obtain an infinite sequence $\mathfrak{C}_1, \cdots, \mathfrak{C}_n, \cdots$ such that (i) is satisfied, and $p_1 + \cdots + p_n + \cdots \subset H = \mathfrak{C}_1 + \cdots + \mathfrak{C}_n + \cdots$. Since $\mathcal{P} = c(p_1 + \cdots + p_n + \cdots) \subset c(H) \subset \mathcal{P}$, the property (ii) is obvious.

II.2.72. CONTINUATION. The following additional facts concerning the sequence $\mathfrak{C}_1, \cdots, \mathfrak{C}_n, \cdots$ of II.2.71 are of importance in the sequel.

(iii) On setting $A_n = \mathfrak{C}_1 + \cdots + \mathfrak{C}_n$, A_n is an A-set. This follows from II.2.71(i) by repeated application of II.2.19.

(iv) The set H of II.2.71(ii) is an H-set. This follows from (iii) and II.2.18.

(v) If C is a proper cyclic element of $\mathcal{P}$, then C is a subset of exactly one of the cyclic chains $\mathfrak{C}_1, \cdots, \mathfrak{C}_n, \cdots$.

PROOF. Since C is connected and nondegenerate and $c(H) = \mathcal{P}$, it follows from (iv) and II.2.68 that $CH = C\mathfrak{C}_1 + \cdots + C\mathfrak{C}_n + \cdots$ is noncountable. Hence at least one of the sets $C\mathfrak{C}_1, \cdots, C\mathfrak{C}_n, \cdots$, say $C\mathfrak{C}_{n_0}$, is nondegenerate. By II.2.45 it follows that $C \subset \mathfrak{C}_{n_0}$. For $n \neq n_0$ we have $C\mathfrak{C}_n \subset \mathfrak{C}_{n_0}\mathfrak{C}_n$. Hence, by II.2.71(i), it follows that $C\mathfrak{C}_n$ is either empty or reduces to a single point. Thus the inclusion $C \subset \mathfrak{C}_n$ does not hold for $n \neq n_0$.

(vi) If the sequence $\mathfrak{C}_n$ is infinite, then $d(\mathfrak{C}_n) \to 0$ for $n \to \infty$. This follows from II.2.71(i) and II.2.47.

(vii) If the sequence $\mathfrak{C}_1, \cdots, \mathfrak{C}_n, \cdots$ is infinite, and δ_n denotes the maximum diameter of the components of $\mathcal{P} - A_n = \mathcal{P} - (\mathfrak{C}_1 + \cdots + \mathfrak{C}_n)$, then $\delta_n \to 0$ for $n \to \infty$. This follows from (iii), II.2.71(ii), II.2.48.

II.2.73. We shall complete presently our information concerning A-sets in a Peano space $\mathcal{P}$. Some relevant facts, already established, will be first recalled.

(a) Every proper cyclic element is an A-set (see II.2.50).

(b) Every cyclic chain is an A-set (see II.2.58).

(c) If x is a cut point and S is a component of $\mathcal{P} - x$, then $S + x$ is an A-set (see II.2.70).

(d) If A is an A-set and S is a component of $\mathcal{P} - A$, then $fr(S)$ is a single point of A, $fr(S)$ is a cut point of $\mathcal{P}$, and $c(S) = S + fr(S)$ is an A-set. Indeed, by II.2.44 the set S is a component of $\mathcal{P} - fr(S)$, and hence $S + fr(S)$ is an A-set by II.2.70.

II.2.74. Let p be a cut point of the Peano space $\mathcal{P}$. Then the components of $\mathcal{P} - p$ are disjoint open sets, and the collection of these components is countable. Let $S_1, \cdots, S_n, \cdots$ be the (finite or infinite) sequence of these components. Since $S_n + p$ is an A-set by II.2.73 and $(S_i + p)(S_j + p) = p$ for $i \neq j$, it follows from II.2.47 that $d(S_n) \to 0$ if the sequence S_n is infinite. Now let the sequence S_n be divided into two subsequences $S_1', \cdots, S_i', \cdots$ and $S_1'', \cdots, S_j'', \cdots$, such that each S_n occurs in one and only one of these subsequences. On setting $A' = p + S_1' + \cdots + S_i' + \cdots$, we assert that A' is an A-set, and the components of $\mathcal{P} - A'$ are precisely $S_1'', \cdots, S_j'', \cdots$.

PROOF. Since $fr(S_i') = p$, $i = 1, 2, \cdots$, we have $A' = c(S_1') + \cdots + c(S_i') + \cdots$. Since each $c(S_i')$ is connected and contains p, it follows that A' is connected. Since $\mathcal{P} - A' = S_1'' + \cdots + S_j'' + \cdots$, and each S_j'' is open, clearly $\mathcal{P} - A'$ is open and hence A' is closed. Since each $c(S_i')$ is an A-set, repeated application of II.2.19 shows that $A_i' = c(S_1') + \cdots + c(S_i')$ is an H-set. Hence $A' = A_1' + \cdots + A_i' + \cdots$ is an H-set by II.2.18. Thus A' is a closed H-set and hence an A-set. Since $\mathcal{P} - A' = S_1'' + \cdots + S_j'' + \cdots$, and S_1'', $\cdots, S_j'', \cdots$ are disjoint connected open sets, it is obvious that the components of $\mathcal{P} - A'$ are precisely $S_1'', \cdots, S_j'', \cdots$ (see I.2.39).

II.2.75. Let A be an A-set in the Peano space $\mathcal{P}$, and let the components of $\mathcal{P} - A$ be divided (without duplication) into two sequences $S_1', \cdots, S_i', \cdots$ and

S_1'', $\cdots$, S_j'', $\cdots$ (cf. II.2.37). Then $A' = A + S_1' + \cdots + S_i' + \cdots$ is an A-set and the components of $\mathcal{P} - A'$ are precisely S_1'', $\cdots$, S_j'', $\cdots$.

PROOF. For each i, we have $fr(S_i') = p_i'$, where p_i' is a single point of A and $S_i' + p_i'$ is an A-set (see II.2.73). By II.2.19 and II.2.18 it follows readily that A' is an H-set (cf. the argument in II.2.74). Since $\mathcal{P} - A' = S_1'' + \cdots + S_j'' + \cdots$, it follows readily that A' is closed and that S_1'', $\cdots$, S_j'', $\cdots$ are the components of $\mathcal{P} - A'$ (cf. the reasoning in II.2.74).

II.2.76. In the Peano space $\mathcal{P}$, let A_1, A_2 be A-sets such that $A_1 + A_2 = \mathcal{P}$ and $A_1A_2 = p$, where p is a single point. Then p is a cut point, and $A_1 - p$, $A_2 - p$ are sums of components of $\mathcal{P} - p$.

PROOF. Since A_2 is nondegenerate and A_1 is closed, $A_2 - p = \mathcal{P} - A_1$ is a nonempty open set. Let S_1'', $\cdots$, S_j'', $\cdots$ be the (finite or countably infinite) sequence of components of $A_2 - p = \mathcal{P} - A_1$ (cf. II.2.37). Then for each j, $fr(S_j'') = p_j''$ is a single point of A_1 (see II.2.35). Since $S_j'' \subset A_2$ and $p_j'' \in c(S_j'')$, it follows that $p_j'' \in c(A_2) = A_2$. Thus $p_j'' \in A_1A_2 = p$, and hence $p_j'' = p$, $j = 1, 2, \cdots$. By II.2.44 it follows that p is a cut point and S_j'' is a component of $\mathcal{P} - p$. Thus $A_2 - p = S_1'' + \cdots + S_j'' + \cdots$ is a sum of components of $\mathcal{P} - p$. Similarly it follows that $A_1 - p$ is a sum of components of $\mathcal{P} - p$.

II.2.77. In the Peano space $\mathcal{P}$, let A_1, $\cdots$, A_n, $\cdots$ be a (finite or infinite) sequence of A-sets, such that $A_1 + \cdots + A_n + \cdots = \mathcal{P}$, and for $i \neq j$ the product A_iA_j is either empty or else reduces to a single point. Then each proper cyclic element C of $\mathcal{P}$ is a subset of exactly one of the sets A_n.

PROOF. Since C is nondegenerate and A_iA_j is empty or a single point for $i \neq j$, clearly C cannot be a subset of more than one of the sets A_n. Now suppose that C is not a subset of any term of the sequence A_n. Then by II.2.45 the product CA_n is either empty or reduces to a single point, $n = 1, 2, \cdots$. Hence $C = C\mathcal{P} = CA_1 + \cdots + CA_n + \cdots$ should be countable, in contradiction with the fact that C is a nondegenerate continuum.

II.2.78. Let $\mathcal{P}'$ be a Peano subspace of the Peano space $\mathcal{P}$ (cf. I.2.9). If C is a proper cyclic element of $\mathcal{P}$, and $C \subset \mathcal{P}'$, then C is also a proper cyclic element of $\mathcal{P}'$ (simple examples show that the converse is generally false).

PROOF. If p and q are points in $\mathcal{P}'$, then let us write $p \bigcirc q(\mathcal{P})$ to express the fact that p and q are conjugate relative to $\mathcal{P}$, and let us write $p \bigcirc q(\mathcal{P}')$ to express the fact that p and q are conjugate relative to $\mathcal{P}'$. Now let p and q be two distinct points of C, and let x' be any point of $\mathcal{P}'$ distinct from p and q. If $C \subset \mathcal{P}' - x'$, then p and q are in the same component of $\mathcal{P}' - x'$ since C is connected. If $x' \in C$, then p and q are still in the same component of $\mathcal{P}' - x'$ since $C - x'$ is connected (cf. II.2.51). Thus $p \bigcirc q(\mathcal{P}')$. Hence by II.2.12 (applied to $\mathcal{P}'$), p and q lie on a proper cyclic element C' of $\mathcal{P}'$. If y is any point of C, then the reasoning used above to show that $p \bigcirc q(\mathcal{P}')$ yields, if applied to p, y and q, y, the relations $p \bigcirc y(\mathcal{P}')$, $q \bigcirc y(\mathcal{P}')$. Thus $C \subset C'$. To derive the complementary inclusion, let y' be any point of C' and x any point of $\mathcal{P}$ distinct from p and y'. If $C' \subset \mathcal{P} - x$, then p and y' are in the same component of $\mathcal{P} - x$ since C' is connected. If $x \in C'$, then p and y' are still in the same component of $\mathcal{P} - x$

since $C' - x$ is connected (cf. II.2.51). Thus $p \bigcirc y'(\mathcal{P})$, and similarly $q \bigcirc y'(\mathcal{P})$. Since C is $\bigcirc$-complete (relative to $\mathcal{P}$), it follows that $y' \in C$. Thus $C' \subset C$.

II.2.79. Let A be an A-set in the Peano space $\mathcal{P}$. Then the proper cyclic elements of A coincide with those proper cyclic elements of $\mathcal{P}$ that are subsets of A (note that A is a Peano subspace of $\mathcal{P}$ by II.2.41).

Proof. Let C be a proper cyclic element of $\mathcal{P}$ that is a subset of A. Then C is also a proper cyclic element of A by II.2.78. Conversely, let C_A be a proper cyclic element of A. Let x_A, y_A be any two distinct points of C_A, and let p be any point of $\mathcal{P}$ distinct from x_A and y_A. If $C_A \subset \mathcal{P} - p$, then x_A, y_A are in the same component of $\mathcal{P} - p$ since C_A is connected. If $p \in C_A$, then x_A, y_A are still in the same component of $\mathcal{P} - p$ since $C_A - p$ is connected (see II.2.51). Thus x_A, y_A are conjugate relative to $\mathcal{P}$. By II.2.12 there follows the existence of a proper cyclic element C of $\mathcal{P}$ such that $x_A + y_A \subset C$. Since A is an A-set, we have $C \subset A$ by II.2.45. By II.2.78 it follows that C is also a proper cyclic element of A. Thus C_A and C are proper cyclic elements of A and $x_A + y_A \subset C_AC$. By II.2.11 (applied to the Peano space A) it follows that $C_A = C$. Thus C_A is a proper cyclic element of $\mathcal{P}$.

II.2.80. Let A be an A-set in the Peano space $\mathcal{P}$, and let p_A, q_A be any two distinct points of A. Then p_A, q_A determine a cyclic chain $C(p_A, q_A)$ relative to $\mathcal{P}$ and a cyclic chain $C_A(p_A, q_A)$ relative to A. We assert that $C(p_A, q_A) = C_A(p_A, q_A)$ (note that A is a Peano space by II.2.41).

Proof. Let γ be a simple arc in $\mathcal{P}$ with end points p_A, q_A. By II.2.60, $C(p_A, q_A)$ consists of γ and of all those proper cyclic elements C of $\mathcal{P}$ that have a nondegenerate intersection with γ. Since A is an A-set and hence $\gamma \subset A$ by II.2.33, it follows that all these proper cyclic elements of $\mathcal{P}$ are subsets of A (cf. II.2.45), and hence they coincide, by II.2.79, with those proper cyclic elements of A that have a nondegenerate intersection with γ. In view of II.2.60 (applied to the Peano space A), there follows the asserted relation $C(p_A, q_A) = C_A(p_A, q_A)$.

II.2.81. Let A_1, A_2 be A-sets in the Peano space $\mathcal{P}$, such that $A_2 \subset A_1 \subset \mathcal{P}$. Then A_2 is an A-set relative to A_1.

Proof. In the first place, A_1 and A_2 are Peano spaces by II.2.41. In particular, A_2 is connected and closed (also relative to A_1). There remains to show that A_2 is $\bigcirc$-complete relative to A_1. If this fact is denied, then we should have two distinct points p_2, q_2 in A_2 and a point x_1 in $A_1 - A_2$ such that the pairs of points p_2, x_1 and q_2, x_1 are conjugate relative to A_1. Then p_2 and x_1 lie on a proper cyclic element of A_1 by II.2.12 (applied to the Peano space A_1), and hence p_2 and x_1 lie on a proper cyclic element of $\mathcal{P}$ by II.2.79. Thus $p_2 \bigcirc x_1$ relative to $\mathcal{P}$, and similarly $q_2 \bigcirc x_1$ relative to $\mathcal{P}$. Since A_2 is an A-set of $\mathcal{P}$, it follows that $x_1 \in A_2$, in contradiction with the assumption that $x_1 \in A_1 - A_2$.

II.2.82. Let $C(p, q)$ be a cyclic chain in the Peano space $\mathcal{P}$. Let γ be a simple arc in $\mathcal{P}$ with end points p, q. By II.2.61, either $C(p, q)$ reduces to γ, or else $C(p, q) = \gamma + C_1 + \cdots + C_n + \cdots$, where $C_1, \cdots, C_n, \cdots$ is the sequence of those proper cyclic elements of $\mathcal{P}$ that have a nondegenerate intersection with γ.

Let us assume that we are dealing with the second case. Let then $C_{n_1}, \cdots, C_{n_k}, \cdots$ be a (finite or infinite) subsequence of the sequence $C_1, \cdots, C_n, \cdots$, and let us put $\mathcal{P}' = \gamma + C_{n_1} + \cdots + C_{n_k} + \cdots$. We assert that $\mathcal{P}'$ is a Peano subspace of $\mathcal{P}$ and the proper cyclic elements of $\mathcal{P}'$ are precisely $C_{n_1}, \cdots, C_{n_k}, \cdots$.

Proof. Let us first show that $\mathcal{P}'$ is a Peano subspace of $\mathcal{P}$. In the first place, $\mathcal{P}'$ is connected and closed, by an argument entirely analogous to that used in II.2.58. We proceed to show that $\mathcal{P}'$ is locally connected.

Case (i). The sequence C_{n_k} is finite. Since γ is a Peano space and each C_{n_k} is a Peano space (see II.2.41, II.2.50), for given $\epsilon > 0$ we can represent γ and also each C_{n_k} as a finite sum of connected sets of diameter less than ϵ. Thus $\mathcal{P}'$ itself appears as a finite sum of this type, and hence $\mathcal{P}'$ is locally connected (see I.2.41).

Case (ii). The sequence C_{n_k} is infinite. Given $\epsilon > 0$, we shall have $d(C_{n_k}) < \epsilon/3$ for k exceeding a certain k_0. Let us subdivide γ into simple arcs $\gamma_1, \cdots, \gamma_N$, such that $d(\gamma_i) < \epsilon/3$ for each $i = 1, \cdots, N$. Let E_i denote the sum of those terms of the sequence C_{n_k} that satisfy the following conditions. (α) $C_{n_k}\gamma_i \neq 0$. (β) $k > k_0$. Then clearly E_i is connected and $d(E_i) < \epsilon$. Then

$$(1) \qquad \mathcal{P}' - (E_1 + \cdots + E_N) \subset \sum C_{n_k}, \qquad k = 1, \cdots, k_0.$$

Since each C_{n_k} is a Peano space and hence can be represented as a finite sum of connected sets of diameter less than ϵ, it follows from (1) that $\mathcal{P}'$ can also be represented in this manner. Hence $\mathcal{P}'$ is locally connected (see I.2.41).

Let us now determine the proper cyclic elements of the Peano space $\mathcal{P}'$. In the first place, by II.2.78 each C_{n_k} is a proper cyclic element of $\mathcal{P}'$. Let us assume that $\mathcal{P}'$ has a proper cyclic element that does not occur in the sequence $C_{n_1}, \cdots, C_{n_k}, \cdots$. By II.2.11 (applied to $\mathcal{P}'$) it follows that $C'C_{n_k}$ is empty or is a single point for every k. Since C' is a nondegenerate, connected and hence noncountable set, it follows that $C'\gamma$ is nondegenerate. Consider now any one of the products $C'C_{n_k}$. If $x' \in \mathcal{P}' - \gamma$, then by II.2.15 (applied to $\mathcal{P}'$ with $E_1 = x'$, $E_2 = \gamma$) it follows that $x' \notin C'C_{n_k}$. Thus $C'C_{n_k} \subset \gamma$. Since $\mathcal{P}' = \gamma + C_{n_1} + \cdots + C_{n_k} + \cdots$, there follows the inclusion $C' \subset \gamma$. By II.2.78, applied to the Peano spaces $\mathcal{P}'$ and γ, C' should be a proper cyclic element of γ, in contradiction with the obvious fact that a simple arc possesses no proper cyclic elements.

II.2.83. Continuation. Let $C'(p, q)$ denote the cyclic chain determined by the points p, q relative to the Peano space $\mathcal{P}'$. Then $C'(p, q) = \mathcal{P}'$. This is an immediate consequence of II.2.82 and II.2.60(1).

II.2.84. A Peano space $\mathcal{P}$ will be said to possess the *property* (π) if and only if every simple arc $\gamma \subset \mathcal{P}$ is a monotone retract of $\mathcal{P}$ (see II.2.39). Let us note that it follows from the theorems in Whyburn [3, Chapter XI] that every unicoherent Peano space has the property (π). On the other hand, examples show that the property (π) does not imply unicoherence. We shall study the property (π) only to the extent needed in the sequel. Clearly, property (π) is invariant under topological transformations.

If $\mathcal{P}$ is a 2-sphere, then $\mathcal{P}$ possesses the property (π). Indeed, let γ be a simple arc in $\mathcal{P}$. We can assume without loss of generality that $\mathcal{P}$ coincides with the sphere $S : x^2 + y^2 + z^2 = 1$, while γ coincides with the arc of circle on S determined by the conditions $x \geqq 0$, $y = 0$. Let p denote a variable point of S, and let $M(p)$ be defined as the point on γ whose z-coordinate agrees with that of p. Clearly, the transformation M is a continuous monotone retraction from S onto γ.

If $\mathcal{P}$ is a 2-cell, then $\mathcal{P}$ possesses the property (π). We subdivide the proof into several steps.

(a) The simple arc $\gamma \subset \mathcal{P}$ is a sub-arc of the boundary of the 2-cell $\mathcal{P}$. We can then assume, without loss of generality, that $\mathcal{P}$ coincides with the square $0 \leqq u \leqq 1$, $0 \leqq v \leqq 1$, while γ coincides with the segment $0 \leqq u \leqq 1$, $v = 0$. Then $M(u, v) = (u, 0)$ is clearly a monotone retraction from $\mathcal{P}$ onto γ.

(b) The end points a, b of the simple arc $\gamma \subset \mathcal{P}$ lie on the boundary of $\mathcal{P}$, but γ is not a sub-arc of the boundary of $\mathcal{P}$. Then γ divides $\mathcal{P}$ into a finite or countably infinite number of 2-cells $\mathcal{P}_\iota$, $\iota = 1, 2, \cdots$. The boundary C_ι of $\mathcal{P}_\iota$ is then a simple closed curve and $\mathcal{P}_\iota \gamma = \gamma_\iota$ is a sub-arc of C_ι. By case (a) above, we have then a monotone retraction $M_i(\mathcal{P}_\iota) = \gamma_\iota$. If we define now a mapping M on $\mathcal{P}$ by $M(p) = M_i(p)$ for $p \in \mathcal{P}_\iota$, then it follows readily that M is a monotone retraction from $\mathcal{P}$ onto γ.

(c) In the general case, we choose two simple arcs α, β in such a manner that $\gamma^* = \alpha + \gamma + \beta$ is a simple arc whose end points lie on the boundary of $\mathcal{P}$. By case (b), we have then a monotone retraction $M^*(\mathcal{P}) = \gamma^*$. Next, we define on γ^* a mapping $m(\gamma^*) = \gamma$ as follows. If $p \in \gamma$, then $m(p) = p$. If $p \in \alpha$, then $m(p)$ is the common end point of α and γ. Similarly, if $p \in \beta$, then $m(p)$ is the common end point of β and γ. Clearly, mM^* is a monotone retraction from $\mathcal{P}$ onto γ.

If $\mathcal{P}$ is a dendrite, then $\mathcal{P}$ has the property (π). Indeed, let γ be a simple arc in $\mathcal{P}$ with end points p, q. By II.2.60(1) it follows that $\gamma = C(p, q)$, and thus γ is an A-set (see II.2.58). Hence γ is a monotone retract of $\mathcal{P}$ by II.2.40.

II.2.85. Continuation. The property (π) is cyclicly reducible. That is, if $\mathcal{P}$ possesses the property (π), then every proper cyclic element C of $\mathcal{P}$ also possesses the property (π), and in fact every A-set in $\mathcal{P}$ possesses the property (π). To prove this statement, let A be an A-set and γ a simple arc in A. By assumption, we have a monotone retraction $\mu(\mathcal{P}) = \gamma$. For clarity, let us use μ_A to refer to μ if thought of as operating from A. Then clearly $\mu_A(A) = \gamma$ is a retraction from A onto γ. If p is any point of γ, then clearly $\mu_A^{-1}(p) = A\mu^{-1}(p)$. Now $\mu^{-1}(p)$ is a continuum since μ is monotone, and hence $A\mu^{-1}(p)$ is a continuum since A is an A-set (see II.2.42). Thus μ_A is monotone.

II.2.86. Continuation. The property (π) is cyclicly extensible. That is, if every proper cyclic element C of $\mathcal{P}$ possesses the property (π), then $\mathcal{P}$ itself also possesses the property (π). We shall make the proof in several steps. The assumption is that every proper cyclic element of $\mathcal{P}$ possesses the property (π).

Case (1). $\mathcal{P}$ is a dendrite. Since $\mathcal{P}$ has no proper cyclic element in this case (cf. II.2.14), our assumption is vacuously satisfied, but $\mathcal{P}$ possesses the property (π) as noted in II.2.84.

Case (2). $\mathcal{P}$ is not a dendrite. Let γ be a simple arc in $\mathcal{P}$ with end points p, q. Let $C(p, q)$ be the cyclic chain determined by the points p, q.

Case (2.a). $\mathcal{P} = C(p, q)$. Then we have $\mathcal{P} = C(p, q) = \gamma + C_1 + \cdots + C_n + \cdots$, where $C_1, \cdots, C_n, \cdots$ is the sequence of the proper cyclic elements of $\mathcal{P}$ (see II.2.61). Using the terminology and the results of II.2.61, let us consider the simple arcs $\gamma_n = \gamma C_n$. By assumption, we have then a monotone retraction $\mu_n(C_n) = \gamma_n$ for each n. Let us now define a transformation $\mu(\mathcal{P}) = \gamma$ as follows: $\mu(x) = x$ if $x \in \gamma$, and $\mu(x) = \mu_n(x)$ if $x \in C_n - \gamma$ We assert that $\mu(x)$ is a monotone retraction from $\mathcal{P}$ onto γ. In the first place, $\mu(x)$ is single-valued. Indeed, if $x \in \gamma$, then $\mu(x) = x$. If $x \in \mathcal{P} - \gamma$, then the inclusion $x \in C_n - \gamma$ holds for precisely one value of n (see II.2.61), and hence $\mu(x) = \mu_n(x)$ is again univocally determined. Clearly $\mu(\mathcal{P}) = \gamma$. To establish the continuity of μ, let us assume that there exists a sequence of points x_k converging to a point x_0 such that $\mu(x_k)$ fails to converge to $\mu(x_0)$. Since $\mathcal{P}$ is compact, we can assume then without loss of generality that $\mu(x_k)$ converges to a point $y_0 \neq \mu(x_0)$. If $x_k \in \gamma$ for infinitely many values of k, then necessarily $x_0 \in \gamma$, and $\mu(x_k) = x_k$ for infinitely many values of k. Thus clearly the sequence $\mu(x_k)$ converges to the same limit as the sequence x_k. That is, $\mu(x_k) \rightarrow x_0 = \mu(x_0)$, in contradiction with the assumption that $\mu(x_k) \rightarrow y_0 \neq \mu(x_0)$. Thus we can assume that $x_k \in \mathcal{P} - \gamma$ if k exceeds a certain k_0. In view of II.2.61, we have then for each k a unique positive integer n_k such that $x_k \in C_{n_k}$. Let us assume first that $d(C_{n_k}) \rightarrow 0$. Since now $\mu(x_k) = \mu_{n_k}(x_k) \in \gamma_{n_k} \subset C_{n_k}$, it follows that $\rho[x_k, \mu(x_k)] \leqq d(C_{n_k}) \rightarrow 0$, and hence $x_0 = \lim x_k = \lim \mu(x_k)$. Thus $x_0 \in \gamma$, and hence $\mu(x_0) = x_0 = \lim \mu(x_k) = y_0$, in contradiction with the assumption that $\mu(x_0) \neq y_0 = \lim \mu(x_k)$. Thus the relation $d(C_{n_k}) \rightarrow 0$ cannot hold. By II.2.52 it follows that some proper cyclic element C_m occurs infinitely often in the sequence C_{n_k}. Thus $x_k \in C_m$ for infinitely many values of k. For simplicity, let us use x_k to denote the infinite subsequence comprised of those points x_k which lie in C_m. Then $x_0 \in C_m$ since C_m is closed, and $\mu(x_k) = \mu_m(x_k)$, $k = 1, 2, \cdots$. Since μ_m is continuous, the relation $x_k \rightarrow x_0$ implies that $\mu(x_k) = \mu_m(x_k) \rightarrow \mu_m(x_0) = \mu(x_0)$, in contradiction with the assumption that $\mu(x_k) \rightarrow y_0 \neq \mu(x_0)$.

The continuity of μ being established, there remains to show that μ is monotone. Let x_0 be a point in γ. Then clearly $\mu^{-1}(x_0) = x_0 + \sum \mu_n^{-1}(x_0)$, where the summation is extended over all those subscripts n for which $x_0 \in C_n$ (in view of II.2.61, the number of such subscripts is either zero, or one, or two). Now if $x_0 \in C_n$, then $\mu_n^{-1}(x_0)$ is a continuum since μ_n is monotone, and $x_0 \in \mu_n^{-1}(x_0)$ since $\mu_n(x_0) = x_0$. Thus $\mu^{-1}(x_0)$ is the sum of x_0 and of at most two continua containing x_0. Hence $\mu^{-1}(x_0)$ is a continuum.

Case (2.b). *The general case.* By II.2.58, $C(p, q)$ is an A-set of $\mathcal{P}$, and hence we have a monotone retraction $\mu^*(\mathcal{P}) = C(p, q)$ (see II.2.40). Now $C(p, q)$ is a Peano space (see II.2.82) for which case (2.a) holds (relative to the given simple arc γ). So by case (2.a) we have a monotone retraction $\mu[C(p, q)] = \gamma$. Clearly, $\mu\mu^*$ is a monotone retraction from $\mathcal{P}$ onto γ (cf. II.1.4).

II.2.87. Given a Peano space $\mathcal{P}$ with the property (π) (see II.2.84), let $C(p, q)$

be a cyclic chain in $\mathcal{P}$, and γ a simple arc with end points p, q. Assuming that $C(p, q)$ does not reduce to γ, we have for $C(p, q)$ the formula $C(p, q) = \gamma + C_1 + \cdots + C_n + \cdots$, as explained in II.2.61. Let us put $\mathcal{P}_n = \gamma + C_1 + \cdots + C_n$. By II.2.82, $\mathcal{P}_n$ is then a Peano subspace of $\mathcal{P}$ whose proper cyclic elements coincide with $C_1, \cdots, C_n$. We assert that there exists a monotone retraction $\mu^n[C(p, q)] = \mathcal{P}_n$.

PROOF. Using the terminology of II.2.61, we define μ^n as follows. If $x \in \mathcal{P}_n$, then we put $\mu^n(x) = x$. If $x \in C(p, q) - \mathcal{P}_n$, then by II.2.61 there exists a unique subscript m such that $x \in C_m$ (clearly $m > n$). Since $\mathcal{P}$ has the property (π), C_m also has the property (π) (see II.2.85), and hence we have a monotone retraction $\mu_m(C_m) = \gamma_m = \gamma C_m$. We define then $\mu^n(x) = \mu_m(x)$. The fact that μ^n is a monotone retraction from $C(p, q)$ onto $\mathcal{P}_n$ follows by an argument entirely analogous to that used in II.2.86, case (2.a).

II.2.88. CONTINUATION. Suppose that the sequence C_n is infinite. Then $d(C_n) \to 0$ by II.2.52. Hence, for each n, the expression $\delta_n = \max d(C_m)$, $m > n$, is meaningful, and clearly $\delta_n \to 0$ for $n \to \infty$. By the definition of μ^n, clearly $\rho[x, \mu^n(x)] \leqq \delta_n$ for every point $x \in C(p, q)$. In other words, the sequence μ^n converges on $C(p, q)$ uniformly to the identity transformation.

II.2.89. Given a Peano space $\mathcal{P}$, let $M(\mathcal{P}) = \mathcal{P}^*$ be a continuous monotone transformation. If every Peano space $\mathcal{P}^*$ related to $\mathcal{P}$ in this manner possesses the property (π) (see II.2.84), then we shall say that $\mathcal{P}$ possesses the *property* (Π). Briefly, $\mathcal{P}$ possesses the property (Π) if and only if every continuous monotone image of $\mathcal{P}$ possesses the property (π). The following remarks will be useful in the sequel.

(a) If $\mathcal{P}$ possesses the property (Π), then $\mathcal{P}$ also possesses the property (π). Indeed, the identity is a continuous monotone transformation from $\mathcal{P}$ onto itself.

(b) If $\mathcal{P}$ possesses the property (Π), then every continuous monotone image of $\mathcal{P}$ also possesses the property (Π). This follows directly from II.1.4.

(c) If $\mathcal{P}$ possesses the property (Π) and $\mathcal{P}^*$ is a continuous monotone image of $\mathcal{P}$, then every A-set (and hence in particular every cyclic chain and every proper cyclic element) of $\mathcal{P}^*$ possesses the property (Π). This follows from (b) and II.2.40.

II.2.90. Let $\mathcal{P}$ be a 2-sphere, and let $M(\mathcal{P}) = \mathcal{P}^*$ be a continuous monotone transformation. Then every proper cyclic element C^* of $\mathcal{P}^*$ is a 2-sphere. Indeed, by II.2.40, II.2.50 we have a monotone retraction $\mu^*(\mathcal{P}^*) = C^*$. Then $\mu^* M(\mathcal{P}) = C^*$ is a continuous monotone transformation. Since C^* is cyclic (see II.2.51), it follows from II.1.35 that C^* is a 2-sphere.

II.2.91. Let $\mathcal{P}$ be a 2-cell, and let $M(\mathcal{P}) = \mathcal{P}^*$ be a continuous monotone transformation. Then every proper cyclic element C^* of $\mathcal{P}^*$ is either a 2-cell or a 2-sphere. The proof is the same as in II.2.90, except that II.1.42 is used instead of II.1.35.

II.2.92. If $\mathcal{P}$ is either a 2-cell or a 2-sphere, then $\mathcal{P}$ possesses the property (Π) (cf. II.2.89). Indeed, let $M(\mathcal{P}) = \mathcal{P}^*$ be any continuous monotone transformation. If C^* is a proper cyclic element of $\mathcal{P}^*$, then C^* is either a 2-cell or a 2-sphere by II.2.90, II.2.91. By II.2.84 it follows that every proper cyclic element of $\mathcal{P}^*$ has the property (π). By II.2.86 it follows that $\mathcal{P}^*$ itself possesses the property (π).

II.2.93. It is obvious that all the conceptions used so far in this chapter are invariant under homeomorphisms. In particular, if $h(\mathcal{P}) = \mathcal{P}^*$ is a topological transformation, and $C_1, \cdots, C_n, \cdots$ is the sequence of the proper cyclic elements of $\mathcal{P}$, then $h(C_1), \cdots, h(C_n), \cdots$ is the sequence of the proper cyclic elements of $\mathcal{P}^*$.

II.2.94. We proceed to apply the theory of cyclic elements of Peano spaces to the study of continuous transformations. Let $T(\mathcal{P}) = \mathcal{P}^*$ be a continuous transformation, where $\mathcal{P}$ and $\mathcal{P}^*$ are Peano spaces (cf. II.1.1). Let $T = LM$, $M(\mathcal{P}) = \mathfrak{M}$, $L(\mathfrak{M}) = \mathcal{P}^*$ be a monotone-light factorization of T (see II.1.17, II.1.18). Let us first assume that the middle-space $\mathfrak{M}$ is not a dendrite. Let then $\mathfrak{C}_1, \cdots, \mathfrak{C}_n, \cdots$ be the (finite or infinite) sequence of the proper cyclic elements of $\mathfrak{M}$ (see II.2.52). By II.2.50, II.2.43 we have a unique monotone retraction $\mu_n(\mathfrak{M}) = \mathfrak{C}_n$. Then $L\mu_nM$ is a continuous transformation from $\mathcal{P}$ into $\mathcal{P}^*$ (that is, $L\mu_nM(\mathcal{P}) \subset \mathcal{P}^*$). Thus we obtain the sequence of transformations $L\mu_1M, \cdots, L\mu_nM, \cdots$. These transformations constitute the cyclic decomposition $\Delta(T)$ of the transformation T. Thus

$$\Delta(T) = [L\mu_1M, \cdots, L\mu_nM, \cdots],$$

where the square brackets are used to indicate that the arrangement of the sequence is immaterial (see I.2.4). If the middle-space $\mathfrak{M}$ is a dendrite, then $\mathfrak{M}$ has no proper cyclic elements, and we define $\Delta(T)$ as the empty sequence, in symbols $\Delta(T) = 0$.

Each one of the transformations $L\mu_nM$ will be referred to as a cyclic partial transformation of T. Since the cyclic decomposition of T is defined in terms of a monotone-light factorization of T, we have to justify yet the notation $\Delta(T)$ by showing that $\Delta(T)$ is independent of the particular choice of the monotone-light factorization. This fact will be established in the next section; in preparation, let us add a remark.

If $\Delta(T) = 0$, then $\mathfrak{M}$ is a dendrite. By II.1.20, the middle-space occurring in any monotone-light factorization of T is then also a dendrite, and hence in this case $\Delta(T)$ is clearly independent of the choice of the monotone-light factorization of T.

II.2.95. CONTINUATION. Let us now consider a second monotone-light factorization $T = L'M'$, $M'(\mathcal{P}) = \mathfrak{M}'$, $L'(\mathfrak{M}') = \mathcal{P}^*$. By II.1.20, there exists then a homeomorphism $h(\mathfrak{M}) = \mathfrak{M}'$, such that $L = L'h$, $M' = hM$. In view of II.2.94 it is sufficient to consider the case when $\mathfrak{M}$ is not a dendrite. If $\mathfrak{C}_1, \cdots, \mathfrak{C}_n, \cdots$ is the sequence of the proper cyclic elements of $\mathfrak{M}$, then on setting $\mathfrak{C}'_n = h(\mathfrak{C}_n)$, $n = 1, 2, \cdots$, we obtain the sequence of the proper cyclic elements of $\mathfrak{M}'$ (cf. II.2.93). We have then (see II.2.50, II.2.43) unique monotone retractions $\mu_n(\mathfrak{M}) = \mathfrak{C}_n$, $\mu'_n(\mathfrak{M}') = \mathfrak{C}'_n$. By definition, the factorizations $T = LM$, $T = L'M'$ yield the cyclic decompositions

(1) $$\Delta(T) = [L\mu_1 M, \cdots, L\mu_n M, \cdots],$$

(2) $$\Delta'(T) = [L'\mu_1' M', \cdots, L'\mu_n' M', \cdots].$$

Now clearly $h\mu_n h^{-1}$ is a monotone retraction from $\mathfrak{M}'$ onto $\mathfrak{C}_n'$. Since there exists precisely one such monotone retraction (see II.2.43), it follows that $\mu_n' = h\mu_n h^{-1}$. Thus $L'\mu_n' M' = Lh^{-1}h\mu_n h^{-1}hM = L\mu_n M$. In other words, the sequences (1), (2) agree term for term, and thus $\Delta(T) = \Delta'(T)$. Hence the cyclic decomposition of T is independent of the choice of a monotone-light factorization for T.

II.2.96. Let there be given two continuous transformations $T_1(\mathcal{P}_1) = \mathcal{P}^*$, $T_2(\mathcal{P}_2) = \mathcal{P}^*$, where $\mathcal{P}_1$, $\mathcal{P}_2$, $\mathcal{P}^*$ are Peano spaces, and $\mathcal{P}_1$, $\mathcal{P}_2$ are homeomorphic. We shall then say that their cyclic decompositions $\Delta(T_1)$, $\Delta(T_2)$ are F-equivalent, in symbols $\Delta(T_1) \sim \Delta(T_2)(F)$, if and only if one of the following two conditions holds:

(i) $\Delta(T_1) = 0 = \Delta(T_2)$.

(ii) $0 \neq \Delta(T_1) = [T_1^1, \cdots, T_1^n, \cdots]$, $0 \neq \Delta(T_2) = [T_2^1, \cdots, T_2^n, \cdots]$, and there exists a rearrangement $\tilde{T}_2^1, \cdots, \tilde{T}_2^n, \cdots$ of the sequence $T_2^1, \cdots, T_2^n, \cdots$ such that $T_1^n \sim \tilde{T}_2^n(F)$ for every n (see II.1.25). This requirement is understood to imply that the sequences $T_1^1, \cdots, T_1^n, \cdots$ and $T_2^1, \cdots, T_2^n, \cdots$ contain the same (finite or infinite) number of terms.

II.2.97. THEOREM. *Let $T_1(\mathcal{P}_1) = \mathcal{P}^*$, $T_2(\mathcal{P}_2) = \mathcal{P}^*$ be a pair of continuous transformations such that $T_1 \sim T_2(F)$ (see II.1.25). Then $\Delta(T_1) \sim \Delta(T_2)(F)$* (cf. II.2.96).

PROOF. Let $T_1 = L_1M_1$, $M_1(\mathcal{P}_1) = \mathfrak{M}_1$, $L_1(\mathfrak{M}_1) = \mathcal{P}^*$, $T_2 = L_2M_2$, $M_2(\mathcal{P}_2) = \mathfrak{M}_2$, $L_2(\mathfrak{M}_2) = \mathcal{P}^*$ be monotone-light factorizations of T_1, T_2. By II.1.28 there exists then a homeomorphism $h(\mathfrak{M}_1) = \mathfrak{M}_2$ with the following properties.

(i) $L_1 = L_2h$.

(ii) If Z_1 is any continuous transformation from $\mathfrak{M}_1$ into $\mathfrak{M}_1$ (that is, $Z_1(\mathfrak{M}_1) \subset \mathfrak{M}_1$), then $L_1Z_1M_1 \sim L_2hZ_1h^{-1}M_2(F)$.

Now if $\mathfrak{M}_1$ is a dendrite, then $\mathfrak{M}_2$ is also a dendrite, and hence $\Delta(T_1) = 0 = \Delta(T_2)$. So we can assume that $\mathfrak{M}_1$ is not a dendrite. Let then $\mathfrak{C}_1^1, \cdots, \mathfrak{C}_1^n, \cdots$ be the sequence of the proper cyclic elements of $\mathfrak{M}_1$. On setting $\mathfrak{C}_2^n = h(\mathfrak{C}_1^n)$, $n = 1, 2, \cdots$, we obtain the sequence of the proper cyclic elements of $\mathfrak{M}_2$. We have then (see II.2.50, II.2.43) unique monotone retractions $\mu_1^n(\mathfrak{M}_1) = \mathfrak{C}_1^n$, $\mu_2^n(\mathfrak{M}_2) = \mathfrak{C}_2^n$, and by definition $\Delta(T_1) = [L_1\mu_1^1M_1, \cdots, L_1\mu_1^nM_1, \cdots]$, $\Delta(T_2) = [L_2\mu_2^1M_2, \cdots, L_2\mu_2^nM_2, \cdots]$. Now clearly $h\mu_1^nh^{-1}$ is a monotone retraction from $\mathfrak{M}_2$ onto $\mathfrak{C}_2^n$. Since there exists only one such monotone retraction, it follows that $\mu_2^n = h\mu_1^nh^{-1}$, and hence $L_2\mu_2^nM_2 = L_2h\mu_1^nh^{-1}M_2$. In view of (ii) (applied with $Z_1 = \mu_1^n$), it follows that $L_2\mu_2^nM_2 \sim L_1\mu_1^nM_1(F)$. Thus $\Delta(T_1) \sim \Delta(T_2)(F)$.

II.2.98. CONTINUATION. The preceding argument yields not merely the existence of a rearrangement as required by II.2.96(ii), but in fact yields a definite mating of the proper cyclic elements of $\mathfrak{M}_1$, $\mathfrak{M}_2$ such that mated elements furnish F-equivalent cyclic partial transformations of T_1, T_2 respectively. A word of caution may be in order. Since $T_1 \sim T_2(F)$, we have, by II.1.29, simultaneous monotone-light factorizations of the form

$$(1) \qquad T_1 = LM_1, \quad M_1(\mathcal{P}_1) = \mathfrak{M}, \quad L(\mathfrak{M}) = \mathcal{P}^*,$$

$$(2) \qquad T_2 = LM_2, \quad M_2(\mathcal{P}_2) = \mathfrak{M}, \quad L(\mathfrak{M}) = \mathcal{P}^*,$$

that is, factorizations with the same middle-space $\mathfrak{M}$ and the same light factor L. Assuming that $\mathfrak{M}$ is not a dendrite, let $\mathfrak{C}^1, \cdots, \mathfrak{C}^n, \cdots$ be the sequence of the proper cyclic elements of $\mathfrak{M}$, and for each n let $\mu^n(\mathfrak{M}) = \mathfrak{C}^n$ be the unique monotone retraction from $\mathfrak{M}$ onto $\mathfrak{C}^n$. By definition, we have then

$$(3) \qquad \Delta(T_1) = [L\mu^1M_1, \cdots, L\mu^nM_1, \cdots],$$

$$(4) \qquad \Delta(T_2) = [L\mu^1M_2, \cdots, L\mu^nM_2, \cdots].$$

By the theorem of II.2.97, there exists a rearrangement of the second sequence such that each element of (3) is F-equivalent to the element occupying the same position in the rearranged sequence (4). In view of the special choice of the factorizations (1), (2) there may be a temptation to expect that a rearrangement of (4) is unnecessary. In other words, one may expect that $L\mu^nM_1 \sim L\mu^nM_2(F)$ for every n if factorizations of the special form (1), (2) are used. Simple examples show that this is generally not the case.

II.2.99. We shall study presently functions of continuous transformations. $\mathcal{P}$ will denote a fixed Peano space, and P^* a fixed metric space (in the applications, P^* will coincide with Euclidean three-space). $\mathfrak{J}$ will denote the class of all continuous transformations from $\mathcal{P}$ into P^* (that is, $T(\mathcal{P}) = \mathcal{P}^* \subset P^*$ for every $T \in \mathfrak{J}$). Thus the spaces $\mathcal{P}$ and P^* will be fixed, while $\mathcal{P}^* = T(\mathcal{P})$ will be a Peano subspace of P^* that depends upon the choice of $T \in \mathfrak{J}$. The image space $\mathcal{P}^*$ may reduce to a single point. This will happen if and only if T is constant on $\mathcal{P}$. If $T_1 \in \mathfrak{J}$, $T_2 \in \mathfrak{J}$, then we put

$$\rho[T_1, T_2] = \max \rho[T_1(x), T_2(x)], \qquad x \in \mathcal{P},$$

where ρ denotes distance in P^*, and the use of the maximum is justified since $\mathcal{P}$ is compact. If $T \in \mathfrak{J}$, $T_n \in \mathfrak{J}$, and $\rho[T, T_n] \to 0$ for $n \to \infty$, then we shall write $T_n \to T$.

If $T \in \mathfrak{J}$, and $T = LM$, $M(\mathcal{P}) = \mathfrak{M}$, $L(\mathfrak{M}) = \mathcal{P}^*$ is a monotone-light factorization of T, then the middle-space $\mathfrak{M}$ is determined by T up to a homeomorphism (see II.1.20). Thus we can speak, for a given $T \in \mathfrak{J}$, of *the* middle-space being a dendrite, or a simple arc, or a 2-cell, and so forth, without specifying a particular monotone-light factorization.

II.2.100. Continuation. Let $T \in \mathfrak{J}$. The cyclic decomposition $\Delta(T)$ of T depends only upon T. If $\Delta(T) \neq 0$, then every cyclic partial transformation of T belongs to $\mathfrak{J}$. Indeed, if $\tilde{T}$ is such a cyclic partial transformation, then clearly $\tilde{T}(\mathcal{P}) \subset T(\mathcal{P}) = \mathcal{P}^* \subset P^*$.

We shall have to use concepts involving T that depend upon the particular choice of a monotone-light factorization for T, and special care will be needed in such cases. Let $T \in \mathfrak{I}$, and let

$$(1) \qquad T = LM, \qquad M(\mathcal{P}) = \mathfrak{M}, \qquad L(\mathfrak{M}) = \mathcal{P}^*$$

be a definite monotone-light factorization of T. Let $\mathfrak{A}$ be an A-set of $\mathfrak{M}$ (note that $\mathfrak{M}$ is a Peano space by I.2.43). By II.2.43 there exists a unique monotone retraction $\alpha(\mathfrak{M}) = \mathfrak{A}$. In terms of α, we define the symbol $T \mid \mathfrak{A}$ by the formula

$$(2) \qquad T \mid A = L\alpha M.$$

Thus $T \mid \mathfrak{A}$ is a continuous transformation from $\mathcal{P}$ into the space $\mathcal{P}^* = T(\mathcal{P})$ (that is, the image of $\mathcal{P}$ under $T \mid \mathfrak{A}$ is a subset of $\mathcal{P}^*$). Thus clearly $T \mid \mathfrak{A} \in \mathfrak{I}$. Since α and M are both monotone, the product αM is also monotone (see II.1.4). We agree to use for $T \mid \mathfrak{A}$ the monotone-light factorization $L \cdot \alpha M$. In other words, we agree to use αM as the monotone factor, $\mathfrak{A}$ as the middle-space, and L (thought of as operating from $\mathfrak{A}$) as the light factor. Thus, after a definite monotone-light factorization (1) has been selected for T, each A-set $\mathfrak{A}$ of the middle-space $\mathfrak{M}$ gives rise to a continuous transformation $T \mid \mathfrak{A}$ for which a definite monotone-light factorization $T \mid \mathfrak{A} = L \cdot \alpha M$ is assigned.

Let now $\mathfrak{A}_1$, $\mathfrak{A}_2$ be two A-sets of $\mathfrak{M}$ such that $\mathfrak{A}_2 \subset \mathfrak{A}_1$. Then $T \mid \mathfrak{A}_1 \in \mathfrak{I}$ as noted above, and the monotone-light factorization assigned to $T \mid \mathfrak{A}_1$ is given by the formula $T \mid \mathfrak{A}_1 = L \cdot \alpha_1 M$, where α_1 is the unique monotone retraction from $\mathfrak{M}$ onto $\mathfrak{A}_1$. The middle-space occurring in this factorization is $\mathfrak{A}_1$, and by II.2.81 (applied to $\mathfrak{M}$), $\mathfrak{A}_2$ is an A-set relative to $\mathfrak{A}_1$. Hence we can consider the transformation

$$(3) \qquad (T \mid \mathfrak{A}_1) \mid \mathfrak{A}_2 = L\alpha_{12}\alpha_1 M,$$

where α_{12} is the unique monotone retraction from $\mathfrak{A}_1$ onto $\mathfrak{A}_2$. Now let α_2 be the unique monotone retraction from $\mathfrak{M}$ onto $\mathfrak{A}_2$. By definition

$$(4) \qquad T \mid \mathfrak{A}_2 = L\alpha_2 M.$$

Now clearly $\alpha_{12}\alpha_1$ is a monotone retraction from $\mathfrak{M}$ onto $\mathfrak{A}_2$. Since there exists only one such retraction, it follows that $\alpha_{12}\alpha_1 = \alpha_2$. Hence (3) and (4) yield the fundamental formula

$$(5) \qquad (T \mid \mathfrak{A}_1) \mid \mathfrak{A}_2 = T \mid \mathfrak{A}_2 .$$

Suppose that $\mathfrak{M}$ is not a dendrite, and let $\mathfrak{C}^1, \cdots, \mathfrak{C}^n, \cdots$ be the sequence of the proper cyclic elements of $\mathfrak{M}$. Then $\mathfrak{C}^n$ contributes to the cyclic decomposition of T the transformation $L\mu^n M$, where μ^n is the monotone retraction from $\mathfrak{M}$ onto $\mathfrak{C}^n$. According to our present notations, we have $L\mu^n M = T \mid \mathfrak{C}^n$ (note that every proper cyclic element is an A-set). Hence we have the formula (cf. II.2.94)

$$(6) \qquad \Delta(T) = [T \mid \mathfrak{C}^1, \cdots, T \mid \mathfrak{C}^n, \cdots].$$

II.2.101. Continuation. Let $\mathfrak{A}_1, \cdots, \mathfrak{A}_n, \cdots$ be a (finite or infinite) sequence of A-sets in $\mathfrak{M}$, such that $\mathfrak{M} = \mathfrak{A}_1 + \cdots + \mathfrak{A}_n + \cdots$, and for each n

the product $(\mathfrak{A}_1 + \cdots + \mathfrak{A}_n)\mathfrak{A}_{n+1}$ is either empty or else reduces to a single point. By II.2.77 (applied to $\mathfrak{M}$), it follows that each proper cyclic element $\mathfrak{C}$ of $\mathfrak{M}$ is a subset of precisely one set $\mathfrak{A}_n$, and by II.2.79 we know that each proper cyclic element of $\mathfrak{A}_n$ is also a proper cyclic element of $\mathfrak{M}$. Now consider a proper cyclic element $\mathfrak{C} \subset \mathfrak{A}_n$. Then $\mathfrak{C}$ contributes to $\Delta(T)$ the transformation $T \mid \mathfrak{C}$ and to $\Delta(T \mid \mathfrak{A}_n)$ the transformation $(T \mid \mathfrak{A}_n) \mid \mathfrak{C} = T \mid \mathfrak{C}$ (see (5)). Thus $\mathfrak{C}$ contributes the same transformation to both $\Delta(T)$ and $\Delta(T \mid \mathfrak{A}_n)$. The preceding remarks yield the formula (cf. I.2.4) $\Delta(T) = \Delta(T \mid \mathfrak{A}_1) + \cdots + \Delta(T \mid \mathfrak{A}_n) + \cdots$. If $\mathfrak{M}$ reduces to a dendrite, then clearly all the individual terms in this formula vanish, and thus the formula is trivial in this case.

II.2.102. Continuation. Let $\Gamma_1, \cdots, \Gamma_n, \cdots$ be a sequence of cyclic chains in $\mathfrak{M}$ that approximate to $\mathfrak{M}$ in the sense of the cyclic chain approximation theorem (see II.2.71, II.2.72). The following properties of the sequence Γ_n will be relevant (see loc. cit.).

(a) $(\Gamma_1 + \cdots + \Gamma_n)\Gamma_{n+1}$ is a single point for every n.

(b) On setting $H = \Gamma_1 + \cdots + \Gamma_n + \cdots$, we have $c(H) = \mathfrak{M}$.

(c) On setting $\mathfrak{A}_n = \Gamma_1 + \cdots + \Gamma_n$, $\mathfrak{A}_n$ is an A-set of $\mathfrak{M}$.

(d) If $\mathfrak{C}$ is a proper cyclic element of $\mathfrak{M}$, then $\mathfrak{C}$ is a subset of exactly one of the cyclic chains Γ_n.

(e) If the sequence Γ_n is infinite, then $d(\Gamma_n) \to 0$ for $n \to \infty$.

(f) Let δ_n denote the maximum diameter of the components of $\mathfrak{M} - \mathfrak{A}_n$ (cf. (c) above). If the sequence Γ_n is infinite, then $\delta_n \to 0$ for $n \to \infty$.

(g) Each Γ_n is a Peano subspace of $\mathfrak{M}$, and the proper cyclic elements of Γ_n coincide with those proper cyclic elements of $\mathfrak{M}$ that are subsets of Γ_n (see II.2.82).

(h) Each Γ_n is an A-set of $\mathfrak{M}$ (cf. II.2.58).

In view of (g) and (h), an argument entirely analogous to that used in II.2.101 yields the formula $\Delta(T) = \Delta(T \mid \Gamma_1) + \cdots + \Delta(T \mid \Gamma_n) + \cdots$.

II.2.103. Continuation. Let $\Gamma(\mathfrak{a}_1, \mathfrak{a}_2)$ be a cyclic chain in $\mathfrak{M}$, and let γ be a simple arc in $\mathfrak{M}$ with end points $\mathfrak{a}_1, \mathfrak{a}_2$. By II.2.61, either $\Gamma(\mathfrak{a}_1, \mathfrak{a}_2)$ reduces to γ, or else

$$\Gamma(\mathfrak{a}_1, \mathfrak{a}_2) = \gamma + \mathfrak{C}_1 + \cdots + \mathfrak{C}_n + \cdots,$$

where $\mathfrak{C}_1, \cdots, \mathfrak{C}_n, \cdots$ is the sequence of those proper cyclic elements of $\mathfrak{M}$ that have a nondegenerate intersection with γ. In the first case, $\Gamma(\mathfrak{a}_1, \mathfrak{a}_2)$ has no proper cyclic elements (cf. II.2.82), and thus $\Delta[T \mid \Gamma(\mathfrak{a}_1, \mathfrak{a}_2)] = 0$. In the second case, we assert that $\Delta[T \mid \Gamma(\mathfrak{a}_1, \mathfrak{a}_2)] = [T \mid \mathfrak{C}_1, \cdots, T \mid \mathfrak{C}_n, \cdots]$. Indeed, by II.2.82, the proper cyclic elements of $\Gamma(\mathfrak{a}_1, \mathfrak{a}_2)$ coincide with $\mathfrak{C}_1, \cdots, \mathfrak{C}_n, \cdots$, and hence $\Delta[T \mid \Gamma(\mathfrak{a}_1, \mathfrak{a}_2)]$ is comprised of the transformations (cf. II.2.100(5)) $[T \mid \Gamma(\mathfrak{a}_1, \mathfrak{a}_2)] \mid \mathfrak{C}_n] = T \mid \mathfrak{C}_n$.

II.2.104. Given $\mathfrak{Z}$ as in II.2.99, let $\Phi(T)$ be a real-valued, non-negative function defined for every $T \in \mathfrak{Z}$. The function $\Phi(T)$ is permitted to have the value $+\infty$ for certain transformations $T \in \mathfrak{Z}$.

In terms of $\Phi(T)$, we define $\Phi[\Delta(T)]$ as follows. If the cyclic decomposition

$\Delta(T)$ of $T \in \mathfrak{J}$ is empty, then we put $\Phi[\Delta(T)] = 0$. If $\Delta(T) \neq 0$, then let $\Delta(T) = [T^1, \cdots, T^n, \cdots]$ (see II.2.94). If the terms of the (finite or infinite) series $\Phi(T^1) + \cdots + \Phi(T^n) + \cdots$ are finite and the series converges, then we put $\Phi[\Delta(T)] = \Phi(T^1) + \cdots + \Phi(T^n) + \cdots$. Otherwise we put $\Phi[\Delta(T)] = +\infty$. Since Φ is non-negative, it is clear that $\Phi[\Delta(T)]$ is independent of the arrangement of the elements of $\Delta(T)$.

If $\Phi(T) = \Phi[\Delta(T)]$ for every $T \in \mathfrak{J}$, then we shall say that Φ is *cyclicly additive.*

II.2.105. CONTINUATION. Let $T \in \mathfrak{J}$, and let γ be a continuum in $\mathcal{P}$ such that $\mathcal{P} - \gamma \neq 0$ and T is constant on γ. That is, $T(\gamma)$ is a single point a_0^* of the space $T(\mathcal{P}) = \mathcal{P}^* \subset P^*$. It is not assumed, however, that γ is a maximal continuum of constancy for T. Let $\mathfrak{D}$ be a component of $\mathcal{P} - \gamma$, and let us define the transformations T_1, T_2 as follows.

$$T_1(x) = \begin{cases} T(x) \text{ in } \mathfrak{D}, \\ a_0^* \text{ in } \mathcal{P} - \mathfrak{D}, \end{cases} \qquad T_2(x) = \begin{cases} T(x) \text{ in } \mathcal{P} - \mathfrak{D}, \\ a_0^* \text{ in } \mathfrak{D}. \end{cases}$$

Clearly, $T_1 \in \mathfrak{J}$, $T_2 \in \mathfrak{J}$. If $\Phi(T) = \Phi(T_1) + \Phi(T_2)$ for every choice of T, γ, $\mathfrak{D}$ with the properties just stated, we shall say that Φ is *additive with respect to continua of constancy.*

II.2.106. Suppose that Φ, given as in II.2.104, satisfies the following conditions (in addition to the condition $\Phi \geqq 0$).

(i) Φ is *lower semi-continuous.* That is, if $T_n \to T$ (see II.2.99), then $\Phi(T) \leqq \liminf \Phi(T_n)$.

(ii) Φ is additive with respect to continua of constancy (see II.2.105).

Let $T \in \mathfrak{J}$, and let us choose a definite monotone-light factorization $T = LM$, $M(\mathcal{P}) = \mathfrak{M}$, $L(\mathfrak{M}) = \mathcal{P}^*$. We assert that the following statements hold.

(a) Let $\mathfrak{A}_0$ be an A-set in $\mathfrak{M}$, such that $\mathfrak{M} - \mathfrak{A}_0 \neq 0$, and let $\mathfrak{S}_1, \cdots, \mathfrak{S}_n, \cdots$ be the (finite or infinite) sequence of the components of $\mathfrak{M} - \mathfrak{A}_0$. Then (cf. II.2.100) $\Phi(T) = \Phi(T \mid \mathfrak{A}_0) + \Phi(T \mid c(\mathfrak{S}_1)) + \cdots + \Phi(T \mid c(\mathfrak{S}_n)) + \cdots$

(b) Let $\mathfrak{a}_0$ be a cut point of $\mathfrak{M}$, and let $\mathfrak{S}_1, \cdots, \mathfrak{S}_n, \cdots$ be the components of $\mathfrak{M} - \mathfrak{a}_0$. Then $\Phi(T) = \Phi(T \mid c(\mathfrak{S}_1)) + \cdots + \Phi(T \mid c(\mathfrak{S}_n)) + \cdots$.

(c) Let $\mathfrak{A}_1$, $\mathfrak{A}_2$ be A-sets in $\mathfrak{M}$, such that $\mathfrak{A}_1 + \mathfrak{A}_2 = \mathfrak{M}$ and $\mathfrak{A}_1\mathfrak{A}_2$ is a single point. Then $\Phi(T) = \Phi(T \mid \mathfrak{A}_1) + \Phi(T \mid \mathfrak{A}_2)$.

(d) Let $\mathfrak{A}_1, \cdots, \mathfrak{A}_n, \cdots$ be a (finite or infinite) sequence of A-sets in $\mathfrak{M}$, such that $\mathfrak{A}_1 + \cdots + \mathfrak{A}_n + \cdots = \mathfrak{M}$, and for each n the product $(\mathfrak{A}_1 + \cdots + \mathfrak{A}_n)\mathfrak{A}_{n+1}$ reduces to a single point. Then

$$\Phi(T) = \Phi(T \mid \mathfrak{A}_1) + \cdots + \Phi(T \mid \mathfrak{A}_n) + \cdots$$

(e) Let $\mathfrak{A}_1, \cdots, \mathfrak{A}_n, \cdots$ be a (finite or infinite) sequence of A-sets in $\mathfrak{M}$, such that for every n the product $(\mathfrak{A}_1 + \cdots + \mathfrak{A}_n)\mathfrak{A}_{n+1}$ is either empty or else reduces to a single point. Then

$$\Phi(T) \geqq \Phi(T \mid \mathfrak{A}_1) + \cdots + \Phi(T \mid \mathfrak{A}_n) + \cdots.$$

(f) Let $\Gamma^1, \cdots, \Gamma^n, \cdots$ be a sequence of cyclic chains in $\mathfrak{M}$ that approximate to $\mathfrak{M}$ in the sense of the cyclic chain approximation theorem (see II.2.71, II.2.72). Then $\Phi(T) = \Phi(T \mid \Gamma^1) + \cdots + \Phi(T \mid \Gamma^n) + \cdots$.

II.2.107. CONTINUATION. PROOF OF STATEMENT (a). We make the proof in several steps.

Case (1). The set $\mathfrak{M} - \mathfrak{A}_0$ has exactly one component $\mathfrak{S}_1$. Then the frontier of $\mathfrak{S}_1$ consists of a single point $\mathfrak{a}_0$ of $\mathfrak{A}_0$, and $c(\mathfrak{S}_1) = \mathfrak{S}_1 + \mathfrak{a}_0$ is an A-set. Furthermore, $\mathfrak{a}_0$ is a cut point of $\mathfrak{M}$ and $\mathfrak{S}_1$ is also a component of $\mathfrak{M} - \mathfrak{a}_0$ (cf. II.2.44, II.2.73). On setting $\gamma = M^{-1}(\mathfrak{a}_0)$, $\mathfrak{D} = M^{-1}(\mathfrak{S}_1)$, it follows by II.1.5 that $\mathfrak{D}$ is a component of $\mathcal{P} - \gamma$. Let α be the monotone retraction from $\mathfrak{M}$ onto $\mathfrak{A}_0$, and let β be the monotone retraction from $\mathfrak{M}$ onto $c(\mathfrak{S}_1) = \mathfrak{S}_1 + \mathfrak{a}_0$ (cf. II.2.40). We have then the formulas (cf. II.2.76) $\alpha M(x) = \mathfrak{a}_0$ for $x \in \mathfrak{D}$, $\alpha M(x) = M(x)$ for $x \in \mathcal{P} - \mathfrak{D}$, $\beta M(x) = M(x)$ for $x \in \mathfrak{D}$, $\beta M(x) = \mathfrak{a}_0$ for $x \in \mathcal{P} - \mathfrak{D}$. On setting $a_0^* = L(\mathfrak{a}_0)$, we have $T(\gamma) = LM(\gamma) = L(\mathfrak{a}_0) = a_0^*$, and hence

$$L\alpha M(x) = \begin{cases} a_0^* & \text{in } \mathfrak{D}, \\ T(x) & \text{in } \mathcal{P} - \mathfrak{D}, \end{cases} \qquad L\beta M(x) = \begin{cases} T(x) & \text{in } \mathfrak{D}, \\ a_0^* & \text{in } \mathcal{P} - \mathfrak{D}. \end{cases}$$

Thus $L\alpha M$, $L\beta M$ coincide with the transformations T_2, T_1 of II.2.105, and hence, by II.2.106(ii), $\Phi(T) = \Phi(L\alpha M) + \Phi(L\beta M)$. Clearly $L\alpha M = T \mid \mathfrak{A}_0$, $L\beta M = T \mid c(\mathfrak{S}_1)$, and the formula $\Phi(T) = \Phi(T \mid \mathfrak{A}_0) + \Phi[T \mid c(\mathfrak{S}_1)]$ follows.

Case (2). $\mathfrak{M} - \mathfrak{A}_0$ has a finite number of components $\mathfrak{S}_1, \cdots, \mathfrak{S}_n$. On setting $\mathfrak{A}_0' = \mathfrak{A}_0 + \mathfrak{S}_2 + \cdots + \mathfrak{S}_n$, by II.2.75 the set $\mathfrak{A}_0'$ is an A-set, and $\mathfrak{M} - \mathfrak{A}_0'$ has the single component $\mathfrak{S}_1$. Hence by case (1) we have the formula

$$(1) \qquad \Phi(T) = \Phi(T \mid \mathfrak{A}_0') + \Phi[T \mid c(\mathfrak{S}_1)].$$

Now $T \mid \mathfrak{A}_0'$ is a transformation with the middle-space $\mathfrak{A}_0' = \mathfrak{A}_0 + \mathfrak{S}_2 + \cdots + \mathfrak{S}_n$. By II.2.81, $\mathfrak{A}_0$ is an A-set relative to $\mathfrak{A}_0'$, and the components of $\mathfrak{A}_0' - \mathfrak{A}_0$ are clearly $\mathfrak{S}_2, \cdots, \mathfrak{S}_n$. On setting $\mathfrak{A}_0'' = \mathfrak{A}_0 + \mathfrak{S}_3 + \cdots + \mathfrak{S}_n$, we obtain by the process already used $\Phi(T \mid \mathfrak{A}_0') = \Phi[(T \mid \mathfrak{A}_0') \mid \mathfrak{A}_0''] + \Phi[(T \mid \mathfrak{A}_0') \mid c(\mathfrak{S}_2)]$. By II.2.100(5) there follows, in view of (1), the formula

$$\Phi(T) = \Phi(T \mid \mathfrak{A}_0'') + \Phi[T \mid c(\mathfrak{S}_1)] + \Phi[T \mid c(\mathfrak{S}_2)].$$

It is now clear that after n steps we obtain the formula

$$\Phi(T) = \Phi(T \mid \mathfrak{A}_0) + \Phi[T \mid c(\mathfrak{S}_1)] + \cdots + \Phi[T \mid c(\mathfrak{S}_n)].$$

Case (3). The sequence of the components $\mathfrak{S}_1, \cdots, \mathfrak{S}_n, \cdots$ of $\mathfrak{M} - \mathfrak{A}_0$ is infinite. Then $d(\mathfrak{S}_n) \to 0$ and hence also $d[c(\mathfrak{S}_n)] \to 0$ for $n \to \infty$ (see II.2.37). Let us put $\mathfrak{A}_n = \mathfrak{A}_0 + \mathfrak{S}_1 + \cdots + \mathfrak{S}_n$. Then $\mathfrak{A}_n$ is an A-set by II.2.75. If α_n is the monotone retraction from $\mathfrak{M}$ onto $\mathfrak{A}_n$, then clearly α_n converges on $\mathfrak{M}$ uniformly to the identity transformation as a consequence of the relation $d(\mathfrak{S}_n) \to 0$ (cf. II.2.40). Hence $T \mid \mathfrak{A}_n = L\alpha_n M$ converges uniformly on $\mathcal{P}$ to $LM = T$. By II.2.106(i) we have therefore the inequality

$$(2) \qquad \Phi(T) \leqq \liminf \Phi(T \mid \mathfrak{A}_n), \qquad n \to \infty.$$

On the other hand, $T \mid \mathfrak{A}_n$ is a transformation with the middle-space $\mathfrak{A}_n$. Since $\mathfrak{A}_n - \mathfrak{A}_0$ has a finite number of components, namely $\mathfrak{S}_1, \cdots, \mathfrak{S}_n$, we have, by case (2) and II.2.100(5),

(3) $$\Phi(T \mid \mathfrak{A}_n) = \Phi(T \mid \mathfrak{A}_0) + \Phi[T \mid c(\mathfrak{S}_1)] + \cdots + \Phi[T \mid c(\mathfrak{S}_n)].$$

From (2) and (3) we obtain

(4) $$\Phi(T) \leqq \Phi(T \mid \mathfrak{A}_0) + \Phi[T \mid c(\mathfrak{S}_1)] + \cdots + \Phi[T \mid c(\mathfrak{S}_n)] + \cdots .$$

Let us put now $\mathfrak{A}'_n = \mathfrak{A}_0 + \mathfrak{S}_{n+1} + \mathfrak{S}_{n+2} + \cdots$. Then $\mathfrak{A}'_n$ is an A-set by II.2.75, and $\mathfrak{M} - \mathfrak{A}'_n$ has a finite number of components, namely $\mathfrak{S}_1, \cdots, \mathfrak{S}_n$. Hence we have, by case (2),

(5) $$\Phi(T) = \Phi(T \mid \mathfrak{A}'_n) + \Phi[T \mid c(\mathfrak{S}_1)] + \cdots + \Phi[T \mid c(\mathfrak{S}_n)].$$

Let α'_n denote the monotone retraction from $\mathfrak{M}$ onto $\mathfrak{A}'_n$. Clearly, as a consequence of the relation $d(\mathfrak{S}_n) \rightarrow 0$, α'_n converges on $\mathfrak{M}$ uniformly to the monotone retraction $\alpha_0(\mathfrak{M}) = \mathfrak{A}_0$ (cf. II.2.40), and hence $T \mid \mathfrak{A}'_n = L\alpha'_n M$ converges on $\mathcal{P}$ uniformly to $L\alpha_0 M = T \mid \mathfrak{A}_0$. By II.2.106(i) there follows the inequality

(6) $$\liminf \Phi(T \mid \mathfrak{A}'_n) \geqq \Phi(T \mid \mathfrak{A}_0), \qquad n \rightarrow \infty .$$

From (5) and (6) we obtain

(7) $$\Phi(T) \geqq \Phi(T \mid \mathfrak{A}_0) + \Phi[T \mid c(\mathfrak{S}_1)] + \cdots + \Phi[T \mid c(\mathfrak{S}_n)] + \cdots .$$

(4) and (7) yield statement (a) in II.2.106.

II.2.108. CONTINUATION. PROOF OF STATEMENT II.2.106(b). Put $\mathfrak{A}_0 = \mathfrak{S}_1 + \mathfrak{a}_0$. Then $\mathfrak{A}_0$ is an A-set and the components of $\mathfrak{M} - \mathfrak{A}_0$ are precisely $\mathfrak{S}_2, \cdots, \mathfrak{S}_n, \cdots$; furthermore, $\mathfrak{A}_0 = c(\mathfrak{S}_1)$ (cf. II.2.74). Hence, by II.2.106(a), $\Phi(T) = \Phi(T \mid \mathfrak{A}_0) + \Phi[T \mid c(\mathfrak{S}_2)] + \cdots = \Phi[T \mid c(\mathfrak{S}_1)] + \cdots + \Phi[T \mid c(\mathfrak{S}_n)] + \cdots$.

II.2.109. CONTINUATION. PROOF OF STATEMENT II.2.106(c). Put $\mathfrak{A}_1\mathfrak{A}_2 = \mathfrak{a}_0$. By II.2.76, $\mathfrak{a}_0$ is a cut point of $\mathfrak{M}$, and $\mathfrak{A}_1 - \mathfrak{a}_0$, $\mathfrak{A}_2 - \mathfrak{a}_0$ are sums of components of $\mathfrak{M} - \mathfrak{a}_0$. In other words, the components of $\mathfrak{M} - \mathfrak{a}_0$ can be divided into two sequences $\mathfrak{S}'_1, \cdots, \mathfrak{S}'_i, \cdots$ and $\mathfrak{S}''_1, \cdots, \mathfrak{S}''_j, \cdots$, such that the components of $\mathfrak{M} - \mathfrak{A}_1$ coincide with $\mathfrak{S}''_1, \cdots, \mathfrak{S}''_j, \cdots$ and the components of $\mathfrak{M} - \mathfrak{A}_2$ coincide with $\mathfrak{S}'_1, \cdots, \mathfrak{S}'_i, \cdots$. By II.2.106(a) and (b) we have therefore

$$\Phi(T) = \Phi(T \mid \mathfrak{A}_1) + \Phi[T \mid c(\mathfrak{S}''_1)] + \cdots + \Phi[T \mid c(\mathfrak{S}''_j)] + \cdots ,$$

$$\Phi(T) = \Phi(T \mid \mathfrak{A}_2) + \Phi[T \mid c(\mathfrak{S}'_1)] + \cdots + \Phi[T \mid c(\mathfrak{S}'_i)] + \cdots ,$$

$$\Phi(T) = \{\Phi[T \mid c(\mathfrak{S}'_1)] + \cdots + \Phi[T \mid c(\mathfrak{S}'_i)] + \cdots\}$$
$$+ \{\Phi[T \mid c(\mathfrak{S}''_1)] + \cdots + \Phi[T \mid c(\mathfrak{S}''_j)] + \cdots\}.$$

These formulas yield immediately the result that $\Phi(T) = \Phi(T \mid \mathfrak{A}_1) + \Phi(T \mid \mathfrak{A}_2)$.

II.2.110. CONTINUATION. PROOF OF STATEMENT II.2.106(d).

Case (1). The given sequence of A-sets is a finite sequence $\mathfrak{A}_1, \cdots, \mathfrak{A}_n$. If $n = 1$, then $\mathfrak{A}_1 = \mathfrak{M}$, $T \mid \mathfrak{A}_1 = T \mid \mathfrak{M} = T$, and thus the assertion is obvious. Assume that the statement has already been established for sequences of $n - 1$ terms. Then for a sequence with n terms $\mathfrak{A}_1, \cdots, \mathfrak{A}_n$ we put $\mathfrak{A}_1' = \mathfrak{A}_1 + \cdots + \mathfrak{A}_{n-1}$, $\mathfrak{A}_2' = \mathfrak{A}_n$. Then $\mathfrak{A}_1'$ is an A-set and $\mathfrak{A}_1, \cdots, \mathfrak{A}_{n-1}$ are A-sets relative to $\mathfrak{A}_1'$ (see II.2.19, II.2.81). Since $T \mid \mathfrak{A}_1'$ is a transformation with the middle space $\mathfrak{A}_1'$, and since statement (d) is assumed to hold for sequences of $n - 1$ A-sets, we have (cf. II.2.100)

$$(1) \qquad \begin{aligned} \Phi(T \mid \mathfrak{A}_1') &= \Phi[(T \mid \mathfrak{A}_1') \mid \mathfrak{A}_1] + \cdots + \Phi[(T \mid \mathfrak{A}_1') \mid \mathfrak{A}_{n-1}] \\ &= \Phi(T \mid \mathfrak{A}_1) + \cdots + \Phi(T \mid \mathfrak{A}_{n-1}). \end{aligned}$$

On the other hand, by II.2.106(c) we have

$$(2) \qquad \Phi(T) = \Phi(T \mid \mathfrak{A}_1') + \Phi(T \mid \mathfrak{A}_2') = \Phi(T \mid \mathfrak{A}_1') + \Phi(T \mid \mathfrak{A}_n).$$

(1) and (2) yield the desired result.

Case (2). The sequence $\mathfrak{A}_1, \cdots, \mathfrak{A}_n, \cdots$ is infinite. Put $\mathfrak{A}_1 + \cdots + \mathfrak{A}_n = \mathfrak{A}_n'$, and let d_n' denote the maximum diameter of the components of $\mathfrak{M} - \mathfrak{A}_n'$. Then

$$(3) \qquad d_n' \to 0 \qquad \text{for} \qquad n \to \infty$$

by II.2.48. Since $T \mid \mathfrak{A}_n'$ is a transformation with the middle-space $\mathfrak{A}_n'$, we have by case (1) (cf. II.2.100)

$$(4) \qquad \begin{aligned} \Phi(T \mid \mathfrak{A}_n') &= \Phi[(T \mid \mathfrak{A}_n') \mid \mathfrak{A}_1] + \cdots + \Phi[(T \mid \mathfrak{A}_n') \mid \mathfrak{A}_n] \\ &= \Phi(T \mid \mathfrak{A}_1) + \cdots + \Phi(T \mid \mathfrak{A}_n). \end{aligned}$$

Now let α_n' be the monotone retraction from $\mathfrak{M}$ onto $\mathfrak{A}_n'$ (note that $\mathfrak{A}_n'$ is an A-set by II.2.19). Clearly, α_n' converges on $\mathfrak{M}$ uniformly to the identity transformation as a consequence of (3) (cf. II.2.40, II.2.43). Hence $T \mid \mathfrak{A}_n' = L\alpha_n'M$ converges on $\mathcal{P}$ uniformly to $LM = T$. By II.2.106(i), it follows that

$$(5) \qquad \Phi(T) \leqq \liminf \Phi(T \mid \mathfrak{A}_n'), \qquad n \to \infty.$$

(4) and (5) yield the inequality

$$(6) \qquad \Phi(T) \leqq \Phi(T \mid \mathfrak{A}_1) + \cdots + \Phi(T \mid \mathfrak{A}_n) + \cdots.$$

Now let $\mathfrak{S}_{n1}', \cdots, \mathfrak{S}_{nk}', \cdots$ be the components of $\mathfrak{M} - \mathfrak{A}_n'$. By II.2.106(a) we have then

$$\Phi(T) = \Phi(T \mid \mathfrak{A}_n') + \Phi[T \mid c(\mathfrak{S}_{n1}')] + \cdots + \Phi[T \mid c(\mathfrak{S}_{nk}')] + \cdots.$$

Since Φ is non-negative, there follows in view of (4) the inequality

$$(7) \qquad \Phi(T) \geqq \Phi(T \mid \mathfrak{A}_1) + \cdots + \Phi(T \mid \mathfrak{A}_n).$$

Since n is arbitrary, (6) and (7) yield $\Phi(T) = \Phi(T \mid \mathfrak{A}_1) + \cdots + \Phi(T \mid \mathfrak{A}_n) + \cdots$.

II.2.111. Continuation. Proof of statement II.2.106(e). Suppose first that the given sequence of A-sets reduces to a single term $\mathfrak{A}_0$. If $\mathfrak{S}_1, \cdots, \mathfrak{S}_k, \cdots$ are the components of $\mathfrak{M} - \mathfrak{A}_0$, then $\Phi(T) = \Phi(T \mid \mathfrak{A}_0) + \Phi[T \mid c(\mathfrak{S}_1)] + \cdots$ by II.2.106(a). Since Φ is non-negative, there follows the inequality $\Phi(T) \geqq \Phi(T \mid \mathfrak{A}_0)$. Suppose now that the statement (e) has been already verified for sequences of not more than n terms.. Given then a sequence of $n + 1$ terms $\mathfrak{A}_1, \cdots, \mathfrak{A}_n, \mathfrak{A}_{n+1}$, let us consider two subcases.

Case (1.a). The sets $\mathfrak{A}_1, \cdots, \mathfrak{A}_n, \mathfrak{A}_{n+1}$ are disjoint. Let $\mathfrak{S}_1, \cdots, \mathfrak{S}_k, \cdots$ be the components of $\mathfrak{M} - \mathfrak{A}_{n+1}$. Since each $\mathfrak{A}_i$ is connected and $\mathfrak{A}_i \subset \mathfrak{M} - \mathfrak{A}_{n+1}$ for $i = 1, \cdots, n$, it follows that each $\mathfrak{A}_i$ is a subset of exactly one of the sets $\mathfrak{S}_1, \cdots, \mathfrak{S}_k, \cdots$. Let us put

$$\textstyle\sum_k = \sum \Phi(T \mid \mathfrak{A}_i), \qquad \mathfrak{A}_i \subset \mathfrak{S}_k .$$

Of course, $\sum_k$ may be vacuous for a given k, and then $\sum_k = 0$. Now for each k the summation contains at most n terms, and hence we have $\sum_k \leqq \Phi[T \mid c(\mathfrak{S}_k)]$ (note that $T \mid c(\mathfrak{S}_k)$ is a transformation with the middle-space $c(\mathfrak{S}_k)$). Thus

$$\Phi(T \mid \mathfrak{A}_1) + \cdots + \Phi(T \mid \mathfrak{A}_n) \leqq \Phi[T \mid c(\mathfrak{S}_1)] + \cdots + \Phi[T \mid c(\mathfrak{S}_k)] + \cdots .$$

On the other hand, we have by II.2.106(a)

$$\Phi(T) = \Phi(T \mid \mathfrak{A}_{n+1}) + \Phi[T \mid c(\mathfrak{S}_1)] + \cdots + \Phi[T \mid c(\mathfrak{S}_k)] + \cdots ,$$

and the inequality $\Phi(T) \geqq \Phi(T \mid \mathfrak{A}_1) + \cdots + \Phi(T \mid \mathfrak{A}_{n+1})$ follows.

Case (1.b). The sets $\mathfrak{A}_1, \cdots, \mathfrak{A}_n, \mathfrak{A}_{n+1}$ are not disjoint, say $\mathfrak{A}_1\mathfrak{A}_2 \neq 0$. Then $\mathfrak{A}_1\mathfrak{A}_2$ is a single point, and hence $\mathfrak{A}_1 + \mathfrak{A}_2$ is an A-set (see II.2.19). We have therefore the inequality

$$\Phi(T) \geqq \Phi[T \mid (\mathfrak{A}_1 + \mathfrak{A}_2)] + \Phi(T \mid \mathfrak{A}_3) + \cdots + \Phi(T \mid \mathfrak{A}_{n+1}). \tag{1}$$

By II.2.106(c) (applied to $T \mid (\mathfrak{A}_1 + \mathfrak{A}_2)$) we have, in view of II.2.100,

$$\Phi[T \mid (\mathfrak{A}_1 + \mathfrak{A}_2)] = \Phi(T \mid \mathfrak{A}_1) + \Phi(T \mid \mathfrak{A}_2). \tag{2}$$

(1) and (2) yield the inequality $\Phi(T) \geqq \Phi(T \mid \mathfrak{A}_1) + \cdots + \Phi(T \mid \mathfrak{A}_{n+1})$.

Thus statement (e) is established for finite sequences of A-sets. If the given sequence $\mathfrak{A}_1, \cdots, \mathfrak{A}_n, \cdots$ is infinite, then by what precedes we have $\Phi(T) \geqq \Phi(T \mid \mathfrak{A}_1) + \cdots + \Phi(T \mid \mathfrak{A}_n)$ for every n, and the desired inequality follows for $n \to \infty$.

II.2.112. Continuation. Proof of statement II.2.106(f). By II.2.106(e) we have the inequality $\Phi(T) \geqq \Phi(T \mid \Gamma^1) + \cdots + \Phi(T \mid \Gamma^n) + \cdots$. To establish the complementary inequality, it is sufficient to consider the case when the sequence Γ^n is infinite, since in the finite case the desired result follows directly from II.2.106(d). Setting $\mathfrak{A}_n = \Gamma^1 + \cdots + \Gamma^n$, $\mathfrak{A}_n$ is an A-set and the maximum diameter of the components of $\mathfrak{M} - \mathfrak{A}_n$ converges to zero for $n \to \infty$ by II.2.72. Hence, if α_n is the monotone retraction from $\mathfrak{M}$ onto $\mathfrak{A}_n$, then α_n converges on $\mathfrak{M}$ uniformly to the identity transformation (cf. II.2.40, II.2.43). Consequently $T \mid \mathfrak{A}_n \doteq L\alpha_n M$ converges on $\mathcal{P}$ uniformly to $LM = T$. By II.2.106(i),

$$(1) \qquad \Phi(T) \leqq \liminf \Phi(T \mid \mathfrak{A}_n), \qquad n \to \infty.$$

On the other hand, II.2.106(d) (applied to $T \mid \mathfrak{A}_n$) yields (cf. II.2.100)

$$(2) \qquad \Phi(T \mid \mathfrak{A}_n) = \Phi(T \mid \Gamma^1) + \cdots + \Phi(T \mid \Gamma^n).$$

(1) and (2) yield the desired inequality $\Phi(T) \leqq \Phi(T \mid \Gamma^1) + \cdots + \Phi(T \mid \Gamma^n) + \cdots$.

II.2.113. CYCLIC ADDITIVITY THEOREM. *Let $\mathfrak{J}$ be the class of all continuous transformations from a fixed Peano space $\mathcal{P}$ into a fixed metric space P^* (that is, $T(\mathcal{P}) \subset P^*$ for every $T \in \mathfrak{J}$). Let $\Phi(T)$ be a real-valued function defined for all $T \in \mathfrak{J}$. Suppose that the following conditions hold.*

(α) *$\mathcal{P}$ possesses the property* (Π) (*see* II.2.89). *In particular, $\mathcal{P}$ may be a 2-cell or a 2-sphere.*

(β) *$\Phi(T) \geqq 0$ for $T \in \mathfrak{J}$. For certain transformations $T \in \mathfrak{J}$, $\Phi(T)$ may be equal to $+\infty$.*

(γ) *$\Phi(T)$ is lower semi-continuous* (*see* II.2.106).

(δ) *If T is constant on $\mathcal{P}$, then $\Phi(T) = 0$.*

(ϵ) *If the middle-space occurring in the monotone-light factorization of a $T \in \mathfrak{J}$ is a simple arc, then $\Phi(T) = 0$.*

(ϕ) *$\Phi(T)$ is additive with respect to continua of constancy* (*see* II.2.105).

Then Φ is cyclicly additive (*see* II.2.104).

PROOF. By II.2.106, Φ possesses all the properties (a) to (f) listed there, as a consequence of the assumptions (β), (γ), (ϕ). Hence, as a special case of II.2.106(e), we have $\Phi[\Delta(T)] \leqq \Phi(T)$. Thus it is sufficient to prove

$$(1) \qquad \Phi(T) \leqq \Phi[\Delta(T)].$$

In view of condition (δ), it is sufficient to consider the case when T is not constant on $\mathcal{P}$. Let $T \in \mathfrak{J}$, and let $T = LM$, $M(\mathcal{P}) = \mathfrak{M}$, $L(\mathfrak{M}) = \mathcal{P}^*$ be a definite monotone-light factorization of T. We make the proof of (1) in several steps.

Case (A). $\mathfrak{M}$ reduces to a single cyclic chain $\Gamma(\mathfrak{a}_1, \mathfrak{a}_2)$. Let γ be a simple arc in $\mathfrak{M}$ with end points $\mathfrak{a}_1$, $\mathfrak{a}_2$.

Subcase (A.1). $\mathfrak{M} = \gamma$. Then $\Delta(T) = 0$ and hence $\Phi[\Delta(T)] = 0$. By condition (ϵ), $\Phi(T) = 0$. Thus (1) holds.

Subcase (A.2). $\mathfrak{M}$ has a finite number of proper cyclic elements $\mathfrak{C}_1, \cdots, \mathfrak{C}_n$. By II.2.63, we have then in $\mathfrak{M}$ a finite system of cyclic chains $\Gamma_1, \cdots, \Gamma_N$, such that $\Gamma_i\Gamma_{i+1}$ is a single point, $i = 1, \cdots, N - 1$, $\Gamma_i\Gamma_j = 0$ otherwise, $\Gamma_1 + \cdots + \Gamma_N = \mathfrak{M}$, and each $\mathfrak{C}_k$, $k = 1, \cdots, n$, coincides with a term of the sequence $\Gamma_1, \cdots, \Gamma_N$. In view of II.2.61 it follows that for each i either Γ_i is one of the proper cyclic elements $\mathfrak{C}_1, \cdots, \mathfrak{C}_n$, or else Γ_i reduces to a simple arc. Hence, in view of the condition (ϵ),

$$\Phi(T \mid \Gamma_1) + \cdots + \Phi(T \mid \Gamma_N) = \Phi(T \mid \mathfrak{C}_1) + \cdots + \Phi(T \mid \mathfrak{C}_n) = \Phi[\Delta(T)].$$

On the other hand, $\Phi(T \mid \Gamma_1) + \cdots + \Phi(T \mid \Gamma_N) = \Phi(T)$ by II.2.106(d), and (1) follows.

Subcase (A.3). $\mathfrak{M}$ has infinitely many proper cyclic elements $\mathfrak{C}_1, \cdots, \mathfrak{C}_n, \cdots$ By II.2.87, II.2.88 we have for each n a monotone retraction $\mu_n(\mathfrak{M}) = \gamma + \mathfrak{C}_1 + \cdots + \mathfrak{C}_n$, such that μ_n converges on $\mathfrak{M}$ uniformly to the identity transformation. Hence the sequence $L\mu_n M$ converges on $\mathcal{P}$ uniformly to $LM = T$. By condition (γ) it follows that

$$\Phi(T) \leqq \liminf \Phi(L\mu_n M), \qquad n \to \infty. \tag{2}$$

On the other hand, if we use for $L\mu_n M$ the monotone-light factorization $L \cdot \mu_n M$ (cf. II.1.4), then the middle-space coincides with $\gamma + \mathfrak{C}_1 + \cdots + \mathfrak{C}_n$. Hence (cf. II.2.82, II.2.83) this middle-space reduces to a single cyclic chain with a finite number of proper cyclic elements, namely $\mathfrak{C}_1, \cdots, \mathfrak{C}_n$. Hence by (A.2)

$$\Phi(L\mu_n M) \leqq \Phi[\Delta(L\mu_n M)]. \tag{3}$$

Now let β_i be the monotone retraction from $\gamma + \mathfrak{C}_1 + \cdots + \mathfrak{C}_n$ onto $\mathfrak{C}_i$, $i = 1, \cdots, n$. By definition

$$\Delta(L\mu_n M) = [L\beta_1\mu_n M, L\beta_2\mu_n M, \cdots, L\beta_n\mu_n M]. \tag{4}$$

Clearly $\beta_i\mu_n$ is a monotone retraction from $\mathfrak{M}$ onto $\mathfrak{C}_i$, and since there exists just one such retraction, $L\beta_i\mu_n M = T \mid \mathfrak{C}_i$. Thus $\Delta(L\mu_n M) = [T \mid \mathfrak{C}_1, \cdots, T \mid \mathfrak{C}_n]$, and consequently, in view of (3), (4), and condition (β),

$$\Phi(L\mu_n M) \leqq \Phi(T \mid \mathfrak{C}_1) + \cdots + \Phi(T \mid \mathfrak{C}_n) \leqq \Phi[\Delta(T)]. \tag{5}$$

(2) and (5) imply (1).

Case (B). $\mathfrak{M}$ does not reduce to a single cyclic chain. Let then $\Gamma^1, \cdots, \Gamma^n, \cdots$ be a sequence of cyclic chains in $\mathfrak{M}$ that approximate to $\mathfrak{M}$ in the sense of the cyclic chain approximation theorem (see II.2.71). By II.2.102 we have then $\Delta(T) = \Delta(T \mid \Gamma^1) + \cdots + \Delta(T \mid \Gamma^n) + \cdots$, and hence (see I.2.4)

$$\Phi[\Delta(T)] = \Phi[\Delta(T \mid \Gamma^1)] + \cdots + \Phi[\Delta(T \mid \Gamma^n)] + \cdots. \tag{6}$$

On the other hand, $T \mid \Gamma^n$ is a transformation with the middle-space Γ^n. Hence by case (A) (cf. II.2.82, II.2.83)

$$\Phi(T \mid \Gamma^n) \leqq \Phi[\Delta(T \mid \Gamma^n)], \qquad n = 1, 2, \cdots. \tag{7}$$

By II.2.106(f), we have

$$\Phi(T) = \Phi(T \mid \Gamma^1) + \cdots + \Phi(T \mid \Gamma^n) + \cdots. \tag{8}$$

(8), (7), (6) yield (1), and the proof of the cyclic additivity theorem is complete.

II.2.114. *Let us assume that* Φ, *given as in* II.2.113, *satisfies the following additional condition.*

(ψ) *If the middle-space occurring in the monotone-light factorization of a* $T \in \mathfrak{I}$ *is nondegenerate and cyclic* (see II.2.1), *then* $\Phi(T) > 0$.

THEOREM. $\Phi(T)$ *vanishes if and only if the middle-space occurring in the monotone-light factorization of* T *is a dendrite.*

Proof. *Sufficiency.* If the middle-space is a dendrite, then $\Delta(T) = 0$, and hence $\Phi(T) = \Phi[\Delta(T)] = 0$ by II.2.113.

Necessity. If $\Phi(T) = 0$, then by II.2.113 we have $\Phi[\Delta(T)] = 0$. Suppose that the middle-space $\mathfrak{M}$ is not a dendrite, and let $\mathfrak{C}_1, \cdots, \mathfrak{C}_n, \cdots$ be the proper cyclic elements of $\mathfrak{M}$. Then $0 = \Phi[\Delta(T)] = \Phi(T \mid \mathfrak{C}_1) + \cdots + \Phi(T \mid \mathfrak{C}_n) + \cdots$. Since Φ is non-negative, it follows that $\Phi(T \mid \mathfrak{C}_n) = 0$ for every n. On the other hand, since $T \mid \mathfrak{C}_n$ is a transformation in $\mathfrak{J}$ with a nondegenerate cyclic middle-space (namely $\mathfrak{C}_n$), we should have $\Phi(T \mid \mathfrak{C}_n) > 0$ by condition (ψ). Thus the assumption that $\mathfrak{M}$ is not a dendrite leads to a contradiction.

CHAPTER II.3. CURVES AND SURFACES

II.3.1. The terms *curve* and *surface* are constantly used in many fields of mathematics, but explicit definitions are usually not given. In the type of study we are engaged in, all concepts involved should be fixed with all possible clarity. Accordingly, we shall present and study in this chapter those concepts of curve and surface that seem to be most suitable for our purposes. The motivation and the historical development of these concepts will be discussed in chapter II.5.

II.3.2. The terms curve and surface are widely used in mathematics to refer to certain geometrical objects in the complex domain. *Let us note explicitly that we are interested in the real case only.*

II.3.3. Since *binary relations* will play an important role in our definitions, we begin with a few general remarks. Of the various definitions of a binary relation, the following is adequate for our purposes. Given any class K of elements to be denoted by $a, b, \cdots$, a binary relation $\mathfrak{B}$ in K may be thought of as a statement with two blank spaces, to be filled in by elements of K. For example, if K is the class of the inhabitants of a city, then "$\square$ is father of $\square$" is a binary relation in K, where the squares represent the blank spaces. For convenience, let us use the letters x and y instead of blank spaces. Then the statement "x is father of y" is a binary relation in K. If we use the symbol $x\mathfrak{B}y$ as an abbreviation for the statement "x is father of y", then the statement $a\mathfrak{B}b$, where a, b are definite elements of K, may be true or false. In the first case, we say that the relation $a\mathfrak{B}b$ holds, in the second case we say that the relation $a\mathfrak{B}b$ fails to hold. Further illustrations of the conception of a binary relation will readily come to the mind of the reader. The relation $p \bigcirc q$ of conjugacy in a Peano space (see II.2.3) is an important example of a binary relation. The binary relations $T_1 \sim T_2(ts)$, $T_1 \sim T_2(F)$ (see II.1.25, II.1.26) are further instances of great importance. As suggested by these examples, a variety of notations are being used to refer to individual binary relations.

II.3.4. Continuation. Let $x\mathfrak{B}y$ be a binary relation in a class K. Let $\mathfrak{K}$ be the class of all those ordered pairs (a, b) of elements of K for which the relation $a\mathfrak{B}b$ holds. The use of ordered pairs is motivated by the fact that generally the relation $a\mathfrak{B}b$ does not imply the relation $b\mathfrak{B}a$. Obviously, the relation $a\mathfrak{B}b$ holds if and only if $(a, b) \in \mathfrak{K}$. In other words, the binary relation $x\mathfrak{B}y$ has the same scope as the binary relation determined by the statement "the ordered pair (x, y) belongs to the class $\mathfrak{K}$". This remark suggests a definition of a binary relation in terms of a class of ordered pairs of elements of a class K. This type of definition is of general use at present in the theory of binary relations, and would be equally acceptable for our purposes.

II.3.5. Continuation. A binary relation $x\mathfrak{B}y$ is *reflexive* if $a\mathfrak{B}a$ for every $a \in K$. It is *symmetric* if $a\mathfrak{B}b$ implies $b\mathfrak{B}a$, and it is *transitive* if $a\mathfrak{B}b$, $b\mathfrak{B}c$ imply

$a\mathfrak{B}c$. For example, the binary relation $p \bigcirc q$ of II.2.3 is reflexive, symmetric, but *not* transitive.

A binary relation which is reflexive, symmetric, and transitive is called an *equivalence relation*, or briefly an *equivalence*. We shall use $x\mathfrak{R}y$ (instead of $x\mathfrak{B}y$) to refer to an equivalence relation.

Given an equivalence $\mathfrak{R}$ in a class K, a nonempty subset E of K is called an *equivalence-class* (relative to $\mathfrak{R}$ in K), if and only if the following conditions hold. (i) If $a \in E$, $b \in E$, then $a\mathfrak{R}b$. (ii) E is maximal with respect to the property (i) (that is, E is not a proper subset of a set that has the property (i)). We have then the following statements.

(a) If E_1, E_2 are equivalence-classes, then either $E_1 = E_2$ or else $E_1E_2 = 0$. This is an immediate consequence of the transitivity of $\mathfrak{R}$.

(b) If a_0 is any element of K, then there exists a unique equivalence class that contains a_0. Indeed, let E_0 be the set of all those elements p of K that satisfy the relation $p\mathfrak{R}a_0$. Then E_0 is not empty, since $a_0 \in E_0$. Since $\mathfrak{R}$ is transitive, E_0 obviously possesses the property (i). If E contains E_0 and E possesses the property (i), then clearly $E \subset E_0$. Thus E_0 is maximal with respect to the property (i). The uniqueness follows from (a).

In view of (a) and (b) it follows that the equivalence-classes constitute a *disjoint covering* of K. In other words, if F is the family of the equivalence-classes relative to $\mathfrak{R}$, then every element of K is contained in precisely one set of the family F. Conversely, every disjoint covering of K can be accounted for in this manner. Indeed, if Φ is any family of subsets E of K that constitute a disjoint covering of K, then the statement "x and y are contained in the same set of the family Φ" determines a binary relation in K that is clearly an equivalence, and the corresponding equivalence-classes clearly coincide with the sets of the family Φ.

II.3.6. Given an equivalence $\mathfrak{R}$ in a class K, let Φ be the family of the equivalence-classes relative to $\mathfrak{R}$ in K. Then Φ may be thought of as a new class derived from K by collapsing each equivalence-class in K into a single element of the new class Φ. This procedure of deriving new classes is applied in a variety of important instances in mathematics.

II.3.7. The relation $T_1 \sim T_2(F)$ (see II.1.25) will now be used to define the geometrical objects that will be termed *F-curves*, *F-surfaces*, and more generally *F-varieties*. The letter F is used to refer to the fact that the relation $T_1 \sim T_2(F)$ was introduced by Fréchet.

Let K be the class of continuous transformations of the form $T(\mathcal{P}) = \mathcal{P}^* \subset P^*$, where P^* is a fixed metric space, and $\mathcal{P}$ is any Peano space (in the applications, P^* will be the Euclidean xyz-space). We choose a Peano space $\mathcal{P}_0$ and define $K(\mathcal{P}_0)$ as the subclass of K consisting of all those continuous transformations $T(\mathcal{P}) = \mathcal{P}^* \subset P^*$ for which $\mathcal{P}$ is homeomorphic with $\mathcal{P}_0$. In the class $K(\mathcal{P}_0)$ we consider the binary relation $T_1 \sim T_2(F)$ (see II.1.25). Obviously, this is an equivalence relation. An *F-variety of the type of* $\mathcal{P}_0$ *in* P^* is now defined as an equivalence class in $K(\mathcal{P}_0)$ with respect to the equivalence $T_1 \sim T_2(F)$. Clearly, if $\mathcal{P}_0$ is replaced by any Peano space $\mathcal{P}_0'$ homeomorphic with $\mathcal{P}_0$, then the F-varieties of type $\mathcal{P}_0'$ coincide with the F-varieties of type $\mathcal{P}_0$. On the other hand, the choice of P^* is relevant. The following special cases are of importance in the sequel.

(a) P^* is the Euclidean three-space, $\mathcal{P}_0$ is a 1-cell (see I.2.31). The corresponding F-varieties are called *F-curves of the type of the* 1-*cell* (or of the type of the simple arc) *in Euclidean three-space.*

(b) P^* is the Euclidean three-space, $\mathcal{P}_0$ is a 1-sphere (see I.2.31). We obtain then the *F-curves of the type of the* 1-*sphere* (or of the type of the simple closed curve) *in Euclidean three-space.*

(c) P^* is the Euclidean three-space, $\mathcal{P}_0$ is a 2-cell (see I.2.31). We obtain then the *F-surfaces of the type of the* 2-*cell in Euclidean three-space.*

(d) P^* is the Euclidean three-space, $\mathcal{P}_0$ is a 2-sphere (see I.2.31). We obtain then the *F-surfaces of the type of the* 2-*sphere in Euclidean three-space.*

II.3.8. Using the terminology of II.3.7, let $T_1(\mathcal{P}_1) = \mathcal{P}_1^*$, $T_2(\mathcal{P}_2) = \mathcal{P}_2^*$ be two transformations in the class $K(\mathcal{P}_0)$. Let $H(\mathcal{P}_1) = \mathcal{P}_2$ be a homeomorphism, and let us put (cf. I.2.10)

$$\delta(T_1, T_2, H) = \max \rho[T_1(x_1), T_2H(x_1)], \qquad x_1 \in \mathcal{P}_1,$$

where the use of the maximum is justified since $\mathcal{P}_1$ is compact. We define the Fréchet distance $d_F(T_1, T_2)$ by the formula

$$d_F(T_1, T_2) = \text{gr.l.b.}\ \delta(T_1, T_2, H),$$

where the greatest lower bound is taken with respect to all homeomorphisms $H(\mathcal{P}_1) = \mathcal{P}_2$. Clearly $0 \leqq d_F(T_1, T_2) < \infty$, and $d_F(T_1, T_2) = d_F(T_2, T_1)$. If T_1, T_2, T_3 are any three elements of the class $K(\mathcal{P}_0)$, then there follows readily the triangle inequality

$$d_F(T_1, T_3) \leqq d_F(T_1, T_2) + d_F(T_2, T_3).$$

It should be emphasized, however, that the distance d_F fails to possess an important property that a distance-function in a metric space should possess. Indeed, the relation $d_F(T_1, T_2) = 0$ does not generally imply that $T_1 = T_2$. For example, if $T_1(\mathcal{P}_1) = \mathcal{P}^*$, $T_2(\mathcal{P}_2) = \mathcal{P}^*$, where $\mathcal{P}_1 \neq \mathcal{P}_2$ and $T_1 \sim T_2(ts)$ (cf. II.1.26), then clearly $d_F(T_1, T_2) = 0$ and $T_1 \neq T_2$. As a less trivial but more revealing example, we mention the following situation. Let P^* be the Euclidean three-space, and let $\mathcal{P}_0$ be chosen as a 1-cell. Let $\mathcal{P}$ be the interval $0 \leqq u \leqq 1$, $\mathcal{P}^*$ the interval $0 \leqq x \leqq 1$, $y = 0$, $z = 0$, and let $T_1(\mathcal{P}) = \mathcal{P}^*$, $T_2(\mathcal{P}) = \mathcal{P}^*$ be defined by the formulas

$$T_1 : x = g(u), \qquad y = 0, \qquad z = 0, \qquad 0 \leqq u \leqq 1,$$

$$T_2 : x = u, \qquad y = 0, \qquad z = 0, \qquad 0 \leqq u \leqq 1,$$

where $g(u) = 2u$ if $0 \leqq u \leqq 1/2$ and $g(u) = 1$ if $1/2 \leqq u \leqq 1$. Then $T_1 \neq T_2$, and clearly even the relation $T_1 \sim T_2(ts)$ fails to hold (cf. II.1.26). On the other hand, $d_F(T_1, T_2) = 0$. Indeed, let $\epsilon > 0$ be given. Then the formula $u' = [g(u) + \epsilon u]/(1 + \epsilon) = h(u)$ defines a homeomorphism $h(\mathcal{P}) = \mathcal{P}$, and

$$\delta(T_1, T_2, h) = \max \rho[T_1(u), T_2h(u)] = \frac{\epsilon}{1 + \epsilon} \max |g(u) - u| < \epsilon.$$

Since ϵ was arbitrary, it follows that $d_F(T_1, T_2) = 0$.

II.3.9. Let $T_1(\mathcal{P}_1) = \mathcal{P}_1^*$, $T_2(\mathcal{P}_2) = \mathcal{P}_2^*$ be two elements of the class $K(\mathcal{P}_0)$ (see II.3.7). If $d_F(T_1, T_2) = 0$, then $\mathcal{P}_1^* = \mathcal{P}_2^*$ (the converse being generally false).

PROOF. Let p_1^* be any point of $\mathcal{P}_1^*$. Then we have a point $p_1 \in \mathcal{P}_1$ such that $p_1^* = T_1(p_1)$. Give $\epsilon > 0$. By assumption, we have then a homeomorphism $H(\mathcal{P}_1) = \mathcal{P}_2$ such that $\rho[T_1(x_1), T_2H(x_1)] < \epsilon$ for $x_1 \in \mathcal{P}_1$. In particular, on setting $p_2 = H(p_1)$, $p_2^* = T_2(p_2)$, it follows that $\rho(p_1^*, p_2^*) < \epsilon$. Since ϵ was arbitrary, it follows that $p_1^* \in \mathcal{P}_2^*$. Thus $\mathcal{P}_1^* \subset \mathcal{P}_2^*$, and similarly $\mathcal{P}_2^* \subset \mathcal{P}_1^*$.

II.3.10. CONTINUATION. $d_F(T_1, T_2) = 0$ if and only if $T_1 \sim T_2(F)$. This is an immediate consequence of II.3.8.

II.3.11. Let $\mathfrak{K}(\mathcal{P}_0, P^*)$ be the class of F-varieties of the type $\mathcal{P}_0$ in P^*, and let $V \in \mathfrak{K}(\mathcal{P}_0, P^*)$. Then V is an equivalence-class of elements of the class $K(\mathcal{P}_0)$ (see II.3.7). Each element $T(\mathcal{P}) = \mathcal{P}^*$ of this equivalence-class is called a *representation* of V. From II.3.9, II.3.10 it follows that the space $\mathcal{P}^*$ is the same for all the representations of V On the other hand, the space $\mathcal{P}$ may be chosen as any Peano space homeomorphic with $\mathcal{P}_0$. Indeed, if $T(\mathcal{P}) = \mathcal{P}^*$ is any representation of V, and if $\mathcal{P}_1$ is any Peano space homeomorphic with $\mathcal{P}_0$, then there exists (since $T \in K(\mathcal{P}_0)$) a homeomorphism $H(\mathcal{P}) = \mathcal{P}_1$. Put $T_1 = TH^{-1}$. Then $T_1(\mathcal{P}_1) = \mathcal{P}^*$ is a representation of V. Indeed, $T(x) = TH^{-1}H(x) = T_1H(x)$ for $x \in \mathcal{P}$. Thus $T_1 \sim T_2(ts)$, and hence $T_1 \sim T(F)$ (cf. II.1.26). That is, T_1 and T belong to the same equivalence class, and consequently T_1 is also a representation of V.

The preceding argument yields also the following statement. If $T(\mathcal{P}) = \mathcal{P}^*$ is a representation of V, and if $H(\mathcal{P}_1) = \mathcal{P}$ is a homeomorphism, then $TH(\mathcal{P}_1) = \mathcal{P}^*$ is also a representation of V. Simple examples (of the type discussed at the end of II.3.8) show that generally we cannot expect to obtain in this manner all the representations of V, and *no general process is known at this time that would yield all the representations of V if one representation is given.*

If $T_1(\mathcal{P}_1) = \mathcal{P}_1^*$, $T_2(\mathcal{P}_2) = \mathcal{P}_2^*$ are two representations of V, then $T_1 \sim T_2(F)$ and hence $\mathcal{P}_1^* = \mathcal{P}_2^*$ as noted above. If $\mathcal{P}_1 = \mathcal{P}_2$, then the representations are termed collocal.

We shall write $V : T(\mathcal{P}) = \mathcal{P}^*$ to express the fact that $T(\mathcal{P}) = \mathcal{P}^*$ is a representation of V. Clearly, V is determined by any one of its representations.

II.3.12. CONTINUATION. Let $T(\mathcal{P}) = \mathcal{P}^*$, $T'(\mathcal{P}') = \mathcal{P}^*$ be two representations of V. Then there exists a sequence of representations $T_n(\mathcal{P}) = \mathcal{P}^*$ of V such that $T_n \sim T'(ts)$ and T_n converges on $\mathcal{P}$ uniformly to T.

PROOF. By assumption, $T \sim T'(F)$. Hence there exists, for every positive integer n, a homeomorphism $H_n(\mathcal{P}) = \mathcal{P}'$ such that $\rho[T(x), T'H_n(x)] < 1/n$ for

$x \in \mathcal{P}$. On setting $T_n = T'H_n$, it follows that $\rho[T(x), T_n(x)] < 1/n$ for $x \in \mathcal{P}$. Thus T_n converges on $\mathcal{P}$ uniformly to T, and by II.3.11, $T_n(\mathcal{P}) = \mathcal{P}^*$ is a representation of V. Clearly, $T_n \sim T'(ts)$.

II.3.13. If $V : T(\mathcal{P}) = \mathcal{P}^*$, $V' : T'(\mathcal{P}') = \mathcal{P}^*$ are two F-varieties of the class $\mathfrak{K}(\mathcal{P}_0, P^*)$, and V' admits of a sequence of representations $T_n(\mathcal{P}) = \mathcal{P}^*$ such that T_n converges on $\mathcal{P}$ uniformly to T, then $V = V'$.

PROOF. Give $\epsilon > 0$. Since T_n and T' are representations of V', and hence $T_n \sim T'(F)$, we have a homeomorphism $H_n(\mathcal{P}) = \mathcal{P}'$ such that $\rho[T_n(x), T'H_n(x)] < \epsilon/2$ for $x \in \mathcal{P}$. Since T_n converges on $\mathcal{P}$ uniformly to T, we can choose n so that $\rho[T(x), T_n(x)] < \epsilon/2$ for $x \in \mathcal{P}$. There follows the inequality $\rho[T(x), T'H_n(x)] < \epsilon$, $x \in \mathcal{P}$. Since ϵ was arbitrary, it follows that $T \sim T'(F)$. Thus T is also a representation of V'. In other words, the equivalence-classes V, V' both contain T, and hence are identical.

II.3.14. Given two F-varieties V_1, V_2 of the class $\mathfrak{K}(\mathcal{P}_0, P^*)$ (see II.3.11), let $T_1'(\mathcal{P}_1') = \mathcal{P}_1^*$, $T_1''(\mathcal{P}_1'') = \mathcal{P}_1^*$ be any two representations of V_1, and let $T_2'(\mathcal{P}_2') = \mathcal{P}_2^*$, $T_2''(\mathcal{P}_2'') = \mathcal{P}_2^*$ be any two representations of V_2. Then (cf. II.3.8)

$$d_F(T_1', T_2') = d_F(T_1'', T_2''). \tag{1}$$

PROOF. By the triangle inequality we have

$$d_F(T_1', T_2') \leqq d_F(T_1', T_1'') + d_F(T_1'', T_2'') + d_F(T_2'', T_2'). \tag{2}$$

Since $T_1' \sim T_1''(F)$, we have $d_F(T_1', T_1'') = 0$ (see II.3.10). Similarly $d_F(T_2'', T_2') = 0$. Thus (2) yields $d_F(T_1', T_2') \leqq d_F(T_1'', T_2'')$. The complementary inequality $d_F(T_1'', T_2'') \leqq d_F(T_1', T_2')$ is established in an analogous manner, and (1) follows.

II.3.15. Given two F-varieties V_1, V_2 of the class $\mathfrak{K}(\mathcal{P}_0, P^*)$ (see II.3.11), let $T_1(\mathcal{P}_1) = \mathcal{P}_1^*$ be a representation of V_1, and let $T_2(\mathcal{P}_2) = \mathcal{P}_2^*$ be a representation of V_2. We define the distance $d(V_1, V_2)$ of V_1 and V_2 by the formula

$$d(V_1, V_2) = d_F(T_1, T_2).$$

By II.3.14, $d(V_1, V_2)$ is independent of the particular choice of the representations used, and thus the notation $d(V_1, V_2)$ is justified. By II.3.8, the distance $d(V_1, V_2)$ has the following properties.

(a) $0 \leqq d(V_1, V_2) < \infty$.

(b) $d(V_1, V_2) = d(V_2, V_1)$.

(c) $d(V_1, V_3) \leqq d(V_1, V_2) + d(V_2, V_3)$ (triangle inequality).

We assert finally that $d(V_1, V_2) = 0$ if and only if $V_1 = V_2$. Indeed, let $T_1(\mathcal{P}_1) = \mathcal{P}_1^*$, $T_2(\mathcal{P}_2) = \mathcal{P}_2^*$ be representations of V_1, V_2 respectively. Then $d(V_1, V_2) = d_F(T_1, T_2)$. Suppose first that $d(V_1, V_2) = 0$. Then $T_1 \sim T_2(F)$ by II.3.10. Thus both of the equivalence-classes V_1, V_2 contain T_1, and hence $V_1 = V_2$. Suppose, secondly, that $V_1 = V_2$. Then T_1, T_2 are in the same equivalence-class. Thus $T_1 \sim T_2(F)$, and hence, by II.3.10, $0 = d_F(T_1, T_2) = d(V_1, V_2)$.

Thus the distance $d(V_1, V_2)$ satisfies all the conditions to qualify as a distance-

function in a metric space. The class $\mathfrak{K}(\mathcal{P}_0, P^*)$ becomes a metric space in this manner, and we shall speak from now on of the space $\mathfrak{K}(\mathcal{P}_0, P^*)$ of the F-varieties of type $\mathcal{P}_0$ in the metric space P^*. In accordance with I.2.36, we set up the following definitions. If $V_n \in \mathfrak{K}(\mathcal{P}_0, P^*)$, $V \in \mathfrak{K}(\mathcal{P}_0, P^*)$, then the sequence V_n is said to converge to V if and only if $d(V_n, V) \to 0$. If this is the case, then we shall write $V_n \to V$. According to I.2.12, the following statements hold.

(α) If $V_n \to V'$, $V_n \to V''$, then $V' = V''$.

(β) If $V_n = V$, $n = 1, 2, \cdots$, then $V_n \to V$.

(γ) If $V_n \to V$, then every infinite subsequence of the sequence V_n also converges to V.

(δ) If $V_n' \to V$, $V_n'' \to V$, then the sequence $V_1', V_1'', \cdots, V_n', V_n'', \cdots$ also converges to V.

On the other hand, the space $\mathfrak{K}(\mathcal{P}_0, P^*)$ is obviously not compact.

II.3.16. Given $V_n : T_n(\mathcal{P}_n) = \mathcal{P}_n^*$, $V : T(\mathcal{P}) = \mathcal{P}^*$ in the space $\mathfrak{K}(\mathcal{P}_0, P^*)$, suppose that $V_n \to V$. Then there exist representations $V_n : T^n(\mathcal{P}) = \mathcal{P}_n^*$, such that T^n converges on $\mathcal{P}$ uniformly to T.

PROOF. By assumption

$$d(V_n, V) = d_F(T_n, T) \to 0 \qquad \text{for } n \to \infty. \tag{1}$$

In view of the definition of d_F, we have for each n a homeomorphism $H_n(\mathcal{P}) = \mathcal{P}_n$ such that

$$\rho[T(x), T_nH_n(x)] < d_F(T_n, T) + 1/n, \qquad x \in \mathcal{P}. \tag{2}$$

Let us put $T^n = T_nH_n$. Then $T^n(\mathcal{P}) = \mathcal{P}_n^*$ is a representation of V_n by II.3.11, and by (1) and (2) clearly T^n converges on $\mathcal{P}$ uniformly to T.

II.3.17. Conversely, if $V_n, V \in \mathfrak{K}(\mathcal{P}_0, P^*)$, and V_n, V admit of representations $V_n : T^n(\mathcal{P}) = \mathcal{P}_n^*$, $V : T(\mathcal{P}) = \mathcal{P}^*$, such that T^n converges on $\mathcal{P}$ uniformly to T, then $V_n \to V$. Indeed, $d(V_n, V) = d_F(T^n, T) \leqq \max \rho[T^n(x), T(x)]$, $x \in \mathcal{P}$. Hence $d(V_n, V) \to 0$.

II.3.18. If $V_1, V_2 \in \mathfrak{K}(\mathcal{P}_0, P^*)$, δ is a positive number, and $V_1 : T_1(\mathcal{P}_1) = \mathcal{P}_1^*$ is any representation of V_1, then the inequality $d(V_1, V_2) < \delta$ holds if and only if V_2 has a representation $V_2 : T_2(\mathcal{P}_1) = \mathcal{P}_2^*$ such that $\rho[T_2(x_1), T_1(x_1)] < \delta$ for $x_1 \in \mathcal{P}_1$. The proof is obvious.

II.3.19. If $V \in \mathfrak{K}(\mathcal{P}_0, P^*)$, and $V : T_1(\mathcal{P}_1) = \mathcal{P}^*$, $V : T_2(\mathcal{P}_2) = \mathcal{P}^*$ are two representations of V such that $T_1 \sim T_2(ts)$ (cf. II.1.26), then the representations will be termed *topologically similar*. Using this terminology, the statements in II.3.12, II.3.13 may be reworded in the following obvious manner. If $V_1 : T_1(\mathcal{P}_1) = \mathcal{P}_1^*$, $V_2 : T_2(\mathcal{P}_2) = \mathcal{P}_2^*$ are two F-varieties in $\mathfrak{K}(\mathcal{P}_0, P^*)$, then $V_1 = V_2$ if and only if, for every $\epsilon > 0$, V_2 admits of a representation $V_2 : T_\epsilon^2(\mathcal{P}_1) = \mathcal{P}_2^*$ which is topologically similar to the representation $V_2 : T_2(\mathcal{P}_2) = \mathcal{P}_2^*$ and satisfies the condition $\rho[T_1(x_1), T_\epsilon^2(x_1)] < \epsilon$, $x_1 \in \mathcal{P}_1$.

II.3.20. An F-variety $V \in \mathfrak{K}(\mathcal{P}_0, P^*)$ gives rise to several geometrical objects that are of importance in the sequel. Let

(1) $$V \cdot T(\mathcal{P}) = \mathcal{P}^*$$

be a representation of V. Then $\mathcal{P}$, $\mathcal{P}^*$ are Peano spaces, and hence T admits of a monotone-light factorization (see II.1.18)

$$T = LM, \qquad M(\mathcal{P}) = \mathfrak{M}, \qquad L(\mathfrak{M}) = \mathcal{P}^*.$$

Assuming that $\mathfrak{M}$ is not a dendrite, let

$$\Delta(T) = [T_1, \cdots, T_n, \cdots]$$

be the cyclic decomposition of T (see II.2.94). Then each one of the cyclic partial transformations T_n is of the form $T_n(\mathcal{P}) = \mathcal{P}_n^*$, where $\mathcal{P}_n^*$ is a subset of $\mathcal{P}^*$. Hence the F-variety V_n defined by the representation

$$V_n : T_n(\mathcal{P}) = \mathcal{P}_n^*$$

belongs to $\mathfrak{K}(\mathcal{P}_0, P^*)$. We define the *cyclic decomposition* $\Delta(V)$ *of* V by the formula (cf. I.2.4)

$$\Delta(V) = [V_1, \cdots, V_n, \cdots].$$

Of course, the sequence V_n may consist of a single term. There arises the question as to the dependence of these various entities upon the choice of the representation (1). As regards $\mathcal{P}^*$, we already noted in II.3.11 that it is independent of the representation. The space $\mathcal{P}^*$ will be termed the *end-space* associated with V. Let now

(2) $$V : T^1(\mathcal{P}^1) = \mathcal{P}^*$$

be a second representation of V, and let

$$T^1 = L^1M^1, \qquad M^1(\mathcal{P}^1) = \mathfrak{M}^1, \qquad L^1(\mathfrak{M}^1) = \mathcal{P}^*$$

be a monotone-light factorization of T^1. Then $T \sim T^1(F)$, and hence the middle-spaces $\mathfrak{M}$ and $\mathfrak{M}^1$ are homeomorphic by II.1.28. Furthermore, by II.2.97, $\Delta(T) \sim \Delta(T^1)(F)$. That is, the cyclic decomposition $\Delta(T^1)$ of T^1 can be written in the form

$$\Delta(T^1) = [T_1^1, \cdots, T_n^1, \cdots]$$

in such a manner that $T_n \sim T_n^1(F)$ for every n. Hence if V_n^1 is the F-variety determined by the transformation T_n^1, then $V_n^1 = V_n$ for every n. In other words, the representation (2) yields the same cyclic decomposition for V as the representation (1).

We assumed that the middle-space $\mathfrak{M}$ is not a dendrite. If $\mathfrak{M}$ is a dendrite, then we define $\Delta(V)$ as an empty sequence of F-varieties, and write $\Delta(V) = 0$. Since by II.1.28 the middle-spaces $\mathfrak{M}$, $\mathfrak{M}^1$ are homeomorphic, it follows that $\Delta(V)$ is independent of the choice of a representation for V in all cases.

Summing up, we have the following statements. The end-space $\mathcal{P}^*$ depends only upon V. The initial space $\mathcal{P}$ is homeomorphic with the fixed Peano space $\mathcal{P}_0$, and can be chosen as any Peano space homeomorphic with $\mathcal{P}_0$ (see II.3.11).

The middle-space $\mathfrak{M}$ is determined by V up to a homeomorphism, and if only the topological character of $\mathfrak{M}$ matters in a certain situation, then we can speak of *the middle-space* without specifying a particular representation of V.

II.3.21. A representation $V : T(\mathcal{P}) = \mathcal{P}^*$ is termed *nondegenerate* if the transformation T is light (see II.1.1). An F-variety $V \in \mathfrak{K}(\mathcal{P}_0, P^*)$ is nondegenerate if it admits of a nondegenerate representation.

An F-variety V will be termed *simple* if the associated middle-space (cf. II.3.20) consists of a single proper cyclic element (that is, if the middle-space is a nondegenerate cyclic Peano space). If $\Delta(V) \neq 0$, then clearly the F-varieties occurring in $\Delta(V)$ are simple (cf. II.2.94).

II.3.22. If $\mathcal{P}_0$ is a 1-cell, then the F-varieties $V \in \mathfrak{K}(\mathcal{P}_0, P^*)$ will be termed *F-curves of the type of the* 1-*cell in P^**. Let us then assume that $\mathcal{P}_0$ is a 1-cell, and let $V : T(\mathcal{P}) = \mathcal{P}^*$ be such an F-curve in P^*. Suppose that the end-space $\mathcal{P}^*$ does not reduce to a single point, and let $T = LM$, $M(\mathcal{P}) = \mathfrak{M}$, $L(\mathfrak{M}) = \mathcal{P}^*$ be a monotone-light factorization of T. Then $\mathfrak{M}$ is a simple arc by II.1.33, and hence we have a topological transformation $h(I) = \mathfrak{M}$, where I is the unit interval $0 \leqq u \leqq 1$. Then $Lh \sim T(F)$ by II.1.61, and hence $Lh(I) = \mathcal{P}^*$ is a representation of V. Setting $Lh = \overline{T}$, clearly $\overline{T}$ is light. The following statement is thus established.

THEOREM. *If $V : T(\mathcal{P}) = \mathcal{P}^*$ is an F-curve of the type of the* 1-*cell in the metric space P^*, such that the end-space $\mathcal{P}^*$ does not reduce to a single point, then V is nondegenerate* (see II.3.21), *and admits of a representation $V : \overline{T}(I) = \mathcal{P}^*$, where $\overline{T}$ is light and I is the unit interval* $0 \leqq u \leqq 1$.

II.3.23. Let us now assume that $\mathcal{P}_0$ is a 1-sphere (simple closed curve). Then the F-varieties $V \in \mathfrak{K}(\mathcal{P}_0, P^*)$ will be termed *F-curves of the type of the* 1-*sphere in P^**. An argument entirely analogous to that used in II.3.22 (the reference to II.1.61, II.1.33 being replaced by a reference to II.1.62, II.1.34) yields the following result.

THEOREM. *Let $V : T(\mathcal{P}) = \mathcal{P}^*$ be an F-curve of the type of the* 1-*sphere in the metric space P^*, such that the end-space $\mathcal{P}^*$ does not reduce to a single point. Then V is nondegenerate (see* II.3.21), *and admits of a representation $V : \overline{T}(C) = \mathcal{P}^*$, where C is the unit circle $u^2 + v^2 = 1$, and $\overline{T}$ is light.*

II.3.24. Let us now assume that $\mathcal{P}_0$ is a 2-sphere. Then the F-varieties $V \in \mathfrak{K}(\mathcal{P}_0, P^*)$ will be termed *F-surfaces of the type of the* 2-*sphere in P^**. An argument entirely analogous to that used in II.3.22 (the reference to II.1.61, II.1.33 being replaced by a reference to II.1.63, II.1.35) yields the following result.

THEOREM. *Let $V : T(\mathcal{P}) = \mathcal{P}^*$ be an F-surface of the type of the* 2-*sphere in the metric space P^*. Suppose that V is simple* (*see* II.3.21). *Then V is nondegenerate* (*see* II.3.21) *and admits of a representation $V : \overline{T}(U) = \mathcal{P}^*$, where U is the unit sphere $u^2 + v^2 + w^2 = 1$ and $\overline{T}$ is light.*

II.3.25. THEOREM. *Let* $V : T(\mathcal{P}) = \mathcal{P}^*$ *be an F-surface of the type of the 2-sphere in* P^* *(see* II.3.24*), such that* $0 \neq \Delta(V) = [V_1, \cdots, V_n, \cdots]$ *(see* II.3.20*). Then each* V_n *is a simple F-surface of the type of the 2-sphere in* P^**, and admits of a representation* $V_n : \bar{T}_n(U) = \mathcal{P}_n^* \subset \mathcal{P}^*$*, where U is the unit sphere* $u^2 + v^2 + w^2 = 1$ *and* $\bar{T}_n$ *is light.*

PROOF. Let $T = LM$, $M(\mathcal{P}) = \mathfrak{M}$, $L(\mathfrak{M}) = \mathcal{P}^*$ be a monotone-light factorization of T. The assumption $\Delta(V) \neq 0$ implies that $\mathfrak{M}$ is not a dendrite. Let $\mathfrak{C}_1, \cdots, \mathfrak{C}_n, \cdots$ be the proper cyclic elements of $\mathfrak{M}$. Then $\Delta(T) = [T_1, \cdots, T_n, \cdots]$,where $T_n = T \mid \mathfrak{C}_n$ (see II.2.100). On setting $L(\mathfrak{C}_n) = \mathcal{P}_n^*$, we have $\mathcal{P}_n^* \subset \mathcal{P}^*$, and $T_n(\mathcal{P}) = \mathcal{P}_n^*$. By definition

$$\Delta(V) = [V_1, \cdots, V_n, \cdots],$$

where V_n is given by

$$V_n : T_n(\mathcal{P}) = \mathcal{P}_n^* .$$

Now T_n is a transformation with the middle-space $\mathfrak{C}_n$. Hence V_n is simple. The last part of the theorem follows now from II.3.24.

II.3.26. Let us now assume that $\mathcal{P}_0$ is a 2-cell. Then the F-varieties $V \in \mathfrak{K}(\mathcal{P}_0, P^*)$ will be termed *F-surfaces of the type of the* 2*-cell in* P^*. Let $V : T(\mathcal{P}) = \mathcal{P}^*$ be such an F-surface, and let $T = LM$, $M(\mathcal{P}) = \mathfrak{M}$, $L(\mathfrak{M}) = \mathcal{P}^*$ be a monotone-light factorization of T. By II.2.91, a proper cyclic element $\mathfrak{C}$ of $\mathfrak{M}$ is either a 2-cell or a 2-sphere. As a consequence, no theorem comparable in simplicity to that in II.3.25 is available in this case. For this reason, we state the relevant results in several separate theorems.

II.3.27. THEOREM. *Let* $V : T(\mathcal{P}) = \mathcal{P}^*$ *be an F-surface of the type of the* 2*-cell in the metric space* P^**, and let* $T = LM$, $M(\mathcal{P}) = \mathfrak{M}$, $L(\mathfrak{M}) = \mathcal{P}^*$ *be a monotone-light factorization of T. Suppose that the middle-space* $\mathfrak{M}$ *is a* 2*-cell. Then V is nondegenerate, and there exists a representation* $V : \bar{T}(Q) = \mathcal{P}^*$*, where Q is the unit square* $0 \leqq u \leqq 1$, $0 \leqq v \leqq 1$*, and* $\bar{T}$ *is light.*

The proof is entirely analogous to that in II.3.22, the reference to II.1.61 being replaced by a reference to II.1.64.

II.3.28. THEOREM. *Let* $V : T(\mathcal{P}) = \mathcal{P}^*$ *be an F-surface of the type of the* 2*-cell in the metric space* P^**, and let* $T = LM$, $M(\mathcal{P}) = \mathfrak{M}$, $L(\mathfrak{M}) = \mathcal{P}^*$ *be a monotone-light factorization of T. Suppose that the middle-space* $\mathfrak{M}$ *is a* 2*-sphere. Then there exists a representation* $V : \bar{T}(C) = \mathcal{P}^*$ *with the following properties.*

(a) *C is the unit disc* $u^2 + v^2 \leqq 1$.

(b) $\bar{T}$ *carries the perimeter of C into a single point* x_0^* of $\mathcal{P}^*$.

(c) $\bar{T}$ *is not constant on any nondegenerate continuum in* $u^2 + v^2 < 1$.

PROOF. By II.1.42, M carries the boundary curve of the 2-cell $\mathcal{P}$ into a single point $\mathfrak{a}_0$ of $\mathfrak{M}$. Hence T carries the perimeter of C into the single point $x_0^* = L(\mathfrak{a}_0)$. In view of II.1.21, we can assume without loss of generality that $\mathfrak{M}$

coincides with the unit sphere $x^2 + y^2 + z^2 = 1$ and $\mathfrak{a}_0$ coincides with the north pole (0, 0, 1). Clearly, there exists a continuous transformation $\bar{M}(C) = \mathfrak{M}$ with the following properties: (α) $\bar{M}$ carries the perimeter of C into the single point $\mathfrak{a}_0$. (β) $\bar{M}$ maps the interior of C topologically onto $\mathfrak{M} - \mathfrak{a}_0$. Let us put $\bar{T} = L\bar{M}$. Obviously, the conditions (a), (b), (c) are then satisfied. Since $\bar{T} \sim T(F)$ by II.1.65, it follows that $\bar{T}(C) = \mathcal{P}^*$ is a representation of V with the desired properties.

II.3.29. THEOREM. *Let $V : T(\mathcal{P}) = \mathcal{P}^*$ be an F-surface of the type of the 2-cell in the metric space P^*, and let $T = LM$, $M(\mathcal{P}) = \mathfrak{M}$, $L(\mathfrak{M}) = \mathcal{P}^*$ be a monotone-light factorization of T. Then V is nondegenerate if and only if $\mathfrak{M}$ is a 2-cell.*

PROOF. If $\mathfrak{M}$ is a 2-cell, then V is nondegenerate by II.3.27. Suppose, conversely, that V is nondegenerate. Then V admits of a representation $V : T_1(\mathcal{P}_1) = \mathcal{P}^*$, where T_1 is light and $\mathcal{P}_1$ is a 2-cell. Thus T_1 admits of the monotone-light factorization $T_1 = T_1 I_1$, where I_1 is the identity transformation on $\mathcal{P}_1$. The middle-space in this factorization is $\mathcal{P}_1$. Since $T \sim T_1(F)$, it follows by II.1.28 that $\mathfrak{M}$ and $\mathcal{P}_1$ are homeomorphic. Thus $\mathfrak{M}$ is a 2-cell.

II.3.30. THEOREM. *Let $V : T(\mathcal{P}) = \mathcal{P}^*$ be an F-surface of the type of the 2-cell in the metric space P^*. Let $I_C(T)$ denote the collection of the components of all the sets in $\mathcal{P}$ of the form $T^{-1}(x^*)$, $x^* \in \mathcal{P}^*$ (see* II.1.13(ii)*). Then V is nondegenerate if and only if the following conditions hold.* (i) *T is not constant on the boundary curve of the 2-cell $\mathcal{P}$.* (ii) *If γ is any continuum of the collection $I_C(T)$, then $\mathcal{P} - \gamma$ is connected.*

PROOF. Let $T = LM$, $M(\mathcal{P}) = \mathfrak{M}$, $L(\mathfrak{M}) = \mathcal{P}^*$ be a monotone-light factorization of T. By II.3.29, V is nondegenerate if and only if $\mathfrak{M}$ is a 2-cell. By II.1.19, $I_C(T) = I(M)$. By II.1.44, $\mathfrak{M}$ is a 2-cell if and only if the collection $I(M) = I_C(T)$ satisfies the conditions (i), (ii) stated above.

II.3.31. Let us return to a general class $\mathfrak{K}(\mathcal{P}_0, P^*)$ (see II.3.11). An F-variety $V \in \mathfrak{K}(\mathcal{P}_0, P^*)$ will be termed *topological* if and only if V admits of a representation $V : H(\mathcal{P}) = \mathcal{P}^*$ where H is a homeomorphism. Since $\mathcal{P}$ is homeomorphic with $\mathcal{P}_0$, it follows that the end-space $\mathcal{P}^*$ is also homeomorphic with $\mathcal{P}_0$. If $I^*(\mathcal{P}^*) = \mathcal{P}^*$ is the identity transformation on $\mathcal{P}^*$, then it follows from II.3.11 that V has the representation $V : I^*(\mathcal{P}^*) = \mathcal{P}^*$. The following statements are now obvious.

(a) A topological F-variety $V \in \mathfrak{K}(\mathcal{P}_0, P^*)$ is univocally determined by its end-space (which is necessarily homeomorphic with $\mathcal{P}_0$).

(b) If V is a topological F-variety in the class $\mathfrak{K}(\mathcal{P}_0, P^*)$, and if $\mathcal{P}$ is any Peano space homeomorphic with $\mathcal{P}_0$, then V admits of a representation of the form $V : H(\mathcal{P}) = \mathcal{P}^*$, where H is a homeomorphism.

(c) An F-variety $V \in \mathfrak{K}(\mathcal{P}_0, P^*)$ with end-space $\mathcal{P}^*$ is topological if and only if it admits of the representation $V : I^*(\mathcal{P}^*) = \mathcal{P}^*$, where I^* is the identity transformation on $\mathcal{P}^*$.

II.3.32. THEOREM. *Let $V : T(\mathcal{P}) = \mathcal{P}^*$ be an arbitrary representation of a topological F-variety $V \in \mathfrak{R}(\mathcal{P}_0, P^*)$. Then T is monotone.*

PROOF. By II.3.31, V has the representation $V : I^*(\mathcal{P}^*) = \mathcal{P}^*$, where I^* is the identity transformation on $\mathcal{P}^*$. Since $I^* = I^*I^*$ is a monotone-light factorization of I^* and $T \sim I^*(F)$, it follows (see II.1.28) that T has a monotone-light factorization of the form $T = I^*M$, where M is monotone. Thus $T = M$, and the theorem is proved.

II.3.33. The converse of the preceding result is known to hold only in a few very special but very important cases which we shall consider presently Let us first introduce some further terminological conventions.

Let us consider a topological F-variety $V \in \mathfrak{R}(\mathcal{P}_0, P^*)$ with end-space $\mathcal{P}^*$. Then $\mathcal{P}^*$ is homeomorphic with $\mathcal{P}_0$. Conversely, if $\mathcal{P}^*$ is any Peano subspace of P^* homeomorphic with $\mathcal{P}_0$, and $I^*(\mathcal{P}^*) = \mathcal{P}^*$ is the identity transformation on $\mathcal{P}^*$, then $V : I^*(\mathcal{P}^*) = \mathcal{P}^*$ is the (unique) topological F-variety in $\mathfrak{R}(\mathcal{P}_0, P^*)$ with end-space $\mathcal{P}^*$ (see II.3.31). Thus there is a biunique correspondence between the topological F-varieties $V \in \mathfrak{R}(\mathcal{P}_0, P^*)$ and those Peano subspaces $\mathcal{P}^* \subset P^*$ that are homeomorphic with $\mathcal{P}_0$.

If $\mathcal{P}^*$ is a Peano subspace of P^* that is homeomorphic with $\mathcal{P}_0$, then a continuous transformation $T(\mathcal{P}) = \mathcal{P}^*$ will be termed an *F-representation of* $\mathcal{P}^*$ if and only if $T(\mathcal{P}) = \mathcal{P}^*$ is a representation of the unique topological F-variety in $\mathfrak{R}(\mathcal{P}_0, P^*)$ with end-space $\mathcal{P}^*$. This definition implies that $\mathcal{P}$ is homeomorphic with $\mathcal{P}^*$. In view of II.3.31, we have the following equivalent wording of this definition. A continuous transformation $T(\mathcal{P}) = \mathcal{P}^*$ is an F-representation of $\mathcal{P}^*$ if and only if $T \sim I^*(F)$, where I^* is the identity transformation on $\mathcal{P}^*$. In view of II.3.32, every F-representation of $\mathcal{P}^*$ is monotone, but the converse is known to hold only in a few special cases.

II.3.34. THEOREM. *Let $\mathcal{P}^*$ be a simple arc in a metric space P^*. Then a continuous transformation $T(\mathcal{P}) = \mathcal{P}^*$ is an F-representation of $\mathcal{P}^*$ if and only if $\mathcal{P}$ is a simple arc and T is monotone.*

PROOF. The necessity follows from II.3.32. Suppose now that $\mathcal{P}$ is a simple arc and T is monotone. Let $I^*(\mathcal{P}^*) = \mathcal{P}^*$ be the identity transformation on $\mathcal{P}^*$. Then $I^* \sim I^*T(F)$ by II.1.66 (with I^*, T replacing the T, M_* in II.1.66). Thus $I^*T = T$ is an F-representation of $\mathcal{P}^*$.

II.3.35. THEOREM. *Let $\mathcal{P}^*$ be a simple closed curve in a metric space P^*. Then a continuous transformation $T(\mathcal{P}) = \mathcal{P}^*$ is an F-representation of $\mathcal{P}^*$ if and only if $\mathcal{P}$ is a simple closed curve and T is monotone.*

The proof is the same as in II.3.34.

II.3.36. THEOREM. *Let $\mathcal{P}^*$ be a 2-cell (2-sphere) in a metric space P^*. Then a continuous transformation $T(\mathcal{P}) = \mathcal{P}^*$ is an F-representation of $\mathcal{P}^*$ if and only if $\mathcal{P}$ is a 2-cell (2-sphere) and T is monotone.*

The proof is the same as in II.3.34. We stated the last three theorems separately for convenience of wording and reference. While the theorems in II.3.34, II.3.35 are practically trivial, the twin theorems of this section lie much deeper.

II.3.37. Let $\mathcal{P}_0$ be a fixed Peano space, and P^* a fixed metric space. Let us consider in P^* two Peano subspaces $\mathcal{P}_1^*$, $\mathcal{P}_2^*$ that are homeomorphic with $\mathcal{P}_0$. Then $\mathcal{P}_1^*$, $\mathcal{P}_2^*$ are the end-spaces of (univocally determined) topological F-varieties V_1, V_2 of the class $\mathfrak{R}(\mathcal{P}_0, P^*)$ (see II.3.31). Let us define the Fréchet distance $d_F(\mathcal{P}_1^*, \mathcal{P}_2^*)$ of the subspaces $\mathcal{P}_1^*$, $\mathcal{P}_2^*$ by the formula

$$(1) \qquad d_F(\mathcal{P}_1^*, \mathcal{P}_2^*) = d(V_1, V_2)$$

From II.3.15 it follows that $d_F(\mathcal{P}_1^*, \mathcal{P}_2^*)$ possesses all the properties of a distance-function in a metric space (see I.2.10). On introducing the distance $d_F(\mathcal{P}_1^*, \mathcal{P}_2^*)$, the class of all those Peano subspaces of P^* that are homeomorphic with $\mathcal{P}_0$ becomes a metric space. If $\mathcal{P}_n^*$, $\mathcal{P}^*$ are Peano subspaces of P^* that are homeomorphic with $\mathcal{P}_0$, and $d_F(\mathcal{P}_n^*, \mathcal{P}^*) \to 0$, then we shall say that the sequence $\mathcal{P}_n^*$ converges to $\mathcal{P}^*$ in the Fréchet sense. From II.3.15, II.3.31 we infer that the definition given by formula (1) can be restated as follows. Let $\mathcal{P}_1^*$, $\mathcal{P}_2^*$ be two Peano subspaces that are homeomorphic with $\mathcal{P}_0$. Let $H(\mathcal{P}_1^*) = \mathcal{P}_2^*$ be a homeomorphism. Let us put $\delta(\mathcal{P}_1^*, \mathcal{P}_2^*, H) = \max \rho[x_1^*, H(x_1^*)]$, $x_1^* \in \mathcal{P}_1^*$. Then $d_F(\mathcal{P}_1^*, \mathcal{P}_2^*) = \text{gr.l.b. } \delta(\mathcal{P}_1^*, \mathcal{P}_2^*, H)$, where the greatest lower bound is taken with respect to all homeomorphisms $H(\mathcal{P}_1^*) = \mathcal{P}_2^*$.

Important special cases are obtained by choosing P^* as the Euclidean three-space E_3, and then choosing $\mathcal{P}_0$ as a simple arc, a simple closed curve, a 2-cell, and a 2-sphere respectively. We obtain thus four metric spaces, the first one being the space of all simple arcs in E_3, where the distance of two simple arcs γ_1, γ_2 is defined as their Fréchet distance $d_F(\gamma_1, \gamma_2)$. This important space may be termed the Fréchet space of simple arcs in E_3. Similarly, we can speak in the same sense of the Fréchet space of simple closed curves, of 2-cells, of 2-spheres, respectively, in E_3. In view of the biunique correspondence between topological F-varieties and their end-spaces, these Fréchet spaces are homeomorphic with the space of topological F-varieties of the type of the 1-cell, of the 1-sphere, of the 2-cell, of the 2-sphere, respectively, in E_3.

II.3.38. Let us consider the class $\mathfrak{R}(\mathcal{P}_0, P^*)$, where $\mathcal{P}_0$ is a 2-cell. This class consists of the F-surfaces of the type of the 2-cell in P^* (see II.3.36). Let $V \in \mathfrak{R}(\mathcal{P}_0, P^*)$, and let $V : T(\mathcal{P}) = \mathcal{P}^*$ be a representation of V. Then $\mathcal{P}$ is a 2-cell. Let C be the boundary curve of $\mathcal{P}$, and let us denote, for clarity, by T_C the transformation T thought of as operating from C only. Then $T_C(C) = C^*$ is a representation of an F-curve Γ^* of the type of the 1-sphere in P^* (see II.3.23). Now let $V : T'(\mathcal{P}') = \mathcal{P}^*$ be an arbitrary second representation of V, and let C' be the boundary curve of the 2-cell $\mathcal{P}'$. Let $T'_{C'}$ denote the transformation T' thought of as operating from C' only. Then $T'_{C'}(C') = C'^*$ is a representation of an F-curve Γ'^* of the type of the 1-sphere in P^*. We assert that $\Gamma^* = \Gamma'^*$. Indeed, let $\epsilon > 0$ be given. Since $T \sim T'(F)$, we have a homeomorphism

$H(\mathcal{P}) = \mathcal{P}'$ such that $\rho[T(x), T'H(x)] < \epsilon$ for $x \in \mathcal{P}$. Then $H(C) = C'$ is also a homeomorphism, and $\rho[T_C(x), T'_{C'}H(x)] < \epsilon$ for $x \in C$. Since ϵ was arbitrary, it follows that $T_C \sim T'_{C'}(F)$, and hence $\Gamma^* = \Gamma'^*$.

Thus Γ^* is independent of the representation chosen for V. The F-curve Γ^*, of the type of the 1-sphere, will be termed *the boundary curve of* V. Thus if $V : T(\mathcal{P}) = \mathcal{P}^*$ is an F-surface of the type of the 2-cell, then its boundary curve is determined by the representation $\Gamma^* : T_C(C) = C^*$, where T_C is T thought of as operating from the boundary C of $\mathcal{P}$ only. The set C^* is the end-space of Γ^*, and hence it is independent of the choice of the representation used.

II.3.39. CONTINUATION. An important special case arises if the boundary curve Γ^* of V is a topological F-curve. In view of II.3.35, this happens if and only if T is monotone and not constant on the boundary C of $\mathcal{P}$ (cf. II.1.36). Suppose that this is the case. Then the end-space C^* of Γ^* is a simple closed curve (see II.1.34), and Γ^* is univocally determined by C^* (see II.3.31). Since C^* is independent of the choice of the representation for V, we can therefore say, without danger of misinterpretation, that V is *bounded by the simple closed curve* C^*. This statement serves merely as an abbreviation of the statement that the boundary curve of V is the topological F-curve of the type of the 1-sphere that is univocally determined by the simple closed curve C^*.

A second important special case arises if the end-space C^* reduces to a single point x_0^*. Since C^* is independent of the choice of the representation, we can say in this case that V is *bounded by the single point* x_0^*, as an abbreviation of the statement that the boundary curve of V is the F-curve of the type of the 1-sphere that is univocally determined by the single point x_0^* as its end-space. Clearly, $V : T(\mathcal{P}) = \mathcal{P}^*$ is bounded by a single point if and only if T is constant on the boundary of the 2-cell $\mathcal{P}$.

Summing up: If $V : T(\mathcal{P}) = \mathcal{P}^*$ is an F-surface of the type of the 2-cell, then V is bounded by a simple closed curve if and only if T is monotone and not constant on the boundary of the 2-cell $\mathcal{P}$, and V is bounded by a single point if and only if T is constant on the boundary of $\mathcal{P}$.

II.3.40. Let $V : T(\mathcal{P}) = \mathcal{P}^*$ be an F-surface of the type of the 2-cell that is bounded by a simple closed curve C^* (see II.3.39). Then T is monotone and not constant on the boundary C of $\mathcal{P}$. Let now $T = LM$, $M(\mathcal{P}) = \mathfrak{M}$, $L(\mathfrak{M}) = \mathcal{P}^*$ be a monotone-light factorization of T. We propose to investigate the proper cyclic elements of $\mathfrak{M}$. The following remarks will be helpful.

(a) M is not constant on C. Indeed, if M were constant on C, then $T = LM$ would also be constant on C, in contradiction with our assumptions.

(b) M is monotone on C. Indeed, assume that this is not the case. Using the notations $I(T)$, $I(M)$, $I_C(T)$ in the sense of II.1.1, II.1.13, we would have then a continuum $\gamma \in I(M)$ such that $C\gamma$ is disconnected. Since $I(M) = I_C(T)$ by II.1.19, we have then a set $E \in I(T)$ such that γ is a component of E. Since T is monotone on C, the set CE is connected, and $CE \cdot \gamma \neq 0$ since $C\gamma \neq 0$ and $\gamma \subset E$. Hence $CE + \gamma$ is connected, and $\gamma \subset CE + \gamma \subset E$. As γ is a component of E, it follows that $\gamma = CE + \gamma$, and hence $CE \subset \gamma$. It follows that $CCE \subset C\gamma$, that is, $CE \subset C\gamma$. Since $\gamma \subset E$, we obtain finally $C\gamma = CE$. Since CE is connected, this contradicts the assumption that $C\gamma$ is disconnected.

(c) In view of (a), (b), and II.1.34 (applied to the transformation M considered on C), it follows that the image of C under M is a simple closed curve $\mathfrak{C}_0$ on $\mathfrak{M}$. Clearly, any two points of $\mathfrak{C}_0$ are conjugate in $\mathfrak{M}$, and hence $\mathfrak{C}_0$ is a subset of a certain proper cyclic element Γ_0 of $\mathfrak{M}$ (see II.2.12).

(d) The proper cyclic element Γ_0 of (c) may be the only proper cyclic element of $\mathfrak{M}$. If this is not the case, let $\Gamma_1, \cdots, \Gamma_n, \cdots$ be the remaining proper cyclic elements of $\mathfrak{M}$. Since $\mathfrak{C}_0 \subset \Gamma_0$ and $\Gamma_0\Gamma_n$, $n > 0$, contains at most one point (see II.2.12), it follows that $\mathfrak{C}_0\Gamma_n$, $n > 0$, is either empty or reduces to a single point.

(e) Now let $\alpha_n(\mathfrak{M}) = \Gamma_n$, $n = 0, 1, \cdots$, be the (unique) monotone retraction from $\mathfrak{M}$ onto Γ_n (see II.2.43). Since $\mathfrak{C}_0 \subset \Gamma_0$, we have then $\alpha_0(\mathfrak{x}) = \mathfrak{x}$ for $\mathfrak{x} \in \mathfrak{C}_0$, and hence (see II.2.100)

$$(1) \qquad \alpha_0 M(x) = M(x) \qquad \text{for } x \in C.$$

Now consider an integer $n > 0$. Since $\mathfrak{C}_0\Gamma_n$ is either empty or reduces to a single point, it follows that $\mathfrak{C}_0 - \mathfrak{C}_0\Gamma_n$ is connected, and hence $\mathfrak{C}_0 - \mathfrak{C}_0\Gamma_n$ is a subset of a component $\mathfrak{S}_n$ of $\mathfrak{M} - \Gamma_n$. Then $fr(\mathfrak{S}_n) = \mathfrak{x}_n$ is a single point (see II.2.44), and (see II.2.40) we have $\alpha_n(\mathfrak{x}) = \mathfrak{x}_n$ for $\mathfrak{x} \in \mathfrak{S}_n + \mathfrak{x}_n = c(\mathfrak{S}_n)$. Since $\mathfrak{C}_0$ is contained in $c(\mathfrak{S}_n)$ (note that $\mathfrak{C}_0 - \mathfrak{S}_n$ is empty or reduces to a single point), it follows that $\alpha_n(\mathfrak{C}_0) = \mathfrak{x}_n$. Hence

$$(2) \qquad \alpha_n M(x) = \mathfrak{x}_n \qquad \text{for } x \in C, n = 1, 2, \cdots.$$

II.3.41. CONTINUATION. Since M is monotone, Γ_n is cyclic, $\alpha_n M(\mathcal{P}) = \Gamma_n$, and $\mathcal{P}$ is a 2-cell, it follows from II.1.42 that Γ_0 is a 2-cell, while Γ_n, $n > 1$, is a 2-sphere (II.1.42 being applied to the monotone transformation $\alpha_n M$). Indeed, $\alpha_0 M$ is not constant on C in view of II.3.40(1), and $\alpha_n M$, $n > 1$, is constant on C in view of II.3.40(2) (note that $\alpha_n M$ is not constant on $\mathcal{P}$, since $\alpha_n M(\mathcal{P}) = \Gamma_n$ is a nondegenerate subset of $\mathfrak{M}$).

Summing up: on setting $M(C) = \mathfrak{C}_0$, $\mathfrak{C}_0$ is a simple closed curve. $\mathfrak{C}_0$ is a subset of a proper cyclic element Γ_0 of $\mathfrak{M}$. Γ_0 is a 2-cell, and all other proper cyclic elements of $\mathfrak{M}$, if any, are 2-spheres.

Now let us consider the cyclic decomposition $\Delta(V)$ of V. Let us put $L(\Gamma_n) = \mathcal{P}_n^*$, $n = 0, 1, \cdots$, $T_n = T \mid \Gamma_n = L\alpha_n M$. Let us denote by V_n the F-variety determined by the representation

$$(1) \qquad V_n : T_n(\mathcal{P}) = \mathcal{P}_n^*, \qquad n = 0, 1, \cdots.$$

By definition, $\Delta(V) = [V_0, V_1, \cdots, V_n, \cdots]$. In view of II.3.40(1), II.3.40(2) it follows that

$$(2) \qquad T_0(x) = L\alpha_0 M(x) = LM(x) = T(x) \qquad \text{for } x \in C,$$

$$(3) \qquad T_n(x) = L\alpha_n(\mathfrak{C}_0) = L(\mathfrak{x}_n) = x_n^* \qquad \text{for } x \in C, n > 1,$$

where x_n^* is a single point, since $\mathfrak{x}_n$ is a single point. In view of the definitions stated in II.3.39, the formulas (1), (2), (3) yield the following result.

THEOREM. *Let V be an F-surface of the type of the 2-cell in a metric space P^*. Suppose that V is bounded by a simple closed curve C^*. Then the cyclic decomposition $\Delta(V)$ of V is nonvacuous, and precisely one of the F-surfaces occurring in $\Delta(V)$ is bounded by C^*. All the other F-surfaces, if any, that occur in $\Delta(V)$ are bounded by single points.*

II.3.42. From this point on, we assume that the metric space P^* of II.3.7 coincides with the Euclidean three-space E_3. An F-variety $V \in \mathfrak{R}(\mathcal{P}_0, E_3)$ can then be represented as follows. Let us choose a Cartesian coordinate system xyz in E_3. Let $V : T(\mathcal{P}) = \mathcal{P}^*$ be a representation of V. If p is a point of $\mathcal{P}$, then its image $p^* = T(p)$ is determined by its coordinates x, y, z which are single-valued, real-valued, continuous functions $x(p), y(p), z(p)$ on $\mathcal{P}$. Then T itself is given by the formulas

$$T : x = x(p), \qquad y = y(p), \qquad z = z(p), \qquad p \in \mathcal{P}.$$

Since V is completely determined by T and hence by the functions $x(p), y(p), z(p)$, we shall write

$$V : x = x(p), \qquad y = y(p), \qquad z = z(p), \qquad p \in \mathcal{P}.$$

It will be convenient to condense such formulas by using $\mathfrak{x}$ to refer to the vector with components x, y, z. The preceding formulas appear then in the form

$$(1) \qquad T : \mathfrak{x} = \mathfrak{x}(p), \qquad p \in \mathcal{P},$$

$$(2) \qquad V : \mathfrak{x} = \mathfrak{x}(p), \qquad p \in \mathcal{P}.$$

If convenient, we shall use x_1, x_2, x_3 instead of x, y, z to refer to the Cartesian coordinates. As usual, $|\mathfrak{x}|$ will denote the length of the vector $\mathfrak{x}$, and symbols like $\mathfrak{x}_1 \pm \mathfrak{x}_2$, $c_1\mathfrak{x}_1 + c_2\mathfrak{x}_2$, and so forth, will be used in conformity with general usage.

Suppose now that a new system of Cartesian coordinates $\bar{x}_1, \bar{x}_2, \bar{x}_3$ is introduced in E_3 by means of formulas

$$(3) \qquad \bar{x}_i = c_i + a_{i1}x_1 + a_{i2}x_2 + a_{i3}x_3, \qquad i = 1, 2, 3,$$

where (a_{ij}) is an orthogonal matrix. On setting $\bar{\mathfrak{x}}(p) = [\bar{x}_1(p), \bar{x}_2(p), \bar{x}_3(p)]$, where $\bar{x}_i(p)$ is derived from $x_1(p), x_2(p), x_3(p)$ by means of the equations (3), we obtain representations for T, V in terms of the new coordinates.

On the other hand, we can interpret the formulas (3) as defining a rigid transformation in E_3. In this case, $\bar{x}_1, \bar{x}_2, \bar{x}_3$ denote the coordinates of the image of the point (x_1, x_2, x_3) in the same coordinate system x_1, x_2, x_3. If $\bar{\mathfrak{x}}(p)$ has the same meaning as above, then we obtain a new F-variety $\bar{V}$ and a new transformation $\bar{T}$ given by

$$(4) \qquad \bar{T} : \mathfrak{x} = \bar{\mathfrak{x}}(p), \qquad p \in \mathcal{P},$$

$$(5) \qquad \bar{V} : \mathfrak{x} = \bar{\mathfrak{x}}(p), \qquad p \in \mathcal{P}.$$

Now let $V_1 : T_1(\mathcal{P}_1) = \mathcal{P}_1^*$ be another F-variety in the class $\mathfrak{K}(\mathcal{P}_0, E_3)$, and let $\overline{T}_1, \overline{V}_1$ be derived from T_1, V_1 by means of (3) in the same manner as $\overline{T}, \overline{V}$ were derived from T, V. Since the transformation (3) preserves Euclidean distance, it follows readily that $d_F(T, T_1) = d_F(\overline{T}, \overline{T}_1)$. Hence (see II.3.15)

$$d(V, V_1) = d(\overline{V}, \overline{V}_1). \tag{6}$$

In particular, if $V = V_1$, it follows that $\overline{V} = \overline{V}_1$. In other words, if V and the transformation (3) are given, then $\overline{V}$ is univocally determined (that is, $\overline{V}$ is independent of the choice of a representation for V). We shall say that $\overline{V}$, given by (5), is the transform of V under the rigid transformation (3). Two F-varieties related in this manner will be termed *congruent.*

II.3.43. Consider the F-varieties $V \in \mathfrak{K}(\mathcal{P}_0, E_3)$, where E_3 is the Euclidean three-space. If $\mathcal{P}$ is any Peano space homeomorphic with $\mathcal{P}_0$, then by II.3.11 we have for each $V \in \mathfrak{K}(\mathcal{P}_0, E_3)$ a representation $V : T(\mathcal{P}) = \mathcal{P}^*$. In view of the definitions and conventions stated in II.3.42, we have the following notations at our disposal in the cases to be enumerated presently.

(a) *F-curves of the type of the* 1*-cell.* We can choose $\mathcal{P}$ as any linear interval $a \leqq u \leqq b$, and we have then for such a curve C a representation of the form

$$C : \mathfrak{x} = \mathfrak{x}(u), \qquad a \leqq u \leqq b.$$

More generally, if γ is any simple arc, then we have for C a representation of the form $C : \mathfrak{x} = \mathfrak{x}(p)$, $p \in \gamma$.

(b) *F-curves of the type of the* 1*-sphere.* We can choose $\mathcal{P}$ as any simple closed curve γ, and then we have a representation of the form $C : \mathfrak{x} = \mathfrak{x}(p)$, $p \in \gamma$.

(c) *F-surfaces of the type of the* 2*-cell.* If S is such a surface, and $\mathfrak{R}$ is any bounded, simply-connected Jordan region in a Euclidean uv-plane, then we have a representation of the form

$$S : \mathfrak{x} = \mathfrak{x}(u, v), \qquad (u, v) \in \mathfrak{R}.$$

If it is desirable to display the components of $\mathfrak{x}$, then we shall write

$$S : \mathfrak{x} = [x(u, v), y(u, v), z(u, v)], \qquad (u, v) \in \mathfrak{R}.$$

Using the complex variable $w = u + iv$, we have representations of the alternative form $S : \mathfrak{x} = \mathfrak{x}(w)$, $w \in \mathfrak{R}$, $S : \mathfrak{x} = [x(w), y(w), z(w)]$, $w \in \mathfrak{R}$.

(d) *F-surfaces of the type of the* 2*-sphere.* If S is such a surface, and $\mathcal{P}$ is any 2-sphere, then we have a representation of the form $S : \mathfrak{x} = \mathfrak{x}(p)$, $p \in \mathcal{P}$.

II.3.44. *We shall be concerned especially with F-surfaces S of the type of the 2-cell in Euclidean three-space. To simplify our statements, such a surface will be termed simply a surface, without further qualifications.* Thus the class of surfaces, in this sense, is a metric space, the distance of two surfaces S_1, S_2 being defined as the quantity $d(V_1, V_2)$ of II.3.15, where the general F-varieties V_1, V_2 are replaced by the special F-varieties S_1, S_2. In the same manner, we agree to use the term *topological surface* as an abbreviation for topological F-variety of the type of the 2-cell in Euclidean three-space (cf. II.3.31).

The following remark will be useful. Given a surface

(1) $$S : \mathfrak{x} = \mathfrak{x}(u, v), \qquad (u, v) \in \mathfrak{R},$$

and an $\epsilon > 0$, there exists a simply-connected Jordan region $\mathfrak{R}_* \subset \mathfrak{R}^0$ such that the surface

(2) $$S_* : \mathfrak{x} = \mathfrak{x}(u, v), \qquad (u, v) \in \mathfrak{R}_*$$

satisfies the inequality (see II.3.15)

(3) $$d(S, S_*) < \epsilon.$$

Proof. *Case* (i). $\mathfrak{R}$ is the circular disc $0 \leqq u^2 + v^2 \leqq 1$. For $0 < \rho < 1$, let us denote by S_ρ the surface

(4) $$S_\rho : \mathfrak{x} = \mathfrak{x}(u, v), \qquad 0 \leqq u^2 + v^2 \leqq \rho^2.$$

The formulas $u' = u/\rho$, $v' = v/\rho$ define then a topological transformation from the disc $0 \leqq u^2 + v^2 \leqq \rho^2$ onto the unit disc. By II.3.11 we have therefore for S the representation

(5) $$S : \mathfrak{x} = \mathfrak{x}\,(u/\rho, v/\rho), \qquad 0 \leqq u^2 + v^2 \leqq \rho^2.$$

In view of the continuity of $\mathfrak{x}(u, v)$, clearly $|\, \mathfrak{x}(u, v) - \mathfrak{x}(u/\rho, v/\rho) \,| < \epsilon$ for $0 \leqq u^2 + v^2 \leqq \rho^2$ if $1 - \rho$ is sufficiently small. If ρ is so chosen, then (4) and (5) show that $d(S, S_\rho) < \epsilon$. Thus the disc $0 \leqq u^2 + v^2 \leqq \rho^2$ may be chosen as $\mathfrak{R}_*$ if ρ is sufficiently close to unity.

Case (ii). In the general case, we have by I.2.50 a topological transformation τ that maps $\mathfrak{R}$ onto the unit disc $0 \leqq \alpha^2 + \beta^2 \leqq 1$ in an auxiliary $\alpha\beta$-plane. If Δ_ρ denotes the disc $0 \leqq \alpha^2 + \beta^2 \leqq \rho^2$, then it follows readily from our discussion in case (i) that the region $\mathfrak{R}_* = \tau^{-1}(\Delta_\rho)$ satisfies our requirements if ρ is sufficiently close to unity.

II.3.45. Let $f(u, v)$ be a real-valued continuous function on a bounded, simply-connected Jordan region $\mathfrak{R}$ in the uv-plane. Let $\mathfrak{D}$ be a domain (connected open set) in the interior $\mathfrak{R}^0$ of $\mathfrak{R}$, and let us denote by M, M^0 the maximum of $f(u, v)$ on $c(\mathfrak{D})$, $\mathfrak{D}$ respectively. Similarly, m, m^0 will denote the minimum of $f(u, v)$ on $c(\mathfrak{D})$, $\mathfrak{D}$ respectively. Clearly $m \leqq m^0 \leqq M^0 \leqq M$. If $m = m^0$ and $M = M^0$ for every choice of $\mathfrak{D}$ in $\mathfrak{R}^0$, then $f(u, v)$ is termed *Lebesgue monotone* in $\mathfrak{R}$.

Theorem. *Let* $S : \mathfrak{x} = [x(u, v), y(u, v), z(u, v)]$, $(u, v) \in \mathfrak{R}$ *be a surface (see* II.3.44), *and suppose that the following conditions hold.* (i) *The coordinate functions* $x(u, v)$, $y(u, v)$, $z(u, v)$ *are Lebesgue monotone in* $\mathfrak{R}$. (ii) S *is bounded by a simple closed curve. Then* S *is nondegenerate.*

Proof. Let us denote by T the transformation $T : \mathfrak{x} = [x(u, v), y(u, v), z(u, v)]$, $(u, v) \in \mathfrak{R}$. Let $T = LM$, $M(\mathfrak{R}) = \mathfrak{M}$, $L(\mathfrak{M}) = \mathcal{P}^*$ be a monotone-light factorization of T. If C is the boundary curve of $\mathfrak{R}$, then by II.3.40(c) the image of C under M is a simple closed curve $\mathfrak{C}_0$ on $\mathfrak{M}$, and M is monotone on C. Clearly, no cut point of $\mathfrak{M}$ lies on $\mathfrak{C}_0$. Suppose that $\mathfrak{M}$ has a cut point $\mathfrak{x}_0 \notin \mathfrak{C}_0$, and put $\gamma = M^{-1}(\mathfrak{x}_0)$. Then γ is a continuum in $\mathfrak{R}$ such that $C\gamma = 0$, and by II.1.5 it follows that $\mathfrak{R} - \gamma$ has at least two components. Since $C\gamma = 0$, it follows that C and all points of $\mathfrak{R}$ sufficiently close to C are contained in the same component of $\mathfrak{R} - \gamma$. Hence $\mathfrak{R} - \gamma$ must have a component $\mathfrak{D}$ such that $c(\mathfrak{D}) \subset \mathfrak{R}^0$. Then clearly $fr(\mathfrak{D}) \subset \gamma$, and since obviously T is constant on γ, it follows that $x(u, v)$, $y(u, v)$, $z(u, v)$ are constant on $fr(\mathfrak{D})$. Since these functions are Lebesgue monotone, it follows that they are constant on $\mathfrak{D}$ also. Thus T itself is constant on $\mathfrak{D}$, and hence on $\gamma + \mathfrak{D}$. Since $\gamma + \mathfrak{D}$ is connected, it follows that $\gamma + \mathfrak{D}$ is a subset of a component of a set $E \in I(T)$, and hence $\gamma + \mathfrak{D}$ is a subset of a set $F \in I(M)$ (see II.1.19). Since $\gamma \subset \gamma + \mathfrak{D}$ and $\gamma \in I(M)$, it follows that $\gamma + \mathfrak{D} = \gamma$, in contradiction with the fact that $\mathfrak{D}$ is a component of $\mathfrak{R} - \gamma$. It is thus established that $\mathfrak{M}$ has no cut point. By II.3.41 it follows that $\mathfrak{M}$ is a 2-cell. By II.3.29 it follows that S is nondegenerate.

II.3.46. Inspection of the preceding proof reveals that we used only the following fact: if $\mathfrak{D}$ is a domain in $\mathfrak{R}^0$, and $\mathfrak{x}(u, v)$ is constant on $fr(\mathfrak{D})$, then $\mathfrak{x}(u, v)$ is constant on $\mathfrak{D}$ also. This remark yields the following result.

THEOREM. *Let $S : \mathfrak{x} = \mathfrak{x}(w)$, $w \in \mathfrak{R}$ be a surface such that the following condition holds: if $\mathfrak{D}$ is any domain in $\mathfrak{R}^0$, and $\mathfrak{x}(w)$ is constant on $fr(\mathfrak{D})$, then $\mathfrak{x}(w)$ is constant on $\mathfrak{D}$ also. Then S is nondegenerate.*

REMARK. The theorems of II.3.45, II.3.46 give rise to various interesting questions (cf. McShane [5]). Since we shall use these theorems only for purposes of comparison, there is no need for us to go into further details.

II.3.47. Let $S : \mathfrak{x} = [x(u, v), y(u, v), z(u, v)]$, $(u, v) \in \mathfrak{R}$ be a surface (see II.3.44). The given representation of S may or may not possess a certain desirable property (P). For example, (P) may be the property: the coordinate functions $x(u, v)$, $y(u, v)$, $z(u, v)$ are analytic in $\mathfrak{R}^0$, and the expression

$$W = \left\{\left[\frac{\partial(y, z)}{\partial(u, v)}\right]^2 + \left[\frac{\partial(z, x)}{\partial(u, v)}\right]^2 + \left[\frac{\partial(x, y)}{\partial(u, v)}\right]^2\right\}^{1/2}$$

is different from zero in $\mathfrak{R}^0$. Incidentally, precisely this property (P) is assumed in many texts in Differential Geometry.

Let (P) be a property of representations. If S has *some* representation that possesses the property (P), then S itself will be said to possess the property (P). For example, we called a surface (or more generally an F-variety) nondegenerate if it possesses a nondegenerate representation. If S possesses a certain property (P), then it does *not* follow that every representation of S possesses the property (P); it follows merely that S has *some* representation with the property (P). Such a representation will be termed *typical* (with respect to the given property (P)). For example, a surface S will be termed Lipschitzian if it possesses a representation where the coordinate functions are Lipschitzian (see I.3.14), and such a representation will be termed a typical representation of the Lipschitzian surface S. Similar conventions will be used in dealing with curves.

CHAPTER II.4. THE TOPOLOGICAL INDEX

II.4.1. In the Euclidean xy-plane, let us introduce the complex variable $z = x + iy$. Then x is the *real part* $\Re z$ and y is the *imaginary part* $\Im z$ of z, while the *absolute value* $|z|$ of z is given by the formula $|z| = (x^2 + y^2)^{1/2}$. If $z \neq 0$, then we have for z the so-called *trigonometric representation* $z = |z| (\cos \alpha + i \sin \alpha)$, where α is a real number. If $z = x + iy \neq 0$ is given, then α has to satisfy the equations $r \cos \alpha = x$, $r \sin \alpha = y$, where $r = |z|$. These equations have infinitely many solutions for α. Indeed, since $(x/r)^2 + (y/r)^2 = 1$, an easy discussion of the functions $\cos \alpha$ and $\sin \alpha$ shows that there exists a solution α_0 such that $0 \leqq \alpha_0 < 2\pi$. If n is any integer (positive, negative, or zero), then clearly $\alpha_0 + 2n\pi$ is also a solution. Conversely, if α is a solution, then we have $\cos \alpha = \cos \alpha_0$, $\sin \alpha = \sin \alpha_0$, and hence $\cos (\alpha - \alpha_0) = 1$. Thus $\alpha - \alpha_0 = 2n\pi$, where n is an integer. Every real number α that satisfies the equation $z = |z| (\cos \alpha + i \sin \alpha)$, where $z \neq 0$, is termed an *argument* of z. *No argument is assigned to the complex number* $z = 0$. If α_k is an argument of $z_k \neq 0$, $k = 1, \cdots, m$, then clearly $\alpha_1 + \cdots + \alpha_m$ is an argument of the product $z_1 \cdots z_m$.

II.4.2. Suppose that $\Re(z) > 0$, and let α_0 be the unique argument of z such that $0 \leqq \alpha_0 < 2\pi$. Then $\cos \alpha_0 = x/r > 0$, and hence we have $-\pi/2 < \alpha_0 < \pi/2$. Thus if z lies in the half plane $\Re(z) > 0$, then z has a unique argument between the limits $-\pi/2$ and $\pi/2$. *This particular argument will be denoted by* $\alpha(z)$. If $\Im(z) > 0$, then $\alpha(z)$ is merely the radian measure of the acute angle enclosed by the positive x-axis and the segment that joins the points 0 and z. If $\Im(z) < 0$, then $\alpha(z)$ admits of the same interpretation, except that $\alpha(z) < 0$ in this case. If $\Im(z) = 0$, then $\alpha(z) = 0$ also. Thus $\alpha(z)$ is a single-valued, real-valued, obviously continuous function of z in the half plane $\Re(z) > 0$ that satisfies the relations

$$z = |z| [\cos \alpha(z) + i \sin \alpha(z)], \qquad -\pi/2 < \alpha(z) < \pi/2.$$

Let us note explicitly that $\alpha(z)$ *is defined only for* $\Re z > 0$. Let now δ denote a number such that $0 < \delta < 1$. For $|z - 1| \leqq \delta$ we have then clearly $\sin |\alpha(z)| \leqq \delta$. Since $\phi/\sin \phi < \pi/2$ for $0 < \phi < \pi/2$, there follows the inequality $|\alpha(z)| \leqq \pi\delta/2 < 2\delta$. Hence we have

$$|\alpha(z_2) - \alpha(z_1)| < 4\delta \qquad \text{if } |z_1 - 1| < \delta, |z_2 - 1| < \delta.$$

II.4.3. Now let $\mathcal{P}$ be a Peano space (see I.2.33), and let $f(p)$ denote a continuous, complex-valued function defined for $p \in \mathcal{P}$. If E is a subset of $\mathcal{P}$, then we shall put

$$M(f, E) = \text{l.u.b.}\ |f(p)|, \qquad m(f, E) = \text{gr.l.b.}\ |f(p)|, \qquad p \in E,$$

$$\omega(f, E) = \text{l.u.b.}\ |f(p_2) - f(p_1)|, \qquad p_1 \in E, p_2 \in E.$$

Since f is continuous on $\mathcal{P}$ and $\mathcal{P}$ is compact, the quantities $M(f, E)$, $m(f, E)$, $\omega(f, E)$ are clearly finite for every set $E \subset \mathcal{P}$.

If $f(p) \neq 0$ on E and if there exists on E a single-valued, real-valued, continuous function $\phi(p)$ such that

$$f(p) = | f(p) | [\cos \phi(p) + i \sin \phi(p)], \qquad p \in E,$$

then we shall say that f satisfies the condition (arg, E), and $\phi(p)$ will be termed *a single-valued continuous argument of f on E*. If for every choice of a complex constant ζ, such that $f(p) - \zeta \neq 0$ for $p \in E$, the function $f(p) - \zeta$ satisfies the condition (arg, E), then we shall say that f satisfies the condition (Arg, E).

II.4.4. Continuation. Let $\mathfrak{C}$ be a connected set in $\mathcal{P}$, such that f satisfies the condition (arg, $\mathfrak{C}$). If $\phi_1(p)$, $\phi_2(p)$ are two single-valued continuous arguments of f on $\mathfrak{C}$, then $\phi_2(p) - \phi_1(p) = 2n\pi$ on $\mathfrak{C}$, where n is an integer independent of p (the converse is of course obvious).

Proof. Let us put $\psi(p) = [\phi_2(p) - \phi_1(p)]/2\pi$, $p \in \mathfrak{C}$. Then $\psi(p)$ is single-valued and continuous on $\mathfrak{C}$, and $\psi(p)$ takes on only integral values, since $\phi_2(p)$, $\phi_1(p)$ are arguments of the same complex number $f(p)$. From I.2.45 it follows that ψ is constant on $\mathfrak{C}$, and thus $\psi(p) = n$ on $\mathfrak{C}$, where n is an integer independent of p.

II.4.5. Continuation. If E is a subset of $\mathcal{P}$, and $\Re f > 0$ on E, then f satisfies the condition (arg, E). Indeed, the function $\phi(p) = \alpha[f(p)]$ is clearly a single-valued continuous argument of f on E (see II.4.2). Let us note the following corollary: if $| f(p) - 1 | < 1$ on E, then f satisfies the condition (arg, E). Indeed, if $| f - 1 | < 1$, then clearly $\Re f > 0$.

II.4.6. Continuation. If E is a subset of $\mathcal{P}$, and $\omega(f, E) < m(f, E)$, then f satisfies the condition (arg, E). Indeed, let p_0 be a point of E. Since $m(f, E) > 0$, we have $f(p_0) \neq 0$. Hence the function $g(p) = f(p)/f(p_0)$ is continuous on $\mathcal{P}$, and satisfies the inequality

$$| g(p) - 1 | = \frac{| f(p) - f(p_0) |}{| f(p_0) |} \leqq \frac{\omega(f, E)}{m(f, E)} < 1, \qquad p \in E.$$

Hence $g(p)$ possesses, by II.4.5, a single-valued continuous argument $\psi(p)$ on E. If β_0 is any argument of $f(p_0)$, then clearly $\beta_0 + \psi(p)$ is a single-valued continuous argument of f on E.

II.4.7. Continuation. If $\mathfrak{C}$ is a connected subset of $\mathcal{P}$, such that $f(p)$ is equal to a constant $c_0 \neq 0$ on $\mathfrak{C}$, and if $\phi(p)$ is a single-valued continuous argument of f on $\mathfrak{C}$, then $\phi(p)$ is constant on $\mathfrak{C}$ (the converse is of course obvious). Indeed, if β_0 is an argument of c_0, then by II.4.4 we have $\phi(p) - \beta_0 = 2n\pi$ on $\mathfrak{C}$, where the integer n is independent of p.

II.4.8. Continuation. Let $\mathfrak{C}$ be a connected subset of $\mathcal{P}$, such that $\omega(f, \mathfrak{C}) < \delta m(f, \mathfrak{C})$, where δ is a constant and $0 < \delta < 1$. Then f satisfies the condition (arg, $\mathfrak{C}$). Furthermore, if $\phi(p)$ is a single-valued continuous argument of f on $\mathfrak{C}$, then $\omega(\phi, \mathfrak{C}) \leqq 4\delta$.

PROOF. The first assertion follows directly from II.4.6. To prove the second assertion, take a point $p_0 \in \mathfrak{C}$. Since $m(f, \mathfrak{C}) > 0$, we have $f(p_0) \neq 0$, and hence the function $g(p) = f(p)/f(p_0)$ is continuous on $\mathcal{P}$ and satisfies the inequality

$$| g(p) - 1 | = \frac{| f(p) - f(p_0) |}{| f(p_0) |} \leqq \frac{\omega(f, \mathfrak{C})}{m(f, \mathfrak{C})} < \delta, \qquad p \in \mathfrak{C}.$$

In view of II.4.2, the function $\psi(p) = \alpha[g(p)]$ is a single-valued continuous argument of g on $\mathfrak{C}$ and satisfies the inequality $| \psi(p_2) - \psi(p_1)| < 4\delta$ for every pair of points p_1, p_2 in $\mathfrak{C}$. If β_0 is an argument of $f(p_0)$, then clearly $\lambda(p) = \beta_0 + \psi(p)$ is a single-valued continuous argument of f on $\mathfrak{C}$. If $\phi(p)$ is any single-valued continuous argument of f on $\mathfrak{C}$, then $\phi(p) - \lambda(p)$ is constant on $\mathfrak{C}$ by II.4.4, and hence $| \phi(p_2) - \phi(p_1) | = | \lambda(p_2) - \lambda(p_1) | = | \psi(p_2) - \psi(p_1) | < 4\delta$ for every pair of points p_1, p_2 in $\mathfrak{C}$. Hence $\omega(\phi, \mathfrak{C}) \leqq 4\delta$.

II.4.9. CONTINUATION. Suppose that $f(p)$ satisfies the condition (arg, E). Let $H(\mathcal{P}) = \mathcal{P}^*$ be a homeomorphism, and let us put $f^*(p^*) = f[H^{-1}(p^*)]$, $E^* = H(E)$. Then f^* satisfies the condition (arg, E^*). Indeed, if $\phi(p)$ is a single-valued continuous argument of f on E, then clearly $\phi^*(p^*) = \phi[H^{-1}(p^*)]$ is a single-valued continuous argument of f^* on E^*.

II.4.10. CONTINUATION. Let E be a subset of $\mathcal{P}$. If every function $f(p)$ that is continuous on $\mathcal{P}$ and different from zero on E satisfies the condition (arg, E), then we shall say that E satisfies the condition (arg).

Let $H(\mathcal{P}) = \mathcal{P}^*$ be a homeomorphism, and put $E^* = H(E)$. If E satisfies the condition (arg) relative to $\mathcal{P}$, then clearly E^* satisfies the condition (arg) relative to $\mathcal{P}^*$ (cf. II.4.9).

II.4.11. CONTINUATION. Let E_1, E_2 be two closed subsets of $\mathcal{P}$ such that the intersection E_1E_2 is connected and nonempty. Suppose that $f(p)$ is continuous on $\mathcal{P}$ and satisfies the conditions (arg, E_1), and (arg, E_2). Then f satisfies the condition (arg, $E_1 + E_2$).

PROOF. Let p_0 be a point of E_1E_2 and let β_0 be an argument of $f(p_0)$. By assumption, we have a single-valued continuous argument $\psi_1(p)$ of f on E_1. Then β_0 and $\psi_1(p_0)$ are arguments of the same complex number $f(p_0) \neq 0$, and hence $\beta_0 = \psi_1(p_0) + 2n\pi$, where n is an integer. Then $\phi_1(p) = \psi_1(p) + 2n\pi$ is a single-valued continuous argument of f on E_1, such that $\phi_1(p_0) = \beta_0$. Similarly we have a single-valued continuous argument $\phi_2(p)$ of f on E_2, such that $\phi_2(p_0) = \beta_0$. Then ϕ_1, ϕ_2 are single-valued continuous arguments of f on the connected set E_1E_2, and hence $\phi_2 - \phi_1$ is constant on E_1E_2 by II.4.4. Since $\phi_2(p_0) - \phi_1(p_0) = 0$, it follows that $\phi_1(p) \equiv \phi_2(p)$ on E_1E_2. Hence if we define $\phi(p) = \phi_1(p)$ on E_1, and $\phi(p) = \phi_2(p)$ on E_2, then $\phi(p)$ is single-valued on $E_1 + E_2$. Since $\phi(p)$ is clearly continuous on $E_1 + E_2$ as a consequence of the assumption that E_1 and E_2 are closed, $\phi(p)$ is a single-valued continuous argument of f on $E_1 + E_2$.

II.4.12. Let Q denote the unit square $0 \leqq u \leqq 1$, $0 \leqq v \leqq 1$ in the Euclidean uv-plane. Then Q satisfies the condition (arg) (see II.4.10).

PROOF. Let $f(u, v)$ be a continuous complex-valued function in Q that is different from zero in Q. We have to show that f satisfies the condition (arg, Q).

Let us deny this. Let us divide Q into two congruent rectangles R_1', R_1'' by a vertical segment. From II.4.11 it follows that f fails to satisfy one of the conditions (arg, R_1'), (arg, R_1''), say the condition (arg, R_1'). Let us subdivide R_1' into two congruent rectangles R_2', R_2'' by a horizontal segment. From II.4.11 it follows that f fails to satisfy one of the conditions (arg, R_2'), (arg, R_2''), say (arg, R_2'). Continuation of this process leads to an infinite sequence of rectangles $R_1', \cdots, R_n', \cdots$, such that the diameter of R_n' converges to zero and f fails to satisfy the condition (arg, R_n'). Now since f is continuous and different from zero in Q, we have $\omega(f, R_n') < m(f, Q) \leqq m(f, R_n')$ for n sufficiently large, and hence f must satisfy the condition (arg, R_n') for large enough n by II.4.6. Thus we reached a contradiction. Hence f satisfies the condition (arg, Q).

II.4.13. If $\mathcal{P}$ is a 2-cell (see I.2.31), then $\mathcal{P}$ satisfies the condition (arg). This follows directly from II.4.10 and II.4.12.

II.4.14. If I is a linear interval $a \leqq u \leqq b$, then I satisfies the condition (arg). The proof is entirely analogous to that in II.4.12. As a corollary, it follows that if $\mathcal{P}$ is a simple arc, then $\mathcal{P}$ satisfies the condition (arg) (cf. I.2.31, II.4.10).

II.4.15. Let $\mathfrak{D}$ be a bounded, simply-connected domain in the Euclidean uv-plane, and let $f(u, v)$ be a continuous complex-valued function in $\mathfrak{D}$ that is different from zero in $\mathfrak{D}$. Then f possesses a single-valued continuous argument in $\mathfrak{D}$.

Proof. By I.2.51, we have a sequence of simply-connected Jordan regions $\mathfrak{R}_1, \cdots, \mathfrak{R}_n, \cdots$, such that $\mathfrak{R}_n \subset \mathfrak{R}_{n+1}^0$ and $\sum \mathfrak{R}_n = \mathfrak{D}$. Let (u_0, v_0) be an interior point of $\mathfrak{R}_1$, and let β_0 be an argument of $f(u_0, v_0)$. Since $\mathfrak{R}_n$ is a 2-cell (see I.2.50), we have by II.4.13 a single-valued continuous argument $\psi_n(u, v)$ for $f(u, v)$ in $\mathfrak{R}_n$. Then $\psi_n(u, v)$ and β_0 are arguments of $f(u_0, v_0)$, and hence $\beta_0 = \psi_n(u_0, v_0) + 2k_n\pi$, where k_n is an integer. Thus $\phi_n(u, v) = \psi_n(u, v) + 2k_n\pi$ is a single-valued continuous argument of $f(u, v)$ in $\mathfrak{R}_n$ such that $\phi_n(u_0, v_0) = \beta_0$, $n = 1, 2, \cdots$. Let us define in $\mathfrak{D}$ the function $\phi(u, v)$ by the formula $\phi(u, v) = \phi_n(u, v)$ if $(u, v) \in \mathfrak{R}_n^0$. We assert that $\phi(u, v)$ is single-valued in $\mathfrak{D}$. Indeed, if $(u, v) \in \mathfrak{R}_n^0\mathfrak{R}_m^0$ where $n > m$, then $\phi_n(u_0, v_0) = \beta_0 = \phi_m(u_0, v_0)$ and hence $\phi_n \equiv \phi_m$ in $\mathfrak{R}_m$ by II.4.4 (note that $\mathfrak{R}_m \subset \mathfrak{R}_n$). Thus ϕ is clearly a single-valued, continuous argument of f in $\mathfrak{D}$.

II.4.16. Let $f(p)$ be a continuous complex-valued function on the Peano space $\mathcal{P}$, and let $\mathfrak{D}$ be a domain (connected open set) in $\mathcal{P}$, such that $f(p) \neq 0$ in $\mathfrak{D}$. If f satisfies the condition (arg, C) for every simple closed curve $C \subset \mathfrak{D}$, then f also satisfies the condition (arg, $\mathfrak{D}$) (cf. II.4.3).

Proof Let us choose a point p_0 in $\mathfrak{D}$, and let β_0 be an argument of $f(p_0)$. We shall keep p_0 and β_0 fixed. Let p_1 be a point in $\mathfrak{D}$ different from p_0. By I.2.41 we have in $\mathfrak{D}$ a simple arc γ with end points p_0, p_1 By II.4.14 there exists on γ a single-valued continuous argument $\psi(p)$ for f. Then $\psi(p_0)$ and β_0 are arguments of the same complex number $f(p_0) \neq 0$, and hence $\beta_0 = \psi(p_0) + 2n\pi$, where n is an integer. Then $\phi(p) = \psi(p) + 2n\pi$ is a single-valued continuous argument of f on γ such that $\phi(p_0) = \beta_0$ By II.4.4, $\phi(p)$ is univocally determined by the condition $\phi(p_0) = \beta_0$. This univocally determined function $\phi(p)$ will be denoted by $\phi(p, \gamma)$. Then $\phi(p, \gamma)$ has the following properties.

(i) $\phi(p, \gamma)$ is a single-valued continuous argument of f on γ.

(ii) $\phi(p_0, \gamma) = \beta_0$.

Now let γ', γ'' be any two simple arcs in $\mathfrak{D}$ with end points p_0, p_1. We assert the formula

$$\phi(p_1, \gamma') = \phi(p_1, \gamma''). \tag{1}$$

Indeed, let E denote the set of the common points of γ', γ''. Then E is a closed subset of γ' which contains (at least) the points p_0, p_1. We consider the function

$$\lambda(p) = \frac{\phi(p, \gamma'') - \phi(p, \gamma')}{2\pi}, \qquad p \in E.$$

Then $\lambda(p)$ is defined and continuous on E, and assumes only integral values on E. Indeed, for every point $p \in E$, $\phi(p, \gamma'')$ and $\phi(p, \gamma')$ are arguments of the same complex number $f(p) \neq 0$, and hence $\lambda(p)$ is an integer. We have $\lambda(p_0) = 0$ (see (ii)). Let us assume now, in contradiction with (1), that $\lambda(p_1) \neq 0$. Let us designate by E_0, E_1 the subsets of E on which $\lambda(p) = 0$, $\lambda(p) \neq 0$ respectively. Then $p_0 \in E_0$, and by assumption $p_1 \in E_1$. Clearly $E_0E_1 = 0$, and E_0 is closed since $\lambda(p)$ is continuous on E. We assert that E_1 is also closed. Indeed, since $E_0 + E_1 = E$ and E is closed, we would have otherwise a point $q_0 \in E$ and a sequence of points $q_n \in E$ such that

$$\lambda(q_0) = 0, \qquad \lambda(q_n) \neq 0, \qquad q_n \to q_0. \tag{2}$$

Now since $\lambda(q_n)$ is an integer k_n and λ is continuous on E, we have in view of (2) the relation $k_n \to 0$. Since k_n is an integer, it follows that $|\lambda(q_n)| = |k_n| = 0$ for n large enough, in contradiction with (2). Summing up, E_0, E_1 are disjoint, nonempty, closed sets on the simple arc γ'. It follows that we have a sub-arc γ_0' of γ', such that the end points a_0, a_1 of γ_0' satisfy the relations

$$a_0 \in E_0, \qquad a_1 \in E_1, \qquad E\gamma_0' = a_0 + a_1. \tag{3}$$

For example, we may choose a_0 as the last point of E_0 in proceeding on γ' from p_0 to p_1, and a_1 as the first point of E_1 that we meet in proceeding on γ' from a_0 to p_1 (note that $p_0 \in E_0$, $p_1 \in E_1$ by assumption). These same two points a_0, a_1 are the end points of a sub-arc γ_0'' of γ'', and since $\gamma_0'\gamma_0'' \subset \gamma'\gamma'' = E$ and hence $\gamma_0'\gamma_0'' = \gamma_0'\gamma_0'\gamma_0'' \subset \gamma_0'E$, we have by (3) the relation $\gamma_0'\gamma_0'' = a_0 + a_1$. Hence $\gamma_0' + \gamma_0''$ is a simple closed curve C in $\mathfrak{D}$. By assumption f possesses a single-valued continuous argument $\phi(p)$ on C. By II.4.4 it follows that $\phi(p) - \phi(p, \gamma')$ is constant on γ_0' and $\phi(p) - \phi(p, \gamma'')$ is constant on γ_0''. Hence

$$\phi(a_0) - \phi(a_0, \gamma') = \phi(a_1) - \phi(a_1, \gamma'), \phi(a_0) - \phi(a_0, \gamma'') = \phi(a_1) - \phi(a_1, \gamma'').$$

Subtraction yields $\lambda(a_0) = \lambda(a_1)$. This is however a contradiction, since $a_0 \in E_0$, $a_1 \in E_1$, and hence $\lambda(a_0) = 0 \neq \lambda(a_1)$. Thus (1) is established.

Let us now define in $\mathfrak{D}$ a function $\mu(q)$ as follows. If $q \neq p_0$, then $\mu(q) =$

$\phi(q, \gamma)$, where γ is any simple arc in $\mathfrak{D}$ with end points p_0, q. For $q = p_0$, we put $\mu(p_0) = \beta_0$. Then $\mu(q)$ is simple-valued in $\mathfrak{D}$ by (1), and for every $q \in \mathfrak{D}$, $\mu(q)$ is clearly an argument of $f(q)$. There remains to show that $\mu(q)$ is continuous in $\mathfrak{D}$. In verifying this fact, we shall consider repeatedly situations where two given points a, b are to be joined by a simple arc γ, but the points a, b may coincide. If this happens, we agree that γ will refer to the single point $a = b$. With this understanding, the continuity of μ in $\mathfrak{D}$ may be proved as follows. Let p_1 be a point in $\mathfrak{D}$, and let $\epsilon > 0$ be given. Let γ_{01} be a simple arc in $\mathfrak{D}$ with end points p_0, p_1, and let $\phi_{01}(p)$ denote the single-valued continuous argument of f on γ_{01} that satisfies the condition $\phi_{01}(p_0) = \beta_0$. Since $f(p_1) \neq 0$ and f is continuous, we have an open set $\bar{G}$ such that $p_1 \in \bar{G} \subset \mathfrak{D}$ and $\omega(f, \bar{G}) < (\epsilon/8)m(f, \bar{G})$ (cf. II.4.3). Since p_1 is an end point of γ_{01}, we can choose on γ_{01} a point q_1 such that the sub-arc of γ_{01} with end points q_1, p_1 lies in $\bar{G}$, and thereupon we can choose an open set G such that $p_1 \in G \subset \bar{G}$ and G has no point in common with the sub-arc p_0q_1 of γ_{01}. Since $\mathcal{P}$ is a Peano space, we can assume that G is connected (cf. I.2.33). Now let p_2 be any point in G. By I.2.41 we have then in G a simple arc γ_{12} with end points p_1, p_2. Let p_1^* be the first point of the set $\gamma_{01}\gamma_{12}$ that is met in proceeding on γ_{12} from p_2 to p_1. Let us denote by γ_{01}^*, γ_{12}^* the sub-arcs $p_0p_1^*$, $p_1^*p_2$ of γ_{01}, γ_{12} respectively. Then $\gamma_{02} = \gamma_{01}^* + \gamma_{12}^*$ is a simple arc with end points p_0, p_2. Let now $\phi_{02}(p)$ denote the single-valued continuous argument of f on γ_{02} that satisfies the condition $\phi_{02}(p_0) = \beta_0$. We have then, by definition,

$$\mu(p_1) = \phi_{01}(p_1), \quad \mu(p_2) = \phi_{02}(p_2). \tag{4}$$

By II.4.4, $\phi_{01} = \phi_{02}$ on γ_{01}^*, and hence

$$\phi_{01}(p_1^*) = \phi_{02}(p_1^*). \tag{5}$$

Since the sub-arcs $p_1^*p_1$, $p_1^*p_2$ of γ_{01}, γ_{02} respectively are comprised in $\bar{G}$ (in fact, the second arc lies in G), we have by II.4.8 the inequalities

$$| \phi_{01}(p_1) - \phi_{01}(p_1^*) | \leqq \epsilon/2, \qquad | \phi_{02}(p_2) - \phi_{02}(p_1^*) | \leqq \epsilon/2. \tag{6}$$

From (4), (5), (6) we infer that $| \mu(p_2) - \mu(p_1) | \leqq \epsilon$ if $p_2 \in G$, and thus the continuity of μ at the arbitrary point $p_1 \in \mathfrak{D}$ is proved.

II.4.17. Suppose that the Peano space $\mathcal{P}$ occurring in II.4.16 is a dendrite (see II.2.1). Then the wording of the theorem and its proof admit of obvious simplifications due to the fact that a dendrite contains no simple closed curve. Indeed, if C is a simple closed curve in $\mathcal{P}$ and p, q are any two distinct points of C, then clearly no point x of $\mathcal{P}$ cuts between p and q, in contradiction with the assumption that $\mathcal{P}$ is a dendrite.

II.4.18. Let $f(p)$ be a continuous complex-valued function on the Peano space $\mathcal{P}$. Let $\mathfrak{C}$ be a connected subset of $\mathcal{P}$, such that $f \neq 0$ on $\mathfrak{C}$ and f fails to satisfy the condition (arg, $\mathfrak{C}$). Then there exists in $\mathcal{P}$ a simple closed curve C such that $f \neq 0$ on C and f fails to satisfy the condition (arg, C) (cf. II.4.3).

Proof. Deny the assertion. Since f is continuous and $f \neq 0$ on $\mathfrak{C}$, we have for every point $p \in \mathfrak{C}$ an open set G_p such that $p \in G_p$ and $f \neq 0$ in G_p. Since $\mathcal{P}$ is a Peano space, we can assume that each G_p is connected. Then the set $\mathfrak{D} = \sum G_p$, $p \in \mathfrak{C}$, is connected since $\mathfrak{C}$ is connected. Obviously, $\mathfrak{D}$ is open and $f \neq 0$ on $\mathfrak{D}$. Hence, by assumption, f satisfies the condition (arg, C) for every simple closed curve $C \subset \mathfrak{D}$. By II.4.16 there follows the existence of a single-valued continuous argument $\phi(p)$ for p in $\mathfrak{D}$. Then $\phi(p)$ is also a single-valued continuous argument for f on $\mathfrak{C}$, in contradiction with the assumption that f fails to satisfy the condition (arg, $\mathfrak{C}$).

II.4.19. Let $f(p)$ be a continuous complex-valued function on a subset E of a Peano space $\mathcal{P}$. If $g(p)$ is a continuous complex-valued function on E such that $f(p) = g(p)^k$ for $p \in E$, where k is a positive integer, then $g(p)$ will be termed *a single-valued continuous kth root of f on E.*

If $f \neq 0$ on E and if there exists a single-valued continuous argument $\phi(p)$ for f on E, then clearly the function

$$g(p) = |f(p)|^{1/k}\left[\cos\frac{\phi(p)}{k} + i\sin\frac{\phi(p)}{k}\right]$$

is a single-valued continuous kth root of f on E. However, a single-valued continuous kth root may exist even if these assumptions are not satisfied. For example, let E denote the unit disc $|z| \leqq 1$ in the complex z-plane, and consider the function $f(z) = z^2$. Then z is a single-valued continuous square root of $f(z)$ on E, even though $f(z)$ does not satisfy the condition (arg, E).

II.4.20. Let $f(p)$ be a continuous complex-valued function on a continuum $\mathfrak{C}$ in a Peano space $\mathcal{P}$, such that (i) $f \neq 0$ on $\mathfrak{C}$, and (ii) f has a single-valued continuous kth root $g_0(p)$ on $\mathfrak{C}$. If λ is a kth root of unity (that is, $\lambda^k = 1$), then clearly $\lambda g_0(p)$ is also a single-valued continuous kth root of f on $\mathfrak{C}$. Conversely, let $g(p)$ be any single-valued continuous kth root of f on $\mathfrak{C}$. Since $f \neq 0$ on $\mathfrak{C}$, we have then also $g_0 \neq 0$ on $\mathfrak{C}$, and thus $h = g/g_0$ is single-valued and continuous on $\mathfrak{C}$. Clearly $h^k = 1$ on $\mathfrak{C}$, and hence h cannot assume more than k distinct values on $\mathfrak{C}$. By I.2.45 it follows that h is constant on $\mathfrak{C}$, say $h = c$ on $\mathfrak{C}$, where $c^k = 1$, and hence $g(p) = cg_0(p)$ on $\mathfrak{C}$, where c is a kth root of unity. In particular, if $g(p_0) = g_0(p_0)$ for a single point $p_0 \in \mathfrak{C}$, then it follows that $g \equiv g_0$ on $\mathfrak{C}$.

II.4.21. Let $f(p)$ be a continuous complex-valued function on a continuum $\mathfrak{C}$ in a Peano space $\mathcal{P}$. Suppose that $f(p)$ has a single-valued continuous kth root $g(p)$ on $\mathfrak{C}$. If $\omega(f, \mathfrak{C})$, $\omega(g, \mathfrak{C})$ denote the oscillations on $\mathfrak{C}$ of f and g respectively, then

$$\omega(g, \mathfrak{C}) \leqq 3\omega(f, \mathfrak{C})^{1/k}. \tag{1}$$

PROOF. If $k = 1$, then $f \equiv g$ and the assertion is obvious. So let us assume that the positive integer k exceeds 1. We can also assume that f is not constant on $\mathfrak{C}$, since otherwise (1) is trivial. Let $M(f, \mathfrak{C})$ denote the maximum of $|f|$ on $\mathfrak{C}$.

Case (i). Let us first assume that

$$M(f, \mathfrak{C}) > 2\omega(f, \mathfrak{C}). \tag{2}$$

Let p_0 be a point of $\mathfrak{C}$ where $|f(p_0)| = M(f, \mathfrak{C})$. By (2) we have then the inequality $|f(p) - f(p_0)| \leqq \omega(f, \mathfrak{C}) < (1/2)M(f, \mathfrak{C}) = |f(p_0)|/2$. Let us now consider the transformation $T : z = f(p)$, $p \in \mathfrak{C}$. The preceeding inequalities express the fact that $T(\mathfrak{C})$ is a subset of an open circular disc D in the z-plane whose radius is $\omega(f, \mathfrak{C})$ and whose center is $f(p_0)$ (note that $|f(p_0)| = M(f, \mathfrak{C}) > 0$ by (2)). Thus $z = 0$ is exterior to D by (2). Now let ψ_0 be an argument of $f(p_0)$, and consider the function

$$f^*(p) = e^{-i\psi_0} f(p), \qquad p \in \mathfrak{C}.$$

Let T^* denote the transformation $z = f^*(p)$, $p \in \mathfrak{C}$. Then $T^*(\mathfrak{C})$ is a subset of the open circular disc D^* with center at $z = |f(p_0)| = M(f, \mathfrak{C})$ and radius $\omega(f, \mathfrak{C})$. Thus D^* is comprised in the half plane $\Re z > 0$ by (2), and hence by II.4.2 we have at our disposal the single-valued continuous argument $\alpha(z)$ of z in D^*. The function

$$h(z) = |z|^{1/k}\left[\cos\frac{\alpha(z)}{k} + i\sin\frac{\alpha(z)}{k}\right], \qquad z \in D^*,$$

is then a single-valued analytic function of z in D^*, and we have the formulas

$$(3) \qquad h(z)^k = z, \qquad h'(z) = \frac{1}{kh(z)^{k-1}}, \qquad |h(z)| = |z|^{1/k}, \qquad z \in D^*.$$

Now let z_1, z_2 be any two points in D^*. We have then the formula

$$h(z_2) - h(z_1) = \int_{z_1}^{z_2} h'(z)\, dz,$$

where the path of integration may be taken as the straight segment with end points z_1, z_2. There follows the inequality

$$|h(z_2) - h(z_1)| \leqq \frac{2}{k}\omega(f, \mathfrak{C})^{1/k} \leqq \omega(f, \mathfrak{C})^{1/k},$$

since $k \geqq 2$. Hence we have for the oscillation $\omega(h, D^*)$ of h in D^* the inequality

$$(4) \qquad \omega(h, D^*) \leqq \omega(f, \mathfrak{C})^{1/k}.$$

Let us introduce the function

$$(5) \qquad g_0(p) = e^{i\psi_0/k} h[e^{-i\psi_0} f(p)] = e^{i\psi_0/k} h[f^*(p)], \qquad p \in \mathfrak{C}.$$

Since $f^*(p) \in D^*$ for $p \in \mathfrak{C}$, it follows that $g_0(p)$ is single-valued and continuous on $\mathfrak{C}$, and clearly $\omega(g_0, \mathfrak{C}) \leqq \omega(h, D^*)$. Hence, by (4),

$$(6) \qquad \omega(g_0, \mathfrak{C}) \leqq \omega(f, \mathfrak{C})^{1/k}.$$

From (3) and (5) it follows that $g_0(p)^k = f(p)$ for $p \in \mathfrak{C}$. Thus $g_0(p)$ is a single-valued continuous kth root of $f(p)$ on $\mathfrak{C}$. If $g(p)$ is any single-valued continuous

kth root of f on $\mathfrak{C}$, then $g(p) = cg_0(p)$, where $|c| = 1$ (see II.4.20), and hence $\omega(g, \mathfrak{C}) = \omega(g_0, \mathfrak{C})$. Thus (1) follows from (6).

Case (ii). $M(f, \mathfrak{C}) \leqq 2\omega(f, \mathfrak{C})$. Let $M(g, \mathfrak{C})$ denote the maximum of $|g(p)|$ on $\mathfrak{C}$. Then clearly

$$\omega(g, \mathfrak{C}) \leqq 2M(g, \mathfrak{C}), \qquad M(g, \mathfrak{C}) = M(f, \mathfrak{C})^{1/k}.$$

There follow the inequalities

$$\omega(g, \mathfrak{C}) \leqq 2M(f, \mathfrak{C})^{1/k} \leqq 2^{1+1/k}\omega(f, \mathfrak{C})^{1/k} \leqq 3\omega(f, \mathfrak{C})^{1/k},$$

since $k \geqq 2$. Thus (1) is proved.

II.4.22. Let $f_1(p), \cdots, f_n(p), \cdots$ be a sequence of continuous complex-valued functions on a Peano space $\mathcal{P}$, such that for each n there exists a single-valued continuous kth root $g_n(p)$ of $f_n(p)$ on $\mathcal{P}$, where k is a fixed positive integer. Suppose that the sequence $f_n(p)$ is equicontinuous on $\mathcal{P}$. Then the sequence $g_n(p)$ is also equicontinuous on $\mathcal{P}$ (cf. I.2.44).

Proof. Give $\epsilon > 0$. By assumption there exists $\delta > 0$ such that $|f_n(p_2) - f_n(p_1)| < (\epsilon/3)^k$ if $\rho(p_1, p_2) < \delta$, $n = 1, 2, \cdots$. Since $\mathcal{P}$ is a Peano space, we have an $\eta > 0$ such that if q_1, q_2 are any two points of $\mathcal{P}$ for which $\rho(q_1, q_2) < \eta$, then there exists in $\mathcal{P}$ a continuum with diameter less than δ that contains q_1 and q_2. Now take any two points q_1, q_2 in $\mathcal{P}$ such that $\rho(q_1, q_2) < \eta$, and let $\mathfrak{C}$ be a continuum in $\mathcal{P}$ such that $d(\mathfrak{C}) < \delta$, $q_1 + q_2 \in \mathfrak{C}$. Then $\omega(f_n, \mathfrak{C}) < (\epsilon/3)^k$, and hence $\omega(g_n, \mathfrak{C}) < \epsilon$ by II.4.21. Since $|g_n(q_2) - g_n(q_1)| \leqq \omega(g_n, \mathfrak{C})$, it follows that $|g_n(q_2) - g_n(q_1)| < \epsilon$ if $\rho(q_2, q_1) < \eta$. Since η is independent of q_1, q_2, n, the equicontinuity of the sequence $g_n(p)$ is proved.

II.4.23. On a Peano space $\mathcal{P}$, let $f_n(p)$ be a sequence of continuous complex-valued functions that converge uniformly to a function $f(p)$ on $\mathcal{P}$. Suppose that for each n there exists a single-valued continuous kth root $g_n(p)$ of $f_n(p)$ on $\mathcal{P}$ where k is a fixed positive integer. Then there exists a single-valued continuous kth root of $f(p)$ on $\mathcal{P}$.

Proof. By I.2.44, the sequence f_n is equicontinuous and uniformly bounded on $\mathcal{P}$. Hence, by II.4.22, the sequence g_n is also equicontinuous on $\mathcal{P}$, and since $|g_n|^k = |f_n|$, the sequence g_n is uniformly bounded on $\mathcal{P}$. By I.2.44 there follows the existence of a subsequence g_{n_i}, $n_1 < n_2 < \cdots < n_i < \cdots$, which converges on $\mathcal{P}$ uniformly to a (necessarily continuous) function $g(p)$. Since $g_{n_i}^k = f_{n_i}$ for each i, there follows the relation $g^k = f$. Thus g is a single-valued continuous kth root of f on $\mathcal{P}$.

II.4.24. Let γ be a simple arc (see I.2.31). Then γ admits of two orientations, and we shall use γ and $-\gamma$ to refer to the two oriented simple arcs so obtained. Let the end points of γ be denoted by a, b in such a manner that a is the first and b is the second end point of γ, and hence b is the first and a is the second end point of $-\gamma$. If $f(p)$ is a continuous complex-valued function on γ and $f \neq 0$ on γ, then $f(p)$ possesses a single-valued continuous argument $\phi(p)$ on γ (see II.4.14). *The variation of the argument of $f(p)$ on the oriented simple arc γ* is then defined by the formula $V_\gamma[\arg f(p)] = \phi(b) - \phi(a)$. If $\phi^*(p)$ is any single-valued continuous argument of f on γ, then $\phi - \phi^*$ is constant on γ by II.4.4, and hence $\phi(b) - \phi(a) = \phi^*(b) - \phi^*(a)$. In other words, $V_\gamma[\arg f(p)]$ *is independent of the particular choice of* $\phi(p)$. The following remarks will be useful.

(a) $V_\gamma[\arg f(p)] = -V_{-\gamma}[\arg f(p)]$.

(b) Let c be a point of γ different from a and b. Let γ_1 denote the oriented sub-arc of γ with first end point a and second end point c, and let γ_2 denote the oriented sub-arc of γ with first end point c and second end point b. Then

$$V_\gamma[\arg f(p)] = V_{\gamma_1}[\arg f(p)] + V_{\gamma_2}[\arg f(p)].$$

Indeed, $\phi(p)$ can be used on both γ_1 and γ_2, and thus the assertion is equivalent to the formula $\phi(b) - \phi(a) = [\phi(c) - \phi(a)] + [\phi(b) - \phi(c)]$.

(c) If $f_1(p), f_2(p)$ are continuous and different from zero on the oriented simple arc γ, then

$$V_\gamma[\arg f_1(p)f_2(p)] = V_\gamma[\arg f_1(p)] + V_\gamma \arg f_2[(p)].$$

Indeed, if $\phi_1(p)$, $\phi_2(p)$ are single-valued continuous arguments on γ for $f_1(p)$, $f_2(p)$ respectively, then clearly $\phi_1(p) + \phi_2(p)$ is a single-valued continuous argument for $f_1(p)f_2(p)$ on γ, and the asserted formula follows.

(d) Let $f_n(p)$ be a sequence of continuous complex-valued functions on γ, such that f_n converges on γ uniformly to a (necessarily continuous) function f. Suppose that all the functions f, f_n are different from zero on γ. Then

$$V_\gamma[\arg f_n(p)] \to V_\gamma[\arg f(p)].$$

Proof. Clearly, the functions $g_n(p) = f_n(p)/f(p)$ are continuous and different from zero on γ, and $g_n \to 1$ uniformly on γ. Hence $| g_n(p) - 1 | < 1$ for n sufficiently large, say $n > n_0$. Let us define

$$\psi_n(p) = \alpha[g_n(p)], \qquad p \in \gamma, n > n_0,$$

where $\alpha(z)$ is the single-valued continuous argument of z defined for $\Re z > 0$ in II.4.2. Since $\alpha(1) = 0$ and $g_n(p) \to 1$ uniformly on γ, it follows that $\psi_n(p) \to 0$ uniformly on γ. Clearly $\psi_n(p)$ is a single-valued continuous argument of $g_n = f_n/f$ on γ. Hence, if ϕ is a single-valued continuous argument of f on γ, then $\phi_n = \psi_n + \phi$ is a single-valued continuous argument of $g_n f = f_n$ on γ, and $\phi_n \to \phi$ uniformly on γ. Hence

$$V_\gamma[\arg f_n(p)] = \phi_n(b) - \phi_n(a) \to \phi(b) - \phi(a) = V_\gamma[\arg f(p)].$$

II.4.25. Let C be a simple closed curve. Then C admits of two orientations, and we shall use C, $-C$ to refer to the two oriented simple closed curves obtained in this manner. Let $f(p)$ be a continuous complex-valued function on C, such that $f \neq 0$ on C. Let $a_0, a_1, \cdots, a_n$ be distinct points on C arranged in conformity with the orientation assigned to C. Let γ_i denote the oriented sub-arc of C with first end point a_{i-1} and second end point a_i, $i = 1, \cdots, n$, and let γ_{n+1} denote the oriented sub-arc of C with first end point a_n and second end point a_0, where all these sub-arcs are oriented in accordance with the orientation of C.

Then the arcs $\gamma_1, \cdots, \gamma_n, \gamma_{n+1}$ have no common interior points. *The variation of the argument of f on the oriented simple closed curve C* is defined by the formula

$$V_C[\arg f(p)] = V_{\gamma_1}[\arg f(p)] + \cdots + V_{\gamma_{n+1}}[\arg f(p)]. \tag{1}$$

By II.4.24(b) this quantity is independent of the number and choice of the points $a_0, a_1, \cdots, a_n$. The following statements will be useful.

(a) $V_C[\arg f(p)] = -V_{-C}[\arg f(p)]$. This follows from II.4.24(a).

(b) If $f_1(p), f_2(p)$ are continuous and not equal to 0 on C, then

$$V_C[\arg f_1(p)f_2(p)] = V_C[\arg f_1(p)] + V_C[\arg f_2(p)].$$

This follows from II.4.24(c).

(c) If $f_n(p)$, $f(p)$ are continuous and different from zero on C, and $f_n \to f$ uniformly on C, then $V_C[\arg f_n(p)] \to V_C[\arg f(p)]$. This follows from II.4.24(d).

(d) $V_C[\arg f(p)]$ is an integral multiple of 2π. Indeed, let $\phi_j(p)$ be a single-valued continuous argument of $f(p)$ on γ_j (see formula (1)). Then (cf. II.4.24)

$$\begin{aligned} V_C[\arg f(p)] & \\ &= [\phi_1(a_1) - \phi_1(a_0)] + [\phi_2(a_2) - \phi_2(a_1)] + \cdots + [\phi_{n+1}(a_0) - \phi_{n+1}(a_n)] \\ &= [\phi_{n+1}(a_0) - \phi_1(a_0)] + [\phi_1(a_1) - \phi_2(a_1)] + \cdots + [\phi_n(a_n) - \phi_{n+1}(a_n)]. \end{aligned} \tag{2}$$

Now each pair of square brackets contains the difference of two arguments of the same complex number. Such a difference being an integral multiple of 2π, it follows that $V_C[\arg f(p)]$ itself is an integral multiple of 2π.

(e) $V_C[\arg f(p)] = 0$ if and only if f possesses a single-valued continuous argument on C. Indeed, if such an argument $\phi(p)$ exists, then we can put, in the formula (2), $\phi_j(p) = \phi(p)$ for each j, and (2) yields then $V_C[\arg f(p)] = 0$. Suppose, conversely, that

$$V_C[\arg f(p)] = 0. \tag{3}$$

Using the same notations as in (d) above, let us note that $\phi_1(a_1) = \phi_2(a_1) + 2k_1\pi$, where k_1 is an integer, since $\phi_1(a_1)$, $\phi_2(a_1)$ are arguments of the same complex number $f(a_1)$. Since $\phi_2(p) + 2k_1\pi$ is also a single-valued continuous argument of f on γ_2, it follows that we can assume without loss of generality that $\phi_1(a_1) = \phi_2(a_1)$. Repeated application of this remark shows that $\phi_1(p), \cdots, \phi_{n+1}(p)$ can be so chosen that

$$\phi_j(a_j) = \phi_{j+1}(a_j), \qquad j = 1, 2, \cdots, n. \tag{4}$$

From (2), (3), (4) there follows then the relation

$$\phi_1(a_0) = \phi_{n+1}(a_0). \tag{5}$$

Let us define on C a function $\phi(p)$ as follows: if $p \in \gamma_j$, then $\phi(p) = \phi_j(p)$. By (4) and (5) this function $\phi(p)$ is then single-valued on C. Once this point is established, it is clear that $\phi(p)$ is a single-valued continuous argument of f on C.

II.4.26. On the oriented simple closed curve C, let there be given three continuous complex-valued functions $f(p)$, $G(p)$, $g(p)$, such that $f = G + g$ and $|G| > |g|$ on C. Then (*theorem of Rouché*)

$$V_C[\arg f(p)] = V_C[\arg G(p)]. \tag{1}$$

PROOF. The assumptions imply that $G \neq 0$ and $f \neq 0$ on C. Hence the function $h(p) = f(p)/G(p)$ is continuous and different from zero on C. Furthermore

$$|h(p) - 1| = \frac{|f(p) - G(p)|}{|G(p)|} = \frac{|g(p)|}{|G(p)|} < 1, \qquad p \in C.$$

By II.4.5, II.4.25(e) it follows that $V_C[\arg h(p)] = 0$. Since $f = hG$, the formula (1) follows by II.4.25(b).

II.4.27. Let $\mathfrak{R}$ be a bounded, simply-connected Jordan region in the complex w-plane with boundary curve C. Let C be oriented in the counterclockwise sense. Suppose that $f(w)$ is a continuous complex-valued function of w such that $f(w) \neq 0$ in $\mathfrak{R}$. Then $V_C[\arg f(w)] = 0$.

PROOF. By II.4.13 (cf. I.2.50), $f(w)$ possesses a single-valued continuous argument $\phi(w)$ in $\mathfrak{R}$. Then $\phi(p)$ is also a single-valued continuous argument of $f(w)$ on C, and hence $V_C[\arg f(w)] = 0$ by II.4.25(e).

II.4.28. Let $\mathfrak{R}$ be a bounded, finitely-connected Jordan region in the w-plane. The *oriented boundary* B of $\mathfrak{R}$ is defined as the sum of its boundary curves with orientations assigned as follows. The exterior boundary curve of $\mathfrak{R}$ is oriented in the counterclockwise sense, the others (if any) are oriented in the clockwise sense. The symbol $C \in B$ will mean that C is a boundary curve, oriented in the manner just described, of $\mathfrak{R}$. Let then $f(w)$ be continuous, complex-valued in $\mathfrak{R}$ and different from zero on B. The variation of the argument of f on the oriented boundary B of $\mathfrak{R}$ is defined by the formula

$$V_B[\arg f(w)] = \sum V_C[\arg f(w)], \qquad C \in B.$$

The following facts are important.

(a) If $f(w)$ is different from zero in $\mathfrak{R}$, then $V_B[\arg f(w)] = 0$.

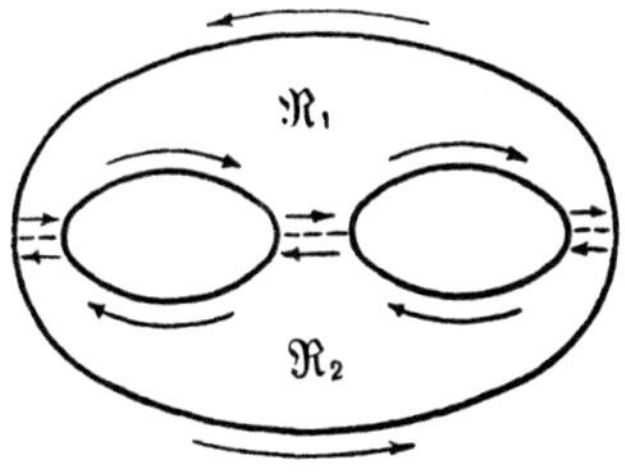

PROOF. If $\mathfrak{R}$ is simply-connected, then the assertion follows from II.4.27. So suppose that $\mathfrak{R}$ is multiply-connected, and let B denote the oriented boundary of $\mathfrak{R}$. The figure indicates the proof clearly. The auxiliary dotted arcs are introduced to decompose $\mathfrak{R}$ into two simply-connected Jordan regions $\mathfrak{R}_1$, $\mathfrak{R}_2$. If C_1 , C_2 are the boundary curves, oriented in the counterclockwise sense, of $\mathfrak{R}_1$, $\mathfrak{R}_2$ respectively, then

$$V_{C_1}[\arg f(w)] = 0, \qquad V_{C_2}[\arg f(w)] = 0,$$

by II.4.27. Addition yields $V_B[\arg f(w)] = 0$, since the contribution of the auxiliary dotted arcs is zero (cf. II.4.24(a)).

(b) Let $f(w)$, $G(w)$, $g(w)$ be continuous complex-valued functions in $\mathfrak{R}$ such that $f = G + g$ in $\mathfrak{R}$ and $|G| > |g|$ on the boundary B of $\mathfrak{R}$. Then $V_B[\arg f(w)] = V_B[\arg G(w)]$. Indeed, if $\sum$ denotes summation over the properly oriented boundary curves $C \in B$, then by II.4.26

$$V_B[\arg f(w)] = \sum V_C[\arg f(w)] = \sum V_C[\arg G(w)] = V_B[\arg G(w)].$$

II.4.29. Let C be an oriented simple closed curve in the complex z-plane, and let z_0 be a point that does not lie on C. We have then the following statements concerning $V_C[\arg (z - z_0)]$.

(a) If z_0 is exterior to C, then $V_C[\arg (z - z_0)] = 0$. Indeed, if $\mathfrak{R}$ denotes the bounded Jordan region determined by C, then $z - z_0 \neq 0$ on $\mathfrak{R}$, and hence the assertion follows directly from II.4.13.

(b) If z_0 is interior to C, and the orientation of C is counterclockwise, then $V_C[\arg (z - z_0)] = 2\pi$. Indeed, the assertion is obvious if C is a circle with center at z_0 . In the general case, let c be a circle, oriented counterclockwise, with center at z_0 and interior to C. Let $\mathfrak{R}$ be the doubly-connected Jordan region bounded by C and c, and let B be the properly oriented boundary of $\mathfrak{R}$. Then $z - z_0 \neq 0$ in $\mathfrak{R}$, and hence by II.4.28(a)

$$0 = V_B[\arg (z - z_0)] = V_C[\arg (z - z_0)] + V_{-c}[\arg (z - z_0)]$$

$$= V_C[\arg (z - z_0)] - 2\pi.$$

Thus $V_C[\arg (z - z_0)] = 2\pi$.

(c) If z_0 is interior to C, and the orientation of C is clockwise, then $V_C[\arg (z - z_0)] = -2\pi$. This follows from (b) and II.4.25(a).

II.4.30. Let σ be an oriented straight segment in the z-plane, and let z_0 be a point that does not lie on σ. Let β denote the radian measure, taken with the positive sign, of the angle subtended by σ at the point z_0 . Let z_1 , z_2 be the first and the second end points of the oriented segment σ. The points z_0 , z_1 , z_2 determine a (rectilinear) triangle Δ. Put $\lambda = 0$ if Δ is degenerate (that is, if z_0 , z_1 , z_2 are collinear), $\lambda = \beta$ if Δ is nondegenerate and the orientation of Δ determined by the arrangement z_0 , z_1 , z_2 of its vertices is counterclockwise, and $\lambda = -\beta$ if Δ is nondegenerate and the orientation just referred to is clockwise. Clearly $V_\sigma[\arg (z - z_0)] = \lambda$. This remark leads to the following generalization.

Let γ be an oriented simple arc, and let $f(p)$ be a continuous complex-valued function on γ. Let p_1 , p_2 denote the first and second end points of γ. Consider the transformation $T : z = f(p)$, $p \in \gamma$, and put $z_1 = f(p_1)$, $z_2 = f(p_2)$. Let z_0

be a point in the z-plane that does not lie on $T(\gamma)$. Suppose that there exists in the z-plane a straight line l through z_0 such that $T(\gamma)$ lies in one of the open half-planes determined by l. Then the oriented segment σ, whose first and second end points are z_1 and z_2, lies in the same open half-plane. If λ is defined, in terms of the point z_0 and the oriented segment σ, in the manner described above, then $V_\gamma[\arg (f(p) - z_0)] = \lambda$. Indeed, using a translation and a rotation, we can reduce the general case to the special case where $z_0 = 0$ and $T(\gamma)$ lies in the open half-plane $\Re z > 0$. Then the single-valued continuous argument $\alpha(z)$ of z is available (see II.4.2). Clearly $\alpha(z)$ is a single-valued continuous argument of z on σ, and $\alpha[f(p)]$ is a single-valued continuous argument of $f(p) - z_0 = f(p)$ on γ. Hence (note that now $z_0 = 0$)

$$V_\gamma[\arg f(p)] = \alpha[f(p_2)] - \alpha[f(p_1)] = \alpha(z_2) - \alpha(z_1) = V_\sigma[\arg z] = \lambda.$$

II.4.31. Let γ be an oriented simple arc. Let $h(\gamma_1) = \gamma$ be a topological transformation from a simple arc γ_1 onto γ. Let γ_1 be oriented in such a manner that its first end point is carried into the first end point of γ, and hence its second end point is carried into the second end point of γ. Suppose that $f(p)$ is a continuous complex-valued function on γ and $f \neq 0$ on γ. Then

$$V_\gamma[\arg f(p)] = V_{\gamma_1}[\arg f(h(p_1))], \tag{1}$$

where p_1 varies on γ_1. Indeed, if $\phi(p)$ is a single-valued continuous argument of $f(p)$ on γ, then $\phi[h(p_1)]$ is a single-valued continuous argument of $f[h(p_1)]$ on γ_1, and the assertion follows.

Let now C be an oriented simple closed curve, and let $h(C_1) = C$ be a topological transformation from a simple closed curve C_1 onto C. Let C_1 be oriented in such a manner that while a point p describes C in the sense assigned on C, the point $p_1 = h^{-1}(p)$ describes C_1 in the sense assigned on C_1. Suppose that $f(p)$ is continuous, complex-valued, and different from zero on C. Then

$$V_C[\arg f(p)] = V_{C_1}[\arg f(h(p_1))]. \tag{2}$$

This follows, in view of II.4.25, directly from the preceding result.

The formulas (1), (2) hold for sense-preserving homeomorphisms h. If h is sense-reversing, then it follows in the same manner that (1) and (2) hold with the minus sign inserted on the right.

II.4.32. Let $\Re$ be a simply-connected, bounded Jordan region in the $w = u + iv$ plane. Let w_0 be a point on the boundary curve of $\Re$, and let $f(w)$ be a complex-valued continuous function in $\Re$ such that $f(w_0) = 0$ and $f(w) \neq 0$ for $w \in \Re - w_0$. Then for every given positive integer k, there exists a single-valued continuous kth root of f in $\Re$.

Proof. Let us first assume that $\Re$ coincides with the region $\Re : v \geqq 0$, $0 \leqq u^2 + v^2 \leqq 1$, and $w_0 = 0$. For every positive integer n, let $\Re_n$ denote the region $\Re_n : v \geqq 0$, $(1/n)^2 \leqq u^2 + v^2 \leqq 1$. Then $\Re_n$ is simply-connected, and $f(w) \neq 0$ in $\Re_n$. Let $w^* = 1$. Then $w^* \in \Re_n$ for every n. Let a^* denote a kth root of $f(w^*)$. Since $\Re_n$ is simply-connected and $f(w) \neq 0$ in $\Re_n$, we have by

II.4.13 a single-valued continuous argument $\phi_n(w)$ of f in $\mathfrak{R}_n$. The function

$$\lambda_n(w) = |f(w)|^{1/k}\left[\cos\frac{\phi_n(w)}{k} + i\sin\frac{\phi_n(w)}{k}\right]$$

is then clearly a single-valued continuous kth root of f in $\mathfrak{R}_n$. Now a^* and $\lambda_n(w^*)$ are kth roots of the same complex number, namely of $f(w^*) \neq 0$, and hence $a^* = c_n\lambda_n(w^*)$, where c_n is a kth root of unity. Then $g_n(w) = c_n\lambda_n(w)$ is also a single-valued continuous kth root of f in $\mathfrak{R}_n$, and $g_n(w^*) = a^*$, $n = 1, 2, \cdots$. If $m \leqq n$, then $R_m \subset R_n$, and thus $g_m(w)$, $g_n(w)$ are two single-valued continuous kth roots of f in $\mathfrak{R}_m$ that agree for $w = w^*$. From II.4.20 it follows that $g_m(w) = g_n(w)$ for $w \in \mathfrak{R}_m$ if $m \leqq n$. Let us now define, in $\mathfrak{R}$, a function $g(w)$ as follows: $g(w) = g_n(w)$ if $w \in \mathfrak{R}_n$, and $g(0) = 0$. In view of the preceeding remark, $g(w)$ is clearly a single-valued continuous kth root of f in $\mathfrak{R}$ (note that $f(0) = 0$, and hence $|g(w)| = |f(w)|^{1/k} \to 0$ for $w \to 0$).

In the general case, we have (cf. I.2.50) a homeomorphism $h(\mathfrak{R}_*) = \mathfrak{R}$, where $\mathfrak{R}_*$ is the region $v_* \geqq 0$, $0 \leqq u_*^2 + v_*^2 \leqq 1$, in the $w_* = u_* + iv_*$ plane, and $h(0) = w_0$. The function $f_*(w_*) = f[h(w_*)]$ admits of a single-valued continuous kth root $g_*(w_*)$ in $\mathfrak{R}_*$ by the preceeding discussion, and clearly $g_*[h^{-1}(w)]$ is then a single-valued continuous kth root of $f(w)$ in $\mathfrak{R}$.

II.4.33. Let $f(w)$ be complex-valued continuous in a bounded, simply-connected Jordan region $\mathfrak{R}$ in the $w = u + iv$ plane. Let w_0 be an interior point of $\mathfrak{R}$, and let C be the boundary curve of $\mathfrak{R}$, where C is oriented in the counterclockwise sense. Suppose that the following conditions are satisfied: (i) $f(w_0) = 0$, (ii) $f(w) \neq 0$ for $w \in \mathfrak{R} - w_0$, (iii) $V_C[\arg f(w)] = \pm 2k\pi$, where k is a positive integer. Then there exists a single-valued continuous kth root of f in $\mathfrak{R}$.

PROOF. A reasoning analogous to that used at the end of II.4.32 shows that there is no loss of generality in assuming that $\mathfrak{R}$ is the unit disc $\mathfrak{R} : |w| \leqq 1$ and $w_0 = 0$ (cf. II.4.31). Let $\mathfrak{R}_1$, $\mathfrak{R}_2$ denote the Jordan regions $\mathfrak{R}_1 : 0 \leqq u^2 + v^2 \leqq 1$, $v \geqq 0$, and $\mathfrak{R}_2 : 0 \leqq u^2 + v^2 \leqq 1$, $v \leqq 0$, respectively. By II.4.32 we have then a single-valued continuous kth root $g_1(w)$ of f in $\mathfrak{R}_1$ and also a single-valued continuous kth root $g_2(w)$ of f in $\mathfrak{R}_2$. Let a_0 be a kth root of $f(1)$. Then $a_0 = c_1g_1(1)$, where c_1 is a kth root of unity. Hence, replacing $g_1(w)$ by $c_1g_1(w)$, we see that we can assume that $g_1(1) = a_0$, and similarly we can assume that $g_2(1) = a_0$. In other words, we can select $g_1(w)$, $g_2(w)$ to satisfy the condition

$$g_1(1) = g_2(1) = a_0. \tag{1}$$

Now let C_1 denote the semicircle $u^2 + v^2 = 1$, $v \geqq 0$, oriented from $w = 1$ toward $w = -1$, and let C_2 denote the semicircle $u^2 + v^2 = 1$, $v \leqq 0$, oriented from $w = -1$ toward $w = 1$. Let ϕ_0 be an argument of the number $a_0 \neq 0$. Since $g_1(w) \neq 0$ on C_1, we have on C_1 a single-valued continuous argument $\phi_1(w, C_1)$ of $g_1(w)$ such that $\phi_1(1, C_1) = \phi_0$ (cf. II.4.14, II.4.4). Similarly, we have a single-valued continuous argument $\phi_2(w, C_2)$ of $g_2(w)$ on C_2 such that $\phi_2(1, C_2) = \phi_0$. Thus

$$\phi_1(1, C_1) = \phi_2(1, C_2) = \phi_0. \tag{2}$$

We have then the formulas

$$(3) \qquad V_{C_1}[\arg g_1(w)] = \phi_1(-1, C_1) - \phi_1(1, C_1),$$

$$(4) \qquad V_{C_2}[\arg g_2(w)] = \phi_2(1, C_2) - \phi_2(-1, C_2).$$

Since $g_1(w)^k = f(w)$ on C_1, and $g_2(w)^k = f(w)$ on C_2, we have (see II.4.24(c))

$$(5) \qquad V_{C_1}[\arg f(w)] = kV_{C_1}[\arg g_1(w)],$$

$$(6) \qquad V_{C_2}[\arg f(w)] = kV_{C_2}[\arg g_2(w)].$$

From (2)—(6) we infer the relations (cf. II.4.25)

$$V_C[\arg f(w)] = V_{C_1}[\arg f(w)] + V_{C_2}[\arg f(w)] = k[\phi_1(-1, C_1) - \phi_2(-1, C_2)].$$

Since $V_C[\arg f(w)] = \pm 2k\pi$ by assumption, it follows that $\phi_1(-1, C_1) = \phi_2(-1, C_2) \pm 2\pi$. Since $\phi_1(-1, C_1)$, $\phi_2(-1, C_2)$ are arguments of $g_1(-1)$, $g_2(-1)$ respectively, and $|g_1(-1)| = |g_2(-1)| = |f(-1)|^{1/k}$, it follows that

$$(7) \qquad g_1(-1) = g_2(-1).$$

Now let u_0 satisfy the inequality $-1 < u_0 < 0$. On the segment $-1 \leqq u \leqq u_0$, g_1 and g_2 are single-valued continuous kth roots of f that agree for $u = -1$ by (7). By II.4.20 it follows that $g_1(u_0) = g_2(u_0)$. Since $g_1(0) = g_2(0) = 0$ (note that $f(0) = 0$), it follows that $g_1(u) = g_2(u)$ for $-1 \leqq u \leqq 0$. Similarly, (1) leads to the conclusion that $g_1(u) = g_2(u)$ for $0 \leqq u \leqq 1$. Hence, if we define $g(w) = g_1(w)$ in $\mathfrak{R}_1$ and $g(w) = g_2(w)$ in $\mathfrak{R}_2$, then $g(w)$ is a single-valued continuous kth root of f in $\mathfrak{R}$.

II.4.34. Let C be an oriented simple closed curve, and let $T: z = f(p)$, $p \in C$, be a continuous transformation from C into the complex z-plane (that is, $f(p)$ is a continuous complex-valued function on C). For each point z_0 in the z-plane, we define an *index* $\mu(z_0, T, C)$ as follows. If $z_0 \in T(C)$, then $\mu(z_0, T, C) = 0$. If $z_0 \notin T(C)$, then $\mu(z_0, T, C) = V_C[\arg (f(p) - z_0)]/2\pi$ (cf. II.4.25). Dropping the subscript zero, we obtain in this manner a function $\mu(z, T, C)$, defined for every z. Let us note the following facts.

(a) Let $\bar{F}$ be the image of C under T, and let $\bar{G}$ be the complement of $\bar{F}$ with respect to the z-plane. Then $\bar{F}$ is a closed bounded set, and hence $\bar{G}$ itself is an open set. The set $\bar{G}$ may be connected or disconnected, but since $\bar{F}$ is bounded, $\bar{G}$ has a unique unbounded component that we denote by Γ_∞ (if $\bar{G}$ is connected, then of course $\Gamma_\infty = \bar{G}$). Each component of $\bar{G}$ is open.

(b) Let Γ be a component of $\bar{G}$ (see (a)). We assert that $\mu(z, T, C)$ is constant on Γ. Indeed, $\mu(z, T, C)$ is clearly a continuous function of z on Γ by II.4.25(c), and $\mu(z, T, C)$ takes on only integral values on Γ by II.4.25(d). Since Γ is connected, it follows that $\mu(z, T, C)$ is constant on Γ (see I.2.45).

(c) $\mu(z, T, C) = 0$ on Γ_∞ (see (a)). Indeed, $\mu(z, T, C)$ is constant on Γ_∞ by (b), and hence it is sufficient to exhibit a single point $z_0 \in \Gamma_\infty$ such that $\mu(z_0, T, C) = 0$. Now the set $\bar{F} = T(C)$ is bounded, and hence there exists a real constant x_0 such that $\mathfrak{R}f(p) > x_0$ for $p \in C$. Then $\mathfrak{R}[f(p) - x_0] > 0$ on C, and hence there exists a single-valued continuous argument for $f(p) - x_0$ on C by II.4.5. By II.4.25(e) it follows that $V_C[\arg(f(p) - x_0)] = 0$, and hence $\mu(x_0, T, C) = 0$. Since clearly $x_0 \in \Gamma_\infty$, the assertion $\mu(z, T, C) = 0$ in Γ_∞ is established.

(d) Summing up: $\mu(z, T, C)$ assumes only integral values, is constant on every component of the complement of the set $\overline{F} = T(C)$, and is equal to zero on $\overline{F}$ and on Γ_∞. Thus $\mu(z, T, C)$ is clearly Borel measurable (see I.3.8).

(e) Let $h(C^*) = C$ be a topological transformation from a simple closed curve C^* onto C, and let C^* be so oriented that h is sense-preserving. Let us put $f^*(p^*) = f[h(p^*)]$, $p^* \in C^*$, and let T^* be the transformation $T^* : z = f^*(p^*)$, $p^* \in C^*$. Then $\mu(z, T, C) \equiv \mu(z, T^*, C^*)$, as an immediate consequence of II.4.31. If h is sense-reversing, then it follows similarly that $\mu(z, T, C) \equiv -\mu(z, T^*, C^*)$.

II.4.35. In the $w = u + iv$ plane, consider a rectangle $R : u_1 \leqq u \leqq u_2$, $v_1 \leqq v \leqq v_2$. Let C be the boundary of R oriented in the counterclockwise sense. Let $f(w)$ be a continuous complex-valued function on C, and let z_0 be a complex number such that the following inequalities hold.

(1) $\Re[f(w) - z_0] > 0$ for $w = u_2 + iv, v_1 \leqq v \leqq v_2$,

(2) $\Im[f(w) - z_0] > 0$ for $w = u + iv_2, u_1 \leqq u \leqq u_2$,

(3) $\Re[f(w) - z_0] < 0$ for $w = u_1 + iv, v_1 \leqq v \leqq v_2$,

(4) $\Im[f(w) - z_0] < 0$ for $w = u + iv_1, u_1 \leqq u \leqq u_2$.

Let T denote the transformation $z = f(w)$, $w \in C$. Then $\mu(z_0, T, C) = 1$.

Indeed, let z_1, z_2, z_3, z_4 denote the images of the vertices (u_1, v_1), (u_2, v_1), (u_2, v_2), (u_1, v_2). In view of (1)-(4), it follows readily from II.4.30 that $V_C[\arg(f(w) - z_0)] = 2\pi$, and hence $\mu(z_0, T, C) = 1$.

If the signs of inequality are reversed in (2) and (4) (or alternatively in (1) and (3)), then it follows similarly that $\mu(z_0, T, C) = -1$.

II.4.36. Let C be a simple closed curve, oriented in the counterclockwise sense, in the $w = u + iv$ plane. Let $w_1, \cdots, w_n$ be distinct points on C, arranged in the counterclockwise sense. We put, for convenience, $w_{n+1} = w_1$. These points divide C into n oriented simple arcs $\gamma_1, \cdots, \gamma_n$, without common interior points, where the end points of γ_k are w_k and w_{k+1} in this order. In the $z = x + iy$ plane, let there be assigned points $\tau_1, \cdots, \tau_n, \tau_{n+1} = \tau_1$ that are not assumed to be distinct. If $\tau_k \neq \tau_{k+1}$, then let σ_k denote the oriented straight segment with first end point τ_k and second end point τ_{k+1}. If $\tau_k = \tau_{k+1}$, then let σ_k denote the single point $\tau_k = \tau_{k+1}$. For each $k = 1, \cdots, n$, let us define on γ_k a function $\tau = h_k(w)$ as follows. If σ_k does not reduce to a single point, then h_k is chosen in such a way that (i) $h(w_k) = \tau_k$, $h(w_{k+1}) = \tau_{k+1}$, and (ii) the equation $\tau = h_k(w)$ defines a topological transformation from γ_k onto σ_k. If σ_k reduces to a single point (that is, if $\tau_k = \tau_{k+1}$), then $h_k(w) = \text{constant} = \tau_k = \tau_{k+1}$ on γ_k. Let us now define on C the function $f(w)$ as follows: $f(w) = h_k(w)$ if $w \in \gamma_k$. Clearly, $f(w)$ is single-valued and continuous on C, and $f(w_k) = \tau_k$, $k = 1, 2, \cdots, n$. Finally, we introduce the transformation $T : \tau = f(w)$, $w \in C$, and we proceed to derive several simple and important facts concerning the index-function $\mu(z, T, C)$ (see II.4.34).

II.4.37. CONTINUATION. If σ_k does not reduce to a single point for a certain k, then the function $h_k(w)$ is not univocally determined. Suppose $g_k(w)$ is a second function satisfying the same conditions as $h_k(w)$. From II.4.30 it follows readily that

$$V_{\gamma_k}[\arg (h_k(w) - z)] = V_{\gamma_k}[\arg (g_k(w) - z)],$$

for every z not on σ_k. By II.4.25, II.4.34 it follows that $\mu(z, T, C)$ is independent of the particular choice of the functions $h_k(w)$ described in II.4.36. From II.4.34(e) we infer that we could assume, without loss of generality, that C is a simple closed convex polygon with vertices $w_1, \cdots, w_n$. In this special case, the function $h_k(w)$ may be chosen, if σ_k does not reduce to a single point, in such a manner that the transformation $\tau = h_k(w)$, $w \in \gamma_k$, maps γ_k linearly onto σ_k. Then each h_k, and hence also the transformation T of II.4.36, is univocally determined.

II.4.38. CONTINUATION. By II.4.34(c), the index-function $\mu(z, T, C)$ vanishes outside of a sufficiently large circular disc in the z-plane. The symbol $\iint \mu(z, T, C)\, dx\, dy$ will denote the integral of $\mu(z, T, C)$ over such a disc. Let ξ_k, η_k be the coordinates of τ_k (that is, $\tau_k = \xi_k + i\eta_k$). Then

$$\iint \mu(z, T, C)\, dx\, dy = \frac{1}{2} \sum_{k=1}^{n} (\xi_k \eta_{k+1} - \eta_k \xi_{k+1}). \tag{1}$$

PROOF. Let us take in the z-plane an auxiliary point τ_0 that does not coincide with any one of the points $\tau_1, \cdots, \tau_n$, and let s_k denote the oriented segment with first end point τ_0 and second end point τ_k. Let us define, if σ_k does not reduce to a single point,

$$V(z, \sigma_k) = V_{\sigma_k}[\arg (\tau - z)],$$

where τ varies on σ_k and $z \notin \sigma_k$. If either σ_k reduces to a single point or z lies on σ_k, we put $V(z, \sigma_k) = 0$. By II.4.34, II.4.25, II.4.30 we have then

$$\mu(z, T, C) = \frac{1}{2\pi} \sum_{k=1}^{n} V(z, \sigma_k) \qquad \text{for } z \notin \sigma_1 + \cdots + \sigma_n. \tag{2}$$

Similarly, we put

$$V(z, s_k) = V_{s_k}[\arg (\tau - z)] \qquad \text{for } z \notin s_k,$$

where τ varies on s_k. If $z \in s_k$, then we put $V(z, s_k) = 0$. Clearly

$$\sum_{k=1}^{n} V(z, \sigma_k) = \sum_{k=1}^{n} [V(z, s_k) + V(z, \sigma_k) - V(z, s_{k+1})] \qquad \text{for } z \notin \sum_{k=1}^{n} s_k + \sum_{k=1}^{n} \sigma_k. \tag{3}$$

For given k, the following cases may arise.

Case (i). The triangle with the vertices τ_0, τ_k, τ_{k+1} is degenerate. From II.4.24(a), (b) it follows immediately that $V(z, s_k) + V(z, \sigma_k) - V(z, s_{k+1}) = 0$ for $z \notin s_k + \sigma_k + s_{k+1}$. Hence in this case

(4) $$\iint [V(z, s_k) + V(z, \sigma_k) - V(z, s_{k+1})]\, dx\, dy = 0,$$

where the integration is extended over any circular disc.

Case (ii). The triangle with vertices τ_0, τ_k, τ_{k+1} is nondegenerate. Let then Γ_k denote the perimeter of this triangle, oriented in the sense τ_0, τ_k, τ_{k+1}. Let ϵ_k be equal to 1 if Γ_k is oriented counterclockwise, and put $\epsilon_k = -1$ if Γ_k is oriented clockwise. By II.4.29 we have then

(5) $$V(z, s_k) + V(z, \sigma_k) - V(z, s_{k+1}) = V_{\Gamma_k}[\arg(\tau - z)] = 2\pi\epsilon_k,$$

where τ varies on Γ_k, if z is interior to the triangle τ_0, τ_k, τ_{k+1}, and

(6) $$V(z, s_k) + V(z, \sigma_k) - V(z, s_{k+1}) = V_{\Gamma_k}[\arg(\tau - z)] = 0,$$

if z is exterior to that triangle. Hence, by a well known formula in Analytic Geometry,

(7) $$\frac{1}{2\pi}\iint [V(z, s_k) + V(z, \sigma_k) - V(z, s_{k+1})]\, dx\, dy = \frac{1}{2}\begin{vmatrix} \xi_0 & \eta_0 & 1 \\ \xi_k & \eta_k & 1 \\ \xi_{k+1} & \eta_{k+1} & 1 \end{vmatrix},$$

where the integration is extended over any circular disc that contains the points τ_0, τ_k, τ_{k+1}.

Now let us note that the determinant in (7) vanishes if the triangle τ_0, τ_k, τ_{k+1} is degenerate. By (4) it follows that (7) holds for every $k = 1, \cdots, n$. The formula (1) follows now from (2), (3), (7) by elementary computations (note that the set $s_1 + \cdots + s_n + \sigma_1 + \cdots + \sigma_n$ is of measure zero).

II.4.39. Continuation. Take a point $z_0 = x_0 + iy_0$ in the z-plane. Let α be any real number. Then the equations $x = x_0 + t\cos\alpha$, $y = y_0 + t\sin\alpha$, $0 \leqq t < +\infty$, represent a ray r issuing from z_0. Suppose that r does not pass through any of the points τ_k. Let m be the number of those subscripts k for which $r\sigma_k \neq 0$. Then

(1) $$|\mu(z_0, T, C)| \leqq m.$$

Proof. If $z_0 \in T(C) = \sigma_1 + \cdots + \sigma_n$, then $\mu(z_0, T, C) = 0$ (see II.4.34), and (1) is obvious. So we can assume that

(2) $$z_0 \notin \sigma_1 + \cdots + \sigma_n.$$

Let r^* be the ray $x = x_0 - t\cos\alpha$, $y = y_0 - t\sin\alpha$, $0 \leqq t < +\infty$, and let us choose on r^* a point $\tau_0 \neq z_0, \tau_1, \cdots, \tau_n$. In terms of this point τ_0, we define the oriented segments s_k and the quantities $V(z, s_k)$, $V(z, \sigma_k)$ as in II.4.38. Let

$\sum'$ denote summation over those nondegenerate triangles $\tau_0, \tau_k, \tau_{k+1}$ that contain z_0 as an interior point. By the formulas (2), (3), (5), (6) in II.4.38 it follows that

$$\mu(z_0, T, C) = \sum{}' \epsilon_k . \tag{3}$$

Clearly, z_0 is interior to a triangle $\tau_0, \tau_k, \tau_{k+1}$ if and only if the ray r intersects the segment σ_k. Hence $\sum' | \epsilon_k | = m$. Thus (1) follows from (3).

II.4.40. Let us now consider a Fréchet curve Γ of the type of the simple closed curve in Euclidean three-space (cf. II.3.7). Let Γ be given in terms of a representation $\Gamma : T(C) = \mathcal{P}^*$, where C is a simple closed curve and $\mathcal{P}^*$ is a subset of Euclidean three-space. By II.3.11 we can assume that C lies in the complex w-plane. Let us now assume that $\mathcal{P}^*$ is a subset of a plane p. If we introduce in p a Cartesian coordinate system ξ, η, and if we put $\tau = \xi + i\eta$, then T may be given in the form $T : \tau = f(w)$, $w \in C$, where $f(w)$ is a complex-valued continuous function on C. Let C be oriented in the counterclockwise sense relative to the $w = u + iv$ plane. There arises an index-function $\mu(\tau, T, C)$, and we propose to determine the manner in which μ depends upon the representation chosen for Γ. So let $T_* : \tau = f_*(w_*)$, $w_* \in C_*$, be a second representation of Γ, where C_* lies in a complex $w_* = u_* + iv_*$ plane, and C_* is oriented in the counterclockwise sense relative to the w_*-plane. We are to compare the index-functions $\mu(\tau, T, C)$ and $\mu(\tau, T_*, C_*)$ (cf. II.4.34). By II.3.12 we can choose for each n a homeomorphism $h_n(C) = C_*$, such that the functions $f_n(w) = f_*[h_n(w)]$ converge on C uniformly to $f(w)$. Let us put $\lambda_n = +1$ if h_n is sense-preserving, and $\lambda_n = -1$ if h_n is sense-reversing. By II.4.34(e) we have

$$\mu(\tau, T_n, C) \equiv \lambda_n \mu(\tau, T_*, C_*), \tag{1}$$

where T_n denotes the transformation $\tau = f_n(w)$, $w \in C$. By II.4.34, II.4.25(c) we have the relation

$$\mu(\tau, T_n, C) \to \mu(\tau, T, C) \qquad \text{for } n \to \infty . \tag{2}$$

If $\lambda_n = +1$ for infinitely many values of n, then (1), (2) imply that $\mu(\tau, T, C) \equiv \mu(\tau, T_*, C_*)$. If $\lambda_n = -1$ for infinitely many values of n, then it follows that $\mu(\tau, T, C) \equiv -\mu(\tau, T_*, C_*)$. Summing up: there exists a constant λ, equal to $+1$ or -1, such that $\mu(\tau, T, C) \equiv \lambda\mu(\tau, T_*, C_*)$.

CHAPTER II.5. GENERAL COMMENTS ON CURVES AND SURFACES

II.5.1. In view of the general occurrence and utility of the terms *curve* and *surface* in various scientific fields and in everyday life, a detailed study of the psychological and technical considerations that lead from the primitive intuitive ideas to the precise formal definitions should be a topic of considerable interest. The book of Menger [1] gives a comprehensive picture as far as the point set theoretical approach is concerned. A brief but illuminating account may be found in G. T. Whyburn [2]. As regards the concepts of curve and surface as used in Chapter II.3, the reader will find a number of interesting and relevant comments in J. W. T. Youngs [2]. We restrict ourselves to remarks meant to prevent misunderstandings on the part of a reader who did not make a special study of the concepts curve and surface, even though he may have used these terms extensively in his work in fields like Differential Geometry, Differential Equations, Complex Variables, Calculus of Variations, and so forth.

II.5.2. For a while, we restrict ourselves to curves, and in fact to curves in Euclidean three-space (only the real case is relevant for our purposes). Disregarding for the moment highly specialized studies, the general interpretation of the term curve in mathematical literature may be described as follows.

(i) Primarily, a curve is thought of as *a point set with certain properties of slenderness.* Thus a circle is a curve, while a circular disc is not a curve. To simplify matters, let us consider only continua in Euclidean three-space E_3. Given then a continuum $\Gamma \subset E_3$ (that is, a bounded, closed, connected set in E_3), in certain simple cases Γ will be classified as a curve, in certain other simple cases Γ will be definitely recognized as not being a curve. From the point of view of most mathematical disciplines, a sharp distinction between curves and non-curves is irrelevant, however. Subject to this vagueness, let us use the term *point curve* to refer to a continuum $\Gamma \subset E_3$ that is admitted as a curve because of its slenderness. The term point curve is used in contradistinction to the term *path curve,* to be considered presently.

(ii) From the analytic point of view, a curve is determined by equations of the form

$$(1) \qquad x = x(u), \quad y = y(u), \quad z = z(u), \qquad u \in I, I : a \leqq u \leqq b,$$

where the functions $x(u)$, $y(u)$, $z(u)$ are continuous in the interval I and possess such further properties of regularity as may be needed for a particular purpose. The points (x, y, z) that correspond to the points u of I by means of the equations (1) constitute a continuum Γ in E_3. In most simple cases, the correspondence between I and Γ will be biunique, but in many simple cases a biunique correspondence cannot be achieved. For example, if Γ is the unit circle, then the representation

$$(2) \qquad x = \cos u, \qquad y = \sin u, \qquad z = 0, \qquad 0 \leqq u \leqq 2\pi,$$

is not biunique, but is still accepted as a useful representation of the unit circle. While (2) presents a minimum amount of duplication, the representation

$$(3) \qquad x = \cos u, \qquad y = \sin u, \qquad z = 0, \qquad 0 \leqq u \leqq 4\pi,$$

is uneconomical from this point of view, since now the unit circle is described twice while u increases from 0 to 4π. On the other hand, we may think of a representation of the form (1) as representing not merely a continuum Γ but rather a *continuous trip* over Γ. Then (2) and (3) represent two entirely different continuous trips around the unit circle. If we consider a representation (1) as describing a continuous trip, then we shall say that (1) determines a *path curve*, where this term will be used for the moment in the same vague manner as the term *point curve* was used in (i) above. The formulation of precise formal definitions for the concepts point curve and path curve is of course a matter of importance, even though it seems that such formal definitions are not absolutely necessary in most mathematical fields. In elementary geometry, the term curve is used, generally, in the sense of point curve. On the other hand, in applications in Analysis the interpretation of a representation (1) in terms of a path curve is more appropriate (cf. the remarks in I.1.5).

(iii) Given a path curve by a representation (1), the locus of the points (x, y, z) obtained by means of the equations (1) is a continuum $\Gamma \subset E_3$ that we shall call, for conciseness, *the trace of the path curve.* Inspection of a lemniscate, for example, reveals that a given point curve may be the trace of several equally plausible path curves. It is then natural to expect that point curves and path curves should be related in the following simple manner: every path curve determines a point curve as its trace, and every point curve is the trace of several path curves. The concept of *continuous traversibility* permits us to condense the preceding statement. A continuum $\Gamma \subset E_3$ is termed continuously traversible if and only if it is possible for a particle, moving continuously according to equations of the form (1), to describe precisely Γ while u (thought of as representing the time) increases from a to b, where the particle is permitted to pass several times through the same point of Γ. Using more technical language, a continuum Γ is continuously traversible if and only if it is the trace of some path curve. The relation between the concepts *path curve* and *point curve,* stated above, may be reworded as follows: point curves are identical with the continuously traversible continua, and the continuously traversible continua are identical with the traces of path curves. In any case, it is probable that this statement represents, except for terminology, the generally accepted view until about 1890, when Peano [1] discovered that *the square is continuously traversible.* From the point of view of mathematical fields of a classical type, this phenomenon is essentially an irrelevant paradox. For instance, if a path curve has finite length, then it follows readily that its trace is a point set without interior points. Hence, a path curve whose coordinate functions $x(u)$, $y(u)$, $z(u)$ satisfy the assumptions usually made in Differential

Geometry, for example, cannot fill a square or a cube. On the other hand, it is apparent that in mathematical fields where a precise formal definition of the term curve is indispensable the discovery of Peano creates an issue that must be thoroughly investigated.

(iv) Returning to the term path curve, we have already noted that this concept is more appropriate in situations where a curve is thought of as being determined by equations of the form (1). In particular, if the representation (1) is not known or is not assumed to be biunique, then the concept of path curve is preferable. Hence we shall use the term path curve in referring to disciplines like Differential Geometry, even though in the literature of that discipline only the term curve is used. Generally speaking, the concept of a path curve (in the vague sense just indicated) is used in mathematical literature without any indication of the intended meaning. On the other hand, the following information is usually given in conection with the use of this concept.

(a) The class of *admissible representations* of the form (1) is described in terms of assumptions concerning the coordinate functions $x(u)$, $y(u)$, $z(u)$. For example, these functions may be required to be analytic, and usually it is also required that the derivatives $x'(u)$, $y'(u)$, $z'(u)$ do not vanish simultaneously. In certain cases, analyticity is replaced by some less restrictive assumption. In any case, a class $K(r)$ of admissible representations r, of the form (1), is defined.

(b) A rule is given for replacing u by a new parameter u^*. For example, u may be required to be a strictly increasing analytic function of u^*. Again, in certain cases analyticity is replaced by some less restrictive requirement. In any case, *a criterion is given to determine whether two admissible representations* r_1, r_2 *belong to the same path curve.* Let us write $r_1 \sim r_2$ to state that r_1, r_2 belong to the same path curve according to the given rule. Then clearly the binary relation $r_1 \sim r_2$ is an *equivalence* (that is, it is symmetric, reflexive, and transitive). Furthermore, *the rule for changing the parameter does not involve the concept of a path curve* but is concerned solely with a relation between admissible representations.

Once the class $K(r)$ and the equivalence $r_1 \sim r_2$ are given, it is clear that the term curve can be eliminated from all statements and proofs. Thus it appears that a formal definition of the concept of a path curve is really unnecessary, while of course the class $K(r)$ and the equivalence $r_1 \sim r_2$ must be identified in a definite manner if the subsequent theory is to have a definite meaning. Alternatively, we may say that the term curve (in the sense of path curve) may be used as an *undefined term,* provided that the class $K(r)$ and the equivalence $r_1 \sim r_2$ are definitely fixed. The situation is analogous to that arising in an axiomatic treatment of geometry where the term *point* is undefined. In our case, the axioms are represented, in a somewhat unusual sense perhaps, by the assignment of the class $K(r)$ and of the equivalence $r_1 \sim r_2$.

The preceding statements may be interpreted as a justification of the general lack of a precise explanation of the concept of a path curve in many mathematical fields. Strictly speaking, the use of the concept of a path curve as an undefined concept should be justified by exhibiting a *mathematical model,* as Analytic Geometry may be used to show the consistency of the axioms of Euclidean geometry. In our case, however, this is an entirely trivial matter. Indeed, given *any* class $K(r)$ of admissible representations r and *any* equivalence $r_1 \sim r_2$ in $K(r)$, let us define a path curve C as an equivalence class, relative to the equivalence $r_1 \sim r_2$ in the class $K(r)$ (see II.3.5, II.3.6). Let us further say that an element r of $K(r)$ is a representation of C if and only if r is an element of the equivalence class C. This mathematical model is obviously adequate for the purpose. Once the existence of a mathematical model is ascertained, one may use this particular model or any other interpretation of the term path curve that is consistent with the class $K(r)$ and the equivalence $r_1 \sim r_2$.

According to this point of view, there are as many different species of path curves as there are possible choices of the class $K(r)$ *and of the equivalence* $r_1 \sim r_2$. In this sense, we should speak of $(K, \sim)$ *path curves* instead of merely path curves. Inspection of the literature shows that in different fields, and in fact on different occasions in the same field, different species of $(K, \sim)$ path curves are used. It also appears that the F-curves of the type of the 1-cell defined in II.3.7 represent merely one particular kind of $(K, \sim)$ path curves.

II.5.3. Remarks analogous to those made in the preceding section apply to surfaces. We turn presently to a review of precise formal definitions. The comments made in II.5.2(iv) suggest that we may expect a variety of formal concepts, each one adjusted to a particular need. As regards point curves, such flexibility does not seem to prevail. According to the generally accepted view of experts (see Menger [1], G. T. Whyburn [2]), a continuum Γ is a curve (a point curve in our sense) if and only if it is one-dimensional in the Menger-Urysohn sense (see Hurewicz-Wallman [1] for an excellent presentation of topological dimension theory). The non-specialist should note the fact that *a point curve in the sense of this definition may lack the property of continuous traversibility* (cf. II.5.2(iii)), as simple examples show. As a consequence, this concept of a general point curve will not play a role in the theory presented in this book. On the other hand, extensive use will be made of certain special point curves (simple arcs, simple closed curves). In this connection, it should be also noted that *topological slenderness,* in the sense of the above formal definition of a point curve, does not generally imply *metric slenderness,* as shown by the so-called Osgood curve (simple closed plane curve of positive two-dimensional measure).

II.5.4. As regards the formal definition of a path curve in terms of an equivalence class (see II.5.2(iv)), it was first used, apparently, by H. Kneser [1] in 1926. Stated for Euclidean three-space, his class $K(r)$ of admissible representations r consists of all representations of the form $x = x(u)$, $y = y(u)$, $z = z(u)$, $0 \leqq u \leqq 1$, where $x(u)$, $y(u)$, $z(u)$ are assumed to be merely continuous. Two such representations are equivalent if and only if one can be obtained from the other by means of a topological mapping of the parameter interval upon itself. However, essentially the same concept has been already adopted by Lebesgue [2] in 1902. Lebesgue uses the same class $K(r)$ and the same equivalence $r_1 \sim r_2$ as Kneser, but instead of using equivalence classes, he proceeds as follows. He first states that a curve C is merely a set of equations $x = x(u)$, $y = y(u)$, $z = z(u)$, $a \leqq u \leqq b$, where $x(u)$, $y(u)$, $z(u)$ are continuous functions in the indicated interval. Two such curves C_1, C_2 are then termed identical if and only if the corresponding sets of equations can be derived from each other by a topological mapping of the parameter intervals. Thus, ultimately, a curve in the sense of Lebesgue is an equivalence class of representations, and hence the difference between the concepts proposed by Lebesgue and Kneser may be considered as purely verbal. Such verbal differences will be disregarded in the following brief historical comments.

Analytically, the equivalence $r_1 \sim r_2$ adopted by Kneser may be described as follows: if the parameter u is replaced by a parameter u^* such that u is a strictly increasing (or strictly decreasing) continuous function of u^*, then the resulting representation, in terms of u^*, is equivalent to the initial representation, and all equivalent representations are obtained in this manner. This wording shows that this equivalence conforms to the generally accepted notion of a change of the parameter, except that only continuity is required instead of the more restrictive assumptions usually made. However, the equivalence $r_1 \sim r_2$ used by Kneser is still too restrictive for several reasons. For example, let r_1 be defined as follows: while u increases from 0 to 1/3, the corresponding point (x, y, z) travels steadily from 0 to 1/2 along the straight segment with end points $(0, 0, 0)$, $(1, 0, 0)$; while u increases from 1/3 to 2/3, the corresponding point rests; and finally, while u increases from 2/3 to 1, the corresponding point travels steadily from 1/2 to 1. If one wishes to introduce the arc length of the resulting path curve as a new paremeter, one finds that the arc length is not a strictly increasing function of u, and hence one is prevented from using the obviously most favorable parameter because of the too restricted character of the equivalence adopted by H. Kneser. Further difficulties arise if one attempts to turn the class of path curves into a metric space, an issue of greatest importance from the point of view of the theory presented in this book. These difficulties disappear if the less restrictive concept of F-equivalence is introduced (see II.3.7). This fundamental improvement is due to Fréchet [3], who introduced his concept of distance in 1924. The general concept of a *variety relative to a binary relation* (see II.3.6; II.3.7) was formulated by J. W. T. Youngs [2].

Let us observe that the concept of F-equivalence admits of a refinement in certain cases. For example, let $T_1(\mathcal{P}_1) = \mathcal{P}^*$, $T_2(\mathcal{P}_2) = \mathcal{P}^*$ be two continuous mappings, where $\mathcal{P}_1$, $\mathcal{P}_2$ are 1-cells. Suppose that $\mathcal{P}_1$, $\mathcal{P}_2$ are assigned definite orientations. For clarity, let us write ${}_0\mathcal{P}_1$, ${}_0\mathcal{P}_2$ to refer to the oriented 1-cells $\mathcal{P}_1$, $\mathcal{P}_2$. The mappings T_1, T_2 are then termed *positively F-equivalent*, in symbols $T_1 \sim T_2(F^+)$, if and only if for every $\epsilon > 0$ there exists a topological mapping $h({}_0\mathcal{P}_1) = {}_0\mathcal{P}_2$, such that (i) the assigned orientation of ${}_0\mathcal{P}_1$ corresponds to the assigned orientation of ${}_0\mathcal{P}_2$ under h, and (ii) $\rho[T_1(x_1), T_2h(x_1)] < \epsilon$ for every point $x_1 \in \mathcal{P}_1$. The relation (F^+) leads to the concept of *oriented F-curves of the type of the 1-cell.* Similar considerations apply if 1-cells are replaced by 1-spheres,

2-cells, 2-spheres, and so forth. Such oriented F-varieties were introduced by McShane [2, 3, 6], in connection with studies in Calculus of Variations. J. W. T. Youngs [3] made a detailed topological study of oriented F-surfaces of the type of the 2-sphere. However, as far as arc length and surface area are concerned, the non-oriented F-varieties seem to be entirely adequate at present.

II.5.5. We turn now to a more specific discussion of the F-varieties defined in II.3.7. Let $\mathcal{P}_0$ be a fixed Peano space, P^* a fixed metric space, and V an F-variety of the type of $\mathcal{P}_0$ in P^*. Then V is given by a representation of the form

$$(1) \qquad V : T(\mathcal{P}) = \mathcal{P}^*, \qquad \mathcal{P}^* \subset P^*,$$

where $\mathcal{P}$ is a Peano space homeomorphic with $\mathcal{P}_0$, and T is a continuous mapping. Now V may possess more favorable representations than the initial representation (1), where the meaning of the term *favorable* depends, of course, upon the particular purpose that one has in mind. For example, in Calculus of Variations one may obtain a minimizing surface in a representation that lacks certain desirable properties of regularity, and one may then look for more favorable representations of the same surface. Generally speaking, it is clear that the following *representation problem* is of fundamental importance: *given V in terms of an initial representation* (1), *find all the representations of V.* In view of the definition of an F-variety (see II.3.7), this problem may be stated in the following equivalent form: *given a continuous mapping $T_1(\mathcal{P}_1) = \mathcal{P}^*$, find all the continuous mappings $T_2(\mathcal{P}_2) = \mathcal{P}^*$ that satisfy the relation $T_1 \sim T_2(F)$* (cf. II.1.24, II.1.25). In this statement it is understood that $\mathcal{P}_1$ and $\mathcal{P}_2$ are homeomorphic Peano spaces, and then $\mathcal{P}^*$ is necessarily a Peano space also (of course, $\mathcal{P}^*$ is generally *not* homeomorphic with $\mathcal{P}_1$, $\mathcal{P}_2$).

In this generality, the *representation problem* seems to be beyond the reach of our present resources in topology. From the point of view of the theory of length and area, presented in this book, the following four cases are of immediate interest. In each of these cases, $\mathcal{P}^*$ may be assumed to be a subspace of Euclidean three-space, a restriction that turns out to be irrelevant as far as the representation problem is concerned.

Case (a). $\mathcal{P}_1$ and $\mathcal{P}_2$ are both 1-cells.
Case (b). $\mathcal{P}_1$ and $\mathcal{P}_2$ are both 1-spheres.
Case (c). $\mathcal{P}_1$ and $\mathcal{P}_2$ are both 2-cells.
Case (d). $\mathcal{P}_1$ and $\mathcal{P}_2$ are both 2-spheres.

For terminology, the reader is referred to I.2.31. For conciseness, these cases will be referred to as the 1-*cell case,* the 1-*sphere case,* the 2-*cell case,* and the 2-*sphere case* respectively. In all these cases, one will naturally ask also for *favorable* representations. While the meaning of the term favorable is of course vague, it has been recognized at a very early stage that *a light representation* (see II.3.21) *is very favorable.* Thus we have the additional problem: *find a light mapping $T_2(\mathcal{P}_2) = \mathcal{P}^*$* (see II.1.1) *such that $T_1 \sim T_2(F)$.* In the following brief historical survey, we shall use present terminology for conciseness, and in particular we shall make use of the concept of a monotone-light factorization (cf. II.1.18), without in any way implying that such concepts were explicitly formulated by all of the authors involved.

II.5.6. Taking first the 1-*cell case,* it is clear that if the mapping $T_1(\mathcal{P}_1) = \mathcal{P}^*$, where $\mathcal{P}_1$ is a 1-cell, is constant on $\mathcal{P}_1$, then there cannot exist any light mapping $T_2(\mathcal{P}_2) = \mathcal{P}^*$ such that $T_1 \sim T_2(F)$. So let us assume that T_1 is not constant. By II.3.22 we have then a light mapping $T_2(\mathcal{P}_2) = \mathcal{P}^*$ such that $T_1 \sim T_2(F)$, where $\mathcal{P}_2$ may be chosen as the unit interval $0 \leqq u \leqq 1$. Except for terminology, this result is due to Fréchet [1, 5]. As regards the representation problem in the 1-cell case, we have the following picture. Let $T_1(\mathcal{P}_1) = \mathcal{P}^*$, $T_2(\mathcal{P}_2) = \mathcal{P}^*$ be continuous mappings, where $\mathcal{P}_1$, $\mathcal{P}_2$ are 1-cells. Suppose that $T_1 \sim T_2(F)$. Then $T_1 \sim T_2(K)$ by II.1.31. Conversely, if $T_1 \sim T_2(K)$, then $T_1 \sim T_2(F)$, as may be readily seen by the following argument. If T_1 is constant on $\mathcal{P}_1$, then the relation $T_1 \sim T_2(K)$ clearly implies that T_2 is constant on $\mathcal{P}_2$, and the relation $T_1 \sim T_2(F)$ is obvious. So let us assume that T_1 is not constant on $\mathcal{P}_1$. By assumption, we have simultaneous monotone-light factorizations of the form

$$(1) \qquad T_1 = LM_1, \qquad M_1(\mathcal{P}_1) = \mathfrak{M}, \qquad L(\mathfrak{M}) = \mathcal{P}^*,$$

$$(2) \qquad T_2 = LM_2, \qquad M_2(\mathcal{P}_2) = \mathfrak{M}, \qquad L(\mathfrak{M}) = \mathcal{P}^*,$$

where the middle space $\mathfrak{M}$ is a simple arc by II.1.33. Hence $T_1 \sim T_2(F)$ by II.1.61. There follows the solution of the representation problem by the following process. Given $T_1(\mathcal{P}_1) = \mathcal{P}^*$, where $\mathcal{P}_1$ is a 1-cell, take any monotone-light factorization (1) of T_1. Take any 1-cell $\mathcal{P}_2$, and let $M_2(\mathcal{P}_2) = \mathfrak{M}$ be any monotone mapping. Then the formula (2) yields a T_2 such that $T_1 \sim T_2(F)$, and conversely, every T_2 such that $T_1 \sim T_2(F)$ can be obtained in this manner (cf. II.1.30). Briefly stated: *in the* 1-*cell case, for given* T_1 *the general solution* T_2 *of the relation* $T_1 \sim T_2(F)$ *is obtained by modifying the monotone factor of* T_1.

In particular, if T_1 is not constant, then we can assume, by a previous remark, that T_1 is light. Then M_1 may be taken as the identity transformation, and M_2 may be taken as any monotone mapping $M_2(\mathcal{P}_2) = \mathcal{P}_1$. The general solution T_2 of the relation $T_1 \sim T_2(F)$ appears then in the form $T_2 = T_1M_2$. In this sense, a light representation T_1 may be termed *a master representation,* since every T_2 satisfying the relation $T_1 \sim T_2(F)$ can be obtained from T_1 by the simple process described by the formula $T_2 = T_1M_2$.

The preceding results, which are due (except for terminology) to Fréchet [1, 5, 7] and Marston Morse [1], yield complete answers to the representation problem in the 1-cell case. Analogous results for the 1-sphere case follow readily from II.3.23, II.1.31, II.1.62, II.1.34. As a consequence of such completely satisfactory solutions of the representation problem in the 1-cell case and in the 1-sphere case, the theory of F-curves corresponding to these cases presents no topological difficulties, as far as arc length is concerned. The situation is entirely different in the 2-sphere case and in the 2-cell case, and in fact a significant discrepancy seems to exist even between these two cases, as we shall presently see.

II.5.7. *The* 2-*sphere case* was first studied by Kerékjártó [1] in 1927. Using present terminology (cf. the concluding remarks in II.5.5), his results may be stated as follows (see however II.5.8).

(i) Given a continuous mapping $T_1(\mathcal{P}_1) = \mathcal{P}^*$, where $\mathcal{P}_1$ is a 2-sphere, there exists a light mapping T_2 such that $T_1 \sim T_2(F)$ if and only if the middle space occurring in the monotone-light factorization of T_1 is a 2-sphere (cf. II.1.28, II.1.20).

(ii) Let $T_1(\mathcal{P}_1) = \mathcal{P}^*$, $T_2(\mathcal{P}_2) = \mathcal{P}^*$ be continuous mappings, where $\mathcal{P}_1$, $\mathcal{P}_2$ are 2-spheres, such that $T_1 \sim T_2(F)$. Then $T_1 \sim T_2(K)$ (cf. II.1.25, II.1.30).

(iii) Conversely, if $T_1(\mathcal{P}_1) = \mathcal{P}^*$, $T_2(\mathcal{P}_2) = \mathcal{P}^*$ are continuous mappings, where $\mathcal{P}_1$, $\mathcal{P}_2$ are 2-spheres, and $T_1 \sim T_2(K)$, then $T_1 \sim T_2(F)$.

(iv) Let us state an important corollary of the preceding statements. Let $\mathcal{P}_1$, $\mathcal{P}_2$ be 2-spheres and let $M_1(\mathcal{P}_1) = \mathcal{P}_2$ be a monotone mapping. Then for every $\epsilon > 0$ there exists a homeomorphism $H_\epsilon(\mathcal{P}_1) = \mathcal{P}_2$ such that $\rho[M_1(x_1), H_\epsilon(x_1)] < \epsilon$ for every point $x_1 \in \mathcal{P}_1$. Indeed, on defining L and M_2 as the identity transformation on $\mathcal{P}_2$, clearly $M_1 = LM_1$, $L = LM_2$, and thus $M_1 \sim L(K)$. Hence, by statement (iii), $M_1 \sim L(F)$, and the existence of the homeomorphism H_ϵ follows directly from the definition of F-equivalence (cf. II.1.25). In other words, *the fundamental approximation theorem of* II.1.57 *is implied by the Kerékjártó statements.*

(v) As a consequence of (iii), it follows (as in the 1-cell case, see II.5.6) that the general solution of the relation $T_1 \sim T_2(F)$, where T_1 is given, is obtained by a *modification of the monotone factor* of T_1.

(vi) Let $T_1(\mathcal{P}_1) = \mathcal{P}^*$, $T_2(\mathcal{P}_2) = \mathcal{P}^*$ be continuous light mappings, where $\mathcal{P}_1$ and $\mathcal{P}_2$ are 2-spheres, such that $T_1 \sim T_2(F)$. Then $T_1 \sim T_2(ts)$ (cf. II.1.67).

II.5.8. C. B. Morrey [2] in 1935, apparently independently of Kerékjártó, took up the study of *the 2-cell case,* and stated a series of results corresponding to the statements (i)-(vi) in II.5.7, where the term 2-sphere is now to be replaced by the term 2-cell throughout. In particular, *both Kerékjártó and Morrey asserted that $T_1 \sim T_2(K)$ implies that $T_1 \sim T_2(F)$, in the 2-sphere case and in the 2-cell case respectively.* Thus it would appear that in both cases the solution of the representation problem follows by the process of modifying the monotone factor (cf. II.5.7(v)). However, in 1940 J. W. T. Youngs discovered that the crucial statement II.5.7(iii) of Kerékjártó and the corresponding statement of Morrey for the 2-cell case were both false. As regards the 2-sphere case, the solution of the representation problem was given subsequently by J. W. T. Youngs [3] in 1944, in terms of an appropriate modification, involving orientation, of the relation $T_1 \sim T_2(K)$. *In terms of this modified relation, the solution of the representation problem is obtained by an appropriate process of modifying the monotone factor.* Since we restrict ourselves, in this book, to the study of surface area in the 2-cell case only, we omit an explicit statement of the beautiful result of J. W. T. Youngs, and we point out only one relevant fact. If $T_1(\mathcal{P}_1) = \mathcal{P}^*$, $T_2(\mathcal{P}_2) = \mathcal{P}^*$ are continuous mappings, where $\mathcal{P}_1$, $\mathcal{P}_2$ are 2-spheres and $T_1 \sim T_2(K)$, and if the common middle space $\mathfrak{M}$ of T_1 and T_2 is either a dendrite or else has at most one proper cyclic element, then the modified relation defined by J W. T. Youngs reduces to the relation $T_1 \sim T_2(K)$. As a consequence, we have the following corollaries of the principal result of J. W. T. Youngs.

(i) *Let* $T_1(\mathcal{P}_1) = \mathcal{P}^*$, $T_2(\mathcal{P}_2) = \mathcal{P}^*$ *be continuous mappings, where* $\mathcal{P}_1$, $\mathcal{P}_2$ *are 2-spheres and* $T_1 \sim T_2(K)$. *If the common middle space of* T_1, T_2 *is a simple arc, then* $T_1 \sim T_2(F)$.

(ii) *Let* T_1, T_2 *satisfy the same assumptions as under* (i), *except that the common middle space of* T_1, T_2 *is a 2-sphere. Then* $T_1 \sim T_2(F)$.

II.5.9. It should be observed that workers in this field seemed to agree that the 2-sphere case and the 2-cell case should be entirely analogous. The 2-sphere case seemed to present certain advantages as far as exposition was concerned, and for this reason in several instances only the 2-sphere case was explicity discussed in the literature. Consequently, the results of J. W. T. Youngs for the 2-sphere case were apparently first assumed to lead, perhaps after some easy adjustments, to an analogous solution of the problems involved in the 2-cell case (cf. C. B. Morrey [4]), even though J. W. T. Youngs [3] himself indicated various difficulties that may arise in the 2-cell case. However, the situation may be more involved than expected. Indeed, T. Radó [28] exhibited simple examples showing that the statements (i), (ii) in II.5.8 fail to hold in the 2-cell case. More precisely, (i) is false, generally, if $\mathcal{P}_1$, $\mathcal{P}_2$ are 2-cells, and (ii) is false, generally, if $\mathcal{P}_1$, $\mathcal{P}_2$ are 2-cells while the common middle space of T_1, T_2 is a 2-sphere (cf. II.1.31). In view of the extreme simplicity of the examples used, it is apparent that the situation in the 1-cell case and the 1-sphere case (see II.5.6) is wholly unreliable as far as analogy with the 2-sphere case and the 2-cell case is concerned. The topological complications in the 2-cell case, suggested by the preceding remarks, indicate the desirability of developing methods in surface area theory that would not depend upon a complete solution of the representation problem in the 2-cell case. This program, initiated by T. Radó [28], is followed in this book.

II.5.10. The development of the program just indicated depends in the first place upon a substantially increased amount of prerequisites with respect to metric results concerning surface area, to compensate for the reduced range of topological tools. The metric results involved will be discussed in subsequent chapters. At this time, we propose to review only the topological issues involved. While we saw above that the relation $T_1 \sim T_2(K)$ does not imply generally the relation $T_1 \sim T_2(F)$, we make essential use of the fact that *in certain special cases the relation* $T_1 \sim T_2(K)$ *does imply the relation* $T_1 \sim T_2(F)$. The relevant special cases (see II.1.60-II.1.65) are due to J. W. T. Youngs [3] and to T. Radó [28]. The fundamental tool in discussing these special cases is *the modification theorem of J. W. T. Youngs* [3], discussed in II.1.47. The method of proof employed by J. W. T. Youngs applies, after plausible adjustments, to several further modification theorems (see II.1.53, II.1.54) that were derived by T. Radó [28]. These modification theorems are used to solve certain special cases of the following *general problem. Let* $\mathcal{P}_1$, $\mathcal{P}_2$ *be homeomorphic Peano spaces, and let* $M(\mathcal{P}_1) = \mathcal{P}_2$ *be a nonconstant continuous monotone mapping. Show that for every* $\epsilon > 0$ *there* *exists a homeomorphism* $H_\epsilon(\mathcal{P}_1) = \mathcal{P}_2$ *such that* $\rho[M(x_1), H_\epsilon(x_1)] < \epsilon$ *for every point* $x_1 \in \mathcal{P}_1$. In this generality, the problem seems to be inaccessible at this time, even though the affirmative answer would represent merely the converse of a well known theorem (see II.1.22) of G. T. Whyburn. If $\mathcal{P}_1$, $\mathcal{P}_2$ are both 1-cells or both 1-spheres, the affirmative answer is practically trivial (see II.1.50, II.1.51). If $\mathcal{P}_1$, $\mathcal{P}_2$ are both 2-spheres, then the affirmative answer is implied by the statements of Kerékjártó (see II.5.7(iv)). For the case when $\mathcal{P}_1$ and $\mathcal{P}_2$ are both 2-cells, the affirmative answer is explicitly stated by C. B. Morrey [2]. However, these results appear in the work of both Kerékjártó and Morrey in a confusing setting (cf. the remarks in II.5.7-II.5.9), and thus the rigorous proofs, based upon the modification theorem of J. W. T. Youngs and its extensions, seem to fill a definite and serious gap. These proofs, even though compact, depend upon advanced tools, including *the topological characterization of the 2-sphere*, indicating that the results lie perhaps deeper than the work of Kerékjártó and Morrey may suggest, and also indicating that the general problem may be quite difficult even for n-cells and n-spheres.

A further important matter is a study of the implications of the relation $T_1 \sim T_2(F)$. Kerékjártó [1] already stated that in the 2-sphere case the relation $T_1 \sim T_2(F)$ implies the relation $T_1 \sim T_2(K)$. For the case when the Peano spaces $\mathcal{P}_1$, $\mathcal{P}_2$ involved in the mappings $T_1(\mathcal{P}_1) = \mathcal{P}^*$, $T_2(\mathcal{P}_2) = \mathcal{P}^*$ are any two homeomorphic Peano spaces, J. W. T. Youngs [3] established the fact that the relation $T_1 \sim T_2(F)$ implies the relation $T_1 \sim T_2(K)$, and he also derived various further important inferences. In particular, the theorem in II.1.28 is due to J. W. T. Youngs, except for the statement concerning partial mappings which is due to T. Radó [28]. This statement concerning partial mappings yields the relevant fact that *the relation* $T_1 \sim T_2(F)$ *implies F-equivalence of the cyclic decompositions of* T_1 *and* T_2 (see II.2.97). As regards details, the lemma of II.1.9, due to T. Radó [28], seems to lead to substantial simplifications as compared with previous literature.

The cyclic additivity theorem (see II.2.113) is partly of metric character, since it is concerned with a functional $\Phi(T)$ where T is a continuous mapping from a fixed Peano space $\mathcal{P}$ into a fixed metric space P^*. A first theorem of this type has been studied by C. B. Morrey [3]. In the case considered by Morrey, $\mathcal{P}$ is a 2-cell, P^* is the Euclidean n-space, and $\Phi(T)$ is the Lebesgue area of the surface determined by the mapping T. However, the proof of Morrey depends upon his erroneous solution of the representation problem (see II.5.8). J. W. T. Youngs established, on the basis of his results concerning the representation problem, the analogous theorem for the case when $\mathcal{P}$ is a 2-sphere and $\Phi(T)$ is again the Lebesgue area (see J. W. T. Youngs [3]). The verification of the original cyclic additivity theorem of Morrey, for the 2-cell case, in an analogous manner seems however to present difficulties, due to the complications mentioned in II.5.9. The cyclic additivity theorem of II.2.113, due to T. Radó [28], covers simultaneously the 2-cell case and the 2-sphere case, and in fact the Peano space $\mathcal{P}$ is required to possess only a certain property (Π) which is shared by the 2-cell and the 2-sphere. The scope of this property (Π) has been determined by G. T. Whyburn who found that for Peano spaces it is equivalent to *unicoherence*. *Thus the cyclic additivity theorem covers such potentially important cases as the n-cell and the n-sphere.* Furthermore, the functional $\Phi(T)$ occurring in the cyclic additivity theorem is required to possess only properties that are common to all lower semi-continuous areas proposed so far. In particular, the theorem applies therefore to *both the Lebesgue area and the lower area* (see part V), an important feature in dealing with some of the fundamental problems in surface area theory. In the applications to surface area, the functional $\Phi(T)$ will possess the property of *F-invariance*. That is, the relation $T_1 \sim T_2(F)$ will imply the relation $\Phi(T_1) = \Phi(T_2)$. Curiously, this property is *not* required in the cyclic additivity theorem of II.2.113. In this respect, the theorem is unnecessarily general; on the other hand, this circumstance explains the fact that *the solution of the representation problem* (see II.5.5) *plays no role in the discussion of the theorem.*

A considerable amount of topology of the plane is needed in the study of *the modified projection principle of Geöcze* (see the remarks in I.1.15). Subsequent studies revealed that *the topological index* (see chapter II.4), a tool not used by Geöcze, is a most efficient device in this connection. Chapter II.4 is devoted therefore to a study of the topological index, a topic familiar to Analysts working in Complex Variables. Except for material that is obviously known generally, most of the topics included are taken from T. Radó and P. V. Reichelderfer [17] and T. Radó [24]. The simple and fundamental estimate for the topological index, given in II.4.39, is due to Schauder [3]. From the topological point of view, the topological index is of course merely a special instance of *the degree of a mapping.* Ample use was made in Chapter II.4 of methods developed in Topology proper, and in particular results and methods of Eilenberg [1] led to substantial simplifications.

From the point of view of applications in other fields, especially in Calculus of Variations, it is important to discuss the concept of *the boundary curve of an F-surface of the type of the* 2-*cell* (see II.3.38), the case where the boundary curve is a simple closed curve being of especial interest. The results discussed in II.3.39-II.3.41 are closely related to the important work of McShane on the structure of surfaces of minimum area (see McShane [5]).

While we are primarily interested in path surfaces (*F*-surfaces in our terminology), certain items included in Chapter II.3 may be considered as establishing a connection with point surfaces. Reference is made here to the brief study, in II.3.31-II.3.37, of *topological F-varieties which may be thought of as the connecting link between path varieties and point varieties* (cf. the remarks in II.5.2 concerning *path curves* and *point curves*). Furthermore, the topological *F*-varieties may be used to achieve a first insight into the nature of the problems that arise in connection with general *F*-varieties. In this connection, it is revealing that the representation problem (see II.5.5) leads to major topological difficulties even in the case of topological *F*-varieties.

II.5.11. The preceding brief review of the major issues discussed in part II indicates that a substantial amount of Topology is involved in what we termed the *analytic* theory of surface area. To ease the task of the reader, we included in part II a certain amount of well known material to maintain the continuity of the presentation. Thus Chapter II.1 is, to a certain limited extent, a self-contained presentation of the theory of continuous mappings of Peano spaces as far as absolutely needed for our purposes. In Chapter II.2, we included a detailed discussion of *cyclic element theory,* subject to analogous limitations. A masterly presentation of this fundamental theory, in its most general aspects, is given in G. T. Whyburn [3]. Since we only need the very special case of Peano spaces, various conceptual simplifications can be achieved in the presentation. In particular, it is possible in the special case of Peano spaces to base the theory upon the binary relation $p \bigcirc q$ of conjugacy (see II.2.3), as has been shown by G. T. Whyburn and Kuratowski [1]. The relation $p \bigcirc q$ is reflexive, symmetric, but in general *not transitive.* And yet, a proper cyclic element is quite analogous to an *equivalence class* in the study of a reflexive, symmetric, and transitive binary relation. In fact, the analogy is not entirely superficial, since the relation $p \bigcirc q$ possesses a certain *weak form of transitivity* (see II.2.5). This observation suggests an approach that was studied by T. Radó and P. V. Reichelderfer [14]. The adoption of this approach in Chapter II.2 may be helpful to the reader who desires to deepen his understanding of the theory by comparing various methods of presentation. The importance of cyclic element theory in the study of surface area was first recognized by C. B. Morrey [3]. As far as F-surfaces of the type of the 2-cell and of the 2-sphere are concerned, the topological information available at this time seems to be adequate. On the other hand, extensions to F-surfaces of other topological types, and more generally to F-varieties of higher dimension, seem to lead to major topological problems. We noted in I.1.3 that the theory of arc length and the theory of surface area may be thought of as merely the special cases $n = 1$ and $n = 2$ of a general theory involving a general positive integer n. The preceding remarks indicate the topological difficulties that may be expected in the case $n > 2$.

The writer wishes to express his gratitude to G. T. Whyburn and J. W. T. Youngs who contributed helpful suggestions concerning several topics in part II.

PART III. ARC-LENGTH AND RELATED TOPICS

CHAPTER III.1. INTERVAL FUNCTIONS

III.1.1. An interval I in Euclidean n-space $x_1 x_2 \cdots x_n$ is a point set determined by inequalities of the form $a_i \leqq x_i \leqq b_i$, where $a_i < b_i$, $i = 1, 2, \cdots, n$. If K is a class of such intervals, and with every $I \in K$ there is associated a finite real number $\phi(I)$, then $\phi(I)$ is termed an *interval function* defined on the class K. Many theorems on arc-length, to be considered in the sequel, are special cases of general theorems on interval functions, and the same remark applies, although to a lesser extent, to surface area. For this reason, a brief discussion of some relevant parts of the theory of interval functions is necessary. The case $n = 1$ is needed in the study of arc-length, and the case $n = 2$ is needed in the study of surface area. However, we shall discuss explicitly only the case $n = 2$, since the analogous results for the case $n = 1$ follow by entirely similar arguments (of course, considerable simplifications are generally possible in the case $n = 1$).

An interval in Euclidean 2-space $x_1 x_2$ will be termed an *oriented rectangle* (a rectangle with sides parallel to the coordinate axes). It will be convenient to use u, v instead of x_1, x_2. An oriented rectangle R may then be given in the form $R : a \leqq u \leqq b, c \leqq v \leqq d$, where $a < b, c < d$. The interior of R will be denoted by R^0. In the case $n = 2$ we shall use the term *rectangle function*, instead of the term interval function. Thus a rectangle function $\phi(R)$ is assumed in the sequel to be defined only on oriented rectangles.

III.1.2. Let $\phi(R)$ be a rectangle function defined for all oriented rectangles R contained in a fixed oriented rectangle R_0. Then $\phi(R)$ will be termed BV (*of bounded variation*) in R_0 if there exists a finite constant M such that $| \phi(R_1) | + \cdots + | \phi(R_n) | < M$ for every finite system of oriented rectangles $R_1, \cdots, R_n$ which satisfy the conditions

$$(1) \qquad R_i \subset R_0, \qquad i = 1, \cdots, n,$$

$$(2) \qquad R_i^0 R_j^0 = 0 \qquad \text{for } i \neq j, i > 0, j > 0.$$

If for every $\epsilon > 0$ there exists an $\eta(\epsilon) > 0$ such that $| \phi(R_1) | + \cdots + | \phi(R_n) | < \epsilon$ for every finite system of oriented rectangles $R_1, \cdots, R_n$ which satisfy, in addition to the conditions (1), (2) just stated, the condition $| R_1 | + \cdots + | R_n | < \eta(\epsilon)$, then $\phi(R)$ will be termed AC (*absolutely continuous*) in R_0. The symbol $| R_i |$ denotes the area of the rectangle R_i.

$\phi(R)$ will be termed *continuous* in R_0 if for every $\epsilon > 0$ there exists a $\delta(\epsilon) > 0$ such that $| \phi(R) | < \epsilon$ for every oriented rectangle R in R_0 that satisfies the condition $| R | < \delta(\epsilon)$.

$\phi(R)$ will be termed *bounded* in R_0 if there exists a finite constant K such that $| \phi(R) | < K$ for every oriented rectangle $R \subset R_0$.

III.1.3. If $\phi(R)$ is AC and bounded in R_0, then $\phi(R)$ is BV in R_0.

PROOF. Let $\eta(\epsilon)$, K be the quantities appearing in the definition of absolute continuity and of boundedness respectively (see III.1.2), and put $\lambda = \eta(1)/2$. Take any finite system $R_1, \cdots, R_m$ of oriented rectangles, without common interior points, in R_0.

Case (1). $| R_j | < \lambda, j = 1, \cdots, m$. Divide the system $R_1, \cdots, R_m$ into groups $G_1, \cdots, G_k$ in such a way that in each group the sum of the areas of the rectangles in that group is less than $\eta(1) = 2\lambda$. This is possible, since $| R_j | < \lambda$, $j = 1, \cdots, m$. Clearly, there exists a grouping of this type where the number k of the groups is a minimum. If $G_1, \cdots, G_k$ is such an extremal grouping, then there is at most one group where the sum of the areas is less than λ, since otherwise two of the groups could be combined into one group, in contradiction with the extremal property of the grouping $G_1, \cdots, G_k$. Hence

$$(k - 1)\lambda \leqq \sum_{j=1}^{m} | R_j | \leqq | R_0 |,$$

and consequently

$$k \leqq 1 + | R_0 |/\lambda. \tag{1}$$

On the other hand, the summation $\sum | \phi(R_j) |$, extended only over the rectangles of a single group G_i, is less than 1 by the definition of $\eta(1)$. Hence, in view of (1), we have

$$\sum_{j=1}^{m} | \phi(R_j) | \leqq 1 + | R_0 |/\lambda. \tag{2}$$

Case (2). $| R_j | \geqq \lambda, j = 1, \cdots, m$. Then clearly $m\lambda \leqq | R_0 |$, and hence

$$\sum_{j=1}^{m} | \phi(R_j) | \leqq mK \leqq K | R_0 |/\lambda. \tag{3}$$

Case (3). In the general case, the system $R_1, \cdots, R_m$ can be split into two subsystems which satisfy the assumptions made in case (1) and case (2) respectively. Hence, by (2) and (3),

$$\sum_{j=1}^{m} | \phi(R_j) | \leqq 1 + | R_0 |/\lambda + K | R_0 |/\lambda.$$

Since $| R_0 |$, λ, K are independent of the choice of the system $R_1, \cdots, R_m$, it follows that $\phi(R)$ is BV in R_0.

III.1.4. Let R be an oriented rectangle in R_0. By a *subdivision* $D(R)$ of R we mean a finite system of oriented rectangles r such that

$$R = \sum r, \qquad | R | = \sum | r |, \qquad r \in D(R).$$

The symbol $r \in D(R)$ is used to state the fact that r occurs in the subdivision $D(R)$. The maximum of the diameters of rectangles $r \in D(R)$ will be denoted by $\| D(R) \|$.

A *special* (or *restricted*) *subdivision* of $R : a \leqq u \leqq b,\ c \leqq v \leqq d$ arises if we take numbers $u_0 = a < u_1 < \cdots < u_{n-1} < u_n = b,\ v_0 = c < v_1 < \cdots < v_m = d$, and subdivide R into the nm rectangles $u_{i-1} \leqq u \leqq u_i,\ v_{j-1} \leqq v \leqq v_j,\ i = 1, \cdots, n,\ j = 1, \cdots, m$. Let us note that unless the contrary is explicitly stated, a subdivision $D(R)$ is not assumed to be of this special type.

III.1.5. If $\phi(R)$ is a rectangle function defined in R_0, then we put (cf. III.1.4)

$$\phi[D(R)] = \sum \phi(r), \qquad r \in D(R).$$

$\phi(R)$ is said to increase (decrease) by subdivision if $\phi(R) \leqq \phi[D(R)]$ ($\phi(R) \geqq \phi[D(R)]$) for every choice of $R \subset R_0$ and of $D(R)$. If $\phi(R) = \phi[D(R)]$ for every choice of $R \subset R_0$ and of $D(R)$, then ϕ is termed *additive* in R_0. We define further, for $R \subset R_0$,

$$U(R, \phi) = \text{l.u.b.}\ \phi[D(R)],$$

where the least upper bound is taken with respect to all subdivisions $D(R)$ of R. Of course, $U(R, \phi)$ may be $+\infty$. If $U(R, \phi)$ is finite for every $R \subset R_0$, then we shall say that ϕ has a finite U-function in R_0. Then $U(R, \phi)$ is again a rectangle function which clearly decreases by subdivision. If $\phi(R)$ itself increases by subdivision, then obviously $U(R, \phi)$ (assumed to be finite) is additive.

III.1.6. Let there be given two rectangle functions $\phi(R)$, $\psi(R)$ in R_0. Let us put

$$\omega(R) = [\phi(R)^2 + \psi(R)^2]^{1/2}.$$

Then $\omega(R)$ is a non-negative rectangle function in R_0. The following statements are obvious.

(i) If $\phi(R)$, $\psi(R)$ are both non-negative and increase by subdivision, then $\omega(R)$ also increases by subdivision.

(ii) If $\phi(R)$, $\psi(R)$ are both non-negative, and both have finite U-functions, then $\omega(R)$ also has a finite U-function.

III.1.7. Let $\phi_n(R)$, $\psi_n(R)$, $n = 0, 1, 2, \cdots$, be a sequence of pairs of rectangle functions in R_0 satisfying the following conditions: (i) $\phi_n(R)$, $\psi_n(R)$ are non-negative, $n = 0, 1, 2, \cdots$, (ii) $\liminf \phi_n(R) \geqq \phi_0(R)$, $\liminf \psi_n(R) \geqq \psi_0(R)$ for every $R \subset R_0$. Then

(1) $$\liminf \omega_n(R) \geqq \omega_0(R),$$

(2) $$\liminf U(R, \phi_n) \geqq U(R, \phi_0),$$

(3) $$\liminf U(R, \psi_n) \geqq U(R, \psi_0),$$

(4) $$\liminf U(R, \omega_n) \geqq U(R, \omega_0),$$

for every $R \subset R_0$, where $\omega_n = (\phi_n^2 + \psi_n^2)^{1/2}$. The proofs are obvious.

III.1.8. Given $\phi_n(R)$, $\psi_n(R)$, $\omega_n(R)$ as in III.1.7, suppose that the following additional conditions are satisfied. (iii) $\phi_n(R)$, $\psi_n(R)$ increase by subdivision, $n = 0, 1, 2, \cdots$. (iv) $\phi_n(R)$, $\psi_n(R)$ have finite U-functions, $n = 0, 1, 2, \cdots$.

(v) $\lim U(R_0, \omega_n) = U(R_0, \omega_0)$. Then $\lim U(R, \omega_n) = U(R, \omega_0)$ for every $R \subset R_0$.

PROOF. Given $R^* \subset R_0$, there exists a subdivision $D(R_0)$ such that $R^* \in D(R_0)$. Let us note that $U(R, \omega_n)$ is additive by conditions (i), (iii), (iv) and by III.1.6, III.1.5. From the condition (v) we obtain therefore, in view of III.1.7(4),

$$(1)\qquad \begin{aligned} U(R_0, \omega_0) &= \lim U(R_0, \omega_n) = \lim \sum U(r, \omega_n) \geqq \sum \liminf U(r, \omega_n) \\ &\geqq \sum U(r, \omega_0) = U(R_0, \omega_0), \end{aligned}$$

where the summations are extended over all rectangles $r \in D(R_0)$. In view of III.1.7(4), it follows from (1) that $\liminf U(r, \omega_n) = U(r, \omega_0)$ for every $r \in D(R_0)$. In particular

$$\liminf U(R^*, \omega_n) = U(R^*, \omega_0).$$

If $n_1 < n_2 < \cdots$ is any infinite sequence of positive integers, then the same reasoning applies to the subsequence $\omega_{n_k}(R)$, and hence

$$(2)\qquad \liminf U(R^*, \omega_{n_k}) = U(R^*, \omega_0).$$

Since (2) holds for every subsequence ω_{n_k}, the relation $\lim U(R^*, \omega_n) = U(R^*, \omega_0)$ follows.

III.1.9. Given $\phi_n(R)$, $\psi_n(R)$, $\omega_n(R)$ as in III.1.7, suppose that the conditions (i)-(v) stated in III.1.7, III.1.8 hold. Then

$$\lim U(R_0, \phi_n) = U(R_0, \phi_0), \qquad \lim U(R_0, \psi_n) = U(R_0, \psi_0).$$

PROOF. Let R be an oriented rectangle in R_0 and let $D(R)$ be any subdivision of R. We have then, in view of the conditions (i), (iii),

$$(\phi_n[D(R)]^2 + \psi_n(R)^2)^{1/2} \leqq (\phi_n[D(R)]^2 + \psi_n[D(R)]^2)^{1/2} \leqq \omega_n[D(R)] \leqq U(R, \omega_n).$$

Since $D(R)$ is arbitrary, there follows the inequality

$$U(R, \phi_n)^2 \leqq U(R, \omega_n)^2 - \psi_n(R)^2.$$

For $n \to \infty$ we obtain, in view of III.1.8 and condition (ii),

$$[\limsup U(R, \phi_n)]^2 \leqq U(R, \omega_0)^2 - \psi_0(R)^2 = U(R, \omega_0)^2 - \omega_0(R)^2 + \phi_0(R)^2.$$

Since $\omega_0(R) \leqq U(R, \omega_0)$, we conclude that

$$(1)\qquad \begin{aligned} \limsup U(R, \phi_n) &\leqq [U(R, \omega_0)^2 - \omega_0(R)^2]^{1/2} + \phi_0(R) \\ &= [U(R, \omega_0) + \omega_0(R)]^{1/2}[U(R, \omega_0) - \omega_0(R)]^{1/2} + \phi_0(R) \\ &\leqq 2^{1/2}U(R, \omega_0)^{1/2}[U(R, \omega_0) - \omega_0(R)]^{1/2} + \phi_0(R). \end{aligned}$$

Now take any subdivision $D(R_0)$. Since $U(R, \phi_n)$ is additive by III.1.5, we obtain from (1), in view of the inequality of Schwarz,

$$\limsup U(R_0, \phi_n) = \limsup U[D(R_0), \phi_n]$$
$$\leqq 2^{1/2}U(R_0, \omega_0)^{1/2}[U(R_0, \omega_0) - \omega_0[D(R_0)]]^{1/2} + U(R_0, \phi_0).$$

Since the difference $U(R_0, \omega_0) - \omega_0[D(R_0)]$ can be made arbitrarily small by proper choice of $D(R_0)$, it follows that $\limsup U(R_0, \phi_n) \leqq U(R_0, \phi_0)$. In view of III.1.7(4), the relation $\lim U(R_0, \phi_n) = U(R_0, \phi_0)$ follows. The relation $\lim U(R_0, \psi_n) = U(R_0, \psi_0)$ is established in the same manner.

III.1.10. Let $\alpha_n(R)$, $\beta_n(R)$, $\gamma_n(R)$, $n = 0, 1, 2, \cdots$, be a sequence of triples of rectangle functions defined in R_0. Set $\lambda_n = (\beta_n^2 + \gamma_n^2)^{1/2}$, $\mu_n = (\gamma_n^2 + \alpha_n^2)^{1/2}$, $\nu_n = (\alpha_n^2 + \beta_n^2)^{1/2}$, $\sigma_n = (\alpha_n^2 + \beta_n^2 + \gamma_n^2)^{1/2}$. Assume that the following conditions are satisfied. (1) The rectangle functions α_n, β_n, γ_n are non-negative, $n = 0, 1, 2, \cdots$. (2) $\liminf \alpha_n(R) \geqq \alpha_0(R)$, $\liminf \beta_n(R) \geqq \beta_0(R)$, $\liminf \gamma_n(R) \geqq \gamma_0(R)$ for every $R \subset R_0$. (3) $\alpha_n, \beta_n, \gamma_n$ increase by subdivision, $n = 0, 1, 2, \cdots$. (4) $\alpha_n, \beta_n, \gamma_n$ have finite U-functions, $n = 0, 1, 2, \cdots$. (5) $\lim U(R_0, \sigma_n) = U(R_0, \sigma_0)$. Then the following relations hold.

$$\text{(1)} \quad \lim U(R_0, \alpha_n) = U(R_0, \alpha_0), \qquad \lim U(R_0, \beta_n) = U(R_0, \beta_0),$$
$$\lim U(R_0, \gamma_n) = U(R_0, \gamma_0),$$

$$\text{(2)} \quad \lim U(R_0, \lambda_n) = U(R_0, \lambda_0), \qquad \lim U(R_0, \mu_n) = U(R_0, \mu_0),$$
$$\lim U(R_0, \nu_n) = U(R_0, \nu_0).$$

PROOF. Since $(\alpha_n^2 + \lambda_n^2)^{1/2} = \sigma_n$, clearly the assumptions made in III.1.9 are satisfied if we set $\phi_n = \alpha_n$, $\psi_n = \lambda_n$, $\omega_n = \sigma_n$. The same remark applies to β_n, μ_n, σ_n and γ_n, ν_n, σ_n. Thus the relations (1) and (2) follow directly from III.1.9.

III.1.11. Suppose that α_n, β_n, γ_n satisfy the assumptions (1)-(5) stated in III.1.10. In view of III.1.8, applied with $\phi_n = \alpha_n$, $\psi_n = \lambda_n$, $\omega_n = \sigma_n$, it follows that the same assumptions (1)-(5) hold if R_0 is replaced by any oriented rectangle $R \subset R_0$. Hence the formulas (1), (2) in III.1.10 remain valid if R_0 is replaced by any $R \subset R_0$.

III.1.12. Let $\phi(R)$ be a rectangle function defined in R_0, and let R be an oriented rectangle in R_0. The *upper Burkill integral* of ϕ over R is defined by the formula

$$\int_R^* \phi = \text{l.u.b.} \limsup \phi[D_n(R)],$$

where the least upper bound is taken with respect to all sequences of subdivisions $D_n(R)$ such that $\| D_n(R) \| \to 0$ for $n \to \infty$. The upper Burkill integral may be equal to $\pm\infty$. Similarly, the *lower Burkill integral* is defined by the formula

$$\int_{R*} \phi = \text{gr.l.b.} \liminf \phi[D_n(R)].$$

If the upper and lower Burkill integrals are finite and equal, then ϕ is termed *Burkill integrable over R,* and its *Burkill integral* $\int_R \phi$ is equal to the common (finite) value of its upper and lower Burkill integrals over R.

III.1.13. Suppose that ϕ is Burkill integrable over R_0. Then for every $\epsilon > 0$ there exists an $\eta = \eta(\epsilon) > 0$, such that

$$\left| \int_{R_0} \phi - \phi[D(R_0)] \right| < \epsilon \qquad \text{if } \| D(R_0) \| < \eta.$$

This is an immediate consequence of the definition of the Burkill integral.

III.1.14. Suppose that ϕ is Burkill integrable over R_0. Then ϕ is Burkill integrable over every oriented rectangle $R \subset R_0$, and $\int_R \phi$ is an additive rectangle function which will be termed the *indefinite Burkill integral* of ϕ in R_0.

PROOF. Let R_1 be any oriented rectangle in R_0, and let $D(R_0)$ be any subdivision of R_0 such that $R_1 \in D(R_0)$. Let $R_1, R_2, \cdots, R_m$ be the rectangles occurring in $D(R_0)$, and for each $i = 1, 2, \cdots, m$, let $D_n(R_i)$ be a sequence of subdivisions such that $\| D_n(R_i) \| \to 0$ for $n \to \infty$. Then the rectangles occurring in $D_n(R_1), \cdots, D_n(R_m)$ constitute, for given n, a subdivision $D_n(R_0)$. Clearly $\| D_n(R_0) \| \to 0$ for $n \to \infty$, and

$$\text{(1)} \qquad \phi[D_n(R_0)] = \phi[D_n(R_1)] + \cdots + \phi[D_n(R_m)].$$

Give $\epsilon > 0$, and let η correspond to ϵ in the sense of III.1.13. Since $\| D_n(R_i) \| \to 0$, $i = 1, 2, \cdots, m$, we have a positive integer N such that $\| D_n(R_i) \| < \eta$ for $n \geqq N$. By III.1.13 there follows the inequality

$$\left| \int_{R_0} \phi - \phi[D_n(R_1)] - \phi[D_N(R_2)] - \cdots - \phi[D_N(R_m)] \right| < \epsilon \qquad \text{for } n \geqq N.$$

Keeping N fixed, and making $n \to \infty$, it follows that the upper and lower Burkill integrals of ϕ over R_1 are both finite and differ by not more than ϵ. Since ϵ was arbitrary, the existence of the Burkill integral of ϕ over R_1 is established. Thus ϕ is Burkill integrable over every oriented rectangle in R_0. The additivity of the indefinite Burkill integral follows now from (1) for $n \to \infty$.

III.1.15. Suppose that ϕ is Burkill integrable over R_0. Then for every $\epsilon > 0$ there exists a $\delta = \delta(\epsilon) > 0$ with the following property. Let $R_1, \cdots, R_m$ be any finite system of oriented rectangles, without common interior points, in R_0, such that the diameter of each R_i, $i = 1, \cdots, m$, is less than δ. Then (cf. III.1.14)

$$\text{(1)} \qquad \left| \sum_{i=1}^{m} \left(\phi(R_i) - \int_{R_i} \phi \right) \right| \leqq \epsilon.$$

PROOF. Given $\epsilon > 0$, we have by III.1.13 an $\eta_1 = \eta(\epsilon/2) > 0$ such that

$$\text{(2)} \qquad \left| \int_{R_0} \phi - \phi[D(R_0)] \right| < \frac{\epsilon}{2} \qquad \text{if } \| D(R_0) \| < \eta_1 .$$

Now let $R_1, \cdots, R_m$ be any system of oriented rectangles, without common interior points, in R_0 such that each R_i, $i = 1, \cdots, m$, has a diameter less than η_1. Clearly, we can choose oriented rectangles $R_{m+1}, \cdots, R_{m+k}$ in R_0, such that the rectangles $R_1, \cdots, R_{m+k}$ constitute a subdivision of R_0 and each of these rectangles has a diameter less than η_1. For each $j = 1, \cdots, k$, let $D_n(R_{m+j})$ be a sequence of subdivisions such that $\| D_n(R_{m+j}) \| \to 0$ for $n \to \infty$. Clearly $\| D_n(R_{m+j}) \| < \eta_1$ for every n, j. By (2) we have therefore (cf. III.1.14)

$$(3) \qquad \left| \sum_{i=1}^{m} \int_{R_i} \phi + \sum_{j=1}^{k} \int_{R_{m+j}} \phi - \sum_{i=1}^{m} \phi(R_i) - \sum_{j=1}^{k} \phi[D_n(R_{m+j})] \right| < \epsilon.$$

By III.1.14 we have the relations

$$\phi[D_n(R_{m+j})] \to \int_{R_{m+j}} \phi, \qquad n \to \infty, j = 1, \cdots, k.$$

Thus (3) yields, for $n \to \infty$, the inequality (1). In other words, the quantity $\delta(\epsilon) = \eta(\epsilon/2)$ satisfies our requirements.

III.1.16. In the special case when the difference

$$\phi(R_i) - \int_{R_i} \phi$$

is either greater than or equal to 0 for $i = 1, \cdots, m$ or less than or equal to 0 for $i = 1, \cdots, m$, the inequality III.1.15(1) can be written in the form

$$\sum_{i=1}^{m} \left| \phi(R_i) - \int_{R_i} \phi \right| < \epsilon.$$

Since the general case may be considered as a combination of two such special cases, we obtain the following theorem.

If ϕ is Burkill integrable on R_0, then for every $\epsilon > 0$ there exists a $\sigma = \sigma(\epsilon) > 0$ with the following property. Let $R_1, \cdots, R_m$ be any finite system of oriented rectangles in R_0, without common interior points, such that each R_i has a diameter less than σ. Then

$$\sum_{i=1}^{m} \left| \phi(R_i) - \int_{R_i} \phi \right| < \epsilon.$$

III.1.17. If $\phi(R)$ is continuous (see III.1.2) and Burkill integrable in R_0, then its indefinite Burkill integral $\beta(R) = \int_R \phi$ is also continuous in R_0.

PROOF. Give $\eta > 0$. Put $\epsilon = \eta/2$, and let $\delta(\epsilon)$ be the quantity associated with ϵ in the sense of the theorem in III.1.15. If R is an oriented rectangle in R_0, then let $d(R)$ denote the diameter of R. We choose a positive integer p that satisfies the inequality $d(R_0)/p < \delta(\epsilon)$. Since $\phi(R)$ is continuous, we have a $\lambda(\epsilon) > 0$ such that $| \phi(R) | < \epsilon/p^2$ if $| R | < \lambda(\epsilon)$. Now take any oriented rectangle

$R^* \subset R_0$, such that $|R^*| < \lambda(\epsilon)$. Let $D_p(R^*)$ denote the subdivision of R^* into p^2 congruent oriented rectangles. Then clearly $\|D_p(R^*)\| \leqq d(R_0)/p < \delta(\epsilon)$, and hence by III.1.15,

$$(1) \qquad |\phi[D_p(R^*)] - \beta[D_p(R^*)]| < \epsilon.$$

If R is any rectangle of the subdivision $D_p(R^*)$, then $|R| < |R^*| < \lambda(\epsilon)$, and hence $|\phi(R)| < \epsilon/p^2$. There follows the inequality

$$(2) \qquad |\phi[D_p(R^*)]| < \epsilon.$$

Since $\beta[D_p(R^*)] = \beta(R^*)$ by III.1.14, we obtain from (1) and (2) the inequality $|\beta(R^*)| < 2\epsilon = \eta$. Since this holds for every $R^* \subset R_0$ such that $|R^*| < \lambda(\epsilon) = \lambda(\eta/2)$, the continuity of $\beta(R)$ is proved.

III.1.18. Suppose that $\phi(R)$ is AC and Burkill integrable in R_0. Then its indefinite Burkill integral $\beta(R) = \int_R \phi$ is also AC in R_0.

PROOF. By assumption, we have for every $\epsilon > 0$ an $\eta = \eta(\epsilon) > 0$ such that the following holds. If $R_1, \cdots, R_n$ is any finite system of oriented rectangles in R_0, without common interior points, such that $|R_1| + \cdots + |R_n| < \eta$, then $|\phi(R_1)| + \cdots + |\phi(R_n)| < \epsilon$. Now let $R_1^*, \cdots, R_m^*$ be any system of oriented rectangles in R_0, without common interior points, such that

$$|R_1^*| + \cdots + |R_m^*| < \eta(\epsilon/2).$$

For each $j = 1, \cdots, m$, let $D_k(R_j^*)$ be a sequence of subdivisions such that $\|D_k(R_j^*)\| \to 0$ for $k \to \infty$. Then $\phi[D_k(R_j^*)] \to \beta(R_j^*)$ for $k \to \infty$, and hence we can choose a positive integer K such that

$$(1) \qquad \sum_{j=1}^{m} |\beta(R_j^*)| < \sum_{j=1}^{m} |\phi[D_K(R_j^*)]| + \frac{\epsilon}{2}.$$

Let $r_1, \cdots, r_N$ denote the rectangles occurring in the subdivisions $D_K(R_1^*), \cdots, D_K(R_m^*)$. Then $|r_1| + \cdots + |r_N| = |R_1^*| + \cdots + |R_m^*| < \eta(\epsilon/2)$, and hence

$$(2) \qquad \sum_{j=1}^{m} |\phi[D_K(R_j^*)]| \leqq \sum_{i=1}^{N} |\phi(r_i)| < \frac{\epsilon}{2}.$$

(1) and (2) yield $\epsilon > \sum |\beta(R_j^*)|, j = 1, \cdots, m$, and the absolute continuity of $\beta(R)$ is proved.

III.1.19. Let $\phi(R)$ be a rectangle function in R_0 with the following properties. (i) $\phi(R)$ is non-negative. (ii) $\phi(R)$ increases by subdivision. (iii) $\phi(R)$ is continuous. (iv) $\phi(R)$ has a finite U-function in R_0 (see III.1.5). Let R_0 be given by

$$R_0 : a \leqq u \leqq b, \qquad c \leqq v \leqq d.$$

Let u_0 satisfy the inequalities $a < u_0 < b$. The segment $u = u_0$, $c \leqq v \leqq d$ will be termed a *light* vertical segment if for every $\epsilon > 0$ there exists a $\delta = \delta(\epsilon, u_0) > 0$, such that $U(R', \phi) < \epsilon$ for every rectangle $R' : a' \leqq u \leqq b', c \leqq v \leqq d$ for which $a' \leqq u_0 \leqq b'$, $b' - a' < \delta$. Otherwise the segment will be termed *heavy*. Light and heavy horizontal segments are defined in a similar way. Since $U(R, \phi)$ is additive (see III.1.5) and non-negative, it follows readily that the class of the heavy segments is countable (possibly empty).

III.1.20. Given $\phi(R)$ as in III.1.19, let R_1, R_2 be two oriented rectangles in R_0, without common interior points, having one side in common, say a vertical side. Then R_1, R_2 are given by formulas

$$R_1 : a_1 \leqq u \leqq l, \qquad \bar{c} \leqq v \leqq \bar{d},$$

$$R_2 : \ l \leqq u \leqq b_2 , \qquad \bar{c} \leqq v \leqq \bar{d}.$$

Let l_1, l_2 be two numbers such that $a_1 < l_1 < l < l_2 < b_2$, and let r, R_1', R_2' denote the rectangles $r : l_1 \leqq l \leqq l_2, \bar{c} \leqq v \leqq \bar{d}$, $R_1' : a_1 \leqq u \leqq l_1, \bar{c} \leqq v \leqq \bar{d}$, $R_2' : l_2 \leqq u \leqq b_2, \bar{c} \leqq v \leqq d$. Given $\eta > 0$, we assert that

$$\text{(1)} \qquad \phi(R_1') + \phi(r) + \phi(R_2') > \phi(R_1) + \phi(R_2) - \eta,$$

if l_1, l_2 are close enough to l. Indeed, if we introduce the auxiliary rectangles $r_1 : l_1 \leqq u \leqq l, \bar{c} \leqq v \leqq \bar{d}$, $r_2 : l \leqq u \leqq l_2, \bar{c} \leqq v \leqq \bar{d}$, then we have by the properties (i) and (ii)

$$\phi(R_1') + \phi(r) + \phi(R_2') = \phi(R_1') + \phi(r_1) + \phi(R_2') + \phi(r_2) - (\phi(r_1) + \phi(r_2) - \phi(r))$$

$$\geqq \phi(R_1) + \phi(R_2) - (\phi(r_1) + \phi(r_2) - \phi(r)).$$

Since ϕ is continuous, we have $\phi(r_1) \to 0$, $\phi(r_2) \to 0$, $\phi(r) \to 0$ for $l_1 \to l$, $l_2 \to l$ and the inequality (1) follows.

III.1.21. Given $\phi(R)$ as in III.1.19, let s be a light segment, and let $D_n(R_0)$ be a sequence of subdivisions such that $\| D_n(R_0) \| \to 0$. Let λ_n denote the summation $\sum\phi(R)$ extended over all rectangles R such that $R \in D_n(R_0)$, $sR \neq 0$. Then $\lambda_n \to 0$. This follows immediately from the fact that $U(R, \phi)$ is non-negative and additive, and $\phi(R) \leqq U(R, \phi)$ for every oriented rectangle $R \subset R_0$.

III.1.22. Given $\phi(R)$ as in III.1.19, and an $\epsilon > 0$, we have by definition a subdivision $D(R_0)$ such that $\phi[D(R_0)] > U(R_0, \phi) - (\epsilon/2)$. If we extend the sides of the rectangles of $D(R_0)$ until they meet the sides of R_0, we obtain a refinement $D'(R_0)$ of $D(R_0)$. Since ϕ increases by subdivision, we have $\phi[D_n'(R_0)] \geqq \phi[D(R_0)]$. Repeated application of the remarks in III.1.20 yields a modification $D''(R_0)$ of $D'(R_0)$ such that $\phi[D''(R_0)] > \phi[D'(R_0)] - (\epsilon/2)$, and such that none of the sides of the rectangles of $D''(R_0)$ lies on a heavy segment. Thus $D''(R_0)$ has the following properties.

(i) $\phi[D''(R_0)] > U(R_0, \phi) - \epsilon$.

(ii) If l is a side of a rectangle $R \in D''(R_0)$, such that l is not a subset of a side of R_0, then l does not lie on a heavy segment.

Now let $D_n(R_0)$ be a sequence of subdivisions such that $\| D_n(R_0) \| \to 0$. Let $D_n^*(R_0)$ be the subdivision obtained by superposition of $D_n(R_0)$ and $D''(R_0)$. Since ϕ increases by subdivision, we have $\phi[D_n^*(R_0)] \geqq \phi[D''(R_0)]$ and hence

$$\text{(1)} \qquad \phi[D_n^*(R_0)] > U(R_0, \phi) - \epsilon.$$

Let α_n denote the summation $\sum \phi(R)$ extended over all rectangles R such that $R \in D_n^*(R_0)$, $R \notin D_n(R_0)$. Since ϕ is non-negative, we have

$$\phi[D_n^*(R_0)] \leqq \phi[D_n(R_0)] + \alpha_n . \tag{2}$$

In view of (ii) above, repeated application of III.1.21 yields $\alpha_n \to 0$. Thus we obtain from (1) and (2) the inequality $\liminf \phi[D_n(R_0)] \geqq U(R_0, \phi) - \epsilon$. Since ϵ was arbitrary, it follows that

$$\liminf \phi[D_n(R_0)] \geqq U(R_0, \phi). \tag{3}$$

In view of the definition of $U(R_0, \phi)$, obviously

$$\limsup \phi[D_n(R_0)] \leqq U(R_0, \phi). \tag{4}$$

Since $D_n(R_0)$ was any sequence such that $\| D_n(R_0) \| \to 0$, (3) and (4) imply that ϕ is Burkill integrable in R_0 and

$$\int_{R_0} \phi = U(R_0, \phi).$$

III.1.23. If $\phi(R)$ satisfies the conditions (i)-(iv) stated in III.1.19, then clearly $\phi(R)$ satisfies the same conditions relative to every oriented rectangle in R_0. Hence III.1.22 yields the following theorem.

Suppose that (i) ϕ is non-negative, (ii) ϕ is continuous, (iii) ϕ increases by subdivision, and (iv) ϕ has a finite U-function in R_0. Then ϕ is Burkill integrable in R_0 and (cf. III.1.5)

$$\int_{R} \phi = U(R, \phi)$$

for every oriented rectangle $R \subset R_0$.

III.1.24. Given a rectangle function $\phi(R)$ in R_0, let (u_0, v_0) be an interior point of R_0. The *upper derivative* of ϕ at (u_0, v_0) is defined as the least upper bound of all those numbers l (including $\pm \infty$) for which there exists a sequence of oriented squares s_n such that $(u_0, v_0) \in \overset{0}{s}_n$, $| s_n | \to 0$, $\phi(s_n)/| s_n | \to l$. Similarly, the *lower derivative* is defined as the greatest lower bound of these same numbers l. We shall use $\overline{D}(u_0, v_0, \phi)$, $\underline{D}(u_0, v_0, \phi)$ to denote the upper and lower derivatives respectively. These derivatives may be equal to $\pm \infty$. If the upper and lower derivatives are finite and equal at a point (u_0, v_0), then their common value is denoted by $D(u_0, v_0, \phi)$ and is called *the derivative* of ϕ at (u_0, v_0). *Note that in defining these derivatives, we use oriented squares, and that we require the inclusion* $(u_0, v_0) \in \overset{0}{s}_n$, *and not merely the inclusion* $(u_0, v_0) \in s_n$ (the symbol $\overset{0}{s}_n$ denotes the interior of s_n). If the rectangle function ϕ is clearly identified by the context, then we shall use the notation $D(u, v)$, with similar conventions for the upper and lower derivatives.

The upper and lower derivatives are Borel measurable. Let us verify this statement for the upper derivative. Give any real number a. Let s be a generic notation for an oriented square. For every pair of positive integers m, n, we define the set E_{amn} by the formula

$$E_{amn} = \sum s^0, \qquad s \subset R_0^0, \qquad |s| < \frac{1}{n}, \qquad \frac{\phi(s)}{|s|} > a + \frac{1}{m}.$$

If E_a denotes the set where $\overline{D}(u, v) > a$, then it follows readily that

$$E_a = \sum_{m=1}^{\infty} \prod_{n=1}^{\infty} E_{amn} .$$

Since E_{amn} is open, it follows that E_a is a Borel set, and hence $\overline{D}(u, v)$ is Borel measurable (see I.3.8).

Let us note that according to our conventions, the upper and lower derivatives are defined only in the interior R_0^0 of R_0. As regards the derivative $D(u, v)$, the set E on which it is defined is the set on which the following conditions hold simultaneously

$$-\infty < \overline{D}(u, v) < +\infty, \qquad -\infty < \underline{D}(u, v) < +\infty, \qquad \overline{D}(u, v) = \underline{D}(u, v).$$

Thus E is a product of three Borel sets and hence E is also a Borel set. Since $D(u, v) = \overline{D}(u, v)$ on E, it follows that $D(u, v)$ is Borel measurable on E (see I.3.8). Of course, the set E may be empty.

III.1.25. Suppose that $\phi(R)$ is non-negative in R_0 and the derivative $D(u, v)$ of ϕ exists a.e. (*almost everywhere*) in R_0^0. Let us put (see IV.2.3)

$$\sigma_j = \sum \phi(s), \qquad s \subset R_0^0, s \in D_{p_j} .$$

If $\sigma_j \to 0$ for $j \to \infty$, then $D(u, v) = 0$ a.e. in R_0^0.

Proof. For each j, let us define a function $g_j(u, v)$ in R_0^0. If (u, v) is an interior point of a square s such that $s \subset R_0^0$, $s \in D_{p_j}$, then $g_j(u, v) = \phi(s)/|s|$. Otherwise $g_j(u, v) = 0$. Clearly

$$\iint_{R_0^0} g_j(u, v)\, du\, dv = \sigma_j, \qquad g_j(u, v) \underset{j \to \infty}{\longrightarrow} D(u, v) \text{ a.e. in } R_0^0$$

By I.3.10 it follows, since $D(u, v) \geqq 0$, that $D(u, v)$ is summable in R_0^0 and

$$(1) \qquad 0 \leqq \iint_{R_0^0} D(u, v)\, du\, dv \leqq \liminf \sigma_j = 0.$$

Since $D(u, v) \geqq 0$ a.e. in R_0^0, (1) implies that $D(u, v) = 0$ a.e. in R_0^0.

III.1.26. Let $f(u, v)$ be a summable function in R_0, and let us put

$$\phi(R) = \iint_R f(u, v)\, du\, dv, \qquad R \subset R_0 .$$

Then the derivative $D(u, v)$ of ϕ exists and is equal to $f(u, v)$ a.e. in R_0^0. This is an immediate consequence of I.3.13.

III.1.27. Suppose that $\phi(R)$ is Burkill integrable on R_0, and let $\beta(R)$ denote its indefinite Burkill integral. Then (cf. III.1.24)

$$\overline{D}(u, v, \phi) = \overline{D}(u, v, \beta), \qquad \underline{D}(u, v, \phi) = \underline{D}(u, v, \beta)$$

a.e. in R_0^0. In particular, if one of the derivatives $D(u, v, \phi)$, $D(u, v, \beta)$ exists a.e. in R_0^0, then the other one also exists a.e. in R_0^0 and $D(u, v, \phi) = D(u, v, \beta)$ a.e. in R_0^0.

PROOF. Let E be the subset of R_0^0 where $\overline{D}(u, v, \phi) > \overline{D}(u, v, \beta)$. Suppose that $| E | > 0$. Then we should have a positive integer k such that the set E_k where $\overline{D}(u, v, \phi) - \overline{D}(u, v, \beta) > (1/k)$ is also of positive measure. If $(u_0, v_0) \in E_k$, then we have, as a consequence of the definition of $\overline{D}(u_0, v_0, \phi)$, a sequence of oriented squares s_n such that $(u_0, v_0) \in s_n^0$, $s_n \subset R_0^0$, $| s_n | \to 0$, $\phi(s_n)/| s_n | \to \overline{D}(u_0, v_0, \phi)$. For this same sequence s_n we have then lim sup $\beta(s_n)/| s_n | \leqq \overline{D}(u_0, v_0, \beta)$. There follows the inequality $[\phi(s_n) - \beta(s_n)]/| s_n | > (1/k)$ for n sufficiently large. Now let $\mathfrak{F}_\eta$ denote the family of oriented squares s that satisfy the conditions $s \subset R_0^0$, $[\phi(s) - \beta(s)]/| s | > (1/k)$, $| s_n | < \eta$, where η is a given positive number. In view of the preceding remarks, the squares of the family $\mathfrak{F}_\eta$ cover the set E_k in the manner required in the Vitali covering theorem (see I.3.3). Hence the family $\mathfrak{F}_\eta$ contains a finite system of oriented squares $\sigma_1, \cdots, \sigma_m$, such that $\sigma_i^0\sigma_j^0 = 0$ for $i \neq j$ and $| \sigma_1 | + \cdots + | \sigma_m | > | E_k |/2$. We have then the following inequalities and inclusions.

$$(1) \qquad | \sigma_i | < \eta, \qquad \sigma_i \subset R_0^0, \qquad i = 1, \cdots, m, \qquad \sigma_i^0\sigma_j^0 = 0 \qquad \text{for } i \neq j.$$

$$(2) \qquad \sum_{i=1}^{m} [\phi(\sigma_i) - \beta(\sigma_i)] > \frac{1}{k} \sum_{i=1}^{m} | \sigma_i | > \frac{| E_k |}{2k}.$$

Since $| E_k | > 0$ and η is arbitrary, these relations contradict III.1.15. Thus it is established that $\overline{D}(u, v, \phi) \leqq \overline{D}(u, v, \beta)$ a.e. in R_0^0. The complementary inequality $\overline{D}(u, v, \phi) \geqq \overline{D}(u, v, \beta)$ is proved in the same manner, and the same argument applies to the lower derivatives.

III.1.28. We shall say that $\phi(R)$ is of *type* A in R_0 if the following conditions are satisfied. (i) ϕ is non-negative. (ii) Let $R_1, \cdots, R_m, R$ be oriented rectangles in R_0, such that $R_iR_j = 0$ for $i \neq j$ and $R_1 + \cdots + R_m \subset R$. Then $\phi(R_1) + \cdots + \phi(R_m) \leqq \phi(R)$. If ϕ is of type A in R_0, then clearly ϕ is of type A in every oriented rectangle in R_0.

THEOREM. *If ϕ is of type* A *in R_0, then its derivative $D(u, v)$ exists a.e. in R_0 and is summable in R_0. Furthermore, for every oriented rectangle $R \subset R_0$ we have the inequality*

$$\iint_R D(u, v)\, du\, dv \leqq \phi(R)$$

The proof is based on several preliminary statements.

(a) Let α be a positive number, and let E_α be the subset of R_0^0 where $\overline{D}(u, v) > \alpha$. Then $\alpha | E_\alpha | \leqq \phi(R_0)$.

PROOF. Let $\mathfrak{F}$ be the family of those oriented squares s that satisfy the following conditions: $s \subset R_0^0$, $\phi(s)/|s| > \alpha$. Clearly, the squares of $\mathfrak{F}$ cover E_α in the sense required by the Vitali covering theorem (see I.3.3). Hence $\mathfrak{F}$ contains a (finite or infinite) sequence of squares $s_1, \cdots, s_n, \cdots$, such that $s_i s_j = 0$ for $i \neq j$ and $|E_\alpha - \sum s_n| = 0$. Since ϕ is of type A, there follows for every positive integer k the inequality $\phi(R_0) \geqq \phi(s_1) + \cdots + \phi(s_k) > \alpha(|s_1| + \cdots + |s_k|)$. For $k \to \infty$ we obtain the inequality

$$\phi(R_0) \geqq \alpha \sum_{n=1}^{\infty} |s_n| \geqq \alpha |E_\alpha|.$$

(b) Since ϕ is of type A in every oriented rectangle $R \subset R_0$ also, (a) implies the inequality $\alpha |E_\alpha R| \leqq \phi(R)$ for all such rectangles R.

(c) Let E^* be the subset of R_0^0 where $\overline{D}(u, v) = +\infty$. Then $E^* \subset E_\alpha$ for every $\alpha > 0$, and from (a) it follows therefore that $|E^*| = 0$. Thus $\overline{D}(u, v) < +\infty$ a.e. in R_0.

(d) The subset of R_0^0 where $\underline{D}(u, v) < \overline{D}(u, v)$ is of measure zero.

PROOF. If the assertion is denied, then we should have numbers x, y, such that $0 < x < y$ and the subset E_{xy} of R_0^0 where $\underline{D}(u, v) < x < y < \overline{D}(u, v)$ is of positive measure. Choose then any $\epsilon > 0$. Since E_{xy} is measurable by III.1.24, we have an open set G such that $E_{xy} \subset G \subset R_0^0$ and $|G| < |E_{xy}| + \epsilon$. Now let $\mathfrak{F}$ denote the family of those oriented squares s in G that satisfy the condition $\phi(s)/|s| < x$. Clearly, the squares of $\mathfrak{F}$ cover E_{xy} in the manner required by the Vitali covering theorem. Hence $\mathfrak{F}$ contains a (finite or infinite) sequence of squares $s_1, \cdots, s_n, \cdots$, such that $s_i s_j = 0$ for $i \neq j$ and $|E_{xy} - \sum s_n| = 0$. There follow the inequalities

$$\sum \phi(s_n) < x \sum |s_n| \leqq x |G| < x(|E_{xy}| + \epsilon).$$

On the other hand, by (b) above we have

$$\sum \phi(s_n) \geqq y \sum |E_{xy} s_n| = y |E_{xy}|.$$

Since ϵ was arbitrary, it follows that $x |E_{xy}| \geqq y |E_{xy}|$. Since $0 < x < y$, this contradicts the assumption that $|E_{xy}| > 0$.

Clearly, (c) and (d) imply that $D(u, v)$ exists a.e. in R_0^0 (cf. III.1.24). Now let us denote, for each positive integer n, by K_n the collection of all squares $s \subset R_0$ of the form $(i - 1)/n \leqq u \leqq i/n$, $(j - 1)/n \leqq v \leqq j/n$, where i, j are integers (positive, negative, or zero). For given n, the collection K_n is finite. Let us replace each square $s^n \in K_n$ by a somewhat smaller oriented square $\bar{s}^n$, with the same center, such that the summation $\sum |s^n - \bar{s}^n|$, extended over all squares $s^n \in K_n$, is less than $1/n$, and let us denote by G_n the set of the interior points of all the squares $\bar{s}^n$ for given n. Then G_n is an open set, and clearly $|R_0 - G_n| \to 0$ for $n \to \infty$. We have therefore a sequence of positive integers $n_1 < n_2 < \cdots$, such that the series $\sum |R_0 - G_{n_k}|$, $k = 1, 2, \cdots$, is convergent. Let us put $F_m = \prod G_{n_k}$, $k = m, m + 1, \cdots$. Then clearly $|R_0 - F_m| \to 0$ for $m \to \infty$. Now let us define, for each positive integer k, a function $g_k(u, v)$ in R_0

as follows. If (u, v) is an interior point of some square $\bar{s}^{nk}$ (see above), then $g_k(u, v) = \phi(\bar{s}^{nk})/|\bar{s}^{nk}|$. Otherwise $g_k(u, v) = 0$ in R_0. Clearly, since ϕ is of type A,

$$\iint_{R_0} g_k(u, v)\, du\, dv \leqq \phi(R_0). \tag{1}$$

Let m be a positive integer, and let (u, v) be a point of F_m such that $D(u, v)$ exists. Then $(u, v) \in G_{n_k}$ for $k \geqq m$, and hence $g_k(u, v)$ is equal to a quotient of the form $\phi(s)/|s|$, where s is one of the squares $\bar{s}^{nk}$ and (u, v) is interior to s. Hence $g_k(u, v) \to D(u, v)$ for $k \to \infty$. Since $D(u, v)$ exists a.e. in R_0, it follows that $g_k(u, v) \to D(u, v)$ a.e. on F_m, $m = 1, 2, \cdots$. Since $|R_0 \dot{-} F_m| \to 0$ for $m \to \infty$, it follows finally that

$$g_k(u, v) \underset{k\to\infty}{\longrightarrow} D(u, v) \text{ a.e. in } R_0 . \tag{2}$$

As $g_k(u, v) \geqq 0$, the relations (1) and (2) imply, by the lemma of Fatou (see I.3.10), the summability of $D(u, v)$ in R_0 and the inequality

$$\iint_{R_0} D(u, v)\, du\, dv \leqq \phi(R_0). \tag{3}$$

Since ϕ is of type A in every oriented rectangle $R \subset R_0$ also, we can replace, in (3), R_0 by any such rectangle R, and the proof is complete.

III.1.29. As matter of fact, most of the rectangle functions $\phi(R)$ that occur in the sequel satisfy a stronger condition than that required in the definition of type A, namely the condition $\phi(R_1) + \cdots + \phi(R_m) \leqq \phi(R)$ as soon as $R_1 + \cdots + R_m \subset R$, $R_i^0 R_j^0 = 0$ for $i \neq j$ (all these rectangles being oriented). Important special cases of rectangle functions of type A include the following ones (many others will occur in the sequel).

(a) $\phi(R)$ is non-negative and additive in R_0. Then ϕ is clearly of type A in R_0.

(b) Let $\Phi(B)$ be a non-negative, completely additive function of Borel sets $B \subset R_0$ (cf. I.3.16). Then Φ gives rise to a rectangle function $\phi(R) = \Phi(R)$ that is clearly of type A in R_0. We define in terms of $\phi(R)$ the derivatives $\overline{D}(u, v, \Phi)$, $\underline{D}(u, v, \Phi)$, $D(u, v, \Phi)$ as follows:

$$\overline{D}(u, v, \Phi) = \overline{D}(u, v, \phi), \qquad \underline{D}(u, v, \Phi) = \underline{D}(u, v, \phi), \qquad D(u, v, \Phi) = D(u, v, \phi).$$

However, $\Phi(R)$ is defined also for arbitrary non-oriented rectangles R in R_0, and this fact leads to several important remarks. For clarity, let us use R^* to refer to arbitrary, not necessarily oriented, rectangles, while R will denote, as before, an oriented rectangle. Then $\Phi(R^*)$ is a function of arbitrary, non-oriented rectangles in R_0. If we use arbitrary non-oriented rectangles and squares in the definitions of type A and of the derivatives (see III.1.28, III.1.24), then we obtain concepts that will be referred to as type A* and derivatives $\overline{D}^*(u, v, \Phi)$,

$\underline{\underline{D}}^*(u, v, \Phi)$, $D^*(u, v, \Phi)$. Clearly, $\underline{D}^*(u, v, \Phi) \leqq \underline{D}(u, v, \Phi) \leqq \overline{D}(u, v, \Phi) \leqq \overline{D}^*(u, v, \Phi)$. Hence, if $D^*(u, v, \Phi)$ exists at a point (u, v), then $D(u, v, \Phi)$ also exists and $D(u, v, \Phi) = D^*(u, v, \Phi)$. Since clearly $\Phi(R^*)$ is of type A^*, a reasoning entirely analogous to that in III.1.28 yields the result that $D^*(u, v, \Phi)$ exists a.e. and hence is equal to $D(u, v, \Phi)$ a.e. in R_0.

III.1.30. Let $\Phi(B)$ be a non-negative, completely additive function of Borel sets $B \subset R_0$. As noted in III.1.29, $\Phi(B)$ gives rise to a rectangle function $\phi(R) = \Phi(R)$. From this point on, R will refer again exclusively to oriented rectangles. As before, we put $\overline{D}(u, v, \Phi) = \overline{D}(u, v, \phi)$, $\underline{D}(u, v, \Phi) = \underline{D}(u, v, \phi)$, $D(u, v, \Phi) = D(u, v, \phi)$ (cf. III.1.24). As noted in III.1.29, ϕ is of type A in R_0. Hence the derivative $D(u, v, \Phi) = D(u, v, \phi)$ exists a.e. in R_0 and is summable in R_0. We proceed to complete for this special situation the general theory of additive set functions (cf. I.3.16). We shall use $D(u, v)$ to refer to $D(u, v, \Phi) = D(u, v, \phi)$ if Φ is clearly identified by the context.

III.1.31. Given $\Phi(B)$ as in III.1.30, suppose that Φ is singular in R_0. Then its derivative $D(u, v)$ vanishes a.e. in R_0.

PROOF. Since Φ is singular, we have two Borel sets E, e such that $E + e = R_0$, $Ee = 0$, $\Phi(E) = 0$, $| e | = 0$ (see I.3.16). Let E^* be the subset of ER_0^0 where $D(u, v) > 0$. The proof will be made if we can show that $| E^* | = 0$. In turn, this last fact will be proved if we can show that for each positive integer n the subset E_n^* of E^* where $D(u, v) > (1/n)$ is of measure zero. Now E_n^* is a Borel subset of E and hence $\Phi(E_n^*) = 0$. Given $\epsilon > 0$, we have therefore (see I.3.16) an open set G such that $E_n^* \subset G \subset R_0^0$ and $\Phi(G) < \epsilon$. Let $\mathfrak{F}$ denote the family of those oriented squares s in G that satisfy the condition $\Phi(s)/| s | > (1/n)$. Clearly, the squares of $\mathfrak{F}$ cover E_n^* in the manner required by the Vitali covering theorem. Hence $\mathfrak{F}$ contains a (finite or infinite) sequence of squares $s_1, \cdots, s_k, \cdots$, such that $s_i s_j = 0$ for $i \neq j$ and $| E_n^* - \sum s_k | = 0$. There follow the inequalities

$$| E_n^* | \leqq \sum_k | s_k | < n \sum_k \Phi(s_k) \leqq n\Phi(G) < n\epsilon.$$

Since n is fixed and ϵ is arbitrary, it follows that $| E_n^* | = 0$.

III.1.32. Given $\Phi(B)$ as in III.1.30, we have (see I.3.16) for Φ a univocally determined Lebesgue decomposition

$$\Phi(B) = \Phi_a(B) + \Phi_s(B) \qquad B \subset R_0, \tag{1}$$

where Φ_a, Φ_s are non-negative, completely additive functions of Borel sets $B \subset R_0$, and Φ_a is absolutely continuous and Φ_s is singular. By I.3.17 we have in R_0 a non-negative, Borel measurable, summable function $f(u, v)$ such that

$$\Phi_a(B) = \iint_B f(u, v)\, du\, dv \tag{2}$$

for every Borel set $B \subset R_0$. In view of III.1.26, there follows from (2) the formula

$$D(u, v, \Phi_a) = f(u, v) \qquad \text{a.e. in } R_0^0 \tag{3}$$

On the other hand, $D(u, v, \Phi_s) = 0$ a.e. in R_0^0 by III.1.31. We obtain therefore, from (1) and (3), the formula $D(u, v, \Phi) = f(u, v)$ a.e. in R_0. It follows now from (1) and (2) that

$$\Phi(B) = \iint_B D(u, v, \Phi)\, du\, dv + \Phi_s(B), \tag{4}$$

for every Borel set $B \subset R_0'$.

III.1.33. Given $\Phi(B)$ as in III.1.30, $\Phi(B)$ is singular if and only if its derivative $D(u, v)$ vanishes a.e. in R_0. Indeed, the condition is necessary by III.1.31. Conversely, if $D(u, v) = 0$ a.e. in R_0, then $\Phi(B) = \Phi_s(B)$ by III.1.32(4), and thus Φ is singular.

III.1.34. Given $\Phi(B)$ as in III.1.30, $\Phi(B)$ is AC if and only if the rectangle function $\phi(R) = \Phi(R)$ is AC (see I.3.16, III.1.2).

PROOF. Suppose first that $\Phi(B)$ is AC. Then $\Phi_s(B) \equiv 0$, and hence by III.1.32(4) we have, in particular,

$$\phi(R) = \Phi(R) = \iint_R D(u, v, \Phi)\, du\, dv$$

for every oriented rectangle $R \subset R_0$. Hence $\phi(R)$ is clearly AC (cf. I.3.13). Suppose, conversely, that $\phi(R)$ is AC. Let e be any Borel set of measure zero in R_0. Give $\eta > 0$. Then we have a set G, open relative to R_0, such that $e \subset G$ and $| G | < \eta$. There exists then a sequence $R_1, \cdots, R_n, \cdots$ of oriented rectangles such that $R_i^0 R_j^0 = 0$ for $i \neq j$ and $G = \sum R_n$ (cf. I.3.2). Since Φ is non-negative, there follows the inequality $0 \leqq \Phi(e) \leqq \sum \Phi(R_n)$. Now since $\phi(R) = \Phi(R)$ is AC by assumption, for given $\epsilon > 0$ the η of the preceding argument may be chosen so that $\Phi(r_1) + \cdots + \Phi(r_k) < \epsilon$ for every finite system of oriented rectangles $r_1, \cdots, r_k$ in R_0 such that $r_i^0 r_j^0 = 0$ for $i \neq j$ and $| r_1 | + \cdots + | r_k | < \eta$. In particular, we shall have then $\Phi(R_1) + \cdots + \Phi(R_n) < \epsilon$ for every n, and hence $\Phi(R_1) + \cdots + \Phi(R_n) + \cdots \leqq \epsilon$. Hence $0 \leqq \Phi(e) \leqq \epsilon$. Since ϵ is arbitrary, it follows that $\Phi(e) = 0$ for every Borel set $e \subset R_0$ of measure zero. Hence Φ is AC in R_0 (see I.3.16).

III.1.35. Given in R_0 a rectangle function $\phi(R)$, we shall say that ϕ satisfies in R_0 the condition (C) if the following statements hold. (i) $\phi(R)$ is non-negative. (ii) Let $r_1, \cdots, r_k$ be a finite system of oriented rectangles in R_0 such that $r_i r_j = 0$ for $i \neq j$, and let $R_1, \cdots, R_n, \cdots$ be a (finite or infinite) sequence of oriented rectangles in R_0 such that $r_1 + \cdots + r_k \subset R_1 + \cdots + R_n^* + \cdots$. Then $\phi(r_1) + \cdots + \phi(r_k) \leqq \phi(R_1) + \cdots + \phi(R_n) + \cdots$, for every choice of the systems $r_1, \cdots, r_k, R_1, \cdots, R_n, \cdots$.

III.1.36. Given $\Phi(B)$ as in III.1.30, the rectangle function $\phi(R) = \Phi(R)$ satisfies the condition (C) in R_0. Indeed, given systems $r_1, \cdots, r_k, R_1, \cdots, R_n, \cdots$ as described in III.1.35(ii), let us put $B_1 = R_1$, $B_2 = R_2 - R_1, \cdots, B_n = R_n - (R_1 + \cdots + R_{n-1}), \cdots$. Then

$\Phi(r_1) + \cdots + \Phi(r_k)$

$$= \Phi(r_1 + \cdots + r_k) \leqq \Phi(R_1 + \cdots + R_n + \cdots)$$
$$= \Phi(B_1 + \cdots + B_n + \cdots)$$
$$= \Phi(B_1) + \cdots + \Phi(B_n) + \cdots \leqq \Phi(R_1) + \cdots + \Phi(R_n) + \cdots .$$

III.1.37. Suppose that the rectangle function $\phi(R)$, given in R_0, has the following properties. (a) ϕ is non-negative. (b) ϕ is additive (see III.1.5). (c) ϕ is continuous (see III.1.2). Then ϕ satisfies the condition (C) of III.1.35.

PROOF. Choose systems $r_1, \cdots, r_k, R_1, \cdots, R_n, \cdots$ as described in III.1.35(ii).

Case (1). The system $R_1, \cdots, R_n, \cdots$ is finite. Then the inequality $\phi(r_1) + \cdots + \phi(r_k) \leqq \phi(R_1) + \cdots + \phi(R_n) + \cdots$ is an immediate consequence of the properties (a) and (b) of ϕ.

Case (2). The system $R_1, \cdots, R_n, \cdots$ is infinite. Give then $\epsilon > 0$. For each n, let R_n^* be an oriented rectangle that contains R_n in its interior, and let $\bar{R}_n$ be the oriented rectangle $R_0R_n^*$. Then $\bar{R}_n$ is the sum of R_n and of not more than four oriented rectangles, comprised in R_0, each of which is contained in $R_n^* - R_n^0$ and hence has an area less than $|R_n^* - R_n^0|$. Since ϕ is continuous, non-negative and additive, it follows that R_n^* can be so chosen that $\phi(\bar{R}_n) < \phi(R_n) + (\epsilon/2^n)$. Now the rectangles $R_1^*, \cdots, R_n^*, \cdots$ cover the closed set $r_1 + \cdots + r_k$ in the manner required in the Borel covering theorem. Hence we have a finite sequence of positive integers $n_1 < n_2 < \cdots < n_m$ such that $r_1 + \cdots + r_k \subset R_{n_1}^* + \cdots + R_{n_m}^*$. Since $r_1 + \cdots + r_k \subset R_0$ and $\bar{R}_n = R_0R_n^*$, it follows that $r_1 + \cdots + r_k \subset \bar{R}_{n_1} + \cdots + \bar{R}_{n_m}$. By case (1), we have therefore

$$\phi(r_1) + \cdots + \phi(r_k) \leqq \phi(\bar{R}_{n_1}) + \cdots + \phi(\bar{R}_{n_m}) \leqq \sum_{n=1}^{\infty} \phi(\bar{R}_n) < \epsilon + \sum_{n=1}^{\infty} \phi(R_n).$$

Since ϵ was arbitrary, the inequality $\phi(r_1) + \cdots + \phi(r_k) \leqq \phi(R_1) + \cdots + \phi(R_n) + \cdots$ follows.

III.1.38. Suppose that the rectangle function $\phi(R)$ satisfies the condition (C) in R_0. Let $R: a \leqq u \leqq b, c \leqq v \leqq d$ be an oriented rectangle in R_0. Let x be a number such that $a < x < b$, and let us denote by $R_x, \bar{R}_x$ the rectangles $a \leqq u \leqq x, c \leqq v \leqq d$ and $x \leqq u \leqq b, c \leqq v \leqq d$ respectively. Then we can choose x to satisfy the conditions

$$(1) \quad x - a < 3(b - a)/4, \qquad b - x < 3(b - a)/4, \qquad \phi(R_x) + \phi(\bar{R}_x) = \phi(R).$$

A similar statement holds, of course, for subdivisions of R into two rectangles that have a horizontal side in common.

PROOF. For $a < x \leqq b$, the quantity $\phi(R_x)$ is a non-negative, nondecreasing function of x, since ϕ satisfies the condition (C). Hence the points of discontinuity of this function constitute a countable (possible empty) set. We can choose therefore an x such that the inequalities in (1) hold and $\phi(R_x)$ is continuous at x. Now give $\epsilon > 0$. Since $\phi(R_x)$ is continuous at x, we can determine a number y

such that $a < y < x$ and $\phi(R_y) > \phi(R_x) - \epsilon$. Since the rectangles R_y and $\bar{R}_x$ are disjoint, we have then $\phi(R) \geqq \phi(R_y) + \phi(\bar{R}_x) > \phi(R_x) + \phi(\bar{R}_x) - \epsilon$. Since ϵ is arbitrary, it follows that $\phi(R) \geqq \phi(R_x) + \phi(\bar{R}_x)$, while the complementary inequality follows directly from the fact that ϕ satisfies the condition (C). Hence $\phi(R) = \phi(R_x) + \phi(\bar{R}_x)$.

III.1.39. Suppose that ϕ satisfies the condition (C) in R_0, and let R be an oriented rectangle in R_0. Then there exists, for given $\delta > 0$, a subdivision $D(R)$ of R such that $\| D(R) \| < \delta$ and $\phi(R) = \phi[D(R)]$ (cf. III.1.4, III.1.5).

This statement follows by repeated application of III.1.38.

III.1.40. Let $\phi(R)$ be a non-negative rectangle function in R_0 (see III.1.1). Given a set $E \subset R_0$, let $R_1, \cdots, R_n, \cdots$ be a finite or infinite sequence of oriented rectangles in R_0 such that $E \subset \sum R_n$. We define

$$\Gamma(E, \phi) = \text{gr.l.b.} \sum \phi(R_n),$$

where the greatest lower bound is taken with respect to all sequences $R_1, \cdots, R_n, \cdots$ that cover E. For the empty set 0 we define $\Gamma(0, \phi) = 0$. The following statements are obvious.

(a) $\Gamma(R, \phi) \leqq \phi(R)$ for every oriented rectangle $R \subset R_0$.

(b) $0 \leqq \Gamma(E, \phi) \leqq \phi(R_0)$ for every set $E \subset R_0$.

(c) If ϕ_1, ϕ_2 are two non-negative rectangle functions in R_0 such that $\phi_1(R) \leqq \phi_2(R)$ for every oriented rectangle $R \subset R_0$, then $\Gamma(E, \phi_1) \leqq \Gamma(E, \phi_2)$ for every set $E \subset R_0$.

(d) If ϕ satisfies the condition (C), then $\Gamma(R, \phi) = \phi(R)$ for every oriented rectangle $R \subset R_0$. Indeed, in this case $\phi(R) \leqq \sum \phi(R_n)$ for every sequence $R_1, \cdots, R_n, \cdots$ that covers R and hence $\phi(R) \leqq \Gamma(R, \phi)$, while the complementary inequality $\phi(R) \geqq \Gamma(R, \phi)$ holds by (a).

III.1.41. Let $\Phi(B)$ be a non-negative, completely additive function of Borel sets $B \subset R_0$. Then Φ gives rise to a non-negative rectangle function $\phi(R) = \Phi(R)$. We define $\Gamma(E, \Phi) = \Gamma(E, \phi)$. We proceed to verify two preliminary statements concerning $\Gamma(E, \Phi)$.

(a) $\Phi(B) \leqq \Gamma(B, \Phi)$ for every Borel set $B \subset R_0$. Indeed, let $R_1, \cdots, R_n, \cdots$ be any (finite or infinite) sequence of oriented rectangles in R_0, such that $B \subset \sum R_n$. Let us put $B_1 = R_1$, $B_2 = R_2 - R_1, \cdots, B_n = R_n - (R_1 + \cdots + R_{n-1}), \cdots$. Then $\Phi(B) \leqq \Phi(\sum B_n) = \sum \Phi(B_n) \leqq \sum \Phi(R_n)$. Thus always $\Phi(B) \leqq \sum \Phi(R_n)$, and the inequality $\Phi(B) \leqq \Gamma(B, \Phi)$ follows.

(b) $\Phi(R) = \Gamma(R, \Phi)$ for every oriented rectangle $R \subset R_0$. This follows directly from III.1.36, III.1.40(d).

III.1.42. Suppose that the rectangle function $\phi(R)$ satisfies the condition (C) in R_0. Then $\Gamma(E, \phi)$ is a Carathéodory outer measure in R_0 (see I.3.18).

PROOF. We have to show that $\Gamma(E, \phi)$ satisfies the conditions (C_1), (C_2), (C_3) stated in I.3.18. Conditions (C_1), (C_2) are obviously satisfied. As regards condition (C_3), let E_1, E_2 be any two sets in R_0 such that $\rho(E_1, E_2) > 0$ (see I.2.10). Give $\epsilon > 0$. By the definition of $\Gamma(E_1 + E_2, \phi)$ we have then a sequence of oriented rectangles $R_1, \cdots, R_n, \cdots$ in R_0 such that $E_1 + E_2 \subset \sum R_n$ and

$$\sum \phi(R_n) < \Gamma(E_1 + E_2, \phi) + \epsilon.$$

In view of III.1.39, we can replace each R_n by a finite system $r_n^1, r_n^2, \cdots$ of oriented rectangles such that $R_n = r_n^1 + r_n^2 + \cdots$, $\phi(R_n) = \phi(r_n^1) + \phi(r_n^2) + \cdots$, and such that the diameter of each r_n^i is less than $\rho(E_1, E_2)$. Let us arrange the rectangles r_n^i, associated in this manner with the rectangles R_n, $n = 1, 2, \cdots$, into a single sequence $r_1, \cdots, r_k, \cdots$. Then $E_1 + E_2 \subset \sum r_k$, $\sum \phi(r_k) = \sum \phi(R_n) < \Gamma(E_1 + E_2, \phi) + \epsilon$, and $d(r_k) < \rho(E_1, E_2)$ for every k. It follows that no rectangle r_k intersects both E_1 and E_2. Hence, if $\sum^1, \sum^2$ denote summations extended over those rectangles r_k that intersect E_1, E_2 respectively, then $E_1 \subset \sum^1 r_k$, $E_2 \subset \sum^2 r_k$, and hence $\Gamma(E_1, \phi) \leqq \sum^1 \phi(r_k)$, $\Gamma(E_2, \phi) \leqq \sum^2 \phi(r_k)$. Furthermore, since no r_k occurs in both of the summations $\sum^1, \sum^2$, we have $\sum^1 \phi(r_k) + \sum^2 \phi(r_k) \leqq \sum \phi(r_k) < \Gamma(E_1 + E_2, \phi) + \epsilon$. There follows the inequality $\Gamma(E_1, \phi) + \Gamma(E_2, \phi) \leqq \Gamma(E_1 + E_2, \phi) + \epsilon$. Since ϵ was arbitrary, it follows that $\Gamma(E_1, \phi) + \Gamma(E_2, \phi) \leqq \Gamma(E_1 + E_2, \phi)$. Since the complementary inequality $\Gamma(E_1, \phi) + \Gamma(E_2, \phi) \geqq \Gamma(E_1 + E_2, \phi)$ is obvious, the relation $\Gamma(E_1, \phi) + \Gamma(E_2, \phi) = \Gamma(E_1 + E_2, \phi)$ follows.

III.1.43. Let $\phi(R)$ be a rectangle function and $\Phi(B)$ a completely additive function of Borel sets in R_0. If $\Phi(R) = \phi(R)$ for every oriented rectangle $R \subset R_0$, then we shall say that Φ is a completely additive extension of ϕ to Borel sets in R_0.

THEOREM. *Let $\phi(R)$ be a rectangle function in R_0. Then $\phi(R)$ admits of a completely additive, non-negative extension to Borel sets in R_0 if and only if ϕ satisfies the condition* (C) *in R_0.*

PROOF. Condition (C) is necessary by III.1.36. Suppose, conversely, that ϕ satisfies the condition (C) in R_0. Then $\Gamma(E, \phi)$ is a Carathéodory outer measure in R_0 (see III.1.42), and $\Gamma(R, \phi) = \phi(R)$ for every oriented rectangle $R \subset R_0$ by III.1.40(d). Clearly Γ is non-negative. By I.3.18, Γ is completely additive in the class of Borel sets in R_0. Hence $\Phi(B) = \Gamma(B, \phi)$ is a non-negative, completely addition extension of ϕ to Borel sets in R_0.

III.1.44. Let $\Phi(B)$ be a completely additive, non-negative function of Borel sets $B \subset R_0$. Then $\Phi(B) = \Gamma(B, \Phi)$ for every Borel set $B \subset R_0$ (cf. III.1.41).

PROOF. By III.1.36, III.1.42, $\Gamma(E, \Phi)$ is a Carathéodory outer measure in R_0 and hence by I.3.18, $\Gamma(B, \Phi)$ is completely additive (and clearly non-negative) on Borel sets. By III.1.41 we have $\Phi(R) = \Gamma(R, \Phi)$ for every oriented rectangle $R \subset R_0$. Hence, if B is any Borel set in R_0, we have

$$\Phi(R_0) = \Gamma(R_0, \Phi) = \Gamma(B, \Phi) + \Gamma(R_0 - B, \Phi). \tag{1}$$

On the other hand, III.1.41 yields

$$\Gamma(B, \Phi) \geqq \Phi(B), \qquad \Gamma(R_0 - B, \Phi) \geqq \Phi(R_0 - B). \tag{2}$$

In view of (1), these inequalities lead, by addition, to the inequalities

$$\Phi(R_0) = \Gamma(B, \Phi) + \Gamma(R_0 - B, \Phi) \geqq \Phi(B) + \Phi(R_0 - B) = \Phi(R_0).$$

It follows that the sign of equality must hold in (2), and thus the relation $\Phi(B) = \Gamma(B, \Phi)$ follows.

III.1.45. If $\Phi_1(B)$, $\Phi_2(B)$ are non-negative, completely additive functions of Borel sets in R_0, and $\Phi_1(R) \geqq \Phi_2(R)$ for every oriented rectangle $R \subset R_0$, then $\Phi_1(B) \geqq \Phi_2(B)$ for every Borel set $B \subset R_0$.

PROOF. Clearly $\Gamma(B, \Phi_1) \geqq \Gamma(B, \Phi_2)$. Hence, by III.1.44, we have $\Phi_1(B) = \Gamma(B, \Phi_1) \geqq \Gamma(B, \Phi_2) = \Phi_2(B)$.

III.1.46. If $\Phi_1(B)$, $\Phi_2(B)$ are completely additive functions (of arbitrary sign) of Borel sets $B \subset R_0$, and $\Phi_1(R) \geqq \Phi_2(R)$ for every oriented rectangle $R \subset R_0$, then $\Phi_1(B) \geqq \Phi_2(B)$ for every Borel set $B \subset R_0$.

PROOF. By I.3.16 we have $\Phi_1 = \Phi_{11} - \Phi_{12}$, $\Phi_2 = \Phi_{21} - \Phi_{22}$, where Φ_{11}, Φ_{12}, Φ_{21}, Φ_{22} are non-negative, completely additive functions of Borel sets in R_0. By assumption we have $\Phi_{11}(R) - \Phi_{12}(R) \geqq \Phi_{21}(R) - \Phi_{22}(R)$ and hence $\Phi_{11}(R) + \Phi_{22}(R) \geqq \Phi_{21}(R) + \Phi_{12}(R)$ for every oriented rectangle $R \subset R_0$. By III.1.45 it follows that $\Phi_{11}(B) + \Phi_{22}(B) \geqq \Phi_{21}(B) + \Phi_{12}(B)$ and hence $\Phi_1(B) = \Phi_{11}(B) - \Phi_{12}(B) \geqq \Phi_{21}(B) - \Phi_{22}(B) = \Phi_2(B)$ for every Borel set $B \subset R_0$.

III.1.47. If $\Phi(B)$, $\Psi(B)$ are completely additive functions (of arbitrary sign) of Borel sets $B \subset R_0$, and $\Phi(R) = \Psi(R)$ for every oriented rectangle $R \subset R_0$, then $\Phi(B) = \Psi(B)$ for every Borel set $B \subset R_0$. This is an immediate consequence of III.1.46.

III.1.48. We shall discuss presently an important special kind of set functions. Let $\mu(B)$ be a function of Borel sets $B \subset R_0$ with the following properties. (i) μ is non-negative. (ii) μ is completely additive. (iii) μ is continuous on oriented rectangles. That is, for every $\epsilon > 0$ there exists a $\delta = \delta(\epsilon) > 0$ such that $\mu(R) < \epsilon$ for every oriented rectangle $R \subset R_0$ such that $|R| < \delta$. In the following discussion of μ, the letters R, r will denote oriented rectangles in R_0, and P_r will denote the center of r. Let us note the following facts.

(a) If σ is a horizontal or vertical segment in R_0, then $\mu(\sigma) = 0$ (note that σ may be a subset of the perimeter of R_0). Indeed, we have in R_0 a sequence of rectangles R_n such that σ is a side of R_n and $|R_n| \to 0$. We have then $0 \leqq \mu(\sigma) \leqq \mu(R_n) \to 0$, and hence $\mu(\sigma) = 0$.

(b) Since μ is non-negative and completely additive, it follows that if $E \subset R_0$ is a set of the form $E = \sigma_1 + \cdots + \sigma_n + \cdots$, where each σ_n is a horizontal or vertical segment, then $\mu(E) = 0$.

(c) Let G be a subset of R_0 that is open relative to R_0 (cf. I.2.26). Then we have a (finite or infinite) sequence of oriented rectangles $R_1, \cdots, R_n, \cdots$ in R_0 such that $G = \sum R_n$, $R_i^0 R_j^0 = 0$ for $i \neq j$. From (b) it follows that $\mu(G) = \sum \mu(R_n)$.

(d) μ is additive on oriented rectangles. That is, if $D(R)$ is any subdivision of the oriented rectangle $R \subset R_0$, then $\mu(R) = \mu[D(R)]$. This is an immediate consequence of (a).

(e) By I.3.17, μ gives rise to a μ-integral in R_0, this integral to be used only for Borel measurable functions. Let $f(u, v)$ be Borel measurable and μ-summable in R_0. We assert that for every $\epsilon > 0$ there exists a continuous function $g(u, v)$ in R_0 such that

(1) $$\int_{R_0} | f - g | \, d\mu < \epsilon.$$

PROOF. *Case* (α). $f(u, v)$ is the characteristic function of a closed set $F \subset R_0$ (that is, $f(u, v) = 1$ for $(u, v) \in F$ and $f(u, v) = 0$ for $(u, v) \in R_0 - F$). If $F = R_0$, then the assertion is trivial. So we can assume that $R_0 - F \neq 0$. Then $G = R_0 - F$ is a nonempty set which is open relative to R_0. Hence, by I.3.16, we have a closed set $F^* \subset G$ such that $\mu(F^*) > \mu(G) - \epsilon$. We have then the relations

(2) $$F \subset R_0 - F^*, \qquad \mu[R_0 - (F + F^*)] < \epsilon.$$

Thus F, F^* are disjoint closed sets in R_0, and hence they have a positive distance $\rho(F, F^*) > 0$. Let $g(u, v)$ be defined as follows in R_0. If $(u, v) \in F$, then $g(u, v) = 1$; if $(u, v) \in F^*$, then $g(u, v) = 0$. If $(u, v) \in R_0 - (F + F^*)$, then

$$g(u, v) = \frac{\min \{\rho(F, F^*), \rho[(u, v), F^*]\}}{\rho(F, F^*)},$$

where $\rho[(u, v), F^*]$ is the distance of (u, v) from F^*. It follows readily that $g(u, v)$ is continuous on R_0. Clearly $0 < g(u, v) \leqq 1$ for $(u, v) \in R_0 - (F + F^*)$. Hence $| f - g | = 0$ on $F + F^*$, and $| f - g | < 1$ on $R_0 - (F + F^*)$. Hence (1) holds in view of (2).

Case (β). $f(u, v)$ is an arbitrary Borel measurable, μ-summable function in R_0. By I.3.17, we have then a function $h(u, v)$ in R_0 of the form $h = c_1 f_1 + \cdots + c_m f_m$, where c_i is a constant and f_i is the characteristic function of a closed set F_i, $i = 1, \cdots, m$, such that

(3) $$\int_{R_0} | f - h | \, d\mu < \frac{\epsilon}{2}.$$

Let $\eta > 0$ be so chosen that $(| c_1 | + \cdots + | c_m |)\eta < (\epsilon/2)$. For each i, we have then by case (α) a continuous function $g_i(u, v)$ in R_0, such that

(4) $$\int_{R_0} | f_i - g_i | \, d\mu < \eta, \qquad i = 1, \cdots, m.$$

If we put $g = c_1 g_1 + \cdots + c_m g_m$, then g is continuous in R_0, and (1) clearly holds as a consequence of (3) and (4).

(f) If $f(u, v)$ is Borel measurable and μ-summable in R_0, and $D(R)$ is a subdivision of an oriented rectangle $R \subset R_0$, then

$$\int_R f \, d\mu = \sum \int_r f \, d\mu, \qquad r \in D(R).$$

This follows immediately from (a).

(g) Let $g(u, v)$ be a continuous function in R_0. Let $D(R_0)$ be any subdivision of R_0. For $r \in D(R_0)$, let P_r be the center of r, and let M_r, m_r denote the maximum and the minimum respectively of $g(u, v)$ in r. By (f) we have then the inequalities

$$\sum m_r\mu(r) \leqq \int_{R_0} g\, d\mu \leqq \sum M_r\mu(r), \qquad r \in D(R_0),$$

$$\sum m_r\mu(r) \leqq \sum g(P_r)\mu(r) \leqq \sum M_r\mu(r), \qquad r \in D(R_0).$$

There follows the inequality

$$\left| \int_{R_0} g\, d\mu - \sum g(P_r)\mu(r) \right| \leqq \sum (M_r - m_r)\mu(r), \qquad r \in D(R_0).$$

Hence, since g is uniformly continuous in R_0, it follows that if $D_n(R_0)$ is a sequence of subdivisions such that $\| D_n(R_0) \| \to 0$ and $\sum^n$ denotes summation over the rectangles $r \in D_n(R_0)$, we have the relation

$$\sum\nolimits^n g(P_r)\mu(r) \to \int_{R_0} g\, d\mu.$$

III.1.49. Given μ as in III.1.48, let $x_i(u, v)$, $y_i(u, v)$, $i = 1, 2, 3$, denote Borel measurable, μ-summable functions in R_0 such that

$$\int_{R_0} | x_i - y_i | \, d\mu < \epsilon, \qquad i = 1, 2, 3.$$

Then we have the inequalities

$$\int_{R_0} (x_1^2 + x_2^2 + x_3^2)^{1/2}\, d\mu < \int_{R_0} (y_1^2 + y_2^2 + y_3^2)^{1/2}\, d\mu + 3\epsilon,$$

$$\sum \left[\left(\int_r x_1\, d\mu \right)^2 + \left(\int_r x_2\, d\mu \right)^2 + \left(\int_r x_3\, d\mu \right)^2 \right]^{1/2}$$

$$< \sum \left[\left(\int_r y_1\, d\mu \right)^2 + \left(\int_r y_2\, d\mu \right)^2 + \left(\int_r y_3\, d\mu \right)^2 \right]^{1/2} + 3\epsilon,$$

where in the second inequality the summation is taken over the rectangles r of any subdivision $D(R_0)$ of R_0. The proof follows immediately from the inequalities in I.3.10.

III.1.50. Given μ as in III.1.48, let $f_1(u, v)$, $f_2(u, v)$, $f_3(u, v)$ be Borel measurable, μ-summable functions in R_0. Let $D_n(R_0)$ be a sequence of subdivisions such that $\| D_n(R_0) \| \to 0$, and let $\sum^n$ denote summation with respect to the rectangles $r \in D_n(R_0)$. Then

$$\sum^n \left[\left(\int_r f_1\,d\mu\right)^2 + \left(\int_r f_2\,d\mu\right)^2 + \left(\int_r f_3\,d\mu\right)^2\right]^{1/2} \to \int_{R_0} (f_1^2 + f_2^2 + f_3^2)^{1/2}\,d\mu.$$

PROOF. By I.3.10 and III.1.48(f), we have

$$\sum^n \left[\left(\int_r f_1\,d\mu\right)^2 + \left(\int_r f_2\,d\mu\right)^2 + \left(\int_r f_3\,d\mu\right)^2\right]^{1/2}$$

$$\leqq \sum^n \int_r (f_1^2 + f_2^2 + f_3^2)^{1/2}\,d\mu = \int_{R_0} (f_1^2 + f_2^2 + f_3^2)^{1/2}\,d\mu.$$

Hence it is sufficient to establish the inequality

$$\liminf \sum^n \left[\left(\int_r f_1\,d\mu\right)^2 + \left(\int_r f_2\,d\mu\right)^2 + \left(\int_r f_3\,d\mu\right)^2\right]^{1/2}$$

(1)

$$\geqq \int_{R_0} (f_1^2 + f_2^2 + f_3^2)^{1/2}\,d\mu.$$

Case (1). f_1, f_2, f_3 are continuous in R_0. For each n, let the functions f_{1n}, f_{2n}, f_{3n} be defined in R_0 as follows. If (u, v) is interior to a rectangle $r \in D_n(R_0)$, then $f_{1n}(u, v) = f_1(P_r)$, where P_r is the center of r. If (u, v) is not interior to any rectangle $r \in D_n(R_0)$, then $f_{1n}(u, v) = 0$. The functions f_{2n}, f_{3n} are defined in the same manner in terms of f_2, f_3. Clearly, since f_1, f_2, f_3 are continuous in R_0, the quantity

$$\epsilon_n = \int_{R_0} |\, f_1 - f_{1n}\,|\,d\mu + \int_{R_0} |\, f_2 - f_{2n}\,|\,d\mu + \int_{R_0} |\, f_3 - f_{3n}\,|\,d\mu$$

converges to zero for $n \to \infty$, in view of III.1.48(b). By III.1.49 we have the inequality

$$\sum^n \left[\left(\int_r f_{1n}\,d\mu\right)^2 + \left(\int_r f_{2n}\,d\mu\right)^2 + \left(\int_r f_{3n}\,d\mu\right)^2\right]^{1/2}$$

(2)

$$< \sum^n \left[\left(\int_r f_1\,d\mu\right)^2 + \left(\int_r f_2\,d\mu\right)^2 + \left(\int_r f_3\,d\mu\right)^2\right]^{1/2} + 3\epsilon_n\,.$$

On the other hand, clearly (cf. III.1.48(a))

$$\sum^n \left[\left(\int_r f_{1n}\,d\mu\right)^2 + \left(\int_r f_{2n}\,d\mu\right)^2 + \left(\int_r f_{3n}\,d\mu\right)^2\right]^{1/2}$$

(3)

$$= \sum^n [f_1(P_r)^2 + f_2(P_r)^2 + f_3(P_r)^2]^{1/2} \mu(r).$$

Since f_1, f_2, f_3 are continuous, we have by III.1.48(g)

$$(4) \qquad \sum^n [f_1(P_r)^2 + f_2(P_r)^2 + f_3(P_r)^2]^{1/2}\mu(r) \to \int_{R_0} (f_1^2 + f_2^2 + f_3^2)^{1/2}\, d\mu.$$

Since $\epsilon_n \to 0$, the inequality (1) follows from (2), (3), (4).

Case (2). *General case.* Give $\epsilon > 0$. By III.1.48(e), we have then continuous functions g_1, g_2, g_3 in R_0 such that

$$\int_{R_0} | f_i - g_i | \, d\mu < \epsilon, \qquad i = 1, 2, 3.$$

By III.1.49 we have the inequalities

$$(5) \qquad \int_{R_0} (f_1^2 + f_2^2 + f_3^2)^{1/2}\, d\mu < \int_{R_0} (g_1^2 + g_2^2 + g_3^2)^{1/2}\, d\mu + 3\epsilon,$$

$$(6) \qquad \begin{aligned} &\sum^n \left[\left(\int_r g_1\, d\mu\right)^2 + \left(\int_r g_2\, d\mu\right)^2 + \left(\int_r g_3\, d\mu\right)^2\right]^{1/2} \\ &\quad < \sum^n \left[\left(\int_r f_1\, d\mu\right)^2 + \left(\int_r f_2\, d\mu\right)^2 + \left(\int_r f_3\, d\mu\right)^2\right]^{1/2} + 3\epsilon. \end{aligned}$$

Since g_1, g_2, g_3 are continuous, we have by case (1)

$$(7) \qquad \begin{aligned} &\liminf \sum^n \left[\left(\int_r g_1\, d\mu\right)^2 + \left(\int_r g_2\, d\mu\right)^2 + \left(\int_r g_3\, d\mu\right)^2\right]^{1/2} \\ &\qquad \geqq \int_{R_0} (g_1^2 + g_2^2 + g_3^2)^{1/2}\, du. \end{aligned}$$

Since ϵ is arbitrary, (5), (6), (7) imply the inequality (1).

III.1.51. Let $\phi_1(B)$, $\phi_2(B)$, $\phi_3(B)$ be functions of Borel sets $B \subset R_0$, with the following properties. (α) ϕ_i is non-negative. (β) ϕ_i is continuous on oriented rectangles in R_0, in the sense of III.1.2. (γ) ϕ_i is completely additive on Borel sets in R_0 ($i = 1, 2, 3$). Let us define, for oriented rectangles $r \subset R_0$, a rectangle function $\psi(r)$ by the formula $\psi(r) = [\phi_1(r)^2 + \phi_2(r)^2 + \phi_3(r)^2]^{1/2}$. Then $\psi(r)$ is Burkill integrable in R_0.

PROOF. Clearly, ψ is non-negative. Since $\psi(r) \leqq \phi_1(r) + \phi_2(r) + \phi_3(r)$, clearly $\psi(r)$ is continuous on oriented rectangles in R_0. Let now R be any oriented rectangle in R_0, and let $D(R)$ be any subdivision of R. By III.1.48(d) we have then $\phi_i(R) = \phi_i[D(R)]$, $i = 1, 2, 3$, and hence by I.3.10

$$
\begin{aligned}
\psi(R) &= [\phi_1(R)^2 + \phi_2(R)^2 + \phi_3(R)^2]^{1/2} \\
&= (\phi_1[D(R)]^2 + \phi_2[D(R)]^2 + \phi_3[D(R)]^2)^{1/2} \\
&\leqq \sum [\phi_1(r)^2 + \phi_2(r)^2 + \phi_3(r)^2]^{1/2} = \sum \psi(r) = \psi[D(R)],
\end{aligned}
$$

where the summation is extended over all rectangles $r \in D(R)$. Thus ψ increases by subdivision. Finally, since $\psi \leqq \phi_1 + \phi_2 + \phi_3$, we have $\psi[D(R)] \leqq \phi_1[D(R)] + \phi_2[D(R)] + \phi_3[D(R)] = \phi_1(R) + \phi_2(R) + \phi_3(R) \leqq \phi_1(R_0) + \phi_2(R_0) + \phi_3(R_0)$. Thus ψ has a finite U-function in R_0 (see III.1.5). By III.1.23 it follows that ψ is Burkill integrable in R_0.

III.1.52. As noted in III.1.1, the preceding theory applies to interval functions in n-dimensional Euclidean space. Of course, in the case $n = 1$ the discussion can be considerably simplified.

In the sequel, we shall have to work with functions $\phi(R)$ of oriented rectangles R that are defined not in a fixed oriented rectangle R_0, but rather for all oriented rectangles in a given bounded domain $\mathfrak{D}$ (connected open set) in the uv-plane. The terms BV, AC are defined then exactly as in III.1.2, where R_0 is replaced by $\mathfrak{D}$. The derivatives of ϕ are defined as in III.1.24. The type A is defined now as follows: ϕ is of type A in $\mathfrak{D}$ if it is of type A, in the sense of III.1.28, in every oriented rectangle $R \subset \mathfrak{D}$. Many of the results obtained in the preceding theory remain valid, with obvious modifications in the proofs. To illustrate: if $\phi(R)$ is AC and bounded in $\mathfrak{D}$, then it is BV in $\mathfrak{D}$. The proof is the same as in III.1.3. Similarly, if $\phi(R)$ is of type A in $\mathfrak{D}$, then the derivative $D(u, v, \phi)$ exists a.e. in $\mathfrak{D}$, is summable on every oriented rectangle $R \subset \mathfrak{D}$ and its integral over R does not exceed $\phi(R)$. This follows immediately from III.1.28. Furthermore, the results stated in III.1.45, III.1.46 are readily seen to apply to completely additive set functions defined for Borel sets $B \subset \mathfrak{D}$.

CHAPTER III.2. BOUNDED VARIATION AND ABSOLUTE CONTINUITY

III.2.1. Let $f(u)$ be a continuous (real-valued) function in a fundamental interval $I_0 : a_0 \leqq u \leqq b_0$. If u is an interior point of I_0, and if the limit of the difference quotient $[f(u + \Delta u) - f(u)]/\Delta u$, for $0 \neq \Delta u \to 0$, exists and is finite, then this limit is termed the *derivative* $f'(u)$ of $f(u)$. Consider now a subinterval $I : a \leqq u \leqq b$ of I_0, and put $\phi(I) = f(b) - f(a)$. Then $\phi(I)$ is an interval function defined in I_0. If the derivative of $\phi(I)$ (cf. III.1.24, III.1.52) exists at an interior point u of I_0, then it will be denoted by $D(u)$, or by $D(u, \phi)$ if explicit reference to ϕ is desirable.

Lemma. *If $f'(u_0)$ exists at an interior point u_0 of I_0, then $D(u_0)$ exists also, and $f'(u_0) = D(u_0)$. Conversely, if $D(u_0)$ exists at an interior point u_0 of I_0, then $f'(u_0)$ exists also, and $f'(u_0) = D(u_0)$.*

Proof. (i) Suppose that $f'(u_0)$ exists at an interior point u_0 of I_0. Take any subinterval $I : a \leqq u \leqq b$ of I_0 such that $a < u_0 < b$. Then

$$\frac{\phi(I)}{|I|} = \frac{f(b) - f(a)}{b - a} = \frac{f(b) - f(u_0)}{b - u_0} \cdot \frac{b - u_0}{b - a} + \frac{f(a) - f(u_0)}{a - u_0} \cdot \frac{u_0 - a}{b - a}.$$

Give $\epsilon > 0$. Since $f'(u_0)$ exists by assumption, we have a $\delta = \delta(\epsilon, u_0) > 0$ such that

$$f'(u_0) - \epsilon < \frac{f(b) - f(u_0)}{b - u_0} < f'(u_0) + \epsilon,$$

$$f'(u_0) - \epsilon < \frac{f(a) - f(u_0)}{a - u_0} < f'(u_0) + \epsilon,$$

if $|I| = b - a < \delta$. Since

$$\frac{b - u_0}{b - a} > 0, \qquad \frac{u_0 - a}{b - a} > 0, \qquad \frac{b - u_0}{b - a} + \frac{u_0 - a}{b - a} = 1,$$

it follows that

$$f'(u_0) - \epsilon < \frac{\phi(I)}{|I|} < f'(u_0) + \epsilon \qquad \text{if } |I| < \delta.$$

As $\epsilon > 0$ was arbitrary, there follows the existence of $D(u_0)$, as well as the relation $D(u_0) = f'(u_0)$.

(ii) Suppose that $D(u_0)$ exists at an interior point u_0 of I_0. Take in I_0 any sequence b_n such that $b_n > u_0$, $b_n \to u_0$. Since $f(u)$ is continuous, we have for each n a point a_n in I_0 such that

$$(1) \qquad \left| \frac{f(b_n) - f(u_0)}{b_n - u_0} - \frac{f(b_n) - f(a_n)}{b_n - a_n} \right| < \frac{1}{n}, \qquad 0 < u_0 - a_n < \frac{1}{n}.$$

Since $D(u_0)$ exists by assumption, we have the relation

$$(2) \qquad \frac{f(b_n) - f(a_n)}{b_n - a_n} \to D(u_0).$$

(1) and (2) show that

$$(3) \qquad \frac{f(b_n) - f(u_0)}{b_n - u_0} \to D(u_0).$$

An analogous reasoning shows that (3) holds also if $b_n < u_0$, $b_n \to u_0$. Thus (3) holds if $u_0 \neq b_n \to u_0$ in any manner. Hence $f'(u_0)$ exists and is equal to $D(u_0)$.

III.2.2. Let $f(u)$ be given as in III.2.1. A great deal of the following discussion remains valid if the assumption of continuity is dropped, but this is irrelevant for our purposes. Let $\phi(I)$ have the same meaning as in III.2.1.

Definition. $f(u)$ is BV in I_0 (*of bounded variation in* I_0) if and only if the interval function $\phi(I)$ is BV in I_0 (see III.1.2).

Definition. $f(u)$ is AC in I_0 (*absolutely continuous in* I_0) if and only if the interval function $\phi(I)$ is AC in I_0 (see III.1.2).

Remark. Since $f(u)$ is continuous in I_0, clearly $\phi(I)$ is continuous and bounded in I_0, in the sense of III.1.2. By III.1.3 it follows that if $f(u)$ is AC in I_0, then $f(u)$ is also BV in I_0 (cf. III.1.52).

III.2.3. Given $f(u)$ as in III.2.1, let $\omega(I)$ denote the oscillation of $f(u)$ in the subinterval I of I_0. That is, $\omega(I) = M(I) - m(I)$, where $M(I)$, $m(I)$ denote the maximum and the minimum, respectively, of $f(u)$ in I. Let $\phi(I)$ have the same meaning as in III.2.1. Then we have in I an interval I^* such that $\omega(I) = |\phi(I^*)|$. Indeed, if $f(u)$ is constant in I, then we can choose $I^* = I$. Otherwise, we have in I two distinct points u', u'' such that $f(u') = m(I)$, $f(u'') = M(I)$, and we can choose then I^* as the interval with end points u', u''. The following statements are immediate consequences of the preceding remarks (cf. III.1.2).

(i) $f(u)$ is BV in I_0 if and only if the interval function $\omega(I)$ is BV in I_0.

(ii) $f(u)$ is AC in I_0 if and only if the interval function $\omega(I)$ is AC in I_0.

III.2.4. Given $f(u)$ as in III.2.1, we introduce the transformation $T : x = f(u)$, $u \in I_0$, where x is thought of as a point on an x-axis (the number line $-\infty < x < +\infty$). For given x_0, the symbol $T^{-1}(x_0)$ denotes the set of all those points $u \in I_0$ whose image under T is x_0. If S is a set in I_0, then $T(S)$ denotes the image of S under T, and $N(x, S)$ denotes the number of distinct points in the set $ST^{-1}(x)$. Thus $N(x, S)$ may be infinite. If explicit reference to T is desirable, we shall write $N(x, T, S)$ instead of $N(x, S)$. If $T_* : x = f_*(u)$, $u \in I_0$, is a second continuous transformation, then $\rho(T, T_*, S)$ denotes the least upper bound of $|f(u) - f_*(u)|$ on the subset S of I_0. If $T_n : x = f_n(u)$, $u \in I_0$, is a sequence of continuous transformations such that $\rho(T, T_n, I_0) \to 0$, then we shall write $T_n \to T$. Thus $T_n \to T$ if and only if $f_n(u) \to f(u)$ uniformly in I_0.

Let S be a subset of I_0 and let δ be a positive number. Then $\omega(\delta, T, S)$ will denote the least upper bound of $| f(u_2) - f(u_1) |$ for all pairs of points u_1, u_2 such that $u_1 \in S$, $u_2 \in S$, $| u_2 - u_1 | \leqq \delta$.

III.2.5. CONTINUATION. The transformation T, associated with $f(u)$, will enable us to express a number of statements concerning $f(u)$ in a geometrical form. A few simple facts, useful in the sequel, will be discussed presently. We define a subset $E_{\max}$ of I_0 as follows: a point $u_0 \in I_0$ belongs to $E_{\max}$ if and only if the following conditions hold. (i) $a_0 < u_0 < b_0$. (ii) There exists a subinterval $I : a \leqq u \leqq b$ of I_0 such that $a < u_0 < b$ and $f(u) \leqq f(u_0)$ for $u \in I$. In other words, $E_{\max}$ is the set of those interior points of I_0 at which $f(u)$ has a weak local maximum. We assert that the set $T(E_{\max})$ is *countable* (possibly empty).

PROOF. Let $x_0 \in T(E_{\max})$. By the definition of $E_{\max}$, there exists then a u_0 and a positive integer n, such that

$$a_0 < u_0 - 1/n < u_0 < u_0 + 1/n < b_0, \qquad f(u_0) = x_0,$$

$$f(u) \leqq f(u_0) = x_0 \qquad \text{for } u_0 - 1/n \leqq u \leqq u_0 + 1/n.$$

Let us denote by $I(x_0)$, $I^*(x_0)$ the intervals

$$I(x_0) : u_0 - 1/n \leqq u \leqq u_0 + 1/n, \tag{1}$$

$$I^*(x_0) : u_0 - 1/3n \leqq u \leqq u_0 + 1/3n. \tag{2}$$

With each point $x_0 \in T(E_{\max})$ there are associated in this manner two intervals $I(x_0)$, $I^*(x_0)$. Clearly $I^*(x_0') \neq I^*(x_0'')$ if $x_0' \neq x_0''$; indeed, if u_0', u_0'' are the mid-points of these intervals, then $f(u_0') = x_0' \neq x_0'' = f(u_0'')$, and hence $u_0' \neq u_0''$. Thus it is sufficient to prove that the class K^* of the intervals $I^*(x_0)$ associated with the various points of $T(E_{\max})$ is countable. Since the length of each $I^*(x_0)$ is a number of the form $2/(3n)$, where n is a positive integer depending upon x_0, it is sufficient to prove that the class K_n of those intervals $I^*(x_0)$ whose length is equal to $2/(3n)$, for fixed n, is countable. In fact, we shall see presently that the class K_n is finite (possibly empty). Indeed, let $I^*(x_0') \in K_n$, $I^*(x_0'') \in K_n$, where $x_0' \neq x_0''$. Let u_0', u_0'' be the mid-points of $I^*(x_0')$, $I^*(x_0'')$ respectively. We assert that $I^*(x_0')$, $I^*(x_0'')$ have no common point. Suppose indeed that we have some point $u^* \in I^*(x_0')I^*(x_0'')$. Then

$$| u^* - u_0' | \leqq 1/3n, \qquad | u^* - u_0'' | \leqq 1/3n,$$

and hence $| u_0' - u_0'' | < 1/n$. Consequently (cf. (1)), $u_0'' \in I(x_0')$, $u_0' \in I(x_0'')$. Hence $f(u_0'') \leqq f(u_0') = x_0'$, $f(u_0') \leqq f(u_0'') = x_0''$. These relations imply that $x_0' = x_0''$, while by assumption $x_0' \neq x_0''$. Thus K_n is a system of disjoint intervals, with the same length $2/(3n)$, in I_0. Hence K_n is finite.

III.2.6. CONTINUATION. Similarly, if $E_{\min}$ is the set of those interior points of I_0 at which $f(u)$ has a weak local minimum, then the set $T(E_{\min})$ is countable (possibly empty). The proof follows by applying the preceding result to the function $-f(u)$.

III.2.7. Given T as in III.2.4, we shall say that an interval $I : a \leqq u \leqq b$ is an interval of constancy of $f(u)$ if $f(u)$ is constant on I. Let us define a subset E_c of I_0 as follows: a point u_0 belongs to E_c if and only if (i) $a_0 < u_0 < b_0$, and (ii) there exists an interval of constancy that contains u_0. We assert that $T(E_c)$ is *countable* (possibly empty).

PROOF. Let $x_0 \in T(E_c)$. By definition, there exists then a point u_0 and an interval I, such that $a_0 < u_0 < b_0$, $u_0 \in I$, $T(I) = x_0$. With each point $x_0 \in T(E_c)$ we associate an interval $I(x_0)$ in this manner. Then clearly $I(x_0')$, $I(x_0'')$ are disjoint if $x_0' \neq x_0''$. Hence it is sufficient to verify that the class K of the intervals $I(x_0)$, associated with the various points $x_0 \in T(E_c)$, is countable. But this is obvious, since K is a class of disjoint, nondegenerate intervals in I_0.

III.2.8. Given T as in III.2.4, let us put (cf. III.2.5, III.2.6, III.2.7)

$$E = a_0 + b_0 + E_{\max} + E_{\min} + E_c, \tag{1}$$

where a_0, b_0 are interpreted as points. In view of the preceding results, $T(E)$ is countable. Let us now define a subset E^* of I_0 as follows: a point u_0 belongs to E^* if and only if for every $\epsilon > 0$ there exist two points u_1, u_2 with the following properties. (i) $a_0 < u_1 < u_0 < u_2 < b_0$. (ii) $u_2 - u_1 < \epsilon$. (iii) $[f(u_2) - f(u_0)] \cdot [f(u_1) - f(u_0)] < 0$. We assert that

$$E^* = I_0 - E. \tag{2}$$

PROOF. Clearly $E^* \subset I_0 - E$. Thus it is sufficient to show that

$$I_0 - E \subset E^*. \tag{3}$$

Let us take any point $u_0 \in I_0 - E$. Then u_0 is an interior point of I_0 (cf. (1)). Given $\epsilon > 0$, we have therefore two points α, β, such that $a_0 < \alpha < u_0 < \beta < b_0$, $\beta - \alpha < \epsilon$. Since $u_0 \in I_0 - E \subset I_0 - E_{\max}$, we have a point γ such that $\alpha \leqq \gamma \leqq \beta$, $f(\gamma) > f(u_0)$. Since $u_0 \in I_0 - E \subset I_0 - E_{\min}$, we have a point δ such that $\alpha \leqq \delta \leqq \beta$, $f(\delta) < f(u_0)$. Clearly $\gamma \neq u_0$, $\delta \neq u_0$, $\gamma \neq \delta$.

Case 1. γ and δ lie on different sides of u_0, say $\gamma < u_0 < \delta$. Then the points $u_1 = \gamma$, $u_2 = \delta$ clearly satisfy the conditions (i), (ii), (iii).

Case 2. γ and δ lie on the same side of u_0, say $\gamma < u_0$, $\delta < u_0$. Since $u_0 \in I_0 - E \subset I_0 - E_c$, we have a point u^* such that $u_0 < u^* \leqq \beta$, $f(u^*) \neq f(u_0)$. If $f(u^*) > f(u_0)$, we put $u_1 = \delta$, $u_2 = u^*$. If $f(u^*) < f(u_0)$, we put $u_1 = \gamma$, $u_2 = u^*$. In either case, the points u_1, u_2 satisfy the conditions (i), (ii), (iii).

Since $\epsilon > 0$ was arbitrary, $u_0 \in E^*$. Thus (3), and hence (2), is proved.

REMARK. Inspection of the proofs reveals that the results of III.2.5-III.2.8 remain valid if the assumption of the continuity of $f(u)$ is dropped. The result of the present section may be restated in the form: the set $T(I_0 - E^*)$ is countable.

III.2.9. Given T as in III.2.4, let $T_n : x = f_n(u)$, $u \in I_0$, be a sequence of continuous transformations such that $T_n \to T$. We assert that

$$N(x, T, I_0) \leqq \liminf N(x, T_n, I_0) \qquad \text{for } n \to \infty, \tag{1}$$

with the possible exception of a countable set of points x.

PROOF. In view of the remark at the end of III.2.8, it is sufficient to show that (1) holds for every point x_0 such that $x_0 \notin T(I_0 - E^*)$. Then $T^{-1}(x_0) \subset E^*$, and hence

$$(2) \qquad N(x_0, T, I_0) = N(x_0, T, E^*).$$

To treat the cases $N(x_0, T, E^*) < \infty$ and $N(x_0, T, E^*) = \infty$ simultaneously, let us take any non-negative integer k such that

$$(3) \qquad k \leqq N(x_0, T, E^*).$$

We assert that

$$(4) \qquad k \leqq \liminf N(x_0, T_n, I_0) \qquad \text{for } n \to \infty.$$

If $k = 0$, then (4) is obvious. So we can assume that $k > 0$. In view of (3), we can choose then k distinct points $u_1, u_2, \cdots, u_k$ in $E^*T^{-1}(x_0)$. By the definition of E^* (see III.2.8), we have then for each $j = 1, 2, \cdots, k$ two points u_j', u_j'', such that the following conditions hold (note that $f(u_j) = x_0$ for $j = 1, 2, \cdots, k$).

(i) $a_0 < u_j' < u_j < u_j'' < b_0$.

(ii) $u_j'' - u_j'$ is less than half of the minimum distance between any two of the points $u_1, u_2, \cdots, u_k$.

(iii) $[f(u_j'') - x_0][f(u_j') - x_0] < 0$.

Now since $T_n \to T$, we infer from (iii) that

(iv) $[f_n(u_j'') - x_0][f_n(u_j') - x_0] < 0$ for n sufficiently large.

Let us denote by I_j the interval $u_j' \leqq u \leqq u_j''$. By (i) and (ii) the intervals $I_1, I_2, \cdots, I_k$ are then disjoint. By (iv), if n is sufficiently large, $f_n(u) - x_0$ changes its sign in each one of the intervals $I_1, I_2, \cdots, I_k$. Hence, for n sufficiently large, each interval I_j contains a point u_j^n such that $f_n(u_j^n) = x_0$. Thus $N(x_0, T_n, I_0) \geqq k$ if n is sufficiently large, and (4) follows. Since k was any integer satisfying (3), the inequality (1) follows now from (2) and (4).

III.2.10. Given T as in III.2.4, let $I: a \leqq u \leqq b$ be a subinterval of I_0. Then the image $T(I)$ of I is either a single point or else an interval with the end points $m(I)$, $M(I)$, and thus $|T(I)| = \omega(I)$ (see III.2.3). Hence, by III.2.3, we have the following statements.

(i) $f(u)$ is BV in I_0 if and only if the interval function $|T(I)|$ is BV in I_0.

(ii) $f(u)$ is AC in I_0 if and only if the interval function $|T(I)|$ is AC in I_0.

III.2.11. Given T as in III.2.4, the image $T(I_0)$ of I_0 is comprised in a certain interval $-K \leqq x \leqq K$, since $f(u)$ is bounded. For $|x| > K$, the multiplicity function $N(x, I_0)$ is clearly equal to zero. By I.3.9, $N(x, I_0)$ is Borel measurable. Since $N(x, I_0) = 0$ for $|x| > K$, it is clear that summability of $N(x, I_0)$ on the interval $-K \leqq x \leqq K$ is equivalent to summability on the whole line $-\infty < x < \infty$, and

$$\int_{-K}^{K} N(x, I_0)\, dx = \int_{-\infty}^{\infty} N(x, I_0)\, dx.$$

To simplify notations, we agree to write $\int N(x, I_0)\, dx$ for the integral taken from $-\infty$ to ∞ (or, equivalently, from $-K$ to K). The statement that $N(x, I_0)$ is summable will mean that $N(x, I_0)$ is summable on the whole line $-\infty < x < \infty$ (or, equivalently, on the interval $-K \leqq x \leqq K$). The same agreements will be used in connection with any function $g(x)$ that vanishes outside of a certain finite interval.

III.2.12. Using the terminology of III.2.11, we have the following statement.

THEOREM. *$f(u)$ is BV in I_0 if and only if $N(x, I_0)$ is summable.*

PROOF. (i) Suppose that $f(u)$ is BV in I_0. If $I : a \leqq u \leqq b$ is a subinterval of I_0, then we shall denote by $g(x, I)$ the characteristic function of the set $T(I)$. That is, $g(x, I) = 1$ if $x \in T(I)$, and $g(x, I) = 0$ if $x \notin T(I)$. Clearly (see III.2.3, III.2.11, III.2.10)

$$\int g(x, I)\, dx = M(I) - m(I) = \omega(I) = | T(I) |. \tag{1}$$

Let n be a positive integer. We subdivide I_0 into n equal parts, and we denote the resulting subintervals by $I_1^n, I_2^n, \cdots, I_n^n$. Let us put

$$g_n(x) = \sum_{k=1}^{n} g(x, I_k^n). \tag{2}$$

Let us take any point x_0. If m is any integer not exceeding $N(x_0, I_0)$, then clearly $m \leqq g_n(x_0)$ for n sufficiently large. Hence $m \leqq \liminf g_n(x_0)$ for $n \to \infty$. Since m was any integer not exceeding $N(x_0, I_0)$, and x_0 was arbitrary, it follows that

$$N(x, I_0) \leqq \liminf_{n\to\infty} g_n(x), \qquad -\infty < x < \infty. \tag{3}$$

On the other hand, since $\omega(I)$ is BV in I_0 (see III.2.3), we have a (finite) constant G such that

$$\sum_{k=1}^{n} \omega(I_k^n) \leqq G, \qquad n = 1, 2, \cdots.$$

In view of (1), (2), it follows that

$$\int g_n(x)\, dx \leqq G, \qquad n = 1, 2, \cdots. \tag{4}$$

By the theorem of Fatou (see I.3.10), (3) and (4) imply that $N(x, I_0)$ is summable.

(ii) Suppose that $N(x, I_0)$ is summable. Let $I_1, \cdots, I_n$ be any system of intervals such that $I_j^0 I_k^0 = 0$ for $j \neq k$. Using the notations $g(x, I_j)$ in the sense explained above, we have obviously

$$\sum_{j=1}^{n} g(x, I_j) \leqq 2N(x, I_0).$$

Integration yields the inequality

$$\sum_{j=1}^{n} \omega(I_j) \leqq 2 \int N(x, I_0)\, dx.$$

By III.2.3 it follows that $f(u)$ is BV in I_0.

III.2.13. Let $f(u)$ be given as in III.2.1. Let σ be a generic notation for a finite system of intervals in I_0 without common interior points. The *total variation* $V(I_0, f)$ of $f(u)$ in I_0 is defined by the formula

$$V(I_0, f) = \underset{\sigma}{\text{l.u.b.}} \sum | \phi(I) |, \qquad I \in \sigma,$$

where $\phi(I)$ has the same meaning as in III.2.2, and the least upper bound is taken with respect to all systems σ. It follows from III.2.2 immediately that $V(I_0, f)$ is finite if and only if $f(u)$ is BV in I_0.

If $I : a \leqq u \leqq b$ is a subinterval of I_0, then the total variation $V(I, f)$ of $f(u)$ in I is defined in the same manner. Instead of $V(I, f)$ we shall use the symbol $V(a, b, f)$ if it is desirable to display the end points of I. If the function $f(u)$ is clearly identified by the context, then we shall write $V(I)$, $V(a, b)$ instead of $V(I, f)$, $V(a, b, f)$.

III.2.14. Continuation. From the remarks made in III.2.3, III.2.10 we infer immediately the formulas

$$V(I_0) = \underset{\sigma}{\text{l.u.b.}} \sum \omega(I), \qquad I \in \sigma,$$

$$V(I_0) = \underset{\sigma}{\text{l.u.b.}} \sum | T(I) |, \qquad I \in \sigma.$$

III.2.15. Continuation. $V(I_0, f)$ is a *lower semi-continuous functional* of f in the following sense. If $f_n(u)$ is a sequence of continuous functions in I_0 converging uniformly to $f(u)$, then

(1) $$V(I_0, f) \leqq \liminf V(I_0, f_n) \qquad \text{for } n \to \infty.$$

In fact, it will be apparent from the following proof that (1) holds under more general conditions, but this is irrelevant for our purposes.

Proof of (1). *Case* (i). $V(I_0, f) < \infty$. For given $\epsilon > 0$, we have then in I_0 a system $I_1, \cdots, I_m$ of intervals, without common interior points, such that

(2) $$V(I_0, f) - \epsilon < \sum_{j=1}^{m} | \phi(I_j) |,$$

where the interval function ϕ has the same meaning as in III.2.1. If ϕ_n has the analogous meaning relative to f_n, then clearly

(3) $$\sum_{j=1}^{m} | \phi(I_j) | = \lim_{n\to\infty} \sum_{j=1}^{m} | \phi_n(I_j) |.$$

By the definition of the total variation we have

(4) $$\sum_{j=1}^{m} |\phi_n(I_j)| \leqq V(I_0, f_n).$$

Since ϵ was arbitrary, the preceding relations imply (1).

Case (ii). $V(I_0, f) = \infty$. For given $G > 0$, we have then in I_0 a system $I_1, \cdots, I_m$ of intervals, without common interior points, such that

(5) $$G < \sum_{j=1}^{m} |\phi(I_j)|.$$

The relations (3), (4) hold as before. Since G was arbitrary, (1) follows from (3), (4), (5).

III.2.16. Given $f(u)$ as in III.2.1, suppose that there exists in I_0 a system of points $u_0 = a_0 < u_1 < \cdots < u_{j-1} < u_j < \cdots < u_n = b_0$, such that $f(u)$ is linear in each one of the intervals $u_{j-1} \leqq u \leqq u_j$. Then $f(u)$ will be termed *quasi-linear* in I_0. Clearly, we have then the formula

$$V(I_0, f) = \sum_{j=1}^{n} | f(u_j) - f(u_{j-1}) |.$$

III.2.17. Given $f(u)$ as in III.2.1, let D_n be a sequence of subdivisions of I_0 such that $\| D_n \| \to 0$ (see III.1.4). We assert that

(1) $$V(I_0, f) = \lim_{n\to\infty} \sum |\phi(I)|, \qquad I \in D_n,$$

where $\phi(I)$ has the same meaning as in III.2.1.

PROOF. For each n, let $f_n(u)$ denote the quasi-linear function in I_0 that agrees with $f(u)$ at the end points of the intervals of D_n and is linear in each interval of D_n. By III.2.16 we have then

(2) $$V(I_0, f_n) = \sum |\phi(I)|, \qquad I \in D_n.$$

Clearly, $f_n(u) \to f(u)$ uniformly in I_0. Hence, by III.2.15, there follows from (2) the inequality

(3) $$V(I_0, f) \leqq \liminf_{n\to\infty} \sum |\phi(I)|, \qquad I \in D_n.$$

On the other hand, by the definition of $V(I_0, f)$ we have

(4) $$V(I_0, f) \geqq \sum |\phi(I)|, \qquad I \in D_n.$$

Clearly, (3) and (4) imply (1).

III.2.18. CONTINUATION. The preceding result implies the formulas (cf. III.2.3, III.2.10)

(1) $$V(I_0, f) = \lim_{n\to\infty} \sum \omega(I), \qquad I \in D_n,$$

(2) $$V(I_0, f) = \lim_{n\to\infty} \sum |T(I)|, \qquad I \in D_n.$$

Indeed, by III.2.3 we have the inequality

$$V(I_0, f) \geqq \sum \omega(I), \qquad I \in D_n. \tag{3}$$

On the other hand, since $\omega(I) \geqq | \phi(I) |$, it follows from III.2.17 that

$$V(I_0, f) \leqq \liminf_{n\to\infty} \sum \omega(I), \qquad I \in D_n. \tag{4}$$

Clearly, (3) and (4) imply (1). Since $| T(I) | = \omega(I)$, (2) follows from (1).

III.2.19. Given $f(u)$ as in III.2.1, the interval function $| \phi(I) |$ is clearly non-negative, continuous, and increases by subdivision (cf. III.1.2, III.1.5, III.1.52). Furthermore, in view of III.1.5, III.2.13, $V(I)$ is the U-function of the interval function $| \phi(I) |$.

Suppose now that $f(u)$ is BV in I_0. In view of the preceding remarks, the general theory developed in III.1 yields the following facts.

(i) The interval function $| \phi(I) |$ is Burkill integrable in I_0, and its indefinite Burkill integral coincides with $V(I)$ (see III.1.23, III.2.13).

(ii) $V(I)$ is an additive interval function (see III.1.5).

(iii) $V(I)$ is a continuous interval function (see III.1.17). As a consequence, $V(a_0, u)$ is a continuous, and clearly nondecreasing, function of u in I_0.

(iv) Since $V(I)$ is additive and non-negative, it possesses a.e. in I_0 a derivative (see III.1.29(a)) that we denote temporarily by $g(u)$. By III.1.28, $g(u)$ is then summable in I_0 and we have the inequality

$$\int_I g(u)\, du \leqq V(I)$$

for every interval $I \subset I_0$.

III.2.20. Suppose that $f(u)$ is continuous and nondecreasing in $I_0 : a_0 \leqq u \leqq b_0$. Let $\phi(I)$ have the same meaning as in III.2.1. Then $\phi(I)$ is clearly non-negative and additive, and hence, by III.1.29(a), the derivative $D(u)$ of $\phi(I)$ exists a.e. in I_0. By III.2.1 it follows that $f'(u)$ exists and is equal to $D(u)$ a.e. in I_0.

III.2.21. Suppose that $f(u)$ is continuous and BV in $I_0 : a_0 \leqq u \leqq b_0$. Then $V(a_0, u)$ is nondecreasing and continuous in I_0 (see III.2.19). Let us put $h(u) = V(a_0, u) - f(u), u \in I_0$. If $a_0 \leqq u_1 < u_2 \leqq b_0$, then clearly (cf. III.2.19)

$$\begin{aligned} h(u_2) - h(u_1) &= V(a_0, u_2) - V(a_0, u_1) - [f(u_2) - f(u_1)] \\ &= V(u_1, u_2) - [f(u_2) - f(u_1)] \geqq 0. \end{aligned}$$

Thus $h(u)$ is nondecreasing and continuous in I_0. For $f(u)$ we obtain the formula $f(u) = V(a_0, u) - h(u)$. Since $V(a_0, u)$ and $h(u)$ are continuous and nondecreasing in I_0, both of these functions possess a derivative a.e. in I_0 by III.2.20. Hence $f'(u)$ also exists a.e. in I_0.

III.2.22. THEOREM. *Suppose that $f(u)$ is continuous and BV in $I_0 : a_0 \leqq u \leqq b_0$. Then $f'(u)$ exists a.e. in I_0 and is summable in I_0, and*

$$\int_I | f'(u) | \, du \leqq V(I) \tag{1}$$

for every interval $I \subset I_0$. *The derivative of the interval function* $V(I)$ *exists and is equal to* $| f'(u) |$ *a.e. in* I_0. *Finally we have, a.e. in* I_0, *the formula*

$$\frac{dV(a_0, u)}{du} = | f'(u) |. \tag{2}$$

PROOF. (i) $f'(u)$ exists a.e. in I_0 by III.2.21. By III.2.1 it follows that the derivative of the interval function $\phi(I)$ exists and is equal to $f'(u)$ a.e. in I_0, and hence clearly the derivative of $| \phi(I) |$ exists and is equal to $| f'(u) |$ a.e. in I_0. Since $V(I)$ is the Burkill integral of $| \phi(I) |$ (see III.2.19), it follows by III.1.27 that the derivative of $V(I)$ exists and is equal to $| f'(u) |$ a.e. in I_0. The inequality (1) appears thus as a direct consequence of III.2.19(iv). The formula (2) follows directly from III.2.1, applied to the function $V(a_0, u)$.

III.2.23. THEOREM. *Suppose that* $f(u)$ *is continuous and BV in* $I_0 : a_0 \leqq u \leqq b_0$. *Then we have the formula* (*cf.* III.2.12)

$$V(I_0) = \int N(x, I_0) \, dx. \tag{1}$$

PROOF. Using the terminology introduced in the course of the proof in III.2.12, let us first observe that the constant G appearing there may be obviously chosen as $V(I_0)$ (see III.2.14). The relations (3) and (4) in III.2.12 yield then, by the theorem of Fatou (see I.3.10), the inequality

$$\int N(x, I_0) \, dx \leqq V(I_0). \tag{2}$$

Now let S denote the set of all the end points of all the intervals I_k^n, $k = 1, \cdots, n$; $n = 1, 2, \cdots$, introduced in III.2.12. We have then the obvious inequality

$$g_n(x) \leqq N(x, I_0) \qquad \text{for } x \notin T(S). \tag{3}$$

Since S is countable, $T(S)$ is also countable and hence of measure zero. Thus (3) yields by integration

$$\sum_{k=1}^{n} \omega(I_k^n) \leqq \int N(x, I_0) \, dx. \tag{4}$$

By III.2.18 we have

$$\lim_{n \to \infty} \sum_{k=1}^{n} \omega(I_k^n) = V(I_0). \tag{5}$$

Clearly, (2), (4), (5) yield (1).

COROLLARY. *If* $f(u)$ *is continuous and BV in* I_0, *then we have the formula* $V(I) = \int N(x, I) \, dx$ *for every interval* $I \subset I_0$. *Indeed,* $f(u)$ *is then continuous and BV in* I *also.*

III.2.24. Suppose that $f(u)$ is continuous and BV in $I_0 : a_0 \leqq u \leqq b_0$. Then $V(I)$ is a finite-valued, non-negative, additive and continuous interval function (see III.2.19). By III.1.37, III.1.43, III.1.52 it follows that $V(I)$ admits of a non-negative, completely additive extension to Borel sets in I_0. This extension, which is unique by III.1.47, III.1.52, will be denoted by $V(B, f)$, where B is a generic notation for a Borel set $B \subset I_0$. If f is clearly identified by the context, we may write $V(B)$ instead of $V(B, f)$.

III.2.25. CONTINUATION. *We assert that* (see III.2.4)

$$V(B) = \int N(x, B)\, dx \tag{1}$$

for every Borel set $B \subset I_0$.

PROOF. By I.2.46, $N(x, B)$ is a measurable function of x, and clearly $N(x, B) \leqq N(x, I_0)$. Since $N(x, I_0)$ is summable by III.2.12, it follows that $N(x, B)$ is also summable. Let us put $g(B) = \int N(x, B)\, dx$. Let $B_1, \cdots, B_n, \cdots$ be any sequence of disjoint Borel sets in I_0, and let $B = \sum B_n$. Since $N(x, B)$ is summable, we have on the x-axis a set $\bar{e}$ of measure zero such that $N(x, B) < \infty$ for $x \notin \bar{e}$. Clearly

$$N(x, B) = \sum_n N(x, B_n) \qquad \text{for } x \notin \bar{e}.$$

By I.3.11, termwise integration is permissible, and we obtain the formula

$$g(B) = \sum_n g(B_n).$$

On the other hand, by the corollary in III.2.23 we have, for every interval $I \subset I_0$, the formula $V(I) = g(I)$. Thus $V(B)$, $g(B)$ are non-negative, completely additive functions of Borel sets which agree for every interval $I \subset I_0$. By III.1.47, III.1.52 $V(B) = g(B)$ for every Borel set $B \subset I_0$, and (1) is proved.

III.2.26. CONTINUATION. By III.2.22, the derivative of the interval function $V(I)$ exists and is equal to $|f'(u)|$ a.e. in I_0 By III.1.32, III.1.52 there follows for $V(B)$ the Lebesgue decomposition

$$V(B, f) = \int_B |f'(u)|\, du + V^0(B, f), \qquad B \subset I_0,$$

where $V^0(B, f)$ is a non-negative, completely additive, singular function of Borel sets $B \subset I_0$, which will be termed the *singular variation* of $f(u)$. If f is clearly identified by the context, we may write $V^0(B)$ instead of $V^0(B, f)$. The Lebesgue decomposition yields the inequality

$$V(B) \geqq \int_B |f'(u)|\, du \qquad \text{for } B \subset I_0.$$

III.2.27. CONTINUATION. $f(u)$ is AC in I_0 if and only if the interval function $V(I)$ is AC in I_0 (cf. III.2.2, III.1.2, III.1.52).

PROOF. If $I : a \leqq u \leqq b$ is any subinterval of I_0, then $| \phi(I) | = | f(b) - f(a) | \leqq V(I)$. Hence if $V(I)$ is AC in I_0, then clearly $f(u)$ is also AC in I_0. Conversely, suppose that $f(u)$ is AC in I_0. By III.2.19, $V(I)$ is the indefinite Burkill integral of the interval function $| \phi(I) |$. Since $| \phi(I) |$ is AC in I_0 by assumption, it follows by III.1.18, III.1.52 that $V(I)$ is AC in I_0.

III.2.28. CONTINUATION. $f(u)$ is AC in I_0 if and only if $V(B)$ is AC in I_0.

PROOF. By III.1.34, III.1.52, $V(B)$ is AC in I_0 if and only if $V(I)$ is AC in I_0, and thus the assertion follows from III.2.27.

III.2.29. CONTINUATION. *If $f(u)$ is AC in I_0, then*

$$V(B) = \int_B | f'(u) | \, du \tag{1}$$

for every Borel set $B \subset I_0$. The converse holds in the stronger form: if $f(u)$ is BV in I_0, and

$$V(I_0) = \int_{I_0} | f'(u) | \, du, \tag{2}$$

then $f(u)$ is AC in I_0.

PROOF. (i) Suppose that $f(u)$ is AC in I_0. By I.3.16 and III.2.28 it follows that the singular variation $V^0(B)$ vanishes, and thus (1) is a consequence of III.2.26.

(ii) Suppose that (2) holds. By III.2.26 it follows that $V^0(I_0) = 0$, and hence $V^0(B) = 0$ for every Borel set $B \subset I_0$. By I.3.16, $V(B)$ is thus AC in I_0, and hence, by III.2.28, $f(u)$ is also AC in I_0.

REMARK. The preceding results yield directly the following statements.

(a) $f(u)$ is AC in I_0 if and only if $f(u)$ is BV in I_0 and (1) holds for every Borel set $B \subset I_0$.

(b) $f(u)$ is AC in I_0 if and only if $f(u)$ is BV in I_0 and (2) holds.

III.2.30. CONTINUATION. *If $f(u)$ is AC in I_0, then*

$$\int_B | f'(u) | \, du = \int N(x, B) \, dx \tag{1}$$

for every Borel set $B \subset I_0$. The converse holds in the stronger form: if $f(u)$ is BV in I_0, and

$$\int_{I_0} | f'(u) | \, du = \int N(x, I_0) \, dx, \tag{2}$$

then $f(u)$ is AC in I_0.

PROOF. The first part of the assertion follows directly from III.2.29, III.2.2, III.2.25. The second part follows directly from III.2.23, III.2.29.

REMARK. The preceding results yield directly the following statements.

(a) $f(u)$ is AC in I_0 if and only if $f(u)$ is BV in I_0 and (1) holds for every Borel set $B \subset I_0$.

(b) $f(u)$ is AC in I_0 if and only if $f(u)$ is BV in I_0 and (2) holds.

III.2.31. CONTINUATION. *$f(u)$ is AC in I_0 if and only if $| T(e) | = 0$ for every Borel set $e \subset I_0$ of measure zero* (cf. III.2.10; recall that $f(u)$ is continuous and BV by assumption).

PROOF. (i) Suppose that $f(u)$ is AC in I_0, and let e be a Borel set of measure zero in I_0. By III.2.30 we have then

$$0 = \int_e | f'(u) | \, du = \int N(x, e) \, dx \geqq | T(e) |.$$

Hence $| T(e) | = 0$.

(ii) Suppose that $| T(e) | = 0$ for every Borel set $e \subset I_0$ of measure zero. By III.2.25 we have then, for every such set e,

$$V(e) = \int N(x, e) \, dx = \int_{T(e)} N(x, e) \, dx = 0,$$

since $| T(e) | = 0$. Thus $V(e) = 0$ for every Borel set $e \subset I_0$ of measure zero. Hence $V(B)$ is AC in I_0 by I.3.16, and consequently $f(u)$ is AC in I_0 by III.2.28.

III.2.32. *If $f(u)$ is AC in $I_0 : a_0 \leqq u \leqq b_0$, then $| T(e) | = 0$ for every set $e \subset I_0$ of measure zero. Furthermore, for every measurable set $E \subset I_0$, the image set $T(E)$ is measurable* (cf. III.2.4).

PROOF. (i) Let e be a set of measure zero in I_0. By I.3.7 we have then a Borel set e^* of measure zero, such that $e \subset e^* \subset I_0$. By III.2.31 it follows that $0 \leqq | T(e) | \leqq | T(e^*) | = 0$. Hence $| T(e) | = 0$.

(ii) Let E be a measurable set in I_0. By I.3.7 we can write $E = E^* + e$, where E^* is an F_σ and $| e | = 0$. Since $T(E) = T(E^*) + T(e)$, and $| T(e) | = 0$ by (i), the measurability of $T(E)$ follows (cf. I.2.46).

III.2.33. Suppose that $f(u)$ is continuous and BV in $I_0 : a_0 \leqq u \leqq b_0$, and let B_0 be a Borel set in I_0. Then $f(u)$ will be termed AC on B_0 (absolutely continuous on B_0), if and only if $V(B)$ is AC on B_0 (see III.2.24, I.3.16). In case B_0 is an interval, this definition is equivalent to that given in III.2.2, in view of III.2.28.

III.2.34. Given $f(u)$ as in III.2.33, suppose that $f(u)$ is AC on a Borel set $B_0 \subset I_0$. Then

$$V(B) = \int_B | f'(u) | \, du \tag{1}$$

for every Borel set $B \subset B_0$. The converse holds in the stronger form: if $f(u)$ is continuous and BV in $I_0 : a_0 \leqq u \leqq b_0$, and B_0 is a Borel set in I_0 such that

$$V(B_0) = \int_{B_0} | f'(u) | \, du, \tag{2}$$

then $f(u)$ is AC on B_0.

PROOF. By III.2.26 we have the formula

$$V(B) = \int_B |f'(u)|\,du + V^0(B) \tag{3}$$

for every Borel set $B \subset I_0$. Since $V^0(B)$ is singular on I_0, it is also singular on B_0, and thus (3) yields the Lebesgue decomposition of $V(B)$ also on B_0.

(i) Suppose that $f(u)$ is AC on B_0. Then, by definition, $V(B)$ is AC on B_0, hence the singular variation $V^0(B)$ vanishes on B_0. Thus (1) follows from (3).

(ii) Suppose that (2) holds. Comparison with (3), for $B = B_0$, shows that $V^0(B_0) = 0$. Thus $V(B)$ is AC on B_0, and hence, by definition, $f(u)$ is AC on B_0.

REMARK. The preceding results yield the following statements, where it is assumed that $f(u)$ is continuous and BV in I_0, and B_0 is a Borel set in I_0.

(a) $f(u)$ is AC on B_0 if and only if (1) holds for every Borel set $B \subset B_0$.

(b) $f(u)$ is AC on B_0 if and only if (2) holds.

III.2.35. Suppose that $f(u)$ is continuous and BV in $I_0 : a_0 \leqq u \leqq b_0$, and let B_0 be a Borel set in I_0. If $f(u)$ is AC on B_0, then we assert that

$$\int_B |f'(u)|\,du = \int N(x, B)\,dx \tag{1}$$

for every Borel set $B \subset B_0$. The converse holds in the stronger form: if

$$\int_{B_0} |f'(u)|\,du = \int N(x, B_0)\,dx, \tag{2}$$

then $f(u)$ is AC on B_0.

PROOF. The first statement follows directly from III.2.34 and III.2.25. The second statement follows directly from III.2.25 and III.2.34.

REMARK. The preceding results yield the following statements, where it is assumed that $f(u)$ is continuous and BV in I_0, and B_0 is a Borel set in I_0.

(a) $f(u)$ is AC on B_0 if and only if (1) holds for every Borel set $B \subset B_0$.

(b) $f(u)$ is AC on B_0 if and only if (2) holds.

III.2.36. Suppose that $f(u)$ is AC on $I_0 : a_0 \leqq u \leqq b_0$. Then

$$\int_a^b f'(u)\,du = f(b) - f(a) \tag{1}$$

for every interval $I : a \leqq u \leqq b$ in I_0.

PROOF. Note first that $f(u)$ is BV in I_0 by III.2.2, and hence the integral appearing in (1) exists by III.2.22.

Case (i). $f(u)$ is nondecreasing in I_0. Then (1) is equivalent to the formula

$$\int_I |f'(u)|\,du = V(I),$$

which was established in III.2.29.

Case (ii). In the general case, we use the auxiliary function $h(u) = V(a_0, u) - f(u)$, introduced in III.2.21. As noted there, $h(u)$ is continuous and nondecreasing in I_0. Since $V(a_0, u)$ is AC in I_0 by III.2.27, clearly $h(u)$ is also AC in I_0. Hence, by case (i),

$$\int_a^b h'(u)\, du = h(b) - h(a), \qquad \int_a^b \frac{dV(a_0, u)}{du}\, du = V(a_0\ \ b) - V(a_0, a),$$

and (1) follows.

III.2.37. Conversely, suppose that (α) $f(u)$ is continuous in $I_0 : a_0 \leqq u \leqq b_0$, ($\beta$) $f'(u)$ exists a.e. in I_0 and is summable in I_0, and (γ) we have, for every interval $I : a \leqq u \leqq b$ in I_0, the formula

$$\text{(1)} \qquad \int_a^b f'(u)\, du = f(b) - f(a).$$

Then $f(u)$ is AC in I_0.

PROOF. (1) yields (cf. III.2.2)

$$|\, \phi(I) \,| = |\, f(b) - f(a) \,| \leqq \int_a^b |\, f'(u) \,|\, du.$$

Hence the interval function $|\, \phi(I) \,|$ is AC on I_0 by I.3.13, and thus $f(u)$ is AC by definition (see III.2.2).

III.2.38. Let $f(u)$ be continuous in $I_0 : a_0 \leqq u \leqq b_0$. We choose a number $0 < \delta < b_0 - a_0$, and we denote by I_δ the interval $a_0 \leqq u \leqq b_0 - \delta$. For $0 < h < \delta$, we introduce the *integral mean*

$$\text{(1)} \qquad f_h(u) = \frac{1}{h} \int_0^h f(u + \eta)\, d\eta = \frac{1}{h} \int_u^{u+h} f(v)\, dv.$$

For fixed h, the function $f_h(u)$ is clearly continuous in I_δ, and for $h \to 0$, clearly $f_h(u) \to f(u)$ uniformly in I_δ. We proceed to discuss the derivative of $f_h(u)$, for fixed h, with respect to u, in I_δ. For $\Delta u \neq 0$ we have, if u is interior to I_δ and Δu is sufficiently small,

$$f_h(u + \Delta u) - f_h(u) = \frac{1}{h}\left[\int_{u+\Delta u}^{u+\Delta u+h} f(v)\, dv - \int_u^{u+h} f(v)\, dv \right]$$

$$= \frac{1}{h}\left[\int_{u+h}^{u+h+\Delta u} f(v)\, dv - \int_u^{u+\Delta u} f(v)\, dv \right].$$

Now since $f(u)$ is continuous, we have for $\Delta u \to 0$ the relations

$$\lim \frac{1}{\Delta u} \int_{u+h}^{u+h+\Delta u} f(v)\, dv = f(u + h), \qquad \lim \frac{1}{\Delta u} \int_{u}^{u+\Delta u} f(v)\, dv = f(u).$$

It follows that $f_h'(u)$ exists and

$$f_h'(u) = \frac{f(u + h) - f(u)}{h}. \tag{2}$$

Thus $f_h(u)$ has a continuous first derivative in the interior of I_δ.

III.2.39. Continuation. Suppose now that $f(u)$ is AC in I_0. By III.2.36 there follows from III.1.38(2) the formula

$$f_h'(u) = \frac{1}{h}\int_0^h f'(u + \eta)\, d\eta = \frac{1}{h}\int_u^{u+h} f'(v)\, dv.$$

In particular, if $f'(u)$ is continuous, it follows that $f_h''(u)$ exists and is continuous in the interior of I_δ, and so forth.

III.2.40. Continuation. Suppose that $f(u)$ is continuous and BV in I_0. Then we have, for every interval $I : a \leqq u \leqq b$ in I_δ, the relations

$$V(a, b, f_h) \leqq V(a, b + \delta, f) \qquad \text{for } 0 < h < \delta, \tag{1}$$

$$V(a, b, f_h) \to V(a, b, f) \qquad \text{for } 0 < h \to 0. \tag{2}$$

Proof of (1). Let σ denote a finite system of intervals $I_j : a_j \leqq u \leqq b_j$, $j = 1, \cdots, n$, without common interior points, in $I : a \leqq u \leqq b$. Then

$$\begin{aligned} \sum_{j=1}^{n} | f_h(b_j) - f_h(a_j) | &= \sum_{j=1}^{n} \left| \frac{1}{h}\int_0^h [f(b_j + \eta) - f(a_j + \eta)]\, d\eta \right| \\ &\leqq \frac{1}{h}\int_0^h \left[\sum_{j=1}^{n} | f(b_j + \eta) - f(a_j + \eta) | \right] d\eta. \end{aligned} \tag{3}$$

Now clearly the intervals $a_j + \eta \leqq u \leqq b_j + \eta$, $j = 1, 2, \cdots, n$, have no common interior points and are comprised in the interval $a \leqq u \leqq b + \delta$. Hence

$$\sum_{j=1}^{n} | f(b_j + \eta) - f(a_j + \eta) | \leqq V(a, b + \delta, f). \tag{4}$$

(3) and (4) yield the inequality

$$\sum_{j=1}^{n} | f_h(b_j) - f_h(a_j) | \leqq V(a, b + \delta, f).$$

Since the interval system σ was arbitrary, the inequality (1) follows (see III.2.13).

PROOF OF (2). The inequality (1) yields $\limsup V(a, b, f_h) \leqq V(a, b+\delta, f)$ for $h \to 0$. Since $V(a, b+\delta, f) \to V(a, b, f)$ for $\delta \to 0$ (see III.2.19),

$$\limsup V(a, b, f_h) \leqq V(a, b, f) \qquad \text{for } h \to 0. \tag{5}$$

As $f_h(u) \to f(u)$ uniformly in $a \leqq u \leqq b$, we have also (see III.2.15)

$$\liminf V(a, b, f_h) \geqq V(a, b, f) \qquad \text{for } h \to 0. \tag{6}$$

Clearly, (5) and (6) imply (2).

III.2.41. Let us return to III.2.38. Let M be the maximum of $|f(u)|$ in I_0. For fixed h, we have then by III.2.38(2) the inequality

$$|f_h'(u)| \leqq \frac{2M}{h}, \qquad a_0 < u < b_0 - \delta.$$

Let now $I: a \leqq u \leqq b$ be any interval such that $a_0 < a < b < b_0 - \delta$. Since I is then interior to an interval in which $f_h'(u)$ is continuous, we have

$$f_h(b) - f_h(a) = \int_a^b f_h'(u)\,du,$$

and hence

$$|f_h(b) - f_h(a)| \leqq \frac{2M(b-a)}{h} \qquad \text{for } a_0 < a < b < b_0 - \delta.$$

By continuity it follows that

$$|f_h(b) - f_h(a)| \leqq \frac{2M(b-a)}{h} \qquad \text{for } a_0 \leqq a < b \leqq b_0 - \delta.$$

Thus $f_h(u)$ is obviously AC in I_δ (cf. III.2.2). By III.2.29 there follows the formula (cf. III.2.38(2))

$$V(I, f_h) = \int_I |f_h'(u)|\,du = \int_I \left| \frac{f(u+h) - f(u)}{h} \right| du,$$

for every interval $I \subset I_\delta$.

III.2.42. Suppose that $f(u)$ is continuous and BV in $I_0: a_0 \leqq u \leqq b_0$, and let I_δ denote again the interval $a_0 \leqq u \leqq b_0 - \delta$, where $0 < \delta < b_0 - a_0$. Then we have, for every interval $I: a \leqq u \leqq b$ in I_δ, the relations

$$\int_a^b \left| \frac{f(u+h) - f(u)}{h} \right| du \leqq V(a, b+\delta, f) \qquad \text{for } 0 < h < \delta, \tag{1}$$

$$\int_a^b \left| \frac{f(u+h) - f(u)}{h} \right| du \to V(a, b, f) \qquad \text{for } 0 < h \to 0. \tag{2}$$

PROOF. In view of III.2.41, the integral appearing in (1) and (2) is equal to $V(a, b, f_h)$, and thus (1) and (2) follow directly from III.2.40.

III.2.43. CONTINUATION. By III.2.19, the total variation $V(a_0, u)$ is a continuous function of u in I_0. Applying the results of III.2.42 to the function $V(a_0, u)$, and noting that $V(a_0, u)$ is clearly nondecreasing, we obtain

$$\int_a^b \frac{V(u, u + h, f)}{h}\, du \leqq V(a, b + \delta, f) \qquad \text{for } 0 < h < \delta,$$

$$\int_a^b \frac{V(u, u + h, f)}{h}\, du \to V(a, b, f) \qquad \text{for } 0 < h \to 0,$$

where $a \leqq u \leqq b$ is any subinterval of I_δ (note that since $V(a_0, u)$ is nondecreasing, its total variation on any interval is equal to its increment on that interval).

For the singular variation $V^0(a_0, u)$ we have by III.2.26 the formula

$$V^0(a_0, u) = V(a_0, u) - \int_{a_0}^{u} | f'(v) |\, dv.$$

Thus $V^0(a_0, u)$ is also a continuous (and nondecreasing) function of u, and hence the results of III.2.42, applied to $V^0(a_0, u)$, yield the formulas

$$\int_a^b \frac{V^0(u, u + h, f)}{h}\, du \leqq V^0(a, b + \delta, f) \qquad \text{for } 0 < h < \delta,$$

$$\int_a^b \frac{V^0(u, u + h, f)}{h}\, du \to V^0(a, b, f) \qquad \text{for } 0 < h \to 0,$$

where $a \leqq u \leqq b$ is any subinterval of I_δ.

III.2.44. The symmetric integral means

$$f_h(u) = \frac{1}{2h} \int_{-h}^{h} f(u + \eta)\, d\eta$$

give rise to entirely analogous considerations, the necessary modifications in the statements and proofs being obvious throughout. For example, the derivative is given this time by the formula

$$f_h'(u) = \frac{f(u + h) - f(u - h)}{2h}.$$

III.2.45. The study of the concepts of bounded variation and of absolute continuity, presented in the preceding sections of this chapter for functions $f(u)$ of a single (real) variable, will be extended in chapter IV to situations involving two (real) variables. However, certain aspects of this extended theory are so closely related to the topics presented above that they are more conveniently treated in the present chapter, since the theorems involved are more or less direct applications of the results obtained for functions of a single (real) variable. Accordingly, we shall consider presently a (real-valued) function $f(u, v)$, defined and continuous in a rectangle $R_0 : a_0 \leqq u \leqq b_0$, $c_0 \leqq v \leqq d_0$. For fixed v in the interval $c_0 \leqq v \leqq d_0$, $f(u, v)$ is a continuous function of u alone in the interval $a_0 \leqq u \leqq b_0$, and hence in every interval $a \leqq u \leqq b$, where $a_0 \leqq a < b \leqq b_0$, it has a total variation (see III.2.13) which will be denoted by $V_u(a, b, v, f)$, and will be termed the total variation of $f(u, v)$, with respect to u for given v, in the interval $a \leqq u \leqq b$. The quantity $V_u(a, b, v, f)$ may be finite or infinite. In any case, for fixed a, b, f, it is a function of v defined for $c_0 \leqq v \leqq d_0$.

Let now B be any Borel set in R_0. Then B_v will denote the set of those points of B whose v-coordinate has a given value v. Thus B_v is again a Borel set. Suppose now that $V_u(a_0, b_0, v, f) < \infty$ for a given value of v. Then $V_u(a, b, v, f)$, where $a_0 \leqq a < b \leqq b_0$, is a function of the interval $I : a \leqq u \leqq b$. By III.2.24, this interval function admits of a non-negative, completely additive extension to Borel sets which will be denoted by $V_u(\beta, v, f)$. In this symbol, β is a generic notation for a Borel set on the interval $a_0 \leqq u \leqq b_0$, v fixed. In particular, if B_v has the meaning explained above, then the symbol $V_u(B_v, v, f)$ is defined, provided that $V_u(a_0, b_0, v, f) < \infty$. If $V_u(a_0, b_0, v, f) = \infty$ for a certain v, then we agree to put $V_u(B_v, v, f) = \infty$ for every Borel set $B \subset R_0$. For given B and f, the quantity $V_u(B_v, v, f)$ is then a function of v defined for $c_0 \leqq v \leqq d_0$. Note that the definition of $V_u(B_v, v, f)$ depends upon the fundamental rectangle R_0 also. In the sequel, R_0 will be fixed, and thus it is unnecessary to use more complicated notations. In fact, if the function $f(u, v)$ is clearly identified by the context, then we shall use the simpler notations $V_u(a, b, v)$, $V_u(B_v, v)$.

III.2.46. Continuation. The preceding definitions are relative to the variable u, as far as total variation is concerned. In the following sections, we shall first develop concepts and theorems in which the variable u will play a preferred part. It will be obvious all along that u and v may be exchanged throughout.

III.2.47. Using the terminology of III.2.45, we assert that, for fixed a, b, f, the quantity $V_u(a, b, v, f)$ is a lower semi-continuous function of v in the interval $c_0 \leqq v \leqq d_0$.

Proof. Let v_n be a sequence of points in $c_0 \leqq v \leqq d_0$, converging to a point v_0. For clarity, let us put $f(u, v_n) = g_n(u)$, $n = 0, 1, 2, \cdots$. Then

$$(1) \qquad V_u(a, b, v_n, f) = V(a, b, g_n), \qquad n = 0, 1, 2, \cdots,$$

(see III.2.45, III.2.13). Since $f(u, v)$ is continuous, clearly $g_n(u) \to g_0(u)$ uniformly in $a \leqq u \leqq b$. Hence, by III.2.15,

$$(2) \qquad V(a, b, g_0) \leqq \liminf V(a, b, g_n) \qquad \text{for } n \to \infty.$$

(1) and (2) yield the asserted relation

$$V_u(a, b, v_0, f) \leqq \liminf V_u(a, b, v_n, f) \qquad \text{for } n \to \infty.$$

COROLLARY. *For fixed a, b, f, the quantity* $V_u(a, b, v, f)$ *is a measurable function of* v *in the interval* $c_0 \leqq v \leqq d_0$.

III.2.48. CONTINUATION. For fixed a, b, v, the quantity $V_u(a, b, v, f)$ is a lower semi-continuous functional of f, in the following sense: if $f_n(u, v)$ is a sequence of continuous functions in R_0 such that $f_n \to f$ uniformly in R_0, then $V_u(a, b, v, f) \leqq \liminf V_u(a, b, v, f_n)$ for $n \to \infty$. This is a direct consequence of III.2.15.

III.2.49. Using the terminology of III.2.45, $f(u, v)$ will be termed BVTu in R_0 (*of bounded variation, in the Tonelli sense, with respect to u in* R_0), if and only if $V_u(a_0, b_0, v, f)$ is summable in the interval $c_0 \leqq v \leqq d_0$.

$f(u, v)$ will be termed ACTu in R_0 (*absolutely continuous, in the Tonelli sense, with respect to u in* R_0), if and only if $f(u, v)$ is BVTu in R_0 and for a.e. v, in the interval $c_0 \leqq v \leqq d_0$, $f(u, v)$ is AC as a function of u in the interval $a_0 \leqq u \leqq b_0$.

Clearly, if $f(u, v)$ is BVTu (ACTu) in R_0, then $f(u, v)$ is also BVTu (ACTu) in every rectangle $a \leqq u \leqq b$, $c \leqq v \leqq d$ in R_0.

In view of I.3.10, the following statement is obvious: if $f(u, v)$ is BVTu in R_0, then $V_u(a_0, b_0, v, f) < \infty$ for a.e. v in the interval $c_0 \leqq v \leqq d_0$.

III.2.50. CONTINUATION. Let e_v denote the subset of the interval $c_0 \leqq v \leqq d_0$ where $V_u(a_0, b_0, v, f) = \infty$. By the preceding remark, e_v is then of measure zero (possibly empty) if $f(u, v)$ is BVTu in R_0. Suppose now that $c_0 \leqq v \leqq d_0$ and $v \notin e_v$. Then $V_u(a_0, b_0, v, f) < \infty$, and hence, by III.2.21, the partial derivative $f_u = \partial f/\partial u$ exists for a.e. u in $a_0 \leqq u \leqq b_0$. Since f_u is clearly measurable, it follows by I.3.6 that f_u exists a.e. in R_0. By III.2.22 it follows further that

$$\int_{a_0}^{b_0} | f_u(u, v) | \, du \leqq V_u(a_0, b_0, v, f) \qquad \text{for } v \notin e_v.$$

Since $| e_v | = 0$ and $V_u(a_0, b_0, v, f)$ is summable in the interval $c_0 \leqq v \leqq d_0$, there follows by I.3.10 the summability of f_u in R_0, as well as the inequality

$$\iint_{R_0} | f_u(u, v) | \, du \, dv \leqq \int_{c_0}^{d_0} V_u(a_0, b_0, v, f) \, dv.$$

III.2.51. Suppose that $f(u, v)$ is continuous and BVTu in $R_0: a_0 \leqq u \leqq b_0$, $c_0 \leqq v \leqq d_0$. Let $r: a \leqq u \leqq b$, $c \leqq v \leqq d$ be any rectangle in R_0. We define

$$W_u(r, f) = \int_c^d V_u(a, b, v, f) \, dv.$$

In view of III.2.49, $W_u(r, f)$ is then a rectangle function which is non-negative and finite for every $r \subset R_0$. From III.2.19 it follows readily that the rectangle function $W_u(r, f)$ is additive (cf. III.1.5). We assert that $W_u(r, f)$ is continuous, in the sense of III.1.2. For clarity, we give the proof in several steps.

(i) Let us put, for $c_0 \leqq v \leqq d_0$,

$$g(v) = \int_c^v V_u(a_0, b_0, v, f)\, dv.$$

By I.3.13 the function $g(v)$ is then continuous, and in fact absolutely continuous, in $c_0 \leqq v \leqq d_0$. Hence for every $\epsilon > 0$ there exists a $\delta' = \delta'(\epsilon) > 0$, such that $g(v_2) - g(v_1) < \epsilon$ if $0 < v_2 - v_1 < \delta'$, $c_0 \leqq v_1 \leqq d_0$, $c_0 \leqq v_2 \leqq d_0$. Equivalently,

$$\int_{v_1}^{v_2} V_u(a_0, b_0, v, f)\, dv < \epsilon$$

if $0 < v_2 - v_1 < \delta'$, $c_0 \leqq v_1 \leqq d_0$, $c_0 \leqq v_2 \leqq d_0$.

(ii) Let us put, for $a_0 \leqq u \leqq b_0$,

$$h(u) = \int_{c_0}^{d_0} V_u(a_0, u, v, f)\, dv.$$

Let e_v be the subset of the interval $c_0 \leqq v \leqq d_0$ where $V_u(a_0, b_0, v, f) = \infty$. Then e_v is of measure zero (possibly empty). In the interval $a_0 \leqq u \leqq b_0$, take a point u_0 and any sequence $u_n \to u_0$. By III.2.19 we have then

(1) $$V_u(a_0, u_n, v, f) \underset{n\to\infty}{\longrightarrow} V_u(a_0, u_0, v, f) \qquad \text{for } v \notin e_v,\ c_0 \leqq v \leqq d_0.$$

Also, $V_u(a_0, u_n, v, f) \leqq V_u(a_0, b_0, v, f)$ for $c_0 \leqq v \leqq d_0$. Since $V_u(a_0, b_0, v, f)$ is summable in the interval $c_0 \leqq v \leqq d_0$, it follows by I.3.11 that termwise integration of the relation (1) is permissible, and thus

$$\int_{c_0}^{d_0} V_u(a_0, u_n, v, f)\, dv \underset{n\to\infty}{\longrightarrow} \int_{c_0}^{d_0} V_u(a_0, u_0, v, f)\, dv,$$

or equivalently, $h(u_n) \to h(u_0)$ for $n \to \infty$. Thus $h(u)$ is continuous in $a_0 \leqq u \leqq b_0$. Hence for every $\epsilon > 0$ there exists a $\delta'' = \delta''(\epsilon) > 0$, such that $h(u_2) - h(u_1) < \epsilon$ if $0 < u_2 - u_1 < \delta''$, $a_0 \leqq u_1 \leqq b_0$, $a_0 \leqq u_2 \leqq b_0$. Equivalently

$$\int_{c_0}^{d_0} V_u(u_1, u_2, v, f)\, dv < \epsilon$$

if $0 < u_2 - u_1 < \delta''$, $a_0 \leqq u_1 \leqq b_0$, $a_0 \leqq u_2 \leqq b_0$.

(iii) Let us put $\delta = \min(\delta', \delta'')$, and let $r : a \leqq u \leqq b$, $c \leqq v \leqq d$ be a subrectangle of R_0, such that $|\, r \,| < \delta^2$.

Case (1). $d - c < \delta$. Then by (i) above

$$W_u(r, f) \leqq \int_c^d V_u(a_0, b_0, v, f)\, dv < \epsilon.$$

Case (2). $d - c \geqq \delta$. Then $b - a < \delta$, and hence by (ii) above

$$W_u(r, f) \leqq \int_{c_0}^{d_0} V_u(a, b, v, f)\, dv < \epsilon.$$

Thus the continuity of the rectangle function $W_u(r, f)$ is established.

REMARK. Suppose that $f(u, v)$ is merely known to be continuous in $R_0 : a_0 \leqq u \leqq b_0, c_0 \leqq v \leqq d_0$. Let $r : a \leqq u \leqq b, c \leqq v \leqq d$ be any rectangle in R_0, and consider the function $V_u(a, b, v, f)$. The following agreement is convenient. If $V_u(a, b, v, f)$ is summable in the interval $c \leqq v \leqq d$, then we put

$$W_u(r, f) = \int_c^d V_u(a, b, v, f)\, dv$$

as before. If $V_u(a, b, v, f)$ is not summable in the interval $c \leqq v \leqq d$, then we agree to put $W_u(r, f) = +\infty$. In view of III.2.49 it follows that $f(u, v)$ is BVTu in r if and only if $W_u(r, f) < +\infty$.

III.2.52. Assume that $f(u, v)$ is continuous and BVTu in $R_0 : a_0 \leqq u \leqq b_0, c_0 \leqq v \leqq d_0$. The rectangle function $W_u(r, f)$ is then finite, non-negative and additive. By III.1.29(a) its derivative exists a.e. in R_0 and is summable in R_0. Let us denote this derivative by $D(u, v)$. By III.1.28 we have then the inequality

$$\iint_r D(u, v)\, du\, dv \leqq W_u(r, f) \tag{1}$$

for every rectangle $r : a \leqq u \leqq b, c \leqq v \leqq d$ in R_0. We assert that

$$D(u, v) = | f_u(u, v) | \qquad \text{a.e. in } R_0. \tag{2}$$

PROOF. Applying (1) to a rectangle $r : a \leqq u \leqq b, c \leqq v \leqq d$ in R_0, we obtain (cf. III.2.51) the inequality

$$\int_c^d [V_u(a, b, v, f) - \int_a^b D(u, v)\, du]\, dv \geqq 0. \tag{3}$$

If a, b are kept fixed, while c and d vary according to the inequalities $c_0 \leqq c < d \leqq d_0$, then by I.3.13 there follows from (3) the relation

$$V_u(a, b, v, f) \geqq \int_a^b D(u, v)\, du \qquad \text{for } v \notin e(a, b), \tag{4}$$

where $e(a, b)$ is a certain subset of measure zero of the interval $c_0 \leqq v \leqq d_0$. Let e denote the sum of all the sets $e(a, b)$ that correspond to rational values of a, b. Further, let e' denote the subset of $c_0 \leqq v \leqq d_0$ where $V_u(a_0, b_0, v, f) = \infty$. Then e and e' are both of measure zero (cf. III.2.49). Let us now consider a point $v \notin e + e'$, $c_0 \leqq v \leqq d_0$. Then (4) holds for every rational pair a, b such that $a_0 \leqq a < b \leqq b_0$. Furthermore, both sides of the inequality (4) are continuous functions of a, b by I.3.13, III.2.19. Hence (4) holds for all pairs a, b such that $a_0 \leqq a < b \leqq b_0$. Now both sides of (4) are functions of the interval $a \leqq u \leqq b$, and from (4) there follows for the derivatives of these interval functions the inequality (cf. I.3.13, III.2.22)

$$|f_u(u, v)| \geqq D(u, v) \tag{5}$$

for a.e. u in $a_0 \leqq u \leqq b_0$. Thus (5) holds, if $v \notin e + e'$, for $a_0 \leqq u \leqq b_0$ with the possible exception of a set of measure zero in this interval, where this exceptional set depends upon v, of course. Since $f_u(u, v)$ and $D(u, v)$ are measurable in R_0 (cf. III.1.24), it follows that (5) holds a.e. in R_0.

Consider now any rectangle $r : a \leqq u \leqq b, c \leqq v \leqq d$, in R_0. By III.2.22 we have then the inequality

$$W_u(r, f) = \int_c^d V_u(a, b, v, f)\, dv \geqq \int_c^d \left[\int_a^b |f_u(u, v)|\, du\right] dv$$

$$= \iint_r |f_u(u, v)|\, du\, dv.$$

By III.1.26 there follows the inequality $D(u, v) \geqq |f_u(u, v)|$ a.e. in R_0, and (2) follows in view of (5).

III.2.53. CONTINUATION. For fixed $r : a \leqq u \leqq b, c \leqq v \leqq d$, $W_u(r, f)$ is a lower semi-continuous function of f in the following sense: if $f_n(u, v)$ is continuous and BVTu in R_0 and $f_n \to f$ uniformly in R_0, then $W_u(r, f) \leqq \liminf W_u(r, f_n)$ for $n \to \infty$. Indeed, by III.2.15 we have $V_u(a, b, v, f) \leqq \liminf V_u(a, b, v, f_n)$, and integration yields the desired relation, in view of the theorem of Fatou (see I.3.10).

III.2.54. Suppose that $f(u, v)$ is continuous and BVTu in R_0. By III.2.51, III.1.37, III.1.43 it follows that the rectangle function $W_u(r, f)$ admits of a non-negative, completely additive extension to Borel sets $B \subset R_0$. This extension, which is unique by III.1.47, will be denoted by $W_u(B, f)$. In view of III.2.52, III.1.32, the Lebesgue decomposition of $W_u(B, f)$ is of the form

$$W_u(B, f) = \iint_B |f_u(u, v)|\, du\, dv + W_u^0(B, f),$$

where $W_u^0(B, f)$ is a non-negative, completely additive, singular function of Borel sets $B \subset R_0$.

III.2.55. Continuation. *$f(u, v)$ is ACTu in R_0 if and only if the rectangle function $W_u(r, f)$ is AC in R_0* (cf. III.2.49, III.1.2).

Proof. (i) Suppose that $f(u, v)$ is ACTu in R_0. Then we have, by III.2.29, the formula

(1)
$$W_u(r, f) = \int_c^d V_u(a, b, v, f)\, dv = \int_c^d \left[\int_a^b | f_u(u, v) |\, du\right] dv = \iint_r | f_u(u, v) |\, du\, dv,$$

for every rectangle $r : a \leqq u \leqq b,\ c \leqq v \leqq d$. From (1) it follows, in view of I.3.13, that $W_u(r, f)$ is AC in R_0.

(ii) Suppose that $W_u(r, f)$ is AC in R_0. By III.1.34 it follows that $W_u(B, f)$ is AC in R_0, and thus the singular part $W_u^0(B, f)$ vanishes. Hence, by III.2.54. we have the relation

(2)
$$W_u(R_0, f) = \iint_{R_0} | f_u(u, v) |\, du\, dv.$$

On the other hand, by definition,

(3)
$$W_u(R_0, f) = \int_{c_0}^{d_0} V_u(a_0, b_0, v, f)\, dv.$$

(2) and (3) yield

(4)
$$\int_{c_0}^{d_0} [V_u(a_0, b_0, v, f) - \int_{a_0}^{b_0} | f_u(u, v) |\, du]\, dv = 0.$$

By III.2.22, III.2.49 we have

$$V_u(a_0, b_0, v, f) \geqq \int_{a_0}^{b_0} | f_u(u, v) |\, du$$

for a.e. v in the interval $c_0 \leqq v \leqq d_0$. Thus (4) implies that

$$V_u(a_0, b_0, v, f) = \int_{a_0}^{b_0} | f_u(u, v) |\, du \qquad \text{for } v \notin e,$$

where e is a certain subset of measure zero of the interval $c_0 \leqq v \leqq d_0$. By III.2.29 it follows that $f(u, v)$ is AC, as a function of u, for $v \notin e$. Thus $f(u, v)$ is ACTu in R_0.

III.2.56. CONTINUATION. *If $f(u, v)$ is ACTu in R_0, then*

$$W_u(B, f) = \iint_B | f_u(u, v) | \, du \, dv \tag{1}$$

for every Borel set $B \subset R_0$.

PROOF. By III.2.55, the rectangle function $W_u(r, f)$ is AC in R_0, and hence, by III.1.34, $W_u(B, f)$ is also AC in R_0. Thus the singular part $W_u^0(B, f)$ of $W_u(B, f)$ vanishes, and (1) follows from III.2.54.

III.2.57. CONTINUATION. *If*

$$W_u(R_0, f) = \iint_{R_0} | f_u(u, v) | \, du \, dv,$$

then $f(u, v)$ is ACTu in R_0.

PROOF. By III.2.54 it follows that $W_u^0(R_0, f) = 0$, and hence the singular part $W_u^0(B, f)$ of $W_u(B, f)$ vanishes. Thus $W_u(B, f)$ is AC in R_0. By III.1.34 it follows that $W_u(r, f)$ is AC in R_0, and hence, by III.2.55, $f(u, v)$ is ACTu in R_0.

III.2.58. CONTINUATION. The results in III.2.56, III.2.57 yield the following statements, where it is assumed that $f(u, v)$ is continuous and BVTu in R_0: $a_0 \leqq u \leqq b_0$, $c_0 \leqq v \leqq d_0$.

(a) $f(u, v)$ is ACTu in R_0 if and only if

$$W_u(R_0, f) = \iint_{R_0} | f_u(u, v) | \, du \, dv.$$

(b) $f(u, v)$ is ACTu in R_0 if and only if

$$W_u(B, f) = \iint_B | f_u(u, v) | \, du \, dv$$

for every Borel set $B \subset R_0$.

III.2.59. Suppose that $f(u, v)$ is continuous in $R_0 : a_0 \leqq u \leqq b_0$, $c_0 \leqq v \leqq d_0$, and satisfies the following conditions. (i) The partial derivative $f_u(u, v)$ exists a.e. in R_0 and is summable in R_0. (ii) For a.e. v in the interval $c_0 \leqq v \leqq d_0$, $f(u, v)$ is AC as a function of u in the interval $a_0 \leqq u \leqq b_0$. Then $f(u, v)$ is ACTu in R_0 (see III.2.49).

PROOF. By III.2.29 it follows that

$$V_u(a_0, b_0, v, f) = \int_{a_0}^{b_0} | f_u(u, v) | \, du,$$

for a.e. v in the interval $c_0 \leqq v \leqq d_0$. Since f_u is summable in R_0, it follows that $V_u(a_0, b_0, v, f)$ is summable in the interval $c_0 \leqq v \leqq d_0$. In view of condition (ii), it is thus proved that $f(u, v)$ is ACTu in R_0.

III.2.60. Suppose that $f(u, v)$ is continuous and BVTu in $R_0 : a_0 \leqq u \leqq b_0$, $c_0 \leqq v \leqq d_0$. By III.2.49, we have then

$$V_u(a_0, b_0, v, f) < \infty \qquad \text{for } v \notin e, \tag{1}$$

where e is a certain subset of measure zero of the interval $c_0 \leqq v \leqq d_0$. Let B be a Borel set in R_0. Then the quantity $V_u(B_v, v, f)$, defined in III.2.45, is a function of v in the interval $c_0 \leqq v \leqq d_0$, and

$$V_u(B_v, v, f) < \infty \qquad \text{for } v \notin e. \tag{2}$$

We assert that $V_u(B_v, v, f)$ is a measurable function of v for $c_0 \leqq v \leqq d_0$.

PROOF. Let us denote by K the class of all those Borel sets $B \subset R_0$ for which it is true that the function $V_u(B_v, v, f)$ is measurable in the interval $c_0 \leqq v \leqq d_0$. We verify the following facts concerning the class K.

(i) If B is a rectangle $r : a \leqq u \leqq b, c \leqq v \leqq d$, in R_0, then $B \in K$. Indeed, we have then clearly $V_u(B_v, v, f) = V_u(a, b, v, f)$ for $c \leqq v \leqq d$, and $V_u(B_v, v, f) = 0$ elsewhere in $c_0 \leqq v \leqq d_0$. In view of III.2.47, the measurability of $V_u(B_v, v, f)$ follows. Thus $B \in K$.

(ii) If B is open (relative to R_0), then $B \in K$. Indeed, B is then the sum of a sequence of rectangles $r^1, \cdots, r^n, \cdots$ without common interior points (see I.3.2). We have then (cf. III.2.24, III.2.19)

$$V_u(B_v, v, f) = \sum_n V_u(r_v^n, v, f), \tag{3}$$

with the possible exception of a subset e^* of the interval $c_0 \leqq v \leqq d_0$, where e^* consists of the set e appearing in (1), (2), and of the countable set of the points of intersection of the horizontal sides of the rectangles r^n with the v-axis. Thus (3) holds for a.e. v in the interval $c_0 \leqq v \leqq d_0$, and in view of (i) the measurability of $V_u(B_v, v, f)$ follows. Thus $B \in K$.

(iii) If $B \in K$, then the set $B^* = R_0 - B$ also belongs to K. Indeed, we have obviously $V_u(B_v, v, f) + V_u(B_v^*, v, f) = V_u(a_0, b_0, v, f)$ for $v \notin e$ (cf. formulas (1), (2)). Since $V_u(a_0, b_0, v, f)$ is measurable by III.2.47, and $V_u(B_v, v, f)$ is measurable by assumption, the measurability of $V_u(B_v^*, v, f)$ follows (note that $|e| = 0$). Thus $B^* \in K$.

(iv) If $B^1, \cdots, B^n, \cdots$ is a sequence of Borel sets belonging to K, such that $B^jB^k = 0$ for $j \neq k$, then $B = \sum B^n$ also belongs to K. Indeed, we have obviously

$$V_u(B_v, v, f) = \sum_n V_u(B_v^n, v, f)$$

for $v \notin e$ (cf. the formulas (1), (2)). Since each term of the summation is measurable by assumption, and $|e| = 0$, the measurability of B follows. Thus $B \in K$.

In view of I.2.46, it follows from (ii), (iii), (iv) that the class K contains all Borel sets in R_0. In other words, if B is any Borel set in R_0, then $V_u(B_v, v, f)$ is a measurable function of v in the interval $c_0 \leqq v \leqq d_0$.

III.2.61. CONTINUATION. For given $B \subset R_0$, clearly $V_u(B_v, v, f) \leqq V_u(a_0, b_0, v, f)$ for $c_0 \leqq v \leqq d_0$. Since $V_u(a_0, b_0, v, f)$ is summable in the interval $c_0 \leqq v \leqq d_0$, it follows that $V_u(B_v, v, f)$ is also summable there. We assert that (cf. III.2.54)

$$(1) \qquad W_u(B, f) = \int_{c_0}^{d_0} V_u(B_v, v, f)\, dv$$

for every $B \subset R_0$.

PROOF. (i) Let us put

$$(2) \qquad \psi(B) = \int_{c_0}^{d_0} V_u(B_v, v, f)\, dv.$$

Then $\psi(B)$ is a (finite-valued) non-negative function of Borel sets $B \subset R_0$. Let $B^1, \cdots, B^n, \cdots$ be a sequence of Borel sets in R_0, such that $B^j B^k = 0$ for $j \neq k$, and let us put $B = \sum B^n$. Then (cf. III.2.60(iv))

$$(3) \qquad V_u(B_v, v, f) = \sum_n V_u(B_v^n, v, f)$$

for a.e. v in $c_0 \leqq v \leqq d_0$, and $V_u(B_v, v, f)$ is summable in $c_0 \leqq v \leqq d_0$, as noted above. From I.3.11 it follows that termwise integration in (3) is permissible, and we obtain the relation $\psi(B) = \sum \psi(B^n)$. Thus $\psi(B)$ is a non-negative, completely additive function of Borel sets $B \subset R_0$.

(ii) The formula (1) holds if B is a rectangle $r : a \leqq u \leqq b$, $c \leqq v \leqq d$, in R_0. In this case clearly (see III.2.51, III.2.45)

$$\int_{c_0}^{d_0} V_u(B_v, v, f)\, dv = \int_c^d V_u(a, b, v, f)\, dv = W_u(r, f).$$

(iii) Thus $W_u(B, f)$, $\psi(B)$ are non-negative, completely additive functions of Borel sets $B \subset R_0$ which agree on every rectangle $r : a \leqq u \leqq b$, $c \leqq v \leqq d$, in R_0. Thus (1) follows from III.1.47.

III.2.62. CONTINUATION. Let B_0 be a Borel set in R_0. Then $W_u(B, f)$ is AC on B_0 if and only if

$$W_u(B_0, f) = \iint_{B_0} | f_u(u, v) |\, du\, dv.$$

Indeed, in view of III.2.54, $W_u(B, f)$ is AC on B_0 if and only if $W_u^0(B_0, f) = 0$ (cf. I.3.16), and thus our assertion follows directly from the Lebesgue decomposition given in III.2.54.

Suppose now that $W_u(B, f)$ is AC on B_0. By the preceding remarks, the singular part $W_u^0(B, f)$ vanishes then on B_0, and the Lebesgue decomposition yields

$$W_u(B, f) = \iint_B | f_u(u, v) | \, du \, dv$$

for every Borel set $B \subset B_0$.

III.2.63. In the preceding sections III.2.45-III.2.62, the variable u played a preferred part. As noted in III.2.46, the discussion applies obviously if u and v are exchanged throughout. If this is done, we obtain the quantities and concepts $V_v(c, d, u, f)$, $V_v(B_u, u, f)$ (cf. III.2.45), BVTv, ACTv (cf. III.2.49), $W_v(r, f)$ (cf. III.2.51), $W_v(B, f)$ (cf. III.2.54), as well as the body of results that follow from those in III.2.45-III.2.62 by exchanging u and v.

III.2.64. Suppose that $f(u, v)$ is continuous in $R_0 : a_0 \leqq u \leqq b_0$, $c_0 \leqq v \leqq d_0$.

Definition. $f(u, v)$ is BVT in R_0 (*of bounded variation, in the Tonelli sense, in* R_0), if and only if $f(u, v)$ is both BVTu and BVTv in R_0 (see III.2.49, III.2.63).

Definition. $f(u, v)$ is ACT in R_0 (*absolutely continuous in* R_0 *in the sense of Tonelli*), if and only if $f(u, v)$ is both ACTu and ACTv in R_0 (see III.2.49, III.2.63).

If $f(u, v)$ is BVT (ACT) in R_0, then we have at our disposal two sets of results simultaneously: those developed in III.2.45-III.2.62 for the BVTu (ACTu) case, and those that follow by exchanging u and v (cf. III.2.63).

III.2.65. Let $\mathfrak{D}$ be a bounded domain (bounded connected open set) in the uv-plane, and let $f(u, v)$ be a (real-valued) function which is defined a.e. in $\mathfrak{D}$, is measurable in $\mathfrak{D}$, and is summable on every closed set $F \subset \mathfrak{D}$. Let $R : a \leqq u \leqq b$, $c \leqq v \leqq d$, be a rectangle in $\mathfrak{D}$, and let δ be the shortest distance between the perimeter of R and the boundary of $\mathfrak{D}$. We consider the *integral mean*

$$f_h(u, v) = \frac{1}{4h^2} \int_{-h}^{h} \int_{-h}^{h} f(u + \xi, v + \eta) \, d\xi \, d\eta, \qquad 0 < h < \delta, \ (u, v) \in R.$$

For fixed h, $f_h(u, v)$ is then defined and continuous in R (see I.3.13). In fact $f_h(u, v)$ has a number of further valuable properties; we shall discuss briefly only those which are relevant for our own purposes. Let us note the relation

$$(1) \qquad f_h(u, v) \to f(u, v) \qquad \text{a.e. in } R,$$

which is a direct consequence of I.3.13.

III.2.66. Continuation. Let us now assume that $f(u, v)$ is continuous in $\mathfrak{D}$. Then clearly

$$(1) \qquad f_h(u, v) \to f(u, v) \qquad \text{uniformly in } R.$$

We assert that f_h has partial derivatives of the first order in the interior of R, given by the formulas

$$(2) \qquad \frac{\partial f_h(u, v)}{\partial u} = \frac{1}{4h^2} \int_{-h}^{h} [f(u + h, v + \eta) - f(u - h, v + \eta)] \, d\eta,$$

(3) $$\frac{\partial f_h(u, v)}{\partial v} = \frac{1}{4h^2} \int_{-h}^{h} [f(u + \xi, v + h) - f(u + \xi, v - h)]\, d\xi.$$

PROOF. Let v and h be fixed, where $c < v < d$, $0 < h < \delta$, and let u vary in the interval $a \leqq u \leqq b$. Let us introduce the auxiliary function

$$g(u) = \frac{1}{2h} \int_{-h}^{h} f(u, v + \eta)\, d\eta.$$

Since v and h are fixed, $g(u)$ is a function of u alone, and clearly $g(u)$ is continuous for $a \leqq u \leqq b$. Obviously

$$f_h(u, v) = \frac{1}{2h} \int_{-h}^{h} g(u + \xi)\, d\xi.$$

Hence, by III.2.44,

$$\frac{\partial f_h(u, v)}{\partial u} = \frac{g(u + h) - g(u - h)}{2h} \qquad \text{for } a < u < b,$$

and (2) follows. The formula (3) is obtained in a similar manner.

REMARK. The formulas (2) and (3) show that $\partial f_h/\partial u$, $\partial f_h/\partial v$ are continuous in the interior of R and in fact (for fixed h) on a rectangle that contains R in its interior. Thus $\partial f_h/\partial u$, $\partial f_h/\partial v$ are bounded and uniformly continuous in the interior of R.

III.2.67. CONTINUATION. Let us now suppose that $f(u, v)$ is continuous in $\mathfrak{D}$ and ACT in every rectangle $R : a \leqq u \leqq b$, $c \leqq v \leqq d$, in $\mathfrak{D}$. Let us take such a rectangle R, and let again δ denote the shortest distance between the perimeter of R and the boundary of $\mathfrak{D}$. Let us introduce the auxiliary rectangle

(1) $$R^* : a - \delta/2 \leqq u \leqq b + \delta/2, \qquad c - \delta/2 \leqq v \leqq d + \delta/2,$$

and let us restrict h by the inequalities

(2) $$0 < h < \delta/2.$$

Then $f(u, v)$ is ACT, by assumption, in the rectangle R^*, and hence (see III.2.50) the partial derivative $f_u(u, v)$ is summable in R^*. Hence the function

(3) $$\begin{aligned} g(u, v) &= \frac{1}{4h^2} \int_{-h}^{h} \int_{-h}^{h} f_u(u + \xi, v + \eta)\, d\xi\, d\eta \\ &= \frac{1}{4h^2} \int_{-h}^{h} \left[\int_{-h}^{h} f_u(u + \xi, v + \eta)\, d\xi \right] d\eta \end{aligned}$$

is defined for $(u, v) \in R$, $0 < h < (\delta/2)$. Now since $f(u, v)$ is AC, for a.e. v in the interval $c - (\delta/2) \leqq v \leqq d + (\delta/2)$, as a function of u in the interval $a - (\delta/2) \leqq u \leqq b + (\delta/2)$, we have (see III.2.36)

$$g(u, v) = \frac{1}{4h^2} \int_{-h}^{h} [f(u + h, v + \eta) - f(u - h, v + \eta)]\, d\eta. \tag{4}$$

(3) and (4) yield, in view of III.2.66(2), the formula

$$\frac{\partial f_h(u, v)}{\partial u} = \frac{1}{4h^2} \int_{-h}^{h} \int_{-h}^{h} f_u(u + \xi, v + \eta)\, d\xi\, d\eta, \qquad (u, v) \in R,\ 0 < h < \delta/2. \tag{5}$$

An analogous argument yields the formula

$$\frac{\partial f_h(u, v)}{\partial v} = \frac{1}{4h^2} \int_{-h}^{h} \int_{-h}^{h} f_v(u + \xi, v + \eta)\, d\xi\, d\eta, \qquad (u, v) \in R,\ 0 < h < \delta/2. \tag{6}$$

From III.2.65(1), applied to the functions f_u, f_v, we obtain the important relations

$$\frac{\partial f_h}{\partial u} \to \frac{\partial f}{\partial u}, \qquad \frac{\partial f_h}{\partial v} \to \frac{\partial f}{\partial v} \qquad \text{a.e. in } R.$$

CHAPTER III.3. ARC-LENGTH

III.3.1. In the present chapter III.3, we begin the study of the length of a curve

$$(1) \qquad C : x = x(u), \qquad y = y(u), \qquad z = z(u), \qquad u \in I_0 : a_0 \leqq u \leqq b_0 ,$$

where the functions $x(u)$, $y(u)$, $z(u)$ are continuous (and real-valued) in I_0, and x, y, z denote Cartesian coordinates in Euclidean three-space. This study will depend upon concepts and methods that belong partly to Analysis and partly to Topology. In the following presentation, an effort will be made to clarify the part played by each of these two types of concepts and methods.

As long as we are concerned with a fixed representation (1), we are actually studying a triple of continuous functions $x(u)$, $y(u)$, $z(u)$ or, equivalently, a continuous vector function $\mathfrak{x}(u)$ with the components $x(u)$, $y(u)$, $z(u)$. The length of C appears then as a certain functional whose argument is a continuous vector function $\mathfrak{x}(u)$. From this point of view, a large number of facts in the theory of arc-length may be stated and discussed within the framework of Analysis proper, and the study of geometrical issues may be deferred, and in fact greatly simplified, on the basis of previously achieved insight into the purely analytic aspects of the theory.

III.3.2. Accordingly, we shall first concern ourselves with *continuous vector functions*. The following notations will be used. We shall write

$$\mathfrak{x} = \mathfrak{x}(u) = [x(u), y(u), z(u)], \qquad u \in I_0 : a_0 \leqq u \leqq b_0 ,$$

to refer to a vector function $\mathfrak{x}(u)$ with components $x(u)$, $y(u)$, $z(u)$, considered in I_0. If circumstances permit, we shall write only

$$\mathfrak{x} = \mathfrak{x}(u), \qquad u \in I_0 .$$

If $I : a \leqq u \leqq b$ is an interval in I_0, and we are considering $\mathfrak{x}(u)$ only in I, we shall write

$$\mathfrak{x} = \mathfrak{x}(u), \qquad u \in I \subset I_0 .$$

We shall assume throughout that $\mathfrak{x}(u)$ *is continuous,* that is, $x(u)$, $y(u)$, $z(u)$ are continuous in I_0, even though many of our results remain valid if this assumption is dropped. Given an interval $I : a \leqq u \leqq b$ in I_0, we define

$$g_1(I, \mathfrak{x}) = | x(b) - x(a) |, \; g_2(I, \mathfrak{x}) = | y(b) - y(a) |, \; g_3(I, \mathfrak{x}) = | z(b) - z(a) | .$$

We shall denote by $\omega(I, x)$ the oscillation of $x(u)$ in I (see III.2.3). The symbols $\omega(I, y)$, $\omega(I, z)$ have the same meaning relative to $y(u)$, $z(u)$ respectively. We put

$$\omega_1(I, \mathfrak{x}) = \omega(I, x), \qquad \omega_2(I, \mathfrak{x}) = \omega(I, y), \qquad \omega_3(I, \mathfrak{x}) = \omega(I, z).$$

The total variations of $x(u)$, $y(u)$, $z(u)$ in I will be denoted by $V(I, x)$, $V(I, y)$, $V(I, z)$ respectively, in conformity with III.2.13. We put

$$v_1(I, \mathfrak{x}) = V(I, x), \qquad v_2(I, \mathfrak{x}) = V(I, y), \qquad v_3(I, \mathfrak{x}) = V(I, z).$$

Let us note that v_1, v_2, v_3 are not necessarily finite. Finally, we put

$$g(I, \mathfrak{x}) = [g_1(I, \mathfrak{x})^2 + g_2(I, \mathfrak{x})^2 + g_3(I, \mathfrak{x})^2]^{1/2},$$

$$\omega(I, \mathfrak{x}) = [\omega_1(I, \mathfrak{x})^2 + \omega_2(I, \mathfrak{x})^2 + \omega_3(I, \mathfrak{x})^2]^{1/2},$$

$$v(I, \mathfrak{x}) = [v_1(I, \mathfrak{x})^2 + v_2(I, \mathfrak{x})^2 + v_3(I, \mathfrak{x})^2]^{1/2}.$$

It is understood that v is infinite if one or more of the quantities v_1, v_2, v_3 is infinite.

III.3.3. CONTINUATION. In conformity with III.1.4, we use the symbol $D(I)$ as a generic notation for a subdivision of the interval I. Thus $D(I)$ is comprised of a finite number of intervals, without common interior points, whose sum is I. We shall write $i \in D(I)$ to state that the interval i occurs in the subdivision $D(I)$. The maximum length of the intervals $i \in D(I)$ will be denoted by $\| D(I) \|$. In conformity with III.1.5, we put

$$g[D(I), \mathfrak{x}] = \sum g(i, \mathfrak{x}), \qquad i \in D(I),$$

$$U(I, g, \mathfrak{x}) = \text{l.u.b.}\ g[D(I), \mathfrak{x}],$$

where the least upper bound is taken with respect to all subdivisions $D(I)$ of I. Note that $U(I, g, \mathfrak{x})$ may be infinite. The symbols $\omega[D(I), \mathfrak{x}]$, $U(I, \omega, \mathfrak{x})$, $v[D(I), \mathfrak{x}]$, $U(I, v, \mathfrak{x})$ have analogous meanings relative to the interval functions ω, v.

III.3.4. CONTINUATION. If $I : a \leqq u \leqq b$ is an interval in I_0, then clearly $g(I, \mathfrak{x})$ is equal to the length of the vector $\mathfrak{x}(b) - \mathfrak{x}(a)$; in symbols

$$g(I, \mathfrak{x}) = | \mathfrak{x}(b) - \mathfrak{x}(a) |.$$

Thus $U(I, g, \mathfrak{x})$ may be interpreted as the total variation of the vector function $\mathfrak{x}(u)$ in I (cf. III.2.13). Also, $g(I, \mathfrak{x})$ may be interpreted as the length of a chord of the curve given by the equations $x = x(u)$, $y = y(u)$, $z = z(u)$, $u \in I$, and thus $U(I, g, \mathfrak{x})$ may be (and in fact later on will be) interpreted as the length of this curve. While according to our plan of exposition the geometrical issues will be taken up only later, the preceding remarks may be in order to motivate our interest in the quantity $U(I, g, \mathfrak{x})$. In view of the geometrical interpretation just referred to, we introduce the more concise notation

$$L(I, \mathfrak{x}) = U(I, g, \mathfrak{x}).$$

Then $L(I, \mathfrak{x})$ is a non-negative, not necessarily finite, interval function which depends upon $\mathfrak{x}(u)$ and upon a variable interval $I \subset I_0$.

III.3.5. CONTINUATION. *We assert that*

(1) $$U(I, g, \mathfrak{x}) = U(I, \omega, \mathfrak{x}) = U(I, v, \mathfrak{x}).$$

PROOF. Clearly $U(I, g, \mathfrak{x}) \leqq U(I, \omega, \mathfrak{x}) \leqq U(I, v, \mathfrak{x})$, and thus (1) will be established if we show that

(2) $$U(I, v, \mathfrak{x}) \leqq U(I, g, \mathfrak{x}).$$

Let $D(I)$ be any subdivision of I, and let us denote by $D^n(I)$ the subdivision obtained by subdividing each interval of $D(I)$ into n equal parts. If i, i^n are generic notations for intervals occurring in $D(I)$, $D^n(I)$ respectively, then we have (see III.2.17, III.3.2)

$$v_1(i, \mathfrak{x}) = V(i, x) = \lim_{n\to\infty} \sum g_1(i^n, \mathfrak{x}), \qquad i^n \subset i,$$

and similar relations hold for $v_2(i, \mathfrak{x})$, $v_3(i, \mathfrak{x})$. It follows that (cf. I.3.10, III.3.2)

$$v(i, \mathfrak{x}) = \lim_{n\to\infty} \{[\sum g_1(i^n, \mathfrak{x})]^2 + [\sum g_2(i^n, \mathfrak{x})]^2 + [\sum g_3(i^n, \mathfrak{x})]^2\}^{1/2}$$

$$\leqq \limsup_{n\to\infty} \sum g(i^n, \mathfrak{x}), \qquad i^n \subset i,$$

and consequently $v(i, \mathfrak{x}) \leqq U(i, g, \mathfrak{x})$. Hence

(3) $$v[D(I), \mathfrak{x}] \leqq \sum U(i, g, \mathfrak{x}), \qquad i \in D(I).$$

We assert that

(4) $$U(I, g, \mathfrak{x}) \geqq \sum U(i, g, \mathfrak{x}), \qquad i \in D(I).$$

If $U(I, g, \mathfrak{x}) = \infty$, then (4) is obvious. If $U(I, g, \mathfrak{x}) < \infty$, then (4) follows from III.1.5. From (3) and (4) we infer that $v[D(I), \mathfrak{x}] \leqq U(I, g, \mathfrak{x})$ for every subdivision $D(I)$ of I, and the inequality (2) follows.

III.3.6. CONTINUATION. *$L(I, \mathfrak{x})$ is a lower semi-continuous functional of $\mathfrak{x}$* in the following sense: if $\mathfrak{x}_n(u)$ is a sequence of continuous vector functions that converge uniformly to $\mathfrak{x}(u)$ in I_0, then

(1) $$L(I, \mathfrak{x}) \leqq \liminf_{n\to\infty} L(I, \mathfrak{x}_n)$$

for every interval $I \subset I_0$.

PROOF. Let $D(I)$ be any subdivision of I. Then clearly

$$g[D(I), \mathfrak{x}] = \lim_{n\to\infty} g[D(I), \mathfrak{x}_n] \leqq \liminf_{n\to\infty} L(I, \mathfrak{x}_n).$$

Since the subdivision $D(I)$ was arbitrary, the inequality (1) follows.

III.3.7. Let $\mathfrak{x}(u)$ be a continuous vector function in $I_0 : a_0 \leqq u \leqq b_0$. If there exists a subdivision $D^*(I_0)$ such that the components $x(u)$, $y(u)$, $z(u)$ are linear functions of u in each interval of $D^*(I_0)$, then $\mathfrak{x}(u)$ will be termed *quasi-linear* in I_0, and $D^*(I_0)$ will be termed a *typical subdivision* for $\mathfrak{x}(u)$. We shall use the notation $q(u)$ to refer to a quasi-linear vector function. Given $q(u)$ in I_0, there exists by definition a typical subdivision $D^*(I_0)$ for $q(u)$. Clearly, $D^*(I_0)$ is not determined univocally by $q(u)$ (for example, every refinement of $D^*(I_0)$ is also a typical subdivision for $q(u)$). The following statements are obvious (cf. III.3.2, III.3.4).

(i) If $D(I_0)$ is any subdivision, and if $D^*(I_0)$ is a typical subdivision for $q(u)$, then $g[D(I_0), q] \leqq g[D^*(I_0), q]$.

(ii) If $D_1^*(I_0)$, $D_2^*(I_0)$ are any two typical subdivisions for $q(u)$, then $g[D_1^*(I_0). q] = g[D_2^*(I_0), q]$. This follows directly from (i).

(iii) We define $l(I_0, q) = g[D^*(I_0), q]$, where $D^*(I_0)$ is any typical subdivision for $q(u)$. In view of (ii), the value of $l(I_0, q)$ is independent of the particular choice of the typical subdivision $D^*(I_0)$.

(iv) $l(I_0, q) = L(I_0, q)$. This follows readily from (i) and (ii).

(v) If $I : a \leqq u \leqq b$ is any subinterval of I_0, then clearly $q(u)$ is quasi-linear in I also. Hence the preceding remarks apply in every interval $I \subset I_0$.

III.3.8. Let $\mathfrak{x}(u)$ be a continuous vector function in $I_0 : a_0 \leqq u \leqq b_0$, and let $D_n(I_0)$ be a sequence of subdivisions such that $\| D_n(I_0) \| \to 0$ for $n \to \infty$. Then

$$g[D_n(I_0), \mathfrak{x}] \to L(I_0, \mathfrak{x}), \tag{1}$$

$$\omega[D_n(I_0), \mathfrak{x}] \to L(I_0, \mathfrak{x}), \tag{2}$$

$$v[D_n(I_0), \mathfrak{x}] \to L(I_0, \mathfrak{x}). \tag{3}$$

Proof. Clearly $g[D_n(I_0), \mathfrak{x}] \leqq \omega[D_n(I_0), \mathfrak{x}] \leqq v[D_n(I_0), \mathfrak{x}] \leqq L(I_0, \mathfrak{x})$, in view of III.3.4, III.3.5. Thus (2) and (3) are implied by (1). To establish (1), it is sufficient to show that

$$L(I_0, \mathfrak{x}) \leqq \liminf_{n\to\infty} g[D_n(I_0), \mathfrak{x}], \tag{4}$$

since $g[D_n(I_0), \mathfrak{x}] \leqq U(I_0, g, \mathfrak{x}) = L(I_0, \mathfrak{x})$. To prove (4), let us denote by $q_n(u)$ the quasi-linear vector function whose components are linear in each interval of $D_n(I_0)$ and agree with the corresponding components of $\mathfrak{x}(u)$ at the end points of the intervals of $D_n(I_0)$. Clearly, $q_n(u) \to \mathfrak{x}(u)$ uniformly in I_0, and hence

$$L(I_0, \mathfrak{x}) \leqq \liminf_{n\to\infty} L(I_0, q_n) \tag{5}$$

by III.3.6. On the other hand, by III.3.7,

$$L(I_0, q_n) = g[D_n(I_0), q_n] = g[D_n(I_0), \mathfrak{x}]. \tag{6}$$

The relations (5) and (6) imply (4).

III.3.9. The result of III.3.5 may be interpreted as yielding three equivalent definitions for the functional $L(I_0, \mathfrak{x})$ by means of the formulas (cf. III.3.3)

$$L(I_0, \mathfrak{x}) = \text{l.u.b.} \sum g(I, \mathfrak{x}), \qquad I \in D(I_0), \tag{1}$$

$$L(I_0, \mathfrak{x}) = \text{l.u.b.} \sum \omega(I, \mathfrak{x}), \qquad I \in D(I_0), \tag{2}$$

$$L(I_0, \mathfrak{x}) = \text{l.u.b.} \sum v(I, \mathfrak{x}), \qquad I \in D(I_0), \tag{3}$$

where the least upper bound is taken with respect to all subdivisions $D(I_0)$ of I_0. Accordingly, each summation $\sum g(I, \mathfrak{x})$, $\sum \omega(I, \mathfrak{x})$, $\sum v(I, \mathfrak{x})$, $I \in D(I_0)$ is a lower bound for $L(I_0, \mathfrak{x})$. Thus (1), (2), (3) represent equivalent definitions for $L(I_0, \mathfrak{x})$ *from below*, that is, *in terms of lower bounds.* The result of III.3.8 may be interpreted as yielding three equivalent definitions for $L(I_0, \mathfrak{x})$ in terms of limit processes. We propose to add still another formula for $L(I_0, \mathfrak{x})$.

Let $\mathfrak{x}(u)$ be a continuous vector function in $I_0 : a_0 \leqq u \leqq b_0$, and let $q_n(u)$ be a sequence of quasi-linear vector functions such that $q_n(u) \to \mathfrak{x}(u)$ uniformly in I_0. Then (see III.3.7)

$$(4) \qquad L(I_0, \mathfrak{x}) \leqq \liminf_{n \to \infty} l(I_0, q_n).$$

Indeed, $l(I_0, q_n) = L(I_0, q_n)$ by III.3.7, and thus (4) follows from III.3.6. The relation (4) states that every sequence $q_n(u)$ that converges uniformly to $\mathfrak{x}(u)$ on I_0 yields an upper bound for $L(I_0, \mathfrak{x})$. *We assert the formula*

$$(5) \qquad L(I_0, \mathfrak{x}) = \text{gr.l.b.} \liminf_{n \to \infty} l(I_0, q_n),$$

where the greatest lower bound is taken with respect to all sequences of quasi-linear vector functions $q_n(u)$ that converge uniformly to $\mathfrak{x}(u)$ on I_0. To prove (5), let us denote by G the right-hand side of (5). Let $D_n(I_0)$ denote a sequence of subdivisions of I_0 such that $\| D_n(I_0) \| \to 0$, and let $q_n(u)$ be chosen as the quasi-linear vector function whose components are linear on each interval of $D_n(I_0)$ and agree with the corresponding components of $\mathfrak{x}(u)$ at the end points of the intervals of $D_n(I_0)$. Clearly $q_n(u) \to \mathfrak{x}(u)$ uniformly in I_0, and $l(I_0, q_n) = g[D_n(I_0), \mathfrak{x}] \to L(I_0, \mathfrak{x})$ by III.3.8. Thus $L(I_0, \mathfrak{x}) \geqq G$. On the other hand, $L(I_0, \mathfrak{x}) \leqq G$ by (4), and (5) is established.

The formula (5) may be interpreted as yielding for $L(I_0, \mathfrak{x})$ an equivalent definition *from above,* that is, a definition *in terms of upper bounds.*

III.3.10. Let $\mathfrak{x}(u)$ be a continuous vector function in $I_0 : a_0 \leqq u \leqq b_0$, and let c be a constant. Then obviously (cf. III.3.4), for every interval $I \subset I_0$,

$$(1) \qquad L(I, c\mathfrak{x}) = | c | L(I, \mathfrak{x}).$$

In particular, if $c = -1$, then

$$(2) \qquad L(I, -\mathfrak{x}) = L(I, \mathfrak{x}).$$

Now let $\mathfrak{x}_1(u)$, $\mathfrak{x}_2(u)$ be two continuous vector functions in I_0. Clearly (cf. III.3.2), $g_j(I, \mathfrak{x}_1 + \mathfrak{x}_2) \leqq g_j(I, \mathfrak{x}_1) + g_j(I, \mathfrak{x}_2)$, $j = 1, 2, 3$, and hence (see I.3.10) $g(I, \mathfrak{x}_1 + \mathfrak{x}_2) \leqq g(I, \mathfrak{x}_1) + g(I, \mathfrak{x}_2)$, for every interval $I \subset I_0$. In view of III.3.4 it follows that

$$(3) \qquad L(I, \mathfrak{x}_1 + \mathfrak{x}_2) \leqq L(I, \mathfrak{x}_1) + L(I, \mathfrak{x}_2).$$

The relation (3) is referred to as *the inequality of Steiner.* (2) and (3) yield

$$(4) \qquad L(I, \mathfrak{x}_1 - \mathfrak{x}_2) \leqq L(I, \mathfrak{x}_1) + L(I, \mathfrak{x}_2).$$

In view of (1), applied with $c = 1/2$, (3) may be written in the form

$$L\left(I, \frac{\mathfrak{x}_1 + \mathfrak{x}_2}{2}\right) \leqq \frac{L(I, \mathfrak{x}_1) + L(I, \mathfrak{x}_2)}{2} \tag{5}$$

which is more convenient at times.

III.3.11. Let $\mathfrak{x}(u) = [x(u), y(u), z(u)]$ be a continuous vector function in $I_0: a_0 \leqq u \leqq b_0$. Then $\mathfrak{x}(u)$ will be termed BV in I_0 (*of bounded variation in* I_0) if and only if the components $x(u)$, $y(u)$, $z(u)$ are BV in I_0. Similarly, $\mathfrak{x}(u)$ will be termed AC in I_0 (*absolutely continuous in* I_0) if and only if the components $x(u)$, $y(u)$, $z(u)$ are AC in I_0 (cf. III.2.2).

III.3.12. Let $\mathfrak{x}(u)$ be a continuous vector function in $I_0 : a_0 \leqq u \leqq b_0$. Then $L(I_0, \mathfrak{x})$ is finite if and only if $\mathfrak{x}(u)$ is BV in I_0.

Proof. We have the inequalities (cf. III.3.2, III.3.4, III.3.5)

$$V(I_0, x) \leqq v[I_0, \mathfrak{x}] \leqq L(I_0, \mathfrak{x}),$$

and similarly $V(I_0, y) \leqq L(I_0, \mathfrak{x})$, $V(I_0, z) \leqq L(I_0, \mathfrak{x})$. Hence, if $L(I_0, \mathfrak{x}) < \infty$, then $x(u)$, $y(u)$, $z(u)$ are BV in I_0.

Conversely, suppose that $x(u)$, $y(u)$, $z(u)$ are BV in I_0. Let $D(I_0)$ be any subdivision of I_0. Then (cf. III.3.3, III.3.2, III.2.19)

$$\begin{aligned} v[D(I_0), \mathfrak{x}] &\leqq v_1[D(I_0), \mathfrak{x}] + v_2[D(I_0), \mathfrak{x}] + v_3[D(I_0), \mathfrak{x}] \\ &= V(I_0, x) + V(I_0, y) + V(I_0, z). \end{aligned}$$

Since $D(I_0)$ was an arbitrary subdivision, there follows by III.3.4, III.3.5 the inequality

$$L(I_0, \mathfrak{x}) \leqq V(I_0, x) + V(I_0, y) + V(I_0, z) < +\infty.$$

III.3.13. Let $\mathfrak{x}(u)$ be a continuous vector function in $I_0 : a_0 \leqq u \leqq b_0$. Suppose that $L(I_0, \mathfrak{x}) < \infty$. Then clearly $L(I, \mathfrak{x}) \leqq L(I_0, \mathfrak{x}) < \infty$ for every interval $I \subset I_0$. In view of III.3.8, the interval function $L(I, \mathfrak{x})$ is the indefinite Burkill integral of the interval function $g(I, \mathfrak{x})$ (cf. III.1.12, III.1.52). Since $g(I, \mathfrak{x})$ is obviously non-negative and continuous, it follows by III.1.14, III.1.17, III.1.52 that $L(I, \mathfrak{x})$ is a (finite-valued) non-negative, additive, continuous interval function for $I \subset I_0$. By III.3.12, III.2.22 it follows that the derivatives $x'(u)$, $y'(u)$, $z'(u)$ of the components of $\mathfrak{x}(u)$ exist a.e. and are summable in I_0. For conciseness, let us introduce the vector function

$$\mathfrak{x}'(u) = [x'(u), y'(u), z'(u)]$$

which thus is defined a.e. in I_0. The length $|\mathfrak{x}'(u)|$ of $\mathfrak{x}'(u)$ is then given by the formula

$$|\mathfrak{x}'(u)| = [x'(u)^2 + y'(u)^2 + z'(u)^2]^{1/2}.$$

Clearly, the derivative of the interval function $g(I, \mathfrak{x})$ exists and is equal to $|\mathfrak{x}'(u)|$ a.e. in I_0, by III.3.2, III.2.1. Since $L(I, \mathfrak{x})$ is the indefinite Burkill integral of $g(I, \mathfrak{x})$, it follows that the derivative of $L(I, \mathfrak{x})$ exists and is equal to

$| \mathfrak{x}'(u) |$ a.e. in I_0 (see III.1.27, III.1.52). By III.1.29, III.1.52 it follows further that, for every interval $I \subset I_0$,

$$\int_I | \mathfrak{x}'(u) | \, du \leqq L(I, \mathfrak{x}).$$

III.3.14. Continuation. If $I : a \leqq u \leqq b$ is an interval in I_0, we agree to put $L(a, b, \mathfrak{x}) = L(I, \mathfrak{x})$, $L(a, a, \mathfrak{x}) = 0$. Let us then consider the function $L(a_0, u, \mathfrak{x})$, $u \in I_0$. In view of the results stated in III.3.13, the following statements are immediate.

(i) $L(a_0, u, \mathfrak{x})$ is non-negative, nondecreasing, continuous in I_0, and $L(a_0, a_0, \mathfrak{x}) = 0$.

(ii) We have a.e. in I_0 the formula (cf. III.2.1)

$$\frac{dL(a_0, u, \mathfrak{x})}{du} = | \mathfrak{x}'(u) |.$$

(iii) For every interval $I : a \leqq u \leqq b$ in I_0, we have the inequality

$$\int_a^b | \mathfrak{x}'(u) | \, du \leqq L(a, b, \mathfrak{x}).$$

III.3.15. Let $\mathfrak{x}(u)$ be a continuous vector function in $I_0 : a_0 \leqq u \leqq b_0$, and suppose that $L(I_0, \mathfrak{x}) < \infty$. Then the interval function $L(I, \mathfrak{x})$ is AC in I_0 if and only if $\mathfrak{x}(u)$ is AC in I_0.

Proof. (i) Suppose that $\mathfrak{x}(u)$ is AC in I_0. If $I : a \leqq u \leqq b$ is any interval in I_0, then clearly $g(I, \mathfrak{x}) \leqq | x(b) - x(a) | + | y(b) - y(a) | + | z(b) - z(a) |$. Since $x(u)$, $y(u)$, $z(u)$ are AC in I_0 by assumption, it follows that $g(I, \mathfrak{x})$ is AC in I_0. Since $L(I, \mathfrak{x})$ is the indefinite Burkill integral of $g(I, \mathfrak{x})$, it follows finally by III.1.18, III.1.52 that $L(I, \mathfrak{x})$ is AC in I_0.

(ii) Suppose that $L(I, \mathfrak{x})$ is AC in I_0. If $I : a \leqq u \leqq b$ is any interval in I_0, then clearly $| x(b) - x(a) | \leqq g(I, \mathfrak{x}) \leqq L(I, \mathfrak{x})$. Hence $x(u)$ is AC in I_0, and it follows similarly that $y(u)$, $z(u)$ are AC in I_0.

III.3.16. Continuation. Clearly, the function $L(a_0, u, \mathfrak{x})$ (see III.3.14) is AC in I_0 if and only if $L(I, \mathfrak{x})$ is AC in I_0. In view of III.3.15 it follows that $L(a_0, u, \mathfrak{x})$ is AC in I_0 if and only if $\mathfrak{x}(u)$ is AC in I_0. Since $L(a_0, u, \mathfrak{x})$ is continuous and nondecreasing in I_0, and $L(a_0, b_0, \mathfrak{x}) - L(a_0, a_0, \mathfrak{x}) = L(I_0, \mathfrak{x})$, we obtain thus by III.2.22, III.2.29, III.3.14 the following statement.

Theorem. *If $\mathfrak{x}(u)$ is a continuous vector function in $I_0 : a_0 \leqq u \leqq b_0$, such that $L(I_0, \mathfrak{x}) < \infty$, then the derivative $\mathfrak{x}'(u)$ (see* III.3.13*) exists a.e. in I_0, and $| \mathfrak{x}'(u) |$ is summable in I_0. We have the inequality*

$$\int_{I_0} | \mathfrak{x}'(u) | \, du \leqq L(I_0, \mathfrak{x}),$$

where the sign of equality holds if and only if $\mathfrak{x}(u)$ is AC in I_0.

III.3.17. Suppose that the vector function $\mathfrak{x}(u)$ is continuous and BV in $I_0 : a_0 \leqq u \leqq b_0$. By III.3.12, III.3.13, the interval function $L(I, \mathfrak{x})$ is then finite-valued, non-negative, continuous and additive for $I \subset I_0$. By III.1.37, III.1.43, III.1.52, it follows that $L(I, \mathfrak{x})$ admits of a non-negative, completely additive extension to Borel sets $B \subset I_0$. This extension, which is unique by III.1.47, III.1.52, will be denoted by $L(B, \mathfrak{x})$. By I.3.16, $L(B, \mathfrak{x})$ possesses a univocally determined Lebesgue decomposition

$$L(B, \mathfrak{x}) = L_a(B, \mathfrak{x}) + L_s(B, \mathfrak{x}), \qquad B \subset I_0, \tag{1}$$

where $L_a(B, \mathfrak{x})$ is non-negative, completely additive and AC, while $L_s(B, \mathfrak{x})$ is non-negative, completely additive and singular. Since the derivative of $L_s(B, \mathfrak{x})$ vanishes a.e. in I_0 (see III.1.31, III.1.52), it follows by III.3.13 that the derivative of $L_a(B, \mathfrak{x})$ is equal to $| \mathfrak{x}'(u) |$ a.e. in I_0. By III.3.13, III.1.32, III.1.52 we have therefore the formulas

$$L(B, \mathfrak{x}) = \int_B | \mathfrak{x}'(u) | \, du + L_s(B, \mathfrak{x}), \tag{2}$$

$$L_a(B, \mathfrak{x}) = \int_B | \mathfrak{x}'(u) | \, du, \tag{3}$$

for every Borel set $B \subset I_0$.

III.3.18. Suppose that the vector function $\mathfrak{x}(u)$ is continuous and BV in $I_0 : a_0 \leqq u \leqq b_0$. Then $\mathfrak{x}(u)$ is AC in I_0 if and only if $L(B, \mathfrak{x})$ is AC in I_0.

Proof. By III.1.34, III.1.52, $L(B, \mathfrak{x})$ is AC in I_0 if and only if the interval function $L(I, \mathfrak{x})$ is AC in I_0, and thus the assertion follows from III.3.15.

III.3.19. Suppose that the vector function $\mathfrak{x}(u)$ is continuous and BV in $I_0 : a_0 \leqq u \leqq b_0$. We have then the following statements concerning $L(B, \mathfrak{x})$ (see III.3.17) for every Borel set $B \subset I_0$.

(i) If $\mathfrak{a}$ is any constant vector, then $L(B, \mathfrak{x} + \mathfrak{a}) = L(B, \mathfrak{x})$.

(ii) If c is any (scalar) constant, then $L(B, c\mathfrak{x}) = | c | \, L(B, \mathfrak{x})$. In particular, $L(B, -\mathfrak{x}) = L(B, \mathfrak{x})$.

Furthermore, if $\mathfrak{x}_1(u)$, $\mathfrak{x}_2(u)$ are continuous and BV in I_0, then the following statements hold for every Borel set $B \subset I_0$.

(iii) $L(B, \mathfrak{x}_1 + \mathfrak{x}_2) \leqq L(B, \mathfrak{x}_1) + L(B, \mathfrak{x}_2)$.

(iv) $L(B, \mathfrak{x}_1 + \mathfrak{x}_2) \geqq | L(B, \mathfrak{x}_1) - L(B, \mathfrak{x}_2) |$.

Proof. If $\mathfrak{a}$ is a constant vector, then clearly $g(I, \mathfrak{x} + \mathfrak{a}) = g(I, \mathfrak{x})$ and hence obviously $L(I, \mathfrak{x} + \mathfrak{a}) = L(I, \mathfrak{x})$ for every interval $I \subset I_0$ (see III.3.2, III.3.4). Thus $L(B, \mathfrak{x} + \mathfrak{a})$, $L(B, \mathfrak{x})$ are two non-negative, completely additive functions of Borel sets $B \subset I_0$ that agree for every interval $I \subset I_0$. By III.1.47, III.1.52 it follows that $L(B, \mathfrak{x} + \mathfrak{a}) = L(B, \mathfrak{x})$ for every Borel set $B \subset I_0$ and (i) is proved. The proofs of (ii) and (iii) are similar. Finally, (iv) is a formal consequence of (ii) and (iii).

III.3.20. We propose to derive a statement analogous to that in III.3.18 for the case when $L(B, \mathfrak{x})$ is singular. The following remarks will be needed. Let $f(u)$ be a (real-valued, scalar) function which is continuous in I_0 and BV in I_0. Then $f(u)$ will be termed singular in I_0 if and only if $f'(u)$ vanishes a.e. in I_0 (cf. III.2.21).

Let $V(B, f)$ have the meaning explained in III.2.24. We assert that $f(u)$ is singular in I_0 if and only if $V(B, f)$ is singular in I_0.

PROOF. By III.2.26 we have for $V(B, f)$ the Lebesgue decomposition

$$V(B, f) = \int_B |f'(u)| \, du + V^0(B, f), \qquad B \subset I_0. \tag{1}$$

Thus $V(B, f)$ is singular in I_0 if and only if the integral in (1) vanishes for every Borel set $B \dot{\subset} I_0$, and by I.3.12 this is the case if and only if $f'(u)$ vanishes a.e. in I_0.

REMARK. If $f(u)$ is constant in I_0, then clearly $f(u)$ is both AC and singular in I_0. Suppose, conversely, that $f(u)$ is both AC and singular in I_0. Since $f(u)$ is AC, we have (see III.2.36)

$$f(b) - f(a) = \int_a^b f'(u) \, du \qquad \text{for } a_0 \leqq a < b \leqq b_0.$$

Since $f(u)$ is singular, we have $f'(u) = 0$ a.e. in I_0, and hence it follows that $f(b) = f(a)$ for $a_0 \leqq a < b \leqq b_0$. Thus $f(u)$ is constant in I_0.

III.3.21. CONTINUATION. Let $f(u)$ be continuous and BV in $I_0 : a_0 \leqq u \leqq b_0$. Then $f(u)$ admits of a univocally determined representation of the form

$$f(u) = f_a(u) + f_s(u), \qquad u \in I_0, \; f_s(a_0) = 0, \tag{1}$$

where $f_a(u)$ is AC and $f_s(u)$ is singular.

PROOF. By III.2.22, $f'(u)$ exists a.e. in I_0 and is summable in I_0. Let us put, for $u \in I_0$,

$$f_a(u) = f(a_0) + \int_{a_0}^{u} f'(v) \, dv, \qquad f_s(u) = f(u) - f_a(u). \tag{2}$$

Then $f_a(u)$ is AC and $f_a'(u) = f'(u)$ a.e. in I_0 (see I.3.13). Hence, $f_s(u)$ is continuous and BV in I_0 and $f_s'(u) = 0$ a.e. in I_0. Thus $f_s(u)$ is singular in I_0. Clearly $f_s(a_0) = 0$ and $f(u) = f_a(u) + f_s(u)$. Thus the existence of a representation of the type (1) is established.

To prove the uniqueness of the representation (1), suppose that $f(u) = f_1(u) + f_2(u)$, where $f_1(u)$ is AC and $f_2(u)$ is singular in I_0, and $f_2(a_0) = 0$. Then $f_a(u) - f_1(u) = f_2(u) - f_s(u)$. Let us put

$$g(u) = f_a(u) - f_1(u) = f_2(u) - f_s(u).$$

Then $g(u)$ is both AC and singular in I_0. By the remark at the end of III.3.20

it follows that $g(u) \equiv c =$ constant in I_0. Since $g(a_0) = 0$, it follows finally that $c = 0$, and thus $f_a(u) \equiv f_1(u)$, $f_s(u) \equiv f_2(u)$ in I_0.

III.3.22. Continuation. The representation III.3.21(1) will be termed *the normalized Lebesgue decomposition* of $f(u)$ in I_0. Suppose now that $f(u)$ is represented in the form $f(u) = f_1(u) + f_2(u)$, where $f_1(u)$ is AC and $f_2(u)$ is singular in I_0. By the reasoning employed in III.3.21 it follows that there exists a constant c such that

$$f_1(u) = f_a(u) - c, \qquad f_2(u) = f_s(u) + c,$$

where $f_a(u)$, $f_s(u)$ are given by the formulas III.3.21(2).

III.3.23. Given a vector function $\mathfrak{x}(u)$ which is continuous and BV in $I_0 : a_0 \leqq u \leqq b_0$, we shall say that $\mathfrak{x}(u)$ is singular in I_0 if and only if the components of $\mathfrak{x}(u)$ are singular in I_0. In view of III.3.20, III.3.11, III.3.13 this definition may be restated as follows: if the vector function $\mathfrak{x}(u)$ is continuous and BV in I_0, then $\mathfrak{x}(u)$ is singular in I_0 if and only if $\mathfrak{x}'(u) = 0$ a.e. in I_0.

III.3.24. Suppose that the vector function $\mathfrak{x}(u) = [x(u), y(u), z(u)]$ is continuous and BV in $I_0 : a_0 \leqq u \leqq b_0$. Let $x(u) = x_a(u) + x_s(u)$, $y(u) = y_a(u) + y_s(u)$, $z(u) = z_a(u) + z_s(u)$ be the normalized Lebesgue decompositions of the components of $\mathfrak{x}(u)$ in I_0, in the sense of III.3.22. The representation $\mathfrak{x}(u) = \mathfrak{x}_a(u) + \mathfrak{x}_s(u)$, where $\mathfrak{x}_a(u) = [x_a(u), y_a(u), z_a(u)]$, $\mathfrak{x}_s(u) = [x_s(u), y_s(u), z_s(u)]$ will be termed the normalized Lebesgue decomposition of $\mathfrak{x}(u)$ in I_0. The following statements are immediate consequences of the preceding discussion.

(i) $\mathfrak{x}_a(u)$ is AC in I_0, and $\mathfrak{x}_a'(u) = \mathfrak{x}'(u)$ a.e. in I_0 (see III.3.21, III.3.11).

(ii) $\mathfrak{x}_s(u)$ is singular in I_0, and $\mathfrak{x}_s(a_0) = 0$ (see III.3.23).

(iii) If $\mathfrak{x}(u)$ is represented in the form $\mathfrak{x}(u) = \mathfrak{x}_1(u) + \mathfrak{x}_2(u)$, where $\mathfrak{x}_1(u)$ is AC and $\mathfrak{x}_2(u)$ is singular in I_0, then there exists a constant vector $\mathfrak{a}$ such that $\mathfrak{x}_1(u) = \mathfrak{x}_a(u) - \mathfrak{a}$, $\mathfrak{x}_2(u) = \mathfrak{x}_s(u) + \mathfrak{a}$ in I_0 (cf. III.3.22).

III.3.25. Suppose that the vector function $\mathfrak{x}(u)$ is continuous and BV in $I_0 : a_0 \leqq u \leqq b_0$. Then $\mathfrak{x}(u)$ is singular in I_0 if and only if $L(B, \mathfrak{x})$ is singular in I_0.

Proof. By III.3.13, the derivative of $L(B, \mathfrak{x})$ is equal to $|\, \mathfrak{x}'(u) \,|$ a.e. in I_0. By III.1.33, III.1.52, $L(B, \mathfrak{x})$ is therefore singular in I_0 if and only if $\mathfrak{x}'(u) = 0$ a.e. in I_0, which by III.3.23 is also necessary and sufficient for $\mathfrak{x}(u)$ to be singular in I_0.

III.3.26. *Suppose that the vector function* $\mathfrak{x}(u)$ *is continuous and BV in* I_0 : $a_0 \leqq u \leqq b_0$, *and let* $\mathfrak{x}(u) = \mathfrak{x}_a(u) + \mathfrak{x}_s(u)$ *be its normalized Lebesgue decomposition in* I_0 (see III.3.24). *Let* $L(B, \mathfrak{x}_a)$, $L(B, \mathfrak{x}_s)$ *be the set functions that are associated with* $\mathfrak{x}_a(u)$, $\mathfrak{x}_s(u)$ *in the same manner as* $L(B, \mathfrak{x})$ *is associated with* $\mathfrak{x}(u)$ (see III.3.17). *Finally, let* $L(B, \mathfrak{x}) = L_a(B, \mathfrak{x}) + L_s(B, \mathfrak{x})$ *be the Lebesgue decomposition of* $L(B, \mathfrak{x})$ *in* I_0 (see III.3.17). *We assert the formulas*

$$L_a(B, \mathfrak{x}) = L(B, \mathfrak{x}_a), \qquad L_s(B, \mathfrak{x}) = L(B, \mathfrak{x}_s), \qquad B \subset I_0.$$

Proof. $L(B, \mathfrak{x}_s)$ is singular in I_0 by III.3.25. Hence (see I.3.16) we have in I_0 a Borel set e_s of measure zero, such that

$$L(B, \mathfrak{x}_s) = L(e_s B, \mathfrak{x}_s), \qquad B \subset I_0. \tag{1}$$

On the other hand, since $L(B, \mathfrak{x}_a)$ is AC in I_0 by III.3.18, and $L_a(B, \mathfrak{x})$ is AC in I_0 by definition, we have (see I.3.16)

$$L(e_s B, \mathfrak{x}_a) = 0, \qquad L_a(e_s B, \mathfrak{x}) = 0, \qquad B \subset I_0. \tag{2}$$

In view of III.3.19 there follow the relations

$$\begin{aligned} L(B, \mathfrak{x}_a) &= L(B - e_s, \mathfrak{x}_a) + L(Be_s, \mathfrak{x}_a) = L(B - e_s, \mathfrak{x}_a) \\ &= L(B - e_s, \mathfrak{x} - \mathfrak{x}_s) \leqq L(B - e_s, \mathfrak{x}) + L(B - e_s, \mathfrak{x}_s) \\ &= L(B - e_s, \mathfrak{x}) \end{aligned} \tag{3}$$

since $L(B - e_s, \mathfrak{x}_s) = 0$ by (1) and $L(Be_s, \mathfrak{x}_a) = 0$ by (2). Similarly

$$\begin{aligned} L(B, \mathfrak{x}_s) &= L(Be_s, \mathfrak{x}_s) = L(Be_s, \mathfrak{x} - \mathfrak{x}_a) \leqq L(Be_s, \mathfrak{x}) + L(Be_s, \mathfrak{x}_a) \\ &= L(Be_s, \mathfrak{x}). \end{aligned} \tag{4}$$

Since $L(B - e_s, \mathfrak{x}) + L(Be_s, \mathfrak{x}) = L(B, \mathfrak{x})$, (3) and (4) yield by addition

$$L(B, \mathfrak{x}_a) + L(B, \mathfrak{x}_s) \leqq L(B, \mathfrak{x}). \tag{5}$$

On the other hand, since $\mathfrak{x}_a(u) + \mathfrak{x}_s(u) = \mathfrak{x}(u)$, we have by III.3.19

$$L(B, \mathfrak{x}_a) + L(B, \mathfrak{x}_s) \geqq L(B, \mathfrak{x}). \tag{6}$$

(5) and (6) show that

$$L(B, \mathfrak{x}) = L(B, \mathfrak{x}_a) + L(B, \mathfrak{x}_s), \qquad B \subset I_0. \tag{7}$$

Comparison with the Lebesgue decomposition

$$L(B, \mathfrak{x}) = L_a(B, \mathfrak{x}) + L_s(B, \mathfrak{x}), \qquad B \subset I_0, \tag{8}$$

yields $L(B, \mathfrak{x}_a) = L_a(B, \mathfrak{x})$, $L(B, \mathfrak{x}_s) = L_s(B, \mathfrak{x})$. Indeed, as noted above, $L(B, \mathfrak{x}_a)$ is AC and $L(B, \mathfrak{x}_s)$ is singular, and by I.3.16 the representation of $L(B, \mathfrak{x})$ as a sum of an absolutely continuous and of a singular (completely additive) set function is univocally determined.

III.3.27. Suppose that the vector functions $\mathfrak{x}_1(u)$, $\mathfrak{x}_2(u)$ are continuous and BV in $I_0 : a_0 \leqq u \leqq b_0$. If $\mathfrak{x}_1(u)$ is AC and $\mathfrak{x}_2(u)$ is singular in I_0, then

$$L(B, \mathfrak{x}_1 + \mathfrak{x}_2) = L(B, \mathfrak{x}_1) + L(B, \mathfrak{x}_2), \qquad B \subset I_0.$$

Proof. Put $\mathfrak{x}(u) = \mathfrak{x}_1(u) + \mathfrak{x}_2(u)$, $u \in I_0$. Then $\mathfrak{x}(u)$ is continuous and BV in I_0. Let $\mathfrak{x}(u) = \mathfrak{x}_a(u) + \mathfrak{x}_s(u)$ be the normalized Lebesgue decomposition of $\mathfrak{x}(u)$ in I_0, and let $L(B, \mathfrak{x}) = L_a(B, \mathfrak{x}) + L_s(B, \mathfrak{x})$ be the Lebesgue decomposition of $L(B, \mathfrak{x})$ in I_0. By III.3.24(iii) we have then a constant vector $\mathfrak{a}$ such that $\mathfrak{x}_1(u) = \mathfrak{x}_a(u) - \mathfrak{a}$, $\mathfrak{x}_2(u) = \mathfrak{x}_s(u) + \mathfrak{a}$. By III.3.19, III.3.26 we obtain the formulas $L(B, \mathfrak{x}_1) = L(B, \mathfrak{x}_a) = L_a(B, \mathfrak{x})$, $L(B, \mathfrak{x}_2) = L(B, \mathfrak{x}_s) = L_s(B, \mathfrak{x})$.

Addition yields $L(B, \mathfrak{x}_1) + L(B, \mathfrak{x}_2) = L_a(B, \mathfrak{x}) + L_s(B, \mathfrak{x}) = L(B, \mathfrak{x}) = L(B, \mathfrak{x}_1 + \mathfrak{x}_2)$.

III.3.28. *Suppose that the vector function $\mathfrak{x}(u)$ is continuous and BV in $I_0 : a_0 \leqq u \leqq b_0$. Let $\mathfrak{x}(u) = \mathfrak{x}_a(u) + \mathfrak{x}_s(u)$ be the normalized Lebesgue decomposition of $\mathfrak{x}(u)$ in I_0* (see III.3.24). *Then we have the formula*

$$L(B, \mathfrak{x}) - \int_B |\mathfrak{x}'(u)| \, du = L(B, \mathfrak{x}_s), \qquad B \subset I_0 .$$

PROOF. Since $L(B, \mathfrak{x}_s) = L_s(B, \mathfrak{x})$ by III.3.26, this is an immediate consequence of III.3.17.

III.3.29. We shall apply presently the preceding results to vector functions of the particular form

(1) $$\mathfrak{x} = \mathfrak{x}(u) = [u, f(u), 0], \qquad u \in I_0 : a_0 \leqq u \leqq b_0 ,$$

where $f(u)$ is continuous and BV in I_0. We assert the formula (cf. III.2.13)

(2) $$L(I_0 , \mathfrak{x}) = \int_{I_0} [1 + f'(u)^2]^{1/2} \, du + V(I_0 , f_s).$$

PROOF. Since now $\mathfrak{x}'(u) = [1, f'(u), 0]$, we have by III.3.28

(3) $$L(I_0 , \mathfrak{x}) = \int_{I_0} [1 + f'(u)^2]^{1/2} \, du + L(I_0 , \mathfrak{x}_s).$$

Let $f(u) = f_a(u) + f_s(u)$ be the normalized Lebesgue decomposition of $f(u)$ in I_0. Since now $x(u) \equiv u$ is AC in I_0, it follows that $\mathfrak{x}_s(u) = [0, f_s(u), 0]$, and hence obviously $L(I_0 , \mathfrak{x}_s) = V(I_0 , f_s)$. Thus (3) implies (2).

III.3.30. *Given $\mathfrak{x}(u)$ as in* III.3.29, *we assert that*

(1) $$L(I_0 , \mathfrak{x}) \leqq | I_0 | + V(I_0 , f),$$

the sign of equality holding if and only if $f(u)$ is singular in I_0.

PROOF. Put $\mathfrak{x}_1(u) = (u, 0, 0)$, $\mathfrak{x}_2(u) = [0, f(u), 0]$. Then clearly

(2) $$L(I_0 , \mathfrak{x}_1) = | I_0 |, \qquad L(I_0 , \mathfrak{x}_2) = V(I_0 , f),$$

and thus (1) follows by the inequality of Steiner (see III.3.10). To discuss the sign of equality in (1), suppose first that $f(u)$ is singular in I_0. Since then $\mathfrak{x}_1(u)$ is AC and $\mathfrak{x}_2(u)$ is singular in I_0, we have by (2) and III.3.27

(3) $$L(I_0 , \mathfrak{x}) = | I_0 | + V(I_0 , f).$$

Suppose, conversely, that (3) holds. We assert that (3) implies

(4) $$L(I, \mathfrak{x}) = | I | + V(I, f)$$

for every interval $I : a \leqq u \leqq b$ in I_0. Indeed, let I', I'' denote the intervals $a_0 \leqq u \leqq a$, $b \leqq u \leqq b_0$ respectively (if $a_0 = a$, for example, then I' is to be omitted in the following argument). The reasoning employed to derive (1) yields the inequalities $L(I', \mathfrak{x}) \leqq |I'| + V(I', f)$, $L(I, \mathfrak{x}) \leqq |I| + V(I, f)$, $L(I'', \mathfrak{x}) \leqq |I''| + V(I'', f)$. If (4) is denied, then addition of these inequalities leads to the inequality $L(I_0, \mathfrak{x}) < |I_0| + V(I_0, f)$ (see III.3.13, III.2.19), in contradiction with (3). Thus (4) holds for every interval $I \subset I_0$. Taking the derivatives of the interval functions involved in (4), we obtain the relation (see III.3.13, III.2.22)

$$[1 + f'(u)^2]^{1/2} = 1 + |f'(u)| \qquad \text{a.e. in } I_0,$$

which yields, by squaring, $f'(u) = 0$ a.e. in I_0. Thus $f(u)$ is singular in I_0 (see III.3.20).

III.3.31. Given $\mathfrak{x}(u)$ as in III.3.29, let us add the further assumption that $f(u)$ is monotone (either nonincreasing or else nondecreasing) in I_0. Then clearly $V(I_0, f) = |f(b_0) - f(a_0)|$, and thus we infer from III.3.30 the following statement.

Given $\mathfrak{x}(u) = [u, f(u), 0]$, where $f(u)$ is continuous, monotone, and singular in $I_0 : a_0 \leqq u \leqq b_0$, the quantity $L(I_0, \mathfrak{x})$ depends only upon the values of $f(u)$ at the end points of I_0, and in fact $L(I_0, \mathfrak{x}) = b_0 - a_0 + |f(b_0) - f(a_0)|$.

III.3.32. We proceed to a study of the inequality of Steiner (see III.3.10, III.3.19). If $\mathfrak{x}_1(u)$, $\mathfrak{x}_2(u)$ are vector functions, continuous and BV in $I_0 : a_0 \leqq u \leqq b_0$, of the special form $\mathfrak{x}_1(u) = [u, y_1(u), z_1(u)]$, $\mathfrak{x}_2(u) = [u, y_2(u), z_2(u)]$, then we shall write the inequality of Steiner in the form

$$L\left(I, \frac{\mathfrak{x}_1 + \mathfrak{x}_2}{2}\right) \leqq \frac{1}{2}[L(I, \mathfrak{x}_1) + L(I, \mathfrak{x}_2)], \tag{1}$$

since the vector function $(\mathfrak{x}_1 + \mathfrak{x}_2)/2$ is again of the special form

$$\frac{\mathfrak{x}_1(u) + \mathfrak{x}_2(u)}{2} = \left[u, \frac{y_1(u) + y_2(u)}{2}, \frac{z_1(u) + z_2(u)}{2}\right],$$

where the first component reduces to u. If $\mathfrak{x}_1(u)$, $\mathfrak{x}_2(u)$ are given in the general form $\mathfrak{x}_1(u) = [x_1(u), y_1(u), z_1(u)]$, $\mathfrak{x}_2(u) = [x_2(u), y_2(u), z_2(u)]$, then we shall use the inequality of Steiner in the form

$$L(I, \mathfrak{x}_1 + \mathfrak{x}_2) \leqq L(I, \mathfrak{x}_1) + L(I, \mathfrak{x}_2). \tag{2}$$

Let us note that in (1) and (2) the interval $I \subset I_0$ can be replaced by any Borel set $B \subset I_0$ (see III.3.19). We shall be interested particularly in conditions under which the sign of equality holds in the inequality of Steiner, and more generally in the inferences that may be drawn if the two sides of the inequality of Steiner differ slightly from each other. Let us note that the result in III.3.27 is a relevant contribution from this point of view. We shall discuss presently certain elementary inequalities concerning vectors which will be needed in the sequel.

III.3.33. Let $\mathfrak{a}_1 = (x_1, y_1, z_1)$, $\mathfrak{a}_2 = (x_2, y_2, z_2)$ be vectors in Euclidean three-space. Then

$$|\mathfrak{a}_1| = (x_1^2 + y_1^2 + z_1^2)^{1/2}, \qquad |\mathfrak{a}_2| = (x_2^2 + y_2^2 + z_2^2)^{1/2}$$

are the lengths of these vectors, and $\mathfrak{a}_1\mathfrak{a}_2 = x_1x_2 + y_1y_2 + z_1z_2$ is their scalar product. By the triangle inequality we have $|\mathfrak{a}_1| + |\mathfrak{a}_2| - |\mathfrak{a}_1 + \mathfrak{a}_2| \geqq 0$. We assert that the sign of equality holds if and only if $|\mathfrak{a}_1||\mathfrak{a}_2| - \mathfrak{a}_1\mathfrak{a}_2 = 0$. This follows readily from the identity

$$(|\mathfrak{a}_1| + |\mathfrak{a}_2|)^2 - |\mathfrak{a}_1 + \mathfrak{a}_2|^2 = 2(|\mathfrak{a}_1||\mathfrak{a}_2| - \mathfrak{a}_1\mathfrak{a}_2).$$

III.3.34. CONTINUATION. If $\mathfrak{a}_1 \neq 0$, $\mathfrak{a}_2 \neq 0$, then $|\mathfrak{a}_1||\mathfrak{a}_2| - \mathfrak{a}_1\mathfrak{a}_2 = 0$ if and only if $\mathfrak{a}_1/|\mathfrak{a}_1| = \mathfrak{a}_2/|\mathfrak{a}_2|$. This follows from the identity

$$|\mathfrak{a}_1||\mathfrak{a}_2| - \mathfrak{a}_1\mathfrak{a}_2 = \frac{|\mathfrak{a}_1||\mathfrak{a}_2|}{2}\left(\frac{\mathfrak{a}_1}{|\mathfrak{a}_1|} - \frac{\mathfrak{a}_2}{|\mathfrak{a}_2|}\right)^2.$$

III.3.35. CONTINUATION. If $\mathfrak{a}_1 \neq 0$, $\mathfrak{a}_2 \neq 0$, then $|\mathfrak{a}_1|^2|\mathfrak{a}_2|^2 - (\mathfrak{a}_1\mathfrak{a}_2)^2 = 0$ if and only if $\mathfrak{a}_1/|\mathfrak{a}_1| = \pm\mathfrak{a}_2/|\mathfrak{a}_2|$. This follows from the identity

$$|\mathfrak{a}_1|^2|\mathfrak{a}_2|^2 - (\mathfrak{a}_1\mathfrak{a}_2)^2 = \frac{|\mathfrak{a}_1|^2|\mathfrak{a}_2|^2}{4}\left(\frac{\mathfrak{a}_1}{|\mathfrak{a}_1|} + \frac{\mathfrak{a}_2}{|\mathfrak{a}_2|}\right)^2\left(\frac{\mathfrak{a}_1}{|\mathfrak{a}_1|} - \frac{\mathfrak{a}_2}{|\mathfrak{a}_2|}\right)^2.$$

III.3.36. CONTINUATION. For any two vectors $\mathfrak{a}_1$, $\mathfrak{a}_2$ we have the inequality

$$(1) \quad |\mathfrak{a}_1||\mathfrak{a}_2| - \mathfrak{a}_1\mathfrak{a}_2 \leqq (|\mathfrak{a}_1| + |\mathfrak{a}_2|)(|\mathfrak{a}_1| + |\mathfrak{a}_2| - |\mathfrak{a}_1 + \mathfrak{a}_2|).$$

Indeed, we have the identity

$$|\mathfrak{a}_1||\mathfrak{a}_2| - \mathfrak{a}_1\mathfrak{a}_2$$
$$= (|\mathfrak{a}_1| + |\mathfrak{a}_2| + |\mathfrak{a}_1 + \mathfrak{a}_2|)(|\mathfrak{a}_1| + |\mathfrak{a}_2| - |\mathfrak{a}_1 + \mathfrak{a}_2|)/2.$$

Since $|\mathfrak{a}_1 + \mathfrak{a}_2| \leqq |\mathfrak{a}_1| + |\mathfrak{a}_2|$, the inequality (1) follows.

III.3.37. CONTINUATION. For any two vectors $\mathfrak{a}_1$, $\mathfrak{a}_2$ we have the inequality

$$|\mathfrak{a}_1|^2|\mathfrak{a}_2|^2 - (\mathfrak{a}_1\mathfrak{a}_2)^2 \leqq (|\mathfrak{a}_1| + |\mathfrak{a}_2|)^3(|\mathfrak{a}_1| + |\mathfrak{a}_2| - |\mathfrak{a}_1 + \mathfrak{a}_2|)/2.$$

PROOF. The identity

$$|\mathfrak{a}_1||\mathfrak{a}_2| + \mathfrak{a}_1\mathfrak{a}_2 = [(|\mathfrak{a}_1| + |\mathfrak{a}_2|)^2 - (\mathfrak{a}_1 - \mathfrak{a}_2)^2]/2$$

yields the inequality

$$|\mathfrak{a}_1||\mathfrak{a}_2| + \mathfrak{a}_1\mathfrak{a}_2 \leqq (|\mathfrak{a}_1| + |\mathfrak{a}_2|)^2/2.$$

Multiplication by the inequality III.3.36(1) yields the desired result.

III.3.38. CONTINUATION. For any two vectors $\mathfrak{a}_1 = (x_1, y_1, z_1)$, $\mathfrak{a}_2 = (x_2, y_2, z_2)$ we have the Lagrange identity

$$|\mathfrak{a}_1|^2|\mathfrak{a}_2|^2 - (\mathfrak{a}_1\mathfrak{a}_2)^2 = (y_1z_2 - y_2z_1)^2 + (z_1x_2 - z_2x_1)^2 + (x_1y_2 - x_2y_1)^2.$$

Suppose now that $\mathfrak{a}_1$, $\mathfrak{a}_2$ are of the special form $\mathfrak{a}_1 = (1, y_1, z_1)$, $\mathfrak{a}_2 = (1, y_2, z_2)$. Then $\mathfrak{a}_1 - \mathfrak{a}_2 = (0, y_1 - y_2, z_1 - z_2)$ and the Lagrange identity reduces to the form

$$|\mathfrak{a}_1|^2 |\mathfrak{a}_2|^2 - (\mathfrak{a}_1\mathfrak{a}_2)^2 = (y_1z_2 - y_2z_1)^2 + |\mathfrak{a}_1 - \mathfrak{a}_2|^2.$$

There follows the statement: if $\mathfrak{a}_1 = (1, y_1, z_1)$, $\mathfrak{a}_2 = (1, y_2, z_2)$, then

$$|\mathfrak{a}_1 - \mathfrak{a}_2| \leqq [|\mathfrak{a}_1|^2 |\mathfrak{a}_2|^2 - (\mathfrak{a}_1\mathfrak{a}_2)^2]^{1/2}.$$

III.3.39. Let $\mathfrak{a}_1$, $\mathfrak{a}_2$ be vectors of the special form $\mathfrak{a}_1 = (1, y_1, z_1)$, $\mathfrak{a}_2 = (1, y_2, z_2)$. Consider the quantities

$$|\mathfrak{a}_1 - \mathfrak{a}_2|, \qquad |\mathfrak{a}_1||\mathfrak{a}_2| - \mathfrak{a}_1\mathfrak{a}_2, \qquad |\mathfrak{a}_1|^2|\mathfrak{a}_2|^2 - (\mathfrak{a}_1\mathfrak{a}_2)^2.$$

We assert that if one of these quantities vanishes, then the other two vanish also. The proof follows readily from III.3.34, III.3.35, III.3.38.

III.3.40. Suppose that the vector functions $\mathfrak{x}_1(u)$, $\mathfrak{x}_2(u)$ are continuous and BV in $I_0 : a_0 \leqq u \leqq b_0$. Let B_0 be a Borel set in I_0. If

$$(1) \qquad L(B_0, \mathfrak{x}_1 + \mathfrak{x}_2) = L(B_0, \mathfrak{x}_1) + L(B_0, \mathfrak{x}_2),$$

then we have also

$$(2) \qquad L(B, \mathfrak{x}_1 + \mathfrak{x}_2) = L(B, \mathfrak{x}_1) + L(B, \mathfrak{x}_2)$$

for every Borel set $B \subset B_0$ (cf. III.3.17).

PROOF. By III.3.19 we have the inequalities

$$(3) \qquad L(B, \mathfrak{x}_1 + \mathfrak{x}_2) \leqq L(B, \mathfrak{x}_1) + L(B, \mathfrak{x}_2),$$

$$(4) \qquad L(B_0 - B, \mathfrak{x}_1 + \mathfrak{x}_2) \leqq L(B_0 - B, \mathfrak{x}_1) + L(B_0 - B, \mathfrak{x}_2).$$

Suppose that the sign of inequality holds in (3). Addition of (3) and (4) yields then $L(B_0, \mathfrak{x}_1 + \mathfrak{x}_2) < L(B_0, \mathfrak{x}_1) + L(B_0, \mathfrak{x}_2)$ in contradiction with (1). Thus the sign of equality must hold in (3), and (2) is proved.

III.3.41. Suppose that the vector functions $\mathfrak{x}_1(u)$, $\mathfrak{x}_2(u)$ are continuous and BV in $I_0 : a_0 \leqq u \leqq b_0$. Let $\mathfrak{x}_i(u) = \mathfrak{x}_{ia}(u) + \mathfrak{x}_{is}(u)$, $i = 1, 2$, be the normalized Lebesgue decompositions of $\mathfrak{x}_1(u)$, $\mathfrak{x}_2(u)$ in I_0 (see III.3.24). We assert that the relation

$$(1) \qquad L(I_0, \mathfrak{x}_1 + \mathfrak{x}_2) = L(I_0, \mathfrak{x}_1) + L(I_0, \mathfrak{x}_2)$$

holds if and only if $L(I_0, \mathfrak{x}_{1a} + \mathfrak{x}_{2a}) = L(I_0, \mathfrak{x}_{1a}) + L(I_0, \mathfrak{x}_{2a})$, $L(I_0, \mathfrak{x}_{1s} + \mathfrak{x}_{2s}) = L(I_0, \mathfrak{x}_{1s}) + L(I_0, \mathfrak{x}_{2s})$.

PROOF. Put $d_a = L(I_0, \mathfrak{x}_{1a}) + L(I_0, \mathfrak{x}_{2a}) - L(I_0, \mathfrak{x}_{1a} + \mathfrak{x}_{2a})$, $d_s = L(I_0, \mathfrak{x}_{1s}) + L(I_0, \mathfrak{x}_{2s}) - L(I_0, \mathfrak{x}_{1s} + \mathfrak{x}_{2s})$. Then $d_a \geqq 0$, $d_s \geqq 0$ by III.3.10. By III.3.27 it follows that

$$L(I_0, \mathfrak{x}_1 + \mathfrak{x}_2) = L(I_0, \mathfrak{x}_{1a} + \mathfrak{x}_{2a}) + L(I_0, \mathfrak{x}_{1s} + \mathfrak{x}_{2s})$$

$$= [L(I_0, \mathfrak{x}_{1a}) + L(I_0, \mathfrak{x}_{2a}) - d_a]$$

$$+ [L(I_0, \mathfrak{x}_{1s}) + L(I_0, \mathfrak{x}_{2s}) - d_s] = L(I_0, \mathfrak{x}_1) + L(I_0, \mathfrak{x}_2) - (d_a + d_s).$$

Since $d_a \geqq 0$, $d_s \geqq 0$, we conclude that (1) holds if and only if $d_a = 0$ and $d_s = 0$.

III.3.42. Suppose that the vector functions $\mathfrak{x}_1(u)$, $\mathfrak{x}_2(u)$ are continuous and BV in $I_0 : a_0 \leqq u \leqq b_0$, and $L(I_0, \mathfrak{x}_1 + \mathfrak{x}_2) = L(I_0, \mathfrak{x}_1) + L(I_0, \mathfrak{x}_2)$. Then the following relations hold.

(i) $| \mathfrak{x}_1' | + | \mathfrak{x}_2' | - | \mathfrak{x}_1' + \mathfrak{x}_2' | = 0$ a.e. in I_0.

(ii) $| \mathfrak{x}_1' | \, | \mathfrak{x}_2' | - \mathfrak{x}_1'\mathfrak{x}_2' = 0$ a.e. in I_0.

(iii) $| \mathfrak{x}_1' |^2 \, | \mathfrak{x}_2' |^2 - (\mathfrak{x}_1'\mathfrak{x}_2')^2 = 0$ a.e. in I_0.

PROOF. By III.3.41 we have the relation $L(I_0, \mathfrak{x}_{1a} + \mathfrak{x}_{2a}) = L(I_0, \mathfrak{x}_{1a}) + L(I_0, \mathfrak{x}_{2a})$, which by III.3.16, III.3.24 is equivalent to the relation

$$\int_{I_0} (| \mathfrak{x}_1' | + | \mathfrak{x}_2' | - | \mathfrak{x}_1' + \mathfrak{x}_2' |) \, du = 0.$$

Since the integrand is clearly non-negative a.e. in I_0, the relation (i) follows. Now (ii) follows from (i) by III.3.33, and clearly (iii) is a consequence of (ii).

III.3.43. Suppose that the vector functions $\mathfrak{x}_1(u)$, $\mathfrak{x}_2(u)$ are continuous and BV in $I_0 : a_0 \leqq u \leqq b_0$. Let E denote the subset of I_0 where $\mathfrak{x}_1'$, $\mathfrak{x}_2'$ both exist and are both different from zero. For $u \in E$ we define the vectors t_1, t_2 by the formulas

$$t_1 = \frac{\mathfrak{x}_1'}{| \mathfrak{x}_1' |}, \qquad t_2 = \frac{\mathfrak{x}_2'}{| \mathfrak{x}_2' |}.$$

The vectors t_1, t_2 may be interpreted as the unit tangent vectors of the curves given by the equations $\mathfrak{x} = \mathfrak{x}_1(u)$, $\mathfrak{x} = \mathfrak{x}_2(u)$ respectively. Suppose now that $L(I_0, \mathfrak{x}_1 + \mathfrak{x}_2) = L(I_0, \mathfrak{x}_1) + L(I_0, \mathfrak{x}_2)$. We assert that $t_1 = t_2$ a.e. on the set E.

PROOF. Since $\mathfrak{x}_1' \neq 0$ and $\mathfrak{x}_2' \neq 0$ on E, the assertion follows from III.3.42(ii) and III.3.34.

III.3.44. Suppose that the vector functions $\mathfrak{x}_1(u)$, $\mathfrak{x}_2(u)$ are AC in $I_0 : a_0 \leqq u \leqq b_0$. Then $L(I_0, \mathfrak{x}_1 + \mathfrak{x}_2) = L(I_0, \mathfrak{x}_1) + L(I_0, \mathfrak{x}_2)$ if and only if $| \mathfrak{x}_1' | \, | \mathfrak{x}_2' | - \mathfrak{x}_1'\mathfrak{x}_2' = 0$ a.e. in I_0, or equivalently (see III.3.33), if and only if $| \mathfrak{x}_1' | + | \mathfrak{x}_2' | - | \mathfrak{x}_1' + \mathfrak{x}_2' | = 0$ a.e. in I_0.

PROOF. The necessity follows from III.3.42. Since $\mathfrak{x}_1(u)$, $\mathfrak{x}_2(u)$ are AC, the sufficiency follows from the formula (cf. III.3.16)

$$L(I_0, \mathfrak{x}_1) + L(I_0, \mathfrak{x}_2) - L(I_0, \mathfrak{x}_1 + \mathfrak{x}_2) = \int_{I_0} (| \mathfrak{x}_1' | + | \mathfrak{x}_2' | - | \mathfrak{x}_1' + \mathfrak{x}_2' |) \, du.$$

III.3.45. Suppose that the vector functions $\mathfrak{x}_1(u)$, $\mathfrak{x}_2(u)$ are AC in $I_0 : a_0 \leqq u \leqq b_0$. Let E, t_1, t_2 have the same meaning as in III.3.43. Then the relation $L(I_0, \mathfrak{x}_1 + \mathfrak{x}_2) = L(I_0, \mathfrak{x}_1) + L(I_0, \mathfrak{x}_2)$ holds if and only if $t_1 = t_2$ a.e. on E.

PROOF. The necessity follows from III.3.43. To prove the sufficiency, we note that obviously $| \mathfrak{x}_1' | + | \mathfrak{x}_2' | - | \mathfrak{x}_1' + \mathfrak{x}_2' | = 0$ a.e. on $I_0 - E$. On E itself, we have now $\mathfrak{x}_1'/| \mathfrak{x}_1' | = \mathfrak{x}_2'/| \mathfrak{x}_2' |$ a.e. by assumption, and hence (see III.3.34, III.3.33) we have also $| \mathfrak{x}_1' | + | \mathfrak{x}_2' | - | \mathfrak{x}_1' + \mathfrak{x}_2' | = 0$ a.e. on E. Thus $| \mathfrak{x}_1' | + | \mathfrak{x}_2' | - | \mathfrak{x}_1' + \mathfrak{x}_2' | = 0$ a.e. in I_0, and by III.3.44 the relation $L(I_0, \mathfrak{x}_1 + \mathfrak{x}_2) = L(I_0, \mathfrak{x}_1) + L(I_0, \mathfrak{x}_2)$ follows.

III.3.46. Suppose that the vector functions $\mathfrak{x}_1(u)$, $\mathfrak{x}_2(u)$ are continuous, BV and singular in $I_0 : a_0 \leqq u \leqq b_0$. The corresponding set functions $L(B, \mathfrak{x}_1)$, $L(B, \mathfrak{x}_2)$ are then singular in I_0 by III.3.25. We have therefore, by I.3.16, two Borel sets e_1, e_2 of measure zero in I_0, such that $L(B, \mathfrak{x}_1) = L(e_1B, \mathfrak{x}_1)$, $L(B, \mathfrak{x}_2) = L(e_2B, \mathfrak{x}_2)$ for every Borel set $B \subset I_0$. Let us note that the sets e_1, e_2 are not determined univocally by $\mathfrak{x}_1(u)$, $\mathfrak{x}_2(u)$. Suppose now that e_1, e_2 can be so chosen that $e_1e_2 = 0$. We assert then that $L(I_0, \mathfrak{x}_1 + \mathfrak{x}_2) = L(I_0, \mathfrak{x}_1) + L(I_0, \mathfrak{x}_2)$.

PROOF. By III.3.10 we have $L(I_0, \mathfrak{x}_1 + \mathfrak{x}_2) \leqq L(I_0, \mathfrak{x}_1) + L(I_0, \mathfrak{x}_2)$. The complementary inequality may be obtained, in view of III.3.19(iv) and the assumption $e_1e_2 = 0$, as follows:

$$\begin{aligned} L(I_0, \mathfrak{x}_1 + \mathfrak{x}_2) &\geqq L(e_1 + e_2, \mathfrak{x}_1 + \mathfrak{x}_2) = L(e_1, \mathfrak{x}_1 + \mathfrak{x}_2) + L(e_2, \mathfrak{x}_1 + \mathfrak{x}_2) \\ &\geqq [L(e_1, \mathfrak{x}_1) - L(e_1, \mathfrak{x}_2)] + [L(e_2, \mathfrak{x}_2) - L(e_2, \mathfrak{x}_1)] \\ &= L(I_0, \mathfrak{x}_1) - L(e_2e_1, \mathfrak{x}_2) + L(I_0, \mathfrak{x}_2) - L(e_1e_2, \mathfrak{x}_1) \\ &= L(I_0, \mathfrak{x}_1) + L(I_0, \mathfrak{x}_2). \end{aligned}$$

III.3.47. Let $\mathfrak{x}_1(u)$, $\mathfrak{x}_2(u)$ be vector functions of the special form $\mathfrak{x}_1(u) = [u, y_1(u), z_1(u)]$, $\mathfrak{x}_2(u) = [u, y_2(u), z_2(u)]$, where $y_1(u)$, $z_1(u)$, $y_2(u)$, $z_2(u)$ are continuous and BV in $I_0 : a_0 \leqq u \leqq b_0$. Suppose that (cf. III.3.32)

$$L\left(I_0, \frac{\mathfrak{x}_1 + \mathfrak{x}_2}{2}\right) = [L(I_0, \mathfrak{x}_1) + L(I_0, \mathfrak{x}_2)]/2.$$

Then $\mathfrak{x}_1' = \mathfrak{x}_2'$ a.e. in I_0.

PROOF. Let E, t_1, t_2 have the same meaning as in III.3.43. Since now $|\mathfrak{x}_1'| \geqq 1$, $|\mathfrak{x}_2'| \geqq 1$ wherever $\mathfrak{x}_1'$, $\mathfrak{x}_2'$ exist, it follows by III.3.43 that $t_1 = t_2$ a.e. in I_0. Hence $\mathfrak{x}_1' = \mathfrak{x}_2'$ a.e. in I_0 by III.3.34, III.3.39.

III.3.48. Given $\mathfrak{x}_1(u)$, $\mathfrak{x}_2(u)$ as in III.3.47, let us add the assumption that $\mathfrak{x}_1(u)$, $\mathfrak{x}_2(u)$ are AC in I_0. Then the relation

$$L\left(I_0, \frac{\mathfrak{x}_1 + \mathfrak{x}_2}{2}\right) = \frac{1}{2}[L(I_0, \mathfrak{x}_1) + L(I_0, \mathfrak{x}_2)] \tag{1}$$

holds if and only if $\mathfrak{x}_1(u) - \mathfrak{x}_2(u)$ is constant in I_0.

PROOF. (i) Suppose that $\mathfrak{x}_1(u) - \mathfrak{x}_2(u) = \mathfrak{a} =$ constant in I_0. Then (see III.3.19)

$$L\left(I_0, \frac{\mathfrak{x}_1 + \mathfrak{x}_2}{2}\right) = L\left(I_0, \mathfrak{x}_2 + \frac{\mathfrak{a}}{2}\right) = L(I_0, \mathfrak{x}_2),$$

$$L(I_0, \mathfrak{x}_1) = L(I_0, \mathfrak{x}_2 + \mathfrak{a}) = L(I_0, \mathfrak{x}_2),$$

and hence (1) holds.

(ii) Suppose that (1) holds. Then $\mathfrak{x}_1' = \mathfrak{x}_2'$ a.e. in I_0 by III.3.47. Since $\mathfrak{x}_1(u)$, $\mathfrak{x}_2(u)$ are both AC in I_0, it follows that the vector function $\mathfrak{x}_1(u) - \mathfrak{x}_2(u)$ is both AC and singular in I_0. Hence $\mathfrak{x}_1(u) - \mathfrak{x}_2(u)$ is constant in I_0 (see III.3.20).

III.3.49. Suppose that the vector functions $\mathfrak{x}_1(u)$, $\mathfrak{x}_2(u)$ are continuous and BV in $I_0 : a_0 \leqq u \leqq b_0$. Then

$$(1)\quad L(I, \mathfrak{x}_1) + L(I, \mathfrak{x}_2) - L(I, \mathfrak{x}_1 + \mathfrak{x}_2) \geqq \int_I (|\mathfrak{x}_1'| + |\mathfrak{x}_2'| - |\mathfrak{x}_1' + \mathfrak{x}_2'|)\, du$$

for every interval $I \subset I_0$.

Proof. Consider the interval function $\psi(I) = L(I, \mathfrak{x}_1) + L(I, \mathfrak{x}_2) - L(I, \mathfrak{x}_1 + \mathfrak{x}_2)$. Then $\psi(I) \geqq 0$ by III.3.10, and $\psi(I)$ is additive by III.3.13. Furthermore, the derivative of $\psi(I)$ is equal to $|\mathfrak{x}_1'| + |\mathfrak{x}_2'| - |\mathfrak{x}_1' + \mathfrak{x}_2'|$ a.e. in I_0 (cf. III.3.13). Thus (1) follows by III.1.29, III.1.52.

III.3.50. Given $\mathfrak{x}_1(u)$, $\mathfrak{x}_2(u)$ as in III.3.49, we assert the inequalities

$$(1)\quad \left[\int_I (|\mathfrak{x}_1'|\,|\mathfrak{x}_2'| - \mathfrak{x}_1'\mathfrak{x}_2')^{1/2}\, du\right]^2$$
$$\leqq [L(I, \mathfrak{x}_1) + L(I, \mathfrak{x}_2)][L(I, \mathfrak{x}_1) + L(I, \mathfrak{x}_2) - L(I, \mathfrak{x}_1 + \mathfrak{x}_2)],$$

$$(2)\quad \left\{\int_I [|\mathfrak{x}_1'|^2\,|\mathfrak{x}_2'|^2 - (\mathfrak{x}_1'\mathfrak{x}_2')^2]^{1/4}\, du\right\}^4$$
$$\leqq [L(I, \mathfrak{x}_1) + L(I, \mathfrak{x}_2)]^3[L(I, \mathfrak{x}_1) + L(I, \mathfrak{x}_2) - L(I, \mathfrak{x}_1 + \mathfrak{x}_2)]/2.$$

Proof. By III.3.36 we have the inequality

$$(|\mathfrak{x}_1'|\,|\mathfrak{x}_2'| - \mathfrak{x}_1'\mathfrak{x}_2')^{1/2} \leqq (|\mathfrak{x}_1'| + |\mathfrak{x}_2'|)^{1/2}(|\mathfrak{x}_1'| + |\mathfrak{x}_2'| - |\mathfrak{x}_1' + \mathfrak{x}_2'|)^{1/2}$$
$$\text{a.e. in } I_0 .$$

Integration yields, in view of the Hölder inequality with $p = q = 2$ (see I.3.10),

$$\left[\int_I (|\mathfrak{x}_1'|\,|\mathfrak{x}_2'| - \mathfrak{x}_1'\mathfrak{x}_2')^{1/2}\, du\right]^2$$
$$\leqq \left[\int_I |\mathfrak{x}_1'|\, du + \int_I |\mathfrak{x}_2'|\, du\right]\left[\int_I (|\mathfrak{x}_1'| + |\mathfrak{x}_2'| - |\mathfrak{x}_1' + \mathfrak{x}_2'|)\, du\right],$$

and (1) follows by III.3.13, III.3.49. The proof of (2) is similar, starting with the inequality (see III.3.37)

$$[|\mathfrak{x}_1'|^2\,|\mathfrak{x}_2'|^2 - (\mathfrak{x}_1'\mathfrak{x}_2')^2]^{1/4} \leqq 2^{-1/4}(|\mathfrak{x}_1'| + |\mathfrak{x}_2'|)^{3/4}$$
$$(|\mathfrak{x}_1'| + |\mathfrak{x}_2'| - |\mathfrak{x}_1' + \mathfrak{x}_2'|)^{1/4},$$

and then applying the Hölder inequality with $p = 4/3$, $q = 4$.

III.3.51. Continuation. The inequalities derived in III.3.50 yield lower bounds for the quantity $L(I_0, \mathfrak{x}_1) + L(I_0, \mathfrak{x}_2) - L(I_0, \mathfrak{x}_1 + \mathfrak{x}_2)$ in the following manner. Let S be a measurable subset of I_0, and let ϵ, δ, M be constants such that $|S| \geqq \delta \geqq 0$, $|\mathfrak{x}_1'|\,|\mathfrak{x}_2'| - \mathfrak{x}_1'\mathfrak{x}_2' \geqq \epsilon \geqq 0$ on S, and $L(I_0, \mathfrak{x}_1) \leqq M$, $L(I_0, \mathfrak{x}_2) \leqq M$. Then we have the inequality

$$(1) \qquad L(I_0, \mathfrak{x}_1) + L(I_0, \mathfrak{x}_2) - L(I_0, \mathfrak{x}_1 + \mathfrak{x}_2) \geqq \epsilon\delta^2/2M.$$

Indeed, obviously

$$\int_{I_0} (|\mathfrak{x}_1'|\,|\mathfrak{x}_2'| - \mathfrak{x}_1'\mathfrak{x}_2')^{1/2}\,du \geqq \int_S (|\mathfrak{x}_1'|\,|\mathfrak{x}_2'| - \mathfrak{x}_1'\mathfrak{x}_2')^{1/2}\,du \geqq \epsilon^{1/2}\delta.$$

By III.3.50(1) it follows that

$$\epsilon\delta^2 \leqq 2M[L(I_0, \mathfrak{x}_1) + L(I_0, \mathfrak{x}_2) - L(I_0, \mathfrak{x}_1 + \mathfrak{x}_2)],$$

and (1) is proved. Next, let S_* be a measurable subset of I_0, and let ϵ_*, δ_*, M be constants such that $|S_*| \geqq \delta_* \geqq 0$, $|\mathfrak{x}_1'|^2\,|\mathfrak{x}_2'|^2 - (\mathfrak{x}_1'\mathfrak{x}_2')^2 \geqq \epsilon_* \geqq 0$ on S_*, and $L(I_0, \mathfrak{x}_1) \leqq M$, $L(I_0, \mathfrak{x}_2) \leqq M$. An analogous reasoning, using III.3.50(2), yields the inequality

$$(2) \qquad L(I_0, \mathfrak{x}_1) + L(I_0, \mathfrak{x}_2) - L(I_0, \mathfrak{x}_1 + \mathfrak{x}_2) \geqq \epsilon_*\delta_*^4/4M^3.$$

The inequalities (1), (2) may be considered as refinements of the results stated in III.3.42.

III.3.52. Suppose that the vector functions $\mathfrak{x}_1(u)$, $\mathfrak{x}_2(u)$ are of the special form $\mathfrak{x}_1(u) = [u, y_1(u), z_1(u)]$, $\mathfrak{x}_2(u) = [u, y_2(u), z_2(u)]$, where $y_1(u)$, $z_1(u)$, $y_2(u)$, $z_2(u)$ are continuous and BV in $I_0 : a_0 \leqq u \leqq b_0$. Since then $|\mathfrak{x}_1' - \mathfrak{x}_2'| \leqq |\mathfrak{x}_1'|^2\,|\mathfrak{x}_2'|^2 - (\mathfrak{x}_1'\mathfrak{x}_2')^2$ by III.3.38, we obtain from III.3.50, III.3.51 the following statements.

(i) For every interval $I \subset I_0$ we have the inequality

$$(1) \qquad \left[\int_I |\mathfrak{x}_1' - \mathfrak{x}_2'|^{1/2}\,du\right]^4 \leqq [L(I, \mathfrak{x}_1) + L(I, \mathfrak{x}_2)]^3[L(I, \mathfrak{x}_1) + L(I, \mathfrak{x}_2) - L(I, \mathfrak{x}_1 + \mathfrak{x}_2)]/2.$$

(ii) Let G be a measurable subset of I_0, and let μ, η, M be constants such that $|G| \geqq \mu \geqq 0$, $|\mathfrak{x}_1' - \mathfrak{x}_2'| \geqq \eta \geqq 0$ on G, and $L(I_0, \mathfrak{x}_1) \leqq M$, $L(I_0, \mathfrak{x}_2) \leqq M$. Then

$$(2) \qquad L(I_0, \mathfrak{x}_1) + L(I_0, \mathfrak{x}_2) - L(I_0, \mathfrak{x}_1 + \mathfrak{x}_2) \geqq \eta^2\mu^4/4M^3.$$

Remark. In view of III.3.10, the inequalities (1), (2) may be written in the form

$$(3) \qquad \left[\int_I |\mathfrak{x}_1' - \mathfrak{x}_2'|^{1/2}\,du\right]^4 \leqq [L(I, \mathfrak{x}_1) + L(I, \mathfrak{x}_2)]^3\left[\frac{L(I, \mathfrak{x}_1) + L(I, \mathfrak{x}_2)}{2} - L\left(I, \frac{\mathfrak{x}_1 + \mathfrak{x}_2}{2}\right)\right],$$

(4) $$\frac{L(I_0, \mathfrak{x}_1) + L(I_0, \mathfrak{x}_2)}{2} - L\left(I_0, \frac{\mathfrak{x}_1 + \mathfrak{x}_2}{2}\right) \geqq \frac{\eta^2\mu^4}{8M^3}.$$

III.3.53. Let $\mathfrak{x}_n(u)$, $n = 0, 1, 2, \cdots$, be a sequence of vector functions which are continuous and BV in $I_0 : a_0 \leqq u \leqq b_0$. Suppose that $\mathfrak{x}_n(u) \to \mathfrak{x}_0(u)$ uniformly in I_0. Then (see III.3.6) $L(I_0, \mathfrak{x}_0) \leqq \liminf L(I_0, \mathfrak{x}_n)$ for $n \to \infty$. Simple examples show that the relation $L(I_0, \mathfrak{x}_n) \to L(I_0, \mathfrak{x}_0)$ is not implied by the uniform convergence of $\mathfrak{x}_n(u)$ to $\mathfrak{x}_0(u)$.

DEFINITION. We shall say that $\mathfrak{x}_n(u)$ converges *in length* to $\mathfrak{x}_0(u)$, in symbols $\mathfrak{x}_n \to \mathfrak{x}_0(L)$, in I_0, if and only if the following conditions hold. (i) $\mathfrak{x}_n(u)$ is continuous and BV in I_0, $n = 0, 1, 2, \cdots$. (ii) $\mathfrak{x}_n(u) \to \mathfrak{x}_0(u)$ uniformly in I_0. (iii) $L(I_0, \mathfrak{x}_n) \to L(I_0, \mathfrak{x}_0)$.

We shall study presently some of the implications of the relation $\mathfrak{x}_n \to \mathfrak{x}_0(L)$.

III.3.54. CONTINUATION. If $\mathfrak{x}_n \to \mathfrak{x}_0(L)$ in I_0, then $\mathfrak{x}_n \to \mathfrak{x}_0(L)$ in every interval $I \subset I_0$.

PROOF. Given $I \subset I_0$, we can construct a subdivision $D(I_0)$ of I_0, such that I coincides with one of the intervals $I^1, \cdots, I^k$ of $D(I_0)$. Let us put

$$\epsilon_j = L(I^j, \mathfrak{x}_0) - \liminf_{n\to\infty} L(I^j, \mathfrak{x}_n), \qquad j = 1, \cdots, k.$$

We have then, since $\mathfrak{x}_n \to \mathfrak{x}_0(L)$ in I_0,

$$L(I_0, \mathfrak{x}_0) = \lim_{n\to\infty} L(I_0, \mathfrak{x}_n) = \lim_{n\to\infty} \sum_{j=1}^{k} L(I^j, \mathfrak{x}_n) \geqq \sum_{j=1}^{k} [\liminf_{n\to\infty} L(I^j, \mathfrak{x}_n)]$$

$$= \sum_{j=1}^{k} [L(I^j, \mathfrak{x}_0) - \epsilon_j] = L(I_0, \mathfrak{x}_0) - \sum_{j=1}^{k} \epsilon_j .$$

Since $\epsilon_j \leqq 0$ by III.3.6, it follows that $\epsilon_1 = \epsilon_2 = \cdots = \epsilon_k = 0$, and thus, in particular, $L(I, \mathfrak{x}_0) = \liminf L(I, \mathfrak{x}_n)$. Since the same reasoning applies to every infinite subsequence of the sequence $\mathfrak{x}_1(u), \mathfrak{x}_2(u), \cdots$, the relation $L(I, \mathfrak{x}_0) = \lim L(I, \mathfrak{x}_n)$ follows.

III.3.55. Given a sequence of (real-valued, scalar) functions $f_n(u)$, $n = 0, 1, 2, \cdots$, in $I_0 : a_0 \leqq u \leqq b_0$, we shall say that $f_n(u)$ converges *in variation* to $f_0(u)$, in symbols $f_n \to f_0(V)$, in I_0 if and only if the following conditions hold. (i) $f_n(u)$ is continuous and BV in I_0, $n = 0, 1, 2, \cdots$. (ii) $f_n(u) \to f_0(u)$ uniformly in I_0. (iii) $V(I_0, f_n) \to V(I_0, f_0)$ (see III.2.13).

III.3.56. Given a sequence of vector functions $\mathfrak{x}_n(u) = [x_n(u), y_n(u), z_n(u)]$, $n = 0, 1, 2, \cdots$, in $I_0 : a_0 \leqq u \leqq b_0$, we shall say that $\mathfrak{x}_n(u)$ converges in variation to $\mathfrak{x}_0(u)$, in symbols $\mathfrak{x}_n \to \mathfrak{x}_0(V)$, in I_0 if and only if $x_n \to x_0(V)$, $y_n \to y_0(V)$, $z_n \to z_0(V)$ in I_0 (see III.3.55).

III.3.57. Given a vector function $\mathfrak{x}(u) = [x(u), y(u), z(u)]$, we shall use the notations $\mathfrak{x}^x(u)$, $\mathfrak{x}^y(u)$, $\mathfrak{x}^z(u)$ to refer to the vector functions $[0, y(u), z(u)]$, $[x(u), 0, z(u)]$, $[x(u), y(u), 0]$ respectively.

THEOREM. *If* $\mathfrak{x}_n \to \mathfrak{x}_0(L)$ *in* I_0 (*see* III.3.53), *then*

$$(1) \qquad \mathfrak{x}_n^x \to \mathfrak{x}_0^x(L), \qquad \mathfrak{x}_n^y \to \mathfrak{x}_0^y(L), \qquad \mathfrak{x}_n^z \to \mathfrak{x}_0^z(L) \qquad \textit{in } I_0 .$$

PROOF. Let I be a generic notation for an interval in I_0. We put (see III.3.2)

$$\alpha_n(I) = g_1(I, \mathfrak{x}_n), \qquad \beta_n(I) = g_2(I, \mathfrak{x}_n),$$

$$\gamma_n(I) = g_3(I, \mathfrak{x}_n), \qquad \sigma_n(I) = g(I, \mathfrak{x}_n),$$

$$\lambda_n(I) = g(I, \mathfrak{x}_n^x), \qquad \mu_n(I) = g(I, \mathfrak{x}_n^y), \qquad \nu_n(I) = g(I, \mathfrak{x}_n^z).$$

We have then the relations (see III.3.4, III.1.5, III.1.52)

$$L(I, \mathfrak{x}_n) = U(I, \sigma_n), \qquad L(I, \mathfrak{x}_n^x) = U(I, \lambda_n),$$

$$L(I, \mathfrak{x}_n^y) = U(I, \mu_n), \qquad L(I, \mathfrak{x}_n^z) = U(I, \nu_n).$$

Clearly, the interval functions $\alpha_n(I)$, $\beta_n(I)$, $\gamma_n(I)$, $\sigma_n(I)$, $\lambda_n(I)$, $\mu_n(I)$, $\nu_n(I)$ satisfy the assumptions in III.1.10 (cf. III.1.52), and the relations (1) follow.

III.3.58. CONTINUATION. In fact, III.3.10 yields the further relations $U(I_0, \alpha_n) \to U(I_0, \alpha_0)$, $U(I_0, \beta_n) \to U(I_0, \beta_0)$, $U(I_0, \gamma_n) \to U(I_0, \gamma_0)$. Since clearly $U(I_0, \alpha_n) = V(I_0, x_n)$, $U(I_0, \beta_n) = V(I_0, y_n)$, $U(I_0, \gamma_n) = V(I_0, z_n)$ (cf. III.2.13), we obtain the following statement (cf. III.3.56).

THEOREM. *If* $\mathfrak{x}_n \to \mathfrak{x}_0(L)$ *in* I_0, *then* $\mathfrak{x}_n \to \mathfrak{x}_0(V)$ *in* I_0.

REMARK. Simple examples show that the converse is generally false.

III.3.59. If $\mathfrak{x}_n \to \mathfrak{x}_0(L)$ in I_0 (see III.3.53), then

$$(1) \qquad L(I_0, \mathfrak{x}_n) + L(I_0, \mathfrak{x}_0) - L(I_0, \mathfrak{x}_n + \mathfrak{x}_0) \to 0.$$

PROOF. Since $\mathfrak{x}_n(u) \to \mathfrak{x}_0(u)$ uniformly in I_0, we have by III.3.6 the inequality

$$(2) \qquad \liminf L(I_0, \mathfrak{x}_n + \mathfrak{x}_0) \geqq L(I_0, 2\mathfrak{x}_0) = 2L(I_0, \mathfrak{x}_0).$$

By the inequality of Steiner we have

$$(3) \qquad 0 \leqq L(I_0, \mathfrak{x}_n) + L(I_0, \mathfrak{x}_0) - L(I_0, \mathfrak{x}_n + \mathfrak{x}_0).$$

From (2) and (3) we infer, in view of the assumption $\mathfrak{x}_n \to \mathfrak{x}_0(L)$,

$$0 \leqq \limsup\, [L(I_0, \mathfrak{x}_n) + L(I_0, \mathfrak{x}_0) - L(I_0, \mathfrak{x}_n + \mathfrak{x}_0)]$$

$$= 2L(I_0, \mathfrak{x}_0) - \liminf L(I_0, \mathfrak{x}_n + \mathfrak{x}_0) \leqq 0,$$

and (1) follows.

III.3.60. If $\mathfrak{x}_n \to \mathfrak{x}_0(L)$ in I_0 (see III.3.53), then

$$(1) \qquad \int_{I_0} (|\,\mathfrak{x}_n'\,| + |\,\mathfrak{x}_0'\,| - |\,\mathfrak{x}_n' + \mathfrak{x}_0'\,|)\, du \to 0,$$

$$(2)\qquad \int_{I_0} (|\,\mathfrak{x}_n'\,|\,|\,\mathfrak{x}_0'\,| - \mathfrak{x}_n'\mathfrak{x}_0')^{1/2}\,du \quad \to 0,$$

$$(3)\qquad \int_{I_0} [|\,\mathfrak{x}_n'\,|^2\,|\,\mathfrak{x}_0'\,|^2 - (\mathfrak{x}_n'\mathfrak{x}_0')^2]^{1/4}\,du \to 0.$$

PROOF. By III.3.49 we have the inequality

$$\int_{I_0} (|\,\mathfrak{x}_n'\,| + |\,\mathfrak{x}_0'\,| - |\,\mathfrak{x}_n' + \mathfrak{x}_0'\,|)\,du$$

$$\leqq L(I_0\,,\mathfrak{x}_n) + L(I_0\,,\mathfrak{x}_0) - L(I_0\,,\mathfrak{x}_n + \mathfrak{x}_0),$$

and (1) follows in view of III.3.59. The relations (2) and (3) are established in a similar manner, using III.3.50.

III.3.61. If $f_n(u)$ is a sequence of (real-valued scalar) measurable functions in $I_0 : a_0 \leqq u \leqq b_0$, and $f_n(u)$ converges in measure to $f_0(u)$ in I_0, then we shall write: $f_n \to f_0(M)$ in I_0. In view of I.3.12, III.3.60 we have then the following statement: if $\mathfrak{x}_n \to \mathfrak{x}_0(L)$ in I_0, then also $|\,\mathfrak{x}_n'\,| + |\,\mathfrak{x}_0'\,| - |\,\mathfrak{x}_n' + \mathfrak{x}_0'\,| \to 0(M)$, $|\,\mathfrak{x}_n'\,|\,|\,\mathfrak{x}_0'\,| - \mathfrak{x}_n'\mathfrak{x}_0' \to 0(M)$, $|\,\mathfrak{x}_n'\,|^2\,|\,\mathfrak{x}_0'\,|^2 - (\mathfrak{x}_n'\mathfrak{x}_0')^2 \to 0(M)$ in I_0.

III.3.62. Suppose that the vector functions $\mathfrak{x}_n(u)$, $n = 0, 1, 2, \cdots$, satisfy the following conditions in $I_0 : a_0 \leqq u \leqq b_0$. (i) $\mathfrak{x}_n(u)$ is AC in I_0 for $n = 1, 2, \cdots$. (ii) $\mathfrak{x}_0(u)$ is continuous and BV in I_0. (iii) $\mathfrak{x}_n(u) \to \mathfrak{x}_0(u)$ uniformly in I_0. Then

$$(1)\qquad \liminf_{n\to\infty} \int_{I_0} |\,\mathfrak{x}_n' - \mathfrak{x}_0'\,|\,du \geqq L(I_0\,,\mathfrak{x}_0) - \int_{I_0} |\,\mathfrak{x}_0'\,|\,du.$$

PROOF. By III.3.16 and condition (i) we have, for $n = 1, 2, \cdots$,

$$\int_{I_0} |\,\mathfrak{x}_n' - \mathfrak{x}_0'\,|\,du \geqq \int_{I_0} |\,\mathfrak{x}_n'\,|\,du - \int_{I_0} |\,\mathfrak{x}_0'\,|\,du = L(I_0\,,\mathfrak{x}_n) - \int_{I_0} |\,\mathfrak{x}_0'\,|\,du,$$

and (1) follows by III.3.6.

III.3.63. Suppose that the vector functions $\mathfrak{x}_n(u)$, $n = 0, 1, 2, \cdots$, satisfy the following conditions in $I_0 : a_0 \leqq u \leqq b_0$. (i) $\mathfrak{x}_n(u)$ is AC in I_0 for $n = 1, 2, \cdots$. (ii) $\mathfrak{x}_0(u)$ is continuous and BV in I_0. (iii) $\mathfrak{x}_n(u) \to \mathfrak{x}_0(u)$ uniformly in I_0. (iv) We have the relation (cf. III.3.62)

$$(1)\qquad \lim_{n\to\infty} \int_{I_0} |\,\mathfrak{x}_n' - \mathfrak{x}_0'\,|\,du = L(I_0\,,\mathfrak{x}_0) - \int_{I_0} |\,\mathfrak{x}_0'\,|\,du.$$

Then $\mathfrak{x}_n \to \mathfrak{x}_0(L)$ in I_0 (see III.3.53).

PROOF. By III.3.16 and condition (i) we have, for $n = 1, 2, \cdots$,

$$L(I_0\,,\mathfrak{x}_n) = \int_{I_0} |\,\mathfrak{x}_n'\,|\,du \leqq \int_{I_0} |\,\mathfrak{x}_n' - \mathfrak{x}_0'\,|\,du + \int_{I_0} |\,\mathfrak{x}_0'\,|\,du.$$

In view of (1) it follows that lim sup $L(I_0, \mathfrak{x}_n) \leqq L(I_0, \mathfrak{x}_0)$. Since $L(I_0, \mathfrak{x}_0) \leqq$ lim inf $L(I_0, \mathfrak{x}_n)$ by III.3.6, the desired relation $L(I_0, \mathfrak{x}_n) \to L(I_0, \mathfrak{x}_0)$ is established.

III.3.64. Suppose that the vector functions $\mathfrak{x}_n(u)$, $n = 0, 1, 2, \cdots$, satisfy the following conditions in $I_0 : a_0 \leqq u \leqq b_0$. (i) $\mathfrak{x}_n(u)$ is AC in I_0 for $n = 1, 2, \cdots$. (ii) $\mathfrak{x}_0(u)$ is continuous and BV in I_0. (iii) $\mathfrak{x}_n(u) \to \mathfrak{x}_0(u)$ uniformly in I_0. (iv) We have the relation

$$(1) \qquad \lim_{n\to\infty} \int_{I_0} | \mathfrak{x}_n' | \, du = \int_{I_0} | \mathfrak{x}_0' | \, du.$$

Then $\mathfrak{x}_n \to \mathfrak{x}_0(L)$ in I_0, and $\mathfrak{x}_0$ is AC in I_0.

PROOF. In view of III.3.16, the relation (1) yields

$$(2) \qquad \lim_{n\to\infty} L(I_0, \mathfrak{x}_n) = \lim_{n\to\infty} \int_{I_0} | \mathfrak{x}_n' | \, du = \int_{I_0} | \mathfrak{x}_0' | \, du \leqq L(I_0, \mathfrak{x}_0).$$

On the other hand, by III.3.6,

$$(3) \qquad L(I_0, \mathfrak{x}_0) \leqq \liminf_{n\to\infty} L(I_0, \mathfrak{x}_n).$$

From (2) and (3) we infer that $L(I_0, \mathfrak{x}_n) \to L(I_0, \mathfrak{x}_0)$. Furthermore, (1), (2), (3) imply that

$$\int_{I_0} | \mathfrak{x}_0' | \, du = L(I_0, \mathfrak{x}_0).$$

Hence $\mathfrak{x}_0(u)$ is AC in I_0 by III.3.16.

III.3.65. Given a sequence of vector functions $\mathfrak{x}_n(u)$, $n = 0, 1, 2, \cdots$, in $I_0 : a_0 \leqq u \leqq b_0$, we shall say that $\mathfrak{x}_n(u)$ converges *strongly in length* to $\mathfrak{x}_0(u)$, in symbols $\mathfrak{x}_n \to \mathfrak{x}_0(\text{SL})$, in I_0 if and only if the following conditions hold. (i) $\mathfrak{x}_n(u)$ is continuous and BV in I_0 for $n = 0, 1, 2, \cdots$. (ii) $\mathfrak{x}_n(u) \to \mathfrak{x}_0(u)$ uniformly in I_0. (iii) $L(I_0, \mathfrak{x}_n - \mathfrak{x}_0) \to 0$.

REMARK 1. If $\mathfrak{x}_n \to \mathfrak{x}_0(\text{SL})$ in I_0, then $\mathfrak{x}_n \to \mathfrak{x}_0(L)$ in I_0. Indeed, by the inequality of Steiner, $L(I_0, \mathfrak{x}_n) \leqq L(I_0, \mathfrak{x}_0) + L(I_0, \mathfrak{x}_n - \mathfrak{x}_0)$, and thus we have lim sup $L(I_0, \mathfrak{x}_n) \leqq L(I_0, \mathfrak{x}_0)$. On the other hand, $L(I_0, \mathfrak{x}_0) \leqq$ lim inf $L(I_0, \mathfrak{x}_n)$ by III.3.6. Hence $L(I_0, \mathfrak{x}_n) \to L(I_0, \mathfrak{x}_0)$.

REMARK 2. Simple examples show that the converse of the preceding remark is generally false.

III.3.66. Given a sequence of vector functions $\mathfrak{x}_n(u)$, $n = 0, 1, 2, \cdots$, in $I_0 : a_0 \leqq u \leqq b_0$, suppose that the following conditions hold. (i) $\mathfrak{x}_n(u)$ is AC in I_0 for $n = 1, 2, \cdots$. (ii) $\mathfrak{x}_0(u)$ is continuous and BV in I_0. (iii) $\mathfrak{x}_n \to \mathfrak{x}_0$ (SL) in I_0. Then $\mathfrak{x}_0(u)$ is AC in I_0.

PROOF. By III.3.16 we have, for $n = 1, 2, \cdots$,

$$L(I_0, \mathfrak{x}_n) = \int_{I_0} |\mathfrak{x}_n'| \, du \leqq \int_{I_0} |\mathfrak{x}_0'| \, du + \int_{I_0} |\mathfrak{x}_n' - \mathfrak{x}_0'| \, du$$

(1)

$$\leqq \int_{I_0} |\mathfrak{x}_0'| \, du + L(I_0, \mathfrak{x}_n - \mathfrak{x}_0) \leqq L(I_0, \mathfrak{x}_0) + L(I_0, \mathfrak{x}_n - \mathfrak{x}_0).$$

By III.3.65, Remark 1, $L(I_0, \mathfrak{x}_n) \to L(I_0, \mathfrak{x}_0)$. Thus (1) yields, for $n \to \infty$,

$$L(I_0, \mathfrak{x}_0) = \int_{I_0} |\mathfrak{x}_0'| \, du.$$

Hence $\mathfrak{x}_0(u)$ is AC in I_0 by III.3.16.

III.3.67. Given a sequence of vector functions $\mathfrak{x}_n(u)$, $n = 0, 1, 2, \cdots$, in $I_0 : a_0 \leqq u \leqq b_0$, suppose that the following conditions hold. (i) $\mathfrak{x}_n(u)$ is AC in I_0, $n = 0, 1, 2, \cdots$. (ii) $\mathfrak{x}_n \to \mathfrak{x}_0(L)$ in I_0. (iii) $\mathfrak{x}_n' \to \mathfrak{x}_0'$ a.e. in I_0. Then $\mathfrak{x}_n \to \mathfrak{x}_0(\mathrm{SL})$ in I_0.

Proof. By III.3.16 we have the formulas

(1) $$L(I_0, \mathfrak{x}_n - \mathfrak{x}_0) = \int_{I_0} |\mathfrak{x}_n' - \mathfrak{x}_0'| \, du, \qquad L(I_0, \mathfrak{x}_n) = \int_{I_0} |\mathfrak{x}_n'| \, du$$

$$n = 0, 1, 2, \cdots$$

In view of condition (ii), we infer from (1) that

(2) $$\int_{I_0} |\mathfrak{x}_n'| \, du \to \int_{I_0} |\mathfrak{x}_0'| \, du.$$

In view of condition (iii), (2) yields (see I.3.11)

(3) $$\int_{I_0} |\mathfrak{x}_n' - \mathfrak{x}_0'| \, du \to 0.$$

(1) and (3) show that $L(I_0, \mathfrak{x}_n - \mathfrak{x}_0) \to 0$.

III.3.68. Suppose that the vector function $\mathfrak{x}(u)$ is continuous and BV in $I_0 : a_0 \leqq u \leqq b_0$. Let $D_n(I_0)$ be a sequence of subdivisions of I_0 such that $\| D_n(I_0) \| \to 0$ (see III.1.4, III.1.52). For each n, let $q_n(u)$ be the quasi-linear vector function defined as follows in I_0 : the components of $q_n(u)$ are linear in each interval of $D_n(I_0)$ and at the end points of these intervals agree with the corresponding components of $\mathfrak{x}(u)$. Geometrically, one may think of the vector functions $q_n(u)$ as determining inscribed polygons of approximation for the curve given by $\mathfrak{x} = \mathfrak{x}(u)$, $u \in I_0$. In view of this interpretation, it is of interest to study the relationships between the sequence $q_n(u)$ and the vector function $\mathfrak{x}(u)$. We list a few relevant statements.

(i) $q_n(u) \to \mathfrak{x}(u)$ uniformly in I_0. This is an obvious consequences of the continuity of $\mathfrak{x}(u)$.

(ii) $q_n'(u) \to \mathfrak{x}'(u)$ a.e. in I_0. This follows readily from III.2.1.

(iii) $q_n \to \mathfrak{x}(L)$ in I_0. This was shown in III.3.9.

(iv) $q_n \to \mathfrak{x}(V)$ in I_0. This follows from (iii) and III.3.58, or else directly from III.2.17.

(v) If $\mathfrak{x}(u)$ is AC in I_0, then $q_n \to \mathfrak{x}(\mathrm{SL})$ in I_0. Since clearly each $q_n(u)$ is AC in I_0, this follows from III.3.67 in view of (ii) and (iii).

(vi) Conversely, if $q_n \to \mathfrak{x}(\mathrm{SL})$ in I_0, then $\mathfrak{x}(u)$ is AC in I_0, by III.3.66.

III.3.69. We shall consider presently a sequence of vector functions $\mathfrak{x}_n(u)$, $n = 0, 1, 2, \cdots$, of the special form $\mathfrak{x}_n(u) = [u, y_n(u), z_n(u)]$, where $y_n(u), z_n(u)$ are continuous and BV in $I_0 : a_0 \leqq u \leqq b_0$, and $\mathfrak{x}_n \to \mathfrak{x}_0(L)$ in I_0 (see III.3.53). As compared with the general case, a number of more precise statements may be established in this special case. A few examples follow, in the way of illustration.

III.3.70. *Given a sequence* $\mathfrak{x}_n(u)$ *as in* III.3.69, *from* III.3.52, III.3.59, I.3.12 *we infer the relations* (cf. III.3.61)

$$\int_{I_0} |\, \mathfrak{x}_n' - \mathfrak{x}_0' \,|^{1/2}\, du \to 0, \qquad \mathfrak{x}_n' \to \mathfrak{x}_0'(M) \qquad \textit{in } I_0 .$$

where the second relation means that $|\, \mathfrak{x}_n' - \mathfrak{x}_0' \,| \to 0(M)$ *in* I_0.

III.3.71. Continuation. The preceding result admits of the following improvement. *If* $0 < \lambda < 1$, *then*

$$\int_{I_0} |\, \mathfrak{x}_n' - \mathfrak{x}_0' \,|^{\lambda}\, du \to 0 \qquad \text{for } n \to \infty . \tag{1}$$

Proof. Let us put $F_n(u) = |\, \mathfrak{x}_n'(u) - \mathfrak{x}_0'(u) \,|$. We have then, by III.3.16,

$$\int_{I_0} F_n(u)\, du \leqq \int_{I_0} |\, \mathfrak{x}_n' \,|\, du + \int_{I_0} |\, \mathfrak{x}_0' \,|\, du \leqq L(I_0\,,\, \mathfrak{x}_n) + L(I_0\,,\, \mathfrak{x}_0).$$

Since $L(I_0\,,\, \mathfrak{x}_n) \to L(I_0\,,\, \mathfrak{x}_0)$ by assumption, there follows the existence of a constant G such that

$$\int_{I_0} F_n\, du \leqq G < \infty, \qquad n = 1, 2, \cdots . \tag{2}$$

By III.3.70 we have the relation

$$F_n \to 0(M) \qquad \text{in } I_0 . \tag{3}$$

Now give $\epsilon > 0$. Let E_n denote the subset of I_0 where $F_n \geqq \epsilon^{1/\lambda}$. In view of (3) we have then a positive integer N such that

$$|\, E_n \,| \leqq \epsilon^{1/(1-\lambda)} \cdot G^{-\lambda/(1-\lambda)} \qquad \text{for } n > N. \tag{4}$$

The Hölder inequality, applied with $p = 1/\lambda$, $q = 1/(1 - \lambda)$, yields

(5) $$\int_{E_n} F_n^\lambda \, du \leqq \left(\int_{E_n} F_n \, du\right)^\lambda \left(\int_{E_n} du\right)^{1-\lambda}.$$

By (2), (4), (5) we obtain now, for $n > N$,

$$\int_{I_0} F_n^\lambda \, du = \int_{I_0 - E_n} F_n^\lambda \, du + \int_{E_n} F_n^\lambda \, du \leqq \epsilon \mid I_0 \mid + G^\lambda \mid E_n \mid^{1-\lambda} = (\mid I_0 \mid + 1)\epsilon.$$

Since ϵ was arbitrary, the relation (1) follows.

III.3.72. Continuation. For $\lambda = 1$ the relation III.3.71(1) may fail to hold, as the following example shows. Let $f(u)$ be a (scalar) function which is continuous, BV, singular and nonconstant in I_0, and let $q_n(u)$ be a sequence of quasi-linear vector functions associated with the vector function $\mathfrak{x}(u) = [u, f(u), 0]$ in the manner described in III.3.68. Then clearly each $q_n(u)$ is of the form $q_n(u) = [u, f_n(u), 0]$. Thus $q_n'(u) = [1, f_n'(u), 0]$ and $\mathfrak{x}_n'(u) = [1, 0, 0]$ a.e. in I_0 (since $f(u)$ is singular). Since $q_n(u)$ is clearly AC in I_0, we obtain (cf. III.2.29)

(1) $$\int_{I_0} \mid q_n' - \mathfrak{x}' \mid du = \int_{I_0} \mid f_n' \mid du = V(I_0, f_n).$$

Now $V(I_0, f_n) \to V(I_0, f)$ by III.3.68(iv). Thus (1) yields

$$\int_{I_0} \mid q_n' - \mathfrak{x}' \mid du \underset{n\to\infty}{\to} V(I_0, f) > 0,$$

since $f(u)$ is not constant in I_0.

III.3.73. Continuation. In view of the preceding negative result, the following remark is of interest. Given the sequence $\mathfrak{x}_n(u)$ as in III.3.69, we have $\mathfrak{x}_n'(u) = [1, y_n'(u), z_n'(u)]$ a.e. in I_0, and thus clearly

(1) $$\mid \mathfrak{x}_n'(u) \mid \geqq 1 \text{ a.e. in } I_0, \qquad n = 0, 1, 2, \cdots.$$

Wherever (1) holds, we can consider the unit vector $t_n = \mathfrak{x}_n'/\mid \mathfrak{x}_n' \mid$ (see III.3.43). Thus t_n is defined a.e. in I_0. *We assert that*

(2) $$\int_{I_0} \mid t_n - t_0 \mid du \to 0 \qquad \textit{for } n \to \infty.$$

Proof. We start with the identity

$$\mid t_n - t_0 \mid^2 = \left| \frac{\mathfrak{x}_n'}{\mid \mathfrak{x}_n' \mid} - \frac{\mathfrak{x}_0'}{\mid \mathfrak{x}_0' \mid} \right|^2 = \frac{2(\mid \mathfrak{x}_n' \mid \mid \mathfrak{x}_0' \mid - \mathfrak{x}_n'\mathfrak{x}_0')}{\mid \mathfrak{x}_n' \mid \mid \mathfrak{x}_0' \mid},$$

which yields, in view of (1), the inequality

$$\mid t_n - t_0 \mid \leqq 2^{1/2}(\mid \mathfrak{x}_n' \mid \mid \mathfrak{x}_0' \mid - \mathfrak{x}_n'\mathfrak{x}_0')^{1/2} \qquad \text{a.e. in } I_0.$$

By III.3.60(2), the relation (2) follows.

III.3.74. Given a sequence of vector functions $\mathfrak{x}_n(u) = [u, y_n(u), z_n(u)]$ as in III.3.69, let us add the assumption that $\mathfrak{x}_0(u)$ is AC in I_0. Then (cf. III.3.71, III.3.72)

$$\int_{I_0} |\mathfrak{x}_n' - \mathfrak{x}_0'| \, du \to 0. \tag{1}$$

PROOF. (i) Let us first assume that

$$\mathfrak{x}_n' \to \mathfrak{x}_0' \qquad \text{a.e. in } I_0 . \tag{2}$$

By I.3.10 we have then

$$\int_{I_0} |y_0'| \, du \leqq \liminf \int_{I_0} |y_n'| \, du. \tag{3}$$

On the other hand, since $\mathfrak{x}_n \to \mathfrak{x}_0(L)$ in I_0 by assumption, we have the relation $\mathfrak{x}_n \to \mathfrak{x}_0(V)$ in I_0 by III.3.58. Hence, by III.2.22, III.2.29,

$$\int_{I_0} |y_n'| \, du \leqq V(I_0, y_n) \to V(I_0, y_0) = \int_{I_0} |y_0'| \, du, \tag{4}$$

and consequently, in view of (3),

$$\int_{I_0} |y_n'| \, du \to \int_{I_0} |y_0'| \, du. \tag{5}$$

The relations (2) and (5) yield, by I.3.11,

$$\int_{I_0} |y_n' - y_0'| \, du \to 0. \tag{6}$$

A similar argument shows that

$$\int_{I_0} |z_n' - z_0'| \, du \to 0. \tag{7}$$

Since $|\mathfrak{x}_n' - \mathfrak{x}_0'| \leqq |y_n' - y_0'| + |z_n' - z_0'|$ a.e. in I_0, (6) and (7) imply (1).

(ii) Now let us drop the assumption (2). Given $\epsilon > 0$, let us denote by $K(\epsilon)$ the class of all those positive integers n for which

$$\int_{I_0} |\mathfrak{x}_n' - \mathfrak{x}_0'| \, du \geqq \epsilon. \tag{8}$$

Clearly, (1) is proved if we can show that the class $K(\epsilon)$ is finite for every $\epsilon > 0$. Suppose that $K(\epsilon)$ is infinite for a certain $\epsilon > 0$. Then there exists an infinite sequence $n_1 < n_2 < \cdots < n_k < \cdots$, such that (8) holds for $n = n_k$, $k = 1, 2, \cdots$. By III.3.70, I.3.12 it follows that this sequence contains an infinite subsequence $m_1 < m_2 < \cdots$, such that

(9) $$\mathfrak{x}'_{m_k} \to \mathfrak{x}'_0 \qquad \text{a.e. in } I_0 .$$

Since each m_k belongs to $K(\epsilon)$, we have also

(10) $$\int_{I_0} | \mathfrak{x}'_{m_k} - \mathfrak{x}'_0 | \, du \geqq \epsilon, \qquad k = 1, 2, \cdots .$$

Clearly, in view of (9), the sequence $\mathfrak{x}_{m_k}(u)$ satisfies all the assumptions used in part (i), and thus (10) contradicts the result obtained there.

III.3.75. Given a sequence of vector functions as in III.3.69, let us add the assumption that $\mathfrak{x}_0$ is AC in I_0. Then $\mathfrak{x}_n \to \mathfrak{x}_0$(SL) in I_0 (see III.3.65).

PROOF. Let $\mathfrak{x}_n(u) = \mathfrak{x}_{na}(u) + \mathfrak{x}_{ns}(u)$ be the normalized Lebesgue decomposition of $\mathfrak{x}_n(u)$ in I_0 (see III.3.24). Since $\mathfrak{x}_0(u)$ is AC in I_0 by assumption, we have then (see III.3.27)

(1) $$\begin{aligned} L(I_0 , \mathfrak{x}_n - \mathfrak{x}_0) &= L(I_0 , \mathfrak{x}_{na} - \mathfrak{x}_0 + \mathfrak{x}_{ns}) = L(I_0 , \mathfrak{x}_{na} - \mathfrak{x}_0) + L(I_0 , \mathfrak{x}_{ns}) \\ &= L(I_0 , \mathfrak{x}_{na} - \mathfrak{x}_0) + L(I_0 , \mathfrak{x}_n) - L(I_0 , \mathfrak{x}_{na}). \end{aligned}$$

By III.3.16, III.3.24(i), III.3.74 we obtain

(2) $$L(I_0 , \mathfrak{x}_{na} - \mathfrak{x}_0) = \int_{I_0} | \mathfrak{x}'_{na} - \mathfrak{x}'_0 | \, du = \int_{I_0} | \mathfrak{x}'_n - \mathfrak{x}'_0 | \, du \to 0.$$

Thus $\mathfrak{x}_{na} \to \mathfrak{x}_0$(SL) in I_0, and hence by III.3.65

(3) $$L(I_0 , \mathfrak{x}_{na}) \to L(I_0 , \mathfrak{x}_0).$$

Since $L(I_0 , \mathfrak{x}_n) \to L(I_0 , \mathfrak{x}_0)$ by assumption, (1), (2), (3) imply the asserted relation $L(I_0 , \mathfrak{x}_n - \mathfrak{x}_0) \to 0$.

REMARK. Comparison of the preceding theorem with that proved in III.3.67 shows the type of improvement that may be expected in dealing with vector functions of the special form $\mathfrak{x}(u) = [u, y(u), z(u)]$.

III.3.76. Given a continuous vector function of the general form $\mathfrak{x}(u) = [x(u), y(u), z(u)]$ in $I_0 : a_0 \leqq u \leqq b_0$, denote by $\tau, \tau_1 , \tau_2 , \tau_3$ the transformations

$$\begin{aligned} &\tau \ : x = x(u), \quad y = y(u), \quad z = z(u), \quad u \in I_0 , \\ &\tau_1 : x = 0 , \quad y = y(u), \quad z = z(u), \quad u \in I_0 , \\ &\tau_2 : x = x(u), \quad y = 0 , \quad z = z(u), \quad u \in I_0 , \\ &\tau_3 : x = x(u), \quad y = y(u), \quad z = 0 , \quad u \in I_0 . \end{aligned}$$

The set $\tau_3(I_0)$ is then a bounded closed set in the xy-plane. We assert that if one of the components $x(u)$, $y(u)$ is BV in I_0, then the (two-dimensional) measure of $\tau_3(I_0)$ is equal to zero. Suppose indeed that $x(u)$ is BV in I_0. Then the corresponding function $N(x, I_0)$ is summable (see III.2.12), and hence we have on the x-axis a set e_x of (linear) measure zero, such that $N(x_0, I_0) < \infty$ for $x_0 \notin e_x$. In other words, the line $x = x_0$, in the xy-plane, intersects the set $\tau_3(I_0)$ in a finite number of points at most if $x_0 \notin e_x$. By I.3.6 it follows that $\tau_3(I_0)$ is of (two-dimensional) measure zero. Similar statements hold for the sets $\tau_1(I_0)$, $\tau_2(I_0)$. Since these sets are the orthogonal projections of the set $\tau(I_0)$ upon the yz-, zx-, xy-planes respectively, it follows that the set $\tau(I_0)$ is of (three-dimensional) measure zero as soon as one at least of the components $x(u)$, $y(u)$, $z(u)$ is BV in I_0. In view of III.3.12 we infer that if $L(I_0, \mathfrak{x}) < \infty$, then the set $\tau(I_0)$ is of (three-dimensional) measure zero, and the sets $\tau_1(I_0)$, $\tau_2(I_0)$, $\tau_3(I_0)$ are of (two-dimensional) measure zero.

III.3.77. The preceding theory is concerned with vector functions defined in a linear interval. An analogous theory can be developed for vector functions defined on a simple closed curve. We proceed to study this case presently, restricting ourselves to facts actually needed in the sequel. Let C be a simple closed curve (see I.2.31), and let p be a generic notation for a point on C. If $x(p)$, $y(p)$, $z(p)$ are (real-valued) continuous functions on C, then $\mathfrak{x}(p) = [x(p), y(p), z(p)]$ is a continuous vector function on C. The concepts needed in our study of such vector functions will now be listed.

III.3.78. Continuation. Let $\phi(\gamma)$ be a real-valued function which is defined for every simple arc γ on C, and also for C itself. We shall then term $\phi(\gamma)$ an interval function defined on C.

Let γ_1, γ_2 be sub-arcs of C which have one end point or both end points (and nothing else) in common. If $\phi(\gamma_1 + \gamma_2) \leqq \phi(\gamma_1) + \phi(\gamma_2)$ for every choice of γ_1, γ_2, then $\phi(\gamma)$ is said to increase by subdivision.

A subdivision $D(C)$ of C is comprised of a finite number of sub-arcs $\gamma_1, \cdots, \gamma_n$, without common interior points, whose sum is C. We shall write $\gamma \in D(C)$ to express the fact that γ is one of the simple arcs occurring in $D(C)$. The symbol $\| D(C) \|$ will denote the maximum diameter of the arcs $\gamma \in D(C)$.

The symbol $\phi[D(C)]$ is defined by the formula $\phi[D(C)] = \sum \phi(\gamma)$, $\gamma \in D(C)$, and the symbol $U(C, \phi)$ by the formula $U(C, \phi) = \text{l.u.b. } \phi[D(C)]$, where the least upper bound is taken with respect to all subdivisions $D(C)$ of C.

III.3.79. Continuation. Let us take a point p^* on C which will be kept fixed. The symbol $U^*(C, \phi)$ is defined as $U(C, \phi)$, except that we use only subdivisions $D(C)$ for which p^* is a point of division. Clearly $U^*(C, \phi) \leqq U(C, \phi)$. If $\phi(\gamma)$ increases by subdivision, then clearly $U^*(C, \phi) = U(C, \phi)$.

III.3.80. Let $f(p)$ be a real-valued function on C. We associate with $f(p)$ an interval function $\phi(\gamma) = | f(b) - f(a) |$, where a, b are the end points of the simple arc γ. We agree to put $\phi(C) = 0$. We define the total variation $V(C, f)$ of $f(p)$ on C by the formula $V(C, f) = U(C, \phi)$. If $V(C, f) < \infty$, then $f(p)$ will be termed BV on C (of bounded variation on C). Since now $\phi(\gamma)$ clearly increases by subdivision, we have also $V(C, f) = U^*(C, \phi)$ (see III.3.79).

If $f_n(p)$ is a sequence of real-valued functions on C, $n = 0, 1, 2, \cdots$, then we shall say that $f_n(p)$ converges to $f_0(p)$ in variation, in symbols $f_n \to f_0(V)$, on C

if and only if the following conditions hold. (i) $f_n(p)$ is continuous and BV on C. (ii) $f_n(p) \to f_0(p)$ uniformly on C. (iii) $V(C, f_n) \to V(C, f_0)$.

III.3.81. Let $\mathfrak{x}(p) = [x(p), y(p), z(p)]$ be a continuous vector function on C (see III.3.77). We associate with $\mathfrak{x}(p)$ the interval functions

$$g_1(\gamma, \mathfrak{x}) = | x(b) - x(a) |, g_2(\gamma, \mathfrak{x}) = | y(b) - y(a) |, g_3(\gamma, \mathfrak{x}) = | z(b) - z(a) |,$$

$$g(\gamma, \mathfrak{x}) = [g_1(\gamma, \mathfrak{x})^2 + g_2(\gamma, \mathfrak{x})^2 + g_3(\gamma, \mathfrak{x})^2]^{1/2},$$

where a, b are the end points of the sub-arc γ of C. We complete these definitions by setting $g_1(C, \mathfrak{x}) = g_2(C, \mathfrak{x}) = g_3(C, \mathfrak{x}) = g(C, \mathfrak{x}) = 0$. Finally, we define $L(C, \mathfrak{x}) = U(C, g)$. Since $g(\gamma, \mathfrak{x})$ clearly increases by subdivision, we have also $L(C, \mathfrak{x}) = U^*(C, g)$ (see III.3.79).

Given a sequence of continuous vector functions $\mathfrak{x}_n(p)$ on C, $n = 0, 1, 2, \cdots$, we shall say that $\mathfrak{x}_n(p)$ converges in length to $\mathfrak{x}_0(p)$, in symbols $\mathfrak{x}_n \to \mathfrak{x}_0(L)$, on C if and only if the following conditions hold. (i) $L(C, \mathfrak{x}_n) < +\alpha, n = 0, 1, 2, \cdots$. (ii) $\mathfrak{x}_n(p) \to \mathfrak{x}_0(p)$ uniformly on C. (iii) $L(C, \mathfrak{x}_n) \to L(C, \mathfrak{x}_0)$. We shall say that $\mathfrak{x}_n(p)$ converges in variation to $\mathfrak{x}_0(p)$, in symbols $\mathfrak{x}_n \to \mathfrak{x}_0(\mathrm{V})$, on C if and only if the components of $\mathfrak{x}_n(p)$ converge in variation to the corresponding components of $\mathfrak{x}_0(p)$ (see III.3.80).

III.3.82. Let us assume that the simple closed curve C is oriented (see I.2.31). Given two real-valued functions $F(p)$, $G(p)$ on C, the Stieltjes integral $\int FdG$ around C is defined as follows. Let $p_1, \cdots, p_n$ be distinct points on C which follow upon each other according to the given orientation of C, and let $\gamma_1, \gamma_2, \cdots, \gamma_{n-1}, \gamma_n$ be the sub-arcs of C with the end points $p_1, p_2; p_2, p_3; \cdots; p_{n-1}, p_n; p_n, p_1$ respectively. The subdivision of C comprised of $\gamma_1, \cdots, \gamma_n$ will be denoted by $D(C)$. On γ_k, we choose a point q_k, $k = 1, 2, \cdots, n$, and we consider the summation $\sum F(q_k)[G(p_{k+1}) - G(p_k)]$, $k = 1, 2, \cdots, n$, where we have put $p_{n+1} = p_1$ for convenience. If this summation approaches a finite limit as $\| D(C) \| \to 0$, then we put

$$\int_C FdG = \lim \sum_{k=1}^{n} F(q_k)[G(p_{k+1}) - G(p_k)], \qquad \| D(C) \| \to 0.$$

Now let p^* be a fixed point on C, and let us restrict ourselves, in the preceding definition, to subdivisions $D(C)$ for which $p_n = p^*$. The resulting limit, if it exists and is finite, will be denoted by $\int_C^* FdG$. Clearly, if $\int_C FdG$ exists, then $\int_C^* FdG$ also exists, and the two integrals are equal. If $\int_C^* FdG$ exists, and if $F(p)$, $G(p)$ are both continuous at p^*, then clearly $\int_C FdG$ also exists, and the two integrals are equal.

III.3.83. In the study initiated in III.3.77, the simple closed curve C may be reduced to the unit interval $I_0 : 0 \leqq u \leqq 1$ by the following device. Let us choose on C a point p^* which will be kept fixed. Next we choose a continuous transformation $T^*(I_0) = C$ subject to the following conditions. (i) $T^*(0) = T^*(1) = p^*$. (ii) The open interval $0 < u < 1$ is mapped by T^* topologically onto $C - p^*$. (iii) If C is oriented, then we require that the point $p = T(u)$

describes C according to the assigned orientation while u increases from 0 to 1. The transformation T^*, once chosen, will be kept fixed.

Given then a continuous vector function $\mathfrak{x}(p) = [x(p), y(p), z(p)]$ on C, we associate with $\mathfrak{x}(p)$ the vector function $\mathfrak{x}[T^*(u)] = (x[T^*(u)], y[T^*(u)], z[T^*(u)])$ on I_0. Since T^* is fixed, and since it will be sufficiently clear from the context whether p or u is considered as the independent variable, we shall use the letters $\mathfrak{x}, x, y, z$ to refer both to $\mathfrak{x}(p), x(p), y(p), z(p)$ and to $\mathfrak{x}[T^*(u)], x[T^*(u)], y[T^*(u)], z[T^*(u)]$. Thus a biunique correspondence is established between the class of continuous vector functions $\mathfrak{x}(p)$ on C and the class of those continuous vector functions $\mathfrak{x}(u)$ on I_0 which satisfy the additional condition $\mathfrak{x}(0) = \mathfrak{x}(1)$. If $\mathfrak{x}(p)$, $\mathfrak{x}(u)$ are corresponding continuous vector functions in this sense, then the following statements are obvious consequences of the remarks made in III.3.79-III.3.82.

(a) $L(C, \mathfrak{x}) = L(I_0, \mathfrak{x})$ (cf. III.3.4).

(b) $V(C, x) = V(I_0, x)$, $V(C, y) = V(I_0, y)$, $V(C, z) = V(I_0, z)$ (cf. III.2.13).

(c) $L(C, \mathfrak{x})$ is finite if and only if $x(p), y(p), z(p)$ are BV on C (cf. III.3.12).

(d) $\mathfrak{x}_n \to \mathfrak{x}_0(L)$ on C if and only if $\mathfrak{x}_n \to \mathfrak{x}_0(L)$ in I_0 (cf. III.3.53).

(e) $\mathfrak{x}_n \to \mathfrak{x}_0(V)$ on C if and only if $\mathfrak{x}_n \to \mathfrak{x}_0(V)$ in I_0 (cf. III.3.56).

(f) If $\mathfrak{x}_n \to \mathfrak{x}_0(L)$ on C, then $\mathfrak{x}_n \to \mathfrak{x}_0(V)$ on C (cf. III.3.58).

(g) Assuming that C is oriented and $L(C, \mathfrak{x}) < \infty$, we have the formulas

$$\int_C x\,dy = \int_{I_0} x\,dy, \qquad \int_C y\,dz = \int_{I_0} y\,dz, \qquad \int_C z\,dx = \int_{I_0} z\,dx,$$

where the existence of these integrals follows from I.3.15, in view of (b), (c).

III.3.84. We shall consider presently the following important special situation. The simple closed curve C lies in the complex $w = u + iv$ plane, and is oriented counterclockwise. We denote by $\mathfrak{K}_0$ the class of continuous vector functions on C of the special form $\mathfrak{x}(w) = [x(w), y(w), 0]$, $w \in C$. Since the z-component will thus not occur at all, we can and shall use the letter z to refer to the complex variable $z = x + iy$. Then each $[x(w), y(w), 0] = \mathfrak{x}(w) \in \mathfrak{K}_0$ gives rise to a continuous transformation

$$(1) \qquad \tau : z = x(w) + iy(w), \qquad w \in C.$$

Geometrically, we may think of the point $z = x + iy$ as describing a closed continuous curve (with multiple points generally), while w describes C in the counterclockwise sense. In terms of the transformation (1), we associate with $\mathfrak{x}(w)$ the index function $\mu(z, \tau, C)$ in the sense of II.4.34. It will be convenient at times to write $\mu(x, y, \tau, C)$ instead of $\mu(z, \tau, C)$, where $z = x + iy$. By II.4.34, $\mu(x, y, \tau, C)$ is then a Borel measurable, integral-valued function defined in the whole xy-plane. We propose to study the integral, in the Lebesgue sense, of this index function, taken over the whole xy-plane. This integral, if it exists, will be denoted by $\iint \mu(x, y, \tau, C)dxdy$, without displaying the limits of integration (cf. the analogous agreements in II.2.11).

III.3.85. Continuation. The following terminology will prove useful. We choose on C a point w^* that will be kept fixed, and we choose a continuous transformation $T^*(I_0) = C$ as explained in III.3.83 (with w^* replacing p^*). Since the letter u is now needed in connection with the complex variable $w = u + iv$, we use the variable t on the unit interval $I_0 : 0 \leqq t \leqq 1$. The transformation T^*, once chosen, will be kept fixed. With each $[x(w), y(w), 0] = \mathfrak{x}(w) \in \mathfrak{R}_0$ we associate the transformations

$$\tau : z = x(w) + iy(w), \qquad \tau^x : x = x[T^*(t)], \qquad \tau^y : y = y[T^*(t)],$$

where τ operates from C into the $z = x + iy$ plane, while τ^x, τ^y operate from $I_0 : 0 \leqq t \leqq 1$ into the x and y axes of the z-plane respectively. The multiplicity functions $N(x, \tau^x, I_0)$, $N(y, \tau^y, I_0)$ will be used in the sense of III.2.4. In the sequel, w^* and T^* are fixed, while $x(w)$, $y(w)$, τ, τ^x, τ^y depend upon the vector function $\mathfrak{x}(w) \in \mathfrak{R}_0$.

III.3.86. Continuation. Let $n > 1$ be a positive integer. Using the fixed transformation $T^*(I_0) = C$ (see III.3.85), we put

$$w_{nk} = T^*\left(\frac{k-1}{n}\right), \qquad k = 1, \cdots, n+1. \tag{1}$$

Then $w_{n1} = w_{n,n+1}$. The points $w_{n1}, w_{n2}, \cdots, w_{n,n+1} = w_{n1} = w^*$ lie on C, and follow upon each other in the counterclockwise sense around C. Let us put now (see III.3.84(1))

$$\tau(w_{nk}) = \zeta_{nk} = \xi_{nk} + i\eta_{nk}, \qquad k = 1, \cdots, n+1. \tag{2}$$

The points $\zeta_{n1}, \zeta_{n2}, \cdots, \zeta_{nn}, \zeta_{n,n+1} = \zeta_{n1}$ are thought of as lying in the $z = x + iy$ plane. To the curve C and the points w_{nk}, ζ_{nk} we apply now the discussion carried out in II.4.36-II.4.39. The transformation T of II.4.36 will now be denoted by T_n, since the points w_{nk}, ζ_{nk} depend upon n. Clearly

$$T_n \to \tau \qquad \text{on } C \tag{3}$$

in the sense of uniform convergence. Hence, by II.4.25(c), II.4.34,

$$\mu(x, y, T_n, C) \to \mu(x, y, \tau, C) \qquad \text{for } (x, y) \notin \tau(C). \tag{4}$$

Since $\mu(x, y, \tau, C) = 0$ for $(x, y) \in \tau(C)$, we infer from (4) the relation

$$|\mu(x, y, \tau, C)| \leqq \liminf_{n\to\infty} |\mu(x, y, T_n, C)|. \tag{5}$$

For each n, let e_{ny} be the set, on the y-axis in the z-plane, comprised of the points $\eta_{n1}, \eta_{n2}, \cdots, \eta_{nn}, \eta_{n,n+1} = \eta_{n1}$. If $y_0 \notin e_{ny}$, then clearly the line $y = y_0$ intersects the straight segment bounded by the points ζ_{nk} and $\zeta_{n,k+1}$ if and only if $(\eta_{nk} - y_0)(\eta_{n,k+1} - y_0) < 0$. But if this inequality holds, then the difference $y[T^*(t)] - y_0$ is of different sign at the end points of the interval $(k-1)/n \leqq t \leqq k/n$, and hence the function $y[T^*(t)]$ takes on the value y_0 at some interior point of this interval. By II.4.39, III.3.85 there follows the inequality

$$(6) \qquad |\mu(x, y_0, T_n, C)| \leqq N(y_0, \tau^y, I_0) \qquad \text{for } y_0 \notin e_{ny}.$$

Let us put $e_y = \sum e_{ny}$, $n = 2, 3, \cdots$. Then e_y is a countable subset of the y-axis, and from (5) and (6) we infer the inequality

$$(7) \qquad |\mu(x, y_0, \tau, C)| \leqq N(y_0, \tau^y, I_0) \qquad \text{for } y \notin e_y.$$

III.3.87. CONTINUATION. Summing up, we have established the following statement: on the y-axis in the $z = x + iy$ plane, we can choose a countable set e_y such that (see III.3.86(6), (7))

$$(1) \qquad |\mu(x, y, \tau, C)| \leqq N(y, \tau^y, I_0) \qquad \text{for } y_0 \notin e_y,$$

$$(2) \qquad |\mu(x, y, T_n, C)| \leqq N(y, \tau^y, I_0) \qquad \text{for } y \notin e_y,\ n = 2, 3, \cdots.$$

Similar statements hold, of course, with y replaced by x.

III.3.88. Using the terminology of III.3.84, III.3.81, *let* $\mathfrak{x}(w) \in \mathfrak{R}_0$ *be a continuous vector function such that* $L(C, \mathfrak{x}) < \infty$. *We assert that the corresponding index function* $\mu(x, y, \tau, C)$ *is summable, and* (see III.3.83(g))

$$(1) \qquad \iint \mu(x, y, \tau, C)\, dx\, dy = \int_C x\, dy = -\int_C y\, dx = \frac{1}{2}\int_C (x\, dy - y\, dx)$$

PROOF. The assumption $L(C, \mathfrak{x}) < \infty$ implies the following facts.

(i) The components $x(w)$, $y(w)$ of $\mathfrak{x}(w)$ are BV on C (see III.3.83(c)), and hence $N(x, \tau^x, I_0)$, $N(y, \tau^y, I_0)$ are summable (see III.3.83(b), III.2.12, III.3.85).

(ii) The set $\tau(C)$ is of (two-dimensional) measure zero (see III.3.85, III.3.76).

(iii) The Stieltjes integrals involved in (1) exist (see III.3.83(g), I.3.15).

From (ii) we infer, by III.3.86(4), the relation

$$(2) \qquad \mu(x, y, T_n, C) \to \mu(x, y, \tau, C) \qquad \text{a.e. in the } xy\text{-plane.}$$

Clearly, since T_n converges uniformly to τ on C, we have a finite constant M such that the functions $\mu(x, y, T_n, C)$, $\mu(x, y, \tau, C)$ vanish outside of the square $-M \leqq x \leqq M$, $-M \leqq y \leqq M$ (cf. II.4.34(c)), and $N(y, \tau^y, I_0)$ vanishes outside of the interval $-M \leqq y \leqq M$. By I.3.10, it follows therefore from III.3.87(1), (2) that $\mu(x, y, \tau, C)$ is summable, and by (2) it follows further that

$$(3) \qquad \iint \mu(x, y, T_n, C)\, dx\, dy \to \iint \mu(x, y, \tau, C)\, dx\, dy.$$

By II.4.38 we have

$$(4) \qquad \iint \mu(x, y, T_n, C)\, dx\, dy = \frac{1}{2}\sum_{k=1}^{n} (\xi_{nk}\eta_{n,k+1} - \eta_{nk}\xi_{n,k+1}).$$

An elementary rearrangement yields the identities

$$(5) \qquad \begin{aligned} &\frac{1}{2}\sum_{k=1}^{n} (\xi_{nk}\eta_{n,k+1} - \eta_{nk}\xi_{n,k+1}) \\ &\qquad = \frac{1}{2}\sum_{k=1}^{n} \xi_{n,k+1}(\eta_{n,k+1} - \eta_{nk}) - \frac{1}{2}\sum_{k=1}^{n} \eta_{n,k+1}(\xi_{n,k+1} - \xi_{nk}), \end{aligned}$$

$$\text{(6)} \qquad \sum_{k=1}^{n} \eta_{n,k+1}(\xi_{n,k+1} - \xi_{nk}) = -\sum_{k=1}^{n} \xi_{nk}(\eta_{n,k+1} - \eta_{nk}),$$

$$\text{(7)} \qquad \sum_{k=1}^{n} \xi_{n,k+1}(\eta_{n,k+1} - \eta_{nk}) = -\sum_{k=1}^{n} \eta_{nk}(\xi_{n,k+1} - \xi_{nk}).$$

For $n \to \infty$, (3)-(7) yield (1) (cf. III.3.82).

III.3.89. Let $[x_n(w), y_n(w), 0] = \mathfrak{x}_n(w) \in \mathfrak{R}_0$, $n = 0, 1, 2, \cdots$ (see III.3.84), be a sequence such that (i) $L(C, \mathfrak{x}_n) < \infty$, $n = 0, 1, 2, \cdots$, and (ii) $\mathfrak{x}_n(w) \to \mathfrak{x}_0(w)$ uniformly on C. Let $\mu(x, y, \tau_n, C)$ be the index function associated with $\mathfrak{x}_n(w)$, where $\tau_n : z = x_n(w) + iy_n(w)$, $w \in C$ (see III.3.84). By II.4.34, II.4.25 it follows that $\mu(x, y, \tau_n, C) \to \mu(x, y, \tau_0, C)$ for $(x, y) \notin \tau_0(C)$. But $|\tau_0(C)| = 0$ by III.3.76. Hence

$$\text{(1)} \qquad \mu(x, y, \tau_n, C) \to \mu(x, y, \tau_0, C)$$

a.e. in the $z = x + iy$ plane. Condition (i) implies, by III.3.88, that $\mu(x, y, \tau_n, C)$ is summable, $n = 0, 1, 2, \cdots$. In view of (1), there arises the question: under what conditions shall we have the relation

$$\text{(2)} \qquad \iint \mu(x, y, \tau_n, C)\, dx\, dy \to \iint \mu(x, y, \tau_0, C)\, dx\, dy,$$

or the stronger relation

$$\text{(3)} \qquad \iint |\mu(x, y, \tau_n, C) - \mu(x, y, \tau_0, C)|\, dx\, dy \to 0?$$

III.3.90. CONTINUATION. *Given the sequence $\mathfrak{x}_n(w)$ as in* III.3.89, *suppose that there exists a finite constant M such that*

$$\text{(1)} \qquad L(C, \mathfrak{x}_n) < M, \qquad n = 0, 1, 2, \cdots.$$

We assert then the relation

$$\text{(2)} \qquad \iint \mu(x, y, \tau_n, C)\, dx\, dy \to \iint \mu(x, y, \tau_0, C)\, dx\, dy.$$

PROOF. In view of III.3.83(g), III.3.88 we have

$$\text{(3)} \qquad \iint \mu(x, y, \tau_n, C)\, dx\, dy = \int_{I_0} x_n\, dy_n, \qquad n = 0, 1, 2, \cdots.$$

By I.3.15 we can write

$$\text{(4)} \qquad \int_{I_0} x_0\, dy_0 - \int_{I_0} x_n\, dy_n = \int_{I_0} (y_n - y_0)\, dx_0 + \int_{I_0} (x_0 - x_n)\, dy_n .$$

Let us denote by m_n the maximum of $| y_n - y_0 | + | x_n - x_0 |$ in I_0, and let us note that by (1) and III.3.83(b)

(5) $$V(I_0, x_n) < M, \qquad V(I_0, y_n) < M.$$

From (4) and (5) we infer the inequality

(6) $$\left| \int_{I_0} x_0 \, dy_0 - \int_{I_0} x_n \, dy_n \right| \leqq 2m_n M.$$

Since $m_n \to 0$, (3) and (6) yield (2).

III.3.91. In discussing the relation III.3.89(3), the following remarks will be helpful. Let $f(x, y)$, $F(y)$ be real-valued, non-negative functions with the following properties. (i) $f(x, y)$ is summable in the square $Q_K : -K \leqq x \leqq K$, $-K \leqq y \leqq K$. (ii) $F(y)$ is summable in the interval $I_K : -K \leqq y \leqq K$. (iii) There exists a set e_y of (linear) measure zero on the y-axis, such that $f(x, y) \leqq F(y)$ for $y \notin e_y$. Let then G_y be a generic notation for a measurable subset of the interval I_K. Since $F(y)$ is summable in this interval, we have (see I.3.13) for every $\epsilon > 0$ an $\eta = \eta(\epsilon) > 0$, such that

(1) $$\int_{G_y} F(y) \, dy < \epsilon \qquad \text{if } | G_y | < \eta(\epsilon).$$

We assert then the inequality

(2) $$\iint_E f(x, y) \, dx \, dy \leqq \left[2K + \int_{-K}^{K} F(y) \, dy\right] \cdot \epsilon$$

for every measurable set $E \subset Q_K$ such that

(3) $$| E | < \min [\epsilon^2, \eta(\epsilon)^2].$$

Proof. Let $c(x, y)$ be the characteristic function of E (that is, $c(x, y) = 1$ if $(x, y) \in E$ and $c(x, y) = 0$ if $(x, y) \notin E$). Let S_y be the subset of I_K where

$$\int_{-K}^{K} c(x, y) \, dx > | E |^{1/2}.$$

Then clearly (see I.3.10)

$$| E | = \iint_{Q_K} c(x, y) \, dx \, dy \geqq \int_{S_y} \left[\int_{-K}^{K} c(x, y) \, dx \right] dy \geqq | E |^{1/2} \, | S_y |.$$

Thus $| S_y | \leqq | E |^{1/2} < \eta(\epsilon)$ by (3), and hence by (1) $\int_{S_y} F(y) \, dy < \epsilon$. In view of condition (iii), we obtain now the inequality (2) as follows

$$\iint_E f(x, y)\, dx\, dy = \iint_{Q_K} f(x, y)c(x, y)\, dx\, dy \leqq \int_{-K}^{K}\left[\int_{-K}^{K} c(x, y)\, dx\right]F(y)\, dy$$

$$= \int_{S_y}\left[\int_{-K}^{K} c(x, y)\, dx\right]F(y)\, dy + \int_{I_K - S_y}\left[\int_{-K}^{K} c(x, y)\, dx\right]F(y)\, dy$$

$$\leqq 2K \int_{S_y} F(y)\, dy + |\, E\,|^{1/2} \int_{-K}^{K} F(y)\, dy < [2K + \int_{-K}^{K} F(y)\, dy]\cdot \epsilon.$$

III.3.92. *Given a sequence* $\mathfrak{x}_n(w)$ *as in* III.3.89, *suppose that*

(1) $$L(C, \mathfrak{x}_n) \to L(C, \mathfrak{x}_0).$$

We assert then the relation

(2) $$\iint |\, \mu(x, y, \tau_n, C) - \mu(x, y, \tau_0, C)\,|\, dx\, dy \to 0.$$

PROOF. Since $\mathfrak{x}_n(w) \to \mathfrak{x}_0(w)$ uniformly in C, we have a finite constant K such that $\mu(x, y, \tau_n, C)$, $n = 0, 1, 2, \cdots$, vanishes outside of the square $Q_K : -K \leqq x \leqq K, -K \leqq y \leqq K$ (see II.4.34, II.4.25), and $N(y, \tau_n^y, I_0)$, $n = 0, 1, 2, \cdots$, vanishes outside of the interval $I_K : -K \leqq y \leqq K$ (see III.3.85). By I.3.11 and III.3.89(1), it is therefore sufficient to show that the sequence $\mu(x, y, \tau_n, C)$ possesses the property (V) in Q_K. We make the proof in two steps.

(i) The sequence $N(y, \tau_n^y, I_0)$ possesses the property (V) in I_K. Indeed, (1) implies by III.3.83(a), (b), (f) the relation $V(I_0, y_n) \to V(I_0, y_0)$. Hence

(3) $$\int_{-K}^{K} N(y, \tau_n^y, I_0)\, dy \to \int_{-K}^{K} N(y, \tau_0^y, I_0)\, dy$$

by III.3.85, III.2.23. We have also (see III.2.9)

(4) $$N(y, \tau_0^y, I_0) \leqq \liminf N(y, \tau_n^y, I_0) \qquad \text{a.e. in } I_K .$$

By I.3.11, the relations (3) and (4) imply that the sequence $N(y, \tau_n^y, I_0)$ possesses the property (V) in I_K. Hence for every $\epsilon > 0$ we have an $\eta(\epsilon) > 0$ such that

(5) $$\int_{G_y} N(y, \tau_n^y, I_0)\, dy < \epsilon \qquad \text{if } |\, G_y\,| < \eta(\epsilon)$$

where G_y is a generic notation for a measurable subset of I_K, and $\eta(\epsilon)$ is independent of n. From (3) we infer the existence of a finite constant M, such that

(6) $$\int_{-K}^{K} N(y, \tau_n^y, I_0)\, dy < M, \qquad n = 0, 1, 2, \cdots .$$

(ii) By III.3.87 we have, for every n, a countable set e_y^n on the y-axis such that

$$| \mu(x, y, \tau_n, C) | \leqq N(y, \tau_n^y, I_0) \qquad \text{for } y \notin e_y^n .$$

Now give $\epsilon > 0$. Let E be any measurable subset of the square Q_K such that $| E | < \min [\epsilon^2, \eta(\epsilon)^2]$ where $\eta(\epsilon)$ is taken from (5). By III.3.88, III.3.91 there follows the inequality (cf. (6))

$$\iint_E | \mu(x, y, \tau_n, C) | \, dx\, dy \leqq (2K + M)\epsilon, \qquad n = 0, 1, 2, \cdots .$$

Since K and M are independent of n, it follows that the sequence $\mu(x, y, \tau_n, C)$ possesses the property (V) in Q_K.

III.3.93. Continuation. Inspection of the preceding proof reveals that the assumption $L(C, \mathfrak{x}_n) \to L(C, \mathfrak{x}_0)$ could be replaced by the weaker assumption $V(I_0, y_n) \to V(I_0, y_0)$, or alternatively by the assumption $V(I_0, x_n) \to V(I_0, x_0)$. Similar remarks apply to several of our preceding results.

III.3.94. We apply presently the theory, developed so far for vector functions, to curves. Unless the contrary is explicitly stated, the term *curve* will be used to refer to a Fréchet curve of the type of the 1-cell in Euclidean three-space (see II.3.43). Such a curve C admits of a representation of the form (see II.3.31)

$$(1) \qquad C : x = x(u), \quad y = y(u), \quad z = z(u), \qquad u \in I_0 ,$$

where $I_0 : a_0 \leqq u \leqq b_0$ is an arbitrarily assigned interval, and $x(u)$, $y(u)$, $z(u)$ are continuous in I_0. We shall also use the notation

$$(2) \qquad C : \mathfrak{x} = \mathfrak{x}(u), \qquad u \in I_0 ,$$

where $\mathfrak{x}(u)$ denotes the continuous vector function with components $x(u)$, $y(u)$, $z(u)$. Each representation (2) of C gives rise to a quantity $L(I_0, \mathfrak{x})$ in the sense of III.3.4, and we propose to show that $L(I_0, \mathfrak{x})$ is independent of the particular choice of a representation (2) for C. The proof is made in the following steps.

(i) If we take a second representation

$$(3) \qquad C : \mathfrak{x} = \mathfrak{x}^*(u^*), \qquad u^* \in I_0^* ,$$

which is topologically similar to (2) (see II.3.19), then $L(I_0, \mathfrak{x}) = L(I_0^*, \mathfrak{x}^*)$.

Proof. By assumption, we have a homeomorphism $h(I_0) = I_0^*$, such that $\mathfrak{x}(u) = \mathfrak{x}^*[h(u)]$ in I_0. If I is any interval in I_0, then $I^* = h(I)$ is an interval in I_0^*. Since clearly $g(I, \mathfrak{x}) = g(I^*, \mathfrak{x}^*)$ (see III.3.2), the relation $L(I_0, \mathfrak{x}) = L(I_0^*, \mathfrak{x}^*)$ follows from III.3.4.

(ii) Suppose now that (3) is any representation of C. We have then (see II.3.19) a sequence of representations

$$(4) \qquad C : \mathfrak{x} = \mathfrak{x}_n(u), \qquad u \in I_0 ,$$

such that the representations (3) and (4) are topologically similar for every n and $\mathfrak{x}_n(u) \to \mathfrak{x}(u)$ uniformly in I_0. By (i) and III.3.6 we have then the relations $L(I_0, \mathfrak{x}_n) = L(I_0^*, \mathfrak{x}^*)$, $L(I_0, \mathfrak{x}) \leqq \liminf L(I_0, \mathfrak{x}_n)$. Hence $L(I_0, \mathfrak{x}) \leqq L(I_0^*, \mathfrak{x}^*)$, and a similar reasoning yields the complementary inequality $L(I_0^*, \mathfrak{x}^*) \leqq L(I_0, \mathfrak{x})$. Thus $L(I_0, \mathfrak{x}) = L(I_0^*, \mathfrak{x}^*)$.

III.3.95. Continuation. We define now the length $L(C)$ of C by the formula $L(C) = L(I_0, \mathfrak{x})$, where $\mathfrak{x} = \mathfrak{x}(u)$, $u \in I_0$, is any representation of C. In view of III.3.94, $L(C)$ depends solely upon C itself; that is, $L(C)$ is independent of the particular choice of the representation. If $L(C) < +\infty$, then C is termed *rectifiable.* We assert that $L(C)$ *is a lower semi-continuous functional of* C, in the following sense: If C_n, $n = 0, 1, 2, \cdots$, is a sequence of curves such that $C_n \to C_0$ (see II.3.15), then $L(C_0) \leqq \liminf L(C_n)$.

Proof. By II.3.16, we can choose representations $C_n : \mathfrak{x} = \mathfrak{x}_n(u)$, $u \in I_0$, $n = 0, 1, 2, \cdots$, such that $\mathfrak{x}_n(u) \to \mathfrak{x}_0(u)$ uniformly in I_0. By III.3.6 it follows that $L(C_0) = L(I_0, \mathfrak{x}_0) \leqq \liminf L(I_0, \mathfrak{x}_n) = \liminf L(C_n)$.

III.3.96. Continuation. The length $L(C)$ is independent of the choice of the Cartesian coordinate system x, y, z, or equivalently, congruent curves have the same length. This is an immediate consequence of the definitions given in II.3.42, III.3.95.

III.3.97. Given a curve (cf. III.3.94)

$$C : \mathfrak{x} = \mathfrak{x}(u), \qquad u \in I_0 : a_0 \leqq u \leqq b_0, \tag{1}$$

we shall say that C *reduces to a point* if and only if $\mathfrak{x}(u)$ is constant in I_0. Clearly, this property is independent of the choice of the representation. Obviously, $L(C) = 0$ if and only if C reduces to a point.

Suppose now that $0 < L(C) < \infty$, and consider the function $L(a_0, u, \mathfrak{x})$ (see III.3.14). Let us put $s = L(a_0, u, \mathfrak{x})$. Then s is a non-negative, nondecreasing, continuous function of u in I_0, and $s(a_0) = 0$, $s(b_0) = L(C)$ (see III.3.14). We shall say that a representation (1) is *in terms of the arc-length* if and only if $a_0 = 0$ and $L(0, u, \mathfrak{x}) \equiv u$ in I_0. In this case, clearly $b_0 = L(0, b_0, \mathfrak{x}) = L(C)$. Thus a representation in terms of the arc-length is of the form

$$C : \mathfrak{x} = \mathfrak{x}(u), \qquad 0 \leqq u \leqq L(C),$$

with the additional property $L(0, u, \mathfrak{x}) \equiv u$. Assuming that $0 < L(C) < \infty$, we assert that such a representation exists.

Proof. Since $L(C) > 0$, the curve C does not reduce to a point, and hence by II.3.22 it admits of a light representation

$$C : \mathfrak{x} = \mathfrak{x}^*(u^*), \qquad u^* \in I_0^* : 0 \leqq u^* \leqq 1. \tag{2}$$

Consider the corresponding function $L(0, u^*, \mathfrak{x}^*)$. Since the representation (2) is light, $\mathfrak{x}^*(u^*)$ is not constant on any interval in I_0^*, and hence $L(0, u^*, \mathfrak{x}^*)$ is a strictly increasing, continuous function of u^* in I_0^*, and $L(0, 0, \mathfrak{x}^*) = 0$, $L(0, 1, \mathfrak{x}^*) = L(C)$ (see III.3.14, III.3.95). Hence the equation $s = L(0, u^*, \mathfrak{x}^*)$ defines a topological transformation $h(u^*) = s$ from I_0^* onto the interval $0 \leqq s \leqq L(C)$. By II.3.11 we have therefore for C the representation

$$(3) \qquad C : \mathfrak{x} = \mathfrak{z}(s) = \mathfrak{x}^*[h^{-1}(s)], \qquad 0 \leqq s \leqq L(C).$$

By the reasoning employed in III.3.94(i) it follows that $h(u^*) = L(0, u^*, \mathfrak{x}^*) = L[0, h(u^*), \mathfrak{z}]$. On setting $h(u^*) = s$, we obtain the identity $s = L(0, s, \mathfrak{z})$ for $0 \leqq s \leqq L(C)$. Thus the representation (3) is in terms of the arc-length.

III.3.98. CONTINUATION. In a representation in terms of the arc-length, we shall use s to denote the independent variable, in conformity with general usage. Given a curve C such that $0 < L(C) < \infty$, let

$$(1) \qquad C : \mathfrak{x} = \mathfrak{x}(s), \qquad 0 \leqq s \leqq L(C),$$

be a representation in terms of the arc-length (see III.3.97). By assumption, $L(0, s, \mathfrak{x}) \equiv s$, and hence the interval function $L(I, \mathfrak{x})$ satisfies the relation $L(I, \mathfrak{x}) = | I |$ for every interval $I : a \leqq s \leqq b$ in I_0 (see III.3.13). Thus $L(I, \mathfrak{x})$ is obviously AC, and hence $\mathfrak{x}(s)$ is also AC (see III.3.15). Furthermore, obviously the derivative of $L(I, \mathfrak{x})$ is equal to 1 identically, and hence $| \mathfrak{x}'(s) | = 1$ a.e. in the interval $0 \leqq s \leqq L(C)$ (see III.3.13). Since $g_1(I, \mathfrak{x}) \leqq g(I, \mathfrak{x}) \leqq L(I, \mathfrak{x}) = | I |$ (see III.3.2, III.3.4), we see that the x-component $x(s)$ of $\mathfrak{x}(s)$ satisfies the condition $| x(b) - x(a) | \leqq b - a$ for every interval $a \leqq s \leqq b$ in the interval $0 \leqq s \leqq L(C)$. Similar remarks apply to the components $y(s)$, $z(s)$ of $\mathfrak{x}(s)$. In other words, the components $x(s)$, $y(s)$, $z(s)$ are Lipschitzian functions in the interval $0 \leqq s \leqq L(C)$ (see I.3.14).

III.3.99. Given a curve C by a representation

$$(1) \qquad C : \mathfrak{x} = \mathfrak{x}(u), \qquad u \in I_0 : a_0 \leqq u \leqq b_0 ,$$

we shall say that the representation is BV (AC) if and only if the vector function $\mathfrak{x}(u)$ is BV (AC) in I_0 (see III.3.11). Our previous study of vector functions yields then readily the following fundamental results.

(i) *If $L(C) < \infty$, then every representation of C is BV* (see III.3.95, III.3.12).

(ii) *If one representation of C is BV, then every representation of C is BV, and* $L(C) < \infty$ (see III.3.12, III.3.95, and (i)).

(iii) *Given C by* (1), *suppose that $L(C) < \infty$. Then* (see III.3.95, III.3.16) *$\mathfrak{x}'(u)$ exists a.e. in I_0, $| \mathfrak{x}'(u) |$ is summable in I_0, and*

$$\int_{I_0} | \mathfrak{x}'(u) | \, du \leqq L(C),$$

the sign of equality holding if and only if the representation (1) *is AC.*

(iv) *If $L(C) < \infty$, then C admits of an AC representation, and in fact of a representation of the form* (1) *such that the components of $\mathfrak{x}(u)$ are Lipschitzian functions. Indeed, if $L(C) = 0$, then $\mathfrak{x}(u)$ is constant for every representation of C, and the assertion is obvious. If $L(C) > 0$, then the representation in terms of the arc-length possesses the desired properties* (see III.3.98).

III.3.100. Continuation. The following remark throws more light upon the preceding theorem (iii). *Let C be any curve. Then C admits of a representation $\mathfrak{x} = \mathfrak{x}(u)$, $u \in I_0$, such that $\mathfrak{x}'(u)$ exists and is equal to zero a.e. in I_0.*

Proof. Let us start with any representation of the form (see III.3.94)

$$C : \mathfrak{x} = \mathfrak{x}^*(u^*), \qquad u^* \in I_0^* : 0 \leqq u^* \leqq 1.$$

Let $f(u)$ denote the Cantor ternary function in the unit interval $I_0 : 0 \leqq u \leqq 1$. Then $f(u)$ is continuous and nondecreasing in I_0, and $f(0) = 0$, $f(1) = 1$. Let us define, for each positive integer n,

$$f_n(u) = \frac{f(u) + u/n}{1 + 1/n}, \qquad u \in I_0 .$$

Then clearly $f_n(u)$ is continuous and strictly increasing in I_0, and $f_n(0) = 0$, $f_n(1) = 1$. Hence the equation $u^* = f_n(u)$ defines a topological transformation from I_0 onto I_0^*. We have therefore (see II.3.11) for C the representation $\mathfrak{x} = \mathfrak{x}^*[f_n(u)]$, $u \in I_0$. Since clearly $f_n(u) \to f(u)$ uniformly in I_0, by II.3.13 there follows for C the representation $\mathfrak{x} = \mathfrak{x}(u) = \mathfrak{x}^*[f(u)]$, $u \in I_0$. Clearly, if u_0 is interior to an interval of constancy of $f(u)$, then $\mathfrak{x}'(u_0)$ exists and is equal to zero. Since the sum of the lengths of the intervals of constancy of $f(u)$ is equal to 1, it follows that $\mathfrak{x}'(u)$ exists and is equal to zero a.e. in I_0.

III.3.101. Let $\mathfrak{x}(u)$ be a continuous vector function in $I_0 : a_0 \leqq u \leqq b_0$. Then $\mathfrak{x}(u)$ determines a curve $C : \mathfrak{x} = \mathfrak{x}(u)$, $u \in I_0$, whose length is equal to $L(I_0, \mathfrak{x})$ (see III.3.95). These remarks yield a geometrical interpretation, and in fact the motivation, for the study of vector functions which was undertaken earlier in this chapter.

III.3.102. Let now Γ be a Fréchet curve of the type of the 1-sphere. Then Γ admits of a representation of the form $\mathfrak{x} = \mathfrak{x}(p)$, $p \in C$, where C is an arbitrarily assigned simple closed curve (see II.3.23). The length of Γ is then defined by the formula $L(\Gamma) = L(C, \mathfrak{x})$ (see III.3.81). If $L(\Gamma) < +\infty$, then Γ is termed rectifiable. The further discussion is entirely analogous to that for Fréchet curves of the type of the 1-cell. By using a transformation $T(I_0) = C$ as indicated in III.3.83, results analogous to those in III.3.99 may be derived. The task of following up these hints in detail is left to the reader.

CHAPTER III.4. GENERAL COMMENTS ON ARC-LENGTH AND RELATED TOPICS

III.4.1. The reader is requested to read again at this time the introductory remarks on arc-length in I.1.1-I.1.8. While in chapter I the terms *path curve* and *arc-length* were used in a vague sense, we have now precise formal definitions at our disposal. If $x = x(u)$, $y = y(u)$, $z = z(u)$ are any three continuous functions in an interval $I : a \leqq u \leqq b$, then the formulas

$$(1) \qquad C : x = x(u), \qquad y = y(u), \qquad z = z(u), \qquad u \in I : a \leqq u \leqq b,$$

represent a curve C (namely a Fréchet curve of the type of the 1-cell) in a perfectly precise sense (see II.3.7). Furthermore, the length $L(C)$ of C is now a precisely defined quantity (see III.3.95) which is independent of the particular choice of a representation (1) for C and is also independent of the particular choice of the Cartesian coordinate system xyz. Still, it is a matter of considerable interest to study *arc-length as a functional* $L(\mathfrak{x})$ *of a continuous vector function* $\mathfrak{x}(u) = [x(u), y(u), z(u)]$, without reference to an invariant geometrical interpretation (see the remarks in III.3.1). From the point of view of exposition, there results the advantage that the reader is first acquainted with the more familiar analytic aspects of the theory, instead of being compelled to begin with a study of the topological material necessary for an invariant treatment. From the point of view of the development of the theory of surface area in greatest possible analogy with the theory of arc-length, the advantage is very real. Indeed, the profound differences between arc-length and surface area arise mainly on account of the entirely different character of the topological problems involved, while on the other hand the type of Analysis employed in the study of arc-length proves most useful in the study of surface area, both by direct analogy and by plausible generalization. Hence we shall follow, in this brief survey of Part III, the pattern laid down in Chapter III.3, and we shall discuss arc-length mainly as a functional $L(\mathfrak{x})$ of a continuous vector function $\mathfrak{x}(u)$, except for the next section III.4.2 which is concerned with an appraisal of some of the main results in invariant form.

III.4.2. Let C be a Fréchet curve of the type of the 1-cell (see II.3.7), and let $L(C)$ be the length of C (see III.3.95). As noted in III.3.97, $L(C) = 0$ if and only if C reduces to a point, in the sense explained there. It is revealing that even for this entirely trivial statement the corresponding result in surface area theory lies quite deep, and in fact the corresponding question in surface area theory leads to paradoxical phenomena (see V.2.55, V.2.70, V.2.71). The next most immediate question is concerned with conditions for $L(C)$ being finite. According to III.3.99, $L(C) < \infty$ if and only if every representation of C, of the form III.4.1(1), is of bounded variation. A corresponding statement for surface area seems to depend upon the solution of an apparently very difficult open problem (see V.2.65, V.2.66).[1] Let us now consider the case $0 < L(C) < \infty$. According to III.3.98, the arc length of C may then be introduced as a parameter, yielding a representation which is optimal in every relevant respect. No corresponding result, of comparable scope and simplicity, is known for surface area.

Finally, let us consider the point set σ that corresponds, in xyz-space, to the points u of the interval I by means of a representation III.4.1(1) of C. If $L(C) < \infty$, then by III.3.76 the three-dimensional measure of σ is equal to zero. If the z-coordinate $z(u)$ vanishes, then σ lies in the xy-plane, and by III.3.76 the two-dimensional measure of σ is equal to zero. Briefly, if $L(C) < \infty$, then σ is a slender set, both topologically and metrically. No similar statement holds for surface area (see I.1.13-I.1.16, V.2.71).

The preceding remarks suffice to call again attention to the fact, already emphasized in chapter I.1, that *the theory of arc-length cannot be used indiscriminately as a model in constructing a theory of surface area.* And yet, the theory of arc-length yields fundamentally important information from this point of view, besides being of course of independent interest also, especially as one of the most attractive applications of the modern theory of functions of a real variable. The selection of topics in part III has been largely determined by these considerations.

III.4.3. As explained above, we consider now arc-length as a functional $L(\mathfrak{x})$ defined for all continuous vector functions $\mathfrak{x}(u) = [x(u), y(u), z(u)]$ (see III.3.4). For definiteness, let us assume that only continuous vector functions $\mathfrak{x}(u)$ defined in a fixed interval $I : a \leqq u \leqq b$ are considered. The resulting theory admits of a generalization of striking scope to the theory of surface area for *the nonparametric case*, that is, the case of surfaces of the form $z = f(x, y)$ (see V.3). As a matter of fact, many results and methods were first developed for surface area in the nonparametric case. For this reason, we present the material included in part III in a form suitable for application to surface area in the nonparametric case, and in several important instances we study topics unnecessary for the theory of arc-length but so closely related to methods used in this theory that considerable economy results by inclusion in part III.

III.4.4. *Functions of intervals* play an important role in the study of $L(\mathfrak{x})$, and also in surface area theory. Accordingly, the necessary information is collected in Chapter III.1. Since the dimension of the intervals involved is irrelevant, as far as the topics discussed in Chapter III.1 are concerned, we discussed there the two-dimensional case, using the term *oriented rectangle* in the sense of *two-dimensional interval.* Thus the results established in Chapter III.1 are directly applicable in surface area theory, while the easy task of rewording the results for the one-dimensional case is left to the reader. Excellent comprehensive treatments of the theory of interval functions may be found in Saks [6] and Kempisty [4]. Hence Chapter III.1 includes only material that is absolutely

---

[1]This problem has been solved by L. Cesari [6]. See the comments in V.4.8.

needed for our purposes. Several fundamental concepts are due to Banach [1], in particular the concepts of *bounded variation* and *absolute continuity* as applied to interval functions (see III.1.2). The concept of *interval functions of type* A (see III.1.28) is also due, essentially, to Banach [1]. Interval functions of type A exhibit a fundamental analogy with monotone functions of a single variable (see the basic theorem in III.1.28) and are of general occurrence in the theory of length and area. These circumstances account for their exceptional importance for our purposes. *The Burkill integral* (see III.1.2, and Saks [6] for bibliography) is an invaluable tool. A fundamental result concerning the derivative of the Burkill integral (see III.1.27) is due to Saks [4]. An important issue arises in connection with the *completely additive extension of a given interval function.* Let $\Phi(R)$ be a function of oriented rectangles R. In the simplest case, when $\Phi(R)$ is equal to the area $|R|$ of R, the Lebesgue (two-dimensional) measure yields the completely additive extension of the rectangle function $|R|$. The existence of this completely additive extension is of course an altogether fundamental and useful fact. By analogy, it is clear that in dealing with any given rectangle function $\Phi(R)$ the existence or nonexistence of a completely additive extension represents an issue of great importance. Sections III.1.35-III.1.43 are concerned with this issue; except for details of exposition, this material is taken from Reichelderfer and Ringenberg [6]. The necessary and sufficient condition (C) (see III.1.35) is remarkable because in the case of the rectangle function $|R|$ it reduces to one of the most obvious properties of the area in the elementary sense.

III.4.5. Chapter III.2 is concerned, in the main, with the concepts of *bounded variation* and *absolute continuity* for functions $f(u)$ of a single variable and for functions $f(u, v)$ of two variables. As regards the case of functions $f(u, v)$, the fundamental concepts and results are due to Tonelli [5]. Sections III.2.49-III.2.64 are devoted to this topic, which is treated here in a slightly more general form with regard to applications in surface area theory. In view of the results on interval functions, established in Chapter III.1, a great deal of indispensable classical material could be treated economically in Chapter III.2. Sections III.2.38-III.2.44 and III.2.65-III.2.67 are concerned with the useful method of *approximation by integral means.* Initial application of this method in the theory of length and area may be found in W. H. Young [6], H. E. Bray [1], and T. Radó [2]. Further relevant information is contained in Helsel and Young [1], Helsel [2], P. M. Young [1]. Chapter III.2 contains also a geometrical study of fundamental importance, initiated by Banach [2], of the concepts of bounded variation and absolute continuity. Let $f(u)$ be continuous in the interval $I: a \leqq u \leqq b$. The formula $x = f(u)$, $u \in I$, defines then a continuous mapping from I into the x-axis. Let $N(x, I)$ denote the number of distinct inverse points of x in I under the mapping $x = f(u)$. Banach observed that $f(u)$ *is of bounded variation if and only if* $N(x, I)$ *is summable.* This observation, jointly with various other facts relating to the mapping $x = f(u)$ (see III.2.4-III.2.9) led to applications and generalizations of great importance in the theory of surface area. In particular, we shall see later on that *the result in* III.2.9 *reveals one of the most relevant reasons of the essential topological simplicity of the theory of arc length, as compared with the theory of surface area.* The results in III.2.5-III.2.9, which are based on conversations between the writer and P. V. Reichelderfer, will be applied in Chapter V.3 to derive interesting geometrical interpretations of the concepts of absolute continuity and of bounded variation introduced by Tonelli.

III.4.6. Chapter III.3 is devoted to the functional $L(\mathfrak{x})$ (see III.4.3). The classical theorems in III.3.12, III.3.16 are especially impressive as regards comparison with surface area theory. Indeed, *these theorems reveal a perfect adjustment between the concepts of arc-length, integral, derivative, bounded variation, absolute continuity. The development of corresponding concepts for surface area, exhibiting equally perfect adjustment, may be considered as one of the main objectives of the theory presented in this book.* Equally challenging, from the point of view of surface area theory, is the availability of three equivalent definitions for $L(\mathfrak{x})$, one *from below*, one *from above*, and one in terms of a *simple limit process* (see III.3.9). Further topics, of great interest for surface area, include the following. The *inequality of Steiner*, expressed by the formula $L(\mathfrak{r}_1 + \mathfrak{r}_2) \leqq L(\mathfrak{r}_1) + L(\mathfrak{r}_2)$ (see III.3.10 and the remarks in I.1.13) leads to a number of relevant questions. In particular, one may ask for conditions for the sign of equality to hold. More generally, one may ask for implications of the assumption that the difference $L(\mathfrak{r}_1) + L(\mathfrak{r}_2) - L(\mathfrak{r}_1 + \mathfrak{r}_2)$ is small. This question was studied by McShane [1], whose results were further improved by Radó and Reichelderfer [16]. In fact, the work just referred to is concerned with surface area in the nonparametric case, but the rewording for arc-length is entirely obvious. The results concerning the inequality of Steiner are relevant for the study of *convergence in length*. Let $\mathfrak{r}_n(u)$, $n = 0, 1, 2, \cdots$, be a sequence of continuous vector functions in an interval $I : a \leqq u \leqq b$, such that $\mathfrak{x}_n(u) \to \mathfrak{x}_0(u)$ uniformly in I. Then $L(\mathfrak{x}_0) \leqq \liminf L(\mathfrak{x}_n)$ (see III.3.6). If $L(\mathfrak{x}_0) = \lim L(\mathfrak{x}_n)$, then $\mathfrak{x}_n(u)$ is said to converge in length to $\mathfrak{x}_0(u)$, in symbols $\mathfrak{x}_n \to \mathfrak{x}_0(L)$. Thus convergence in length is a strong type of convergence whose implications are both interesting and relevant. A detailed study of convergence in length was first given by Adams and Clarkson [1] and Adams and Lewy [2]. Further studies were made by McShane [1] and Radó and Reichelderfer [20], whose work is concerned, in whole or in part, also with the analogous concept of *convergence in area.* These studies were continued and completed, and the methods were improved as regards economy and elegance, by Miriam Ayer [1], [2]. A significant point, revealed by the investigation of the phenomena connected with convergence in length, is a discrepancy between the parametric and nonparametric case. This discrepancy is of interest for two reasons. First, the parametric and nonparametric cases in surface area theory present a striking contrast as regards topological and analytic difficulties, and thus a study of the corresponding situation for arc-length is instructive. Second, the parametric and nonparametric cases in arc-length theory are generally entirely analogous, and thus it may be hoped that the gap, revealed by the study of convergence in length, will be ultimately bridged somehow. The results of Miriam Ayer seem to justify this hope. The presentation of all these topics in chapter III.3 follows closely that of Miriam Ayer [2]. In particular, the theorem on Lebesgue decomposition (see III.3.26) and the simple vector inequalities in III.3.36-III.3.38 prove most useful in achieving an improved exposition. A complete clarification of the relations between the various convergence types, studied in Miriam Ayer [2], may be of interest, and extension of the results to surface area seems to involve problems of great importance and difficulty.

The important paper of A. P. Morse [1] came to the attention of the writer only after the manuscript of this book was completed. Many of the results discussed in chapter III.3, relating to convergence in length and convergence in variation, are due to A. P. Morse or are more or less direct consequences of his results. Reference should be made here also to the comprehensive researches on arc length from the measure-theoretical point of view, a topic which lies beyond the scope of this book.

III.4.7. The sections III.3.77-III.3.93 are concerned with *closed curves* (Fréchet curves of the type of the 1-sphere). Let us call attention to a set of relevant questions that arise in this connection. For brevity, we shall use a geometrical terminology whose precise meaning the reader will readily determine by comparison with the various sections involved of Chapter III.3. Let K denote the perimeter of the unit circle, and let $x(p)$, $y(p)$ be continuous functions of the point $p \in K$. The formulas $x = x(p)$, $y = y(p)$, $p \in K$, determine then a closed continuous oriented curve C in the xy-plane, where K itself is oriented in the counterclockwise sense. If (x, y) is a point not on C, then let $\mu(x, y)$ denote the topological index of (x, y) with respect to C. If (x, y) is a point on C, then we put $\mu(x, y) = 0$. If C is of a simple character, a polygon for example, then *the signed area enclosed by* C is clearly equal to the double integral of $\mu(x, y)$ (note that $\mu(x, y)$ vanishes outside of a sufficiently large disc, and thus $\iint \mu(x, y)\, dx\, dy$ is thought of as being taken over such a disc). Thus *in simple cases* we have the formula

$$\sigma(C) = \iint \mu(x, y)\, dx\, dy, \tag{1}$$

where $\sigma(C)$ denotes the signed area enclosed by C. Also, by virtue of a well known Calculus formula we have in simple cases

$$\sigma(C) = \frac{1}{2} \int_K (x\, dy - y\, dx) \tag{2}$$

The determination of the range of validity of these formulas is of course an interesting and important problem. W. H. Young [4], [5], [6], in particular, has studied this question in great detail. He defined $\sigma(C)$ as follows: let P_n denote a sequence of approximating polygons, inscribed in C, represented by equations $x = x_n(p)$, $y = y_n(p)$ and let $\mu_n(x, y)$ denote the topological index relative to P_n. Then, *by definition*,

$$\sigma_Y(P_n) = \iint \mu_n(x, y)\, dx\, dy, \tag{3}$$

and also *by definition*

$$\sigma_Y(C) = \lim_{n\to\infty} \iint \mu_n(x, y)\, dx\, dy, \tag{4}$$

provided that the limit exists and is independent of the particular choice of the sequence of inscribed polygons P_n. The notation σ_Y is used to refer to W. H. Young as the author of the definition (4). The definition (4) raises a number of questions, some of which we shall state presently.

(i) In view of (1), which is obvious in simple cases, we may ask whether

$$\sigma_Y(C) = \iint \mu(x, y)\, dx\, dy. \tag{5}$$

Quite precisely: *if $\sigma_Y(C)$ exists, does it follow that $\mu(x, y)$ is summable and* (5) *holds? Conversely, if $\mu(x, y)$ is summable, does it follow that $\sigma_Y(C)$ exists and* (5) *holds?*

(ii) Concerning $\mu(x, y)$ we made the arbitrary agreement that $\mu(x, y) = 0$ if the point (x, y) lies on C. Thus in cases where the points on C constitute a point set S of positive two-dimensional measure, the evidence in favor of (5) may be considered inconclusive. Therefore, let us raise *the same questions as in* (i) *under the assumption that the set S is of two-dimensional measure zero.*

(iii) *Under what conditions does $\sigma_Y(C)$ exist?*

W. H. Young apparently did not raise these questions in a general form. One of his results, stated in a somewhat restricted form, shows that *if C is rectifiable, then $\sigma_Y(C)$ exists and*

$$\sigma_Y(C) = \frac{1}{2} \int_K (x\, dy - y\, dx). \tag{6}$$

However, he did not raise the question whether (5) holds in this simple case. The results in III.3.88-III.3.93, due to T. Radó [10], are concerned with issues that arise in connection with the work of W. H. Young. In particular, the theorem of III.3.88, stating that

$$\iint \mu(x, y)\, dx\, dy = \frac{1}{2} \int_K (x\, dy - y\, dx) \tag{7}$$

whenever C is rectifiable, completes the result of W. H. Young expressed by (6). Also, (7) may be considered as a general justification of the familiar area-formula in Calculus. The theorems in III.3.90 and III.3.92 yield information concerning *the continuity of the signed area enclosed by a curve C, considered as a functional of this curve.* However, further studies along the lines just indicated are clearly desirable.

PART IV. PLANE TRANSFORMATIONS

CHAPTER IV.1. TOPOLOGICAL FOUNDATIONS

IV.1.1. We shall study topological properties of transformations given by formulas of the form

$$T : x = x(u, v), \qquad y = y(u, v), \qquad (u, v) \in E,$$

where $x(u, v)$, $y(u, v)$ are single-valued functions defined on a set E in the uv-plane. For technical reasons it will be convenient to introduce the complex variables $w = u + iv$, $z = x + iy$. If we put $x(u, v) + iy(u, v) = t(w)$, then T is given in the form

$$T : z = t(w), \qquad w \in E.$$

Biuniqueness of the transformation will not be assumed unless the contrary is explicitly stated. That is, every point $w \in E$ has a unique image $z = t(w)$, but a point z may have any number of inverse points in E. In other words, the equation $z = t(w)$ may have any number of solutions in E for given z. The variables u, v and x, y denote Cartesian coordinates in the Euclidean planes uv, xy respectively. We shall term these planes also the w- and z-planes respectively. In agreement with I.2.5, we shall employ the following terminology. If z is any point, then $T^{-1}(z)$ denotes the set of all those points $w \in E$ for which $t(w) = z$ holds. If the set E is not clearly identified by the context, we write more explicitly $ET^{-1}(z)$. If G is any set in the w-plane, then $N(z, G)$ denotes the number (possibly zero or infinite) of the points of the set $GT^{-1}(z)$. Thus $N(z, G)$ may have for its value any non-negative integer or $+\infty$. If the transformation T is not clearly indicated by the context, then we may use the more explicit notation $N(z, T, G)$. Given a second transformation $T^* : z = t^*(w)$, $w \in E$, and any set G in the w-plane, we define $\rho(T, T^*, G)$ as the least upper bound of $| t^*(w) - t(w) |$ for $w \in EG$ if $EG \neq 0$, and we put $\rho(T, T^*, G) = 0$ if $EG = 0$. Clearly, if T_1, T_2, T_3 are three transformations defined on E, then we have the triangle inequality

$$\rho(T_1, T_3, G) \leqq \rho(T_1, T_2, G) + \rho(T_2, T_3, G),$$

for every set G. Obviously, $\rho(T, T^*, G) = 0$ implies that $T \equiv T^*$ on EG, this statement being vacuous if $EG = 0$. If T_n is a sequence of transformations defined on E, and if $\rho(T, T_n, G) \to 0$ for $n \to \infty$, then we shall write $T_n \to T$, $w \in G$. Given $T : z = t(w)$, $w \in E$, and given a number $\delta > 0$ and a set G, we shall use $\omega(\delta, T, G)$ to denote the least upper bound of $| t(w_1) - t(w_2) |$ for all pairs of points w_1, w_2 in EG such that $| w_1 - w_2 | \leqq \delta$, it being understood that

$\omega(\delta, T, G) = 0$ if $EG = 0$. Of course notations and concepts introduced in I.2 will be used freely.

IV.1.2. We shall use $\mathfrak{R}$, with superscripts and subscripts as needed, as a generic notation to refer to a bounded, finitely-connected Jordan region in the w-plane. The set of interior points of $\mathfrak{R}$ will be denoted by $\mathfrak{R}^0$ (in a general way, if E is any set in a topological space, then E^0 will denote the set of its interior points). Then $\mathfrak{R} - \mathfrak{R}^0$ is the boundary of $\mathfrak{R}$; it consists of a finite number of simple closed curves (see I.2.47). We shall be especially concerned with continuous transformations from $\mathfrak{R}$ into the z-plane (cf. I.2.29). Such a transformation may be given (cf. IV.1.1) either in the real form $T : x = x(u, v)$, $y = y(u, v)$, $(u, v) \in \mathfrak{R}$, or else equivalently in the complex form $T : z = t(w)$, $w \in \mathfrak{R}$. In the first case, $x(u, v)$ and $y(u, v)$ are single-valued, real-valued, continuous functions in $\mathfrak{R}$, in the second case $t(w)$ is a single-valued, complex-valued, continuous function in $\mathfrak{R}$. Since $\mathfrak{R}$ is compact (see I.2.17), T is necessarily bounded in $\mathfrak{R}$; that is, there exists a circular disc Δ in the z-plane, such that $T(\mathfrak{R}) \subset \Delta$. Clearly, $N(z, T, \mathfrak{R}) = 0$ for $z \notin T(\mathfrak{R})$ and hence *a fortiori* for $z \notin \Delta$.

IV.1.3. Given T as in IV.1.2, let k be a non-negative integer. We define in the z-plane a set $\mathfrak{K}(k, T, \mathfrak{R})$, to be termed the *kernel of order k of the image of $\mathfrak{R}$ under T*, as follows. A point z_0 belongs to $\mathfrak{K}(k, T, \mathfrak{R})$ if and only if there exists an $\epsilon = \epsilon(k, T, \mathfrak{R}, z_0) > 0$, such that if a continuous transformation $T^* : z = t^*(w)$, $w \in \mathfrak{R}$, satisfies the inequality $\rho(T^*, T, \mathfrak{R}) < \epsilon$, then it also satisfies the inequality $N(z_0, T^*, \mathfrak{R}) \geqq k$ (cf. IV.1.1). Briefly: for every continuous transformation T^*, defined on $\mathfrak{R}$ and sufficiently close to T, the point z_0 has at least k distinct inverse points under T^* in $\mathfrak{R}$. How close T^* should be to T for this purpose depends upon the point z_0 and of course also upon T, $\mathfrak{R}$, and k. The notation $\epsilon(k, T, \mathfrak{R}, z_0)$ is meant to stress this fact which must be kept in mind if errors are to be avoided. Clearly $\mathfrak{K}(k, T, \mathfrak{R}) \supset \mathfrak{K}(k + 1, T, \mathfrak{R})$, $k = 0, 1, 2, \cdots$. The set $\mathfrak{K}(0, T, \mathfrak{R})$ coincides with the whole (finite) z-plane. For $k > 0$ the set $\mathfrak{K}(k, T, \mathfrak{R})$ may be empty. We proceed to define $\mathfrak{K}(\infty, T, \mathfrak{R})$. *The definition given for finite k remains meaningful for $k = +\infty$, but we use, for important technical reasons, a different definition in this case, a fact that should be kept in mind.* We define

$$\mathfrak{K}(\infty, T, \mathfrak{R}) = \prod_{k=0}^{\infty} \mathfrak{K}(k, T, \mathfrak{R}).$$

The notation on the right conforms to standard practice, according to which it means that k assumes all non-negative, integral and finite values, but of course the substitution $k = +\infty$ is not permitted on the right. Otherwise our definition of $\mathfrak{K}(\infty, T, \mathfrak{R})$ would be of course meaningless.

IV.1.4. Continuation. We define now in the z-plane a function $\kappa(z, T, \mathfrak{R})$ as follows: If $z \in \mathfrak{K}(\infty, T, \mathfrak{R})$, then $\kappa(z, T, \mathfrak{R}) = +\infty$. Otherwise there exists a largest non-negative integer l such that $z \in \mathfrak{K}(l, T, \mathfrak{R})$. We put then $\kappa(z, T, \mathfrak{R}) = l$. That is, $\kappa(z, T, \mathfrak{R}) = l$ if and only if

$$z \in \mathfrak{K}(l, T, \mathfrak{R}) - \mathfrak{K}(l + 1, T, \mathfrak{R}),$$

where l is any non-negative integer, and $\kappa(z, T, \mathfrak{R}) = +\infty$ if and only if $z \in \mathfrak{K}(l, T, \mathfrak{R})$ for every non-negative integer l. We assert the inequality $\kappa(z, T, \mathfrak{R}) \leqq N(z, T, \mathfrak{R})$. Indeed, this is obvious if $N(z, T, \mathfrak{R}) = +\infty$. So we can assume that $N(z, T, \mathfrak{R}) = k$, where k is a non-negative integer. Then for every $\epsilon > 0$, $\rho(T, T, \mathfrak{R}) = 0 < \epsilon$ and $N(z, T, \mathfrak{R}) < k + 1$. Hence $z \notin \mathfrak{K}(k + 1, T, \mathfrak{R})$, and thus $\kappa(z, T, \mathfrak{R}) \leqq k = N(z, T, \mathfrak{R})$.

$\kappa(z, T, \mathfrak{R})$ will be termed *the essential multiplicity* of the point z in the image of $\mathfrak{R}$ under T.

IV.1.5. CONTINUATION. In the sequel we shall have to verify frequently an inequality of the type

(1) $$k \leqq \kappa(z_0, T, \mathfrak{R}),$$

where k is a non-negative integer and z_0 is a given point, and hence we insert here the following obvious remark. By the definition of $\kappa(z_0, T, \mathfrak{R})$, (1) is equivalent to the relation $z_0 \in \mathfrak{K}(k, T, \mathfrak{R})$, which is equivalent to the statement that

(2) $$N(z_0, T^*, \mathfrak{R}) \geqq k \qquad \text{if } \rho(T^*, T, \mathfrak{R}) < \epsilon(k, T, \mathfrak{R}, z_0),$$

where T^* is assumed to be defined and continuous in $\mathfrak{R}$. Thus to establish (1), it is sufficient to verify (2), and conversely (1) implies (2). This remark enables us to treat the cases $\kappa(z_0, T, \mathfrak{R}) = +\infty$ and $\kappa(z_0, T, \mathfrak{R}) < +\infty$ in a uniform manner.

IV.1.6. Given two continuous transformations (cf. IV.1.2)

$$T_1 : z = t_1(w_1), \qquad w_1 \in \mathfrak{R}_1,$$

$$T_2 : z = t_2(w_2), \qquad w_2 \in \mathfrak{R}_2,$$

where the w_1- and w_2-planes may or may not coincide, let us suppose that $T_1 \sim T_2(ts)$ (see II.1.26). In other words, we assume the existence of a homeomorphism $\tau(\mathfrak{R}_1) = \mathfrak{R}_2$, such that $t_1(w_1) = t_2\tau(w_1)$ for every point $w_1 \in \mathfrak{R}_1$. We have then also $t_2(w_2) = t_1\tau^{-1}(w_2)$ for $w_2 \in \mathfrak{R}_2$. Clearly $N(z, T_1, \mathfrak{R}_1) \equiv N(z, T_2, \mathfrak{R}_2)$. Let now $T_1^* : z = t_1^*(w_1)$, $w_1 \in \mathfrak{R}_1$, be a continuous transformation, and let us put $T_2^* = T_1^*\tau^{-1}$. Then T_2^* may be given in the form $T_2^* : z = t_2^*(w_2)$, $w_2 \in \mathfrak{R}_2$, where $t_2^*(w_2) = t_1^*\tau^{-1}(w_2)$. It follows that

$$| t_1^*(w_1) - t_1(w_1) | = | t_2^*\tau(w_1) - t_2\tau(w_1) | = | t_2^*(w_2) - t_2(w_2) |,$$

where $w_2 = \tau(w_1)$. Thus clearly (see IV.1.1)

$$\rho(T_1, T_1^*, \mathfrak{R}_1) = \rho(T_2, T_2^*, \mathfrak{R}_2), \qquad N(z, T_1^*, \mathfrak{R}_1) = N(z, T_2^*, \mathfrak{R}_2).$$

As an immediate consequence of these relations, it follows that (cf. IV.1.3, IV.1.4)

$$\mathfrak{K}(k, T_1, \mathfrak{R}_1) \equiv \mathfrak{K}(k, T_2, \mathfrak{R}_2), \qquad \mathfrak{K}(\infty, T_1, \mathfrak{R}_1) \equiv \mathfrak{K}(\infty, T_2, \mathfrak{R}_2),$$

$$\kappa(z, T_1, \mathfrak{R}_1) \equiv \kappa(z, T_2, \mathfrak{R}_2), \qquad \epsilon(k, T_1, \mathfrak{R}_1, z) = \epsilon(k, T_2, \mathfrak{R}_2, z).$$

The last equation is to be interpreted as follows: if an $\epsilon > 0$ can be used as $\epsilon(k, T_1, \mathfrak{R}_1, z)$ in the sense of IV.1.3, then the same ϵ can also be used as $\epsilon(k, T_2, \mathfrak{R}_2, z)$.

Briefly: *topologically similar transformations give rise to identical essential multiplicity functions* κ. In particular, κ is unchanged if a new Cartesian coordinate system is introduced in the w-plane, since this operation is formally equivalent to passing from the given transformation T to a topologically similar transformation. The fact that κ is unchanged if we change the coordinate system in the z-plane is obvious.

IV.1.7. Given T as in IV.1.2, let z_0 be a point in the z-plane such that $\kappa(z_0, T, \mathfrak{R}) < +\infty$. Given then any $\eta > 0$, there exists a continuous transformation $T^* : z = t^*(w)$, $w \in \mathfrak{R}$, such that

$$\rho(T^*, T, \mathfrak{R}) < \eta, \qquad N(z_0, T^*, \mathfrak{R}) = \kappa(z_0, T, \mathfrak{R}).$$

This transformation T^* depends upon T, $\mathfrak{R}$, z_0, η.

Proof. Put $\kappa(z_0, T, \mathfrak{R}) = k < +\infty$. Consider the class K of all continuous transformations $T_* : z = t_*(w)$, $w \in \mathfrak{R}$, which satisfy the conditions (cf. IV.1.3)

$$\rho(T_*, T, \mathfrak{R}) < \eta, \qquad \rho(T_*, T, \mathfrak{R}) < \epsilon(k, T, \mathfrak{R}, z_0).$$

This class K is not empty, since clearly $T \in K$. By the definition of $\epsilon(k, T, \mathfrak{R}, z_0)$ (see IV.1.3) we have $N(z_0, T_*, \mathfrak{R}) \geqq k$ for $T_* \in K$. We assert that the sign of equality holds for some $T_* \in K$. Indeed, if this were not true, then the number min $[\eta, \epsilon(k, T, \mathfrak{R}, z_0)]$ would serve as an $\epsilon(k + 1, T, \mathfrak{R}, z_0)$ in the sense of IV.1.3. This would imply that $z_0 \in \mathfrak{K}(k + 1, T, \mathfrak{R})$, and hence $\kappa(z_0, T, \mathfrak{R}) \geqq k + 1$, while by assumption $\kappa(z_0, T, \mathfrak{R}) = k$. Hence we have a transformation $T^* \in K$ such that $N(z_0, T^*, \mathfrak{R}) = k = \kappa(z_0, T, \mathfrak{R})$, and (since $T^* \in K$) simultaneously $\rho(T^*, T, \mathfrak{R}) < \eta$.

IV.1.8. Given T as in IV.1.2, let z_0 be a point in the z-plane, and let k be a non-negative (finite) integer such that $\kappa(z_0, T, \mathfrak{R}) \geqq k$ (observe that $\kappa(z_0, T, \mathfrak{R})$ itself may be infinite). Let T^* be a continuous transformation in $\mathfrak{R}$ that satisfies the inequality (cf. IV.1.3)

$$\rho(T^*, T, \mathfrak{R}) < \epsilon(k, T, \mathfrak{R}, z_0). \tag{1}$$

Observe that $\epsilon(k, T, \mathfrak{R}, z_0)$ is available, since $\kappa(z_0, T, \mathfrak{R}) \geqq k$. By IV.1.3, the inequality (1) implies that $N(z_0, T^*, \mathfrak{R}) \geqq k$. We assert that we have also

$$\kappa(z_0, T^*, \mathfrak{R}) \geqq k. \tag{2}$$

Proof. Let us put $\epsilon^* = \epsilon(k, T, \mathfrak{R}, z_0) - \rho(T^*, T, \mathfrak{R})$. Then $\epsilon^* > 0$ by (1). Let now T_* be any continuous transformation in $\mathfrak{R}$ such that $\rho(T^*, T_*, \mathfrak{R}) < \epsilon^*$. In view of (1) it follows that

$$\rho(T_*, T, \mathfrak{R}) \leqq \rho(T_*, T^*, \mathfrak{R}) + \rho(T^*, T, \mathfrak{R}) < \epsilon(k, T, \mathfrak{R}, z_0).$$

Hence, by the definition of $\epsilon(k, T, \mathfrak{R}, z_0)$,

$$N(z_0, T_*, \mathfrak{R}) \geqq k. \tag{3}$$

Thus $\rho(T^*, T_*, \mathfrak{R}) < \epsilon^*$ implies (3). In other words, ϵ^* serves as an $\epsilon(k, T^*, \mathfrak{R}, z_0)$ in the sense of IV.1.3. Hence $z_0 \in \mathfrak{K}(k, T^*, \mathfrak{R})$ and (2) follows (see IV.1.4).

IV.1.9. Given T as in IV.1.2, we assert that for $k < +\infty$ the set $\mathfrak{K}(k, T, \mathfrak{R})$ is open (possibly empty). Consequently, $\mathfrak{K}(\infty, T, \mathfrak{R})$ is a G_δ (see IV.1.3, I.2.46).

PROOF. The assertion is obvious if $\mathfrak{K}(k, T, \mathfrak{R}) = 0$. So assume that $\mathfrak{K}(k, T, \mathfrak{R}) \neq 0$. Let z_0 be any point of $\mathfrak{K}(k, T, \mathfrak{R})$. Then $\epsilon(k, T, \mathfrak{R}, z_0)$ is available (cf. IV.1.3). Let z_* be any point that satisfies the inequality

$$(1) \qquad | z_0 - z_* | < \epsilon(k, T, \mathfrak{R}, z_0).$$

Let us put $\lambda = z_0 - z_*$, $\epsilon_* = \epsilon(k, T, \mathfrak{R}, z_0) - |\lambda|$. Then $\epsilon_* > 0$ in view of (1). Let $T_* : z = t_*(w)$, $w \in \mathfrak{R}$, be any continuous transformation that satisfies the inequality

$$(2) \qquad \rho(T_*, T, \mathfrak{R}) < \epsilon_* .$$

Consider the auxiliary transformation $T^* : z = t^*(w) = t_*(w) + \lambda$, $w \in \mathfrak{R}$. Then the equation $z_0 = t^*(w)$ is equivalent to the equation $z_* = t_*(w)$, and thus

$$(3) \qquad N(z_0, T^*, \mathfrak{R}) = N(z_*, T_*, \mathfrak{R}).$$

On the other hand

$$\rho(T^*, T, \mathfrak{R}) \leqq \rho(T^*, T_*, \mathfrak{R}) + \rho(T_*, T, \mathfrak{R}) < |\lambda| + \epsilon_* = \epsilon(k, T, \mathfrak{R}, z_0).$$

Hence, by the definition of $\epsilon(k, T, \mathfrak{R}, z_0)$, it follows that $N(z_0, T^*, \mathfrak{R}) \geqq k$, and hence, in view of (3),

$$(4) \qquad N(z_*, T_*, \mathfrak{R}) \geqq k.$$

Thus (2) implies (4). In other words, ϵ_* serves as an $\epsilon(k, T, \mathfrak{R}, z_*)$ in the sense of IV.1.3. Hence every point z_* that satisfies (1) belongs to $\mathfrak{K}(k, T, \mathfrak{R})$, and consequently $\mathfrak{K}(k, T, \mathfrak{R})$ is open.

IV.1.10. Given T as in IV.1.2, let us consider the set $T(\mathfrak{R})$. Suppose that this set has no interior points. Then $\kappa(z, T, \mathfrak{R}) \equiv 0$.

PROOF. Assume that there exists a point z_0 such that $\kappa(z_0, T, \mathfrak{R}) \neq 0$. Then $z_0 \in \mathfrak{K}(1, T, \mathfrak{R})$ (cf. IV.1.4). Hence, by IV.1.9, $\mathfrak{K}(1, T, \mathfrak{R})$ contains an open circular disc Δ with center z_0 and radius $\epsilon(1, T, \mathfrak{R}, z_0) > 0$. By IV.1.4, we have then $1 \leqq \kappa(z, T, \mathfrak{R}) \leqq N(z, T, \mathfrak{R})$ for $z \in \Delta$, and hence $\Delta \subset T(\mathfrak{R})$. Thus z_0 would be an interior point of $T(\mathfrak{R})$, in contradiction to the assumption that $T(\mathfrak{R})$ has no interior points.

IV.1.11. Given T as in IV.1.2, $\kappa(z, T, \mathfrak{R})$ is a lower semi-continuous function of z. That is, if $z_n \to z_0$, then

$$(1) \qquad \kappa(z_0, T, \mathfrak{R}) \leqq \liminf \kappa(z_n, T, \mathfrak{R}).$$

PROOF. Let k be any (finite) non-negative integer such that $k \leqq \kappa(z_0, T, \mathfrak{R})$. Then $z_0 \in \mathfrak{K}(k, T, \mathfrak{R})$. By IV.1.9, $\mathfrak{K}(k, T, \mathfrak{R})$ is then a nonempty open set, and hence $z_n \in \mathfrak{K}(k, T, \mathfrak{R})$ for n large, say $n > n_0$. Consequently $\kappa(z_n, T, \mathfrak{R}) \geqq k$ for $n > n_0$, and hence

$$\liminf_{n\to\infty} \kappa(z_n, T, \mathfrak{R}) \geqq k.$$

Since k was restricted only by the conditions $k < +\infty$, $k \leqq \kappa(z_0, T, \mathfrak{R})$, the inequality (1) follows.

IV.1.12. Given T as in IV.1.2, let $T_n : z = t_n(w)$, $w \in \mathfrak{R}$, be a sequence of continuous transformations such that $\rho(T_n, T, \mathfrak{R}) \to 0$ for $n \to \infty$. We assert that

$$(1) \qquad \kappa(z, T, \mathfrak{R}) \leqq \liminf \kappa(z, T_n, \mathfrak{R}) \qquad \text{for } n \to \infty.$$

Proof. Let z_0 be any point, and let k be any non-negative (finite) integer such that $k \leqq \kappa(z_0, T, \mathfrak{R})$. Then $z_0 \in \mathfrak{K}(k, T, \mathfrak{R})$, and hence $\epsilon(k, T, \mathfrak{R}, z_0)$ is available (see IV.1.3). For n large, say $n > n_0$, we shall have $\rho(T_n, T, \mathfrak{R}) < \epsilon(k, T, \mathfrak{R}, z_0)$. By IV.1.8 it follows that $\kappa(z_0, T_n, \mathfrak{R}) \geqq k$ for $n > n_0$. Hence for $n \to \infty$ we obtain $\liminf \kappa(z_0, T_n, \mathfrak{R}) \geqq k$. Since k was restricted only by the conditions $k < +\infty$, $k \leqq \kappa(z_0, T, \mathfrak{R})$, the inequality (1) follows.

IV.1.13. We observed in IV.1.4 that always $\kappa(z, T, \mathfrak{R}) \leqq N(z, T, \mathfrak{R})$. To get a preliminary view of the possibilities involved, let us consider the following example. Let $\phi(u)$, $\psi(u)$ be any two continuous functions in the interval $0 \leqq u \leqq 1$. For each positive integer n, let us subdivide this interval into n equal parts, and let us denote by $\phi_n(u)$ the function that agrees with $\phi(u)$ at the points of division and is linear on each one of the n subintervals. Let $\psi_n(u)$ be defined in the same manner in terms of $\psi(u)$. Let us now consider the transformations

$$T : x = x(u, v) = \phi(u), \qquad y = y(u, v) = \psi(u), \qquad (u, v) \in Q,$$

$$T_n : x = x_n(u, v) = \phi_n(u), \qquad y = y_n(u, v) = \psi_n(u), \qquad (u, v) \in Q,$$

where Q is the unit square $Q : 0 \leqq u \leqq 1, 0 \leqq v \leqq 1$. Clearly $\rho(T_n, T, Q) \to 0$ for $n \to \infty$ (cf. IV.1.1), and hence $\kappa(z, T, Q) \leqq \liminf \kappa(z, T_n, Q)$. On the other hand, the set $T_n(Q)$ consists clearly of a finite number of straight segments (some or even all of which may reduce to single points). Thus $T_n(Q)$ is a point set without interior points, and hence $\kappa(z, T_n, Q) \equiv 0$ by IV.1.10. Since $\kappa(z, T, Q) \leqq \liminf \kappa(z, T_n, Q)$, it follows that $\kappa(z, T, Q) \equiv 0$. Now we can choose the functions $\phi(u)$, $\psi(u)$ in such a manner that the point $x = \phi(u)$, $y = \psi(u)$ passes through every point of a given square $\bar{s}$ in the z-plane while u varies from 0 to 1. Then $T(Q) \supset \bar{s}$, and $N(z, T, Q) \geqq 1$ for $z \in \bar{s}$, while $\kappa(z, T, Q) \equiv 0$. This example shows that the converse of the result in IV.1.10 is generally false. It also shows that caution is justified in dealing with the essential multiplicity κ.

IV.1.14. Given T as in IV.1.2, and a (finitely-connected) Jordan region $\mathfrak{R}^* \subset \mathfrak{R}$, we have the inequality $\kappa(z, T, \mathfrak{R}^*) \leqq \kappa(z, T, \mathfrak{R})$.

Proof. Let z_0 be any point. If $\kappa(z_0, T, \mathfrak{R}) = +\infty$, then the assertion is obvious. So we can assume that $\kappa(z_0, T, \mathfrak{R}) = k < +\infty$. Assume that $\kappa(z_0, T, \mathfrak{R}^*) > k$. Then

$$(1) \qquad z_0 \in \mathfrak{K}(k + 1, T, \mathfrak{R}^*),$$

and hence $\epsilon(k + 1, T, \mathfrak{R}^*, z_0)$ is available (see IV.1.3). By IV.1.7, applied with $\eta = \epsilon(k + 1, T, \mathfrak{R}^*, z_0)$, we have a continuous transformation $T_* : z = t_*(w)$, $w \in \mathfrak{R}$, such that

$$\rho(T_* , T, \mathfrak{R}) < \epsilon(k + 1, T, \mathfrak{R}^*, z_0), \qquad N(z_0 , T_* , \mathfrak{R}) = \kappa(z_0 , T, \mathfrak{R}) = k.$$

We have then *a fortiori*

$$\text{(2)} \qquad \rho(T_* , T, \mathfrak{R}^*) < \epsilon(k + 1, T, \mathfrak{R}^*, z_0), \qquad N(z_0 , T, \mathfrak{R}^*) \leqq k.$$

In view of the definition of $\epsilon(k + 1, T, \mathfrak{R}^*, z_0)$, the relations (1) and (2) contradict each other.

IV.1.15. Given T as in IV.1.2, let z_0 be any point and k a non-negative (finite) integer such that $k \leqq \kappa(z_0 , T, \mathfrak{R})$. Let $T^* : z = t^*(w)$, $w \in \mathfrak{R}$, be a continuous transformation such that

$$\text{(1)} \qquad \rho(T^*, T, \mathfrak{R}) < \epsilon(k, T, \mathfrak{R}, z_0).$$

Note that the assumption $k \leqq \kappa(z_0 , T, \mathfrak{R})$ implies that $z_0 \in \mathfrak{K}(k, T, \mathfrak{R})$, and hence $\epsilon(k, T, \mathfrak{R}, z_0)$ is available (cf. IV.1.3, IV.1.4). By the definition of $\epsilon(k, T, \mathfrak{R}, z_0)$, (1) implies that $N(z_0 , T^*, \mathfrak{R}) \geqq k$. We improve this statement by showing that (see IV.1.2)

$$\text{(2)} \qquad N(z_0 , T^*, \mathfrak{R}^0) \geqq k.$$

PROOF. By I.2.50, $\mathfrak{R}$ can be transformed, by a topological transformation, into a region whose boundary curves are circles. Hence, in view of IV.1.6, we can assume without loss of generality that $\mathfrak{R}$ itself is bounded by circles. Let $C_1 , \cdots , C_m$ be these boundary circles, the notation being so chosen that C_m encloses $C_1 , \cdots , C_{m-1}$ (if $m = 1$, then this requirement is of course vacuous). For each positive integer i, let C_{li} , $l = 1, 2, \cdots , m$, be a circle concentric with C_l , contained in $\mathfrak{R}^0$, and having a radius that differs from that of C_l by $1/i$. For i sufficiently large, say $i \geqq i_0$, the circles $C_{1i} , \cdots , C_{mi}$ will not intersect each other and will bound a Jordan region $\mathfrak{R}_i \subset \mathfrak{R}^0$. We assume in the sequel that $i \geqq i_0$. We have the obvious relations

$$\mathfrak{R}_i \subset \mathfrak{R}^0_{i+1} , \qquad \sum_{i=i_0}^{\infty} \mathfrak{R}_i = \mathfrak{R}^0.$$

Let us denote by γ_{li} the doubly-connected region bounded by the concentric circles C_l and C_{li} , $l = 1, 2, \cdots , m$. Then

$$\mathfrak{R} = \mathfrak{R}^0_i + \gamma_{1i} + \cdots + \gamma_{mi}$$

We define now in $\mathfrak{R}$, for $i \geqq i_0$, a continuous transformation

$$\tau_i : w' = \psi_i(w), \qquad w \in \mathfrak{R},$$

as follows: For $w \in \gamma_{li}$, w' is the point on C_{li} nearest to w. For $w \in \mathfrak{R}^0_i$, $w' = w$. Thus τ_i is a (clearly continuous) retraction from $\mathfrak{R}$ onto $\mathfrak{R}_i$ (see II.2.39). Clearly

(3) $$| \psi_i(w) - w | \leqq 1/i \qquad \text{for } w \in \mathfrak{R}.$$

Let us now return to the transformation T^*, and let us assume that we have, in contradiction to (2), the inequality

(4) $$N(z_0, T^*, \mathfrak{R}^0) < k.$$

For i large enough, say $i > i_0 + i_1$, $\mathfrak{R}_i^0$ will include then all the points of the set $T^{*-1}(z_0) \cdot \mathfrak{R}^0$ (observe that this set is finite by (4)). Thus we have, for $i > i_0 + i_1$,

(5) $$N(z_0, T^*, \mathfrak{R}_i) = N(z_0, T^*, \mathfrak{R}^0),$$

(6) $$t^*(w) - z_0 \neq 0 \qquad \text{for } w \in \mathfrak{R}_i - \mathfrak{R}_i^0.$$

Consider now the auxiliary transformation

(7) $$T_i^* : z = t_i^*(w) = t^*[\psi_i(w)], \qquad w \in \mathfrak{R}.$$

Clearly, T_i^* is continuous in $\mathfrak{R}$, and (cf. (4), (5), (6), (7))

(8) $$N(z_0, T_i^*, \mathfrak{R}) = N(z_0, T^*, \mathfrak{R}_i) = N(z_0, T^*, \mathfrak{R}^0) < k, \qquad i > i_0 + i_1.$$

On the other hand, (3), (7) yield (cf. IV.1.1)

$$| t_i^*(w) - t(w) | \leqq | t^*[\psi_i(w)] - t[\psi_i(w)] | + | t[\psi_i(w)] - t(w) |$$

$$\leqq \rho(T^*, T, \mathfrak{R}) + \omega(1/i, T, \mathfrak{R}).$$

Since this holds for every point $w \in \mathfrak{R}$, it follows that

$$\rho(T_i^*, T, \mathfrak{R}) \leqq \rho(T^*, T, \mathfrak{R}) + \omega(1/i, T, \mathfrak{R}), \qquad i > i_0 + i_1.$$

Since the ω-term converges to zero for $i \to \infty$, it follows that we shall have (see (1))

(9) $$\rho(T_i^*, T, \mathfrak{R}) < \epsilon(k, T, \mathfrak{R}, z_0)$$

for i sufficiently large. In view of the definition of $\epsilon(k, T, \mathfrak{R}, z_0)$, the inequalities (8) and (9) contradict each other, and the proof is complete.

IV.1.16. Given T as in IV.1.2, let z_0 be a point such that $\kappa(z_0, T, \mathfrak{R}) < +\infty$, and let $\eta > 0$ be arbitrarily assigned. By IV.1.7 we have then a continuous transformation $T^* : z = t^*(w)$, $w \in \mathfrak{R}$, such that $\rho(T^*, T, \mathfrak{R}) < \eta$, $N(z_0, T^*, \mathfrak{R}) = \kappa(z_0, T, \mathfrak{R})$. We want to show that we can select T^* to satisfy the additional condition $N(z_0, T^*, \mathfrak{R}) = N(z_0, T^*, \mathfrak{R}^0)$.

Proof. Let us put $\kappa(z_0, T, \mathfrak{R}) = k < +\infty$. Then $z_0 \in \mathfrak{K}(k, T, \mathfrak{R})$, and hence $\epsilon(k, T, \mathfrak{R}, z_0)$ is available (see IV.1.3). We define

(1) $$\eta_1 = \min [\eta, \epsilon(k, T, \mathfrak{R}, z_0)].$$

Then $\eta_1 > 0$. By IV.1.7 we have a continuous transformation

$$T^* : z = t^*(w), \qquad w \in \mathfrak{R},$$

such that

(2) $$\rho(T^*, T, \mathfrak{R}) < \eta_1, \qquad N(z_0, T^*, \mathfrak{R}) = \kappa(z_0, T, \mathfrak{R}) = k.$$

(1), (2) yield

(3) $$\rho(T^*, T, \mathfrak{R}) < \eta,$$

(4) $$\rho(T^*, T, \mathfrak{R}) < \epsilon(k, T, \mathfrak{R}, z_0).$$

By IV.1.15 it follows from (4) that

(5) $$N(z_0, T^*, \mathfrak{R}^0) \geqq k.$$

Since $N(z_0, T^*, \mathfrak{R}^0) \leqq N(z_0, T^*, \mathfrak{R})$, we obtain from (2), (5)

(6) $$N(z_0, T^*, \mathfrak{R}) = N(z_0, T^*, \mathfrak{R}^0) = \kappa(z_0, T, \mathfrak{R}).$$

In view of (3), (6), T^* satisfies all of our requirements.

IV.1.17. Given T as in IV.1.2, there exists a sequence of (finitely-connected) Jordan regions $\mathfrak{R}_i$ with the following properties:

(i) $$\mathfrak{R}_i \subset \mathfrak{R}_{i+1}^0, \qquad i = 1, 2, \cdots.$$

(ii) $$\sum_{i=1}^{\infty} \mathfrak{R}_i = \mathfrak{R}^0.$$

(iii) $$\kappa(z, T, \mathfrak{R}_i) \to \kappa(z, T, \mathfrak{R}) \qquad \text{for } i \to \infty.$$

(Let us observe that this is merely a preliminary lemma; a stronger statement will be proved in the next section.)

PROOF. In view of IV.1.6, we can assume without loss of generality that $\mathfrak{R}$ is bounded by circles. This being assumed, we assert that the regions $\mathfrak{R}_i$ defined in IV.1.15 in the course of the proof satisfy our requirements. In fact, those regions $\mathfrak{R}_i$ were only defined for i exceeding a certain i_0. Let us define regions $\mathfrak{R}_i$, for i up to i_0, subject only to the conditions $\mathfrak{R}_i \subset \mathfrak{R}_{i+1}^0$, $\mathfrak{R}_i \subset \mathfrak{R}_{i_0+1}^0$. The fact that (i) and (ii) hold has been already noted in IV.1.15. Let now z_0 be any point. By IV.1.14 we have then $\kappa(z_0, T, \mathfrak{R}_i) \leqq \kappa(z_0, T, \mathfrak{R}_{i+1}) \leqq \kappa(z_0, T, \mathfrak{R})$. Hence, if (iii) is denied (for $z = z_0$), it follows that the sequence $\kappa(z_0, T, \mathfrak{R}_i)$ is a nondecreasing, bounded, and hence convergent sequence with a finite limit that we denote by l, and

(1) $$l < \kappa(z_0, T, \mathfrak{R}),$$

(2) $$\kappa(z_0, T, \mathfrak{R}_i) \leqq l < +\infty, \qquad i = 1, 2, \cdots.$$

(1) implies that $z_0 \in \mathfrak{K}(l + 1, T, \mathfrak{R})$. Hence $\epsilon(l + 1, T, \mathfrak{R}, z_0)$ is available (cf. IV.1.3). Let us apply IV.1.16 to T in $\mathfrak{R}_i$, with

$$\eta = \epsilon(l + 1, T, \mathfrak{R}, z_0)/2.$$

There follows the existence of a continuous transformation

$$T_i \cdot z = t_i(w), \qquad w \in \Re_i,$$

such that (cf. (2))

(3)
$$\rho(T_i, T, \Re_i) < \epsilon(l + 1, T, \Re, z_0)/2,$$
$$N(z_0, T_i, \Re_i) = N(z_0, T_i, \Re_i^0) = \kappa(z_0, T, \Re_i) < +\infty.$$

It follows that (cf. (2))

(4)
$$t_i(w) \neq z_0 \qquad \text{for } w \in \Re_i - \Re_i^0,$$

(5)
$$N(z_0, T_i, \Re_i) \leqq l.$$

From this point on, we assume that i exceeds the i_0 of IV.1.15. Then $\Re_i$ is bounded by circles, and the retraction $\tau_i : w' = \psi_i(w)$, $w \in \Re$, defined in IV.1.15, is therefore available. Let us introduce the transformations $T_i^* : z = t_i^*(w) = t_i[\psi_i(w)]$, $w \in \Re$. We have then, by (4), (5),

(6)
$$N(z_0, T_i^*, \Re) = N(z_0, T_i, \Re_i) \leqq l.$$

On the other hand, since $| \psi_i(w) - w | \leqq 1/i$ in $\Re$, we have in $\Re$ (cf. IV.1.1)

$$| t_i^*(w) - t(w) | \leqq | t_i[\psi_i(w)] - t[\psi_i(w)] | + | t[\psi_i(w)] - t(w)$$
$$\leqq \rho(T_i, T, \Re_i) + \omega(1/i, T, \Re).$$

Consequently, in view of (3),

$$\rho(T_i^*, T, \Re) \leqq \rho(T_i, T, \Re_i) + \omega(1/i, T, \Re)$$
$$< \epsilon(l + 1, T, \Re, z_0)/2 + \omega(1/i, T, \Re).$$

Since the ω-term converges to zero for $i \to \infty$, it follows that, for i sufficiently large, we shall have $\rho(T_i^*, T, \Re) < \epsilon(l + 1, T, \Re, z_0)$, and hence, by the definition of $\epsilon(l + 1, T, \Re, z_0)$,

(7)
$$N(z_0, T_i^*, \Re) \geqq l + 1.$$

(6) and (7) contradict each other, and the proof is complete.

IV.1.18. Given T as in IV.1.2, let $\Re_n^*$ be a sequence of finitely-connected Jordan regions with the following properties:

(a) $\Re_n^* \subset \Re$, $\qquad n = 1, 2, \cdots.$

(b) For every closed set $F \subset \Re^0$, there exists an integer $\nu(F)$ such that $F \subset \Re_n^*$ for $n > \nu(F)$.

Under these conditions we have the relation

(1)
$$\kappa(z, T, \Re_n^*) \to \kappa(z, T, \Re) \qquad \text{for } n \to \infty.$$

Proof. Consider a sequence of Jordan regions $\Re_i$ with the properties listed in IV.1.17. For fixed i, $\Re_i$ is a closed set in $\Re^0$, and hence by condition (b) we have an integer n_i such that $\mathfrak{R}_i \subset \mathfrak{R}_n^*$ for $n > n_i$. By IV.1.14 it follows that $\kappa(z, T, \mathfrak{R}_i) \leqq \kappa(z, T, \mathfrak{R}_n^*)$ for $n > n_i$. For fixed i and $n \to \infty$ we obtain thus

$$\kappa(z, T, \mathfrak{R}_i) \leqq \liminf_{n\to\infty} \kappa(z, T, \mathfrak{R}_n^*).$$

For $i \to \infty$ we obtain by IV.1.17

$$(2) \qquad \kappa(z, T, \mathfrak{R}) \leqq \liminf_{n\to\infty} \kappa(z, T, \mathfrak{R}_n^*).$$

Since $\mathfrak{R}_n^* \subset \mathfrak{R}$, we have by IV.1.14

$$(3) \qquad \kappa(z, T, \mathfrak{R}_n^*) \leqq \kappa(z, T, \mathfrak{R}), \qquad n = 1, 2, \cdots.$$

(2) and (3) imply (1).

IV.1.19. CONTINUATION. Returning to the preliminary lemma in IV.1.17, we note the following improvement. Given T as in IV.1.2, let us consider any sequence $\mathfrak{R}_i$ of finitely-connected Jordan regions such that

(α) $\mathfrak{R}_i \subset \mathfrak{R}_{i+1}^0$, $i = 1, 2, \cdots$.

(β) $\sum_{i=1}^{\infty} \mathfrak{R}_i = \mathfrak{R}^0$.

In IV.1.17 we asserted that amongst all the sequences $\mathfrak{R}_i$ satisfying (α), (β) there exist some that satisfy the further condition:

(γ) $\kappa(z, T, \mathfrak{R}_i) \to \kappa(z, T, \mathfrak{R})$ for $i \to \infty$.

We can now assert that every sequence $\mathfrak{R}_i$ with the properties (α), (β) also satisfies condition (γ). Indeed, $\mathfrak{R}_i \subset \mathfrak{R}$ by (α), (β), and hence by IV.1.18 we have only to show that if F is any closed set in $\mathfrak{R}^0$, then $F \subset \mathfrak{R}_i$ for i sufficiently large. Now let w be any point of F. By (β) we have $w \in \mathfrak{R}_i$ for some i, and hence $w \in \mathfrak{R}_{i+1}^0$ by (α). We have therefore an open circular disc $\Delta(w, i)$ such that $w \in \Delta(w, i) \subset \mathfrak{R}_{i+1}^0$. Since F is bounded and closed, it follows by the Borel covering theorem that there exists a finite system of such discs, say $\Delta(w_1, i_1), \cdots, \Delta(w_m, i_m)$, such that $F \subset \Delta(w_1, i_1) + \cdots + \Delta(w_m, i_m)$. In view of ($\alpha$), it follows that $F \subset \mathfrak{R}_i$ for $i > i_1 + \cdots + i_m$.

IV.1.20. We observed in IV.1.4 that $\kappa(z, T, \mathfrak{R}) \leqq N(z, T, \mathfrak{R})$. In fact, we have the stronger inequality

$$(1) \qquad \kappa(z, T, \mathfrak{R}) \leqq N(z, T, \mathfrak{R}^0).$$

Indeed, if k is any finite non-negative integer such that $k \leqq \kappa(z, T, \mathfrak{R})$, then by IV.1.15 (applied to $T^* = T$) it follows that $N(z, T, \mathfrak{R}^0) \geqq k$. Since this holds for every finite $k \leqq \kappa(z, T, \mathfrak{R})$, the inequality (1) follows.

IV.1.21. The statement in IV.1.14 may be generalized as follows: Given T as in IV.1.2, let $\mathfrak{R}_1, \cdots, \mathfrak{R}_n$ be finitely-connected Jordan regions such that $\mathfrak{R}_i \subset \mathfrak{R}$, $i = 1, 2, \cdots, n$, $\mathfrak{R}_i^0\mathfrak{R}_j^0 = 0$ for $i \neq j$. Then

$$(1) \qquad \sum_{i=1}^{n} \kappa(z, T, \mathfrak{R}_i) \leqq \kappa(z, T, \mathfrak{R}).$$

(Note that the regions $\mathfrak{R}_1, \cdots, \mathfrak{R}_n$ are permitted to have common boundary points.)

PROOF. Take a point z_0, and let $k_1, \cdots, k_n$ be non-negative (finite) integers such that

(2) $$k_i \leqq \kappa(z_0, T, \mathfrak{R}_i), \qquad i = 1, 2, \cdots, n.$$

Then $z_0 \in \mathfrak{K}(k_i, T, \mathfrak{R}_i)$, and hence $\epsilon(k_i, T, \mathfrak{R}_i, z_0)$ is available (cf. IV.1.3). Take now an ϵ_0 such that $0 < \epsilon_0 < \epsilon(k_i, T, \mathfrak{R}_i, z_0)$, $i = 1, 2, \cdots, n$. Let $T^* : z = t^*(w)$, $w \in \mathfrak{R}$, be any continuous transformation such that

(3) $$\rho(T^*, T, \mathfrak{R}) < \epsilon_0 .$$

Then *a fortiori* $\rho(T^*, T, \mathfrak{R}_i) < \epsilon(k_i, T, \mathfrak{R}_i, z_0)$, $i = 1, 2, \cdots, n$. By IV.1.15 (applied to T, T^* in $\mathfrak{R}_i$) it follows that $N(z_0, T^*, \mathfrak{R}_i^0) \geqq k_i$, and hence (note that $\mathfrak{R}_1^0, \cdots, \mathfrak{R}_n^0$ are disjoint)

(4) $$N(z_0, T^*, \mathfrak{R}) \geqq k_1 + \cdots + k_n .$$

Thus (3) implies (4), and hence ϵ_0 serves as an $\epsilon(k_1 + \cdots + k_n, T, \mathfrak{R}, z_0)$ in the sense of IV.1.3. Consequently

$$z_0 \in \mathfrak{K}(k_1 + \cdots + k_n, T, \mathfrak{R}),$$

and hence

(5) $$\kappa(z_0, T, \mathfrak{R}) \geqq k_1 + \cdots + k_n .$$

Since (5) holds for every choice of the non-negative integers $k_1, \cdots, k_n$, such that (2) is satisfied, the inequality (1) follows.

IV.1.22. Given $T : z = t(w)$, $w \in \mathfrak{R}$, as in IV.1.2, take a point z_0 and a non-negative (finite) integer $k \leqq \kappa(z_0, T, \mathfrak{R})$. Then $\epsilon(k, T, \mathfrak{R}, z_0)$ is available (see IV.1.3). Let now $T_n : z = t_n(w)$, $w \in \mathfrak{R}$, be a sequence of continuous transformations such that (cf. IV.1.1)

(1) $$\rho(T_n, T, \mathfrak{R}) \to 0 \qquad \text{for } n \to \infty .$$

We assert that any number η satisfying

(2) $$0 < \eta < \epsilon(k, T, \mathfrak{R}, z_0)$$

will serve as an $\epsilon(k, T_n, \mathfrak{R}, z_0)$ for n sufficiently large. Indeed, let an η satisfying (2) be chosen. In view of (1) we shall have then, for a properly chosen n_0,

(3) $$\rho(T_n, T, \mathfrak{R}) < \epsilon(k, T, \mathfrak{R}, z_0) - \eta \qquad \text{if } n > n_0 .$$

Let then T^* be any continuous transformation, defined in $\mathfrak{R}$, such that

(4) $$\rho(T^*, T_n, \mathfrak{R}) < \eta.$$

It follows that

$$\rho(T^*, T, \mathfrak{R}) \leqq \rho(T^*, T_n, \mathfrak{R}) + \rho(T_n, T, \mathfrak{R}) < \epsilon(k, T, \mathfrak{R}, z_0).$$

Hence, by the definition of $\epsilon(k, T, \mathfrak{R}, z_0)$,

(5) $$N(z_0, T^*, \mathfrak{R}) \geqq k.$$

Thus, for $n > n_0$, (4) implies (5), and hence η serves as an $\epsilon(k, T_n, \mathfrak{R}, z_0)$.

IV.1.23. A continuous transformation $T : z = t(w)$, $w \in \mathfrak{R}$, will be termed *polyhedral* if there exists a curvilinear triangulation of $\mathfrak{R}$, made up of curvilinear triangles $\delta_1, \cdots, \delta_m$, such that T maps each δ_i topologically onto a *rectilinear* triangle in the z-plane, vertices being carried into vertices. Clearly, if T is polyhedral and $T^* : z = t^*(w^*)$, $w^* \in \mathfrak{R}^*$, is topologically similar to T (cf. IV.1.6), then T^* is also polyhedral.

Given $T : z = t(w)$, $w \in \mathfrak{R}$, as in IV.1.2, there exists a sequence of continuous transformations $T_n : z = t_n(w)$, $w \in \mathfrak{R}$, such that each T_n is polyhedral and $\rho(T_n, T, \mathfrak{R}) \to 0$.

Proof. By I.2.50, IV.1.6 we can assume, without loss of generality, that $\mathfrak{R}$ is bounded by rectilinear polygons. Let then $\mathfrak{J}_n$ be a rectilinear triangulation of $\mathfrak{R}$, such that every triangle of $\mathfrak{J}_n$ has a diameter less than $1/n$. Let $w_1^n, \cdots, w_N^n$ be the vertices in the triangulation $\mathfrak{J}_n$, and let us choose in the z-plane an equal number of points $z_1^n, \cdots, z_N^n$ subject to the following conditions: (i) The points $z_1^n, \cdots, z_N^n$ are distinct. (ii) No three of these points are collinear. (iii) $| t(w_i^n) - z_i^n | < 1/n$ for $i = 1, 2, \cdots, N$. Clearly these conditions are compatible. Let the points $z_1^n, \cdots, z_N^n$ be joined by straight segments in the same manner as the points $w_1^n, \cdots, w_N^n$ are joined in the triangulation $\mathfrak{J}_n$. Define now a continuous transformation $T_n : z = t_n(w)$, $w \in \mathfrak{R}$, as follows: T_n carries the vertices w_i^n of $\mathfrak{J}_n$ into the corresponding points z_i^n, and in each triangle δ^n of $\mathfrak{J}_n$ the transformation T_n is an affine transformation. Clearly T_n is polyhedral, and $\rho(T_n, T, \mathfrak{R}) \to 0$ as an immediate consequence of the uniform continuity of T in $\mathfrak{R}$.

IV.1.24. Given $T : z = t(w)$, $w \in \mathfrak{R}$, as in IV.1.2, we define in the z-plane an *index function* $\mu(z, T, \mathfrak{R})$ as follows. Let $C_1, \cdots, C_m$ be the boundary curves of $\mathfrak{R}$, each C_i being oriented positively with respect to $\mathfrak{R}$. That is, if C_i is the exterior boundary curve of $\mathfrak{R}$, then the orientation of C_i agrees with the counterclockwise orientation, and if C_i is an interior boundary curve of $\mathfrak{R}$, then the orientation of C_i agrees with the clockwise orientation. For each point z, we have then an index $\mu(z, T, C_i)$ in the sense of II.4.34. Let us recall the convention stated there to the effect that $\mu(z, T, C_i) = 0$ whenever $z \in T(C_i)$. We define now $\mu(z, T, \mathfrak{R})$ as follows:

(i) If $z \notin T(C_1 + \cdots + C_m)$, then

$$\mu(z, T, \mathfrak{R}) = \sum_{i=1}^{m} \mu(z, T, C_i).$$

(ii) If $z \in T(C_1 + \cdots + C_m)$, then $\mu(z, T, \mathfrak{R}) = 0$.

The convention (ii) *is arbitrary, and can be justified only on grounds of technical convenience.* The following remarks should be kept in mind. If $\mu(z_0, T, \mathfrak{R}) \neq 0$ for a point z_0, then by (ii)

$$z_0 \notin T(C_1 + \cdots + C_m) = T(\mathfrak{R} - \mathfrak{R}^0).$$

The converse, however, is generally false. Indeed, if $z_0 \notin T(\mathfrak{R} - \mathfrak{R}^0)$, then $\mu(z_0, T, \mathfrak{R})$ is given by (i), and the indices $\mu(z_0, T, C_\iota)$ may very well cancel each other.

IV.1.25. Continuation. Let us put

$$\delta(z_0, T, \mathfrak{R}) = \min |t(w) - z_0|, \qquad w \in \mathfrak{R} - \mathfrak{R}^0.$$

In other words, $\delta(z_0, T, \mathfrak{R})$ is the distance of z_0 from the image of the boundary of $\mathfrak{R}$. The following statements are useful in the sequel:

(a) If $\mu(z_0, T, \mathfrak{R}) \neq 0$, then $\delta(z_0, T, \mathfrak{R}) > 0$, while the converse is generally false. This follows directly from IV.1.24.

(b) If $\delta(z_0, T, \mathfrak{R}) = 0$, then $\mu(z_0, T, \mathfrak{R}) = 0$. Indeed, the assumption implies that $z_0 \in T(\mathfrak{R} - \mathfrak{R}^0)$, and hence $\mu(z_0, T, \mathfrak{R}) = 0$ by definition.

(c) If $\delta(z_0, T, \mathfrak{R}) > 0$, and if the point z_1 satisfies the inequality $|z_1 - z_0| < \delta(z_0, T, \mathfrak{R})$, then $\mu(z_1, T, \mathfrak{R}) = \mu(z_0, T, \mathfrak{R})$.

Proof. Let C_ι be a boundary curve of $\mathfrak{R}$. We can write

$$t(w) - z_1 = [t(w) - z_0] + (z_0 - z_1).$$

We have $|t(w) - z_0| \geqq \delta(z_0, T, \mathfrak{R}) > |z_0 - z_1|$ for $w \in \mathfrak{R} - \mathfrak{R}^0$ and hence *a fortiori* for $w \in C_\iota$. By the theorem of Rouché (see II.4.26, II.4.34) it follows that $\mu(z_1, T, C_\iota) = \mu(z_0, T, C_\iota)$, $\iota = 1, 2, \cdots, m$. To infer that $\mu(z_1, T, \mathfrak{R}) = \mu(z_0, T, \mathfrak{R})$ we must make sure that $z_0 \notin T(\mathfrak{R} - \mathfrak{R}^0)$, $z_1 \notin T(\mathfrak{R} - \mathfrak{R}^0)$ (cf. IV.1.24). However, $z_0 \notin T(\mathfrak{R} - \mathfrak{R}^0)$ since $\delta(z_0, T, \mathfrak{R}) > 0$, and $z_1 \notin T(\mathfrak{R} - \mathfrak{R}^0)$ since $|t(w) - z_1| \geqq |t(w) - z_0| - |z_1 - z_0| \geqq \delta(z_0, T, \mathfrak{R}) - |z_1 - z_0| > 0$ for $w \in \mathfrak{R} - \mathfrak{R}^0$.

(d) If $\delta(z_0, T, \mathfrak{R}) > 0$ and if the continuous transformation $T^* : z = t^*(w)$, $w \in \mathfrak{R}$, satisfies the inequality $\rho(T^*, T, \mathfrak{R} - \mathfrak{R}^0) < \delta(z_0, T, \mathfrak{R})$, then $\mu(z_0, T^*, \mathfrak{R}) = \mu(z_0, T, \mathfrak{R})$.

Proof. The assumption $\delta(z_0, T, \mathfrak{R}) > 0$ implies that $z_0 \notin T(\mathfrak{R} - \mathfrak{R}^0)$. For $w \in \mathfrak{R} - \mathfrak{R}^0$ we have $|t^*(w) - z_0| \geqq |z_0 - t(w)| - |t^*(w) - t(w)| \geqq \delta(z_0, T, \mathfrak{R}) - \rho(T^*, T, \mathfrak{R} - \mathfrak{R}^0) > 0$, and hence $z_0 \notin T^*(\mathfrak{R} - \mathfrak{R}^0)$. Hence, by IV.1.24,

$$\mu(z_0, T, \mathfrak{R}) = \sum_{\iota=1}^{m} \mu(z_0, T, C_\iota), \qquad \mu(z_0, T^*, \mathfrak{R}) = \sum_{\iota=1}^{m} \mu(z_0, T^*, C_\iota).$$

Hence it is sufficient to show that

$$\text{(1)} \qquad \mu(z_0, T, C_\iota) = \mu(z_0, T^*, C_\iota), \qquad \iota = 1, 2, \cdots, m.$$

Now for $w \in \mathfrak{R} - \mathfrak{R}^0$ we have $|t(w) - z_0| \geqq \delta(z_0, T, \mathfrak{R}) > \rho(T^*, T, \mathfrak{R} - \mathfrak{R}^0) \geqq |t^*(w) - t(w)|$. Since $t^*(w) - z_0 = [t(w) - z_0] + [t^*(w) - t(w)]$, formula (1) follows from the theorem of Rouché (see II.4.26, II.4.34).

(e) If $\mu(z_0, T, \mathfrak{R}) \neq 0$, then $N(z_0, T, \mathfrak{R}^0) \geqq 1$. Equivalently, if $z_0 \notin T(\mathfrak{R}^0)$, then $\mu(z_0, T, \mathfrak{R}) = 0$.

Proof. Since $\mu(z_0, T, \mathfrak{R}) \neq 0$, we have (see IV.1.24) $z_0 \notin T(\mathfrak{R} - \mathfrak{R}^0)$, and hence $t(w) - z_0 \neq 0$ for $w \in \mathfrak{R} - \mathfrak{R}^0$. Furthermore, by IV.1.24(i),

$$(2) \qquad \mu(z_0, T, \mathfrak{R}) = \sum_{i=1}^{m} \mu(z_0, T, C_i).$$

Assume, in contradiction to our assertion, that $N(z_0, T, \mathfrak{R}^0) = 0$. Then $t(w) - z_0 \neq 0$ in $\mathfrak{R}^0$, and hence also in $\mathfrak{R}$, since $t(w) - z_0 \neq 0$ on $\mathfrak{R} - \mathfrak{R}^0$. By II.4.28 it follows that the summation in (2) is equal to zero, in contradiction to the assumption that $\mu(z_0, T, \mathfrak{R}) \neq 0$.

(f) Let there be given two continuous transformations $T : z = t(w)$, $w \in \mathfrak{R}$, $T^* : z = t^*(w^*)$, $w^* \in \mathfrak{R}^*$, where the w- and w^*-planes may or may not coincide. Suppose that T and T^* are topologically similar; that is, there exists a homeomorphism $\tau(\mathfrak{R}) = \mathfrak{R}^*$ such that $t(w) = t^*[\tau(w)]$ for $w \in \mathfrak{R}$. Then

$$(3) \qquad \mu(z, T, \mathfrak{R}) \equiv \mu(z, T^*, \mathfrak{R}^*) \qquad \text{or} \qquad \mu(z, T, \mathfrak{R}) \equiv -\mu(z, T^*, \mathfrak{R}^*).$$

Proof. Let $C_1, \cdots, C_m$ be the boundary curves of $\mathfrak{R}$, positively oriented with respect to $\mathfrak{R}$ (cf. IV.1.24), and let $C_1^*, \cdots, C_m^*$ be the boundary curves of $\mathfrak{R}^*$, positively oriented with respect to $\mathfrak{R}^*$ (since τ is a homeomorphism, $\mathfrak{R}$ and $\mathfrak{R}^*$ have the same connectivity). Now τ carries each boundary curve C_i into a boundary curve $C_{k_i}^*$ of $\mathfrak{R}^*$, but the resulting orientation may or may not agree with that assigned to $C_{k_i}^*$ relative to $\mathfrak{R}^*$. To stress this point, let us write $\tau(C_i) = \lambda_i C_{k_i}^*$, where $\lambda_i = +1$ if τ carries the oriented curve C_i into $C_{k_i}^*$ with the proper orientation relative to $\mathfrak{R}^*$, and $\lambda_i = -1$ if this is not the case. Since τ is a homeomorphism, we have either $\lambda_i = +1$ for $i = 1, 2, \cdots, m$, or $\lambda_i = -1$ for $i = 1, 2, \cdots, m$. Now let z_0 be any point. If $z_0 \in T(C_1 + \cdots + C_m)$, then clearly $z_0 \in T^*(C_1^* + \cdots + C_m^*)$ and $\mu(z_0, T, \mathfrak{R}) = 0 = \mu(z_0, T^*, \mathfrak{R}^*)$. In the general case, we have by II.2.40

$$\mu(z_0, T, C_i) = \lambda_i \mu(z_0, T^*, C_{k_i}^*), \qquad i = 1, 2, \cdots, m,$$

and summation yields the formula (3), since λ_i is independent of i and also independent of z_0.

IV.1.26. Given T as in IV.1.2, let z_0 be a point such that $\mu(z_0, T, \mathfrak{R}) \neq 0$. Then $\kappa(z_0, T, \mathfrak{R}) > 0$ (cf. IV.1.4, IV.1.24).

Proof. The assumption implies that $\delta(z_0, T, \mathfrak{R}) > 0$ (see IV.1.25). Now let $T^* : z = t^*(w)$, $w \in \mathfrak{R}$, be any continuous transformation such that $\rho(T^*, T, \mathfrak{R}) < \delta(z_0, T, \mathfrak{R})$. By IV.1.25(d) we have then $\mu(z_0, T^*, \mathfrak{R}) = \mu(z_0, T, \mathfrak{R}) \neq 0$, and hence $N(z_0, T^*, \mathfrak{R}) \geqq N(z_0, T^*, \mathfrak{R}^0) > 0$ by IV.1.25(e), applied to T^*. Thus $\delta(z_0, T, \mathfrak{R})$ serves as an $\epsilon(1, T, \mathfrak{R}, z_0)$ (see IV.1.3), and hence $z_0 \in \mathfrak{K}(1, T, \mathfrak{R})$ and finally $\kappa(z_0, T, \mathfrak{R}) \geqq 1$.

IV.1.27. Given T as in IV.1.2, let z_0 be a point in the z-plane. A Jordan region $\mathfrak{R}^* \subset \mathfrak{R}$ will be termed an *indicator region* for z_0 under T if and only if $\mu(z_0, T, \mathfrak{R}^*) \neq 0$ (see IV.1.24). For brevity we shall say that $\mathfrak{R}^*$ is an *indicator region* (z_0, T). Note that the condition implies that $z_0 \notin T(\mathfrak{R}^* - \mathfrak{R}^{*0})$, or equivalently that $\delta(z_0, T, \mathfrak{R}^*) > 0$ (cf. IV.1.25). The converse does not hold; that is, the condition $\delta(z_0, T, \mathfrak{R}^*) > 0$ does not guarantee that $\mathfrak{R}^*$ is an indicator region (z_0, T). An *indicator system* (z_0, T) *of order* k is defined as a system of k indicator regions $\mathfrak{R}_1, \cdots, \mathfrak{R}_k$ in $\mathfrak{R}$, such that $\mathfrak{R}_i^0 \mathfrak{R}_j^0 = 0$ for $i \neq j$.

The point z_0 being kept fixed, consider the set $T^{-1}(z_0)$. This set is closed and possibly empty. If $T^{-1}(z_0) \neq 0$, then each component of $T^{-1}(z_0)$ is a continuum which will be termed a *maximal model continuum* for z_0 under T in $\mathfrak{R}$. Such a continuum γ will be termed *essential* if the following conditions are satisfied: (i) $\gamma \subset \mathfrak{R}^0$. (ii) If O is any open set that contains γ, then there exists an indicator region (z_0, T) in O which contains γ in its interior. Since $T(\gamma) = z_0$, it is clear that if $\mathfrak{R}^*$ is an indicator region (z_0, T) such that $\gamma \subset \mathfrak{R}^*$, then necessarily $\gamma \subset \mathfrak{R}^{*0}$. Indeed, as noted above, $z_0 \notin T(\mathfrak{R}^* - \mathfrak{R}^{*0})$. *Let us note emphatically that if a maximal model continuum γ has a nonempty intersection with the boundary of $\mathfrak{R}$, then γ cannot be essential relative to $\mathfrak{R}$.*

IV.1.28. Given T as in IV.1.2, let z_0 be any point, and let K be the number (possibly $+\infty$) of the essential maximal model continua for z_0 under T in $\mathfrak{R}^0$. Then $K \leqq \kappa(z_0, T, \mathfrak{R})$.

PROOF. If $\kappa(z_0, T, \mathfrak{R}) = +\infty$ or $K = 0$, then the assertion is obvious. So we can assume that $K > 0$ and $\kappa(z_0, T, \mathfrak{R}) < +\infty$. Let us put $m = 1 + \kappa(z_0, T, \mathfrak{R})$, and let us assume, in contradiction to our assertion, that $K \geqq m$. Then we have (at least) m distinct essential maximal model continua $\gamma_1, \cdots, \gamma_m$ for z_0 under T in $\mathfrak{R}^0$. Let $\eta > 0$ be smaller than the distance of any two of these continua, and let O_i, $i = 1, 2, \cdots, m$, denote the set of those points w whose distance from γ_i is less than $\eta/2$. Then $O_i\mathfrak{R}^0$ is an open set containing γ_i, and the sets $O_1\mathfrak{R}^0, \cdots, O_m\mathfrak{R}^0$ are disjoint. By definition (see IV.1.27), we have for each i a Jordan region $\mathfrak{R}_i$ such that (i) $\gamma_i \subset \mathfrak{R}_i^0$, (ii) $\mathfrak{R}_i \subset O_i\mathfrak{R}^0$, and (iii) $\mathfrak{R}_i$ is an indicator region (z_0, T). The regions $\mathfrak{R}_1, \cdots, \mathfrak{R}_m$ are clearly disjoint. As a consequence of (iii), we have $\mu(z_0, T, \mathfrak{R}_i) \neq 0$, and hence, by IV.1.26, $\kappa(z_0, T, \mathfrak{R}_i) > 0$, $i = 1, 2, \cdots, m$. By IV.1.21 it follows that

$$\kappa(z_0, T, \mathfrak{R}) \geqq \sum_{i=1}^{m} \kappa(z_0, T, \mathfrak{R}_i) \geqq m = 1 + \kappa(z_0, T, \mathfrak{R}).$$

Thus the assumption that $K > \kappa(z_0, T, \mathfrak{R})$ leads to a contradiction.

We proceed to establish a series of lemmas which will yield the fundamental theorem that $\kappa(z, T, \mathfrak{R})$ is actually equal to the number of essential maximal model continua for z under T in $\mathfrak{R}^0$. Some of these lemmas could be stated in far greater generality, but we shall restrict ourselves to the simplest statements adequate for our purposes.

IV.1.29. Let $\mathfrak{R}$ be a simply-connected, bounded Jordan region in the w-plane, and let $f(w)$ be a continuous complex-valued function defined on the boundary C of $\mathfrak{R}$ such that $f(w) \neq 0$ on C and (cf. II.4.25)

$$(1) \qquad V_C\,[\arg f(w)] = 0.$$

Then there exists in $\mathfrak{R}$ a continuous complex-valued function $F(w)$ with the following properties: (i) $F(w) = f(w)$ on C. (ii) $F(w) \neq 0$ for $w \in \mathfrak{R}$. (iii) The maximum of $| F(w) |$ in $\mathfrak{R}$ does not exceed the maximum of $| f(w) |$ on C.

PROOF. *Case* (a). $\mathfrak{R}$ is the circular disc $| w | \leqq 1$. In view of (1) we have (see II.4.25(e)) a single-valued, real-valued, continuous function $\phi(w)$ on C, such that $f(w) = |f(w)| [\cos \phi(w) + i \sin \phi(w)]$ for $w \in C$. Since $f(w) \neq 0$ on C, the (real) logarithm of $|f(w)|$ is continuous on C. Let us put $g(w) = \log |f(w)| + i\phi(w)$, $w \in C$. Then $g(w)$ is single-valued and continuous on C, and

$$(2) \qquad f(w) = e^{g(w)} \qquad \text{for } w \in C.$$

Let us put

$$(3) \qquad M = \max |f(w)|, \qquad w \in C,$$

and let us define

$$G(w) = \begin{cases} (1 - |w|) \log M + |w| g(w/|w|) & \text{for } 0 < |w| \leqq 1, \\ \log M & \text{for } w = 0. \end{cases}$$

Then $G(w)$ is single-valued and continuous in $\mathfrak{R}$ (even at $w = 0$), and

$$(4) \qquad \begin{aligned} G(w) &= g(w) && \text{for } |w| = 1, \\ \mathfrak{R}G(w) &= (1 - |w|) \log M + |w| \log |f(w/|w|)| && \text{for } 0 < |w| \leqq 1. \end{aligned}$$

There follows the estimate

$$\mathfrak{R}G(w) \leqq (1 - |w|) \log M + |w| \log M = \log M \qquad \text{for } 0 < |w| \leqq 1.$$

Since $\mathfrak{R}G(0) = \log M$, it follows that

$$(5) \qquad \mathfrak{R}G(w) \leqq \log M \qquad \text{for } |w| \leqq 1.$$

From (2), (3), (4), (5) it follows that the function

$$F(w) = e^{G(w)}, \qquad |w| \leqq 1,$$

satisfies our requirements.

Case (b). $\mathfrak{R}$ is a general simply-connected, bounded Jordan region. Let then $\mathfrak{R}^*$ denote the disc $|w^*| \leqq 1$ in an auxiliary w^*-plane, and let $w^* = \psi(w)$ be a topological transformation that maps $\mathfrak{R}$ onto $\mathfrak{R}^*$ (see I.2.50). In view of II.4.31, the function $f^*(w^*) = f[\psi^{-1}(w^*)]$, $|w^*| \leqq 1$, satisfies then the assumptions made in case (a) above. If $F^*(w^*)$ is the corresponding function in $|w^*| \leqq 1$ whose existence we established in case (a), then the function $F(w) = F^*[\psi(w)]$, $w \in \mathfrak{R}$, clearly satisfies our requirements.

IV.1.30. On the boundary C of a simply-connected, bounded Jordan region $\mathfrak{R}$ in the w-plane, let there be given a complex-valued continuous function $f(w)$ which vanishes at exactly one point w_0 on C. Then there exists in $\mathfrak{R}$ a continuous function $F(w)$ with the following properties: (i) $F(w) = f(w)$ on C. (ii) $F(w) \neq 0$ in $\mathfrak{R}^0$. (iii) The maximum of $|F(w)|$ in $\mathfrak{R}$ does not exceed the maximum of $|f(w)|$ on C.

PROOF. As in IV.1.29, case (b), we can verify that it is sufficient to consider the special case when $\mathfrak{R}$ is the circular disc $|w| \leqq 1$. For $w \neq w_0$, let w^* denote the second point of intersection of the line through w and w_0 with the circle

$|w| = 1$. Then the function, defined for $|w| \leqq 1$ by the formulas

$$F(w) = \begin{cases} \dfrac{w - w_0}{w^* - w_0} f(w^*) & \text{for } w \neq w_0 , \\ 0 = f(w_0) & \text{for } w = w_0 , \end{cases}$$

clearly satisfies our requirements.

IV.1.31. Let $f(w)$ be a continuous complex-valued function that is defined and different from zero on the boundary C of a bounded, simply-connected Jordan region $\mathfrak{R}$ in the w-plane, and let w^* be an interior point of $\mathfrak{R}$. Then there exists in $\mathfrak{R}$ a continuous function $F(w)$ with the following properties: (i) $F(w) = f(w)$ on the boundary of $\mathfrak{R}$. (ii) $F(w^*) = 0$. (iii) $F(w) \neq 0$ in $\mathfrak{R} - w^*$. (iv) The maximum of $|F(w)|$ in $\mathfrak{R}$ does not exceed the maximum of $|f(w)|$ on the boundary of $\mathfrak{R}$.

PROOF. Using an auxiliary topological transformation as in IV.1.29, we can assume without loss of generality that $\mathfrak{R}$ coincides with the circular disc $|w| \leqq 1$ and that $w^* = 0$. Then the function

$$F(w) = \begin{cases} |w| f(w/|w|) & \text{for } w \neq 0, \\ 0 & \text{for } w = 0. \end{cases}$$

clearly satisfies our requirements.

IV.1.32. Given $T : z = t(w)$, $w \in \mathfrak{R}$, as in IV.1.2, let z_0 be a point in the z-plane such that $N(z_0, T, \mathfrak{R}) < +\infty$, $\kappa(z_0, T, \mathfrak{R}) > 0$. Since $\kappa(z_0, T, \mathfrak{R}) \leqq N(z_0, T, \mathfrak{R}^0)$ (see IV.1.20), $\kappa(z_0, T, \mathfrak{R})$ is also finite. Let k be a positive integer such that $k \leqq \kappa(z_0, T, \mathfrak{R})$. Then we have the inequalities

$$0 < k \leqq \kappa(z_0, T, \mathfrak{R}) \leqq N(z_0, T, \mathfrak{R}^0) < +\infty . \tag{1}$$

Furthermore, $z_0 \in \mathfrak{K}(k, T, \mathfrak{R})$, and hence $\epsilon(k, T, \mathfrak{R}, z_0)$ is available (*see* IV.1.3).

LEMMA. *Under the conditions just described, take any number τ such that*

$$0 < \tau < \epsilon(k, T, \mathfrak{R}, z_0)/2. \tag{2}$$

Then there exists in $\mathfrak{R}^0$ an indicator system (z_0, T) of order k (see IV.1.27) *comprised of Jordan regions $\mathfrak{R}_1, \cdots, \mathfrak{R}_k$, such that*

$$|t(w) - z_0| > \tau/2 \qquad \text{for } w \in \mathfrak{R}_i - \mathfrak{R}_i^0, \; i = 1, 2, \cdots, k,$$

$$|t(w) - z_0| < \tau \qquad \text{for } w \in \mathfrak{R}_i, \; i = 1, 2, \cdots, k.$$

PROOF. Let G denote the set of those points $w \in \mathfrak{R}^0$ where $|t(w) - z_0| < \tau$. Then G contains the set $T^{-1}(z_0) \cdot \mathfrak{R}^0$, and hence G is not empty. Indeed, $T^{-1}(z_0) \cdot \mathfrak{R}^0$ contains at least $k > 0$ distinct points (cf. (1)). The set G is clearly open, and hence each one of its components is also open. As $T^{-1}(z_0)$ is a finite set on account of the assumption $N(z_0, T, \mathfrak{R}) < +\infty$, the set G has only a finite number of components that contain some point of $T^{-1}(z_0)$. Let $G_1, \cdots, G_m$ be these components of G. Then each G_i is a domain (connected open set) in $\mathfrak{R}^0$, and

$$(3)\qquad \begin{aligned} t(w) - z_0 &\neq 0 \qquad &&\text{for } w \in \mathfrak{R}^0 - (G_1 + \cdots + G_m), \\ G_iG_j &= 0 &&\text{for } i \neq j. \end{aligned}$$

Let w_i be a boundary point of G_i. Since $| t(w) - z_0 | < \tau$ in G and hence in G_i, clearly $| t(w_i) - z_0 | \leqq \tau$, and if the sign of equality does not hold, then w_i must obviously lie on the boundary of $\mathfrak{R}$.

IV.1.33. CONTINUATION. We shall now modify $t(w)$ in each one of the components $G_1, \cdots, G_m$. We take one of these components G_i, and proceed according to the following instructions:

Case (α). $| t(w) - z_0 | = \tau$ on the whole boundary of G_i. Let then $\mathfrak{R}_{in}$, $n = 1, 2, \cdots$, be a sequence of finitely-connected Jordan regions that fill up G_i in the following sense:

$$(1)\qquad \mathfrak{R}_{i,n} \subset \mathfrak{R}^0_{i,n+1}, \; n = 1, 2, \cdots, \qquad \sum_{n=1}^{\infty} \mathfrak{R}^0_{in} = G_i.$$

Such a sequence $\mathfrak{R}_{in}$ exists (see I.2.48). For n sufficiently large we shall have then

$$(2)\qquad | t(w) - z_0 | > \tau/2 \qquad \text{for } w \in G_i - \mathfrak{R}^0_{in}.$$

Indeed, if E_i is the subset of G_i where $| t(w) - z_0 | \leqq \tau/2$, then E_i is a closed bounded set in G_i, since $| t(w) - z_0 | = \tau$ on the boundary of G_i. From (1) it follows readily, by means of the Borel covering theorem, that $E_i \subset \mathfrak{R}^0_{in}$ for n sufficiently large (cf. the argument used in IV.1.19), and hence (2) holds. We pick a region $\mathfrak{R}_{in}$ for which (2) holds and denote it by $\mathfrak{R}_i$. Then

$$(3)\qquad \mathfrak{R}_i \subset G_i, \qquad 0 < \tau/2 < | t(w) - z_0 | < \tau \qquad \text{for } w \in G_i - \mathfrak{R}^0_i.$$

Since $N(z_0, T, \mathfrak{R}) < +\infty$ by assumption, $\mathfrak{R}_i$ contains only a finite number of points of the set $T^{-1}(z_0)$ in its interior and none on its boundary (cf. (3)). We can take therefore a simply-connected Jordan region $\mathfrak{R}^*_i \subset \mathfrak{R}^0_i$ such that $\mathfrak{R}_i \cdot T^{-1}(z_0) \subset \mathfrak{R}^{*0}_i$. In the Jordan region $\mathfrak{R}'_i = \mathfrak{R}_i - \mathfrak{R}^{*0}_i$ the function $t(w) - z_0$ is then different from zero. Hence by IV.1.25(e) (applied to $\mathfrak{R}'_i$) we have

$$0 = \mu(z_0, T, \mathfrak{R}'_i) = \mu(z_0, T, \mathfrak{R}_i) - \mu(z_0, T, \mathfrak{R}^*_i),$$

and hence

$$(4)\qquad \mu(z_0, T, \mathfrak{R}_i) = \mu(z_0, T, \mathfrak{R}^*_i).$$

We have to consider two subcases.

Case (α_1). $\mu(z_0, T, \mathfrak{R}_i) = 0$. Then $\mu(z_0, T, \mathfrak{R}^*_i) = 0$ by (4). Since $\mathfrak{R}^*_i$ is simply-connected, the lemma of IV.1.29 applies to the function $f(w) = t(w) - z_0$ and the region $\mathfrak{R}^*_i$. There follows the existence of a continuous function $F_i(w)$ with the following properties: (1) $F_i(w) = t(w) - z_0$ on the boundary of $\mathfrak{R}^*_i$. (2) $F_i(w) \neq 0$ in $\mathfrak{R}^*_i$. (3) The maximum of $| F_i(w) |$ in $\mathfrak{R}^*_i$ does not exceed the maximum of $| f(w) | = | t(w) - z_0 |$ on the boundary of $\mathfrak{R}^*_i$, and hence $| F_i(w) | <$

τ in $\mathfrak{R}_i^*$ (observe that $\mathfrak{R}_i^* \subset G_\iota$). We define now $t_i(w) = F_i(w) + z_0$ for $w \in \mathfrak{R}_i^*$. Then $t_i(w)$ possesses the following properties:

(a) $t_i(w)$ is continuous in $\mathfrak{R}_i^*$.

(b) $t_i(w) = t(w)$ on the boundary of $\mathfrak{R}_i^*$.

(c) $| t_i(w) - t(w) | = | F_i(w) + z_0 - t(w) | \leqq | F_i(w) | + | t(w) - z_0 | < 2\tau < \epsilon(k, T, \mathfrak{R}, z_0)$ in $\mathfrak{R}_i^*$ (cf. IV.1.32(2)).

(d) $t_i(w) \neq z_0$ in $\mathfrak{R}_i^*$.

Case (α_2). $\mu(z_0, T, \mathfrak{R}_\iota) \neq 0$. Let us observe first that $\mathfrak{R}_\iota$ is in this case an indicator region (z_0, T)·(cf. IV.1.27), and by (3),

$$| t(w) - z_0 | > \tau/2 \qquad \text{on } \mathfrak{R}_\iota - \mathfrak{R}_\iota^0,$$

$$| t(w) - z_0 | < \tau \qquad \text{in } \mathfrak{R}_\iota$$

Let us take now a point w_i interior to $\mathfrak{R}_i^*$. Since $t(w) - z_0 \neq 0$ on the boundary of $\mathfrak{R}_i^*$, we have by IV.1.31 a continuous function $F_i(w)$ with the following properties: (1) $F_i(w) = t(w) - z_0$ on the boundary of $\mathfrak{R}_i^*$. (2) $F_i(w_i) = 0$. (3) $F_i(w) \neq 0$ in $\mathfrak{R}_i^* - w_i$. (4) The maximum of $| F_i(w) |$ in $\mathfrak{R}_i^*$ does not exceed the maximum of $| t(w) - z_0 |$ on the boundary of $\mathfrak{R}_i^*$, and hence $| F_i(w) | < \tau$ in $\mathfrak{R}_i^*$, since $\mathfrak{R}_i^* \subset \mathfrak{R}_\iota \subset G_\iota$. We define $t_i(w) = F_i(w) + z_0$ for $w \in \mathfrak{R}_i^*$. Then $t_i(w)$ possesses the following properties:

(a) $t_i(w)$ is continuous in $\mathfrak{R}_i^*$.

(b) $t_i(w) = t(w)$ on the boundary of $\mathfrak{R}_i^*$.

(c) $| t_i(w) - t(w) | < \epsilon(k, T, \mathfrak{R}, z_0)$ in $\mathfrak{R}_i^*$ (cf. the argument in the case (α_1)).

(d) The equation $t_i(w) = z_0$ has the unique solution $w = w_i$ in $\mathfrak{R}_i^*$.

IV.1.34. Continuation. *Case* (β). There exists a point w_i^* on the boundary of G_ι such that $| t(w_i^*) - z_0 | < \tau$. Then w_i^* is necessarily a boundary point of $\mathfrak{R}$, as noted in IV.1.32, and thus w_i^* lies on a boundary curve of $\mathfrak{R}$. On this boundary curve we have then a short arc σ with the following properties:

(i) w_i^* is an interior point of σ.

(ii) $t(w) \neq z_0$ at the end points of σ.

(iii) $| t(w) - z_0 | < \tau$ on σ (observe that $| t(w_i^*) - z_0 | < \tau$).

As a consequence of (iii), it is clear that every point in $\mathfrak{R}^0$ that lies sufficiently near to σ belongs to G_ι. Since the set $T^{-1}(z_0)$ is finite by assumption, it follows that we can join the end points of σ by a simple arc γ which lies in G_ι except for its end points, in such a manner that the simple closed curve $C = \sigma + \gamma$ contains in its interior all the points of the set $G_\iota \cdot T^{-1}(z_0)$ and the interior of C is contained in G_ι. On C we define a continuous function $f(w)$ as follows. We set $f(w) = t(w) - z_0$ on γ, and note that we have then, in view of (ii),

$$f(w) = t(w) - z_0 \neq 0 \qquad \text{on } \gamma. \tag{1}$$

As regards σ, we observe that $f(w)$ is already defined at the end points of σ by (1), and that $0 < | f(w) | < \tau$ at the end points of σ (see (iii)). Clearly, we can extend the definition of $f(w)$ to the interior points of σ in such a way that

(2) $$| f(w) | < \tau \text{ on } \sigma, \qquad f(w_i^*) = 0,$$

(3) $$f(w) \neq 0 \text{ on } \sigma - w_i^* .$$

In view of (1), (2), (3) and of the inclusion $\gamma \subset G_i$ (except for end points), it follows that $| f(w) | < \tau$ on $C = \sigma + \gamma$, $f(w) \neq 0$ on $C - w_i^*$, and $f(w_i^*) = 0$. If we denote by $\mathfrak{R}_i^*$ the simply-connected Jordan region bounded by C, then we have therefore by IV.1.30 a continuous function $F_i(w)$ with the following properties: (1) $F_i(w) = f(w)$ on C. (2) $F_i(w) \neq 0$ in $\mathfrak{R}_i^{*0}$, and hence $F_i(w) \neq 0$ in $\mathfrak{R}_i^* - w_i^*$, since $f(w) \neq 0$ on $C - w_i^*$. (3) The maximum of $| F_i(w) |$ in $\mathfrak{R}_i^*$ does not exceed the maximum of $| f(w) |$ on C, and hence $| F_i(w) | < \tau$ in $\mathfrak{R}_i^*$. We define $t_i(w) = F_i(w) + z_0$, $w \in \mathfrak{R}_i^*$. Then $t_i(w)$ possesses the following properties:

(a) $t_i(w)$ is continuous in $\mathfrak{R}_i^*$.

(b) $t_i(w) = t(w)$ on the simple arc γ (cf. formula (1)).

(c) $| t_i(w) - t(w) | < \epsilon(k, T, \mathfrak{R}, z_0)$ in $\mathfrak{R}_i^*$ (cf. the argument in case (α_1)).

(d) The equation $t_i(w) = z_0$ has no solution in $\mathfrak{R}_i^* - \sigma$.

IV.1.35. Continuation. We define now in $\mathfrak{R}$ a continuous transformation T^* as follows

$$T^* : z = t^*(w) = \begin{cases} t(w) & \text{for } w \in \mathfrak{R} - (\mathfrak{R}_1^* + \cdots + \mathfrak{R}_m^*), \\ t_i(w) & \text{for } w \in \mathfrak{R}_i^*,\ i = 1, \cdots, m. \end{cases}$$

By IV.1.33, IV.1.34 the function $t^*(w)$ is clearly continuous in $\mathfrak{R}$ and satisfies the inequality $| t^*(w) - t(w) | < \epsilon(k, T, \mathfrak{R}, z_0)$. Thus (cf. IV.1.1)

$$\rho(T^*, T, \mathfrak{R}) < \epsilon(k, T, \mathfrak{R}, z_0).$$

By IV.1.32(1), IV.1.15 it follows that

(1) $$k \leqq N(z_0, T^*, \mathfrak{R}^0).$$

On the other hand, $N(z_0, T^*, \mathfrak{R}^0)$ is clearly equal to the number ν of those components G_i for which the case (α_2) occurs (see IV.1.33, IV.1.34, IV.1.32(3)). Thus $\nu \geqq k$ in view of (1). Now in the case (α_2) we observed that G_i contains an indicator region (z_0, T) for which the inequalities specified in the lemma of IV.1.32 hold. Since $\nu \geqq k$ and the components $G_1, \cdots, G_m$ are disjoint, the lemma of IV.1.32 is proved.

IV.1.36. Given T as in IV.1.2, let z_0 be a point in the z-plane, and let k be a non-negative integer such that $k \leqq \kappa(z_0, T, \mathfrak{R})$. Then $z_0 \in \mathfrak{K}(k, T, \mathfrak{R})$ and hence $\epsilon(k, T, \mathfrak{R}, z_0)$ is available. Let τ be a number such that

(1) $$0 < \tau < \epsilon(k, T, \mathfrak{R}, z_0)/2.$$

Then there exists in $\mathfrak{R}^0$ an indicator system (z_0, T) of order k, comprised of Jordan regions $\mathfrak{R}_1, \cdots, \mathfrak{R}_k$ such that $| t(w) - z_0 | < 2\tau$ for $w \in \mathfrak{R}_i$, $i = 1, \cdots, k$.

Proof. If $k = 0$, then the assertion is vacuously true. So we can assume that $k \geqq 1$. Let us observe that our present assertion is an immediate consequence of IV.1.32 if we assume that $N(z_0, T, \mathfrak{R}) < +\infty$. Our point is precisely to dispense with this assumption.

By IV.1.23 we have a sequence of continuous transformations $T_n : z = t_n(w)$, $w \in \mathfrak{R}$, each of which is polyhedral in $\mathfrak{R}$, such that

$$\rho(T_n, T, \mathfrak{R}) \to 0 \qquad \text{for } n \to \infty. \tag{2}$$

Since T_n is polyhedral, surely $N(z_0, T_n, \mathfrak{R}) < +\infty$. Let us put

$$\sigma = \epsilon(k, T, \mathfrak{R}, z_0) - 2\tau.$$

Then $\sigma > 0$ by (1). Define

$$\eta = 2\tau + \sigma/2. \tag{3}$$

Then $0 < \eta < \epsilon(k, T, \mathfrak{R}, z_0)$. By (2) and IV.1.22, it follows that η will serve as an $\epsilon(k, T_n, \mathfrak{R}, z_0)$ for n sufficiently large, say $n > n_0$. Thus we can choose

$$\epsilon(k, T_n, \mathfrak{R}, z_0) = \eta \qquad \text{for } n > n_0. \tag{4}$$

In view of (3), (4), we have then $0 < \tau < \epsilon(k, T_n, \mathfrak{R}, z_0)/2$ for $n > n_0$. In view of (2), we shall have $\rho(T_n, T, \mathfrak{R}) < \tau/2$ for n large. By IV.1.8 it follows from (2) that $\kappa(z_0, T_n, \mathfrak{R}) \geqq k$ for n large. Summing up, we can choose a large n, say $n = \nu$, such that the following conditions hold simultaneously:

$$\begin{aligned} &\kappa(z_0, T_\nu, \mathfrak{R}) \geqq k > 0. \\ &\rho(T_\nu, T, \mathfrak{R}) < \tau/2. \\ &0 < \tau < \epsilon(k, T_\nu, \mathfrak{R}, z_0)/2. \end{aligned} \tag{5}$$

Since T_ν is polyhedral, we have of course $N(z_0, T_\nu, \mathfrak{R}) < +\infty$. Thus the lemma of IV.1.32 applies to T_ν, and hence there exists in $\mathfrak{R}^0$ an indicator system (z_0, T_ν) of order k, comprised of Jordan regions $\mathfrak{R}_1, \cdots, \mathfrak{R}_k$ such that

$$| t_\nu(w) - z_0 | > \tau/2 \qquad \text{for } w \in \mathfrak{R}_i - \mathfrak{R}_i^0, \; i = 1, \cdots, k, \tag{6}$$

$$| t_\nu(w) - z_0 | < \tau \qquad \text{for } w \in \mathfrak{R}_i, \; i = 1, \cdots, k. \tag{7}$$

We assert that each $\mathfrak{R}_i$ is also an indicator region (z_0, T). Indeed, since $\mathfrak{R}_i$ is an indicator region (z_0, T_ν), we have (see IV.1.27)

$$\mu(z_0, T_\nu, \mathfrak{R}_i) \neq 0. \tag{8}$$

(5), (6) yield (cf. IV.1.25) $\rho(T, T_\nu, \mathfrak{R}_i - \mathfrak{R}_i^0) < \delta(z_0, T_\nu, \mathfrak{R}_i)$. Hence, by IV.1.25(d),

$$\mu(z_0, T, \mathfrak{R}_i) = \mu(z_0, T_\nu, \mathfrak{R}_i). \tag{9}$$

(8), (9) yield $\mu(z_0, T, \mathfrak{R}_i) \neq 0$. Hence $\mathfrak{R}_i$ is an indicator region (z_0, T), by IV.1.27. From (5), (7) we infer, for $w \in \mathfrak{R}_i$,

$$| t(w) - z_0 | \leqq | t(w) - t_\nu(w) | + | t_\nu(w) - z_0 | < \tau/2 + \tau < 2\tau,$$

and the proof is complete.

IV.1.37. Given T as in IV.1.2, let z_0 be a point in the z-plane, and let $\mathfrak{R}_1$ be an indicator region (z_0, T) in $\mathfrak{R}^0$. Given any number $\eta > 0$, there exists a Jordan region $\mathfrak{R}_2$ with the following properties: (i) $\mathfrak{R}_2 \subset \mathfrak{R}_1^0$. (ii) $\mathfrak{R}_2$ is an indicator region (z_0, T). (iii) The maximum of $| t(w) - z_0 |$ in $\mathfrak{R}_2$ is less than η.

PROOF. Since $\mathfrak{R}_1$ is an indicator region (z_0, T), we have $\mu(z_0, T, \mathfrak{R}_1) \neq 0$ by IV.1.27. Hence $\kappa(z_0, T, \mathfrak{R}_1) \geqq 1$ by IV.1.26 (applied to T in $\mathfrak{R}_1$). Thus we can apply IV.1.36 to T in $\mathfrak{R}_1$ with $k = 1$ and any τ that satisfies the inequality

$$(1) \qquad 0 < \tau < \epsilon(1, T, \mathfrak{R}_1, z_0)/2.$$

We choose τ to satisfy both (1) and

$$(2) \qquad \tau < \eta/2.$$

By IV.1.36, we have then in $\mathfrak{R}_1^0$ an indicator region (z_0, T), say $\mathfrak{R}_2$, such that $| t(w) - z_0 | < 2\tau$ for $w \in \mathfrak{R}_2$. In view of (2), $\mathfrak{R}_2$ satisfies our requirements.

IV.1.38. Given T as in IV.1.2, let z_0 be a point in the z-plane, and let $\mathfrak{R}_*$ be an indicator region (z_0, T) in $\mathfrak{R}^0$. Then there exists in $\mathfrak{R}_*^0$ an essential maximal model continuum for z_0 under T (cf. IV.1.27).

PROOF. Let η_n be a sequence of positive numbers converging to zero. By successive applications of IV.1.37, we obtain a sequence of Jordan regions $\mathfrak{R}_n$ with the following properties:

(i) $\mathfrak{R}_{n+1} \subset \mathfrak{R}_n^0 \subset \mathfrak{R}_*^0$; $n = 1, 2, \cdots$.

(ii) $\mathfrak{R}_n$ is an indicator region (z_0, T), $n = 1, 2, \cdots$.

(iii) The maximum of $| t(w) - z_0 |$ in $\mathfrak{R}_n$ is less than η_n, $n = 1, 2, \cdots$.

Let us define

$$(1) \qquad \gamma = \prod_{n=1}^{\infty} \mathfrak{R}_n .$$

Since each $\mathfrak{R}_n$ is a continuum, it follows from (i) (cf. I.2.39) that γ is a nonempty continuum (which may reduce to a single point). Clearly $\gamma \subset \mathfrak{R}_*^0$. As a consequence of (iii), $t(w) = z_0$ on γ. Hence γ is contained in a maximal model continuum Γ for z_0 under T. We assert that $\Gamma = \gamma$. Assume indeed that there exists a point $w_0 \in \Gamma$, $w_0 \notin \gamma$. Then (cf. (1)), we shall have $w_0 \notin \mathfrak{R}_n$ for some n. Now Γ is connected and $\Gamma\mathfrak{R}_n^0 \neq 0$, since $\Gamma \supset \gamma$ and $\gamma \subset \mathfrak{R}_n^0$. Hence (see I.2.40) $\Gamma \cdot (\mathfrak{R}_n - \mathfrak{R}_n^0) \neq 0$. But this is a contradiction, since $t(w) = z_0$ on Γ, and $t(w) \neq z_0$ on $\mathfrak{R}_n - \mathfrak{R}_n^0$ since $\mathfrak{R}_n$ is an indicator region (z_0, T) (cf. IV.1.27). Hence $\gamma = \Gamma$. In other words, γ itself is a maximal model continuum for z_0 under T. To show that γ is essential, let O be any open set that contains γ. In view of (1) and (i), we shall have $\mathfrak{R}_n \subset O$ for n large. Then $\mathfrak{R}_n$ is an indicator region (z_0, T) that contains γ and is contained in O. Hence γ is essential (see IV.1.27).

IV.1.39. THEOREM. *Given T as in IV.1.2, let z_0 be a point in the z-plane. Then $\kappa(z_0, T, \mathfrak{R})$ is equal to the number (possibly $+\infty$) of the essential maximal model continua for z_0 under T in $\mathfrak{R}^0$.*

Proof. Let K denote the number (possibly $+\infty$) of the essential maximal model continua for z_0 under T in $\mathfrak{R}^0$. By IV.1.28, $\kappa(z_0, T, \mathfrak{R}) \geqq K$. The complementary inequality $\kappa(z_0, T, \mathfrak{R}) \leqq K$ will be established if we can show that every non-negative (finite) integer k that satisfies the inequality $k \leqq \kappa(z_0, T, \mathfrak{R})$ also satisfies the inequality $k \leqq K$. In verifying this fact it is clearly sufficient to consider the case when $k > 0$. Now if $0 < k \leqq \kappa(z_0, T, \mathfrak{R})$, then by IV.1.36 we have in $\mathfrak{R}^0$ an indicator system (z_0, T) of order k. Let $\mathfrak{R}_1, \cdots, \mathfrak{R}_k$ be the regions of this system. Then $\mathfrak{R}_i^0\mathfrak{R}_j^0 = 0$ for $i \neq j$, and by IV.1.38 each region $\mathfrak{R}_i$ contains in its interior an essential maximal model continuum γ_i for z_0 under T. Thus we obtain a system of k distinct continua $\gamma_1, \cdots, \gamma_k$, and hence $k \leqq K$.

IV.1.40. Let $T: z = t(w)$, $w \in \mathfrak{R}$, $T^*: z = t^*(w^*)$, $w^* \in \mathfrak{R}^*$, be two continuous transformations, where the w- and w^*-planes may or may not coincide. Suppose that T and T^* are topologically similar (cf. IV.1.6). That is, there exists a homeomorphism $\tau(\mathfrak{R}) = \mathfrak{R}^*$ such that $t(w) = t^*[\tau(w)]$ for $w \in \mathfrak{R}$. Let z_0 be any point in the z-plane, and let $\mathfrak{R}_1$ be any indicator region (z_0, T) in $\mathfrak{R}^0$. Then $\mathfrak{R}_1^* = \tau(\mathfrak{R}_1)$ is an indicator region (z_0, T^*) as an immediate consequence of IV.1.25(f), IV.1.27. In view of IV.1.27 it is now clear that if γ is an essential maximal model continuum for z_0 under T in $\mathfrak{R}^0$, then $\gamma^* = \tau(\gamma)$ is an essential maximal model continuum for z_0 under T^* in $\mathfrak{R}^{*0}$.

IV.1.41. The following definitions will be used constantly in the sequel. Let $\mathfrak{R}$ be a (bounded, finitely-connected) Jordan region in the w-plane. A sequence of Jordan regions $\mathfrak{R}_n$ will be said to fill up $\mathfrak{R}$ from the interior if the following conditions hold: (i) $\mathfrak{R}_n \subset \mathfrak{R}$, $n = 1, 2, \cdots$. (ii) If F is any closed set in $\mathfrak{R}^0$, then $F \subset \mathfrak{R}_n^0$ for n sufficiently large.

Let $\mathfrak{D}$ be a bounded domain (connected open set) in the w-plane. A sequence $\mathfrak{R}_n$ of Jordan regions will be said to fill up $\mathfrak{D}$ from the interior if the following conditions hold: (i) $\mathfrak{R}_n \subset \mathfrak{D}$, $n = 1, 2, \cdots$. (ii) If F is any closed set contained in $\mathfrak{D}$, then $F \subset \mathfrak{R}_n^0$ for n sufficiently large. The term closed set is used here in the absolute sense (relative to the w-plane, and not relative to $\mathfrak{D}$). Since $\mathfrak{D}$ is bounded, a closed set $F \subset \mathfrak{D}$ is then compact (see I.2.38).

IV.1.42. We shall now consider transformations T given in the form

$$T: z = t(w), \qquad w \in \mathfrak{D},$$

where the following standard conditions will be assumed to hold: (i) $\mathfrak{D}$ is a bounded domain (connected open set). (ii) $t(w)$ is continuous and bounded in $\mathfrak{D}$. Note that the boundedness of $t(w)$ is now a separate requirement, since $\mathfrak{D}$ is not compact. We shall use simultaneously the real representation

$$T: x = x(u, v), \quad y = y(u, v), \qquad (u, v) \in \mathfrak{D},$$

where $x + iy = z$, $u + iv = w$, $x(u, v) + iy(u, v) = t(w)$.

IV.1.43. Given T as in IV.1.42, let us take a point z. Let $\mathfrak{R}_n$ be a sequence of Jordan regions that fill up $\mathfrak{D}$ from the interior, in the sense of IV.1.41 (such sequences exist by I.2.48). We assert that the sequence $\kappa(z, T, \mathfrak{R}_n)$ is convergent (its limit being possibly $+\infty$). Indeed, since $\mathfrak{R}_k$ is itself a closed set in $\mathfrak{D}$, we shall have $\mathfrak{R}_k \subset \mathfrak{R}_j$ for fixed k and for j sufficiently large. Hence by IV.1.14 $\kappa(z, T, \mathfrak{R}_k) \leqq \kappa(z, T, \mathfrak{R}_j)$ for fixed k and large j. For $j \to \infty$ it follows that

$$\kappa(z, T, \mathfrak{R}_k) \leqq \liminf_{j \to \infty} \kappa(z, T, \mathfrak{R}_j).$$

For $k \to \infty$ we obtain now

$$\limsup_{k \to \infty} \kappa(z, T, \mathfrak{R}_k) \leqq \liminf_{j \to \infty} \kappa(z, T, \mathfrak{R}_j).$$

Since the notation used for the subscripts is of course immaterial, the last relation proves the convergence of the sequence $\kappa(z, T, \mathfrak{R}_n)$. If $\mathfrak{R}_n^*$ is a second sequence of Jordan regions filling up $\mathfrak{D}$ from the interior, then by the same argument the sequence $\kappa(z, T, \mathfrak{R}_n^*)$ is also convergent. Since $\mathfrak{R}_1, \mathfrak{R}_1^*, \cdots, \mathfrak{R}_n, \mathfrak{R}_n^*, \cdots$, is also a sequence of Jordan regions filling up $\mathfrak{D}$ from the interior, it follows that the sequences $\kappa(z, T, \mathfrak{R}_n)$, $\kappa(z, T, \mathfrak{R}_n^*)$ have the same limit. In other words, the limit of $\kappa(z, T, \mathfrak{R}_n)$ is independent of the particular choice of the sequence $\mathfrak{R}_n$. We can therefore define

$$\kappa(z, T, \mathfrak{D}) = \lim_{n \to \infty} \kappa(z, T, \mathfrak{R}_n).$$

The function $\kappa(z, T, \mathfrak{D})$ will be termed *the essential multiplicity of the point z in the image of $\mathfrak{D}$ under T.*

IV.1.44. Continuation. If $\mathfrak{R}$ is any Jordan region in $\mathfrak{D}$, then $\kappa(z, T, \mathfrak{R}) \leqq \kappa(z, T, \mathfrak{D})$. Indeed, let $\mathfrak{R}_n$ be a sequence of Jordan regions that fill up $\mathfrak{D}$ from the interior (see IV.1.41). Then $\mathfrak{R} \subset \mathfrak{R}_n$ for n sufficiently large, and hence, by IV.1.14, $\kappa(z, T, \mathfrak{R}) \leqq \kappa(z, T, \mathfrak{R}_n)$. For $n \to \infty$ we obtain the inequality $\kappa(z, T, \mathfrak{R}) \leqq \kappa(z, T, \mathfrak{D})$.

IV.1.45. Continuation. Let z_0 be a point in the z-plane. If k is any non-negative integer such that $k \leqq \kappa(z_0, T, \mathfrak{D})$, and if $\mathfrak{R}_n$ is a sequence of Jordan regions that fill up $\mathfrak{D}$ from the interior, then we have $k \leqq \kappa(z_0, T, \mathfrak{R}_n)$ for n sufficiently large. Indeed, if we put $\kappa_n = \kappa(z_0, T, \mathfrak{R}_n)$, then by definition $\kappa_n \to \kappa(z_0, T, \mathfrak{D})$. If $\kappa(z_0, T, \mathfrak{D}) = +\infty$, then it follows immediately that $\kappa_n > k$ for n large enough. If $\kappa(z_0, T, \mathfrak{D}) < +\infty$, then $\kappa_n < +\infty$ also (see IV.1.44), and since κ_n and $\kappa(z_0, T, \mathfrak{D})$ are integers, the relations $\kappa_n \to \kappa(z_0, T, \mathfrak{D})$, $k \leqq \kappa(z_0, T, \mathfrak{D})$ clearly imply that $\kappa_n = \kappa(z_0, T, \mathfrak{D})$ and hence $\kappa_n \geqq k$ for n sufficiently large.

IV.1.46. Continuation. Given a point z_0 in the z-plane, consider the set $T^{-1}(z_0)$ in $\mathfrak{D}$. Assume that $T^{-1}(z_0) \neq 0$. If γ is a component of $T^{-1}(z_0)$, then γ is connected, but γ may not be closed (since $\mathfrak{D}$ itself is open). It may happen however that γ is closed. In this case, since $\mathfrak{D}$ is bounded, γ is a compact connected set, and hence a continuum that may reduce to a single point. We shall say then that γ is a *maximal model continuum for z_0 under T in $\mathfrak{D}$.* If for every open set O, that contains the maximal model continuum γ, there exists a (finitely-connected) Jordan region $\mathfrak{R}$ such that $\gamma \subset \mathfrak{R}^0$, $\mathfrak{R} \subset O\mathfrak{D}$, $\mu(z_0, T, \mathfrak{R}) \neq 0$ (cf. IV.1.24), then γ will be termed *essential* (for z_0 under T in $\mathfrak{D}$). According to the terminology introduced in IV.1.27, we shall term a Jordan region $\mathfrak{R}$ with the properties $\mathfrak{R} \subset \mathfrak{D}$, $\mu(z_0, T, \mathfrak{R}) \neq 0$ an indicator region (z_0, T) in $\mathfrak{D}$. If γ is an essential maximal model continuum for z_0 under T in $\mathfrak{D}$, and if $\mathfrak{D}_0$ is any sub-domain of $\mathfrak{D}$ that contains γ, then clearly γ is also an essential maximal model continuum for z_0 under T in $\mathfrak{D}_0$, and the converse is equally obvious.

IV.1.47. THEOREM. *Given a continuous transformation* $T: z = t(w)$, $w \in \mathfrak{D}$, *as in* IV.1.42, $\kappa(z, T, \mathfrak{D})$ *is equal to the number (possibly* $+\infty$*) of the essential maximal model continua for* z *under* T *in* $\mathfrak{D}$ (*cf.* IV.1.46, IV.1.43).

PROOF. Let $\mathfrak{R}_n$ be a sequence of Jordan regions that fill up $\mathfrak{D}$ from the interior, and let K be the number (possibly $+\infty$) of the essential maximal model continua for z under T in $\mathfrak{D}$, and let K_n be the corresponding number for $\mathfrak{R}_n^0$ (cf. IV.1.27). By IV.1.39 we have $K_n = \kappa(z, T, \mathfrak{R}_n)$, and by definition $\kappa(z, T, \mathfrak{D}) = \lim \kappa(z, T, \mathfrak{R}_n)$. Hence $K_n \to \kappa(z, T, \mathfrak{D})$. If γ is an essential maximal model continuum for z under T in $\mathfrak{R}_n^0$, then clearly γ is also an essential maximal model continuum for z under T in $\mathfrak{D}$. Thus $K_n \leqq K$. On the other hand, if $\gamma_1, \cdots, \gamma_m$ is any finite system of essential maximal model continua for z under T in $\mathfrak{D}$, then $\gamma_1 + \cdots + \gamma_m$ is a closed set, and hence $\gamma_1, \cdots, \gamma_m$ will be contained in $\mathfrak{R}_n^0$ for n sufficiently large. Thus $\liminf K_n \geqq K$. It follows that $K_n \to K$. The relations $K_n \to K$, $K_n = \kappa(z, T, \mathfrak{R}_n)$, $\kappa(z, T, \mathfrak{R}_n) \to \kappa(z, T, \mathfrak{D})$ yield $\kappa(z, T, \mathfrak{D}) = K$.

IV.1.48. CONTINUATION. If $\mathfrak{D}_0$ is any sub-domain of $\mathfrak{D}$, then $\kappa(z, T, \mathfrak{D}_0) \leqq \kappa(z, T, \mathfrak{D})$. This is an immediate corollary of IV.1.47. A second corollary is concerned with the invariance of $\kappa(z, T, \mathfrak{D})$ under topological similarity. Let $T: z = t(w)$, $w \in \mathfrak{D}$, and $T^*: z = t^*(w^*)$, $w^* \in \mathfrak{D}^*$, be two transformations as in IV.1.42, and suppose that there exists a homeomorphism $\tau(\mathfrak{D}) = \mathfrak{D}^*$ such that $t(w) = t^*[\tau(w)]$ for $w \in \mathfrak{D}$. If γ is an essential maximal model continuum for a point z under T in $\mathfrak{D}$, then the same argument as that used in IV.1.40 shows that $\gamma^* = \tau(\gamma)$ is an essential maximal model continuum for z under T^* in $\mathfrak{D}^*$. In view of IV.1.47 it follows that $\kappa(z, T, \mathfrak{D}) = \kappa(z, T^*, \mathfrak{D}^*)$ if T and T^* are topologically similar.

IV.1.49. CONTINUATION. If T is topological (that is, if $w_1 \neq w_2$ implies that $t(w_1) \neq t(w_2)$), then $\kappa(z, T, \mathfrak{D}) = N(z, T, \mathfrak{D})$ (cf. IV.1.1). Indeed, in this case for every point $w_0 \in \mathfrak{D}$ we have $w_0 = T^{-1}(z_0)$, where $z_0 = t(w_0)$. Thus w_0 is a maximal model continuum for z_0 under T in $\mathfrak{D}$. Let now $\mathfrak{R}$ denote a small circular disc with center w_0. Then clearly $\mu(z_0, T, \mathfrak{R}) = \pm 1$, and hence w_0 is an essential maximal model continuum for z_0 under T in $\mathfrak{D}$. Thus $\kappa(z, T, \mathfrak{D}) = 1$ if $N(z, T, \mathfrak{D}) = 1$, and clearly (cf. IV.1.4) $\kappa(z, T, \mathfrak{D}) = 0$ if $N(z, T, \mathfrak{D}) = 0$.

IV.1.50. Given a continuous transformation $T: z = t(w)$, $w \in \mathfrak{R}$, as in IV.1.2, we defined in IV.1.4 the essential multiplicity function $\kappa(z, T, \mathfrak{R})$. On the other hand, we can consider T as operating from the domain $\mathfrak{D} = \mathfrak{R}^0$, and then we have the essential multiplicity function $\kappa(z, T, \mathfrak{R}^0)$ as defined in IV.1.43. We assert that $\kappa(z, T, \mathfrak{R}) = \kappa(z, T, \mathfrak{R}^0)$. This follows from the definition given in IV.1.43 and from the result in IV.1.17, or more immediately from the theorems in IV.1.39 and IV.1.47.

IV.1.51. Given T as in IV.1.42, $\kappa(z, T, \mathfrak{D})$ *is a lower semi-continuous function of* z.

PROOF. Let z_j be a sequence of points converging to z_0. Let $\mathfrak{R}_n$ be a sequence of Jordan regions filling up $\mathfrak{D}$ from the interior. For fixed n we have then, by IV.1.11, IV.1.44,

$$\kappa(z_0, T, \mathfrak{R}_n) \leqq \liminf_{j\to\infty} \kappa(z_j, T, \mathfrak{R}_n) \leqq \liminf_{j\to\infty} \kappa(z_j, T, \mathfrak{D}).$$

For $n \to \infty$ we obtain, in view of IV.1.43,

$$\kappa(z_0, T, \mathfrak{D}) \leqq \liminf_{j\to\infty} \kappa(z_j, T, \mathfrak{D}).$$

IV.1.52. Let there be given continuous transformations

$$T : z = t(w), \qquad w \in \mathfrak{D},$$

$$T_j : z = t_j(w), \qquad w \in \mathfrak{D}_j, j = 1, 2, \cdots,$$

as in IV.1.42. Suppose that the following conditions hold:

(i) If F is any closed set contained in $\mathfrak{D}$, then $F \subset \mathfrak{D}_j$ for j sufficiently large.

(ii) If F is any closed set contained in $\mathfrak{D}$, then $\rho(T_j, T, F) \to 0$ for $j \to \infty$ (cf. IV.1.1). Note that this requirement is meaningful, since T_j is defined on F for j sufficiently large, by condition (i).

Under these conditions, we have the inequality

$$\kappa(z, T, \mathfrak{D}) \leqq \liminf \kappa(z, T_j, \mathfrak{D}_j) \qquad \text{for } j \to \infty. \tag{1}$$

PROOF. Let $\mathfrak{R}_n$ be a sequence of Jordan regions that fill up $\mathfrak{D}$ from the interior. By condition (i), T_j will be defined on $\mathfrak{R}_n$ for fixed n and for j sufficiently large. In view of condition (ii), it follows therefore from IV.1.12, IV.1.44 that

$$\kappa(z, T, \mathfrak{R}_n) \leqq \liminf \kappa(z, T_j, \mathfrak{R}_n) \leqq \liminf \kappa(z, T_j, \mathfrak{D}_j) \qquad \text{for } j \to \infty.$$

For $n \to \infty$ we obtain, in view of IV.1.43, the inequality (1).

IV.1.53. We shall say that a sequence $\mathfrak{D}_j$ of domains fills up the domain $\mathfrak{D}$ if $\mathfrak{D}_j \subset \mathfrak{D}$, $j = 1, 2, \cdots$, and for every closed set $F \subset \mathfrak{D}$ we have $F \subset \mathfrak{D}_j$ for j sufficiently large. Using this terminology, we have the following corollaries to IV.1.52.

COROLLARY 1. *Given T as in* IV.1.42, *let $\mathfrak{D}_j$ be a sequence of domains that fill up $\mathfrak{D}$. Then $\kappa(z, T, \mathfrak{D}) = \lim \kappa(z, T, \mathfrak{D}_j)$ for $j \to \infty$.*

Indeed, we have $\kappa(z, T, \mathfrak{D}_j) \leqq \kappa(z, T, \mathfrak{D})$ by IV.1.48, and hence the assertion follows from IV.1.52, applied with $T_j : z = t(w)$, $w \in \mathfrak{D}_j$.

COROLLARY 2. *Given the continuous transformations $T : z = t(w)$, $w \in \mathfrak{D}$, and $T_j : z = t_j(w)$, $w \in \mathfrak{D}$, $j = 1, 2, \cdots$, as in* IV.1.42, *suppose that $\rho(T, T_j, F) \to 0$ for every closed set contained in $\mathfrak{D}$. Then $\kappa(z, T, \mathfrak{D}) \leqq \liminf \kappa(z, T_j, \mathfrak{D})$ for $j \to \infty$.*

This is merely the special case $\mathfrak{D}_j = \mathfrak{D}$ of IV.1.52.

Corollary 1 states a continuity property of $\kappa(z, T, \mathfrak{D})$ with respect to the third argument $\mathfrak{D}$, while Corollary 2 states a property of lower semicontinuity of $\kappa(z, T, \mathfrak{D})$ with respect to the second argument T.

IV.1.54. Given $T : z = t(w)$, $w \in \mathfrak{D}$, as in IV.1.42, let $\mathfrak{D}_i$ denote a (finite or infinite) sequence of domains in $\mathfrak{D}$, such that $\mathfrak{D}_i\mathfrak{D}_j = 0$ for $i \neq j$. Then

$$\sum_i \kappa(z, T, \mathfrak{D}_i) \leqq \kappa(z, T, \mathfrak{D}).$$

The proof follows directly from IV.1.47, IV.1.46.

IV.1.55. Continuation. Let us put $E = \mathfrak{D} - \sum \mathfrak{D}_i$. We have then

$$\kappa(z, T, \mathfrak{D}) = \sum_i \kappa(z, T, \mathfrak{D}_i) \qquad \text{for } z \notin T(E).$$

Proof. In view of IV.1.54, it is sufficient to show that

$$\kappa(z, T, \mathfrak{D}) \leqq \sum_i \kappa(z, T, \mathfrak{D}_i) \qquad \text{for } z \notin T(E). \tag{1}$$

Now let z_0 be any point not in $T(E)$, and let γ_0 be any essential maximal model continuum for z_0 under T in $\mathfrak{D}$. Since $T(\gamma_0) = z_0$, and $z_0 \notin T(E)$, it follows that $\gamma_0 \subset \mathfrak{D} - E = \sum \mathfrak{D}_i$. Since γ_0 is connected, γ_0 must be a subset of a component of $\sum \mathfrak{D}_i$. By I.2.39, the components of $\sum \mathfrak{D}_i$ are precisely the domains $\mathfrak{D}_i$. It follows that every essential maximal model continuum for z_0 under T in $\mathfrak{D}$ is a subset of some domain $\mathfrak{D}_i$. The inequality (1) follows now directly from IV.1.47, IV.1.46.

IV.1.56. Given $T : z = t(w)$, $w \in \mathfrak{D}$, as in IV.1.42, we introduce certain subsets of $\mathfrak{D}$ that will play an important role in the sequel.

(i) *The set* $\mathcal{E}(T, \mathfrak{D})$. A point $w_0 \in \mathfrak{D}$ belongs to $\mathcal{E}(T, \mathfrak{D})$ if w_0, taken by itself, is an essential maximal model continuum, under T in $\mathfrak{D}$, for its image $z_0 = t(w_0)$ (cf. IV.1.46).

(ii) *The set* $\mathcal{E}^*(T, \mathfrak{D})$. A point $w_0 \in \mathfrak{D}$ belongs to $\mathcal{E}^*(T, \mathfrak{D})$ if and only if w_0 lies on an essential maximal model continuum, under T in $\mathfrak{D}$, for its image $z_0 = t(w_0)$.

(iii) *The set* $\mathcal{N}(T, \mathfrak{D})$. A point $w_0 \in \mathfrak{D}$ belongs to $\mathcal{N}(T, \mathfrak{D})$ if and only if (α) $w_0 \in \mathcal{E}(T, \mathfrak{D})$, and ($\beta$) there exists an open set O such that $w_0 \in O$ and no essential maximal model continuum, under T in $\mathfrak{D}$, of the point $z_0 = t(w_0)$ intersects the set $O - w_0$.

(iv) *The set* $\mathcal{I}(T, \mathfrak{D})$. A point $w_0 \in \mathfrak{D}$ belongs to $\mathcal{I}(T, \mathfrak{D})$ if and only if there exists an open set O containing w_0 such that $t(w) \neq t(w_0)$ for $w \in O\mathfrak{D} - w_0$.

Observe that one or more of the sets just defined may be empty. Clearly $\mathcal{E}^*(T, \mathfrak{D}) \supset \mathcal{E}(T, \mathfrak{D}) \supset \mathcal{N}(T, \mathfrak{D})$. If $\mathfrak{D}_0$ is any subdomain of $\mathfrak{D}$, then the definitions apply to T considered in $\mathfrak{D}_0$, and yield the sets $\mathcal{E}(T, \mathfrak{D}_0)$, $\mathcal{E}^*(T, \mathfrak{D}_0)$, $\mathcal{N}(T, \mathfrak{D}_0)$, $\mathcal{I}(T, \mathfrak{D}_0)$. Clearly (cf. IV.1.46) $\mathcal{E}(T, \mathfrak{D}_0) = \mathfrak{D}_0\mathcal{E}(T, \mathfrak{D})$, $\mathcal{N}(T, \mathfrak{D}_0) = \mathfrak{D}_0 \mathcal{N}(T, \mathfrak{D})$, $\mathcal{I}(T, \mathfrak{D}_0) = \mathfrak{D}_0\mathcal{I}(T, \mathfrak{D})$. On the other hand, the formula $\mathcal{E}^*(T, \mathfrak{D}_0) = \mathfrak{D}_0\mathcal{E}^*(T, \mathfrak{D})$ is generally false. Indeed, suppose that there exists a point z_0 that possesses an essential maximal model continuum γ under T in $\mathfrak{D}$, where γ does not reduce to a single point (examples for this may be easily constructed). Let w_0 be a point of γ, and let $\mathfrak{D}_0$ be a sub-domain of $\mathfrak{D}$ such that $w_0 \in \mathfrak{D}_0$, $\gamma(\mathfrak{D} - \mathfrak{D}_0) \neq 0$. For example, we may choose $\mathfrak{D}_0$ as an open circular disc with center at w_0 and with a radius less than half of the diameter of γ. Then $w_0 \in \mathfrak{D}_0\mathcal{E}^*(T, \mathfrak{D})$, but $w_0 \notin \mathcal{E}^*(T, \mathfrak{D}_0)$. While the formula $\mathcal{E}^*(T, \mathfrak{D}_0) = \mathfrak{D}_0\mathcal{E}^*(T, \mathfrak{D})$ is generally not available, the inclusion $\mathcal{E}^*(T, \mathfrak{D}_0) \subset \mathfrak{D}_0\mathcal{E}^*(T, \mathfrak{D})$ is obvious (cf. IV.1.46).

If there is no danger of misunderstanding, then we shall use the more concise notations $\mathcal{E}(T)$, $\mathcal{E}(\mathfrak{D})$ or simply $\mathcal{E}$ instead of $\mathcal{E}(T, \mathfrak{D})$, with similar conventions concerning $\mathcal{E}^*$, $\mathcal{N}$, $\mathcal{J}$. Then clearly (cf. IV.1.1, IV.1.47)

$$(1) \qquad N(z, T, \mathcal{N}) \leqq N(z, T, \mathcal{E}) \leqq \kappa(z, T, \mathfrak{D}) \leqq N(z, T, \mathcal{E}^*).$$

Suppose now that we have $\kappa(z, T, \mathfrak{D}) = N(z, T, \mathcal{E}) < +\infty$ for a certain point z. This means that z has a finite number of essential maximal model continua under T in $\mathfrak{D}$, each of which reduces to a single point. Clearly it follows that the sign of equality holds throughout in (1).

Let us next assume that we have $N(z, T, \mathcal{E}^*) < +\infty$ for a certain point z. If γ is an essential maximal model continuum for z under T in $\mathfrak{D}$, then γ must reduce to a single point. Indeed, if this were not the case, then γ would contain infinitely many points (observe that γ is connected), and we should have $N(z, T, \mathcal{E}^*) = +\infty$, while we assumed that $N(z, T, \mathcal{E}^*) < +\infty$. It follows now immediately that the sign of equality holds throughout in (1). Thus we see that for the sign of equality to hold throughout in (1), for a certain point z, either one of the following two conditions is sufficient: (a) $N(z, T, \mathcal{E}) = \kappa(z, T, \mathfrak{D}) < +\infty$, or (b) $N(z, T, \mathcal{E}^*) < +\infty$.

IV.1.57. Continuation. Let n be a positive integer and m any nonzero integer. Define a subset E_{mn} of $\mathfrak{D}$ as follows. A point $w_0 \in \mathfrak{D}$ belongs to E_{mn} if and only if there exists a (bounded, finitely-connected) Jordan region $\mathfrak{R}$ with the following properties: (α) $w_0 \in \mathfrak{R}^0$, $\mathfrak{R} \subset \mathfrak{D}$. ($\beta$) $\mathfrak{R}$ is contained in the open circular disc $| w - w_0 | < 1/n$. (γ) $\mu[t(w_0), T, \mathfrak{R}] = m$ (cf. IV.1.24). We assert that E_{mn} is an open (possibly empty) set.

Proof. If $E_{mn} = 0$, then there is nothing to prove. If $E_{mn} \neq 0$, then let w_0 be any point of E_{mn}, and let $\mathfrak{R}$ be a corresponding Jordan region with the properties (α), (β), (γ). We assert that $\mathfrak{R}$ has these same properties (α), (β), (γ) with respect to every point w_1 that is sufficiently close to w_0 (once this is verified, it will be established that E_{mn} is open). As regards the properties (α), (β), our assertion is obvious. As regards (γ), let us note that $\mu[t(w_0), T, \mathfrak{R}] = m \neq 0$, and hence $\delta[t(w_0), T, \mathfrak{R}] > 0$ by IV.1.25(a). Since $t(w)$ is continuous, it follows that we have an $\eta > 0$ such that $| t(w_1) - t(w_0) | < \delta[t(w_0), T, \mathfrak{R}]$ for $| w_1 - w_0 | < \eta$. By IV.1.25(c) it follows that $\mu[t(w_1), T, \mathfrak{R}] = \mu[t(w_0), T, \mathfrak{R}] = m$ for $| w_1 - w_0 | < \eta$. Thus $\mathfrak{R}$ possesses the property (γ) with respect to every point w_1 that is sufficiently close to w_0.

IV.1.58. Continuation. The set $\mathcal{E}$ (see IV.1.56) is a Borel set (see I.2.46).

Proof. Since the set E_{mn} (see IV.1.57) is open (or empty), our assertion follows from the formula

(1) $$\mathcal{E} = \prod_{n=1}^{\infty} \sum_{m=-\infty}^{+\infty} E_{mn}, \qquad m \neq 0,$$

which we shall verify presently.

(i) Take any point $w_0 \in \mathcal{E}$, and let n be any positive integer. Then w_0 is an essential maximal model continuum (under T in $\mathfrak{D}$) for the point $z_0 = t(w_0)$, and hence (cf. IV.1.46) we have in the open circular disc $| w - w_0 | < 1/n$ a (finitely-connected) Jordan region $\mathfrak{R}$ that contains w_0 in its interior and is an indicator region $(t(w_0), T)$ in $\mathfrak{D}$. Then $\mu[t(w_0), T, \mathfrak{R}] \neq 0$. Hence, if we put $m = \mu[t(w_0), T, \mathfrak{R}]$, then $m \neq 0$, and $w_0 \in E_{mn}$ (cf. IV.1.57). In other words, for every positive integer n the point w_0 belongs to E_{mn} for some integer $m \neq 0$. Thus the set on the left in (1) is a subset of the set on the right.

(ii) Suppose conversely that $w_0 \in \mathfrak{D}$ is a point of the set on the right in (1). Let O be any open set such that $w_0 \in O \subset \mathfrak{D}$. Choose a positive integer n so large that the open disc $| w - w_0 | < 1/n$ is contained in O. By assumption

$$w_0 \in \sum_{m=-\infty}^{+\infty} E_{mn}, \qquad m \neq 0.$$

Hence we have some integer $m \neq 0$ such that $w_0 \in E_{mn}$ (observe that m depends upon n). By the definition of E_{mn} we have then a Jordan region $\mathfrak{R}_n$ with the properties (α), (β), (γ) listed in IV.1.57. Note that since m depends upon n, the notation $\mathfrak{R}_n$ is justified. Then $w_0 \in \mathfrak{R}_n^0$, $\mathfrak{R}_n \subset O$, $\mu[t(w_0), T, \mathfrak{R}_n] = m \neq 0$. Thus $\mathfrak{R}_n$ is an indicator region $(t(w_0), T)$ containing w_0 and contained in O. Furthermore, the diameter of $\mathfrak{R}_n$ is less than $2/n$. Now let γ_0 be the component of $T^{-1}[t(w_0)]$ that contains w_0. Since $T(\gamma_0) = t(w_0)$ and $t(w) \neq t(w_0)$ on the boundary of $\mathfrak{R}_n$ as a consequence of the relations $\mu[t(w_0), T, \mathfrak{R}_n] = m \neq 0$ (cf. IV.1.24), it follows that $\gamma_0 \subset \mathfrak{R}_n^0$. Thus the diameter of γ_0 does not exceed $2/n$. Since this holds for all sufficiently large values of n, γ_0 reduces to the point w_0 itself. Thus $w_0 \in \mathcal{E}$ (cf. IV.1.56). In other words, the set on the right in (1) is a subset of the set on the left.

IV.1.59. Continuation. The set $\mathcal{E}^*$ of IV.1.56 is a Borel set. Since the argument is quite analogous to that used for $\mathcal{E}$ in IV.1.58, we shall merely sketch the proof. For every positive integer n, define a subset E_n of $\mathfrak{D}$ as follows. A point $w_0 \in \mathfrak{D}$ belongs to E_n if and only if there exists a (finitely-connected) Jordan region $\mathfrak{R}$ with the following properties: (α) $w_0 \in \mathfrak{R}^0$, $\mathfrak{R} \subset \mathfrak{D}$. (β) $T(\mathfrak{R})$ is contained in the open circular disc $| z - t(w_0) | < 1/n$. (γ) $\mu[t(w_0), T, \mathfrak{R}] \neq 0$. Then E_n is an open (possibly empty) set and $\mathcal{E}^* = \prod E_n$, $n = 1, 2, \cdots$. Hence $\mathcal{E}^*$ is a Borel set.

IV.1.60. Continuation. The set $\mathcal{N}$ of IV.1.56 is a Borel set. Again, we shall merely indicate the proof. Using the notations of IV.1.3, IV.2.3, IV.1.56, the first step consists of verifying the formula

$$\mathcal{N} = \mathcal{E} \cdot \sum_{j=1}^{\infty} \sum s^0 \cdot \{T^{-1}[\mathfrak{K}(1, T, s)] - T^{-1}[\mathfrak{K}(2, T, s)]\}, \qquad s \in D_{p_j}, s \subset \mathfrak{D}.$$

Now the sets $\mathfrak{K}(1, T, s)$, $\mathfrak{K}(2, T, s)$ are open (cf. IV.1.9), and hence the corresponding inverse sets are also open. Since $\mathcal{E}$ is a Borel set by IV.1.58, it follows that $\mathcal{N}$ is a Borel set.

IV.1.61. Continuation. The set $\mathcal{G}$ of IV.1.56 is a Borel set.

Proof. Let m, n be positive integers, and let us define a subset G_{mn} of $\mathfrak{D}$ as follows. A point $w_0 \in \mathfrak{D}$ belongs to G_{mn} if and only if the following conditions hold: (α) The closed circular disc $|w - w_0| \leqq 1/m$ is comprised in $\mathfrak{D}$. (β) $t(w) \neq t(w_0)$ for $1/(m + n) \leqq |w - w_0| \leqq 1/m$. Clearly G_{mn} is an open set, and

$$\mathcal{G} = \sum_{m=1}^{\infty} \prod_{n=1}^{\infty} G_{mn} .$$

Thus $\mathcal{G}$ is a Borel set.

IV.1.62. Continuation. Let w_0 be any point of the set $\mathcal{N}$ (see IV.1.56), and let O be any open set such that $w_0 \in O$. Then there exists a Jordan region $\mathfrak{R}$ with the following properties. (a) $w_0 \in \mathfrak{R}^0$, $\mathfrak{R}^0 \subset O\mathfrak{D}$. (b) $\mathfrak{R}$ is simply-connected. (c) $t(w) \neq t(w_0)$ for $w \in \mathfrak{R} - \mathfrak{R}^0$. (d) $\mu[t(w_0), T, \mathfrak{R}] \neq 0$. (e) $\kappa[t(w_0), T, \mathfrak{R}] = 1$.

Proof. Let us denote by $U(w_0, \rho)$ the open circular disc $|w - w_0| < \rho$. Since $w_0 \in \mathcal{N} \subset \mathfrak{D}$, we can choose ρ so small that $U(w_0, \rho) \subset O\mathfrak{D}$ and no essential maximal model continuum of the point $z_0 = t(w_0)$ intersects the set $U(w_0, \rho) - w_0$. By the definition of $\mathcal{N}$, we have then a finitely-connected Jordan region $\mathfrak{R}^*$ such that

$$(1) \qquad w_0 \in \mathfrak{R}^{*0}, \qquad \mu[t(w_0), T, \mathfrak{R}^*] \neq 0, \qquad \mathfrak{R}^* \subset U(w_0, \rho).$$

If $\mathfrak{R}^*$ happens to be simply-connected, then $\mathfrak{R}^*$ satisfies all of our requirements. Indeed, (a), (b), (d) are then clearly satisfied. (c) is a consequence of (d) (see IV.1.24), and (e) is a consequence of the fact that no essential maximal model continuum of the point $t(w_0)$ intersects $U(w_0, \rho) - w_0$ (note that an essential maximal model continuum for $t(w_0)$ under T in $\mathfrak{R}^*$ is also an essential maximal model continuum for $t(w_0)$ under T in $\mathfrak{D}$). If $\mathfrak{R}^*$ is multiply-connected, then let C_0 be the exterior boundary curve of $\mathfrak{R}^*$, and let $C_1, \cdots, C_m$ be the other boundary curves of $\mathfrak{R}^*$. Let $\mathfrak{R}_0, \mathfrak{R}_1, \cdots, \mathfrak{R}_m$ be the bounded, simply-connected Jordan regions bounded by $C_0, C_1, \cdots, C_m$ respectively. Since $C_i \subset U(w_0, \rho) \subset O\mathfrak{D}$, $i = 0, 1, \cdots, m$, it follows that

$$\mathfrak{R}_i \subset O\mathfrak{D}, \qquad i = 0, 1, \cdots, m, \; w_0 \in \mathfrak{R}_0^0,$$

$$w_0 \notin \mathfrak{R}_i, \; i = 1, 2, \cdots, m, \qquad \mathfrak{R}_i \subset U(w_0, \rho) - w_0, \; i = 1, 2, \cdots, m,$$

$$(2) \qquad \mu[t(w_0), T, \mathfrak{R}^*] = \mu[t(w_0), T, \mathfrak{R}_0] - \sum_{i=1}^{m} \mu[t(w_0), T, \mathfrak{R}_i].$$

$\mathfrak{R}_0$ clearly satisfies the conditions (a), (b), (e). We assert that

$$(3) \qquad \mu[t(w_0), T, \mathfrak{R}_i] = 0, \qquad i = 1, 2, \cdots, m.$$

Indeed, if $\mu[t(w_0), T, \mathfrak{R}_i] \neq 0$ for some i, $1 \leq i \leq m$, then $\mathfrak{R}_i$ is an indicator region $(t(w_0), T)$, and hence (see IV.1.38) $\mathfrak{R}_i^0$ contains an essential maximal model continuum γ for $t(w_0)$ under T. Then $\gamma \subset \mathfrak{R}_i \subset U(w_0, \rho) - w_0$, in contradiction to the fact that no such continuum intersects $U(w_0, \rho) - w_0$. Now (1), (2), (3) imply that $\mu[t(w_0), T, \mathfrak{R}_0] \neq 0$. Thus $\mathfrak{R}_0$ satisfies (d) and hence also (c) (cf. IV.1.24).

IV.1.63. Continuation. Let w_0 be a point in $\mathcal{N}$, and let $\mathfrak{R}_1$, $\mathfrak{R}_2$ be two bounded, simply-connected Jordan regions with the following properties: (a) $w_0 \in \mathfrak{R}_i^0$, $\mathfrak{R}_i \subset \mathfrak{D}$, $i = 1, 2$. (b) $\mu[t(w_0), T, \mathfrak{R}_i] \neq 0$, $i = 1, 2$. (c) $\kappa[t(w_0), T, \mathfrak{R}_i] = 1$, $i = 1, 2$. Then $\mu[t(w_0), T, \mathfrak{R}_1] = \mu[t(w_0), T, \mathfrak{R}_2]$.

Proof. By IV.1.62, we have a Jordan region $\mathfrak{R}$ with the properties (a)-(e) listed there, where we choose now $O = \mathfrak{R}_1^0\mathfrak{R}_2^0$. Let C_1, C_2, C be the boundary curves of $\mathfrak{R}_1$, $\mathfrak{R}_2$, $\mathfrak{R}$ respectively, and let $\mathfrak{R}_1^*$ be the doubly-connected Jordan region bounded by C_1 and C. Clearly $\mathfrak{R}_1^* \subset \mathfrak{R}_1$, $w_0 \in \mathfrak{R}_1 - \mathfrak{R}_1^*$. In view of property (c), w_0 is the only essential maximal model continuum for $t(w_0)$ in $\mathfrak{R}_1$, and hence we have by IV.1.38 (applied to $\mathfrak{R}_1^*$)

$$(1) \qquad \mu[t(w_0), T, \mathfrak{R}_1^*] = 0.$$

On the other hand, clearly

$$(2) \qquad \mu[t(w_0), T, \mathfrak{R}_1^*] = \mu[t(w_0), T, \mathfrak{R}_1] - \mu[t(w_0), T, \mathfrak{R}].$$

(1) and (2) yield $\mu[t(w_0), T, \mathfrak{R}_1] = \mu[t(w_0), T, \mathfrak{R}]$. Similarly it follows that $\mu[t(w_0), T, \mathfrak{R}_2] = \mu[t(w_0), T, \mathfrak{R}]$. Hence $\mu[t(w_0), T, \mathfrak{R}_1] = \mu[t(w_0), T, \mathfrak{R}_2]$.

IV.1.64. Given T as in IV.1.42, we define for each point $w \in \mathfrak{D}$ an *essential local index* $i_e(w, T)$ as follows:

(i) For $w \in \mathfrak{D} - \mathcal{N}$, we put $i_e(w, T) = 0$ (cf. IV.1.56).

(ii) Consider next a point $w_0 \in \mathcal{N}$. By IV.1.62 we can choose a simply-connected Jordan region $\mathfrak{R}$ such that (a) $w_0 \in \mathfrak{R}^0$, $\mathfrak{R} \subset \mathfrak{D}$, (b) $\mu[t(w_0), T, \mathfrak{R}] \neq 0$, (c) $\kappa[t(w_0), T, \mathfrak{R}] = 1$. By IV.1.63, $\mu[t(w_0), T, \mathfrak{R}]$ has the same value for all simply-connected Jordan regions $\mathfrak{R}$ with these properties (a), (b), (c). Hence we can define $i_e(w_0, T) = \mu[t(w_0), T, \mathfrak{R}]$, where any simply-connected Jordan region $\mathfrak{R}$ with the properties (a), (b), (c) can be used to evaluate $i_e(w_0, T)$.

The function $i_e(w, T)$ is now defined for every point $w \in \mathfrak{D}$, and we see that (see property (b))

$$i_e(w, T) = 0 \text{ for } w \in \mathfrak{D} - \mathcal{N}, \qquad i_e(w, T) \neq 0 \text{ for } w \in \mathcal{N}.$$

By IV.1.62, for a point $w \in \mathcal{N}$ we can choose a Jordan region $\mathfrak{R}$, suitable for the evaluation of $i_e(w, T)$, in any assigned open set O that contains the point w. Now let w_0 be a point in $\mathfrak{D}$, and let $\mathfrak{D}_0$ be a subdomain of $\mathfrak{D}$ that contains w_0. Let us put $\mathcal{N}_0 = \mathfrak{D}_0\mathcal{N} = \mathfrak{D}_0\mathcal{N}(T, \mathfrak{D})$. By IV.1.56 we have then $\mathcal{N}_0 = \mathcal{N}(T, \mathfrak{D})$. Thus the inclusions $w_0 \in \mathcal{N}(T, \mathfrak{D})$, $w_0 \in \mathcal{N}(T, \mathfrak{D}_0)$ hold or fail to hold simultaneously. These remarks show that the value of $i_e(w, T)$ at a point $w \in \mathfrak{D}$ depends solely upon the behavior of T in the vicinity of that point. It also follows that if

$T_1 : z = t_1(w)$, $w \in \mathfrak{D}_1$, $T_2 : z = t_2(w)$, $w \in \mathfrak{D}_2$, are two continuous transformations and w_0 is a common point of $\mathfrak{D}_1$ and $\mathfrak{D}_2$ such that $T_1 \equiv T_2$ in a certain vicinity of w_0, then $i_e(w_0, T_1) = i_e(w_0, T_2)$. These observations justify the term *local index*, and the lack of reference to $\mathfrak{D}$ in the notation $i_e(w, T)$. The term *essential* is used to distinguish this local index from a *strong local index* to be defined in the sequel.

If the transformation T is clearly identified by the context, we shall use the notation $i_e(w)$ instead of $i_e(w, T)$.

IV.1.65. CONTINUATION. Take a point $w_0 \in \mathcal{N} = \mathcal{N}(T, \mathfrak{D})$. Let $\mathfrak{R}_n$ be a sequence of Jordan regions with the following properties: (i) $w_0 \in \mathfrak{R}_n^0$, $\mathfrak{R}_n \subset \mathfrak{D}$. (ii) $\mathfrak{R}_n$ is simply-connected. (iii) The diameter of $\mathfrak{R}_n$ converges to zero as $n \to \infty$. (iv) $t(w) \neq t(w_0)$ on $\mathfrak{R}_n - \mathfrak{R}_n^0$. Then $\mu[t(w_0), T, \mathfrak{R}_n] = i_e(w_0)$ for n sufficiently large.

PROOF. Since $w_0 \in \mathcal{N}$, we have a simply-connected Jordan region $\mathfrak{R}$ with the properties (a), (b), (c) described in IV.1.64(ii). Since the diameter of $\mathfrak{R}_n$ converges to zero, we shall have $\mathfrak{R}_n \subset \mathfrak{R}^0$ for n large. Let us then denote by $\mathfrak{R}_n^*$ the doubly-connected Jordan region $\mathfrak{R} - \mathfrak{R}_n^0$. Since w_0 is the only essential maximal model continuum for $t(w_0)$ in $\mathfrak{R}$ (note that $\kappa[t(w_0), T, \mathfrak{R}] = 1$ and $w_0 \in \mathcal{N}$), we have, by IV.1.38, $\mu[t(w_0), T, \mathfrak{R}_n^*] = 0$. Since $\mu[t(w_0), T, \mathfrak{R}] \neq 0$, we have $t(w) \neq t(w_0)$ on $\mathfrak{R} - \mathfrak{R}^0$. By assumption, $t(w) \neq t(w_0)$ on $\mathfrak{R}_n - \mathfrak{R}_n^0$. By IV.1.24, it follows that $0 = \mu[t(w_0), T, \mathfrak{R}_n^*] = \mu[t(w_0), T, \mathfrak{R}] - \mu[t(w_0), T, \mathfrak{R}_n]$. Since $\mu[t(w_0), T, \mathfrak{R}] = i_e(w_0)$ by IV.1.64 (ii), it follows that $\mu[t(w_0), T, \mathfrak{R}_n] = i_e(w_0)$ as soon as $\mathfrak{R}_n \subset \mathfrak{R}^0$.

IV.1.66. CONTINUATION. The function $i_e(w)$ is Borel measurable in $\mathfrak{D}$. The argument is entirely analogous to that used in IV.1.57, IV.1.58, and hence we shall merely outline the proof. Let m be an integer different from zero (m may be positive or negative), and let F_m denote the set of those points $w \in \mathcal{N}$ where $i_e(w) = m$. The first step of the proof consists in verifying the formula

$$F_m = \mathcal{N} \prod_{n=1}^{\infty} E_{mn} ,$$

where E_{mn} is the set defined in IV.1.57. Since each set E_{mn} is open (see IV.1.57), and since $\mathcal{N}$ is a Borel set (see IV.1.60), it follows that F_m is a Borel set, $m = \pm 1, \pm 2, \cdots$. Of course, F_m may be empty. Given any real number a, let E_a denote the set of those points $w \in \mathfrak{D}$ where $i_e(w) \geqq a$. Since $i_e(w)$ takes on only integral values, clearly (cf. IV.1.64) $E_a = \sum F_m$, $m \geqq a$, if $a > 0$, $E_a = (\mathfrak{D} - \mathcal{N}) + \sum F_m$, $m \geqq a$, if $a \leqq 0$. Since F_m, $\mathfrak{D}$, $\mathcal{N}$ are Borel sets, it follows that $i_e(w)$ is Borel measurable.

IV.1.67. CONTINUATION. The function $i_e(w)$ assumes only integral values. On the set $\mathfrak{D} - \mathcal{N}$, $i_e(w)$ vanishes by definition. On the set $\mathcal{N}$ itself $i_e(w)$ may have any integral values different from zero. Consider, for example, the transformation

(1) $$T_1 : z = w^m, \qquad |w| < 1,$$

where m is a positive integer. It is easy to verify that $i_e(0, T_1) = m$. On the other hand, if we consider the transformation

(2) $$T_2 : z = \overline{w}^m, \qquad |w| < 1,$$

where m is a positive integer and $\overline{w} = u - iv$ is the conjugate of $w = u + iv$, then we find that $i_e(0, T_2) = -m$. An easy discussion shows that $i_e(w, T_1) = +1$ for $0 < |w| < 1$, and $i_e(w, T_2) = -1$ for $0 < |w| < 1$. Thus in these two very special cases the inequality $|i_e(w)| > 1$ holds only at isolated points. It would seem unreasonable to expect a statement of comparable simplicity for a general continuous transformation $T : z = t(w)$, $w \in \mathfrak{D}$. And yet, we shall see presently that the inequality $|i_e(w, T)| > 1$ can hold at most for a countable set of values of w. The proof requires a series of lemmas. Several of these lemmas involve a positive integer k and are trivially true if $k = 1$. Hence we shall consider only the case $k > 1$.

IV.1.68. Given a continuous transformation $T : z = t(w)$, $w \in \mathfrak{R}$, as in IV.1.2, suppose that the following conditions hold:

(i) $\mathfrak{R}$ is simply-connected.

(ii) The point $z = 0$ possesses precisely one essential maximal model continuum under T in $\mathfrak{R}$, and this continuum reduces to a single point w_0.

(iii) $i_e(w_0) = \pm k$, where k is a positive integer greater than 1.

Then $t(w)$ possesses a single-valued continuous kth root in $\mathfrak{R}$ (cf. II.4.19).

Proof. Let us observe that $t(w)$ may vanish at points other than w_0, as far as our assumptions are concerned. This fact accounts for the cautious treatment of the following argument.

By assumption, $\kappa(0, T, \mathfrak{R}) = 1$ (cf. IV.1.39), and hence (see IV.1.16) we have for each positive integer n a continuous transformation $T_n : z = t_n(w)$, $w \in \mathfrak{R}$, such that

(1) $$\rho(T_n, T, \mathfrak{R}) < 1/n, \qquad N(0, T_n, \mathfrak{R}) = N(0, T_n, \mathfrak{R}^0) = 1.$$

The condition (ii) implies that $w_0 \in \mathcal{N}(T, \mathfrak{R}^0)$, and hence by IV.1.62, IV.1.64 we have a Jordan region $\mathfrak{R}^*$ with the following properties: (α) $w_0 \in \mathfrak{R}^{*0}$, $\mathfrak{R}^* \subset \mathfrak{R}^0$. ($\beta$) $\mathfrak{R}^*$ is simply-connected. (γ) $\mu(0, T, \mathfrak{R}^*) = i_e(w_0, T) = \pm k$. By IV.1.25(a) it follows that $\delta(0, T, \mathfrak{R}^*) > 0$, and hence (1) yields, in view of IV.1.25(d),

(2) $$\mu(0, T_n, \mathfrak{R}^*) = \mu(0, T, \mathfrak{R}^*) = \pm k \qquad \text{for } n \text{ large}.$$

Since $k \neq 0$, it follows from (2) by IV.1.25(e) that

(3) $$N(0, T_n, \mathfrak{R}^{*0}) > 0 \qquad \text{for } n \text{ large}.$$

(1) and (3) show that

(4) $$N(0, T_n, \mathfrak{R} - \mathfrak{R}^{*0}) = 0, \qquad N(0, T_n, \mathfrak{R}^{*0}) = 1 \qquad \text{for } n \text{ large}.$$

In view of IV.1.25(e), (4) implies that

(5) $$\mu(0, T_n, \mathfrak{R} - \mathfrak{R}^{*0}) = 0 \qquad \text{for } n \text{ large}.$$

By (1) and (2) we have $t_n(w) \neq 0$ on $\mathfrak{R} - \mathfrak{R}^0$ and also on $\mathfrak{R}^* - \mathfrak{R}^{*0}$ for n large. By IV.1.24 it follows that $\mu(0, T_n, \mathfrak{R} - \mathfrak{R}^{*0}) = \mu(0, T_n, \mathfrak{R}) - \mu(0, T_n, \mathfrak{R}^*)$. Hence, by (5) and (2),

$$\mu(0, T_n, \mathfrak{R}) = \pm k \qquad \text{for } n \text{ large.} \tag{6}$$

(1) and (6) imply, by II.4.33, II.4.34, that $t_n(w)$ possesses a single-valued continuous kth root in $\mathfrak{R}$. Since $t_n(w) \to t(w)$ uniformly in $\mathfrak{R}$ by (1), it follows that $t(w)$ itself possesses a single-valued continuous kth root in $\mathfrak{R}$ (see II.4.23).

IV.1.69. Let there be given two continuous transformations $T : z = t(w)$, $w \in \mathfrak{D}$, $T^* : z = t^*(w)$, $w \in \mathfrak{D}$ (cf. IV.1.42), such that the following conditions hold: (i) $t(w) = t^*(w)^k$, where k is an integer greater than 1. (ii) $\kappa(0, T, \mathfrak{D}) > 0$. Then $\kappa(0, T^*, \mathfrak{D}) > 0$.

Proof. The assumption implies (see IV.1.47, IV.1.46) that there exists an indicator region $\mathfrak{R}$ for the point $z = 0$ under T in $\mathfrak{D}$. We have then $\mu(0, T, \mathfrak{R}) \neq 0$. By II.4.25 it follows that $\mu(0, T, \mathfrak{R}) = k\mu(0, T^*, \mathfrak{R})$, and hence $\mu(0, T^*, \mathfrak{R}) \neq 0$. By IV.1.26 we obtain $\kappa(0, T^*, \mathfrak{R}) > 0$, and hence $\kappa(0, T^*, \mathfrak{D}) > 0$ by IV.1.44.

IV.1.70. Continuation. In view of IV.1.51 the inequality $\kappa(0, T^*, \mathfrak{D}) > 0$ implies that $\kappa(\zeta, T^*, \mathfrak{D}) > 0$ if $|\zeta|$ is sufficiently small. Let ζ be any point in the z-plane such that $\zeta \neq 0$ and $\kappa(\zeta, T^*, \mathfrak{D}) > 0$. We have then (see IV.1.47) some essential maximal model continuum γ^* for ζ under T^* in $\mathfrak{D}$. We assert that γ^* is an essential maximal model continuum for the point ζ^k under T in $\mathfrak{D}$.

Proof. The assumptions imply that $t^*(w) = \zeta$ for $w \in \gamma^*$. Hence $t(w) = t^*(w)^k = \zeta^k$ for $w \in \gamma^*$. Thus $\gamma^* \subset T^{-1}(\zeta^k)$. Since γ^* is connected, it follows that γ^* is contained in a component γ of $T^{-1}(\zeta^k)$. We assert that $\gamma^* = \gamma$. Indeed, we have $t(w) = \zeta^k$ for $w \in \gamma$. Hence, in view of the relation $t(w) = t^*(w)^k$, the function $t^*(w)$ cannot assume more than k distinct values on γ. Since $t^*(w)$ is continuous and γ is connected, it follows that $t^*(w)$ is constant on γ (see I.2.45). But $t^*(w) = \zeta$ for $w \in \gamma^* \subset \gamma$. Hence $t^*(w) = \zeta$ on γ. Since γ^* is a maximal model continuum for ζ under T^* in $\mathfrak{D}$ and $\gamma^* \subset \gamma$, it follows that $\gamma^* = \gamma$.

Thus it is established that γ^* is a maximal model continuum for ζ^k under T in $\mathfrak{D}$. We have yet to show that γ^* is essential for ζ^k under T. Let now O be any open set such that $\gamma^* \subset O$. In view of the relation $t(w) = t^*(w)^k$ we have the identity

$$t(w) - \zeta^k = t^*(w)^k - \zeta^k = [t^*(w) - \zeta]g(w), \tag{1}$$

where we have put

$$g(w) = t^*(w)^{k-1} + \zeta t^*(w)^{k-2} + \cdots + \zeta^{k-2}t^*(w) + \zeta^{k-1}.$$

The function $g(w)$ is continuous in $\mathfrak{D}$, and clearly $g(w) = k\zeta^{k-1} \neq 0$ for $w \in \gamma^*$. Hence we have an open set O^* such that

$$\gamma^* \subset O^* \subset \mathfrak{D}, \qquad g(w) \neq 0 \text{ for } w \in O^*. \tag{2}$$

Since $\gamma^* \subset O$, we have $\gamma^* \subset OO^*$. Let us recall that γ^* is an essential maximal model continuum for ζ under T^* in $\mathfrak{D}$. By IV.1.46 we have therefore a (finitely-connected) Jordan region $\mathfrak{R}$ such that

(3) $$\gamma^* \subset \mathfrak{R}^0, \qquad \mathfrak{R} \subset OO^* \subset \mathfrak{D}, \qquad \mu(\zeta, T^*, \mathfrak{R}) \neq 0.$$

(1), (2), (3) imply, by II.4.28, that $\mu(\zeta^k, T, \mathfrak{R}) = \mu(\zeta, T^*, \mathfrak{R}) \neq 0$. Thus $\mathfrak{R}$ is an indicator region (ζ^k, T) containing γ^* and contained in the arbitrarily assigned open set $O \supset \gamma^*$. Hence γ^* is an essential maximal model continuum for ζ^k under T (see IV.1.46).

IV.1.71. Given T as in IV.1.42, let z_0 be a point such that $\kappa(z_0, T, \mathfrak{D}) = 1$. Then z_0 possesses a unique essential maximal model continuum under T in $\mathfrak{D}$ (see IV.1.47). Let us further assume that this continuum reduces to a single point w_0 such that $i_e(w_0, T) = \pm k$, where k is an integer greater than 1. Under these conditions we assert the existence of a $\delta > 0$, such that $\kappa(z, T, \mathfrak{D}) \geqq k$ for $0 < | z - z_0 | < \delta$.

PROOF. Clearly we can assume, without loss of generality, that $z_0 = 0$. Let $\mathfrak{R}$ be a simply-connected region such that $w_0 \in \mathfrak{R}^0$, $\mathfrak{R} \subset \mathfrak{D}$. Then T, considered in $\mathfrak{R}$, satisfies all the assumptions of IV.1.68, and hence we have in $\mathfrak{R}$ a single-valued continuous function $t^*(w)$ such that $t(w) = t^*(w)^k$ in $\mathfrak{R}$. Consider now the transformations $T : z = t(w)$, $w \in \mathfrak{R}^0$, $T^* : z = t^*(w)$, $w \in \mathfrak{R}^0$. Then the assumptions of IV.1.69 are satisfied (with $\mathfrak{D}$ replaced by $\mathfrak{R}^0$), and hence $\kappa(0, T^*, \mathfrak{R}^0) > 0$. By IV.1.51 there follows the existence of a $\rho > 0$ such that

(1) $$\kappa(\zeta, T^*, \mathfrak{R}^0) > 0 \qquad \text{for } 0 < | \zeta | < \rho.$$

Let us put $\delta = \rho^k$, and let us take any point z such that $0 < | z | < \delta$. Then z possesses k distinct kth roots $\zeta_1, \cdots, \zeta_k$, and clearly $0 < | \zeta_i | < \rho$, $i = 1, \cdots, k$. Hence, by (1), $\kappa(\zeta_i, T^*, \mathfrak{R}^0) > 0$, $i = 1, \cdots, k$. By IV.1.47 it follows that for each i we have an essential maximal model continuum γ_i^* for ζ_i under T^* in $\mathfrak{R}^0$. The continua $\gamma_1^*, \cdots, \gamma_k^*$ are distinct, since $t^*(w) = \zeta_i$ on γ_i^* and the points $\zeta_1, \cdots, \zeta_k$ are distinct. By IV.1.70, γ_i^* is an essential maximal model continuum for $\zeta_i^k = z$ under T in $\mathfrak{R}^0$, $i = 1, \cdots, k$. Hence $\kappa(z, T, \mathfrak{R}^0) \geqq k$ by IV.1.47, and thus *a fortiori* $\kappa(z, T, \mathfrak{D}) \geqq k$ (see IV.1.44).

IV.1.72. Given T as in IV.1.42, let k be an integer greater than 1, and let $\mathfrak{D}_0$ be a subdomain of $\mathfrak{D}$. We define an auxiliary set $\bar{E}(k, T, \mathfrak{D}_0)$ in the z-plane as follows. A point z_0 belongs to $\bar{E}(k, T, \mathfrak{D}_0)$ if and only if (i) z_0 has a unique essential maximal model continuum under T in $\mathfrak{D}_0$, and (ii) this continuum reduces to a single point where $i_e(w, T) = \pm k$. We assert that $\bar{E}(k, T, \mathfrak{D}_0)$ is an isolated and hence countable (possibly empty) set.

PROOF. If $z_0 \in \bar{E}(k, T, \mathfrak{D}_0)$, then by IV.1.71, applied to T in $\mathfrak{D}_0$, there exists a $\delta > 0$ such that $\kappa(z, T, \mathfrak{D}_0) \geqq k > 1$ for $0 < | z - z_0 | < \delta$. Since clearly $\kappa(z, T, \mathfrak{D}_0) = 1$ if $z \in \bar{E}(k, T, \mathfrak{D}_0)$ (cf. IV.1.47), it follows that the domain $0 < | z - z_0 | < \delta$ is clear of points of $\bar{E}(k, T, \mathfrak{D}_0)$. Thus every point of $\bar{E}(k, T, \mathfrak{D}_0)$ is an isolated point.

IV.1.73. Given T as in IV.1.42, and an integer $k \geqq 2$, let $\mathcal{N}(k, T, \mathfrak{D})$ denote the set of those points $w \in \mathfrak{D}$ where $i_e(w, T) = \pm k$. Then $\mathcal{N}(k, T, \mathfrak{D})$ is a countable (possibly empty) set.

PROOF. Using the notations of IV.1.72 and IV.2.3, we first verify the formula

$$(1) \qquad T[\mathcal{N}(k, T, \mathfrak{D})] = \sum_{j=1}^{\infty} \sum_{s} \bar{E}(k, T, s^0), \quad s \in D_{pj}, s^0 \subset \mathfrak{D}.$$

Suppose first that $z_0 \in T[\mathcal{N}(k, T, \mathfrak{D})]$. Then there exists a point w_0 such that $z_0 = t(w_0)$, $w_0 \in \mathcal{N}(k, T, \mathfrak{D})$. Since $k \neq 0$, it follows by IV.1.64 that $w_0 \in \mathcal{N}$. By the definition of $\mathcal{N}$ (see IV.1.56), w_0 is then an essential maximal model continuum for z_0 under T in $\mathfrak{D}$, and there exists an open set O containing w_0 such that the set $O - w_0$ is not intersected by any essential maximal model continuum for z_0 under T in $\mathfrak{D}$. For j large, we shall have a square s such that $w_0 \in s^0 \subset O$, $s \in D_{pj}$. Clearly $z_0 \in \bar{E}(k, T, s^0)$. Thus the set on the left in (1) is a subset of the set on the right. Conversely, let z_0 be a point such that for some j and for some $s \in D_{pj}$ we have the inclusion $z_0 \in \bar{E}(k, T, s^0)$. Then z_0 has a unique essential maximal model continuum under T in s^0, this continuum reduces to a single point w_0, and $i_e(w_0, T) = \pm k$ (see IV.1.72). Clearly $w_0 \in \mathcal{N}(k, T, \mathfrak{D})$, and hence $t(w_0) = z_0 \in T[\mathcal{N}(k, T, \mathfrak{D})]$.

Thus (1) is verified. Since each set $\bar{E}(k, T, s^0)$ is countable by IV.1.72, it follows that $T[\mathcal{N}(k, T, \mathfrak{D})]$ is countable. We proceed to show that $\mathcal{N}(k, T, \mathfrak{D})$ itself is countable. If $T[\mathcal{N}(k, T, \mathfrak{D})] = 0$, then $\mathcal{N}(k, T, \mathfrak{D}) = 0$, and the assertion is obvious. So we can assume that $T[\mathcal{N}(k, T, \mathfrak{D})] \neq 0$. Clearly

$$(2) \qquad \mathcal{N}(k, T, \mathfrak{D}) = \sum_{z} \mathcal{N}(k, T, \mathfrak{D})T^{-1}(z), \qquad z \in T[\mathcal{N}(k, T, \mathfrak{D})].$$

Now consider a point $z_0 \in T[\mathcal{N}(k, T, \mathfrak{D})]$, and let w_0 be a point such that $w_0 \in \mathcal{N}(k, T, \mathfrak{D})T^{-1}(z_0)$. Then $i_e(w_0, T) = \pm k$, and hence (cf. IV.1.64), $w_0 \in \mathcal{N}T^{-1}(z_0)$. Thus $\mathcal{N}(k, T, \mathfrak{D})T^{-1}(z_0) \subset \mathcal{N}T^{-1}(z_0)$. On the other hand, for fixed z_0 the set $\mathcal{N}T^{-1}(z_0)$ is clearly countable, in view of the definition of $\mathcal{N}$. Hence $\mathcal{N}(k, T, \mathfrak{D})T^{-1}(z_0)$ is countable for every $z_0 \in T[\mathcal{N}(k, T, \mathfrak{D})]$. Since $T[\mathcal{N}(k, T, \mathfrak{D})]$ is countable, (2) shows that $\mathcal{N}(k, T, \mathfrak{D})$ is also countable.

IV.1.74. THEOREM. *Given T as in* IV.1.42, *the set of those points $w \in \mathfrak{D}$ where $i_e(w, T) \neq 0, \pm 1$ is countable (possibly empty).*

PROOF. If this set is denoted by E, then clearly $E = \sum \mathcal{N}(k, T, \mathfrak{D})$, $k = 2, 3, \cdots$ (see IV.1.73). Since each set $\mathcal{N}(k, T, \mathfrak{D})$, $k \geqq 2$, is countable by IV.1.73, the theorem follows.

IV.1.75. Given T as in IV.1.42, we define a *strong local index function* $i_s(w, T)$ as follows:

(i) If $w \in \mathfrak{D} - \mathcal{I}(T, \mathfrak{D})$ (cf. IV.1.56), then $i_s(w, T) = 0$.

(ii) If $w_0 \in \mathcal{I}(T, \mathfrak{D})$, then we proceed as follows. By the definition of $\mathcal{I}(T, \mathfrak{D})$, there exists an open set O such that $w_0 \in O \subset \mathfrak{D}$ and $t(w) \neq t(w_0)$ for $w \in O - w_0$. As a consequence, there exist Jordan regions $\mathfrak{R}$ with the following properties: (α) $w_0 \in \mathfrak{R}^0$, $\mathfrak{R} \subset \mathfrak{D}$. ($\beta$) $\mathfrak{R}$ is simply-connected. (γ) $t(w) \neq t(w_0)$ for $w \in \mathfrak{R} - w_0$. We define

$$i_s(w_0, T) = \mu[t(w_0), T, \mathfrak{R}].$$

It remains to show, of course, that $i_s(w_0, T)$ is independent of the particular choice of $\mathfrak{R}$, provided that $\mathfrak{R}$ satisfies the conditions (α), (β), (γ). Let $\mathfrak{R}_n$ denote the closed circular disc $|w - w_0| \leqq 1/n$. Given any Jordan region $\mathfrak{R}$ that satisfies the conditions (α), (β), (γ), we shall have $\mathfrak{R}_n \subset \mathfrak{R}^0$ and $t(w) \neq t(w_0)$ on $\mathfrak{R}_n - \mathfrak{R}_n^0$ for n sufficiently large. If $\mathfrak{R}_n^*$ denotes the doubly-connected Jordan region $\mathfrak{R} - \mathfrak{R}_n^0$, then $t(w) \neq t(w_0)$ in $\mathfrak{R}_n^*$, and hence, by IV.1.25(e), IV.1.24,

$$0 = \mu[t(w_0), T, \mathfrak{R}_n^*] = \mu[t(w_0), T, \mathfrak{R}] - \mu[t(w_0), T, \mathfrak{R}_n].$$

Hence $\mu[t(w_0), T, \mathfrak{R}] = \mu[t(w_0), T, \mathfrak{R}_n]$ for n sufficiently large. Thus $\mu[t(w_0), T, \mathfrak{R}]$ is independent of the choice of $\mathfrak{R}$ (subject to the conditions (α), (β), (γ)).

The preceding remarks also show that $i_s(w, T)$ depends only upon the behavior of T in the vicinity of the point w. If T is clearly identified by the context, then we shall write $i_s(w)$ instead of $i_s(w, T)$.

IV.1.76. Continuation. $i_s(w) = i_e(w)$ for $w \in \mathcal{G}$ (cf. IV.1.56, IV.1.64). Indeed, if $i_s(w) \neq 0$, then clearly $w \in \mathcal{N}$, and $i_s(w) = i_e(w)$ as a direct consequence of IV.1.65. If $w \in \mathcal{G}$ and $i_s(w) = 0$, then clearly $w \notin \mathcal{N}$, and hence $i_e(w) = 0$ (see IV.1.64).

IV.1.77. Continuation. $i_s(w)$ is Borel measurable.

Proof. By IV.1.75, IV.1.76 we have

$$i_s(w) = \begin{cases} i_e(w) & \text{for } w \in \mathcal{G}, \\ 0 & \text{for } w \in \mathfrak{D} - \mathcal{G}. \end{cases}$$

Since $\mathfrak{D}$ and $\mathcal{G}$ are Borel sets and $i_e(w)$ is Borel measurable (see IV.1.61, IV.1.66), it follows that $i_s(w)$ is Borel measurable.

IV.1.78. Continuation. Let us denote by $\mathcal{G}_0(T, \mathfrak{D})$ the subset of $\mathcal{G}(T, \mathfrak{D})$ where $i_s(w, T) = 0$. If there is no danger of ambiguity, we shall write $\mathcal{G}_0(T)$, $\mathcal{G}_0(\mathfrak{D})$, or simply $\mathcal{G}_0$ instead of $\mathcal{G}_0(T, \mathfrak{D})$. In view of IV.1.76, we have $\mathcal{G}_0 = \mathcal{G} - \mathcal{N}$, and hence $\mathcal{G}_0$ is also a Borel set (cf. IV.1.60, IV.1.61, IV.1.64). As a formal consequence of the relation $\mathcal{G}_0 = \mathcal{G} - \mathcal{N}$, we obtain the inclusion $(\mathfrak{D} - \mathcal{N})(\mathfrak{D} - \mathcal{G}_0) \subset \mathfrak{D} - \mathcal{G}$.

IV.1.79. Continuation. $i_s(w, T) = 0, \pm 1$ *except at the points of a countable set* (which may be empty). Indeed, let E be the subset of $\mathfrak{D}$ on which $|i_s(w, T)| > 1$. Since $i_s(w, T) = 0$ for $w \in \mathfrak{D} - \mathcal{G}$, it follows that $E \subset \mathcal{G}$. By IV.1.76 it follows that $|i_e(w, T)| > 1$ on E, and hence E is countable by IV.1.74.

IV.1.80. We shall now discuss, to the extent needed in the sequel, transformations of the special form

$$T : x = f(u, v), \qquad y = v, \qquad (u, v) \in R_0,$$

where $R_0 : a_0 \leqq u \leqq b_0, c_0 \leqq v \leqq d_0$ is an oriented rectangle and $f(u, v)$ is continuous in R_0 (and not merely in R_0^0). We shall use, if convenient, the compact notations $\mathcal{E}$, $\mathcal{N}$, $\cdots$ instead of $\mathcal{E}(T, R_0^0)$, $\mathcal{N}(T, R_0^0)$, $\cdots$, and we shall write $N(x, y, E)$ instead of $N(z, E)$, with analogous conventions for other functions that we shall have occasion to consider. Since $f(u, v)$ is continuous in R_0, it is also bounded there, and hence $T(R_0)$ is comprised in a certain oriented rectangle of the form $\bar{R}_0 : -M \leqq x \leqq M,\ c_0 \leqq y \leqq d_0$. Let λ be a number such that

$$(1) \qquad c_0 < \lambda < d_0 .$$

We shall then denote by g_λ, $\bar{g}_\lambda$ the segments $a_0 \leqq u \leqq b_0$, $v = \lambda$ and $-M \leqq x \leqq M$, $y = \lambda$ respectively. Clearly $T(g_\lambda) \subset \bar{g}_\lambda$, $T^{-1}(\bar{g}_\lambda) = g_\lambda$. For fixed λ, satisfying (1), let us define a subset $\mathcal{E}_\lambda$ of g_λ as follows. A point (u_0, λ) of g_λ belongs to $\mathcal{E}_\lambda$ if and only if for every $\epsilon > 0$ there exist two points (u_1, λ), (u_2, λ) such that the following conditions hold. (i) $a_0 < u_1 < u_0 < u_2 < b_0$. (ii) $u_2 - u_1 < \epsilon$. (iii) $[f(u_1, \lambda) - f(u_0, \lambda)][f(u_2, \lambda) - f(u_0, \lambda)] < 0$. Of course u_1, u_2 depend on ϵ. We assert that (cf. IV.1.56)

$$(2) \qquad \mathcal{E}_\lambda \subset \mathcal{E} = \mathcal{E}(T, R_0^0).$$

PROOF. Take any point $(u_0, \lambda) \in \mathcal{E}_\lambda$. Give any $\epsilon > 0$, and let (u_1, λ), (u_2, λ) be a pair of corresponding points with the properties (i), (ii), (iii). Since $f(u, v)$ is continuous, we can then select two numbers v_1, v_2 such that the following conditions hold. (α) $c_0 < v_1 < \lambda < v_2 < d_0$. ($\beta$) $v_2 - v_1 < \epsilon$. (γ) $[f(u_1, v) - f(u_0, \lambda)][f(u_2, v) - f(u_0, \lambda)] < 0$ for $v_1 \leqq v \leqq v_2$. Let us now consider the oriented rectangle $r : u_1 \leqq u \leqq u_2,\ v_1 \leqq v \leqq v_2$. Clearly $f(u, v) \neq f(u_0, \lambda)$ on the vertical sides of r and $v \neq \lambda$ on the horizontal sides of r. The point (u_0, λ) is interior to r, and the diameter of r is less than $2^{1/2}\epsilon$. If we put $w_0 = u_0 + i\lambda$, $z_0 = f(u_0, \lambda) + i\lambda$, then by II.4.35 it follows that $\mu(z_0, T, r) = \pm 1$. More precisely (see II.4.35),

$$\mu(z_0, T, r) = +1 \qquad \text{if } f(u_1, \lambda) < f(u_0, \lambda) \text{ and } f(u_2, \lambda) > f(u_0, \lambda),$$

$$\mu(z_0, T, r) = -1 \qquad \text{if } f(u_1, \lambda) > f(u_0, \lambda) \text{ and } f(u_2, \lambda) < f(u_0, \lambda).$$

Thus r is an indicator region for the point $z_0 = T(w_0)$ under T in R_0^0 (cf. IV.1.46). Since the diameter of r is arbitrarily small, it follows by IV.1.56 that $w_0 \in \mathcal{E}$.

IV.1.81. CONTINUATION. For fixed λ, satisfying $c_0 < \lambda < d_0$, we have

$$(1) \qquad N(x, \lambda, R_0) = N(x, \lambda, R_0^0) = N(x, \lambda, \mathcal{E}),$$

with the possible exception of a countable set of points (x, λ).

PROOF. λ being fixed, T may be considered as a continuous transformation from the segment g_λ into the segment $\bar{g}_\lambda$ (see IV.1.80). The set $\mathcal{E}_\lambda$ coincides then with the set E^* of III.2.8, and hence the set $g_\lambda - \mathcal{E}_\lambda$ coincides with the set E of III.2.8. Therefore, by III.2.8, the set $T(g_\lambda - \mathcal{E}_\lambda)$ is countable (possibly empty). Hence it is sufficient to verify (1) for points (x_0, λ) such that

$$(2) \qquad (x_0, \lambda) \notin T(g_\lambda - \mathcal{E}_\lambda).$$

Now if $N(x_0, \lambda, R_0) = 0$, then all the terms involved in (1) vanish. So we can assume that $T^{-1}(x_0, \lambda) \neq 0$. Let us take any point (u_0, λ) such that

$$(3) \qquad (u_0, \lambda) \in T^{-1}(x_0, \lambda).$$

(2) and (3) imply that $(u_0, \lambda) \in \mathcal{E}_\lambda$. Hence $(u_0, \lambda) \in \mathcal{E}$ by IV.1.80(2). Since (u_0, λ) was any point of $T^{-1}(x_0, \lambda)$, it follows that $T^{-1}(x_0, \lambda) \subset \mathcal{E}$, and hence

$$(4) \qquad N(x_0, \lambda, R_0) \leqq N(x_0, \lambda, \mathcal{E}).$$

Since obviously $N(x_0, \lambda, R_0) \geqq N(x_0, \lambda, R_0^0) \geqq N(x_0, \lambda, \mathcal{E})$, (4) yields $N(x_0, \lambda, R_0) = N(x_0, \lambda, R_0^0) = N(x_0, \lambda, \mathcal{E})$.

IV.1.82. Continuation. The set $g_\lambda \cdot (R_0 - R_0^0)$ consists of precisely two points (recall that $c_0 < \lambda < d_0$), and hence the set $T[g_\lambda \cdot (R_0 - R_0^0)]$ contains at most two points. The same remark applies if R_0 is replaced by any oriented rectangle $R \subset R_0$, given by $R : a \leqq u \leqq b, c \leqq v \leqq d$, and if λ is restricted by the inequalities $c < \lambda < d$. Since obviously $N(x, y, \mathcal{E}R^0) \leqq \kappa(x, y, R^0) \leqq N(x, y, \mathcal{E}^*R^0) \leqq N(x, y, R^0)$ (cf. IV.1.56), it follows from IV.1.81 (applied to R) that $N(x, \lambda, \mathcal{E}R^0) = \kappa(x, \lambda, R^0) = N(x, \lambda, \mathcal{E}^*R^0) = N(x, \lambda, R^0) = N(x, \lambda, R)$ for fixed λ satisfying $c < \lambda < d$, with the possible exception of a countable set of points (x, λ). In fact, the same conclusion holds if the condition $c < \lambda < d$ is replaced by the condition $\lambda \neq c, d$. Indeed, if either $\lambda < c$ or $\lambda > d$, then all the terms involved vanish, and thus the assertion is obvious.

IV.1.83. Continuation. Assume again that $c_0 < \lambda < d_0$, and let (u_0, λ) be an interior point of the segment g_λ (see IV.1.80). Suppose that the partial derivative $f_u(u_0, \lambda)$ exists and is different from zero. We assert that $i_s(u_0, \lambda) = \operatorname{sgn} f_u(u_0, \lambda)$ (cf. IV.1.75). That is, $i_s(u_0, \lambda) = +1$ if $f_u(u_0, \lambda) > 0$, and $i_s(u_0, \lambda) = -1$ if $f_u(u_0, \lambda) < 0$.

Proof. The cases $f_u(u_0, \lambda) > 0, f_u(u_0, \lambda) < 0$ can be treated in the same manner, and we discuss the proof only for the case when $f_u(u_0, \lambda) > 0$. We have, by the definition of f_u, two points (u_1, λ), (u_2, λ) such that (i) $a_0 < u_1 < u_0 < u_2 < b_0$, (ii) $f(u, \lambda) < f(u_0, \lambda)$ for $u_1 \leqq u < u_0$, (iii) $f(u, \lambda) > f(u_0, \lambda)$ for $u_0 < u \leqq u_2$. Let v_1, v_2 be any two numbers such that $c_0 < v_1 < \lambda < v_2 < d_0$, and let us consider the rectangle $R : u_1 \leqq u \leqq u_2, v_1 \leqq v \leqq v_2$. We assert that

$$(1) \qquad T(u, v) \neq [f(u_0, \lambda), \lambda] = T(u_0, \lambda) \qquad \text{for } (u, v) \in R - (u_0, \lambda).$$

Indeed, the conditions $f(u, v) = f(u_0, \lambda)$, $y = \lambda$, $(u, v) \in R - (u_0, \lambda)$ imply that $v = \lambda$ (see IV.1.80) and hence these conditions yield the relations $f(u, \lambda) = f(u_0, \lambda)$, $u \neq u_0$, $u_1 \leqq u \leqq u_2$ which contradict (ii) or (iii). Thus (1) holds. It follows that (u_0, λ) is a point of the set $\mathcal{I}$ defined in IV.1.56. Hence by IV.1.75, $i_s(u_0, \lambda) = \mu[T(u_0, \lambda), T, R]$. On the other hand, by II.4.35 the conditions (i), (ii), (iii) imply that $\mu[T(u_0, \lambda), T, R] = +1$. Thus $i_s(u_0, \lambda) = +1 = \operatorname{sgn} f_u(u_0, \lambda)$.

IV.1.84. Transformations of the form $T : x = u, y = f(u, v), (u, v) \in R_0$, give rise to results entirely analogous to those discussed in IV.1.80 to IV.1.83. Of course, this case may be considered as differing from that in IV.1.80 merely in the notations for the variables.

CHAPTER IV.2. METRIC FOUNDATIONS

IV.2.1. As in Chapter IV.1, we shall be concerned with transformations $T: z = t(w)$, $w \in \mathfrak{D}$, where $\mathfrak{D}$ is a bounded domain, and $t(w)$ is single-valued, continuous and bounded in $\mathfrak{D}$. Under certain conditions it will be convenient to give T in terms of real equations of the form

$$T: x = x(u, v), \qquad y = y(u, v), \qquad (u, v) \in \mathfrak{D},$$

where $x + iy = z$, $u + iv = w$, $x(u, v) + iy(u, v) = t(w)$. Then $x(u, v)$, $y(u, v)$ are single-valued, real-valued, bounded, continuous functions in $\mathfrak{D}$. It will be imperative to condense notations to the extent compatible with clarity. If $f(u, v)$ is a summable function on some measurable set $E \subset \mathfrak{D}$, then we shall also write $f(w)$ instead of $f(u, v)$, and to denote the (Lebesgue) integral of $f(u, v)$ on E we shall use any one of the symbols

$$\iint_E f(u, v)\, du\, dv, \qquad \iint_E f(u, v), \qquad \iint_E f, \qquad \iint_E f(w)\, du\, dv, \qquad \iint_E f(w).$$

Similar conventions will be used concerning integration in the $z = x + iy$ plane, with w, u, v, du, dv replaced by z, x, y, dx, dy. As regards integration in the z-plane, in most cases in the sequel the integrand, say $h(x, y)$ or $h(z)$, will be defined in the whole plane and will vanish outside of some bounded measurable set E. In such cases we shall write

$$\iint h \quad \text{for} \quad \iint_E h,$$

and it will be understood that if the range of integration is not displayed, then the range is the set on which $h \neq 0$ or equivalently any bounded measurable set outside of which h vanishes. To illustrate, let $H(z)$ be defined and measurable in the whole z-plane, and let $N(z, T, E\mathfrak{D})$ denote again the number (possibly $+\infty$) of distinct points in the set $T^{-1}(z)E\mathfrak{D}$, where E is any set in the w-plane (cf. IV.1.1). Since T is bounded, $N(z, T, E\mathfrak{D}) = 0$ for z outside of a certain circular disc Δ and hence the same holds for $H(z)N(z, T, E\mathfrak{D})$. According to our agreement, the symbol

$$\iint H(z)N(z, T, E\mathfrak{D}) \tag{1}$$

designates then the integral of $H(z)N(z, T, E\mathfrak{D})$ extended over Δ or equivalently over any measurable set outside of which the integrand vanishes (the existence of the integral being assumed). Since the function $\kappa(z, T, \mathfrak{D}^*)$, where $\mathfrak{D}^*$ is any subdomain of $\mathfrak{D}$, vanishes outside of the set $T(\mathfrak{D})$ (see IV.1.47), analogous remarks apply to integrals of the form

$$(2) \qquad \iint H(z)\kappa(z, T, \mathfrak{D}^*).$$

It will happen that $H(z)$ is defined and measurable only on some measurable set $\bar{E}$ that contains the set $T(\mathfrak{D})$. In such a case, we may extend the definition of $H(z)$ to the whole plane, by setting $H(z) = 0$ outside of $\bar{E}$. Clearly, integrals of the form (1) or (2) are entirely independent of the values of $H(z)$ outside of $T(\mathfrak{D})$.

IV.2.2. Continuation. If $\bar{E}$ is a set in the z-plane, then $g(z, \bar{E})$ will denote its characteristic function. That is, $g(z, \bar{E}) = 1$ if $z \in \bar{E}$ and $g(z, \bar{E}) = 0$ if $z \notin \bar{E}$. If E is a set in the w-plane, then $\bar{A}(E\mathfrak{D}, k)$ will denote the set of those points z where $N(z, E\mathfrak{D}) = k$. We have used here the simplified notation $N(z, E\mathfrak{D})$ instead of $N(z, T, E\mathfrak{D})$ (cf. IV.1.1) and k is permitted to be infinite. Thus $\bar{A}(\mathfrak{D}, +\infty)$, for example, is the set of those points z for which $N(z, \mathfrak{D}) = +\infty$. Clearly

$$g[z, T(E)] \leqq N(z, T, E\mathfrak{D}).$$

Obviously, if $g[z_0, T(E)] = 0$ for a certain z_0 then $N(z_0, T, E) = 0$. The symbol $\bar{A}(E\mathfrak{D}, k)$ depends also upon T and we shall use instead the symbol $\bar{A}(E\mathfrak{D}, T, k)$ if explicit reference to T is desirable. If $E_1, \cdots, E_m$ are disjoint subsets of a set $E \subset \mathfrak{D}$, then clearly

$$\sum_{\iota=1}^{m} N(z, E_\iota) \leqq N(z, E),$$

and *a fortiori*

$$\sum_{\iota=1}^{m} g[z, T(E_\iota)] \leqq N(z, E).$$

IV.2.3. The following remarks will be useful in the sequel. The domain $\mathfrak{D}$ being bounded, we can choose a square Q, with sides parallel to the u- and v-axes respectively, such that $\mathfrak{D} \subset Q$. Let $p_1, p_2, \cdots, p_j, \cdots$ be the sequence of positive primes. Thus $p_1 = 2, p_2 = 3, p_3 = 5, \cdots$. Let D_{p_j} denote the subdivision of Q into p_j^2 congruent smaller squares. To express the fact that a square s occurs in the subdivision D_{p_j} we shall write $s \in D_{p_j}$. The convenience of these subdivisions D_{p_j} results from the following remark. Let $w_0 = u_0 + iv_0$ be any point in $\mathfrak{D}$, and suppose that w_0 is a boundary point of a certain square $s \in D_{p_i}$. If l is the side length of Q and (α, β) is the lower left vertex of Q then it follows that we have either

$$(1) \qquad u_0 = \alpha + \frac{ln_\iota}{p_i}, \qquad n_\iota \text{ an integer, } 0 < n_\iota < p_i,$$

or else

$$(2) \qquad v_0 = \beta + \frac{lm_\iota}{p_i}, \qquad m_\iota \text{ an integer, } 0 < m_\iota < p_i.$$

Suppose (1) happens for two different primes p_j, p_i, $i < j$. Then we should have $n_j p_i = n_i p_j$, and hence p_j, being a prime, should be a divisor of either n_j or of p_i which is impossible since $0 < n_j < p_j$, $0 < p_i < p_j$. Thus (1) can happen for at most one prime p_j and the same holds of course for (2). It follows that for given $w_0 \in \mathfrak{D}$ it can happen for at most two values of j that w_0 is a boundary point of some square $s \in D_{p_j}$. Hence, for given $w_0 \in \mathfrak{D}$, we have a $j(w_0)$ such that for $j > j(w_0)$, the point w_0 is an *interior* point of precisely one square $s \in D_{p_j}$. Now let us consider any finite system of points $w_1, \cdots, w_m$ in $\mathfrak{D}$. For j large enough, no square $s \in D_{p_j}$ will contain more than one of these points, and by the preceding remark none of these points will be a boundary point of any square $s \in D_{p_j}$. Hence, for j large enough, the number of those squares $s \in D_{p_j}$ that contain some point of the system $w_1, \cdots, w_m$ will be precisely equal to m.

In the sequel, we shall use rectangles in the w-plane with sides parallel to the u- and v-axes respectively. For brevity we shall use *oriented rectangle* to refer to a rectangle of this type. Thus an oriented rectangle R is given in the form

$$R : a \leqq u \leqq b, \qquad c \leqq v \leqq d.$$

The symbol R^0 will denote the set of the interior points of R.

IV.2.4. Given T as in IV.2.1, let E be any set in the w-plane. Then

$$N(z, E\mathfrak{D}) \geqq \sum g[z, T(s^0E)], \tag{1}$$

$$N(z, E\mathfrak{D}) = \lim_{j\to\infty} \sum g[z, T(s^0E)], \tag{2}$$

$$N(z, E\mathfrak{D}) \geqq \sum N(z, s^0E), \tag{3}$$

$$N(z, E\mathfrak{D}) = \lim_{j\to\infty} \sum N(z, s^0E), \tag{4}$$

where the summations are extended over all squares s that satisfy the conditions (cf. IV.2.3) $s \in D_{p_j}$, $s^0 \subset \mathfrak{D}$.

Proof. (3) follows directly from IV.2.2, and by IV.2.2 the inequality (1) is a consequence of (3). Furthermore, by (3) and IV.2.2, the relation (4) is a consequence of (2). Thus it is sufficient to verify (2). Now (2) is obvious, in view of (1), if $N(z_0, E\mathfrak{D}) = 0$ for a given point z_0. Hence we can assume that $N(z_0, E\mathfrak{D}) > 0$. Let then m be any positive integer not exceeding $N(z_0, E\mathfrak{D})$. Then we can pick m distinct points $w_1, \cdots, w_m$ in the set $T^{-1}(z_0)\cdot E\mathfrak{D}$. By IV.2.3, for j sufficiently large there will be precisely m squares $s \in D_{p_j}$ such that $w_i \in s^0$ for some $i = 1, \cdots, m$. Hence, for j large enough, we shall have

$$m \leqq \sum g[z_0, T(s^0E)], \qquad s \in D_{p_j}, s^0 \subset \mathfrak{D}.$$

Hence

$$m \leqq \liminf_{j\to\infty} \sum g[z_0, T(s^0E)], \qquad s \in D_{p_j}, s^0 \subset \mathfrak{D}.$$

Since m was any positive integer not exceeding $N(z_0, E\mathfrak{D})$ the relation (2) follows in view of (1).

REMARK. If $N(z_0, E\mathfrak{D}) < +\infty$ then we can choose $m = N(z_0, E\mathfrak{D})$. The preceding argument is worded to cover the case $N(z_0, E\mathfrak{D}) = +\infty$ also.

IV.2.5. CONTINUATION. If E is any set in the w-plane and O is any open set in $\mathfrak{D}$, then

$$N(z, EO) \geqq \sum g[z, T(s^0E)],$$

$$N(z, EO) = \lim_{j\to\infty} \sum g[z, T(s^0E)],$$

$$N(z, EO) \geqq \sum N(z, s^0E),$$

$$N(z, EO) = \lim_{j\to\infty} \sum N(z, s^0E),$$

where the summation is extended over all squares s such that $s \in D_{p_j}$, $s^0 \subset O$. The proof is entirely analogous to that in IV.2.4.

IV.2.6. Given T as in IV.2.1, let B be a Borel set in $\mathfrak{D}$. Then $N(z, B)$ is measurable.

PROOF. By IV.2.4

(1) $$N(z, B) = \lim_{j\to\infty} \sum g[z, T(s^0B)], \qquad s \in D_{p_j}, s^0 \subset \mathfrak{D}.$$

Now $g[z, T(s^0B)]$ is a measurable function, since s^0B is a Borel set (see I.2.46). Thus for each j the summation in (1) is a measurable function of z and hence $N(z, B)$ is also measurable.

IV.2.7. Given T as in IV.2.1, we choose in the w-plane a Borel set $\mathfrak{B}$, not necessarily a subset of $\mathfrak{D}$, which will be kept fixed and will be referred to as *the base set.* If B is any Borel set in $\mathfrak{D}$ then the set $T(B\mathfrak{B})$ is measurable by I.2.46 and the function $N(z, B\mathfrak{B})$ is measurable by IV.2.6.

IV.2.8. CONTINUATION. Suppose that T satisfies the following condition: if e is any Borel set of measure zero in $\mathfrak{D}$ then $| T(e\mathfrak{B}) | = 0$. Then we assert that $N(z, E\mathfrak{B})$ is measurable for every measurable set $E \subset \mathfrak{D}$.

PROOF. *Case* (i). $| E | = 0$. Then there exists (see I.3.7) a Borel set E^* of measure zero such that $E \subset E^*$. Since $E^*\mathfrak{D}\mathfrak{B}$ is then also a Borel set of measure zero, we have $| T(E^*\mathfrak{D}\mathfrak{B}) | = 0$. As $N(z, E^*\mathfrak{D}\mathfrak{B}) = 0$ for $z \notin T(E^*\mathfrak{D}\mathfrak{B})$ it follows that $N(z, E^*\mathfrak{D}\mathfrak{B}) = 0$ a.e. Since $E \subset E^*$, *a fortiori* $N(z, E\mathfrak{B}) = 0$ a.e., and hence $N(z, E\mathfrak{B})$ is measurable.

Case (ii). E is any measurable set in $\mathfrak{D}$. Then (see I.3.7) we have a Borel set B such that $E = B + E_0$, $BE_0 = 0$, $| E_0 | = 0$, and hence

$$N(z, E\mathfrak{B}) = N(z, B\mathfrak{B}) + N(z, E_0\mathfrak{B}).$$

Thus the measurability of $N(z, E\mathfrak{B})$ follows by IV.2.7 and case (i).

IV.2.9. CONTINUATION. Under the conditions stated in IV.2.8, $| E | = 0$,

$E \subset \mathfrak{D}$ imply that $|T(E\mathfrak{B})| = 0$, and for every measurable set $E \subset \mathfrak{D}$ the set $T(E\mathfrak{B})$ is measurable.

PROOF. The first assertion follows from the proof in IV.2.8, case (i). Since $T(E\mathfrak{B})$ is the set where $N(z, E\mathfrak{B}) > 0$, the second assertion follows from the measurability of $N(z, E\mathfrak{B})$.

IV.2.10. CONTINUATION. Given T as in IV.2.1, we define for oriented rectangles $R \subset \mathfrak{D}$ (cf. IV.2.3)

$$G(R) = |T(R^0\mathfrak{B})|.$$

Then $G(R)$ is a non-negative, finite rectangle function. In view of IV.2.2, IV.2.1 we have the equivalent formula

$$G(R) = \iint g[z, T(R^0\mathfrak{B})].$$

Note that $G(R)$ depends also upon T and the base set $\mathfrak{B}$, and more explicit notations like $G(R, T)$, $G(R, \mathfrak{B})$, $G(R, T, \mathfrak{B})$ may be used if desirable.

IV.2.11. CONTINUATION. Given T as in IV.2.1, T will be termed BV$\mathfrak{B}$ in $\mathfrak{D}$ (*of bounded variation with respect to the base set* $\mathfrak{B}$ *in* $\mathfrak{D}$) if the rectangle function $G(R)$ of IV.2.10 is BV in $\mathfrak{D}$ (see III.1.52).

IV.2.12. CONTINUATION. The following remarks are useful in dealing with the definitions given in IV.2.10, IV.2.11. If R is an oriented rectangle such that $R^0 \subset \mathfrak{D}$ but R itself is not a subset of $\mathfrak{D}$, then the definition of $G(R)$ still applies. Let r_n be a sequence of oriented rectangles in R^0 such that $r_1 + \cdots + r_n + \cdots = R^0$, $r_n \subset r_{n+1}$. Then $T(r_n^0\mathfrak{B}) \subset T(R^0\mathfrak{B})$, and $T(R^0\mathfrak{B}) = T(r_1^0\mathfrak{B}) + \cdots + T(r_n^0\mathfrak{B}) + \cdots$. Hence clearly $g[z, T(r_n^0\mathfrak{B})] \to g[z, T(R^0\mathfrak{B})]$ and consequently $G(r_n) \to G(R)$. Now let $R_1^0, \cdots, R_m^0$ be any system of oriented open rectangles in $\mathfrak{D}$ without common interior points, and let $R_1^*, \cdots, R_m^*$ be oriented rectangles such that $R_i^* \subset R_i^0$ for $i = 1, \cdots, m$. Suppose that T is BV$\mathfrak{B}$ in $\mathfrak{D}$.Then we have a finite constant M, depending upon T, $\mathfrak{B}$, and $\mathfrak{D}$ only, such that

$$G(R_1^*) + \cdots + G(R_m^*) \leqq M.$$

By the preceding remark, we can choose $R_1^*, \cdots, R_m^*$ in such a way that the summation on the left differs as little as we please from the summation obtained by replacing $R_1^*, \cdots$ by $R_1, \cdots$. Hence we have also

$$G(R_1) + \cdots + G(R_m) \leqq M.$$

In other words, we can use oriented rectangles R that are not in $\mathfrak{D}$ provided that $R^0 \subset \mathfrak{D}$.

IV.2.13. CONTINUATION. *T is BV$\mathfrak{B}$ in $\mathfrak{D}$ if and only if $N(z, \mathfrak{D}\mathfrak{B})$ is summable.*

PROOF. Suppose first that T is BV$\mathfrak{B}$ in $\mathfrak{D}$. Then there exists a finite constant M such that

$$G(r_1) + \cdots + G(r_m) \leqq M \tag{1}$$

for every finite system of oriented rectangles $r_1, \cdots, r_m$, without common interior points, such that $r_i^0 \subset \mathfrak{D}$ (cf. IV.2.12). Define, for $j = 1, 2, \cdots,$

$$\psi_j(z) = \sum g[z, T(s^0\mathfrak{B})], \qquad s \in D_{p_j}, s^0 \subset \mathfrak{D}.$$

By IV.2.4 we have then the relations

$$\text{(2)} \qquad \psi_j(z) \leqq N(z, \mathfrak{D}\mathfrak{B}), \qquad \psi_j(z) \to N(z, \mathfrak{D}\mathfrak{B}).$$

Clearly (cf. IV.2.10, IV.2.12)

$$\text{(3)} \qquad \iint \psi_j(z) = \sum G(s), \qquad s \in D_{p_j}, s^0 \subset \mathfrak{D}.$$

(1), (3) yield the inequality

$$\text{(4)} \qquad \iint \psi_j(z) \leqq M, \qquad j = 1, 2, \cdots.$$

Since $\psi_j(z)$ and $N(z, \mathfrak{D}\mathfrak{B})$ are non-negative functions, (2) and (4) imply the summability of $N(z, \mathfrak{D}\mathfrak{B})$ by I.3.10.

Suppose conversely that $N(z, \mathfrak{D}\mathfrak{B})$ is summable. Let $r_1, \cdots, r_m$ be any finite system of oriented rectangles, without common interior points, such that $r_i^0 \subset \mathfrak{D}$, $i = 1, \cdots, m$. Using IV.2.2, IV.2.10, IV.2.12, we obtain the inequality

$$\sum_{i=1}^{m} G(r_i) \leqq \iint N(z, \mathfrak{D}\mathfrak{B}).$$

Since the integral on the right exists by assumption, it follows that T is BV$\mathfrak{B}$ in $\mathfrak{D}$.

IV.2.14. Continuation. Corollary. *If T is BV$\mathfrak{B}$ in $\mathfrak{D}$, then the set $\bar{A}(\mathfrak{D}\mathfrak{B}, +\infty)$ is of measure zero* (*cf.* IV.2.2). *Indeed, $N(z, \mathfrak{D}\mathfrak{B})$ is summable by* IV.2.13 *and hence $N(z, \mathfrak{D}\mathfrak{B}) < \infty$ a.e.*

IV.2.15. Suppose that T, given as in IV.2.1, is BV$\mathfrak{B}$ in $\mathfrak{D}$. If $\Phi(w)$ is a finite-valued, real-valued function in $\mathfrak{D}$, then we define in the z-plane a function $\sigma(z, \Phi)$ as follows (cf. IV.2.2).

$$\sigma(z, \Phi) = \begin{cases} 0 & \text{if } z \in \bar{A}(\mathfrak{D}\mathfrak{B}, +\infty), \\ \sum \Phi(w), \quad w \in T^{-1}(z) \cdot \mathfrak{D}\mathfrak{B} & \text{if } z \notin \bar{A}(\mathfrak{D}\mathfrak{B}, +\infty). \end{cases}$$

In other words, $\sigma(z, \Phi) = 0$ if z has infinitely many inverse points in the set $\mathfrak{D}\mathfrak{B}$. If z has a finite number of inverse points $w_1, \cdots, w_m$ in $\mathfrak{D}\mathfrak{B}$ then $\sigma(z, \Phi) = \Phi(w_1) + \cdots + \Phi(w_m)$. If z has no inverse points in the set $\mathfrak{D}\mathfrak{B}$, that is, if $z \notin T(\mathfrak{D}\mathfrak{B})$, then $\sigma(z, \Phi) = 0$. Clearly $\sigma(z, \Phi) = 0$ for $z \notin T(\mathfrak{D})$. Since T is bounded, it follows that $\sigma(z, \Phi)$ vanishes outside of a certain circular disc. Thus the remarks made in IV.2.1 apply to integrations involving $\sigma(z, \Phi)$. The following remarks should be kept in mind in dealing with $\sigma(z, \Phi)$.

(i) $\sigma(z, \Phi)$ depends also upon T, $\mathfrak{D}$, $\mathfrak{B}$ and more explicit notations like $\sigma(z, \Phi,$

T), and so on, may be used if desirable. A completely explicit notation would be $\sigma(z, \Phi, T, \mathfrak{D}, \mathfrak{B})$.

(ii) Independently of Φ, $\sigma(z, \Phi) = 0$ for $z \notin T(\mathfrak{D})$ and for $z \in \bar{A}(\mathfrak{D}\mathfrak{B}, +\infty)$.

(iii) Since $|\bar{A}(\mathfrak{D}\mathfrak{B}, +\infty)| = 0$ (see IV.2.14) the arbitrary convention of setting $\sigma(z, \Phi) = 0$ for $z \in \bar{A}(\mathfrak{D}\mathfrak{B}, +\infty)$ is irrelevant as far as integration is concerned.

IV.2.16. CONTINUATION. The following properties of $\sigma(z, \Phi)$ are listed for convenient reference:

(i) If $\Phi(w) \equiv 0$ in $\mathfrak{D}$ then $\sigma(z, \Phi) \equiv 0$.

(ii) If $\Phi(w) \geqq 0$ in $\mathfrak{D}$ then $\sigma(z, \Phi) \geqq 0$.

(iii) If $c_1, \cdots, c_m$ are constants, then

$$\sigma(z, c_1\Phi_1 + \cdots + c_m\Phi_m) = c_1\sigma(z, \Phi_1) + \cdots + c_m\sigma(z, \Phi_m).$$

(iv) If $\Phi_1(w) \geqq \Phi_2(w)$ in $\mathfrak{D}$ then $\sigma(z, \Phi_1) \geqq \sigma(z, \Phi_2)$.

(v) If $\Phi_n(w) \to \Phi(w)$ in $\mathfrak{D}$ then $\sigma(z, \Phi_n) \to \sigma(z, \Phi)$.

(vi) If $\Phi(w)$ is the characteristic function of a set $E \subset \mathfrak{D}$, then $\sigma(z, \Phi) = N(z, E\mathfrak{B})$ a.e. in the z-plane.

These statements are obvious consequences of the definition of $\sigma(z, \Phi)$. As regards (vi), the exceptional set is easily seen to be a subset of $\bar{A}(\mathfrak{D}\mathfrak{B}, +\infty)$ which is of measure zero by IV.2.14. As regards (i) to (v), these statements are worded to apply for every choice of $\mathfrak{B}$.

IV.2.17. CONTINUATION. We proceed to list conditions for $\sigma(z, \Phi)$ to be measurable. It is assumed that T is BV$\mathfrak{B}$ in $\mathfrak{D}$.

(i) If $\Phi(w)$ is the characteristic function of a Borel set $B \subset \mathfrak{D}$, then $\sigma(z, \Phi)$ is measurable. This follows directly from IV.2.16(vi), IV.2.7.

(ii) Let $\Phi_i(w)$, $i = 1, 2, \cdots, m$ be the characteristic function of a Borel set $B_i \subset \mathfrak{D}$, and let $c_1, \cdots, c_m$ be constants. Then $\sigma(z, c_1\Phi_1 + \cdots + c_m\Phi_m)$ is measurable. This follows directly from (i) above and IV.2.16(iii).

(iii) If $\Phi(w)$ is Borel measurable and bounded in $\mathfrak{D}$, then $\sigma(z, \Phi)$ is measurable.

PROOF. By assumption we have a constant M such that $|\Phi(w)| < M$ in $\mathfrak{D}$. Given a positive integer $n > 1$, define B_j^n as the set of those points $w \in \mathfrak{D}$ where

$$-M + \frac{2Mj}{n} \leqq \Phi(w) < -M + \frac{2M(j+1)}{n}, \qquad j = 0, 1, \cdots, n-1.$$

Then $B_0^n, B_1^n, \cdots, B_{n-1}^n$ are disjoint Borel sets whose sum is $\mathfrak{D}$. Let $\Phi_j^n(w)$ be the characteristic function of the set B_j^n, and define

$$\Phi_n(w) = \sum_{j=0}^{n-1}\left(-M + \frac{2Mj}{n}\right)\Phi_j^n(w).$$

By (ii) above, $\sigma(z, \Phi_n)$ is measurable. Clearly

$$|\Phi(w) - \Phi_n(w)| < 2M/n \qquad \text{in } \mathfrak{D},$$

and hence $\Phi_n(w) \to \Phi(w)$ in $\mathfrak{D}$. Thus $\sigma(z, \Phi_n) \to \sigma(z, \Phi)$ by IV.2.16(v), and hence $\sigma(z, \Phi)$ is measurable.

(iv) If $\Phi(w)$ is finite-valued and Borel measurable in $\mathfrak{D}$, then $\sigma(z, \Phi)$ is measurable.

PROOF. Define, for each positive integer n,

$$\Phi_n(w) = \begin{cases} n & \text{if } \Phi(w) \geqq n, \\ \Phi(w) & \text{if } |\Phi(w)| < n, \\ -n & \text{if } \Phi(w) \leqq -n. \end{cases}$$

Then (see I.3.8) $\Phi_n(w)$ is bounded and Borel measurable, and hence $\sigma(z, \Phi_n)$ is measurable by (iii) above. Clearly $\Phi_n(w) \to \Phi(w)$ in $\mathfrak{D}$, and hence $\sigma(z, \Phi_n) \to \sigma(z, \Phi)$ by IV.2.16(v). Thus $\sigma(z, \Phi)$ is measurable.

IV.2.18. Given T as in IV.2.1, let $H(z)$ be a finite-valued, real-valued function in the z-plane. Define

$$F(w, H) = H[t(w)], \qquad w \in \mathfrak{D}.$$

The function $F(w, H)$ depends also upon T and $\mathfrak{D}$ and more explicit notations like $F(w, H, T)$, and so on, may be used if desirable. The following properties are obvious:

(i) If $H(z) \equiv 0$, then $F(w, H) \equiv 0$ in $\mathfrak{D}$.

(ii) If $H(z) \geqq 0$, then $F(w, H) \geqq 0$ in $\mathfrak{D}$.

(iii) If $c_1, \cdots, c_m$ are constants, then $F(w, c_1H_1 + \cdots + c_mH_m) = c_1F(w, H_1) + \cdots + c_mF(w, H_m)$.

(iv) If $H_1(z) \geqq H_2(z)$, then $F(w, H_1) \geqq F(w, H_2)$.

(v) If $H_n(z) \to H(z)$ everywhere in the z-plane, then

$$F(w, H_n) \to F(w, H) \qquad \text{in } \mathfrak{D}.$$

(vi) If $H(z)$ is the characteristic function of a set $\bar{E}$ in the z-plane, then $F(w, H)$ agrees with the characteristic function of the set in $T^{-1}(\bar{E})$ in $\mathfrak{D}$.

IV.2.19. CONTINUATION. To obtain information about the measurability of $F(w, H)$, we need the following remark. If $\bar{B}$ is a Borel set in the z-plane, then $T^{-1}(\bar{B})$ is a Borel set in $\mathfrak{D}$.

PROOF. Let K denote the class of all those sets $\bar{E}$ in the z-plane for which the set $T^{-1}(\bar{E})$ is a Borel set. Then K possesses the following properties:

(i) If $\bar{E}$ is closed, then $\bar{E} \in K$. Indeed, the set $T^{-1}(\bar{E})$ is then closed relative to $\mathfrak{D}$ (see I.2.26) and hence $T^{-1}(\bar{E})$ is a Borel set.

(ii) If $\bar{E} \in K$ then the complement $\bar{F}$ of $\bar{E}$ is also in K. Indeed, $T^{-1}(\bar{F}) = \mathfrak{D} - T^{-1}(\bar{E})$. Since $T^{-1}(\bar{E})$ is a Borel set by assumption, the assertion follows.

(iii) If $\bar{E}_1, \cdots, \bar{E}_m, \cdots$ belong to K, then their sum $\bar{E}$ also belongs to K since (cf. I.2.5)

$$T^{-1}(\bar{E}) = \sum_m T^{-1}(\bar{E}_m).$$

As a consequence of these properties (i), (ii), (iii), K contains all Borel sets (see I.2.46).

IV.2.20. CONTINUATION. We now obtain readily the following statements:

(i) If $H(z)$ is the characteristic function of a Borel set $\overline{B}$ in the z-plane, then $F(w, H)$ is Borel measurable. This follows directly from IV.2.18(vi) and IV.2.19.

(ii) If $H_i(z)$, $i = 1, 2, \cdots, n$ is the characteristic function of a Borel set $\overline{B}_i$ in the z-plane, and if $c_1, \cdots, c_m$ are constants, then $F(w, c_1 H_1 + \cdots + c_m H_m)$ is Borel measurable. This follows from (i) above and IV.2.18(iii).

(iii) If $H(z)$ is Borel measurable and bounded in the z-plane, then $F(w, H)$ is Borel measurable. The proof is entirely analogous to that in IV.2.17(iii).

(iv) If $H(z)$ is finite-valued, Borel measurable in the z-plane, then $F(w, H)$ is Borel measurable. The proof is entirely analogous to that in IV.2.17(iv).

REMARK. Note that we do not now need the assumption that T is BV$\mathfrak{B}$ in $\mathfrak{D}$, and that in the conclusions we obtain Borel measurability. In the sequel it will happen that the function $H(z)$ is defined only on some subset $\overline{E}$ of the z-plane. In such cases we shall agree to set $H(z) = 0$ for $z \notin \overline{E}$. If $H(z)$ is Borel measurable on a Borel set $\overline{E}$, then the function so extended is clearly Borel measurable in the z-plane.

IV.2.21. Let us now assume that T, given as in IV.2.1, is BV$\mathfrak{B}$ in $\mathfrak{D}$. For every Borel set $B \subset \mathfrak{D}$, the function $N(z, B\mathfrak{B})$ is then summable by IV.2.7, IV.2.13 (note that $0 \leqq N(z, B\mathfrak{B}) \leqq N(z, \mathfrak{D}\mathfrak{B})$). We can therefore introduce the set function (cf. IV.2.1)

$$\mu(B) = \iint N(z, B\mathfrak{B}),$$

defined, finite, and non-negative for all Borel sets $B \subset \mathfrak{D}$. We assert that $\mu(B)$ is completely additive on Borel sets in $\mathfrak{D}$. Indeed, if $B_1, \cdots, B_n, \cdots$ is a sequence of disjoint Borel sets in $\mathfrak{D}$, and if B is their sum, then clearly

$$N(z, B\mathfrak{B}) = \sum_{n=1}^{\infty} N(z, B_n\mathfrak{B}).$$

Since all the functions involved are summable and non-negative, termwise integration is permissible, and we obtain the desired relation

$$\mu(B) = \sum_{n=1}^{\infty} \mu(B_n).$$

Thus $\mu(B)$ is a measure on Borel sets in $\mathfrak{D}$, and consequently $\mu(B)$ gives rise to a (μ)-integral (see I.3.17). Since μ is defined only for Borel sets in $\mathfrak{D}$, only Borel measurable functions will be integrated with respect to μ. The measure $\mu(B)$ depends also upon T, $\mathfrak{D}$ and $\mathfrak{B}$ and more explicit notations like $\mu(B, T, \mathfrak{B})$, and so on, may be used if desirable.

IV.2.22. CONTINUATION. Let $\Phi(w)$ be a finite-valued, Borel measurable function in $\mathfrak{D}$. We propose to establish the transformation formula (cf. IV.2.21, IV.2.1)

$$\iint_{\mathfrak{D}} \Phi \, d\mu = \iint \sigma(z, \Phi) \tag{1}$$

in the greatest possible generality. Let us repeat that μ and σ depend upon T, $\mathfrak{D}$ and $\mathfrak{B}$ too. As will become apparent in the sequel, this formula is of fundamental importance for the study of the transformation of double integrals. We shall proceed by proving (1) for a sequence of special cases of increasing generality.

IV.2.23. Continuation. (1) in IV.2.22 holds if $\Phi(w)$ agrees with the characteristic function of a Borel set B in $\mathfrak{D}$. Indeed, in this case we have the obvious relations (cf. IV.2.21, IV.2.7, IV.2.17, IV.2.16(vi)),

$$\iint_{\mathfrak{D}} \Phi \, d\mu = \iint_{B} d\mu = \mu(B) = \iint N(z, B\mathfrak{B}), \qquad \iint \sigma(z, \Phi) = \iint N(z, B\mathfrak{B}).$$

IV.2.24. Continuation. Let $c_1, \cdots, c_m$ be constants, and let $\Phi_i(w)$ agree with the characteristic function of a Borel set B_i in $\mathfrak{D}$, $i = 1, 2, \cdots, m$. Then IV.2.22(1) holds for $\Phi = c_1\Phi_1 + \cdots + c_m\Phi_m$, as a direct consequence of IV.2.23, IV.2.16(iii).

IV.2.25. Continuation. Suppose that $\Phi(w)$ is Borel measurable and bounded in $\mathfrak{D}$. Then IV.2.22(1) holds.

Proof. By assumption we have a constant M such that $| \Phi(w) | < M$ in $\mathfrak{D}$. For each positive integer n, let us introduce the function $\Phi_n(w)$ as in IV.2.17(iii). Clearly

$$\Phi_n \leqq \Phi \leqq \Phi_n + 2M/n. \tag{1}$$

By IV.2.16 it follows that

$$\sigma(z, \Phi_n) \leqq \sigma(z, \Phi) \leqq \sigma(z, \Phi_n) + \frac{2M}{n} N(z, \mathfrak{D}\mathfrak{B}). \tag{2}$$

By IV.2.24 we have

$$\iint_{\mathfrak{D}} \Phi_n \, d\mu = \iint \sigma(z, \Phi_n). \tag{3}$$

(1), (2), (3) yield successively (cf. IV.2.21)

$$\iint_{\mathfrak{D}} \Phi_n \, d\mu \leqq \iint_{\mathfrak{D}} \Phi \, d\mu \leqq \iint_{\mathfrak{D}} \Phi_n \, d\mu + \frac{2M}{n} \mu(\mathfrak{D}),$$

$$\iint \sigma(z, \Phi_n) \leqq \iint \sigma(z, \Phi) \leqq \iint \sigma(z, \Phi_n) + \frac{2M}{n} \mu(\mathfrak{D}),$$

$$\left| \iint_{\mathfrak{D}} \Phi \, d\mu - \iint \sigma(z, \Phi) \right| \leqq \frac{2M}{n} \mu(\mathfrak{D}).$$

Since n is arbitrary, IV.2.22(1) follows.

IV.2.26. Continuation. *Now let $\Phi(w)$ be finite-valued, non-negative and Borel measurable in $\mathfrak{D}$. We assert that if one of the two integrals*

$$(1) \qquad \iint_{\mathfrak{D}} \Phi \, d\mu, \qquad \iint \sigma(z, \Phi)$$

exists, then the other one exists also, and IV.2.22(1) *holds.*

PROOF. For each positive integer n define

$$\Psi_n(w) = \begin{cases} n & \text{if } \Phi(w) \geqq n, \\ \Phi(w) & \text{if } \Phi(w) < n. \end{cases}$$

By IV.2.25 we have then

$$(2) \qquad \iint_{\mathfrak{D}} \Psi_n \, d\mu = \iint \sigma(z, \Psi_n).$$

Clearly, since $\Phi \geqq 0$,

$$(3) \qquad 0 \leqq \Psi_1 \leqq \cdots \leqq \Psi_n \leqq \cdots, \qquad \Psi_n \to \Phi \qquad \text{in } \mathfrak{D},$$

and hence, by IV.2.16,

$$(4) \qquad 0 \leqq \sigma(z, \Psi_1) \leqq \cdots \leqq \sigma(z, \Psi_n) \leqq \cdots, \qquad \sigma(z, \Psi_n) \to \sigma(z, \Phi)$$

everywhere in the z-plane. Hence, by the lemma of Fatou, $\Phi(w)$ is (μ)-summable in $\mathfrak{D}$ if and only if the sequence of the integrals of the functions Ψ_n, with respect to μ in $\mathfrak{D}$, is bounded, and $\sigma(z, \Phi)$ is summable if and only if the sequence of the integrals of the functions $\sigma(z, \Psi_n)$ is bounded. In view of (2) it follows that the integrals (1) exist or fail to exist simultaneously. If they exist, then by I.3.11 we obtain from (3) and (4) the relations

$$\iint_{\mathfrak{D}} \Psi_n \, d\mu \to \iint_{\mathfrak{D}} \Phi \, d\mu, \qquad \iint \sigma(z, \Psi_n) \to \iint \sigma(z, \Phi),$$

and IV.2.22(1) follows in view of (2).

IV.2.27. THEOREM. *Suppose that T, given as in* IV.2.1, *is $BV\mathfrak{B}$ in $\mathfrak{D}$. If $\Phi(w)$ is finite-valued, Borel measurable in $\mathfrak{D}$, then the formula (cf.* IV.2.21, IV.2.15, IV.2.1)

$$(1) \qquad \iint_{\mathfrak{D}} \Phi \, d\mu = \iint \sigma(z, \Phi)$$

holds as soon as the integral on the left exists. If $\Phi(w) \geqq 0$ in $\mathfrak{D}$, then the formula holds as soon as one of the two integrals involved exists.

PROOF. The second half of the theorem is merely a restatement of IV.2.26. To prove the first half, assume that $\Phi(w)$ is summable (μ). Then $|\Phi(w)|$ is also summable (μ) (see I.3.17), and hence by IV.2.26

$$\iint_{\mathfrak{D}} |\Phi| \, d\mu = \iint \sigma(z, |\Phi|).$$

Furthermore, $|\Phi| - \Phi$ is also summable (μ) and non-negative. Hence by IV.2.26

$$\iint_{\mathfrak{D}} (|\Phi| - \Phi) \, d\mu = \iint \sigma(z, |\Phi| - \Phi)$$

Subtraction yields, in view of IV.2.16(iii), the formula (1).

IV.2.28. To derive an important application of the preceding result we need the following remark. Let T be BV$\mathfrak{B}$ in $\mathfrak{D}$, and let $H(z)$ be a finite-valued function defined in the whole z-plane. Then (cf. IV.2.15, IV.2.18)

(1) $$\sigma[z, F(w, H)] = H(z)N(z, \mathfrak{D}\mathfrak{B})$$

a.e. in the z-plane.

PROOF. Since T is BV$\mathfrak{B}$ in $\mathfrak{D}$, the set $\overline{A}(\mathfrak{D}\mathfrak{B}, +\infty)$ is of measure zero (see IV.2.14). Thus it is sufficient to show that (1) holds if $z \notin \overline{A}(\mathfrak{D}\mathfrak{B}, +\infty)$. Then z has at most a finite number of inverse points in the set $\mathfrak{D}\mathfrak{B}$. If no such inverse points exist, then (1) holds since then both sides are equal to zero. So we can assume that there exist such points; let us denote them by $w_1, \cdots, w_k$. Then $k = N(z, \mathfrak{D}\mathfrak{B})$. By definition

$$\sigma[z, F(w, H)] = \sum_{i=1}^{k} F(w_i, H) = \sum_{i=1}^{k} H[t(w_i)]$$

$$= \sum_{i=1}^{k} H(z) = kH(z) = N(z, \mathfrak{D}\mathfrak{B})H(z),$$

and (1) is proved.

IV.2.29. THEOREM. *Suppose that T, given as in* IV.2.1, *is BV$\mathfrak{B}$ in $\mathfrak{D}$. Let $H(z)$ be a finite-valued, Borel measurable function in the z-plane. Then the formula* (cf. IV.2.21, IV.2.1)

(1) $$\iint_{\mathfrak{D}} H[t(w)] \, d\mu = \iint H(z)N(z, \mathfrak{D}\mathfrak{B})$$

holds as soon as one of the two integrals involved exists.

PROOF. (i) Suppose that the integral on the left exists. In view of IV.2.28 the formula (1) is then merely a special case of IV.2.27, IV.2.20, for $\Phi(w) = H[t(w)] = F(w, H)$.

(ii) $H(z) \geqq 0$, and the integral on the right in (1) exists. In view of IV.2.28 the assertion is again a special case of the second half of the theorem in IV.2.27.

(iii) The integral on the right in (1) exists. Since $N(z, \mathfrak{D}\mathfrak{B}) \geqq 0$, it follows that the integrals

$$\iint | H(z) | N(z, \mathfrak{D}\mathfrak{B}), \qquad \iint [| H(z) | - H(z)]N(z, \mathfrak{D}\mathfrak{B})$$

also exist. Hence, by (ii) above

$$\iint_{\mathfrak{D}} | H[t(w)] | \, d\mu = \iint | H(z) | N(z, \mathfrak{D}\mathfrak{B}),$$

$$\iint_{\mathfrak{D}} \{| H[t(w)] | - H[t(w)]\} \, d\mu = \iint [| H(z) | - H(z)]N(z, \mathfrak{D}\mathfrak{B})$$

Subtraction yields the formula (1).

IV.2.30. Given T as in IV.2.1, let $E_1, \cdots, E_n$ be disjoint subsets of a set $E \subset \mathfrak{D}$. Then (cf. IV.2.1, IV.2.2)

$$\text{(1)} \qquad \sum_{j=1}^{n} [N(z, E_j) - g(z, T(E_j))] \leqq N(z, E) - g(z, T(E)).$$

Proof. Let z_0 be a point in the z-plane. Then (1) is obvious if $N(z_0, E) = +\infty$. Hence we can assume that $N(z_0, E) < +\infty$. If $g[z_0, T(E_j)] = 0$ for a certain j, then $N(z_0, E_j) = 0$ for that j (see IV.2.2). Hence, if $g[z_0, T(E_j)] = 0$ for every j, then every term of the summation in (1) is equal to zero, and (1) is obvious. So we can assume $g[z_0, T(E_j)] = 1$ for at least one j. Then (cf. IV.2.2)

$$\sum_{j=1}^{n} [N(z_0, E_j) - g(z_0, T(E_j))] \leqq \left[\sum_{j=1}^{n} N(z_0, E_j)\right] - 1 \leqq N(z_0, E) - 1$$

$$\leqq N(z_0, E) - g[z_0, T(E)]$$

IV.2.31. Suppose that T, given as in IV.2.1, is BV$\mathfrak{B}$ in $\mathfrak{D}$. We define, for every oriented rectangle R such that $R^0 \subset \mathfrak{D}$ (cf. IV.2.21),

$$G^*(R) = \mu(R^0) = \iint N(z, R^0\mathfrak{B}).$$

Clearly (cf. IV.2.10)

$$G(R) \leqq G^*(R).$$

If $R_1, \cdots, R_m$ are oriented rectangles, without common interior points, in an oriented rectangle R such that $R^0 \subset \mathfrak{D}$ then (see IV.2.2)

$$\sum_{j=1}^{m} N(z, R_j^0\mathfrak{B}) \leqq N(z, R^0\mathfrak{B}).$$

Integration yields the inequality

$$\sum_{j=1}^{m} G^*(R_j) \leqq G^*(R).$$

Hence the rectangle function $G^*(R)$ is of type A in $\mathfrak{D}$ (cf. III.1.52). As a consequence the derivative of $G^*(R)$ exists a.e. in $\mathfrak{D}$. We shall denote this derivative by $D^*(w)$. This derivative depends also upon T and $\mathfrak{B}$ and more explicit notations like $D^*(w, \mathfrak{B})$, and so on, will be used if desirable. By III.1.52, $D^*(w)$ is summable on every oriented rectangle $R \subset \mathfrak{D}$ and

$$(1) \qquad \iint_R D^*(w) \leqq G^*(R).$$

In fact, this inequality holds as soon as $R^0 \subset \mathfrak{D}$ as we shall verify presently. Indeed, suppose that $R^0 \subset \mathfrak{D}$. Let r_n be a sequence of oriented rectangles such that $r_n \subset R^0$, $r_1 \subset r_2 \subset \cdots$, $r_1 + \cdots + r_n + \cdots = R^0$. Then $r_n \subset \mathfrak{D}$ and hence

$$(2) \qquad \iint_{r_n} D^*(w) \leqq G^*(r_n) = \iint N(z, r_n^0\mathfrak{B}) \leqq \iint N(z, R^0\mathfrak{B}) = G^*(R).$$

Note that $G^*(R) < +\infty$ since T is BV$\mathfrak{B}$ in $\mathfrak{D}$ (cf. IV.2.13). Since (2) holds for every n, the inequality (1) follows. We proceed to show that $D^*(w)$ is summable in $\mathfrak{D}$ itself. Indeed, since $\mathfrak{D}$ is a bounded domain, we have a sequence of oriented rectangles $R_1, \cdots, R_k, \cdots$, such that

$$(3) \qquad \mathfrak{D} = \sum_{k=1}^{\infty} R_k, \qquad R_i^0 R_j^0 = 0,\ i \neq j.$$

(3) implies, in view of IV.2.21, the inequalities

$$(4) \qquad \sum_{k=1}^{\infty} G^*(R_k) = \sum_{k=1}^{\infty} \mu(R_k^0) \leqq \mu(\mathfrak{D}) < +\infty.$$

(1) and (4) yield

$$(5) \qquad \sum_{k=1}^{\infty} \iint_{R_k} D^*(w) \leqq \mu(\mathfrak{D}) < +\infty.$$

By I.3.10 the summability of $D^*(w)$ in $\mathfrak{D}$ follows. In fact (5) yields, in view of I.3.10, the inequality

$$\iint_{\mathfrak{D}} D^*(w) \leqq \mu(\mathfrak{D}),$$

which will be generalized in the sequel.

IV.2.32. Continuation. Let again $R_1, \cdots, R_m$ be oriented rectangles, without common interior points, comprised in an oriented rectangle R such that $R^0 \subset \mathfrak{D}$. By IV.2.30 we have then the inequality

$$\sum_{j=1}^{m} [N(z, R_j^0\mathfrak{B}) - g(z, T(R_j^0\mathfrak{B}))] \leqq N(z, R^0\mathfrak{B}) - g[z, T(R^0\mathfrak{B})].$$

Since T is BV$\mathfrak{B}$ in $\mathfrak{D}$, all the functions involved are summable (see IV.2.13). Integration yields (cf. IV.2.10, IV.2.31)

$$\sum_{j=1}^{m} [G^*(R_j) - G(R_j)] \leqq G^*(R) - G(R).$$

Since clearly $G^*(R) \geqq G(R)$, it follows that the rectangle function $G'(R) = G^*(R) - G(R)$ is also of type A in $\mathfrak{D}$ (cf. III.1.52). As a consequence, $G'(R)$ has a derivative $D'(w)$ a.e. in $\mathfrak{D}$, and furthermore $D'(w)$ is summable in every oriented rectangle $R \subset \mathfrak{D}$ and satisfies the inequality

$$(1) \qquad \iint_R D'(w) \leqq G'(R) = G^*(R) - G(R).$$

Consequently, $G(R) = G^*(R) - G'(R)$ also has a derivative $D(w)$ a.e. in $\mathfrak{D}$, and clearly $D(w) = D^*(w) - D'(w)$ a.e. in $\mathfrak{D}$. Since $G^*(R) \geqq G(R)$ for every $R \subset \mathfrak{D}$,

$$(2) \qquad D^*(w) \geqq D(w) \qquad \text{a.e. in } \mathfrak{D}.$$

The inequality (1) can now be rewritten in the form

$$(3) \qquad \iint_R [D^*(w) - D(w)] \leqq G^*(R) - G(R).$$

The derivative $D(w)$ will also be denoted by $D(w, \mathfrak{B})$ if explicit reference to the base set $\mathfrak{B}$ is desirable.

IV.2.33. Continuation. We assert the inequality (cf. IV.2.21)

$$(1) \qquad \iint_B D^*(w) \leqq \mu(B)$$

for every Borel set $B \subset \mathfrak{D}$.

Proof. By IV.2.31, $D^*(w)$ is summable in $\mathfrak{D}$. Thus we can define a set function $\phi(B)$ by the formula

$$\phi(B) = \iint_B D^*(w)$$

for all Borel sets (in fact, for all measurable sets) $B \subset \mathfrak{D}$. Clearly $\phi(B)$ is non-negative and completely additive on Borel sets in $\mathfrak{D}$ (see I.3.16). By IV.2.31(1) we have

$$\phi(R) \leqq \mu(R^0) \leqq \mu(R)$$

for every oriented rectangle $R \subset \mathfrak{D}$. By III.1.45, III.1.52 the inequality (1) follows.

IV.2.34. Continuation. We assert that $D(w) = D^*(w)$ a.e. in $\mathfrak{D}$.

Proof. Take any oriented rectangle $R \subset \mathfrak{D}$. Let us put, for $j = 1, 2, \cdots$ (cf. IV.2.3),

$$(1) \qquad \psi_j(z) = \sum g[z, T(s^0\mathfrak{B})], \qquad s \in D_{p_j}, s^0 \subset R^0.$$

By IV.2.5 we have then the relations

$$0 \leqq \psi_j(z) \leqq N(z, R^0\mathfrak{B}), \ \psi_j(z) \to N(z, R^0\mathfrak{B}).$$

Since $N(z, R^0\mathfrak{B})$ is summable (cf. IV.2.13), it follows that (cf. I.3.11, IV.2.1, IV.2.10)

(2) $$\iint \psi_j(z) = \sum G(s) \underset{j\to\infty}{\longrightarrow} \iint N(z, R^0\mathfrak{B}) = \mu(R^0),$$

where the summation is extended over the same squares s as in (1). Similarly, if we put

$$\lambda_j(z) = \sum N(z, s^0\mathfrak{B}), \qquad s \in D_{p_j}, s^0 \subset R^0,$$

then we have by IV.2.5 the relations

$$0 \leqq \lambda_j(z) \leqq N(z, R^0\mathfrak{B}), \lambda_j(z) \to N(z, R^0\mathfrak{B}),$$

and termwise integration yields this time the relations (cf. IV.2.31)

(3) $$\iint \lambda_j(z) = \sum G^*(s) \underset{j\to\infty}{\longrightarrow} \iint N(z, R^0\mathfrak{B}) = \mu(R^0)$$

(2) and (3) yield

(4) $$0 = \lim_{j\to\infty} \sum [G^*(s) - G(s)], \qquad s \in D_{p_j}, s^0 \subset R^0.$$

Now the rectangle function $G^*(R) - G(R)$ is non-negative and has a derivative a.e. in $\mathfrak{D}$ (see IV.2.32, IV.2.31). Hence, by III.1.25, it follows from (4) that $D^*(w) = D(w)$ a.e. in R. Since R was any oriented rectangle in $\mathfrak{D}$, it follows that $D^*(w) = D(w)$ a.e. in $\mathfrak{D}$.

IV.2.35. Continuation. $D(w) = 0$ a.e. on $D - \mathfrak{B}$.

Proof. Since $\mathfrak{B}$ is a Borel set, we have by IV.2.33, IV.2.34

$$\iint_{\mathfrak{D}-\mathfrak{B}} D(w) \leqq \mu(\mathfrak{D} - \mathfrak{B}) = \iint N[z, (\mathfrak{D} - \mathfrak{B})\mathfrak{B}] = 0.$$

Since $D(w)$ is non-negative, the assertion follows.

IV.2.36. Suppose that T, given as in IV.2.1, is BV$\mathfrak{B}$ in $\mathfrak{D}$. Let $\overline{E}$ be a set of measure zero in the z-plane. Then $D(w) = 0$ a.e. on the set $T^{-1}(\overline{E})$.

Proof. Since T is bounded, it is clearly sufficient to consider the case when $\overline{E}$ is bounded.

Case (i). $\overline{E}$ is a Borel set of measure zero. Then $T^{-1}(\overline{E})$ is a Borel set (see IV.2.19). Hence, if we put $E = T^{-1}(\overline{E})$, we have by IV.2.33, IV.2.34

$$\iint_E D(w) \leqq \mu(E) = \iint N(z, E\mathfrak{B}).$$

Since $N(z, E\mathfrak{B}) = 0$ for $z \notin T(E)$ and $T(E) \subset \overline{E}$, it follows that $N(z, E\mathfrak{B}) = 0$ a.e. in the z-plane. Hence

$$\iint_E D(w) = 0.$$

Since $D(w)$ is non-negative, it follows that $D(w) = 0$ a.e. on $E = T^{-1}(\overline{E})$.

Case (ii). $\overline{E}$ is any bounded set of measure zero. By I.3.7 we have then a Borel set $\overline{B}$ of measure zero, such that $\overline{E} \subset \overline{B}$. By case (i) we have $D(w) = 0$ a.e. on $T^{-1}(\overline{B})$. Since $T^{-1}(\overline{E}) \subset T^{-1}(\overline{B})$, *a fortiori* $D(w) = 0$ a.e. on $T^{-1}(\overline{E})$.

IV.2.37. Corollary 1. *$D(w) = 0$ a.e. on the set $T^{-1}[\overline{A}(\mathfrak{D}\mathfrak{B}, +\infty)]$ (see* IV.2.2). *Indeed, the set $\overline{A}(\mathfrak{D}\mathfrak{B}, +\infty)$ is of measure zero by* IV.2.14.

Corollary 2. *If E is a subset of $\mathfrak{D}$ such that $| T(E) | = 0$, then $D(w) = 0$ a.e. on E. Indeed, if we put $\overline{E} = T(E)$, then clearly $E \subset T^{-1}(\overline{E})$. Since $D(w) = 0$. a.e. on $T^{-1}(\overline{E})$ by* IV.2.36, *a fortiori $D(w) = 0$ a.e. on E.*

IV.2.38. Given T as in IV.2.1, we shall work in the sequel with several different base sets simultaneously, and the following remarks will be useful. Let $\mathfrak{B}$, $\mathfrak{B}_1$, $\mathfrak{B}_2$ be Borel sets in the w-plane, such that $\mathfrak{B} = \mathfrak{B}_1 + \mathfrak{B}_2$.

(i) T is BV$\mathfrak{B}$ if and only if T is both BV$\mathfrak{B}_1$ and BV$\mathfrak{B}_2$ in D.

Proof. Let us put, for every oriented rectangle $R \subset \mathfrak{D}$,

$$G(R) = | T(R^0\mathfrak{B}) |, \; G_1(R) = | T(R^0\mathfrak{B}_1) |, \; G_2(R) = | T(R^0\mathfrak{B}_2) |.$$

Since $\mathfrak{B} = \mathfrak{B}_1 + \mathfrak{B}_2$ and hence $T(R^0\mathfrak{B}) = T(R^0\mathfrak{B}_1) + T(R^0\mathfrak{B}_2)$, we have the inequalities

$$(1) \qquad G_1(R) \leqq G(R), \; G_2(R) \leqq G(R), \; G(R) \leqq G_1(R) + G_2(R),$$

and the assertion follows in view of the definitions given in IV.2.11.

(ii) Suppose that T is BV$\mathfrak{B}$ in $\mathfrak{D}$. By IV.2.32 the derivatives $D(w, \mathfrak{B}_1)$, $D(w, \mathfrak{B}_2)$, $D(w, \mathfrak{B})$ exist then a.e. in $\mathfrak{D}$ (cf. (i)). The inequalities (1) imply that

$$(2) \qquad \begin{aligned} &D(w, \mathfrak{B}_1) \leqq D(w, \mathfrak{B}), \; D(w, \mathfrak{B}_2) \leqq D(w, \mathfrak{B}), \\ &D(w, \mathfrak{B}) \leqq D(w, \mathfrak{B}_1) + D(w, \mathfrak{B}_2) \qquad \text{a.e. in } \mathfrak{D}. \end{aligned}$$

By IV.2.35 we have

$$(3) \qquad D(w, \mathfrak{B}_1) = 0 \qquad \text{a.e. on } \mathfrak{D} - \mathfrak{B}_1,$$

$$(4) \qquad D(w, \mathfrak{B}_2) = 0 \qquad \text{a.e. on } \mathfrak{D} - \mathfrak{B}_2.$$

From (2), (3), (4) we infer the relations

$$(5) \qquad D(w, \mathfrak{B}_1) = D(w, \mathfrak{B}) \qquad \text{a.e. on } \mathfrak{D} - \mathfrak{B}_2,$$

$$(6) \qquad D(w, \mathfrak{B}_2) = D(w, \mathfrak{B}) \qquad \text{a.e. on } \mathfrak{D} - \mathfrak{B}_1.$$

(iii) Using the notations and assumptions of (ii), suppose that $\mathfrak{B}_1\mathfrak{B}_2 = 0$. Then

$$(7) \qquad D(w, \mathfrak{B}) = D(w, \mathfrak{B}_1) + D(w, \mathfrak{B}_2) \qquad \text{a.e. in } \mathfrak{D}.$$

This is a formal consequence of (3), (4), (5), (6).

(iv) Using the assumptions and notations of (ii), we assert that

$$(8) \qquad D(w, \mathfrak{B}) = D(w, \mathfrak{B}_1) \qquad \text{a.e. on } \mathfrak{B}_1.$$

Indeed, by (iii) (applied to the base sets $\mathfrak{B}_1$ and $\mathfrak{B} - \mathfrak{B}_1$) we have $D(w, \mathfrak{B}) = D(w, \mathfrak{B}_1) + D(w, \mathfrak{B} - \mathfrak{B}_1)$ a.e. in $\mathfrak{D}$. By IV.2.35, applied to the base set $\mathfrak{B} - \mathfrak{B}_1$, we have $D(w, \mathfrak{B} - \mathfrak{B}_1) = 0$ a.e. on $\mathfrak{B}_1$, and (8) follows.

(v) Using the assumptions and notations of (ii), we have

$$(9) \qquad D(w, \mathfrak{B}) = D(w, \mathfrak{B}_1) = D(w, \mathfrak{B}_2) \qquad \text{a.e. on } \mathfrak{B}_1\mathfrak{B}_2 .$$

Indeed, (8) implies that $D(w, \mathfrak{B}) = D(w, \mathfrak{B}_1)$ a.e. on $\mathfrak{B}_1$, and hence *a fortiori* a.e. on $\mathfrak{B}_1\mathfrak{B}_2$. Interchanging $\mathfrak{B}_1$ and $\mathfrak{B}_2$, we have similarly $D(w, \mathfrak{B}) = D(w, \mathfrak{B}_2)$ a.e. on $\mathfrak{B}_1\mathfrak{B}_2$, and (9) follows.

IV.2.39. Given T and a base set $\mathfrak{B}$ (see IV.2.7), we shall say that T is AC$\mathfrak{B}$ in $\mathfrak{D}$ (*absolutely continuous in* $\mathfrak{D}$ *with respect to the base set* $\mathfrak{B}$) if and only if the rectangle function $G(R)$ of IV.2.10 is AC in $\mathfrak{D}$ (see III.1.52). Explicitly, T is AC$\mathfrak{B}$ in $\mathfrak{D}$ if for every $\epsilon > 0$ there exists an $\eta(\epsilon) > 0$ such that the following statement holds. Let $R_1, \cdots, R_m$ be any system of oriented rectangles in $\mathfrak{D}$ without common interior points, such that

$$(1) \qquad |R_1| + \cdots + |R_m| \leqq \eta(\epsilon).$$

Then we have the inequality

$$(2) \qquad G(R_1) + \cdots + G(R_m) \leqq \epsilon.$$

According to the definition given in III.1.52, it is assumed that $R_i \subset \mathfrak{D}$, $i = 1, 2, \cdots, m$. We want to verify that (1) implies (2) even if we only require that $R_i^0 \subset \mathfrak{D}$, $i = 1, 2, \cdots, m$. Indeed, let us select, for each $i = 1, 2, \cdots, m$, an oriented rectangle $R_i^* \subset R_i^0$. Then in view of (1) we have $|R_1^*| + \cdots + |R_m^*| \leqq \eta(\epsilon)$, and hence by assumption $G(R_1^*) + \cdots + G(R_m^*) \leqq \epsilon$. By the argument used in IV.2.12, we can choose R_i^* in such a manner that $G(R_i^*)$ differs as little as we please from $G(R_i)$, and hence the assertion follows. This reasoning assumes that T is BV$\mathfrak{B}$ in $\mathfrak{D}$, and this is indeed the case, because the boundedness of T implies obviously the boundedness of the rectangle function $G(R)$, and hence (see III.1.52) T is BV$\mathfrak{B}$ if it is AC$\mathfrak{B}$ in $\mathfrak{D}$. Summing up: in dealing with (1) and (2) we can use oriented rectangles R such that $R^0 \subset \mathfrak{D}$. Furthermore, if T is AC$\mathfrak{B}$ in $\mathfrak{D}$, then it is also BV$\mathfrak{B}$ in $\mathfrak{D}$. As a consequence, the results derived previously, under the assumption that T is BV$\mathfrak{B}$ in $\mathfrak{D}$, apply if T is AC$\mathfrak{B}$ in $\mathfrak{D}$.

IV.2.40. Suppose that T, given as in IV.2.1, is BV$\mathfrak{B}$ in $\mathfrak{D}$. Then T is AC$\mathfrak{B}$ in $\mathfrak{D}$ if and only if the associated set function $\mu(B)$ (see IV.2.21) is AC on Borel sets in $\mathfrak{D}$ (cf. I.3.16).

Proof. Suppose first that $\mu(B)$ is AC on Borel sets in $\mathfrak{D}$. If R^0 is any oriented open rectangle in $\mathfrak{D}$ then clearly

$$G(R) = |T(R^0\mathfrak{B})| \leqq \iint N(z, R^0\mathfrak{B}) = \mu(R^0).$$

Hence, if $R_1^0, \cdots, R_m^0$ is any disjoint system of oriented open rectangles in $\mathfrak{D}$, then

$$\sum_{i=1}^m G(R_i) \leqq \sum_{i=1}^m \mu(R_i^0) = \mu\Big(\sum_{i=1}^m R_i^0\Big).$$

By I.3.16 it follows that T is AC$\mathfrak{B}$ in $\mathfrak{D}$.

Suppose, conversely, that T is AC$\mathfrak{B}$ in $\mathfrak{D}$. Then T is also BV$\mathfrak{B}$ in $\mathfrak{D}$ (see IV.2.39). Take any open set O in $\mathfrak{D}$. Let us put, for $j = 1, 2, \cdots$ (cf. IV.2.2, IV.2.3),

$$\psi_j(z) = \sum g[z, T(s^0\mathfrak{B})], \qquad s \in D_{p_j}, s^0 \subset O.$$

By IV.2.5 we have then

$$(1) \qquad 0 \leqq \psi_j(z) \leqq N(z, O\mathfrak{B}), \psi_j(z) \to N(z, O\mathfrak{B}).$$

Since $N(z, O\mathfrak{B})$ is summable (cf. IV.2.13), termwise integration of the sequence $\psi_j(z)$ is permissible by (1) and I.3.11, and we obtain the relation

$$(2) \qquad \iint \psi_j(z) = \sum G(s) \underset{j\to\infty}{\longrightarrow} \iint N(z, O\mathfrak{B}) = \mu(O),$$

where the summation is extended over all squares s such that $s \in D_{p_j}$, $s^0 \subset O$. Now give $\epsilon > 0$. Since T is AC$\mathfrak{B}$ in $\mathfrak{D}$, we have an $\eta(\epsilon) > 0$ with the property explained in IV.2.39. Suppose now that the open set $O \subset \mathfrak{D}$ satisfies the inequality

$$(3) \qquad | O | \leqq \eta(\epsilon).$$

Then, for $j = 1, 2, \cdots$, clearly

$$\eta(\epsilon) \geqq | O | \geqq \sum | s |, \qquad s \in D_{p_j}, s^0 \subset O.$$

Hence, by the definition of $\eta(\epsilon)$,

$$(4) \qquad \epsilon \geqq \sum G(s), \qquad s \in D_{p_j}, s^0 \subset O.$$

(2), (4) yield the inequality

$$(5) \qquad \mu(O) \leqq \epsilon.$$

Thus (3) implies (5). Now let B be any Borel set in $\mathfrak{D}$ such that

$$(6) \qquad | B | \leqq \eta(\epsilon)/2.$$

By I.3.7 we have then an open set $O_\epsilon \subset \mathfrak{D}$ such that $B \subset O_\epsilon$ and $| O_\epsilon | < \eta(\epsilon)$. By the preceding argument, we have then $\mu(O_\epsilon) \leqq \epsilon$ and hence *a fortiori*

$$(7) \qquad \mu(B) \leqq \epsilon.$$

Thus (6) implies (7), and hence μ is AC on Borel sets in $\mathfrak{D}$ (cf. I.3.16).

IV.2.41. Given T as in IV.2.1, T is AC$\mathfrak{B}$ in $\mathfrak{D}$ if and only if the rectangle function $G^*(R)$ (see IV.2.31) is AC in $\mathfrak{D}$ (see III.1.52).

PROOF. Since $G(R) \leqq G^*(R) = \mu(R^0)$, the assertion follows immediately from IV.2.40.

IV.2.42. Suppose that T, given as in IV.2.1, is BV$\mathfrak{B}$ in $\mathfrak{D}$. Then T is AC$\mathfrak{B}$ in $\mathfrak{D}$ if and only if $| T(e\mathfrak{B}) | = 0$ for every set $e \subset \mathfrak{D}$ of measure zero.

PROOF. Suppose that $| T(e\mathfrak{B}) | = 0$ for every set $e \subset \mathfrak{D}$ of measure zero.

Let us choose then in $\mathfrak{D}$ any Borel set e of measure zero. Since $N(z, e\mathfrak{B}) = 0$ for $z \notin T(e\mathfrak{B})$, it follows that

$$\mu(e) = \iint N(z, e\mathfrak{B}) = 0.$$

Hence μ is AC by I.3.16, and hence, by IV.2.40, T is AC$\mathfrak{B}$ in $\mathfrak{D}$.

Suppose conversely that T is AC$\mathfrak{B}$ in $\mathfrak{D}$. Let e be any set of measure zero in $\mathfrak{D}$. Since by IV.2.40 the set function μ is AC on the Borel sets in $\mathfrak{D}$ we have for every $\epsilon > 0$ an $\eta(\epsilon) > 0$ such that $\mu(B) \leqq \epsilon$ for every Borel set $B \subset \mathfrak{D}$ such that $|B| \leqq \eta(\epsilon)$. Since $|e| = 0$, we have in $\mathfrak{D}$ an open set O_ϵ such that $e \subset O_\epsilon$, and $|O_\epsilon| < \eta(\epsilon)$. It follows that

$$\iint N(z, O_\epsilon\mathfrak{B}) = \mu(O_\epsilon) \leqq \epsilon.$$

Since $T(e\mathfrak{B}) \subset T(O_\epsilon\mathfrak{B})$ and $|T(O_\epsilon\mathfrak{B})| \leqq \iint N(z, O_\epsilon\mathfrak{B})$, it follows that the exterior measure of $T(e\mathfrak{B})$ does not exceed ϵ. Since ϵ was arbitrary, the assertion $|T(e\mathfrak{B})| = 0$ follows.

IV.2.43. Continuation. In the sufficiency part of the proof, we used actually the assumption that $|T(e\mathfrak{B})| = 0$ if $|e| = 0$ only for Borel sets $e \subset \mathfrak{D}$. Hence we have the following statement. If T is BV$\mathfrak{B}$ in $\mathfrak{D}$, and if $|T(e\mathfrak{B})| = 0$ for every Borel set $e \subset \mathfrak{D}$ of measure zero, then T is AC$\mathfrak{B}$ in $\mathfrak{D}$.

IV.2.44. Suppose that T, given as in IV.2.1, is AC$\mathfrak{B}$ in $\mathfrak{D}$. Then (cf. IV.2.32, IV.2.39, IV.2.21)

$$\mu(B) = \iint_B D(w) \tag{1}$$

for every Borel set $B \subset \mathfrak{D}$.

Proof. Since μ is AC by IV.2.40, we have (see I.3.17) a non-negative, Borel measurable, summable function $\delta(w)$ in $\mathfrak{D}$, such that

$$\mu(B) = \iint_B \delta(w) \tag{2}$$

for every Borel set B in $\mathfrak{D}$. In particular, we have for every oriented open rectangle $R^0 \subset \mathfrak{D}$

$$\mu(R^0) = \iint_{R^0} \delta(w) = \iint_R \delta(w),$$

and hence (see IV.2.31)

$$G^*(R) = \iint_R \delta(w).$$

By I.3.13 and IV.2.31 it follows that $D^*(w) = \delta(w)$ a.e. in $\mathfrak{D}$, and hence, by IV.2.34,

(3) $$D(w) = \delta(w) \qquad \text{a.e. in } \mathfrak{D}.$$

(2) and (3) imply (1).

IV.2.45. THEOREM. *Suppose that* T, *given as in* IV.2.1, *is* $BV\mathfrak{B}$ *in* $\mathfrak{D}$. *Then* (*cf.* IV.2.32, IV.2.13)

(1) $$\iint_{\mathfrak{D}} D(w) \leqq \iint N(z, \mathfrak{D}\mathfrak{B}),$$

and the sign of equality holds if and only if T *is* $AC\mathfrak{B}$ *in* $\mathfrak{D}$.

PROOF. (1) holds by IV.2.34, IV.2.33. Suppose now that T is AC$\mathfrak{B}$ in $\mathfrak{D}$. Then the sign of equality holds in (1) by IV.2.44. Suppose conversely that

(2) $$\iint_{\mathfrak{D}} D(w) = \iint N(z, \mathfrak{D}\mathfrak{B}).$$

By IV.2.33, IV.2.34, IV.2.21 we have, for every Borel set $B \subset \mathfrak{D}$, the inequalities

(3) $$\iint_{\mathfrak{D}-B} D(w) \leqq \iint N[z, (\mathfrak{D} - B)\mathfrak{B}],$$

(4) $$\iint_{B} D(w) \leqq \iint N(z, B\mathfrak{B}).$$

We assert that the sign of equality holds in both (3) and (4). Indeed, we would obtain otherwise, by addition, the inequality

$$\iint_{\mathfrak{D}} D(w) < \iint N(z, \mathfrak{D}\mathfrak{B}),$$

in contradiction to (2). Thus we have

(5) $$\iint_{B} D(w) = \iint N(z, B\mathfrak{B}) = \mu(B)$$

for every Borel set $B \subset \mathfrak{D}$. Now (5) clearly implies that $\mu(B) = 0$ if $|B| = 0$, and hence μ is AC on Borel sets B in $\mathfrak{D}$ (cf. I.3.16). By IV.2.40 it follows that T is AC$\mathfrak{B}$ in $\mathfrak{D}$.

IV.2.46. Assume that T, given as in IV.2.1, is AC$\mathfrak{B}$ in $\mathfrak{D}$. Then $\mu(B)$ is AC on Borel sets in $\mathfrak{D}$ by IV.2.40. Hence there exists a non-negative, Borel measurable, summable function $\delta(w)$ in $\mathfrak{D}$, such that

(1) $$\iint_{\mathfrak{D}} \Phi(w)\, d\mu = \iint_{\mathfrak{D}} \Phi(w)\delta(w)$$

for every Borel measurable function $\Phi(w)$ as soon as one of the two integrals involved exists (see I.3.17). Choosing $\Phi(w)$ as the characteristic function of a Borel set $B \subset \mathfrak{D}$, we have by (1)

$$\mu(B) = \iint_B \delta(w). \tag{2}$$

By the argument used in IV.2.44, we infer from (2) that $\delta(w) = D(w)$ a.e. in $\mathfrak{D}$. Thus (1) can be rewritten in the form

$$\iint_{\mathfrak{D}} \Phi(w)\, d\mu = \iint_{\mathfrak{D}} \Phi(w) D(w), \tag{3}$$

and this formula holds for every Borel measurable function $\Phi(w)$, as soon as one of the two integrals involved exists. In view of (3), the (μ)-integrals appearing in the transformation formulas of IV.2.27, IV.2.29 can be replaced by Lebesgue integrals if T is AC$\mathfrak{B}$ in $\mathfrak{D}$. The formulas obtained in this manner can however be further generalized. We proceed to discuss this point.

IV.2.47. Assume that T, given as in IV.2.1, is AC$\mathfrak{B}$ in $\mathfrak{D}$. Let $\Phi(w)$ be a finite-valued measurable function in $\mathfrak{D}$. We assert that the function $\sigma(z, \Phi)$ of IV.2.15 is measurable in the z-plane.

Proof. Since $\Phi(w)$ is finite-valued and measurable in $\mathfrak{D}$, we have (see I.3.8) a finite-valued, Borel measurable function $\Phi_1(w)$ in $\mathfrak{D}$, such that

$$\Phi(w) = \Phi_1(w) \qquad \text{a.e. in } \mathfrak{D}. \tag{1}$$

Let E be the set of those points $w \in \mathfrak{D}$ where (1) fails to hold. Then $|E| = 0$ and hence $|T(E\mathfrak{B})| = 0$ by IV.2.42. Clearly $\sigma(z, \Phi_1) = \sigma(z, \Phi)$ for $z \notin T(E\mathfrak{B})$. Since $\sigma(z, \Phi_1)$ is measurable by IV.2.17, the measurability of $\sigma(z, \Phi)$ follows.

IV.2.48. Assume that T, given as in IV.2.1, is AC$\mathfrak{B}$ in $\mathfrak{D}$. If E is any measurable set in $\mathfrak{D}$, then $N(z, E\mathfrak{B})$ is measurable.

Proof. Let $\Phi(w)$ be the characteristic function of E. Then $\Phi(w)$ is defined, finite-valued and measurable in $\mathfrak{D}$, and hence $\sigma(z, \Phi)$ is measurable by IV.2.47. Since $N(z, E\mathfrak{B}) = \sigma(z, \Phi)$ a.e. by IV.2.16(vi), IV.2.39, the measurability of $N(z, E\mathfrak{B})$ follows.

IV.2.49. Assume that T, given as in IV.2.1, is AC$\mathfrak{B}$ in $\mathfrak{D}$. Let $\Phi_1(w)$, $\Phi_2(w)$ be two finite-valued functions in $\mathfrak{D}$ such that $\Phi_1 = \Phi_2$ a.e. in $\mathfrak{D}$. Then $\sigma(z, \Phi_1) = \sigma(z, \Phi_2)$ a.e. in the z-plane.

Proof. Let E be the subset of $\mathfrak{D}$ where $\Phi_1 \neq \Phi_2$. Then $|E| = 0$, and hence $|T(E\mathfrak{B})| = 0$ by IV.2.42. Since clearly $\sigma(z, \Phi_1) = \sigma(z, \Phi_2)$ for $z \notin T(E\mathfrak{B})$, the assertion follows.

IV.2.50. Theorem. *Assume that T, given as in* IV.2.1, *is AC$\mathfrak{B}$ in $\mathfrak{D}$. If $\Phi(w)$ is a finite-valued, measurable function in $\mathfrak{D}$, then*

$$\iint_{\mathfrak{D}} \Phi(w) D(w) = \iint \sigma(z, \Phi), \tag{1}$$

as soon as the integral on the left exists. If $\Phi \geqq 0$, *then the formula holds as soon as one of the two integrals involved exists.*

PROOF. *Case* (i). $\Phi(w)$ is finite-valued and Borel measurable in $\mathfrak{D}$. Then the theorem is an immediate consequence of IV.2.27, IV.2.46.

Case (ii). $\Phi(w)$ is finite-valued, measurable, and $\Phi(w)D(w)$ is summable. By I.3.8 we have a finite-valued, Borel-measurable function $\Phi_1(w)$ in $\mathfrak{D}$, such that $\Phi = \Phi_1$ a.e. in $\mathfrak{D}$. Since $\Phi(w)D(w)$ is summable in $\mathfrak{D}$, it follows that $\Phi_1(w)D(w)$ is also summable in $\mathfrak{D}$. Hence, by case (i), $\sigma(z, \Phi_1)$ is summable, and

$$(2) \qquad \iint_{\mathfrak{D}} \Phi_1(w)D(w) = \iint \sigma(z, \Phi_1).$$

But $\sigma(z, \Phi_1) = \sigma(z, \Phi)$ a.e. in the z-plane by IV.2.49, and $\Phi_1(w)D(w) = \Phi(w)D(w)$ a.e. in $\mathfrak{D}$. Hence (2) implies (1).

Case (iii). $\Phi(w)$ is finite-valued, measurable, non-negative, and $\sigma(z, \Phi)$ is summable. Let again $\Phi_1(w)$ be a finite-valued, Borel-measurable function in $\mathfrak{D}$, such that $\Phi_1 = \Phi$ a.e. in $\mathfrak{D}$. Since $\Phi \geqq 0$, we can assume that $\Phi_1 \geqq 0$ also. Then $\sigma(z, \Phi_1) = \sigma(z, \Phi)$ a.e. by IV.2.49, and hence $\sigma(z, \Phi_1)$ is also summable. By case (i) it follows that $\Phi_1(w)D(w)$ is summable in $\mathfrak{D}$ and (2) holds. Since $\sigma(z, \Phi) = \sigma(z, \Phi_1)$ a.e. in the z-plane and $\Phi(w)D(w) = \Phi_1(w)D(w)$ a.e. in $\mathfrak{D}$, the formula (1) follows.

IV.2.51. Assume that T, given as in IV.2.1, is BV$\mathfrak{B}$ in $\mathfrak{D}$. If $H(z)$, $H^*(z)$ are finite-valued functions in the z-plane such that $H(z) = H^*(z)$ a.e., then

$$(1) \qquad H[t(w)]D(w) = H^*[t(w)]D(w) \qquad \text{a.e. in } \mathfrak{D}.$$

PROOF. Let $\bar{E}$ denote the set in the z-plane where $H(z) \neq H^*(z)$. Since $|\bar{E}| = 0$ by assumption, we have by IV.2.36

$$(2) \qquad D(w) = 0 \qquad \text{a.e. on } T^{-1}(\bar{E}),$$

and clearly

$$(3) \qquad H[t(w)] = H^*[t(w)] \qquad \text{for } w \notin T^{-1}(\bar{E}).$$

Finally, by IV.2.32, $D(w)$ exists (and hence is finite) a.e. in $\mathfrak{D}$. Hence formula (1) holds a.e. on $\mathfrak{D} - T^{-1}(\bar{E})$ in view of (3), and it holds a.e. on $T^{-1}(\bar{E})$ in view of (2).

IV.2.52. Assume that T, given as in IV.2.1, is BV$\mathfrak{B}$ in $\mathfrak{D}$. If $H(z)$ is finite-valued and measurable in the z-plane, then $H[t(w)]D(w)$ is measurable in $\mathfrak{D}$.

PROOF. By I.3.8 we have a finite-valued, Borel-measurable function $H^*(z)$ such that $H(z) = H^*(z)$ a.e. in the z-plane. By IV.2.51, we have then $H[t(w)]D(w) = H^*[t(w)]D(w)$ a.e. in $\mathfrak{D}$. Since $H^*[t(w)]D(w)$ is measurable by IV.2.20, the measurability of $H[t(w)]D(w)$ follows.

IV.2.53. THEOREM. *Assume that* T, *given as in* IV.2.1, *is* AC$\mathfrak{B}$ *in* $\mathfrak{D}$. *If* $H(z)$ *is a finite-valued, measurable function in the z-plane, then*

(1) $$\iint_{\mathfrak{D}} H[t(w)]D(w) = \iint H(z)N(z, \mathfrak{D}\mathfrak{B}),$$

as soon as one of the two integrals involved exists.

PROOF. By I.3.8, we have a finite-valued, Borel measurable function $H^*(z)$ such that $H^*(z) = H(z)$ a.e. in the z-plane. Since $N(z, \mathfrak{D}\mathfrak{B})$ is finite a.e. in the z-plane (see IV.2.39, IV.2.14), it follows that

(2) $$H^*(z)N(z, \mathfrak{D}\mathfrak{B}) = H(z)N(z, \mathfrak{D}\mathfrak{B})$$

a.e. in the z-plane. By IV.2.51, IV.2.39 it follows further that

(3) $$H^*[t(w)]D(w) = H[t(w)]D(w) \qquad \text{a.e. in } \mathfrak{D}.$$

Now suppose that one of the two integrals involved in (1) exists. By (2) and (3) there follows the existence of one of the integrals involved in the formula

(4) $$\iint_{\mathfrak{D}} H^*[t(w)]D(w) = \iint H^*(z)N(z, \mathfrak{D}\mathfrak{B}),$$

and hence (4) holds by IV.2.29, IV.2.39, IV.2.46. In view of (2), (3), (4), formula (1) follows.

IV.2.54. Assume that T, given as in IV.2.1, is BV$\mathfrak{B}$ in $\mathfrak{D}$. The corresponding set function $\mu(B)$ (see IV.2.21) gives rise to a Lebesgue decomposition (see I.3.16) which we shall study presently in more detail than absolutely necessary, in order to clarify its geometrical implications in this particular situation. Let K denote the class of all Borel sets of measure zero in $\mathfrak{D}$. By the general theory (see I.3.16), there exists a set $B_0 \in K$ such that

(1) $$\mu(B) \leqq \mu(B_0) \qquad \text{for } B \in K.$$

Let us put

(2) $$\mathfrak{B}_1 = B_0\mathfrak{B}.$$

Then $\mathfrak{B}_1$ is a Borel set of measure zero, and (cf. IV.2.21)

(3) $$\mu(\mathfrak{B}_1) = \iint N(z, \mathfrak{B}_1\mathfrak{B}) = \iint N(z, B_0\mathfrak{B}) = \mu(B_0).$$

(1) and (3) show that

(4) $$\mu(B) \leqq \mu(\mathfrak{B}_1) \qquad \text{for } B \in K.$$

Let now B be any Borel set such that $B \in K$, $B \subset \mathfrak{D} - \mathfrak{B}_1$. Then $B + \mathfrak{B}_1 \in K$, and by (4) it follows that

$$\mu(B) + \mu(\mathfrak{B}_1) = \mu(B + \mathfrak{B}_1) \leqq \mu(\mathfrak{B}_1).$$

Hence $\mu(B) = 0$. Thus we see that

(5) $$\mu(B) = 0 \qquad \text{if } B \in K,\ B \subset \mathfrak{D} - \mathfrak{B}_1.$$

Let us put now

$$(6) \qquad \mathfrak{B}_2 = \mathfrak{B} - \mathfrak{B}_1 .$$

Then $\mathfrak{B}_2$ is a Borel subset of $\mathfrak{B}$, and we have the relations

$$\mathfrak{B} = \mathfrak{B}_1 + \mathfrak{B}_2 , \; \mathfrak{B}_1\mathfrak{B}_2 = 0, \; | \mathfrak{B}_1 | = 0.$$

We assert that T is BV$\mathfrak{B}_1$ and AC$\mathfrak{B}_2$. Indeed, by IV.2.38, T is both BV$\mathfrak{B}_1$ and BV$\mathfrak{B}_2$. In view of IV.2.43 the proof will be complete if we can show that $| T(e\mathfrak{B}_2) | = 0$ for every Borel set $e \subset \mathfrak{D}$ of measure zero. Now let e be such a set. By (5) and (6) we have then

$$0 = \mu(e\mathfrak{B}_2) = \iint N(z, e\mathfrak{B}_2\mathfrak{B}) = \iint N(z, e\mathfrak{B}_2) \geqq | T(e\mathfrak{B}_2) |.$$

Hence $| T(e\mathfrak{B}_2) | = 0$.

IV.2.55. The sets $\mathfrak{B}_1$, $\mathfrak{B}_2$ of the preceding section are not determined univocally by their properties listed there. For convenience of reference, we state the result obtained in the following form.

THEOREM. *Suppose that T, given as in* IV.2.1, *is BV$\mathfrak{B}$ in $\mathfrak{D}$. Then there exists a decomposition of the base set $\mathfrak{B}$ into two Borel sets $\mathfrak{B}^a$, $\mathfrak{B}^s$ such that the following holds:* (i) $\mathfrak{B} = \mathfrak{B}^a + \mathfrak{B}^s$, $\mathfrak{B}^a\mathfrak{B}^s = 0$. (ii) $| \mathfrak{B}^s | = 0$. (iii) *T is BV$\mathfrak{B}^s$ and AC$\mathfrak{B}^a$ in $\mathfrak{D}$.*

Indeed, it follows from IV.2.54 that the sets $\mathfrak{B}^s = \mathfrak{B}_1$, $\mathfrak{B}^a = \mathfrak{B}_2$ satisfy our requirements. A little later we shall discuss another way of obtaining a decomposition with the properties (i), (ii), (iii).

IV.2.56. CONTINUATION. In terms of any decomposition $\mathfrak{B} = \mathfrak{B}^a + \mathfrak{B}^s$, with the properties described in IV.2.55, we can construct the Lebesgue decomposition of the set function $\mu(B)$ as follows (cf. I.3.16, IV.2.21). Let us use the explicit notations $G^*(R, \mathfrak{B})$, $\mu(B, \mathfrak{B})$, $D(w, \mathfrak{B})$ to refer to $G^*(R)$, $\mu(B)$, $D(w)$ (cf. IV.2.31, IV.2.21, IV.2.32). We shall consider simultaneously the base sets $\mathfrak{B}^a$, $\mathfrak{B}^s$ also, using corresponding notations $G^*(R, \mathfrak{B}^a)$, $\mu(B, \mathfrak{B}^a)$, $D(w, \mathfrak{B}^a)$, $G^*(R, \mathfrak{B}^s)$, $\mu(B, \mathfrak{B}^s)$, $D(w, \mathfrak{B}^s)$. We have then the relations

$$(1) \qquad D(w, \mathfrak{B}) = D(w, \mathfrak{B}^a) \qquad \text{a.e. in } \mathfrak{D},$$

$$(2) \qquad D(w, \mathfrak{B}^s) = 0 \qquad \text{a.e. in } \mathfrak{D}.$$

Indeed, $D(w, \mathfrak{B}^s) = 0$ a.e. on $\mathfrak{D} - \mathfrak{B}^s$ by IV.2.35, and $| \mathfrak{B}^s | = 0$. Thus (2) follows. On the other hand, $D(w, \mathfrak{B}) = D(w, \mathfrak{B}^a)$ a.e. on $\mathfrak{D} - \mathfrak{B}^s$ by IV.2.38, and (1) follows since $| \mathfrak{B}^s | = 0$. Furthermore (cf. IV.2.55, IV.2.21)

$$(3) \qquad \begin{aligned} \mu(B, \mathfrak{B}) &= \iint N(z, B\mathfrak{B}) = \iint N(z, B\mathfrak{B}^a) + \iint N(z, B\mathfrak{B}^s) \\ &= \mu(B, \mathfrak{B}^a) + \mu(B, \mathfrak{B}^s). \end{aligned}$$

Since T is AC$\mathfrak{B}^a$, the set function $\mu(B, \mathfrak{B}^a)$ is absolutely continuous on Borel sets in $\mathfrak{D}$, by IV.2.40. On the other hand, $\mu(B, \mathfrak{B}^s)$ is singular. Indeed, if B is any Borel set in $\mathfrak{D} - \mathfrak{B}^s$, then $N(z, B\mathfrak{B}^s) \equiv 0$, and hence

$$\mu(B, \mathfrak{B}^s) = \iint N(z, B\mathfrak{B}^s) = 0.$$

Since $|\mathfrak{B}^s| = 0$, it follows that $\mu(B, \mathfrak{B}^s)$ is singular (see I.3.16). Thus (3) yields the Lebesgue decomposition of $\mu(B, \mathfrak{B})$. As we know, this decomposition of $\mu(B, \mathfrak{B})$ is univocally determined (see I.3.16).

IV.2.57. THEOREM. *Assume that T, given as in* IV.2.1, *is* $BV\mathfrak{B}$ *in* $\mathfrak{D}$. *Then the corresponding set function* $\mu(B)$ (*see* IV.2.21) *admits of a univocally determined Lebesgue decomposition*

$$\mu(B) = \mu_a(B) + \mu_s(B)$$

for Borel sets $B \subset \mathfrak{D}$, *where* μ_a *is absolutely continuous, and* μ_s *is singular. Let* $\Phi(w)$ *be any finite-valued, Borel measurable function in* $\mathfrak{D}$, *and let* $H(z)$ *be any finite-valued Borel measurable function defined in the whole z-plane. Then* (*cf.* IV.2.32, IV.2.15)

$$\iint_{\mathfrak{D}} \Phi(w)D(w, \mathfrak{B}) + \iint_{\mathfrak{D}} \Phi(w)\, d\mu_s = \iint \sigma(z, \Phi, \mathfrak{B}), \tag{1}$$

as soon as the integrals on the left exist. If $\Phi \geqq 0$, *then the formula holds as soon as one of the two sides of* (1) *exists. Furthermore*

$$\iint_{\mathfrak{D}} H[t(w)]D(w, \mathfrak{B}) + \iint_{\mathfrak{D}} H[t(w)]\, d\mu_s = \iint H(z)N(z, \mathfrak{D}\mathfrak{B}), \tag{2}$$

as soon as one of the two sides of (2) *exists.*

PROOF. Let $\mathfrak{B}^a + \mathfrak{B}^s = \mathfrak{B}$ be a decomposition as in IV.2.55. By IV.2.56 we have then the formulas $\mu_a(B) = \mu(B, \mathfrak{B}^a)$, $\mu_s(B) = \mu(B, \mathfrak{B}^s)$. Assume that the integrals

$$\iint_{\mathfrak{D}} \Phi(w)D(w, \mathfrak{B}), \qquad \iint_{\mathfrak{D}} \Phi(w)\, d\mu_s$$

exist. By IV.2.56(1) we have then

$$\iint_{\mathfrak{D}} \Phi(w)D(w, \mathfrak{B}) = \iint_{\mathfrak{D}} \Phi(w)D(w, \mathfrak{B}^a). \tag{3}$$

Since T is AC$\mathfrak{B}^a$ in $\mathfrak{D}$, we have by IV.2.46

$$\iint_{\mathfrak{D}} \Phi(w)D(w, \mathfrak{B}^a) = \iint_{\mathfrak{D}} \Phi(w)\, d\mu_a\,. \tag{4}$$

Hence (see IV.2.27, IV.2.56(3))

$$\iint_{\mathfrak{D}} \Phi(w) D(w, \mathfrak{B}) + \iint_{\mathfrak{D}} \Phi(w)\, d\mu_s = \iint_{\mathfrak{D}} \Phi(w)\, d\mu_a + \iint_{\mathfrak{D}} \Phi(w)\, d\mu_s$$

$$= \iint_{\mathfrak{D}} \Phi(w)\, d\mu = \iint \sigma(z, \Phi, \mathfrak{B}).$$

Thus (1) is established in the general case.

Let us assume next that $\Phi \geqq 0$ and $\sigma(z, \Phi, \mathfrak{B})$ is summable. By IV.2.27, we have then (cf. IV.2.38)

$$\iint \sigma(z, \Phi, \mathfrak{B}) = \iint_{\mathfrak{D}} \Phi(w)\, d\mu = \iint_{\mathfrak{D}} \Phi(w)\, d\mu_a + \iint_{\mathfrak{D}} \Phi(w)\, d\mu_s\, .$$

Since T is AC$\mathfrak{B}^a$ in $\mathfrak{D}$ the formulas (3), (4) are available and (1) follows.

Taking up (2), assume first that the integrals

$$\iint_{\mathfrak{D}} H[t(w)] D(w, \mathfrak{B}), \qquad \iint_{\mathfrak{D}} H[t(w)]\, d\mu_s$$

exist. By IV.2.56(1) we have then

$$\iint_{\mathfrak{D}} H[t(w)] D(w, \mathfrak{B}) = \iint_{\mathfrak{D}} H[t(w)] D(w, \mathfrak{B}^a). \tag{5}$$

Since T is AC$\mathfrak{B}^a$ in $\mathfrak{D}$, we have by IV.2.46

$$\iint_{\mathfrak{D}} H[t(w)] D(w, \mathfrak{B}^a) = \iint_{\mathfrak{D}} H[t(w)]\, d\mu_a\, . \tag{6}$$

By IV.2.56(3), IV.2.29 we obtain, in view of (5), (6),

$$\iint_{\mathfrak{D}} H[t(w)] D(w, \mathfrak{B}) + \iint_{\mathfrak{D}} H[t(w)]\, d\mu_s = \iint_{\mathfrak{D}} H[t(w)]\, d\mu_a + \iint_{\mathfrak{D}} H[t(w)]\, d\mu_s$$

$$= \iint_{\mathfrak{D}} H[t(w)]\, d\mu = \iint H(z) N(z, \mathfrak{D}\mathfrak{B}).$$

Assume finally that $H(z)N(z, \mathfrak{D}\mathfrak{B})$ is summable. By IV.2.29 we have then

$$\iint H(z) N(z, \mathfrak{D}\mathfrak{B}) = \iint_{\mathfrak{D}} H[t(w)]\, d\mu = \iint_{\mathfrak{D}} H[t(w)]\, d\mu_a + \iint_{\mathfrak{D}} H[t(w)]\, d\mu_s\, .$$

Since T is AC$\mathfrak{B}^a$ in $\mathfrak{D}$, the formulas (5) and (6) are available, and (2) follows.

IV.2.58. Given T as in IV.2.1, let us denote by $\overline{D}(w, \mathfrak{B})$ the upper derivative of the rectangle function $\underline{G}(R)$ defined in IV.2.10. Let w_0 be any point in $\mathfrak{D}$. By definition (cf. III.1.24), $\overline{D}(w_0, \mathfrak{B})$ is then equal to the least upper bound of

$$\limsup_{n\to\infty} \frac{| T(s_n^0\mathfrak{B}) |}{| s_n |}$$

for all possible choices of the sequence of oriented squares s_n such that $w_0 \in s_n^0$, $s_n \subset \mathfrak{D}$, $| s_n | \to 0$. Let us note that $\overline{D}(w_0, \mathfrak{B})$ may be infinite. Now let σ_n be any sequence of oriented squares such that $w_0 \in \sigma_n \subset \mathfrak{D}$, $| \sigma_n | \to 0$. We assert that

$$\limsup_{n\to\infty} \frac{| T(\sigma_n\mathfrak{B}) |}{| \sigma_n |} \leqq \overline{D}(w_0, \mathfrak{B}). \tag{1}$$

PROOF. The assertion would be obvious if we had assumed that $w_0 \in \sigma_n^0$. Since we only assumed that $w_0 \in \sigma_n$, some explanation is needed. Since $\sigma_n \subset \mathfrak{D}$, we have for each n an oriented square s_n with the following properties:

(i) $\sigma_n \subset s_n^0$, $s_n \subset \mathfrak{D}$.

(ii) $| s_n | < (1 + 1/n) | \sigma_n |$.

Then $w_0 \in s_n^0$, $s_n \subset \mathfrak{D}$, $| s_n | \to 0$, and hence, by the definition of $\overline{D}(w_0, \mathfrak{B})$,

$$\limsup_{n\to\infty} \frac{| T(s_n^0\mathfrak{B}) |}{| s_n |} \leqq \overline{D}(w_0, \mathfrak{B}).$$

Clearly, (i) and (ii) imply that $| T(\sigma_n\mathfrak{B}) | \leqq | T(s_n^0\mathfrak{B}) |$, $| s_n |/| \sigma_n | \to 1$, and (1) follows.

IV.2.59. Given T as in IV.2.1, choose any base set $\mathfrak{B}$. Let e be a subset of $\mathfrak{B}\mathfrak{D}$ such that

$$| e | = 0, \overline{D}(w, \mathfrak{B}) < M \qquad \text{for } w \in e, \tag{1}$$

where M is a finite constant. Then $| T(e) | = 0$.

PROOF. Let us introduce a network of oriented squares in the w-plane (see I.3.4). Let $\epsilon > 0$ be given. Since $| e | = 0$, we can choose an open set O such that $e \subset O \subset \mathfrak{D}$, $| O | < \epsilon$. Let us denote by $\mathfrak{F}$ the family of those network squares σ for which the following conditions hold:

(i) $\sigma \subset O$.

(ii) $| T(\sigma\mathfrak{B}) |/| \sigma | < M$.

Let w_0 be any point in e. Then $w_0 \in O$, and hence we have a sequence of network squares σ_n such that $w_0 \in \sigma_n \subset O$, $| \sigma_n | \to 0$ (note that we could not generally assert the inclusion $w_0 \in \sigma_n^0$). (1) implies, by IV.2.58, that

$$\limsup_{n\to\infty} \frac{| T(\sigma_n\mathfrak{B}) |}{| \sigma_n |} < M.$$

Hence we shall have, for n sufficiently large,

$$\frac{| T(\sigma_n\mathfrak{B}) |}{| \sigma_n |} < M, \qquad \sigma_n \subset O.$$

Thus σ_n belongs to the family $\mathfrak{F}$ for n sufficiently large. In other words, the squares of the family $\mathfrak{F}$ cover e. Hence (see I.3.4), we have a (finite or countably infinite) sequence of oriented squares s, such that $s_j \in \mathfrak{F}$, $e \subset \sum s_j$, $s_j^0 s_k^0 = 0$ for $j \neq k$. Since $e \subset \mathfrak{B}$, it follows that $e = e\mathfrak{B} \subset \sum s_j\mathfrak{B}$, and hence

$$(2) \qquad T(e) = T(e\mathfrak{B}) \subset \sum T(s_j\mathfrak{B}).$$

On the other hand,

$$(3) \qquad \sum | T(s_j\mathfrak{B}) | < \sum M | s_j | \leqq M | O | < M\epsilon.$$

Since ϵ was arbitrary, (2) and (3) imply that $| T(e) | = 0$.

IV.2.60. Given T as in IV.2.1, choose any base set $\mathfrak{B}$. Let E be a subset of $\mathfrak{B}\mathfrak{D}$ such that

$$(1) \qquad | E | = 0, \overline{D}(w, \mathfrak{B}) < +\infty \qquad \text{for } w \in E.$$

Then $| T(E) | = 0$.

PROOF. For each positive integer n, let e_n be the subset of E on which $\overline{D}(w, \mathfrak{B}) < n$. In view of (1) we have then $E = \sum e_n$ and hence $T(E) \subset \sum T(e_n)$. Since $| T(e_n) | = 0$, $n = 1, 2, \cdots$, by IV.2.59, it follows that $| T(E) | = 0$.

IV.2.61. Suppose that T, given as in IV.2.1, is BV$\mathfrak{B}$ in $\mathfrak{D}$. Let $\mathfrak{B}_\infty$ be the subset of $\mathfrak{B}$ on which $\overline{D}(w, \mathfrak{B}) = +\infty$. We assert that the sets

$$\mathfrak{B}^a = \mathfrak{B} - \mathfrak{B}_\infty, \ \mathfrak{B}^s = \mathfrak{B}_\infty$$

satisfy the conditions stated in the theorem of IV.2.55.

PROOF. In the first place, $\mathfrak{B}_\infty$ is a Borel set (see III.1.24), and hence $\mathfrak{B}^a$, $\mathfrak{B}^s$ are both Borel sets. Since $| \mathfrak{B}_\infty | = 0$ by IV.2.32, and since T is surely BV$\mathfrak{B}^a$ and BV$\mathfrak{B}^s$ by IV.2.38, we have to show only that T is AC$\mathfrak{B}^a$. For this purpose it is sufficient to verify (cf. IV.2.42) that the conditions

$$(1) \qquad e \subset \mathfrak{D}, \ | e | = 0$$

imply that

$$(2) \qquad | T[e \cdot (\mathfrak{B} - \mathfrak{B}_\infty)] | = 0.$$

Now if we put $E = e \cdot (\mathfrak{B} - \mathfrak{B}_\infty)$, then E is a subset of measure zero of $\mathfrak{D}\mathfrak{B}$ (cf. (1)), and $\overline{\mathfrak{D}}(w, \mathfrak{B}) < +\infty$ on E by the definition of $\mathfrak{B}_\infty$. Hence, by IV.2.60, $| T(E) | = 0$, and (2) is established.

IV.2.62. THEOREM. *Suppose that T, given as in* IV.2.1, *is BV$\mathfrak{B}$ in $\mathfrak{D}$. Let $\mathfrak{B}_\infty$ be the subset of $\mathfrak{B}$ on which $\overline{D}(w, \mathfrak{B}) = +\infty$. Then the Lebesgue decomposition $\mu(B) = \mu_a(B) + \mu_s(B)$ of the corresponding set function* (*cf.* IV.2.21)

$$\mu(B) = \iint N(z, B\mathfrak{B})$$

is given by the formulas

$$\mu_a(B) = \iint N[z, B \cdot (\mathfrak{B} - \mathfrak{B}_\infty)],$$

$$\mu_s(B) = \iint N[z, B\mathfrak{B}_\infty],$$

where μ_a is absolutely continuous and μ_s is singular on Borel sets in $\mathfrak{D}$.

PROOF. By IV.2.61, the sets $\mathfrak{B}^a = \mathfrak{B} - \mathfrak{B}_\infty$, $\mathfrak{B}^s = \mathfrak{B}_\infty$ satisfy the conditions stated in IV.2.55, and hence the assertion follows directly from IV.2.56.

REMARK. Let us recall that the Lebesgue decomposition of μ is univocally determined (see I.3.16) and the preceding theorem merely indicates a particular way of determining this decomposition.

IV.2.63. THEOREM. *Suppose that T, given as in* IV.2.1, *is $BV\mathfrak{B}$ in $\mathfrak{D}$. Let $\mathfrak{B}_\infty$ be the subset of $\mathfrak{B}$ on which $\overline{D}(w, \mathfrak{B}) = +\infty$. Then T is $AC\mathfrak{B}$ in $\mathfrak{D}$ if and only if* $| T(\mathfrak{B}_\infty) | = 0$.

PROOF. Since $| \mathfrak{B}_\infty | = 0$ by IV.2.32, the necessity of the condition follows from IV.2.42. To prove the sufficiency, assume that $| T(\mathfrak{B}_\infty) | = 0$. If B is any Borel set in $\mathfrak{D}$, then *a fortiori* $| T(B\mathfrak{B}_\infty) | = 0$, and hence $N(z, B\mathfrak{B}_\infty) = 0$ a.e. in the z-plane, since $N(z, B\mathfrak{B}_\infty) = 0$ for $z \notin T(B\mathfrak{B}_\infty)$. By IV.2.62 it follows that the singular set function $\mu_s(B)$, in the Lebesgue decomposition of $\mu(B)$, vanishes identically, and hence $\mu(B)$ is absolutely continuous on Borel sets in $\mathfrak{D}$. By IV.2.40 it follows that T is AC$\mathfrak{B}$ in $\mathfrak{D}$.

IV.2.64. In presenting the preceding theory, we operated with *oriented rectangles*, that is rectangles with sides parallel to the axes u, v in the w-plane. There arises the question whether the concepts and the results of the theory are independent of the choice of the coordinate system. The affirmative answer, as far as the fundamental concepts and the principal results are concerned, is suggested by a glance at the main transformation formulas obtained. For example the formula (see IV.2.29)

$$\iint_{\mathfrak{D}} H[t(w)]\, d\mu = \iint H(z)N(z, \mathfrak{D}\mathfrak{B})$$

contains only quantities that are clearly invariant under changes of the Cartesian coordinate systems in both the uv- and the xy-planes. To obtain precise statements concerning the invariant character of the theory, let us make a preliminary remark. Suppose that T, given as in IV.2.1, is BV$\mathfrak{B}$ in $\mathfrak{D}$. Then we have at our disposal the corresponding non-negative, completely additive function $\mu(B)$ of Borel sets $B \subset \mathfrak{D}$. Let w be a point in $\mathfrak{D}$ and let s_n be a sequence of squares, not necessarily oriented, such that $w \in \overset{0}{s}_n$, $s_n \subset \mathfrak{D}$, $| s_n | \to 0$. By III.1.29, we have in $\mathfrak{D}$ a set e of measure zero, such that for $w \notin e$ the limit of the ratio $\mu(\overset{0}{s}_n)/| s_n |$ exists and is independent of the choice of the sequence s_n. Let us denote this limit by $d(w)$. In particular, we can use oriented squares s_n in computing $d(w)$, and by IV.2.31, IV.2.32, IV.2.34 it follows that $d(w) = D(w) = D^*(w)$ a.e. in $\mathfrak{D}$.

Let us observe now that the summability of the function $N(z, \mathfrak{D}\mathfrak{B})$ is clearly a feature of the transformation T itself and is independent of the choice of the

Cartesian coordinate systems in the uv- and xy-planes respectively. By IV.2.13, T is BV$\mathfrak{B}$ in $\mathfrak{D}$ if and only if $N(z, \mathfrak{D}\mathfrak{B})$ is summable. Hence the property of being BV$\mathfrak{B}$ in $\mathfrak{D}$ is also independent of the choice of the Cartesian coordinate system. Now by IV.2.40, T is AC$\mathfrak{B}$ in $\mathfrak{D}$ if and only if it is BV$\mathfrak{B}$ in $\mathfrak{D}$ and if the corresponding set function $\mu(B)$ is absolutely continuous on Borel sets in $\mathfrak{D}$. This criterion is also independent, in view of the previous remarks, of the choice of the Cartesian coordinate system. Finally, from our preliminary remark concerning the derivatives d, D, D^* it follows that D and D^* are also invariant if sets of measure zero are disregarded, which is of course all we need in studying formulas involving double integrals. The invariant character of our theory is thus established, as far as the main concepts and results are concerned.

CHAPTER IV.3. DERIVATIVES AND JACOBIANS

IV.3.1. We shall combine presently the metrical and topological information obtained in IV.1 and IV.2. In particular, we shall investigate the relationships between derivatives corresponding to various choices of a base set $\mathfrak{B}$ (see IV.2.7). The quantities to be studied will depend upon arguments like w, T, $\mathfrak{D}$, $\mathfrak{B}$, z, and notations threaten to become excessively involved at times. For example, to be quite explicit, we should have to use a symbol like $D[w, T, \mathcal{E}^*(T, \mathfrak{D})]$ (cf. IV.2.32, IV.1.56). We shall write instead $D(w, T, \mathcal{E}^*)$ or even more concisely $D(w, \mathcal{E}^*)$ if no misunderstandings are likely to arise.

IV.3.2. Given T as in IV.2.1, we introduce in the z-plane a multiplicity function $\nu(z, T, \mathfrak{D}_*)$, where $\mathfrak{D}_*$ is any subdomain of $\mathfrak{D}$, as follows:

(i) If $N(z_0, \mathfrak{D}_*\mathcal{N}) = +\infty$, then we put $\nu(z_0, T, \mathfrak{D}_*) = 0$ (cf. IV.2.1, IV.1.56).

(ii) If $N(z_0, \mathfrak{D}_*\mathcal{N}) < +\infty$, then the set $T^{-1}(z_0)\mathfrak{D}_*\mathcal{N}$ is finite, and we put (cf. IV.1.64) $\nu(z_0, T, \mathfrak{D}_*) = \sum i_e(w)$, $w \in T^{-1}(z_0)\cdot\mathfrak{D}_*\mathcal{N}$.

If the summation is vacuous (that is, if the set $T^{-1}(z_0)\cdot\mathfrak{D}_*\mathcal{N}$ is empty), then it is understood that $\nu(z_0, T, \mathfrak{D}_*) = 0$. Clearly, $\nu(z, T, \mathfrak{D}_*)$ coincides with the function $\sigma(z, \Phi)$ defined in IV.2.15 with $\mathfrak{D}$, $\mathfrak{B}$, Φ replaced by $\mathfrak{D}_*$, $\mathcal{N}$, i_e respectively. Since $i_e(w)$ is Borel measurable by IV.1.66, it follows (see IV.2.17) that $\nu(z, T, \mathfrak{D}_*)$ is measurable as soon as T is BV$\mathcal{N}$ in $\mathfrak{D}_*$ and hence *a fortiori* if T is BV$\mathcal{N}$ in $\mathfrak{D}$. Since T is bounded, clearly $\nu(z, T, \mathfrak{D}_*)$ vanishes outside of a sufficiently large circular disc.

IV.3.3. Continuation. We assert that $|\nu(z, T, \mathfrak{D})| \leqq \kappa(z, T, \mathfrak{D})$, with the possible exception of a countable set of points z.

Proof. Let E be the subset of $\mathfrak{D}$ where $|i_e(w)| > 1$. Then E is a countable (possibly empty) set by IV.1.74. Hence the set $T(E)$ is also countable. Hence our assertion will be proved if we show that

$$(1) \qquad |\nu(z_0, T, \mathfrak{D})| \leqq \kappa(z_0, T, \mathfrak{D}) \qquad \text{if } z_0 \notin T(E).$$

If $\nu(z_0, T, \mathfrak{D}) = 0$, then (1) is obvious. So we can assume that $\nu(z_0, T, \mathfrak{D}) \neq 0$. But then (see IV.3.2) the set $T^{-1}(z_0)\cdot\mathfrak{D}\mathcal{N}$ is finite and nonempty. Let $w_1, \cdots, w_m$ be the points of this set. By definition

$$(2) \qquad \nu(z_0, T, \mathfrak{D}) = i_e(w_1) + \cdots + i_e(w_m).$$

Clearly (cf. IV.1.56), $m \leqq \kappa(z_0, T, \mathfrak{D})$.. Since $z_0 \notin T(E)$, it follows that $w_j \notin E$, $j = 1, 2, \cdots, m$. Hence $|i_e(w_j)| \leqq 1, j = 1, 2, \cdots, m$. Thus $|\nu(z_0, T, \mathfrak{D})| \leqq m$ by (2), and (1) follows.

IV.3.4. Continuation. Let $\mathfrak{R}$ be a finitely-connected Jordan region in $\mathfrak{D}$, and let z_0 be a point satisfying the following conditions:

(i) $N(z_0, \mathcal{E}^*) < +\infty$ (cf. IV.1.56).

(ii) $z_0 \notin T(\mathfrak{R} - \mathfrak{R}^0)$.

Then $\nu(z_0, T, \mathfrak{R}^0) = \mu(z_0, T, \mathfrak{R})$ (cf. IV.1.24).

PROOF. *Case* (a). $\kappa(z_0, T, \mathfrak{R}) = 0$. By IV.3.2, IV.1.26 it follows immediately that $\nu(z_0, T, \mathfrak{R}^0) = 0 = \mu(z_0, T, \mathfrak{R}^0)$.

Case (b). $\kappa(z_0, T, \mathfrak{R}) > 0$. Then condition (i) implies that the point z_0 has, under T in $\mathfrak{R}^0$, a finite number of essential maximal model continua, each of which reduces to a single point (cf. IV.1.56). Let $w_1, \cdots, w_m$ be these points. Clearly (see IV.1.56)

$$(1) \qquad w_j \in N, j = 1, 2, \cdots, m; \qquad \nu(z_0, T, \mathfrak{R}^0) = \sum_{j=1}^{m} i_e(w_j).$$

By IV.1.64, we have then Jordan regions $\mathfrak{R}_j$, $j = 1, 2, \cdots, m$, satisfying the following conditions:

(α) $w_j \in \mathfrak{R}_j^0$, $\mathfrak{R}_j \subset \mathfrak{R}^0$.

(β) $\mathfrak{R}_j$ is simply connected.

(γ) $t(w) \neq z_0$ for $w \in \mathfrak{R}_j - \mathfrak{R}_j^0$.

(δ) $\mathfrak{R}_j\mathfrak{R}_k = 0$ for $j \neq k$.

(ϵ) $\mu(z_0, T, \mathfrak{R}_j) = i_e(w_j)$.

Let us denote by $\mathfrak{R}_*$ the Jordan region $\mathfrak{R} - (\mathfrak{R}_1^0 + \cdots + \mathfrak{R}_m^0)$. Since $\mathfrak{R}_*$ does not contain any of the points $w_1, \cdots, w_m$, we have (cf. IV.1.47) $\kappa(z_0, T, \mathfrak{R}_*^0) = 0$, and hence by IV.1.26, IV.1.50

$$0 = \mu(z_0, T, \mathfrak{R}_*) = \mu(z_0, T, \mathfrak{R}) - \sum_{j=1}^{m} \mu(z_0, T, \mathfrak{R}_j).$$

In writing the last term of this formula, we used (ii) and (γ). In view of (1) and (ϵ), the formula $\nu(z_0, T, \mathfrak{R}^0) = \mu(z_0, T, \mathfrak{R})$ follows.

IV.3.5. Let $T : z = t(w)$, $w \in \mathfrak{R}$, be a continuous transformation defined in a finitely-connected, bounded Jordan region $\mathfrak{R}$ (and not merely in $\mathfrak{R}^0$). Then the argument used in IV.3.4 applies, as inspection shows, not only to Jordan subregions of $\mathfrak{R}^0$ but also to $\mathfrak{R}$ itself, and we have the following result. Let z_0 be a point such that $N(z_0, \mathcal{E}^*) < +\infty$ and $z_0 \notin T(\mathfrak{R} - \mathfrak{R}^0)$. Then $\nu(z_0, T, \mathfrak{R}^0) = \mu(z_0, T, \mathfrak{R})$. In this statement, $\mathcal{E}^*$ denotes the set $\mathcal{E}^*(T, \mathfrak{R}^0)$ (see IV.1.56).

IV.3.6. Given T as in IV.2.1, we introduce the following notations, where R^0 denotes any oriented open rectangle in $\mathfrak{D}$.

$g_1(z, R^0)$ denotes the characteristic function of the set in the z-plane where $\nu(z, T, R^0) \neq 0$ (cf. IV.3.2).

$g_2(z, R^0)$ denotes the characteristic function of the set $T(R^0N)$ (cf. IV.1.56).

$g_3(z, R^0)$ denotes the characteristic function of the set $T(R^0\mathcal{E})$ (cf. IV.1.56).

$g_4(z, R^0)$ denotes the characteristic function of the set in the z-plane where $\kappa(z, T, R^0) \neq 0$ (cf. IV.1.43).

$g_5(z, R^0)$ denotes the characteristic function of the set $T(R^0\mathcal{E}^*)$ (cf. IV.1.56).

$g_6(z, R^0)$ denotes the characteristic function of the set $T(R^0)$.

Since T is bounded in $\mathfrak{D}$, the functions $g_i(z, R^0)$, $i = 1, 2, \cdots, 6$, all vanish outside of a sufficiently large circular disc in the z-plane. The functions $g_i(z, R^0)$, $i = 2, 3, 5, 6$, are measurable by I.2.46 and IV.1.58, IV.1.59, IV.1.60. The function $g_4(z, R^0)$ is measurable since $\kappa(z, T, R^0)$ is measurable (cf. IV.1.51). As regards $g_1(z, R^0)$ we can only assert that $g_1(z, R^0)$ is measurable if T is BV$\mathcal{N}$ in $\mathfrak{D}$ (cf. IV.3.2). As an immediate consequence of the definitions given, we have the inequalities

$$g_i(z, R^0) \leqq g_{i+1}(z, R^0), \qquad i = 1, 2, \cdots, 5.$$

IV.3.7. Continuation. We introduce now the rectangle functions (cf. IV.2.1)

$$G_i(R) = \iint g_i(z, R^0), \qquad i = 1, \cdots, 6,$$

where R is any oriented rectangle in $\mathfrak{D}$. In view of the remarks in IV.3.6, $G_1(R)$ is defined only if T is BV$\mathcal{N}$ in $\mathfrak{D}$, while $G_i(R)$, $i = 2, \cdots, 6$, is defined without any further restriction upon T. Assuming the existence of $G_1(R)$ we have by IV.3.6 the inequalities

$$G_i(R) \leqq G_{i+1}(R), \qquad i = 1, 2, \cdots, 5.$$

For $i = 2, 3, 5, 6$, clearly $G_i(R)$ coincides with the rectangle function $G(R)$ of IV.2.10 if the base set $\mathfrak{B}$ is chosen as $\mathcal{N}$, $\mathcal{E}$, $\mathcal{E}^*$, $\mathfrak{D}$ respectively. It seems that no such simple interpretation is available for $G_1(R)$ and $G_4(R)$.

IV.3.8. Continuation. Let us now consider the functions $N(z, R^0\mathcal{N})$, $N(z, R^0\mathcal{E})$, $\kappa(z, T, R^0)$, $N(z, R^0\mathcal{E}^*)$, $N(z, R^0)$, where R^0 is any oriented open rectangle in $\mathfrak{D}$ (cf. IV.2.1, IV.1.56). All of these functions are measurable. Indeed, $\kappa(z, T, R^0)$ is measurable by IV.1.51 while the remaining four functions are measurable by I.2.46, IV.1.58, IV.1.59, IV.1.60, IV.2.6. Since T is bounded in $\mathfrak{D}$ all these functions vanish outside of a sufficiently large circular disc in the z-plane. Without further restrictions upon T we cannot assert that these functions are summable. In defining now five more rectangle functions $G_i^*(R)$, $i = 2, \cdots, 6$, it is understood that any particular one of these definitions can be used only if the function involved in that definition is summable. The definitions are as follows:

$$G_2^*(R) = \iint N(z, R^0\mathcal{N}).$$

$$G_3^*(R) = \iint N(z, R^0\mathcal{E}).$$

$$G_4^*(R) = \iint \kappa(z, T, R^0).$$

$$G_5^*(R) = \iint N(z, R^0\mathcal{E}^*).$$

$$G_6^*(R) = \iint N(z, R^0).$$

In view of IV.1.56, we have the inequalities

$$G_i^*(R) \leqq G_{i+1}^*(R), \qquad i = 2, 3, 4, 5,$$

as soon as the quantities involved exist, it being obvious that the existence of $G_{i+1}^*(R)$ implies the existence of $G_i^*(R)$. Clearly (cf. IV.3.7)

$$G_i(R) \leqq G_i^*(R), \qquad i = 2, \cdots, 6,$$

as soon as $G_i^*(R)$ exists. Clearly, $G_i^*(R)$ coincides, for $i = 2, 3, 5, 6$, with the rectangle function $G^*(R)$ of IV.2.31 if the base set $\mathfrak{B}$ be chosen as $\mathcal{N}$, $\mathcal{E}$, $\mathcal{E}^*$, $\mathfrak{D}$ respectively. No such simple interpretation seems to be available for $G_4^*(R)$.

IV.3.9. Continuation. If the derivative (cf. III.1.24) of the rectangle function $G_i(R)$ exists at a point $w \in \mathfrak{D}$ then this derivative will be denoted by $D_i(w)$, $i = 1, \cdots, 6$. It will be convenient to use, along with these notations, the more suggestive alternative notations $D_\nu(w)$, $D(w, \mathcal{N})$, $D(w, \mathcal{E})$, $D_\kappa(w)$, $D(w, \mathcal{E}^*)$, $D(w, \mathfrak{D})$, corresponding to the cases $i = 1, 2, 3, 4, 5, 6$ in order. For $i = 2, 3, 5, 6$ these alternative notations conform to those already agreed upon in IV.2.31, IV.2.32 (cf. the remarks in IV.3.7). The notations $D_\nu(w)$, $D_\kappa(w)$ are self-explanatory. Similarly, the derivatives of the rectangle functions $G_i^*(R)$ will be denoted by $D_i^*(w)$ or alternatively by $D^*(w, \mathcal{N})$, $D^*(w, \mathcal{E})$, $D_\kappa^*(w)$, $D^*(w, \mathcal{E}^*)$, $D^*(w, \mathfrak{D})$, corresponding to the cases $i = 2, 3, 4, 5, 6$, in this order. For $i = 2, 3, 5, 6$, there is again agreement with the notations adopted in IV.2.31, IV.2.32. In view of IV.3.7, IV.3.8 we have the inequalities $D_i(w) \leqq D_{i+1}(w)$, $D_i^*(w) \leqq D_{i+1}^*(w)$, $D_i(w) \leqq D_i^*(w)$ as soon as the derivatives involved exist. We proceed to study the relationships between these various derivatives.

IV.3.10. Assume that T, given as in IV.2.1, is BV$\mathcal{N}$ in $\mathfrak{D}$ (cf. IV.2.11, IV.1.56). By IV.2.31, IV.2.32, IV.2.34, IV.3.9 we have then

$$D(w, \mathcal{N}) = D^*(w, \mathcal{N}) \qquad \text{a.e. in } \mathfrak{D}.$$

We assert further that $D_\nu(w)$ (see IV.3.9) exists a.e. in $\mathfrak{D}$ and

$$(1) \qquad D_\nu(w) = D(w, \mathcal{N}) \qquad \text{a.e. in } \mathfrak{D}.$$

Proof. Let R^0 be any oriented open rectangle in $\mathfrak{D}$, and let $R_1^0, \cdots, R_m^0$ be any finite system of oriented open rectangles such that $R_i^0 R_j^0 = 0$ for $i \neq j$. $R_i^0 \subset R^0$, $i = 1, 2, \cdots, m$. We proceed to show that (cf. IV.3.6)

$$(2) \qquad \sum_{i=1}^{m} [N(z, R_i^0 \mathcal{N}) - g_1(z, R_i^0)] \leqq N(z, R^0 \mathcal{N}) - g_1(z, R^0).$$

Clearly, in view of the position of the rectangles R_i^0,

$$\sum_{i=1}^{m} N(z, R_i^0 \mathcal{N}) \leqq N(z, R^0 \mathcal{N}),$$

and hence (2) is obvious if either $g_1(z, R^0) = 0$ or else if $g_1(z, R_i^0) = 1$ for some i. There remains therefore the task of verifying (2) for points such that simultaneously

(3) $$g_1(z, R^0) = 1, \; g_1(z, R_\imath^0) = 0, \qquad \imath = 1, 2, \cdots, m.$$

Since (2) is obvious if $N(z, R^0\mathcal{N}) = +\infty$ we can further assume that $N(z, R^0\mathcal{N}) < +\infty$. Now (3) implies that $N(z, R^0\mathcal{N}) \geqq 1$ (cf. IV.3.2). Thus the set $T^{-1}(z)\cdot R^0\mathcal{N}$ is finite and nonempty. Let this set consist of the points $w_1, \cdots, w_k$, where $k \geqq 1$. If all these points lie in $R_1^0 + \cdots + R_m^0$, then clearly

(4) $$\nu(z, T, R^0) = \sum_{\imath=1}^{m} \nu(z, T, R_\imath^0).$$

But (3) implies that $\nu(z, T, R_\imath^0) = 0$, $\imath = 1, 2, \cdots, m$, and hence (4) implies that $\nu(z, T, R^0) = 0$, in contradiction to (3). Thus we see that at least one of the points $w_1, \cdots, w_k$ lies in $R^0 - (R_1^0 + \cdots + R_m^0)$. Hence

$$N(z, R^0\mathcal{N}) \geqq 1 + \sum_{\imath=1}^{m} N(z_0, R_\imath^0\mathcal{N}),$$

and consequently

$$N(z, R^0\mathcal{N}) - g_1(z, R^0) \geqq N(z, R^0\mathcal{N}) - 1 \geqq \sum_{\imath=1}^{m} N(z, R_\imath^0\mathcal{N}).$$

In view of (3), the inequality (2) follows. Now since T is BV$\mathcal{N}$ by assumption, all the functions involved in (2) are measurable and summable (cf. IV.3.6, IV.2.31). Integrating (2) we obtain (cf. IV.3.7, IV.3.8) the inequality

$$\sum_{\imath=1}^{m} [G_2^*(R_\imath) - G_1(R_\imath)] \leqq G_2^*(R) - G_1(R).$$

Thus the rectangle function $\Psi(R) = G_2^*(R) - G_1(R)$ is of type A in $\mathfrak{D}$ (cf. III.1.28, III.1.52). Hence the derivative of $\Psi(R)$ exists a.e. in $\mathfrak{D}$. Since $G_2^*(R)$ also has a derivative a.e. in $\mathfrak{D}$ (note that T is BV$\mathcal{N}$ in $\mathfrak{D}$ by assumption), it follows that $G_1(R) = G_2^*(R) - \Psi(R)$ also has a derivative a.e. in $\mathfrak{D}$. Thus $D_1(w) = D_\nu(w)$ exists a.e. in $\mathfrak{D}$ (cf. IV.3.9).

Now let R^0 be any oriented rectangle contained in $\mathfrak{D}$. Let us put (cf. IV.2.3, IV.3.6)

(5) $$\psi_j(z) = \sum g_1(z, s^0), \; \chi_j(z) = \sum g_2(z, s^0), \qquad \jmath = 1, 2, \cdots,$$

where the summation is extended over all oriented squares s such that

$$s \in D_{p_j}, \; s^0 \subset R^0.$$

From the definition of the quantities involved (cf. IV.3.6) it follows immediately that

(6) $$0 \leqq \psi_j(z) \leqq \chi_j(z) \leqq N(z, R^0\mathcal{N}).$$

We want to verify the relation

(7) $$\psi_j(z) \to N(z, R^0\mathcal{N}) \text{ for } \jmath \to \infty, \qquad \text{a.e. in the } z\text{-plane}.$$

Now let us observe that T is BV$\mathcal{N}$ by assumption, and hence $N(z, \mathfrak{D}\mathcal{N})$ is summable (see IV.2.13). Thus $N(z, \mathfrak{D}\mathcal{N}) < +\infty$ and hence *a fortiori* $N(z, R^0\mathcal{N}) < +\infty$ a.e. in the z-plane. Consequently, in verifying (6), it is sufficient to show that

$$(8) \qquad \psi_j(z_0) \to N(z_0, R^0\mathcal{N}) \qquad \text{if } N(z_0, R^0\mathcal{N}) < +\infty.$$

Now (8) is obvious if $N(z_0, R^0\mathcal{N}) = 0$ (cf. (6)). Thus we can assume that $0 < N(z_0, R^0\mathcal{N}) < +\infty$. Then the set $T^{-1}(z_0)\cdot R^0\mathcal{N}$ is finite and nonempty. Let $w_1^*, \cdots, w_n^*$ be the points of this set. For j large enough, say $j > j_0$, we shall have then the following situation (cf. IV.2.3). In the subdivision $D_{\nu j}$ we shall have n squares $s_1, \cdots, s_n$ such that $w_i^* \in s_i^0$, $s_i^0 \subset R^0$. Clearly, since each s_i^0 contains precisely one point of the set $T^{-1}(z_0)\cdot\mathcal{N}$, we have (cf. IV.3.2)

$$\nu(z_0, T, s_i^0) = i_e(w_i^*) \neq 0.$$

Hence

$$g_1(z_0, s_i^0) = 1, \qquad i = 1, 2, \cdots, n.$$

Thus (cf. (5)), $\psi_j(z_0) \geqq N(z_0, R^0\mathcal{N})$ for $j > j_0$, and (8) follows in view of (6). Summing up, we have now the following facts at our disposal:

(i) $N(z, R^0\mathcal{N})$ is summable, as a consequence of the assumption that T is BV$\mathcal{N}$.

(ii) $0 \leqq \psi_j(z) \leqq \chi_j(z) \leqq N(z, R^0\mathcal{N})$, $j = 1, 2, \cdots$.

(iii) $\psi_j(z) \to N(z, R^0\mathcal{N})$ and hence also (by (ii)) $\chi_j(z) \to N(z, R^0\mathcal{N})$ for $j \to \infty$ a.e. in the z-plane.

By I.3.11, it follows that

$$\lim \iint \psi_j(z) = \lim \iint \chi_j(z) = \iint N(z, R^0\mathcal{N}) \qquad \text{for } j \to \infty.$$

In view of IV.3.7, IV.3.8, (5), these relations may be rewritten in the form

$$G_2^*(R) = \lim_{j\to\infty} \sum G_1(s), \qquad s \in D_{\nu j}, s^0 \subset R^0,$$

$$G_2^*(R) = \lim_{j\to\infty} \sum G_2(s), \qquad s \in D_{\nu j}, s^0 \subset R^0.$$

Subtraction yields

$$\lim_{j\to\infty} \sum [G_2(s) - G_1(s)] = 0, \qquad s \in D_{\nu j}, s^0 \subset R^0.$$

Now the rectangle function $G_2(r) - G_1(r)$ is non-negative (cf. IV.3.7), and has a.e. in $\mathfrak{D}$ a derivative, namely $D(w, \mathcal{N}) - D_\nu(w)$; indeed, we verified earlier in the present section that the derivatives $D(w, \mathcal{N})$, $D_\nu(w)$ of $G_2(r)$, $G_1(r)$ respectively exist a.e. in $\mathfrak{D}$. Thus (1) follows from III.1.25.

IV.3.11. Let us assume now that T, given as in IV.2.1, is BV$\mathcal{E}$ in $\mathfrak{D}$ (cf. IV.1.56, IV.2.11). By IV.2.34, IV.3.9 the derivatives $D(w, \mathcal{E})$, $D^*(w, \mathcal{E})$ exist and are equal to each other a.e. in $\mathfrak{D}$. Since $\mathcal{N} \subset \mathcal{E}$ (cf. IV.1.56), T is also BV$\mathcal{N}$ in $\mathfrak{D}$, and hence in view of IV.3.10, IV.3.9 we obtain the result that the derivatives

$D_\nu(w)$, $D(w, \mathcal{N})$, $D^*(w, \mathcal{N})$, $D(w, \mathcal{E})$, $D^*(w, \mathcal{E})$ exist a.e. in $\mathfrak{D}$ and satisfy the relations $D_\nu(w) = D^*(w, \mathcal{N}) = D(w, \mathcal{N}) \leqq D(w, \mathcal{E}) = D^*(w, \mathcal{E})$ a.e. in $\mathfrak{D}$. It is not known whether the sign of equality holds throughout without further assumptions.

IV.3.12. Given T as in IV.2.1, let us assume now that $\kappa(z, T, \mathfrak{D})$ is summable (cf. IV.1.43). Take any oriented open rectangle $R^0 \subset \mathfrak{D}$, and let $R_1, \cdots, R_m$ be any finite system of oriented rectangles in R without common interior points. By IV.1.54 we have then the inequality

$$\sum_{\iota=1}^{m} \kappa(z, T, R_\iota^0) \leqq \kappa(z, T, R^0).$$

Integration yields (cf. IV.3.8)

$$\sum_{\iota=1}^{m} G_4^*(R_\iota) \leqq G_4^*(R).$$

Thus the rectangle function $G_4^*(R)$ is of type A (see III.1.28, III.1.52), and hence its derivative $D_4^*(w)$ exists a.e. in $\mathfrak{D}$. Since clearly $N(z, \mathcal{N}) \leqq N(z, \mathcal{E}) \leqq \kappa(z, T, \mathfrak{D})$ (cf. IV.1.56), the summability of $\kappa(z, T, \mathfrak{D})$ implies the summability of $N(z, \mathcal{N})$, $N(z, \mathcal{E})$. Thus T is BV$\mathcal{N}$ and BV$\mathcal{E}$ in $\mathfrak{D}$ (cf. IV.2.13). Hence, by IV.3.10, IV.3.11, the derivatives $D_1(w)$, $D_2(w)$, $D_3(w)$, $D_2^*(w)$, $D_3^*(w)$ exist a.e. in $\mathfrak{D}$, and we have just proved that $D_4^*(w)$ exists a.e. in $\mathfrak{D}$. On the other hand, we have not accounted as yet for the derivative $D_4(w)$. We proceed to show that this also exists a.e. in $\mathfrak{D}$ and that all the derivatives $D_1(w), \cdots, D_4^*(w)$ are equal to each other a.e. in $\mathfrak{D}$. Let us put (cf. IV.2.3) for $j = 1, 2, \cdots,$

$$\psi_j(z) = \sum g_1(z, s^0), \qquad s \in D_{\nu_j}, s^0 \subset R^0.$$

We have then the relations (cf. IV.3.10, IV.1.56)

$$0 \leqq \psi_j(z) \leqq N(z, R^0\mathcal{N}) \leqq \kappa(z, T, \mathfrak{D}),$$

$$(1) \qquad \psi_j(z) \to N(z, R^0\mathcal{N}) \qquad \text{for } j \to \infty, \text{ a.e. in the } z\text{-plane.}$$

Since $\kappa(z, T, \mathfrak{D})$ is summable, it follows that termwise integration of the sequence $\psi_j(z)$ is permissible (cf. I.3.11), and in view of IV.3.7, IV.3.8 we obtain thus

$$(2) \qquad G_2^*(R) = \lim_{j\to\infty} \sum G_1(s), \qquad s \in D_{\nu_j}, s^0 \subset R^0.$$

Let us put next, for $j = 1, 2, \cdots,$

$$(3) \qquad \lambda_j(z) = \sum \kappa(z, T, s^0), \qquad s \in D_{\nu_j}, s^0 \subset R^0.$$

Clearly (cf. IV.1.54)

$$(4) \qquad 0 \leqq \lambda_j(z) \leqq \kappa(z, T, R^0) \leqq \kappa(z, T, \mathfrak{D}).$$

We assert that

$$(5) \qquad \lambda_j(z) \to N(z, R^0\mathcal{N}) \qquad \text{for } j \to \infty, \text{ a.e. in the } z\text{-plane.}$$

Indeed, since $\kappa(z, T, \mathfrak{D})$ is summable, $\kappa(z, T, \mathfrak{D}) < +\infty$ a.e. in the z-plane. Hence it is sufficient to show that

$$(6) \qquad \lambda_j(z_0) \to N(z_0, R^0\mathcal{N}) \qquad \text{if } \kappa(z_0, T, \mathfrak{D}) < +\infty.$$

Now this is obvious if $\kappa(z_0, T, R^0) = 0$, since $0 \leqq \lambda_j(z_0) \leqq \kappa(z_0, T, R^0)$, $0 \leqq N(z_0, R^0\mathcal{N}) \leqq \kappa(z_0, T, R^0)$. So we can assume that $0 < \kappa(z_0, T, R^0) < +\infty$. Let us put $\kappa(z_0, T, R^0) = k$. Then we know, by IV.1.47, that z_0 has exactly k essential maximal model continua $\gamma_1, \cdots, \gamma_k$ in R^0. Let us put $\delta = 1$ if all these continua reduce to single points; otherwise let δ be a positive number less than the smallest one of the diameters of those of the continua $\gamma_1, \cdots, \gamma_k$ that do not reduce to single points. For j large enough, the diagonal of the squares s of the subdivision D_{p_j} will be less than δ, and hence no square of D_{p_j} will contain in its interior one of the continua $\gamma_1, \cdots, \gamma_k$ that does not reduce to a single point. Hence, for j large, we shall have $\kappa(z_0, T, s^0) = N(z_0, s^0\mathcal{N})$ for every $s \in D_{p_j}$, $s^0 \subset R^0$. Now since $N(z_0, R^0\mathcal{N}) \leqq \kappa(z_0, T, R^0) < +\infty$, the set $T^{-1}(z_0)\cdot R^0\mathcal{N}$ is finite (possibly empty). Hence, for j large, no square s of D_{p_j} will contain more than one point of this set, and every point of this set will be interior to a square of D_{p_j} (see IV.2.3). It follows that for j sufficiently large we shall have $\lambda_j(z_0) = N(z_0, R^0\mathcal{N})$, and (6) is proved. Since $\kappa(z, T, \mathfrak{D})$ is summable, (4) and (5) imply (see I.3.11) that termwise integration of the sequence $\lambda_j(z)$ is permissible. From (3), (5) we obtain thus (cf. IV.3.8)

$$(7) \qquad G_2^*(R) = \lim_{j\to\infty} \sum G_4^*(s), \qquad s \in D_{p_j}, s^0 \subset R^0.$$

(2), (7) yield by subtraction

$$(8) \qquad 0 = \lim_{j\to\infty} \sum [G_4^*(s) - G_1(s)], \qquad s \in D_{p_j}, s^0 \subset R^0.$$

Now we have already proved that the derivatives $D_4^*(w)$ and $D_1(w)$ exist a.e.in $\mathfrak{D}$. Thus the rectangle function $G_4^*(r) - G_1(r)$ has a derivative a.e. in $\mathfrak{D}$, and by IV.3.8 this rectangle function is non-negative. Thus (8) implies, in view of III.1.25, the relation $D_4^*(w) = D_1(w)$ a.e. in R^0. Since R was arbitrary, it follows that $D_4^*(w) = D_1(w)$ a.e. in $\mathfrak{D}$. Now let R be any oriented rectangle in $\mathfrak{D}$. By IV.3.7, IV.3.8, we have the inequalities $G_1(R) \leqq G_4(R) \leqq G_4^*(R)$. Since the derivatives $D_4^*(w)$ and $D_1(w)$ exist and are equal to each other a.e. in $\mathfrak{D}$, it follows that the derivative $D_4(w)$ exists a.e. in $\mathfrak{D}$, and $D_1(w) = D_4(w) = D_4^*(w)$ a.e. in $\mathfrak{D}$. In view of IV.3.11, we can summarize our present information as follows. The derivatives $D_1(w), \cdots, D_4^*(w)$ exist a.e. in $\mathfrak{D}$, and satisfy the inequalities (cf. IV.3.9)

$$D_1(w) = D_2^*(w) = D_2(w) \leqq D_3(w) = D_3^*(w) \leqq D_4^*(w) = D_4(w) = D_1(w) \quad \text{a.e. in } \mathfrak{D}.$$

It follows that all these derivatives are equal to each other a.e. in $\mathfrak{D}$.

IV.3.13. Using the alternative notations explained in IV.3.9, the preceding result may be restated as follows. If $\kappa(z, T, \mathfrak{D})$ is summable, then the derivatives $D_\nu(w)$, $D(w, \mathcal{N})$, $D^*(w, \mathcal{N})$, $D(w, \mathcal{E})$, $D^*(w, \mathcal{E})$, $D_\kappa(w)$, $D_\kappa^*(w)$ exist and are equal to each other a.e. in $\mathfrak{D}$. We assert further that $D_\kappa(w)$ is summable in $\mathfrak{D}$ and

$$(1) \qquad \iint_{\mathfrak{D}} D_\kappa(w) \leqq \iint \kappa(z, T, \mathfrak{D}).$$

Indeed, since $0 \leqq N(z, \mathcal{E}) \leqq \kappa(z, T, \mathfrak{D})$, the summability of $\kappa(z, T, \mathfrak{D})$ implies the summability of $N(z, \mathcal{E})$. By IV.2.33, IV.2.34 it follows that $D(w, \mathcal{E})$ is summable in $\mathfrak{D}$ and

$$\iint_{\mathfrak{D}} D(w, \mathcal{E}) \leqq \iint N(z, \mathcal{E}).$$

Since $D(w, \mathcal{E}) = D_\kappa(w)$ a.e. in $\mathfrak{D}$, and $N(z, \mathcal{E}) \leqq \kappa(z, T, \mathfrak{D})$, the inequality (1) follows.

IV.3.14. Let us assume now that T, given as in IV.2.1, is BV$\mathcal{E}^*$ in $\mathfrak{D}$ (see IV.1.56, IV.2.11). By IV.2.34, IV.3.9 the derivatives $D_5(w)$, $D_5^*(w)$ exist and are equal to each other a.e. in $\mathfrak{D}$. Furthermore, since $\kappa(z, T, \mathfrak{D}) \leqq N(z, \mathcal{E}^*)$, it follows by IV.2.13 that $\kappa(z, T, \mathfrak{D})$ is summable, and hence all the results stated in IV.3.13 hold.

Since $N(z, \mathcal{E}^*)$ is summable by IV.2.13, we have $N(z, \mathcal{E}^*) < +\infty$ a.e. in the z-plane. If $\mathfrak{D}_0$ is any subdomain of $\mathfrak{D}$, and if we put $\mathcal{E}_0^* = \mathcal{E}^*(T, \mathfrak{D}_0)$, then *a fortiori* $N(z, \mathcal{E}_0^*) < +\infty$ a.e. in the z-plane. By IV.1.56(3) it follows that $\kappa(z, T, \mathfrak{D}_0) = N(z, \mathcal{E}_0^*)$ a.e. in the z-plane and hence

$$(1) \qquad \iint \kappa(z, T, \mathfrak{D}_0) = \iint N(z, \mathcal{E}_0^*).$$

Now at a point z where $N(z, \mathcal{E}^*) < +\infty$, we have clearly $N(z, \mathcal{E}_0^*) = N(z, \mathfrak{D}_0\, \mathcal{E}^*)$. Since $N(z, \mathcal{E}^*) < +\infty$ a.e., it follows that (1) can be rewritten in the form

$$\iint \kappa(z, T, \mathfrak{D}_0) = \iint N(z, \mathfrak{D}_0\, \mathcal{E}^*).$$

Choosing $\mathfrak{D}_0$ as an oriented open rectangle $R^0 \subset \mathfrak{D}$, we obtain (cf. IV.3.8) $G_4^*(R) = G_5^*(R)$. Since this holds for every $R \subset \mathfrak{D}$, and since the derivatives $D_4^*(w)$, $D_5^*(w)$ are already known to exist a.e. in $\mathfrak{D}$, it follows that $D_4^*(w) = D_5^*(w)$ a.e. in $\mathfrak{D}$. In view of the result in IV.3.13 and the inequalities in IV.3.9, it follows that all the derivatives $D_1(w), \cdots, D_5^*(w)$ exist and are equal to each other a.e. in $\mathfrak{D}$.

IV.3.15. Suppose that T, given as in IV.2.1, is BV$\mathcal{E}^*$ in $\mathfrak{D}$. Then we have in the z-plane a set $\bar{E}$ of measure zero, such that for every finitely-connected Jordan region $\mathfrak{R} \subset \mathfrak{D}$ the conditions (cf. IV.1.24, IV.3.2)

$$(1) \qquad \mu(z, T, \mathfrak{R}) \neq 0 \text{ and } z \notin \bar{E}$$

imply that

$$(2) \qquad \mu(z, T, \mathfrak{R}) = \nu(z, T, \mathfrak{R}^0).$$

Proof. Let us choose $\bar{E}$ as the set of those points z where $N(z, \mathcal{E}^*) = +\infty$.

Then $|\bar{E}| = 0$ by IV.2.14. Let now $\mathfrak{R}$ be any finitely-connected Jordan region in $\mathfrak{D}$ and let z be a point that satisfies (1). Then (cf. IV.1.24) $z \notin T(\mathfrak{R} - \mathfrak{R}^0)$ and $N(z, \mathcal{E}^*) < +\infty$. The relation (2) follows now directly from IV.3.4.

IV.3.16. Let $T : z = t(w)$, $w \in \mathfrak{R}$, be a continuous transformation defined in a bounded, finitely-connected Jordan region $\mathfrak{R}$ (and not merely in $\mathfrak{R}^0$). Suppose that T is BV$\mathcal{E}^*$ in $\mathfrak{R}^0$. Then we have in the z-plane a set $\bar{E}$ of measure zero, such that the conditions $\mu(z, T, \mathfrak{R}) \neq 0$, $z \notin \bar{E}$, imply that $\mu(z, T, \mathfrak{R}) = \nu(z, T, \mathfrak{R}^0)$.

The proof is entirely analogous to that in IV.3.15 except that IV.3.5 is used instead of IV.3.4.

IV.3.17. Let $T : z = t(w)$, $w \in \mathfrak{R}$, be a continuous transformation defined in a bounded, finitely-connected Jordan region $\mathfrak{R}$ which is BV$\mathcal{E}^*$ in $\mathfrak{R}^0$. Then

$$(1) \qquad |\mu(z, T, \mathfrak{R})| \leqq \kappa(z, T, \mathfrak{R}^0)$$

a.e. in the z-plane.

Proof. By IV.3.3 we have

$$(2) \qquad |\nu(z, T, \mathfrak{R}^0)| \leqq \kappa(z, T, \mathfrak{R}^0) \qquad \text{for } z \notin \bar{G},$$

where $\bar{G}$ is a countable (possibly empty) set. Let $\bar{E}$ be the set of measure zero that appears in IV.3.16. Let z be any point such that

$$(3) \qquad z \notin \bar{G} + \bar{E}.$$

Since $|\bar{G} + \bar{E}| = 0$, the proof will be made by showing that (3) implies (1). Now if $\mu(z, T, \mathfrak{R}) = 0$, then (1) is obvious. On the other hand, if $\mu(z, T, \mathfrak{R}) \neq 0$, then by (3) and IV.3.16 we have

$$(4) \qquad \mu(z, T, \mathfrak{R}) = \nu(z, T, \mathfrak{R}^0).$$

(2) and (4) imply (1).

IV.3.18. Assume now that T, given as in IV.2.1, is BV$\mathfrak{D}$ in $\mathfrak{D}$ (cf. IV.2.11). By IV.2.34 (cf. IV.3.9) the derivatives $D_6(w) = D(w, \mathfrak{D})$ and $D_6^*(w) = D^*(w, \mathfrak{D})$ exist and are equal to each other a.e. in $\mathfrak{D}$. Since clearly $N(z, \mathcal{E}^*) \leqq N(z, \mathfrak{D})$, T is also BV$\mathcal{E}^*$ in $\mathfrak{D}$ (cf. IV.2.13), and hence the results stated in IV.3.14 hold. Hence all the derivatives $D_1(w)$, $D_2(w)$, $D_2^*(w)$, $\cdots$, $D_6(w)$, $D_6^*(w)$ exist a.e. in $\mathfrak{D}$. Furthermore, using the alternative notations and inequalities of IV.3.9, we have a.e. in $\mathfrak{D}$

$$D_\nu(w) = D^*(w, \mathcal{N}) = D(w, \mathcal{N}) = D^*(w, \mathcal{E}) = D(w, \mathcal{E}) = D_\kappa^*(w)$$
$$= D_\kappa(w) = D^*(w, \mathcal{E}^*) = D(w, \mathcal{E}^*) \leqq D(w, \mathfrak{D}) = D^*(w, \mathfrak{D}).$$

However, we are unable to prove that the sign of equality holds throughout without further restricting T. We proceed to consider an important special case where the equality of all these derivatives can be established a.e. in $\mathfrak{D}$.

IV.3.19. Continuation. Let us assume that T is BV$\mathfrak{D}$ in $\mathfrak{D}$. Let us consider the set $\mathcal{I}_0 = \mathcal{I}_0(T, \mathfrak{D})$ defined in IV.1.78. If $|T(\mathcal{I}_0)| = 0$, then all the derivatives $D_\nu(w)$, $\cdots$, $D^*(w, \mathfrak{D})$ are equal to each other a.e. in $\mathfrak{D}$.

PROOF. The derivatives $D(w, \mathcal{N})$ and $D(w, \mathfrak{D})$ coincide with the derivative $D(w, \mathcal{B})$ in IV.2.32 for $\mathcal{B} = \mathcal{N}$ and $\mathcal{B} = \mathfrak{D}$ respectively. Hence, by IV.2.38,

$$(1) \qquad D(w, \mathfrak{D}) = D(w, \mathcal{N}) \qquad \text{a.e. on } \mathcal{N}.$$

We can write

$$(2) \qquad \mathfrak{D} - \mathcal{N} = (\mathfrak{D} - \mathcal{N})\mathcal{J}_0 + (\mathfrak{D} - \mathcal{N})(\mathfrak{D} - \mathcal{J}_0).$$

Now $| T(\mathcal{J}_0) | = 0$ by assumption. By IV.2.37 it follows that

$$(3) \qquad D(w, \mathfrak{D}) = 0 \qquad \text{a.e. on } \mathcal{J}_0 .$$

On the other hand (see IV.1.78)

$$(4) \qquad (\mathfrak{D} - \mathcal{N})(\mathfrak{D} - \mathcal{J}_0) \subset \mathfrak{D} - \mathcal{J}.$$

Clearly (cf. IV.1.56, IV.2.2)

$$T(\mathfrak{D} - \mathcal{J}) \subset \overline{A}(\mathfrak{D}, +\infty).$$

Since T is BV$\mathfrak{D}$, the set $\overline{A}(\mathfrak{D}, +\infty)$ is of measure zero (see IV.2.14). Thus $| T(\mathfrak{D} - \mathcal{J}) | = 0$, and by IV.2.37 it follows that

$$(5) \qquad D(w, \mathfrak{D}) = 0 \qquad \text{a.e. on } \mathfrak{D} - \mathcal{J}.$$

(2), (3), (4), (5) imply that $D(w, \mathfrak{D}) = 0$ a.e. on $\mathfrak{D} - \mathcal{N}$. Since $0 \leqq D(w, \mathcal{N}) \leqq D(w, \mathfrak{D})$ a.e., it follows that

$$(6) \qquad D(w, \mathcal{N}) = D(w, \mathfrak{D}) = 0 \qquad \text{a.e. on } \mathfrak{D} - \mathcal{N}.$$

(1), (6) show that $D(w, \mathcal{N}) = D(w, \mathfrak{D})$ a.e. in $\mathfrak{D}$. In view of the relations listed in IV.3.18, it follows that all the derivatives $D_v(w), \cdots, D^*(w, \mathfrak{D})$ are equal to each other a.e. in $\mathfrak{D}$.

IV.3.20. CONTINUATION. Assuming again that T is BV$\mathfrak{D}$ in $\mathfrak{D}$ and $| T(\mathcal{J}_0) | = 0$, let us consider any subdomain $\mathfrak{D}_0$ of $\mathfrak{D}$. We assert that $\kappa(z, T, \mathfrak{D}_0) = N(z, T, \mathfrak{D}_0)$ a.e. in the z-plane. In particular, $\kappa(z, T, \mathfrak{D}) = N(z, T, \mathfrak{D})$ a.e. in the z-plane.

PROOF. Let us put $\overline{E} = \overline{A}(\mathfrak{D}, +\infty) + T(\mathcal{J}_0)$ (cf. IV.2.2). The set $\overline{A}(\mathfrak{D}, +\infty)$ is of measure zero by IV.2.14, while $T(\mathcal{J}_0)$ is of measure zero by assumption. Thus $\overline{E}$ is of measure zero, and hence it is sufficient to show that

$$\kappa(z, T, \mathfrak{D}_0) = N(z, T, \mathfrak{D}_0) \qquad \text{if } z \notin \overline{E}.$$

So let us take a point $z_0 \notin \overline{E}$. If $N(z_0, T, \mathfrak{D}_0) = 0$, then $\kappa(z_0, T, \mathfrak{D}_0) = 0$ too, by IV.1.47. Hence we can assume that $0 < N(z_0, T, \mathfrak{D}_0) < +\infty$. Then the set $T^{-1}(z_0) \cdot \mathfrak{D}_0$ is finite and nonempty. Let $w_1, \cdots, w_m$ be the points of this set, where $m = N(z_0, T, \mathfrak{D}_0)$. Since $z_0 \notin \overline{E}$ and hence $z_0 \notin T(\mathcal{J}_0)$, we have $w_i \in \mathcal{J} - \mathcal{J}_0$, $i = 1, 2, \cdots, m$ (cf. IV.1.56). But $\mathcal{J} - \mathcal{J}_0 \subset \mathcal{N}$ (cf. IV.1.78). Hence $N(z_0, T, \mathfrak{D}_0\mathcal{N}) \geqq m = N(z_0, T, \mathfrak{D}_0)$. Since $N(z_0, T, \mathfrak{D}_0\mathcal{N}) \leqq \kappa(z_0, T, \mathfrak{D}_0) \leqq N(z_0, T, \mathfrak{D}_0)$ (cf. IV.1.56), it follows that $\kappa(z_0, T, \mathfrak{D}_0) = N(z_0, T, \mathfrak{D}_0)$.

IV.3.21. Given T as in IV.2.1, we shall introduce presently certain quantities that will be termed *generalized Jacobians*. If the derivative $D(w, \mathfrak{D})$ exists at a point $w \in \mathfrak{D}$, then we define (cf. IV.3.9, IV.1.75)

$$\mathcal{J}_s(w) = i_s(w)D(w, \mathfrak{D}).$$

The quantity $\mathcal{J}_s(w)$ will be termed the *strong generalized Jacobian*. If the derivative $D_\kappa(w)$ exist at a point $w \in \mathfrak{D}$ then we define (cf. IV.3.9, IV.1.64)

$$\mathcal{J}_e(w) = i_e(w)D_\kappa(w).$$

The quantity $\mathcal{J}_e(w)$ will be termed the *essential generalized Jacobian*. Let us now write T in the real form $x = x(u, v)$, $y = y(u, v)$, $(u, v) \in \mathfrak{D}$ (cf. IV.2.1). If the partial derivatives x_u, x_v, y_u, y_v exist at a point $(u, v) \in \mathfrak{D}$ then we shall put

$$J(w) = x_u y_v - x_v y_u .$$

$J(w)$ will be termed the *ordinary Jacobian*. These Jacobians depend of course upon T and if desirable we shall denote them also by $\mathcal{J}_s(w, T)$, $\mathcal{J}_e(w, T)$, $J(w, T)$. We shall study presently the relationships between these three types of Jacobians.

IV.3.22. Assume that T, given as in IV.2.1, is BV$\mathfrak{D}$ in $\mathfrak{D}$. Then $\mathcal{J}_s(w)$ and $\mathcal{J}_e(w)$ exist and are equal to each other a.e. in $\mathfrak{D}$.

Proof. By IV.3.18, $D(w, \mathfrak{D})$ and $D_\kappa(w)$ exist a.e. in $\mathfrak{D}$, but we do not know whether they are equal to each other a.e. in $\mathfrak{D}$. At any rate, $\mathcal{J}_s(w)$ and $\mathcal{J}_e(w)$ exist a.e. in $\mathfrak{D}$ (see IV.3.21). Let us write (cf. IV.1.56) $\mathfrak{D} = \mathcal{G} + (\mathfrak{D} - \mathcal{G})$.

Case (i). $w \in \mathcal{G}$. By IV.1.76, we have

$$(1) \qquad i_s(w) = i_e(w) \qquad \text{for } w \in \mathcal{G}.$$

We proceed to show that (cf. IV.1.78)

$$(2) \qquad D(w, \mathfrak{D}) = D_\kappa(w) \qquad \text{a.e. on } \mathcal{G} - \mathcal{G}_0 .$$

Indeed (see IV.1.56, IV.1.78), $\mathcal{G} - \mathcal{G}_0 \subset N \subset \mathcal{E}^*$, and by IV.2.38, IV.3.14 we have $D(w, \mathfrak{D}) = D(w, \mathcal{E}^*)$ a.e. on $\mathcal{E}^*$ and $D(w, \mathcal{E}^*) = D_\kappa(w)$ a.e. in $\mathfrak{D}$, and (2) follows. In view of IV.3.21, (1) and (2) imply that

$$\mathcal{J}_s(w) = \mathcal{J}_e(w) \qquad \text{a.e. on } \mathcal{G} - \mathcal{G}_0 .$$

On the other hand, $i_s(w) = i_e(w) = 0$ for $w \in \mathcal{G}_0$, by IV.1.76, IV.1.78. Hence

$$\mathcal{J}_s(w) = \mathcal{J}_e(w) = 0 \qquad \text{a.e. on } \mathcal{G}_0 .$$

Thus $\mathcal{J}_s(w) = \mathcal{J}_e(w)$ a.e. on $\mathcal{G}$.

Case (ii). $w \in \mathfrak{D} - \mathcal{G}$. Clearly (cf. IV.2.2), $T(\mathfrak{D} - \mathcal{G}) \subset \overline{A}(\mathfrak{D}, +\infty)$. Since T is BV$\mathfrak{D}$ in $\mathfrak{D}$, it follows by IV.2.14 that $| T(\mathfrak{D} - \mathcal{G}) | = 0$. Hence, by IV.2.37, $D(w, \mathfrak{D}) = 0$ a.e. on $\mathfrak{D} - \mathcal{G}$. Since $0 \leqq D_\kappa(w) \leqq D(w, \mathfrak{D})$ a.e., it follows that $D_\kappa(w) = D(w, \mathfrak{D}) = 0$ and hence $\mathcal{J}_s(w) = \mathcal{J}_e(w) = 0$ a.e. on $\mathfrak{D} - \mathcal{G}$.

IV.3.23. Assume that T, given as in IV.2.1, is BV$\mathfrak{D}$ in $\mathfrak{D}$. Then $\mathcal{J}_s(w)$ exists a.e. in $\mathfrak{D}$, but since $i_s(w)$ may be different from ± 1, it is not *a priori* clear whether the formula $| \mathcal{J}_s(w) | = D(w, \mathfrak{D})$ holds a.e. in $\mathfrak{D}$. In fact, no general information is available at present on this point. On the other hand, if we add the assumption that $| T(\mathcal{G}_0) | = 0$ (cf. IV.1.78), then it follows that $| \mathcal{J}_s(w) | = D(w, \mathfrak{D})$ a.e. in $\mathfrak{D}$.

PROOF We have, by definition,

$$|\mathfrak{J}_s(w)| = |i_s(w)|\, D(w, \mathfrak{D}) \qquad \text{a.e. in } \mathfrak{D}.$$

As we observed in IV.3.22(ii), $D(w, \mathfrak{D}) = 0$ a.e. on $\mathfrak{D} - \mathcal{G}$. Hence

$$(1) \qquad |\mathfrak{J}_s(w)| = D(w, \mathfrak{D}) = 0 \qquad \text{a.e. on } \mathfrak{D} - \mathcal{G}.$$

In $\mathcal{G} - \mathcal{G}_0$ we have $i_s(w) \neq 0$ (cf. IV.1.78) and $|i_s(w)| \leqq 1$ except on a countable set (cf. IV.1.79). Thus $i_s(w) = \pm 1$ a.e. on $\mathcal{G} - \mathcal{G}_0$, and hence

$$(2) \qquad |\mathfrak{J}_s(w)| = D(w, \mathfrak{D}) \qquad \text{a.e. on } \mathcal{G} - \mathcal{G}_0 .$$

By assumption, $|T(\mathcal{G}_0)| = 0$, and hence $D(w, \mathfrak{D}) = 0$ a.e. on $\mathcal{G}_0$ by IV.2.37. Thus

$$(3) \qquad |\mathfrak{J}_s(w)| = D(w, \mathfrak{D}) = 0 \qquad \text{a.e. on } \mathcal{G}_0 .$$

(1), (2), (3) show that $|\mathfrak{J}_s(w)| = D(w, \mathfrak{D})$ a.e. in $\mathfrak{D}$.

IV.3.24. Assume that T, given as in IV.2.1, is BV$\mathfrak{D}$ in $\mathfrak{D}$. Then T is also BV$\mathcal{G}$ where $\mathcal{G}$ is the set $\mathcal{G} = \mathcal{G}(T, \mathfrak{D})$ defined in IV.1.56. Since $\mathcal{G}$ is a Borel set (see IV.1.61) we can use $\mathcal{G}$ as a base set, and by IV.2.32 the derivative $D(w, \mathcal{G})$ exists a.e. in $\mathfrak{D}$. In a sense, on the set $\mathcal{G}$ the transformation behaves almost like a biunique transformation, and it is interesting to observe that the Jacobian $\mathfrak{J}_s(w)$ can be expressed equally well in terms of the derivative $D(w, \mathcal{G})$. Indeed, we assert that $\mathfrak{J}_s(w) = i_s(w)D(w, \mathcal{G})$ a.e. in $\mathfrak{D}$ (assuming of course that T is BV$\mathfrak{D}$ in $\mathfrak{D}$).

PROOF. By IV.2.38 we have $D(w, \mathfrak{D}) = D(w, \mathcal{G})$ a.e. on $\mathcal{G}$. We observed in IV.3.22(ii) that $D(w, \mathfrak{D}) = 0$ a.e. on $\mathfrak{D} - \mathcal{G}$. Since clearly $0 \leqq D(w, \mathcal{G}) \leqq D(w, \mathfrak{D})$ a.e., it follows that $D(w, \mathfrak{D}) = D(w, \mathcal{G}) = 0$ a.e. on $\mathfrak{D} - \mathcal{G}$. Thus $D(w, \mathfrak{D}) = D(w, \mathcal{G})$ a.e. in $\mathfrak{D}$, and hence also $i_s(w)D(w, \mathcal{G}) = i_s(w)D(w, \mathfrak{D}) = \mathfrak{J}_s(w)$ a.e. in $\mathfrak{D}$.

IV.3.25. Given T as in IV.2.1, assume that $\kappa(z, T, \mathfrak{D})$ is summable. Then $|\mathfrak{J}_e(w)| = D_\kappa(w)$ a.e. in $\mathfrak{D}$.

PROOF. By definition we have $|\mathfrak{J}_e(w)| = |i_e(w)|\, D_\kappa(w)$ a.e. in $\mathfrak{D}$. By IV.1.67, IV.1.64, IV.1.74, $|i_e(w)| = 1$ a.e. on $\mathcal{N}$ and hence

$$(1) \qquad |\mathfrak{J}_e(w)| = D_\kappa(w) \qquad \text{a.e. on } \mathcal{N}.$$

By IV.1.64, $i_e(w) = 0$ on $\mathfrak{D} - \mathcal{N}$, and hence

$$(2) \qquad |\mathfrak{J}_e(w)| = 0 \qquad \text{a.e. on } \mathfrak{D} - \mathcal{N}.$$

By IV.2.35, $D(w, \mathcal{N}) = 0$ a.e. on $\mathfrak{D} - \mathcal{N}$, and hence by IV.3.13 we have also

$$(3) \qquad D_\kappa(w) = 0 \qquad \text{a.e. on } \mathfrak{D} - \mathcal{N}.$$

(1), (2), (3) show that $|\mathfrak{J}_e(w)| = D_\kappa(w)$ a.e. in $\mathfrak{D}$.

IV.3.26. Assume that T, given as in IV.2.1, is BV$\mathcal{E}^*$ in $\mathfrak{D}$ (cf. IV.1.56). Then $|\mathfrak{J}_e(w)| = D_\kappa(w) = D(w, \mathcal{E}^*)$ and $\mathfrak{J}_e(w) = i_e(w)D(w, \mathcal{E}^*)$ a.e. in $\mathfrak{D}$.

PROOF. Since $\kappa(z, T, \mathfrak{D}) \leqq N(z, \mathcal{E}^*)$, it follows by IV.2.13 that $\kappa(z, T, \mathfrak{D})$ is summable. Thus the assertions follow directly from IV.3.25, IV.3.14, IV.3.21.

IV.3.27. In studying the ordinary Jacobian $J(w) = x_u y_v - x_v y_u$, we shall need some facts about affine transformations. If a, b, c, d are (real) constants, then the formulas

$$T^* : \begin{cases} x = x_0 + a(u - u_0) + b(v - v_0), \\ y = y_0 + c(u - u_0) + d(v - v_0) \end{cases}$$

define an affine transformation that carries the point $w_0 = u_0 + iv_0$ into the point $z_0 = x_0 + iy_0$. Using complex notation, we may write equivalently

$$T^* : z = t^*(w) = z_0 + (a + ic)(u - u_0) + (b + id)(v - v_0).$$

We shall put $ad - bc = \Delta$. If $\Delta \neq 0$ then T^* is biunique and maps the w-plane onto the whole z-plane. If $\Delta = 0$, then T^* maps the w-plane either onto a single straight line or else onto a single point.

IV.3.28. Continuation. Let us first assume that

$$(1) \qquad \Delta = ad - bc \neq 0.$$

Then there exists a positive constant k, depending only upon a, b, c, d, such that

$$(2) \qquad | t^*(w'') - t^*(w') | \geqq k | w'' - w' |$$

for every choice of the points w', w''.

Proof. In view of (1), T^* has a single-valued inverse defined in the whole z-plane. By solving the equations defining T^* for u, v, we obtain for the inverse transformation formulas of the type

$$u = u_0 + \alpha(x - x_0) + \beta(y - y_0),$$

$$v = v_0 + \gamma(x - x_0) + \delta(y - y_0),$$

where α, β, γ, δ depend only upon a, b, c, d, and the existence of a constant k satisfying (2) follows. The exact (that is, the largest) constant k satisfying (2) is found, by an elementary discussion, to be given by the formula

$$2k^2 = a^2 + b^2 + c^2 + d^2 - [(a^2 + b^2 + c^2 + d^2)^2 - 4(ad - bc)^2]^{1/2}.$$

IV.3.29. Continuation. Given T^* as in IV.3.27, we find the estimate

$$| t^*(w) - z_0 | \leqq K | w - w_0 |,$$

where K is given by the formula

$$K = (a^2 + b^2 + c^2 + d^2)^{1/2}.$$

The assumption $\Delta \neq 0$ is not needed in deriving this estimate.

IV.3.30. Continuation. Let us assume again that

$$(1) \qquad \Delta = ad - bc \neq 0.$$

In the sequel we shall have occasion to consider an oriented square s with center at the point w_0. In view of (1), $T^*(s)$ is a parallelogram $\bar{p}$ with center at the point z_0 and we have the well known formula

$$|\bar{p}|/|s| = |\Delta|.$$

Since T^* is biunique in view of (1), it is easily seen that (cf. IV.1.24)

$$\mu(z, T^*, s) = \begin{cases} \operatorname{sgn} \Delta & \text{for } z \in \bar{p}^0, \\ 0 & \text{for } z \notin \bar{p}, \end{cases}$$

where sgn $\Delta = +1$ if $\Delta > 0$ and sgn $\Delta = -1$ if $\Delta < 0$.

IV.3.31. Given a transformation $T : z = t(w)$, $w \in \mathfrak{D}$, as in IV.2.1, let $w_0 = u_0 + iv_0$ be a point in $\mathfrak{D}$, and let us put $t(w_0) = z_0 = x_0 + iy_0$. Let us consider the affine transformation

$$T^* : z = t^*(w) = z_0 + (a + ic)(u - u_0) + (b + id)(v - v_0),$$

where a, b, c, d are for the moment arbitrary real constants. Then $t^*(w_0) = t(w_0) = z_0$. We propose to study the relation between T and T^* for fixed w_0, z_0, a, b, c, d. We define

$$\zeta(w) = \begin{cases} \dfrac{t(w) - t^*(w)}{|w - w_0|} & \text{for } w \in \mathfrak{D}, w \neq w_0, \\ 0 & \text{for } w = w_0. \end{cases}$$

Clearly, $\zeta(w)$ is continuous on the set $\mathfrak{D} - w_0$, but it may be discontinuous at the point w_0 itself. We have then the formula (note that $t(w_0) = t^*(w_0) = z_0$)

$$t(w) = t^*(w) + |w - w_0| \zeta(w) \qquad \text{for } w \in \mathfrak{D}.$$

Let s be an oriented square in $\mathfrak{D}$ with center at w_0. The function $\zeta(w)$ is continuous on the perimeter $s - s^0$ of s and hence we can define

$$m(s) = \max |\zeta(w)|, \qquad w \in s - s^0.$$

The function $\zeta(w)$ may be discontinuous at w_0, and hence it may be unbounded in s. Still we can define

$$M(s) = \text{l.u.b. } |\zeta(w)|, \qquad w \in s,$$

it being understood that possibly $M(s) = +\infty$. Clearly $\zeta(w)$, $m(s)$, $M(s)$ depend upon T, z_0, w_0, a, b, c, d.

IV.3.32. CONTINUATION. Let us now suppose that $\Delta = ad - bc \neq 0$. Then the constant k of IV.3.28 is available. Let us assume that for a certain choice of s the inequality $m(s) < k/2$ holds. Let l be the side length of s. We introduce two auxiliary oriented squares s', s'' with center at w_0, whose side lengths are given by the formulas

$$l' = \left(1 - \frac{2m(s)}{k}\right)l, \qquad l'' = \left(1 + \frac{2m(s)}{k}\right)l.$$

Note that s' can be constructed in view of the inequality $m(s) < k/2$.

IV.3.33. CONTINUATION. Let us assume now that

$$\Delta = ad - bc \neq 0, M(s) < +\infty.$$

We can now introduce an auxiliary square S'', with center at w_0, whose side length L'' is given by the formula

$$L'' = \left(1 + \frac{2^{1/2}M(s)}{k}\right)l.$$

Since $\Delta \neq 0$, $T^*(S'')$ is a parallelogram $\bar{P}''$ with center at z_0. We assert that

$$(1) \qquad | t^*(w) - z^* | > | t(w) - t^*(w) | \qquad \text{for } w \in s, z^* \notin \bar{P}''.$$

PROOF. Since z^* is exterior to $\bar{P}'' = T^*(S'')$ and T^* is biunique, we have a point w^* exterior to S'' such that $z^* = t^*(w^*)$. Since w^* is exterior to S'' we have the inequality

$$| w - w^* | > \frac{(L'' - l)}{2} = \frac{2^{1/2}M(s)l}{2k} \qquad \text{for } w \in s.$$

Hence, by IV.3.28,

$$(2) \; | t^*(w) - z^* | = | t^*(w) - t^*(w^*) | \geqq k | w - w^* | > \frac{2^{1/2}M(s)l}{2} \text{ for } w \in s.$$

On the other hand (cf. IV.3.31)

$$(3) \qquad | t(w) - t^*(w) | \leqq | w - w_0 || \zeta(w) | \leqq \frac{2^{1/2}M(s)l}{2} \qquad \text{for } w \in s.$$

(2) and (3) imply (1).

IV.3.34. CONTINUATION. Let us now assume only that

$$\Delta \neq 0, m(s) < k/2.$$

Let us consider the squares s, s', s'' (see IV.3.32), and let us put

$$T^*(s') = \bar{p}', T^*(s) = \bar{p}, T^*(s'') = \bar{p}''.$$

Then $\bar{p}', \bar{p}, \bar{p}''$ are similar parallelograms with center at z_0 and $\bar{p}' \subset \bar{p} \subset \bar{p}''$. Let us denote by $\bar{\sigma}$ the doubly-connected (closed) region bounded by the perimeters of $\bar{p}'$ and $\bar{p}''$. We assert that

$$(1) \qquad | t^*(w) - z^* | > | t(w) - t^*(w) | \qquad \text{for } w \in s - s^0, z^* \notin \bar{\sigma}^0.$$

PROOF. *Case* (i). z^* is not interior to $\bar{p}''$. Since T^* is biunique, we have then a point w^* not interior to s'', such that $t^*(w^*) = z^*$. If w is any point on the perimeter of s, then (cf. IV.3.28)

$$(2) \quad | t^*(w) - z^* | = | t^*(w) - t^*(w^*) | \geqq k | w - w^* | \geqq k \cdot \frac{l'' - l}{2} = m(s)l,$$

and further, by IV.3.31,

(3) $$| t(w) - t^*(w) | \leqq | w - w_0 || \zeta(w) | \leqq \frac{2^{1/2} m(s) l}{2}.$$

(2) and (3) imply (1).

Case (ii). $z^* \in \bar{p}'$. We have then a point $w^* \in s'$ such that $t^*(w^*) = z^*$. If w is any point on the perimeter of s, then we have by IV.3.28, IV.3.31,

(4) $$| t^*(w) - z^* | = | t^*(w) - t^*(w^*) | \geqq k | w - w^* | \geqq k \cdot \frac{l - l'}{2} = m(s) l,$$

(5) $$| t(w) - t^*(w) | \leqq | w - w_0 || \zeta(w) | \leqq \frac{2^{1/2} m(s) l}{2}.$$

(4) and (5) imply (1).

IV.3.35. Continuation. We assume again that

$$\Delta \neq 0, \; m(s) < k/2.$$

We assert that

(1) $$\mu(z, T, s) = \begin{cases} 0 & \text{for } z \text{ not interior to } \bar{p}'', \\ \operatorname{sgn} \Delta & \text{for } z \in \bar{p}'. \end{cases}$$

Proof. Let $z^* \notin \bar{\sigma}^0$ (cf. IV.3.34). We write

$$t(w) - z^* = [t^*(w) - z^*] + [t(w) - t^*(w)].$$

By IV.3.34 we have then

$$| t^*(w) - z^* | > | t(w) - t^*(w) | \qquad \text{for } w \in s - s^0.$$

Hence, by the theorem of Rouché (see II.4.26), $\mu(z^*, T, s) = \mu(z^*, T^*, s)$. In view of IV.3.30 the formulas (1) follow.

IV.3.36. Continuation. Given T, T^*, s as in IV.3.35, such that

$$\Delta \neq 0, \; m(s) < k/2,$$

suppose that we are also given a sequence of transformations $T_n : z = t_n(w)$, $w \in \mathfrak{D}_n$, such that the following conditions hold: (i) $s \subset \mathfrak{D}_n$ for n large, say $n > n_0$. (ii) $\rho(T_n, T, s) \to 0$ for $n \to \infty$ (cf. IV.1.1). Then there exists an integer n_1 such that

(1) $$\mu(z, T_n, s) = \mu(z, T, s) \qquad \text{for } n > n_1, \; z \in \bar{p}',$$

where n_1 is independent of z.

Proof. Let $z_0 \in \bar{p}'$ and $n > n_0$. Since $\Delta \neq 0$ and $m(s) < k/2$, we have, by IV.3.35, $\mu(z_0, T, s) \neq 0$. By IV.1.25, we have therefore a closed circular disc $U(z_0)$ with center z_0 such that

(2) $$\mu(z, T, s) = \mu(z_0, T, s) \neq 0 \qquad \text{for } z \in U(z_0).$$

(2) implies (cf. IV.1.24) that $U(z_0) \cdot T(s - s^0) = 0$. Hence condition (ii) implies, by IV.1.25, that

$$\mu(z, T_n, s) = \mu(z, T, s) \quad \text{for } z \in U(z_0),\ n > n(z_0),$$

where $n(z_0)$ is some sufficiently large integer depending upon z_0. Clearly we can assume that $n(z_0) > n_0$. Now the discs $U(z_0)$, associated in this manner with the various points $z_0 \in \bar{p}'$, cover $\bar{p}'$ in the sense required by the Borel covering theorem. Hence we have a finite number of points $z_0^1, z_0^2, \cdots, z_0^m$ in $\bar{p}'$ such that

$$\bar{p}' \subset U(z_0^1) + \cdots + U(z_0^m).$$

Clearly, if we put $n_1 = n(z_0^1) + \cdots + n(z_0^m)$, then (1) holds.

IV.3.37. Continuation Let us now assume that

$$\Delta = ad - bc \neq 0,\ m(s) < k/2,\ M(s) < +\infty.$$

We assert that $\bar{p}' \subset T(s) \subset \bar{P}''$ (see IV.3.33, IV.3.34, IV.3.31).

Proof. By IV.3.35 we have $\mu(z, T, s) = \operatorname{sgn} \Delta \neq 0$ for $z \in \bar{p}'$, and hence every point $z \in \bar{p}'$ is comprised in $T(s)$ by IV.1.25. Let us now consider a point $z^* \notin \bar{P}''$. Let us write

$$t(w) - z^* = [t^*(w) - z^*] + [t(w) - t^*(w)]. \tag{1}$$

Let w be any point in s. By IV.3.33 we have then the inequality

$$|\, t^*(w) - z^* \,| > |\, t(w) - t^*(w) \,|. \tag{2}$$

(1) and (2) imply that $t(w) - z^* \neq 0$ for $w \in s$, and thus $z^* \notin T(s)$.

IV.3.38. Continuation. Let us now assume that

$$\Delta = ad - bc \neq 0,\ m(s) < k/2.$$

We assert that $T(s - s^0) \subset \bar{\sigma}^0$ (cf. IV.3.34).

Proof. Let w be any point of $s - s^0$ and z^* any point not in $\bar{\sigma}^0$. By IV.3.34 we have the inequality

$$|\, t^*(w) - z^* \,| > |\, t(w) - t^*(w) \,|.$$

Hence $t(w) - z^* = [t^*(w) - z^*] + [t(w) - t^*(w)] \neq 0$.

IV.3.39. Continuation. Let us assume that

$$\Delta = ad - bc = 0,\ M(s) < +\infty.$$

We assert the inequality (cf. IV.3.31, IV.3.29)

$$|\, T(s) \,| < 2[K + M(s)]M(s) \,|\, s \,|. \tag{1}$$

Proof. Since $\Delta = 0$, the transformation T^* maps the whole w-plane either onto the single point z_0, or else onto a straight line passing through z_0. In either case we have a straight line g through z_0 such that

$$t^*(w) \in g \qquad \text{for every } w.$$

Let now $\overline{S}$ denote the segment on g with length $2^{1/2}Kl$ and center z_0. Since $| t^*(w) - z_0 | \leqq K | w - w_0 |$ by IV.3.29, it follows that

$$(2) \qquad T^*(s) \subset \overline{S}.$$

On the other hand, by IV.3.31,

$$(3) \qquad | t(w) - t^*(w) | = | w - w_0 || \zeta(w) | \leqq \frac{2^{1/2}M(s)l}{2} \qquad \text{for } w \in s.$$

(2) and (3) imply that $T(s)$ is comprised in a rectangle with sides equal to $2^{1/2}M(s)l$ and $2^{1/2}Kl + 2^{1/2}M(s)l$, and the inequality (1) follows.

IV.3.40. Continuation. Let us now assume only that

$$\Delta = ad - bc = 0.$$

We have then (cf. IV.3.39) a straight line g passing through z_0 such that $t^*(w) \in g$ for every w. Now let $\bar{r}$ be a rectangle in the z-plane defined as follows:

(i) The center of $\bar{r}$ is at z_0.

(ii) Two of the sides of $\bar{r}$ are parallel to g and have the length $2^{1/2}l[K + m(s)]$.

(iii) The other two sides of $\bar{r}$ are perpendicular to g and have the length $2^{1/2}m(s)l$. Since $t^*(w) \in g$ for every w and $t^*(w_0) = z_0$, it follows by IV.3.29 that $T^*(s - s^0)$ is comprised in the segment $\overline{S}$ on g with center at z_0 and length $2^{1/2}Kl$. By IV.3.31 we have

$$| t(w) - t^*(w) | \leqq 2^{1/2}m(s)l/2 \qquad \text{for } w \in s - s^0.$$

It follows that $T(s - s^0) \subset \bar{r}$. By II.4.34 it follows that $\mu(z, T, s) = 0$ for $z \notin \bar{r}$.

IV.3.41. Continuation. Using the assumptions and notations of IV.3.40, let z^* be a point such that (i) $z^* \notin \bar{r}$, and (ii) $N[z^*, T, \mathcal{E}^*(T, \mathfrak{D})] < +\infty$ (cf. IV.1.56). Then $\nu(z^*, T, s^0) = 0$ (cf. IV.3.2).

Proof. By IV.3.40, $z^* \notin T(s - s^0)$. In view of condition (ii) it follows by IV.3.4 that $\nu(z^*, T, s^0) = \mu(z^*, T, s)$. By IV.3.40 it follows that $\nu(z^*, T, s^0) = 0$.

IV.3.42. Theorem. *Given T as in* IV.2.1, *assume that T is $BV\mathcal{E}^*$ in $\mathfrak{D}$ and that the ordinary Jacobian $J(w)$ exists a.e. in $\mathfrak{D}$* (cf. IV.1.56, IV.2.11, IV.3.21). *Then*

$$(1) \qquad J(w) = \mathcal{J}_e(w) \qquad \textit{a.e. in } \mathfrak{D},$$

$$(2) \qquad | J(w) | = D_\kappa(w) = D(w, \mathcal{E}^*) = D(w, \mathcal{N}) \qquad \textit{a.e. in } \mathfrak{D}.$$

Proof. Let us note first that $\mathcal{J}_e(w)$, $D_\kappa(w)$, $D(w, \mathcal{E}^*)$, $D(w, \mathcal{N})$ exist a.e. in $\mathfrak{D}$ (cf. IV.3.14). For convenience, let us introduce the following auxiliary sets:

(1) E_1 is the set of those points $w \in \mathfrak{D}$ where $J(w)$ fails to exist. Then $| E_1 | = 0$ by assumption.

(2) E_2 is the set of those points $w \in \mathfrak{D}$ where one at least of $D_\kappa(w)$, $D(w, \mathcal{E}^*)$, $D(w, \mathcal{N})$, $D_\nu(w)$ fails to exist. Then $| E_2 | = 0$ by IV.3.14.

(3) E_3 is the set of those points $w \in \mathfrak{D}$ where $D_\kappa(w)$, $D(w, \mathcal{E}^*)$, $D(w, \mathcal{N})$, $D_\nu(w)$ all exist but fail to have the same value. Then $| E_3 | = 0$ by IV.3.14.

(4) E_4 is the set of those points $w \in \mathfrak{D}$ where $D(w, \mathcal{E}^*)$ and $\mathcal{J}_e(w)$ exist but $D(w, \mathcal{E}^*) \neq | \mathcal{J}_e(w) |$. Then $| E_4 | = 0$ by IV.3.26.

(5) E_5 is the set of those points $w \in \mathfrak{D} - \mathcal{N}$ where $D(w, \mathcal{E}^*)$ exists and is different from zero. Then $| E_5 | = 0$. Indeed, $D(w, \mathcal{E}^*) = D(w, \mathcal{N})$ a.e. in $\mathfrak{D}$ by IV.3.14, and $D(w, \mathcal{N}) = 0$ a.e. on $\mathfrak{D} - \mathcal{N}$ by IV.2.35. Clearly if w_0 is a point in $\mathfrak{D} - E_5$ where $D(w_0, \mathcal{E}^*)$ exists and is different from zero, then necessarily $w_0 \in \mathcal{N}$.

(6) E_6 is defined as follows. Let w_0 be a point where $J(w_0)$ exists. Let us associate with the point w_0 the affine transformation

$$T^* : \begin{cases} x = x_0 + a(u - u_0) + b(v - v_0), \\ y = y_0 + c(u - u_0) + d(v - v_0), \end{cases}$$

where we have put

$$x_0 + iy_0 = z_0 = t(w_0),\ w_0 = u_0 + iv_0 ,$$

$$a = x_u(u_0, v_0),\ b = x_v(u_0, v_0),\ c = y_u(u_0, v_0),\ d = y_v(u_0, v_0).$$

According to the notations adopted in IV.3.27 we have then

$$\Delta = ad - bc = J(w_0).$$

We define now first a set E_6^* as follows. A point $w_0 \in \mathfrak{D}$ belongs to E_6^* if and only if there exists a sequence s_n of oriented squares, with center at w_0, such that $| s_n | \to 0$ and $m(s_n) \to 0$ for $n \to \infty$, where $m(s)$ is defined as in IV.3.31 and is relative to the transformation T^* associated with w_0 in the manner just explained. Finally we put $E_6 = \mathfrak{D} - E_6^*$. Then $| E_6 | = 0$ by I.3.14.

(7) Finally, $E_7 = E_1 + \cdots + E_6$. Then $| E_7 | = 0$. We proceed to show that (1) and (2) hold for every point $w_0 \in \mathfrak{D} - E_7$.

IV.3.43. Continuation. Let $w_0 \in \mathfrak{D} - E_7$. We shall treat the cases $\Delta \neq 0$ and $\Delta = 0$ separately.

Case (i). $\Delta = J(w_0) \neq 0$, $w_0 \in \mathfrak{D} - E_7$. Then we have (see IV.3.42(6)) a sequence of oriented squares s_n, with center at w_0, such that

$$(1) \qquad | s_n | \to 0,\ m(s_n) \to 0.$$

Discarding, if necessary, a finite number of these squares, we shall have $s_n \subset \mathfrak{D}$, $m(s_n) < k/2$, where k is the constant defined in IV.3.28. Let us note that k depends only upon the point w_0 and is independent of the sequence s_n. Let l_n denote the side length of s_n and let us introduce the auxiliary oriented squares s_n', s_n'' with center at w_0 and side lengths l_n', l_n'' given by

$$(2) \qquad l_n' = \left(1 - \frac{2m(s_n)}{k}\right)l_n, \qquad l_n'' = \left(1 + \frac{2m(s_n)}{k}\right)l_n.$$

Let us denote (cf. IV.3.34) by $\bar{p}_n'$, $\bar{p}_n$, $\bar{p}_n''$ the parallelograms $T^*(s_n')$, $T^*(s_n)$,

$T^*(s_n'')$ and by $\bar{\sigma}_n$ the doubly-connected closed region bounded by the perimeters of $\bar{p}_n'$, $\bar{p}_n''$ (see IV.3.42(6) concerning T^*). By IV.3.35

(3)
$$\mu(z, T, s_n) = \begin{cases} 0 & \text{for } z \text{ not interior to } \bar{p}_n'', \\ \operatorname{sgn} \Delta & \text{for } z \in \bar{p}_n'. \end{cases}$$

Since T is BV$\mathcal{E}^*$, we have by IV.2.14

(4)
$$N(z, T, \mathcal{E}^*) < +\infty \qquad \text{a.e. in the } z\text{-plane.}$$

By IV.3.38

(5)
$$T(s_n - s_n^0) \subset \bar{\sigma}_n^0.$$

(3), (4), (5) imply, by IV.3.4,

$$\nu(z, T, s_n^0) = \begin{cases} 0 & \text{a.e. outside of } \bar{p}_n'', \\ \operatorname{sgn} \Delta = \pm 1 & \text{a.e. in } \bar{p}_n'. \end{cases}$$

Hence (cf. IV.3.7)

$$| \bar{p}_n' | \leqq G_1(s_n) \leqq | \bar{p}_n'' |.$$

In view of IV.3.30 we obtain the inequalities

(6)
$$| \Delta | \cdot \frac{| s_n' |}{| s_n |} \leqq \frac{G_1(s_n)}{| s_n |} \leqq | \Delta | \cdot \frac{| s_n'' |}{| s_n |}.$$

Since $D_\nu(w_0)$ exists (cf. IV.3.42(2), (7)), for $n \to \infty$ we obtain from (1), (2), (6)

(7)
$$D_\nu(w_0) = | \Delta | = | J(w_0) |.$$

By IV.3.42(3), (7) it follows that

(8)
$$| J(w_0) | = D_\kappa(w_0) = D(w_0, \mathcal{E}^*) = D(w_0, \mathcal{N}).$$

It remains to show that $J(w_0) = \mathcal{J}_e(w_0)$. Let us note that $D(w_0, \mathcal{E}^*) \neq 0$ by (7) and (8). Hence, in view of IV.3.42(5), (7), it follows that $w_0 \in \mathcal{N}$. By (3) we have $\mu(z_0, T, s_n) = \operatorname{sgn} \Delta = \pm 1$. Hence, by IV.1.65, we have $\mu(z_0, T, s_n) = i_e(w_0)$ for n large, and consequently

(9)
$$i_e(w_0) = \operatorname{sgn} \Delta = \operatorname{sgn} J(w_0).$$

(8) and (9) yield (cf. IV.3.21)

$$J(w_0) = | J(w_0) | \operatorname{sgn} J(w_0) = i_e(w_0) D_\kappa(w_0) = \mathcal{J}_e(w_0).$$

IV.3.44. CONTINUATION. *Case* (ii). $J(w_0) = \Delta = 0$, $w_0 \in \mathfrak{D} - E_7$. By IV.3.42(6), (7) we have a sequence of oriented sqaures s_n, with center at w_0, such that

(1)
$$| s_n | \to 0, \; m(s_n) \to 0.$$

For each n, let $\bar{r}_n$ be the rectangle associated with the square s_n in the sense of IV.3.40. Since T is BV$\mathcal{E}^*$, we have $N(z, T, \mathcal{E}^*) < +\infty$ a.e. in the z-plane. By IV.3.40 we have $\mu(z, T, s_n) = 0$ for $z \notin \bar{r}_n$. Hence, by IV.3.41, $\nu(z, T, s^0) = 0$ a.e. outside of $\bar{r}_n$. By IV.3.7, IV.3.40 it follows that

$$(2) \qquad G_1(s_n) \leqq |\bar{r}_n| = 2[K + m(s_n)]m(s_n) \, |s_n|.$$

Since $D_\nu(w_0)$ exists (see IV.3.42(2), (7)), it follows from (1) and (2) that $D_\nu(w_0) = 0$. By IV.3.42(3), (7) it follows that $D_\kappa(w_0) = 0$ and hence $\mathcal{J}_e(w_0) = 0$ (cf. IV.3.21). Thus $J(w_0) = 0 = \mathcal{J}_e(w_0)$, and the proof is complete (observe that $D_\kappa(w_0) = D(w_0, \mathcal{E}^*) = D(w_0, \mathcal{N})$ by IV.3.42(3), (7)).

IV.3.45. CONTINUATION. We have just proved that $J(w) = \mathcal{J}_e(w)$ a.e. in $\mathfrak{D}$ under the conditions described in IV.3.42. The following remarks are of interest from the point of view of the geometrical interpretation of the ordinary Jacobian $J(w)$.

(i) Under the conditions described in IV.3.42, we have

$$(1) \qquad J(w) = 0 \qquad \text{a.e. on } \mathfrak{D} - \mathcal{N}.$$

Indeed, $D(w, \mathcal{N}) = 0$ a.e. on $\mathfrak{D} - \mathcal{N}$ by IV.2.35 and thus (1) is a direct consequence of IV.3.42.

(ii) Under the conditions described in IV.3.42, we have the formulas (cf. IV.3.42, IV.3.14, IV.3.21)

$$(2) \qquad J(w) = i_e(w)D(w, \mathcal{N}), \ |J(w)| = D(w, \mathcal{N}) \qquad \text{a.e. in } \mathfrak{D}.$$

Since the existence, a.e. in $\mathfrak{D}$, of $D(w, \mathcal{N})$ is guaranteed as soon as T is BV$\mathcal{N}$ in $\mathfrak{D}$, there arises the question whether the formulas (1), (2) could be derived under conditions weaker than those used in IV.3.42. The formulas (1) and (2) also suggest the use of $\mathcal{N}$ as the fundamental base set. The questions suggested by these remarks have not been investigated as yet.

IV.3.46. Given T as in IV.2.1, let us assume that the following conditions hold:

(i) $\kappa(z, T, \mathfrak{D})$ is summable.

(ii) The ordinary Jacobian $J(w) = x_u y_v - x_v y_u$ exists a.e. in $\mathfrak{D}$. Then we have the inequality (cf. IV.3.21)

$$(1) \qquad |J(w)| \leqq |\mathcal{J}_e(w)| = D_\kappa(w) \qquad \text{a.e. in } \mathfrak{D}.$$

PROOF. Let us introduce the following auxiliary sets.

(1) E_1 is the set of those points $w \in \mathfrak{D}$ where $J(w)$ fails to exist. Then $|E_1| = 0$ by assumption.

(2) E_2 is the set of those points $w \in \mathfrak{D}$ where $D_\kappa(w)$ fails to exist. Then $|E_2| = 0$ by IV.3.13.

(3) E_3 is the set of those points $w \in \mathfrak{D}$ where $D_\kappa(w)$ exists but the equation $D_\kappa(w) = |\mathcal{J}_e(w)|$ fails to hold. Then $|E_3| = 0$ by IV.3.25.

(4) Finally, E_6 is defined as in IV.3.42(6). Then $|E_6| = 0$, as we have observed there. Then (1) will be proved if we show that

(2) $$| J(w_0) | \leqq | \mathfrak{J}_e(w_0) | = D_\kappa(w_0) \qquad \text{if } w_0 \notin E_1 + E_2 + E_3 + E_6 .$$

Since (2) clearly holds if $J(w_0) = 0$, we can further assume that

(3) $$J(w_0) \neq 0.$$

Let w_0 be a point such that $w_0 \notin E_1 + E_2 + E_3 + E_6$, $J(w_0) \neq 0$. Then we have (see the definition of E_6) a sequence of oriented squares s_n, with center at w_0, such that

(4) $$| s_n | \to 0, \; m(s_n) \to 0.$$

Discarding, if necessary, a finite number of these squares, we shall have (cf. IV.3.28)

(5) $$s_n \subset \mathfrak{D}, \; m(s_n) < k/2.$$

Let l_n be the side length of s_n, and let s_n' be the oriented square with center w_0 and side length l_n' given by

(6) $$l_n' = \left(1 - \frac{2m(s_n)}{k}\right) l_n .$$

Let us denote again by $\bar{p}_n$, $\bar{p}_n'$ the parallelograms $T^*(s_n)$, $T^*(s_n')$ (cf. the definition of E_6). By IV.3.35 we have then, in view of (3),

$$\mu(z, T, s_n) = \operatorname{sgn} J(w_0) \neq 0 \qquad \text{for } z \in \bar{p}_n',$$

and hence, by IV.1.26, IV.1.50, $\kappa(z, T, s_n^0) \neq 0$ for $z \in \bar{p}_n'$. By IV.3.7, IV.3.30 we obtain the inequalities

$$G_4(s_n) \geqq | \bar{p}_n' | = | J(w_0) | \, | s_n' |,$$

$$\frac{G_4(s_n)}{| s_n |} \geqq | J(w_0) | \cdot \frac{| s_n' |}{| s_n |}.$$

Since $D_\kappa(w_0)$ exists (cf. the definition of E_2), and since $| s_n' | / | s_n | \to 1$ by (5) and (6), the relations (2) follow (cf. the definition of E_3).

IV.3.47. *Given T as in* IV.2.1, *let us assume that T is BV$\mathfrak{D}$ in $\mathfrak{D}$ and the ordinary Jacobian $J(w) = x_u y_v - x_v y_u$ exists a.e. in $\mathfrak{D}$. Then* (cf. IV.3.21)

$$J(w) = \mathfrak{J}_s(w) \qquad \textit{a.e. in } \mathfrak{D}.$$

Proof. We have $\mathfrak{J}_s(w) = \mathfrak{J}_e(w)$ a.e. in $\mathfrak{D}$ by IV.3.22, and $\mathfrak{J}_e(w) = J(w)$ a.e. in $\mathfrak{D}$ by IV.2.38 and IV.3.42.

Remark. Comparison of IV.3.47, IV.3.42 reveals that we failed to state, in analogy with IV.3.42, the formula $| J(w) | = D(w, \mathfrak{D})$ a.e. in $\mathfrak{D}$. As a matter of fact, it is not known whether this formula is true if T is only known to be BV$\mathfrak{D}$, and we shall prove it presently only under further restrictions upon T.

IV.3.48. Given T as in IV.2.1, suppose that the following assumptions hold:

(i) T is BV$\mathfrak{D}$ in $\mathfrak{D}$.

(ii) The ordinary Jacobian $J(w) = x_u y_v - x_v y_u$ exists a.e. in $\mathfrak{D}$.

(iii) $| T(\mathscr{I}_0) | = 0$ (see IV.1.78).

Then $| J(w) | = D(w, \mathfrak{D})$ a.e. in $\mathfrak{D}$.

PROOF. By IV.2.38 and IV.3.42, we have $| J(w) | = D(w, \mathcal{E}^*)$ a.e. in $\mathfrak{D}$ and by (i), (iii) and IV.3.19 we have $D(w, \mathcal{E}^*) = D(w, \mathfrak{D})$ a.e. in $\mathfrak{D}$.

REMARK. It seems to be a matter of interest to decide whether the condition (iii) is really necessary. In the absence of adequate information on this point, we proceed to consider a very special but quite important case in which the conditions (i), (ii), (iii) hold.

IV.3.49. Given $T : z = t(w)$, $w \in \mathfrak{D}$, as in IV.2.1, let us use simultaneously the real equations $T : x = x(u, v)$, $y = y(u, v)$, $(u, v) \in \mathfrak{D}$. If both $x(u, v)$ and $y(u, v)$ are totally differentiable at a point $u_0 + iv_0 = w_0 \in \mathfrak{D}$ (see I.3.14), then we shall say that T is totally differentiable at that point. Let us note that if T is totally differentiable at w_0, then the ordinary Jacobian $J(w_0)$ exists (see I.3.14).

IV.3.50. Suppose that T, given as in IV.2.1, satisfies the following conditions at a point $w_0 \in \mathfrak{D}$.

(i) T is totally differentiable at w_0 (see IV.3.49).

(ii) $J(w_0) = 0$.

(iii) The derivative $D(w_0, \mathfrak{D})$ exists.

Then $D(w_0, \mathfrak{D}) = 0$.

PROOF. Let s be any oriented square, with center at w_0, comprised in $\mathfrak{D}$. Let us consider the auxiliary affine transformation

$$T^* : \begin{cases} x = x_0 + a(u - u_0) + b(v_0 - v), \\ \\ y = y_0 + c(u - u_0) + d(v - v_0), \end{cases}$$

where $x_0 + iy_0 = z_0 = t(w_0)$, $a = x_u(u_0, v_0)$, $b = x_v(u_0, v_0)$, $c = y_u(u_0, v_0)$, $d = y_v(u_0, v_0)$. Using the notations of IV.3.31, we see readily that condition (i) implies (cf. I.3.14) the relations

$$(1) \qquad M(s) < +\infty \text{ and } M(s) \to 0 \qquad \text{for } | s | \to 0,$$

while condition (ii) yields $J(w_0) = \Delta = 0$. By IV.3.39 it follows that

$$(2) \qquad | T(s) |/| s | < 2[K + M(s)]M(s),$$

where the constant K is independent of s (see IV.3.29). Since $D(w_0, \mathfrak{D})$ exists by assumption, (2) and (1) yield, for $| s | \to 0$, the desired formula $D(w_0, \mathfrak{D}) = 0$.

REMARK. A somewhat closer discussion of the auxiliary transformation T^* would show that condition (iii) is unnecessary, but this is irrelevant for our purposes.

IV.3.51. Assume that T, given as in IV.2.1, satisfies the following conditions:

(i) T is AC$\mathfrak{D}$ in $\mathfrak{D}$.

(ii) T is totally differentiable a.e. in $\mathfrak{D}$ (see IV.3.49). Then $| J(w) | = D(w, \mathfrak{D})$ a.e. in $\mathfrak{D}$.

PROOF. In view of IV.2.39 and IV.3.48, it is sufficient to show that

(1) $$| T(\mathcal{S}_0) | = 0.$$

Let us observe that by IV.2.39, IV.3.47, IV.3.22, IV.3.14

(2) $$J(w) = \mathcal{J}_e(w) = i_e(w)D(w, \mathcal{E}^*) \quad \text{a.e. in } \mathfrak{D}.$$

By IV.1.76, IV.1.78

(3) $$i_e(w) = i_s(w) = 0 \quad \text{on } \mathcal{S}_0 .$$

(2), (3) imply that

(4) $$J(w) = 0 \quad \text{a.e. on } \mathcal{S}_0 .$$

Let now w_0 be a point in $\mathcal{S}_0$ where T is totally differentiable, (4) holds, and $D(w_0, \mathfrak{D})$ exists (note that these conditions hold a.e. on $\mathcal{S}_0$). By IV.3.50 we have then $D(w_0, \mathfrak{D}) = 0$. Thus $D(w, \mathfrak{D}) = 0$ a.e. on $\mathcal{S}_0$. Since $\mathcal{S}_0$ is a Borel set (see IV.1.78, IV.1.77), and since T is AC$\mathfrak{D}$ in $\mathfrak{D}$, it follows by IV.2.21, IV.2.44 (applied with $\mathcal{B} = \mathfrak{D}$, $B = \mathcal{S}_0$) that

$$\iint N(z, \mathcal{S}_0) = \iint_{\mathcal{S}_0} D(w, \mathfrak{D}) = 0.$$

Thus $N(z, \mathcal{S}_0) = 0$ a.e. in the z-plane. Since $N(z, \mathcal{S}_0) \geqq 1$ on $T(\mathcal{S}_0)$, (1) follows.

CHAPTER IV.4. SPECIAL CLASSES OF TRANSFORMATIONS

IV.4.1. Given a transformation $T: z = t(w)$, $w \in \mathfrak{D}$, as in IV.2.1, we introduce the following definitions:

(i) T is sBV in $\mathfrak{D}$ (*strongly of bounded variation in* $\mathfrak{D}$) if T is BV$\mathfrak{D}$ in $\mathfrak{D}$ (see IV.2.11).

(ii) T is sAC in $\mathfrak{D}$ (*strongly absolutely continuous in* $\mathfrak{D}$) if T is AC$\mathfrak{D}$ in $\mathfrak{D}$ (see IV.2.39).

(iii) T is eBV in $\mathfrak{D}$ (*essentially of bounded variation in* $\mathfrak{D}$) if $\kappa(z, T, \mathfrak{D})$ is summable (see IV.1.4, IV.2.1, IV.1.43).

(iv) T is eAC in $\mathfrak{D}$ (*essentially absolutely continuous in* $\mathfrak{D}$) if T is AC$\mathcal{E}^*(T, \mathfrak{D})$ in $\mathfrak{D}$ (see IV.2.39, IV.1.56).

The following statements are obvious.

(a) If T is sAC in $\mathfrak{D}$, then T is also sBV in $\mathfrak{D}$ (cf. IV.2.39).

(b) If T is sAC in $\mathfrak{D}$, and if $\mathfrak{D}^*$ is any subdomain of $\mathfrak{D}$, then T is also sAC in $\mathfrak{D}^*$ (cf. IV.2.39).

(c) If T is sBV in $\mathfrak{D}$, and if $\mathfrak{D}^*$ is any subdomain of $\mathfrak{D}$, then T is also sBV in $\mathfrak{D}^*$ (cf. IV.2.11).

(d) Statements analogous to (a), (b), (c), with sAC, sBV replaced by eAC, eBV (cf. IV.1.56).

(e) If T is sAC in $\mathfrak{D}$, then T is also eAC in $\mathfrak{D}$ (cf. IV.2.39).

(f) If T is sBV in $\mathfrak{D}$, then T is also eBV in $\mathfrak{D}$.

Let us stress the lack of analogy between the definitions of sBV and eBV. The former is defined as bounded variation with respect to the base-set $\mathfrak{D}$, but the latter is *not* defined as bounded variation with respect to the base-set $\mathcal{E}^*(T, \mathfrak{D})$ as the definition of eAC may lead one to expect.

IV.4.2. Suppose that T is eAC in $\mathfrak{D}$. By IV.2.39, T is then also BV$\mathcal{E}^*(T, \mathfrak{D})$ in $\mathfrak{D}$, and hence (see IV.3.14) the derivatives $D_\nu(w, T^*)$, $D(w, T, \mathcal{N})$, $D(w, T, \mathcal{E})$, $D_\kappa(w, T)$, $D(w, T, \mathcal{E}^*)$, $D^*(w, T, \mathcal{N})$, $D^*(w, T, \mathcal{E})$, $D_\kappa(w, T)$, $D^*(w, T, \mathcal{E}^*)$ exist and are equal to each other a.e. in $\mathfrak{D}$. As a matter of fact, we should write, more explicitly, $\mathcal{N}(T, \mathfrak{D})$, $\mathcal{E}(T, \mathfrak{D})$, $\mathcal{E}^*(T, \mathfrak{D})$ instead of $\mathcal{N}$, $\mathcal{E}$, $\mathcal{E}^*$ in the preceding statements, but the more concise notation is sufficiently clear if T and $\mathfrak{D}$ remain fixed in a given situation. Let us note that the essential generalized Jacobian $\mathcal{J}_e(w) = i_e(w)D_\kappa(w)$ satisfies, by IV.3.26, a.e. in the z-plane the equations

$$(1) \qquad |\mathcal{J}_e(w)| = D_\kappa(w) = D(w, \mathcal{E}^*), \quad \mathcal{J}_e(w) = i_e(w)D(w, \mathcal{E}^*).$$

Let us agree again that we shall use the more concise notations $\mathcal{J}_e(w)$, $i_e(w)$ $D_\kappa(w)$, and so on, instead of $\mathcal{J}_e(w, T)$, $i_e(w, T)$, $D_\kappa(w, T)$, and so on, if the context is such that no misunderstandings can arise.

IV.4.3. *Suppose that T is* eAC *in $\mathfrak{D}$. Let $H(z)$ be a finite-valued Borel measurable function in the z-plane. Then* (cf. IV.2.1)

$$\iint_{\mathfrak{D}} H[t(w)]\,\mathcal{J}_e(w) = \iint H(z)\nu(z, \mathfrak{D}), \tag{1}$$

as soon as the integral on the left exists.

PROOF. $H[t(w)]$ is Borel measurable by IV.2.20, and $i_e(w)$ is Borel measurable by IV.1.66. Hence the function

$$\Phi(w) = H[t(w)]i_e(w) \tag{2}$$

is Borel measurable (and finite-valued) in $\mathfrak{D}$. We propose to compute the function $\sigma(z, \Phi)$ corresponding to the base-set $\mathfrak{B} = \mathcal{E}^*$ (cf. IV.2.15). We assert that (cf. IV.3.2)

$$\sigma(z, \Phi) = H(z)\nu(z, \mathfrak{D}) \qquad \text{a.e. in the } z\text{-plane.} \tag{3}$$

Indeed, since T is BV$\mathcal{E}^*$ in $\mathfrak{D}$ (cf. IV.4.2), we have $N(z, \mathcal{E}^*) < +\infty$ a.e. in the z-plane, by IV.2.13. Hence it is sufficient to show that

$$\sigma(z_0, \Phi) = H(z_0)\nu(z_0, \mathfrak{D}) \qquad \text{if } N(z_0, \mathcal{E}^*) < +\infty. \tag{4}$$

Let then z_0 be a point such that $N(z_0, \mathcal{E}^*) < +\infty$, and let us consider the set $E = T^{-1}(z_0)\mathcal{E}^*$. Then E is a finite set, and we assert that

$$E = T^{-1}(z_0)\mathcal{N}. \tag{5}$$

Indeed, the assumption $N(z_0, \mathcal{E}^*) < +\infty$ implies that z_0 has, under T in $\mathfrak{D}$, at most a finite number of essential maximal model continua, each of which reduces to a single point. Each of these points belongs then, by definition, to $\mathcal{N}$ (cf. IV.1.56), and (5) follows. We have (using IV.2.15 with $\mathfrak{B} = \mathcal{E}^*$)

$$\sigma(z_0, \Phi) = \sum \Phi(w) = \sum H[t(w)]i_e(w), \qquad w \in T^{-1}(z_0)\mathcal{E}^*. \tag{6}$$

Now $H[t(w)] = H(z_0)$ for $w \in T^{-1}(z_0)$, and thus (5) and (6) imply that

$$\sigma(z_0, \Phi) = \sum H(z_0)i_e(w) = H(z_0) \sum i_e(w), \quad w \in T^{-1}(z_0)\mathcal{N}. \tag{7}$$

On the other hand, by definition (cf. IV.3.2).

$$\nu(z_0, \mathfrak{D}) = \sum i_e(w), \qquad w \in T^{-1}(z_0)\mathcal{N}. \tag{8}$$

(7) and (8) imply (4), and thus (3) is verified. Since $\Phi(w)$ (cf. (2)) is finite-valued and Borel measurable, we have by IV.2.50 and in view of (2) and (3),

$$\iint_{\mathfrak{D}} H[t(w)]i_e(w)D(w, \mathcal{E}^*) = \iint H(z)\nu(z, \mathfrak{D})$$

as soon as the integral on the left exists. The formula (1) follows by IV.4.2(1).

IV.4.4. *Suppose that T is eAC in $\mathfrak{D}$. If $H(z)$ is a finite-valued measurable function in the z-plane, then $H[t(w)]\mathcal{J}_e(w)$ is measurable in $\mathfrak{D}$, and* (cf. IV.2.1)

$$\iint_{\mathfrak{D}} H[t(w)]\,\mathcal{J}_e(w) = \iint H(z)\nu(z, \mathfrak{D}), \tag{1}$$

as soon as the integral on the left exists.

PROOF. We have (see I.3.8) a finite-valued, Borel measurable function $H^*(z)$ such that $H^*(z) = H(z)$ a.e. in the z-plane. Noting that T is BV$\mathcal{E}^*$ (cf. IV.4.2), we infer from IV.2.51 that $H[t(w)]D(w, \mathcal{E}^*) = H^*[t(w)]D(w, \mathcal{E}^*)$ a.e. in $\mathfrak{D}$, and hence (cf. IV.4.2)

$$H[t(w)]\mathcal{J}_e(w) = H^*[t(w)]\mathcal{J}_e(w) \qquad \text{a.e. in } \mathfrak{D}. \tag{2}$$

Since $H(z) = H^*(z)$ a.e. in the z-plane, we also have

$$H(z)\nu(z, \mathfrak{D}) = H^*(z)\nu(z, \mathfrak{D}) \qquad \text{a.e. in the } z\text{-plane}. \tag{3}$$

Let us observe that the functions involved in (2) are measurable in $\mathfrak{D}$ by IV.2.52 and IV.1.66. Suppose now that $H[t(w)]\mathcal{J}_e(w)$ is summable in $\mathfrak{D}$. Then by (2) the function $H^*[t(w)]\mathcal{J}_e(w)$ is also summable in $\mathfrak{D}$, and hence by IV.4.3

$$\iint_{\mathfrak{D}} H^*[t(w)]\,\mathcal{J}_e(w) = \iint H^*(z)\nu(z, \mathfrak{D}). \tag{4}$$

(2), (3), (4) imply (1).

IV.4.5. CONTINUATION. In applications, there arise cases where the function $H(z)$, known to be measurable, is equal to $\pm\infty$ on a certain set of measure zero. In such a case, let us replace $H(z)$ by a function $H^0(z)$ defined as follows: $H^0(z) = H(z)$ if $H(z)$ is finite, and $H^0(z) = 0$ if $H(z) = \pm\infty$. Then $H^0(z)$ is measurable and finite-valued, and the preceding result applies, but there is a little point that should be cleared up. Suppose we have, in a certain application, a set of measure zero on which $H(z)$ is not known to be finite, but it is not certain either that $H(z) = \pm\infty$ on this set. The set in question is then not univocally determined by the function $H(z)$, and there arises the following question. Let $\bar{E}_1$, $\bar{E}_2$ be two sets of measure zero in the z-plane, such that $H(z)$ is measurable and finite-valued for z not in $\bar{E}_1$ and also for z not in $\bar{E}_2$. Let $H_i^0(z)$, $i = 1, 2$, be the function that is equal to $H(z)$ for z not in $\bar{E}_i$, and equal to zero on $\bar{E}_i$. Then $H_1^0(z)$, $H_2^0(z)$ are finite-valued and measurable, and $H_1^0(z) = H_2^0(z)$ a.e. in the z-plane. The argument used in IV.4.4 (based on IV.4.2, IV.2.51) yields that $H_1^0[t(w)]\mathcal{J}_e(w) = H_2^0[t(w)]\mathcal{J}_e(w)$ a.e. in $\mathfrak{D}$. Hence, if one of these functions is summable in $\mathfrak{D}$, then the other one is also summable, and we obtain, since $H_1^0(z) = H_2^0(z) = H(z)$ a.e. in the z-plane, the relations

$$\iint_{\mathfrak{D}} H_1^0[t(w)]\,\mathcal{J}_e(w) = \iint_{\mathfrak{D}} H_2^0[t(w)]\,\mathcal{J}_e(w) = \iint H_1^0(z)\nu(z, \mathfrak{D})$$

$$= \iint H_2^0(z)\nu(z, \mathfrak{D}) = \iint H(z)\nu(z, \mathfrak{D}).$$

Thus, if a function $H(z)$ is measurable and finite-valued, with the exception of a doubtful set of measure zero, and if we define $H(z)$ to be equal to zero on this doubtful set, then our possibly imperfect knowledge of this set does not lead to any trouble as far as the transformation formula (1) in IV.4.4 is concerned. In this sense, the formula (1) may be used for functions $H(z)$ that fail to be (or are merely not known to be) finite on a certain set of measure zero. Similar remarks apply to all the transformation formulas considered in the sequel.

IV.4.6. *Suppose that T is eAC in $\mathfrak{D}$. If $H(z)$ is a finite-valued, measurable function in the z-plane, then*

$$(1) \qquad \iint_{\mathfrak{D}} H[t(w)] \mid \mathfrak{J}_e(w) \mid = \iint H(z)\kappa(z, T, \mathfrak{D}),$$

as soon as one of the two integrals involved exists.

Proof. By IV.2.53 (applied with $\mathfrak{B} = \mathcal{E}^*$) we have

$$(2) \qquad \iint_{\mathfrak{D}} H[t(w)]D(w, \mathcal{E}^*) = \iint H(z)N(z, T, \mathcal{E}^*),$$

as soon as one of the two integrals involved exists. By IV.1.56, $N(z, T, \mathcal{E}^*) = \kappa(z, T, \mathfrak{D})$ for every point z such that $N(z, T, \mathcal{E}^*) < +\infty$. Since T is BV$\mathcal{E}^*$ (cf. IV.4.2), it follows by IV.2.14 that $N(z, T, \mathcal{E}^*) = \kappa(z, T, \mathfrak{D})$ a.e. in the z-plane. By IV.4.2 we have $\mid \mathfrak{J}_e(w) \mid = D(w, \mathcal{E}^*)$ a.e. in $\mathfrak{D}$. Thus the formulas (1) and (2) are equivalent, and the proof is complete.

IV.4.7. Continuation. For $H(z) \equiv 1$ we obtain the following formulas, valid as soon as T is eAC in $\mathfrak{D}$, from IV.4.4, IV.4.6.

$$\iint_{\mathfrak{D}} \mathfrak{J}_e(w) = \iint \nu(z, T, \mathfrak{D}), \qquad \iint_{\mathfrak{D}} \mid \mathfrak{J}_e(w) \mid = \iint \kappa(z, T, \mathfrak{D}).$$

The expressions on the right-hand sides of these formulas may be interpreted as the *essential signed area* and the *essential absolute area* respectively of the image of $\mathfrak{D}$ under T.

IV.4.8. *Given T as in* IV.2.1, *suppose that $\kappa(z, T, \mathfrak{D})$ is summable. By* IV.3.25, IV.3.13 *we have then the inequality*

$$\iint_{\mathfrak{D}} \mid \mathfrak{J}_e(w) \mid \leqq \iint \kappa(z, T, \mathfrak{D}).$$

We assert that the sign of equality holds if and only if T is eAC *in $\mathfrak{D}$.*

Proof. The sufficiency of the condition follows from IV.4.7. To show the necessity, let us assume that

$$(1) \qquad \iint_{\mathfrak{D}} \mid \mathfrak{J}_e(w) \mid = \iint \kappa(z, T, \mathfrak{D}).$$

By IV.3.25, IV.3.13, IV.2.34, IV.2.33, IV.2.21, IV.1.56 we obtain, using (1),

$$\iint_{\mathfrak{D}} |\mathcal{J}_e(w)| = \iint_{\mathfrak{D}} D_\kappa(w) = \iint_{\mathfrak{D}} D(w, \mathcal{E}) \leqq \iint N(z, T, \mathcal{E})$$

$$\leqq \iint \kappa(z, T, \mathfrak{D}) = \iint_{\mathfrak{D}} |\mathcal{J}_e(w)|.$$

It follows that the sign of equality holds throughout. In particular

$$\iint N(z, T, \mathcal{E}) = \iint \kappa(z, T, \mathfrak{D}). \tag{2}$$

Since $0 \leqq N(z, T, \mathcal{E}) \leqq \kappa(z, T, \mathfrak{D})$ (cf. IV.1.56), (2) implies that $N(z, T, \mathcal{E}) = \kappa(z, T, \mathfrak{D}) < +\infty$ a.e. in the z-plane. By IV.1.56 it follows that $N(z, T, \mathcal{E}^*) = \kappa(z, T, \mathfrak{D})$ a.e. in the z-plane. Since $\kappa(z, T, \mathfrak{D})$ is summable by assumption, it follows that $N(z, T, \mathcal{E}^*)$ is also summable. But then, by IV.3.26, IV.2.13, $D(w, \mathcal{E}^*) = |\mathcal{J}_e(w)|$ a.e. in $\mathfrak{D}$. Thus (1) yields

$$\iint_{\mathfrak{D}} D(w, \mathcal{E}^*) = \iint N(z, T, \mathcal{E}^*).$$

Hence, by IV.2.45, T is AC$\mathcal{E}^*$ in $\mathfrak{D}$, and the proof is complete (cf. IV.4.1).

IV.4.9. Let $\mathfrak{R}$ be a bounded, finitely-connected Jordan region in the w-plane, and let $T : z = t(w)$, $w \in \mathfrak{R}$, be a continuous transformation defined in $\mathfrak{R}$ (and not merely in $\mathfrak{R}^0$). Suppose that (i) T is eAC in $\mathfrak{R}^0$ and (ii) $|T(\mathfrak{R} - \mathfrak{R}^0)| = 0$. If $H(z)$ is a finite-valued, measurable function in the z-plane, then (cf. IV.1.24)

$$\iint_{\mathfrak{R}^0} H[t(w)]\mathcal{J}_e(w) = \iint H(z)\mu(z, T, \mathfrak{R}) \tag{1}$$

as soon as the integral on the left exists. In particular (taking $H(z) \equiv 1$), we have the formula

$$\iint_{\mathfrak{R}^0} \mathcal{J}_e(w) = \iint \mu(z, T, \mathfrak{R}). \tag{2}$$

PROOF. Since T is BV$\mathcal{E}^*$ in $\mathfrak{R}^0$ (cf. IV.4.2), we have $N(z, T, \mathcal{E}^*) < +\infty$ a.e. in the z-plane, by IV.2.14. Hence, by IV.3.5, condition (ii) implies that $\nu(z, \mathfrak{R}^0) = \mu(z, T, \mathfrak{R})$ a.e. in the z-plane. Thus (1) and (2) appear as direct consequences of IV.4.3, IV.4.7.

IV.4.10. CONTINUATION. Let us assume that the image, under T, of each one of the boundary curves of $\mathfrak{R}$ is rectifiable (see III.3.95). We have then (see IV.1.24, III.3.76, III.3.88)

$$\iint \mu(z, T, \mathfrak{R}) = \int_B x\, dy = -\int_B y\, dx = \frac{1}{2}\int_B x\, dy - y\, dx,$$

where B denotes the boundary of $\mathfrak{R}$, each boundary curve being taken with the proper orientation. In view of IV.4.9 there follow the formulas

$$\iint_{\mathfrak{R}^\circ} \mathcal{J}_e(w) = \iint \mu(z, T, \mathfrak{R}) = \int_B x\, dy = -\int_B y\, dx = \frac{1}{2}\int_B x\, dy - y\, dx.$$

IV.4.11. Given a transformation $T : z = t(w)$, $w \in \mathfrak{D}$, as in IV.2.1, suppose there exists a sequence of transformations $T_n : z = t_n(w)$, $w \in \mathfrak{D}_n$ with the following properties:

(i) The domains $\mathfrak{D}_n$ fill up $\mathfrak{D}$ from the interior (see IV.1.53).

(ii) $\rho(T_n, T, F) \to 0$ for $n \to \infty$, for every closed set $F \subset \mathfrak{D}$ (cf. IV.1.1).

(iii) T_n is eAC in $\mathfrak{D}_n$, $n = 1, 2, \cdots$.

(iv) $$\liminf_{n\to\infty} \iint_{\mathfrak{D}_n} | \mathcal{J}_e(w, T_n) | < +\infty.$$

Then T is eBV in $\mathfrak{D}$, and

(1) $$\iint_{\mathfrak{D}} | \mathcal{J}_e(w, T) | \leqq \iint \kappa(z, T, \mathfrak{D}) \leqq \liminf_{n\to\infty} \iint_{\mathfrak{D}_n} | \mathcal{J}_e(w, T_n) |.$$

Proof. Let us put (cf. condition (iv))

(2) $$l = \liminf_{n\to\infty} \iint_{\mathfrak{D}_n} | \mathcal{J}_e(w, T_n) | < +\infty.$$

Let $\mathfrak{R}_j$, $j = 1, 2, \cdots$, be a sequence of Jordan regions that fill up $\mathfrak{D}$ from the interior. By IV.1.43 we have then

(3) $$\kappa(z, T, \mathfrak{D}) = \lim_{j\to\infty} \kappa(z, T, \mathfrak{R}_j).$$

By condition (i), we have for each j an $n(j)$ such that $\mathfrak{R}_j \subset \mathfrak{D}_n$ for $n > n(j)$. By IV.1.44 it follows that

(4) $$\kappa(z, T_n, \mathfrak{R}_j) \leqq \kappa(z, T_n, \mathfrak{D}_n) \qquad \text{for } n > n(j).$$

By IV.4.7 and condition (iii) we have

$$\iint_{\mathfrak{D}_n} | \mathcal{J}_e(w, T_n) | = \iint \kappa(z, T_n, \mathfrak{D}_n).$$

Hence, by (2) and condition (iv),

(5) $$\liminf_{n\to\infty} \iint \kappa(z, T_n, \mathfrak{D}_n) = l < +\infty.$$

By condition (ii) and IV.1.12 we have

(6) $$\kappa(z, T, \mathfrak{R}_j) \leqq \liminf_{n\to\infty} \kappa(z, T_n, \mathfrak{R}_j), \qquad j = 1, 2, \cdots.$$

(4), (5), (6) imply, by the Fatou lemma (see I.3.10), that $\kappa(z, T, \mathfrak{R}_j)$ is summable, and

(7) $$\iint \kappa(z, T, \mathfrak{R}_j) \leqq l < +\infty, \qquad j = 1, 2, \cdots .$$

Applying again the Fatou lemma, we infer from (3) and (7) that $\kappa(z, T, \mathfrak{D})$ is summable and

(8) $$\iint \kappa(z, T, \mathfrak{D}) \leqq l < +\infty .$$

Thus T is eBV in $\mathfrak{D}$. The inequalities (1) follow now directly from (8), (2) and IV.3.13, IV.3.25.

IV.4.12. Closure theorem for eAC transformations. *Given a transformation* $T : z = t(w)$, $w \in \mathfrak{D}$, *as in* IV.2.1, *suppose that* (a) $\mathfrak{J}_e(w, T)$ *exists a.e. in* $\mathfrak{D}$ *and is summable there, and* (b) *there exists a sequence of transformations* $T_n : z = t_n(w)$, $w \in \mathfrak{D}_n$, *with the following properties.*

(i) T_n *is eAC in* $\mathfrak{D}_n$, $n = 1, 2, \cdots$.
(ii) *The domains* $\mathfrak{D}_n$ *fill up* $\mathfrak{D}$ *from the interior* (*see* IV.1.53).
(iii) $\rho(T_n, T, F) \to 0$ *for every closed set* $F \subset \mathfrak{D}$ (*cf.* IV.1.1).
(iv) *For every oriented rectangle* $R \subset \mathfrak{D}$ *we have*

$$\iint_R | \mathfrak{J}_e(w, T_n) | \underset{n\to\infty}{\longrightarrow} \iint_R | \mathfrak{J}_e(w, T) | .$$

Then T *is eAC in* $\mathfrak{D}$.

IV.4.13. Continuation. The proof will be made in two steps. Let $\mathfrak{R}$ be any finitely-connected Jordan region such that

(1) $$\mathfrak{R} \subset \mathfrak{D}, \ | \mathfrak{R} - \mathfrak{R}^0 | = 0.$$

For example, $\mathfrak{R}$ may be taken as a polygonal region to insure that the second condition in (1) holds. We assert that

(2) $$\iint_{\mathfrak{R}^0} | \mathfrak{J}_e(w, T) | = \iint \kappa(z, T, \mathfrak{R}^0).$$

Indeed, give $\epsilon > 0$. Since $\mathfrak{J}_e(w, T)$ is summable in $\mathfrak{D}$, and hence in $\mathfrak{R}$, by assumption, we can choose an open set O such that

(3) $$\mathfrak{R} \subset O \subset \mathfrak{D},$$

(4) $$\iint_O | \mathfrak{J}_e(w, T) | < \iint_{\mathfrak{R}} | \mathfrak{J}_e(w, T) | + \epsilon.$$

Since $| \mathfrak{R} - \mathfrak{R}^0 | = 0$, we can rewrite (4) in the form

(5) $$\iint_O | \mathcal{J}_e(w, T) | < \iint_{\mathfrak{R}^0} | \mathcal{J}_e(w, T) | + \epsilon.$$

Next we can choose a finite system of oriented rectangles $R_1, \cdots, R_m$, such that

(6) $$R_i^0 R_j^0 = 0 \qquad \text{for } i \neq j.$$

(7) $$\mathfrak{R} \subset R_1 + \cdots + R_m \subset O.$$

(5), (6), (7) yield

(8) $$\sum_{j=1}^{m} \iint_{R_j} | \mathcal{J}_e(w, T) | < \iint_{\mathfrak{R}^0} | \mathcal{J}_e(w, T) | + \epsilon,$$

(9) $$\iint_{\mathfrak{R}^0} | \mathcal{J}_e(w, T_n) | \leqq \sum_{j=1}^{m} \iint_{R_j} | \mathcal{J}_e(w, T_n) |.$$

By condition (iv) we have

(10) $$\sum_{j=1}^{m} \iint_{R_j} | \mathcal{J}_e(w, T_n) | \underset{n\to\infty}{\longrightarrow} \sum_{j=1}^{m} \iint_{R_j} | \mathcal{J}_e(w, T) |.$$

Now let us apply the result of IV.4.11 to the transformations T, T_n considered in the domain $\mathfrak{R}^0$. We obtain the inequalities

(11) $$\iint_{\mathfrak{R}^0} | \mathcal{J}_e(w, T) | \leqq \iint \kappa(z, T, \mathfrak{R}^0) \leqq \liminf_{n\to\infty} \iint_{\mathfrak{R}^0} | \mathcal{J}_e(w, T_n) |.$$

(11), (9), (10), (8) yield

$$\iint_{\mathfrak{R}^0} | \mathcal{J}_e(w, T) | \leqq \iint \kappa(z, T, \mathfrak{R}^0) < \iint_{\mathfrak{R}^0} | \mathcal{J}_e(w, T) | + \epsilon.$$

Since ϵ was arbitrary, (2) follows.

IV.4.14. Continuation. We proceed to prove the theorem of IV.4.12. Let us take a sequence of finitely-connected Jordan regions $\mathfrak{R}_j$ that fill up $\mathfrak{D}$ from the interior (cf. IV.1.91) and satisfy the condition $| \mathfrak{R}_j - \mathfrak{R}_j^0 | = 0, j = 1, 2, \cdots$. For example, we may take each $\mathfrak{R}_j$ as a polygonal region. By IV.1.43, IV.1.44, IV.1.50 we have then

(1) $$\kappa(z, T, \mathfrak{D}) = \lim_{j\to\infty} \kappa(z, T, \mathfrak{R}_j^0),$$

(2) $$\kappa(z, T, \mathfrak{R}_j^0) \leqq \kappa(z, T, \mathfrak{D}), \qquad j = 1, 2, \cdots.$$

By IV.4.13 we have (cf. condition (a) in IV.4.12)

(3) $$\iint \kappa(z, T, \mathfrak{R}_j^0) = \iint_{\mathfrak{R}_j^0} | \mathcal{J}_e(w, T) | \leqq \iint_{\mathfrak{D}} | \mathcal{J}_e(w, T) | < +\infty.$$

(1), (2) and (3) imply (cf. I.3.10) that $\kappa(z, T, \mathfrak{D})$ is summable, and that the relation (1) can be integrated termwise. In view of (3) there follows the inequality

$$\iint \kappa(z, T, \mathfrak{D}) \leqq \iint_{\mathfrak{D}} | \mathcal{J}_e(w, T) |. \tag{4}$$

In view of IV.4.8, it follows that the sign of equality holds in (4), and hence T is eAC in $\mathfrak{D}$ by IV.4.8.

IV.4.15. Corollary to IV.4.12. *Condition* (iv) *in* IV.4.12 *can be replaced by the condition*

(iv*) *For every oriented rectangle* $R \subset \mathfrak{D}$ *we have*

$$\iint_R | \mathcal{J}_e(w, T) - \mathcal{J}_e(w, T_n) | \underset{n \to \infty}{\longrightarrow} 0. \tag{1}$$

Proof. Let us put $f_n(w) = \mathcal{J}_e(w, T_n)$, $f(w) = \mathcal{J}_e(w, T)$. We have then the obvious inequalities

$$| f(w) | \leqq | f_n(w) | + | f(w) - f_n(w) |, \tag{2}$$

$$| f_n(w) | \leqq | f(w) | + | f_n(w) - f(w) |. \tag{3}$$

(1) and (2) yield

$$\iint_R | f(w) | \leqq \liminf \iint_R | f_n(w) |, \tag{4}$$

while (1) and (3) yield

$$\limsup \iint_R | f_n(w) | \leqq \iint_R | f(w) |. \tag{5}$$

(4) and (5) yield

$$\iint_R | \mathcal{J}_e(w, T_n) | \to \iint_R | \mathcal{J}_e(w, T) |.$$

Thus condition (iv*) implies condition (iv) in IV.4.12, and the fact that T is eAC in $\mathfrak{D}$ follows.

IV.4.16. Modified closure theorem for eAC transformations. *Given a transformation* $T : z = t(w)$, $w \in \mathfrak{D}$, *as in* IV.2.1, *suppose that* (a) $\mathcal{J}_e(w, T)$ *exists a.e. in* $\mathfrak{D}$ *and is summable in* $\mathfrak{D}$, *and* (b) *there exists a sequence of transformations* $T_n : z = t_n(w)$, $w \in \mathfrak{D}_n$, *with the following properties*:

(i) T_n *is eAC in* $\mathfrak{D}_n$.

(ii) *The domains* $\mathfrak{D}_n$ *fill up* $\mathfrak{D}$ *from the interior* (*see* IV.1.53).

(iii) $\rho(T_n, T, F) \to 0$ *for* $n \to \infty$, *for every closed set* $F \subset \mathfrak{D}$ (*cf.* IV.1.1).

(iv)
$$\liminf_{n\to\infty} \iint_{\mathfrak{D}_n} | \mathcal{J}_e(w, T_n) | \leqq \iint_{\mathfrak{D}} | \mathcal{J}_e(w, T) |.$$

Then T is eAC in $\mathfrak{D}$.

Proof. By IV.4.11 we obtain, in view of condition (iv),

$$\iint_{\mathfrak{D}} | \mathcal{J}_e(w, T) | \leqq \iint \kappa(z, T, \mathfrak{D}) \leqq \liminf_{n\to\infty} \iint_{\mathfrak{D}_n} | \mathcal{J}_e(w, T_n) | \leqq \iint_{\mathfrak{D}} | \mathcal{J}_e(w, T) |.$$

Consequently

$$\iint_{\mathfrak{D}} | \mathcal{J}_e(w, T) | = \iint \kappa(z, T, \mathfrak{D}).$$

By IV.4.8 it follows that T is eAC in $\mathfrak{D}$.

Remark. It may be of interest to note that the conditions assumed above imply the conditions assumed in IV.4.12 as can be shown easily. In other words, the preceding theorem is weaker than that in IV.4.12 but it seems to be more convenient in some applications.

IV.4.17. Special closure theorem for eAC transformations. *Given a transformation $T : z = t(w)$, $w \in \mathfrak{D}$, as in* IV.2.1, *suppose that* (a) *the ordinary Jacobian $J(w, T)$* (*see* IV.3.21) *exists a.e. in $\mathfrak{D}$ and is summable in $\mathfrak{D}$, and* (b) *there exists a sequence of transformations $T_n : z = t_n(w)$, $w \in \mathfrak{D}_n$, with the properties*:

(i) *T_n is eAC in $\mathfrak{D}_n$, and the ordinary Jacobian $J(w, T_n)$ exists a.e. in $\mathfrak{D}_n$.*

(ii) *The domains $\mathfrak{D}_n$ fill up $\mathfrak{D}$ from the interior* (*cf.* IV.1.53).

(iii) *$\rho(T_n, T, F) \to 0$ for $n \to \infty$ for every closed set $F \subset \mathfrak{D}$* (*cf.* IV.1.1).

(iv) *For every oriented rectangle $R \subset \mathfrak{D}$ we have*

$$\iint_R | J(w, T_n) | \to \iint_R | J(w, T) |.$$

Then T is eAC in $\mathfrak{D}$.

Proof. By IV.3.42 and IV.2.39, $J(w, T_n) = \mathcal{J}_e(w, T_n)$ a.e. in $\mathfrak{D}$. Hence, if we can prove that $\mathcal{J}_e(w, T)$ exists a.e. in $\mathfrak{D}$ and $J(w, T) = \mathcal{J}_e(w, T)$ a.e. in $\mathfrak{D}$, then the present closure theorem will follow directly from that in IV.4.12. Furthermore, it is clearly sufficient to show that $J(w, T) = \mathcal{J}_e(w, T)$ a.e. in every oriented rectangle $R \subset \mathfrak{D}$ (since the same is then true for $\mathfrak{D}$ itself). Finally, by IV.3.42, it is sufficient for this purpose to show that T is BV$\mathcal{E}^*(T, R^0)$ in every oriented open rectangle R^0 such that $R \subset \mathfrak{D}$, and we proceed to establish this last fact. So let R be an oriented rectangle in $\mathfrak{D}$. From condition (iv) it follows that

(1)
$$\lim_{n\to\infty} \iint_{R^0} | J(w, T_n) | = \iint_{R^0} | J(w, T) | < +\infty.$$

In view of condition (iii) we have (cf. IV.1.52)

$$\kappa(z, T, R^0) \leqq \liminf_{n\to\infty} \kappa(z, T_n, R^0). \tag{2}$$

Since T_n is eAC in $\mathfrak{D}_n$, it is also eAC in R^0 (cf. IV.4.1) for n sufficiently large (see condition (ii) and note that $R \subset \mathfrak{D}$). Hence we have, for n sufficiently large (see IV.3.42, IV.2.39, IV.4.7),

$$\iint_{R^0} | J(w, T_n) | = \iint_{R^0} | \mathcal{J}_e(w, T_n) | = \iint \kappa(z, T_n, R^0). \tag{3}$$

(1), (2), (3) imply, by the lemma of Fatou (see I.3.10), that $\kappa(z, T, R^0)$ is summable and

$$\iint \kappa(z, T, R^0) \leqq \iint_{R^0} | J(w, T) |. \tag{4}$$

By IV.3.46 (applied to the domain R^0), we have (since $\kappa(z, T, R^0)$ is now known to be summable)

$$\iint_{R^0} | J(w, T) | \leqq \iint_{R^0} | \mathcal{J}_e(w, T) |. \tag{5}$$

(4), (5) imply, by IV.4.8, that

$$\iint \kappa(z, T, R^0) = \iint_{R^0} | \mathcal{J}_e(w, T) |.$$

By IV.4.8 it follows that T is eAC in R^0 and *a fortiori* BV$\mathcal{E}^*(T, R^0)$ in R^0 (cf. IV.2.39, IV.4.1).

IV.4.18. Corollary to IV.4.17. Condition (iv) in IV.4.17 can be replaced by the following condition:

(iv*) For every oriented rectangle $R \subset \mathfrak{D}$ we have

$$\iint_R | J(w, T) - J(w, T_n) | \underset{n\to\infty}{\longrightarrow} 0.$$

The proof is entirely similar to that in IV.4.15.

IV.4.19. A modification of the special closure theorem for eAC transformations. *Given a transformation* $T : z = t(w), w \in \mathfrak{D}$, *as in* IV.2.1, *suppose that* (a) *the ordinary Jacobian* $J(w, T)$ *exists a.e. in* $\mathfrak{D}$ *and is summable in* $\mathfrak{D}$, *and* (b) *there exists a sequence of transformations* $T_n : z = t_n(w)$, $w \in \mathfrak{D}_n$, *with the following properties*:

(i) T_n *is eAC in* $\mathfrak{D}_n$, *and the ordinary Jacobian* $J(w, T_n)$ *exists a.e. in* $\mathfrak{D}_n$.

(ii) *The domains* $\mathfrak{D}_n$ *fill up* $\mathfrak{D}$ *from the interior* (cf. IV.1.53).

(iii) $\rho(T_n, T, F) \to 0$ *for* $n \to \infty$, *for every closed set* $F \subset \mathfrak{D}$ (*cf.* IV.1.1).

(iv) $$\liminf_{n\to\infty} \iint_{\mathfrak{D}_n} | J(w, T_n) | \leqq \iint_{\mathfrak{D}} | J(w, T) |.$$

Then T is eAC in $\mathfrak{D}$.

PROOF. By IV.3.42, $J(w, T_n) = \mathcal{J}_e(w, T_n)$ a.e. in $\mathfrak{D}_n$. Hence condition (iv) yields

(1) $$\liminf_{n\to\infty} \iint_{\mathfrak{D}_n} | \mathcal{J}_e(w, T_n) | \leqq \iint_{\mathfrak{D}} | J(w, T) |.$$

By IV.4.11 it follows that $\kappa(z, T, \mathfrak{D})$ is summable. Hence, by IV.3.46 and IV.3.13, $\mathcal{J}_e(w, T)$ is summable in $\mathfrak{D}$, and

(2) $$\iint_{\mathfrak{D}} | J(w, T) | \leqq \iint_{\mathfrak{D}} | \mathcal{J}_e(w, T) |.$$

(1) and (2) yield

(3) $$\liminf_{n\to\infty} \iint_{\mathfrak{D}_n} | \mathcal{J}_e(w, T_n) | \leqq \iint_{\mathfrak{D}} | \mathcal{J}_e(w, T) |.$$

By IV.4.16 it follows that T is eAC in $\mathfrak{D}$.

IV.4.20. COROLLARY TO IV.4.19. *Given a transformation* $T : z = t(w)$, $w \in \mathfrak{D}$, *as in* IV.2.1, *suppose that* (a) *the ordinary Jacobian* $J(w, T)$ *exists a.e. in* $\mathfrak{D}$ *and is summable in* $\mathfrak{D}$, *and* (b) *there exists a sequence of finitely-connected Jordan regions* $\mathfrak{R}_n$ *with the following properties*:

(i) *The regions* $\mathfrak{R}_n$ *fill up* $\mathfrak{D}$ *from the interior* (*cf.* IV.1.41).

(ii) *T is eAC in* $\mathfrak{R}_n^0$, $n = 1, 2, \cdots$.

Then T is eAC in $\mathfrak{D}$. *This is merely a trivial but quite useful corollary of* IV.4.19 ($t_n(w) = t(w)$, $\mathfrak{D}_n = \mathfrak{R}_n^0$).

IV.4.21. Given a transformation $T : z = t(w)$, $w \in \mathfrak{D}$, as in IV.2.1, suppose that T is eAC in $\mathfrak{D}$ and the ordinary Jacobian $J(w, T)$ exists a.e. in $\mathfrak{D}$. Then $J(w, T) = \mathcal{J}_e(w, T)$ a.e. in $\mathfrak{D}$ by IV.4.1, IV.3.42, IV.2.39. *As a consequence, we can replace, in the various transformation formulas previously derived for eAC transformations,* $\mathcal{J}_e(w, T)$ *by* $J(w, T)$. We shall now list, for convenience of reference, the formulas obtained in this manner from IV.4.4, IV.4.6, IV.4.7, IV.4.9, IV.4.10. For conciseness, we write $J(w)$ for $J(w, T)$ in these formulas.

(i_1) $$\iint_{\mathfrak{D}} H[t(w)]J(w) = \iint H(z)\nu(z, T, \mathfrak{D}).$$

(i_2) $$\iint_{\mathfrak{D}} H[t(w)] \, | J(w) | = \iint H(z)\kappa(z, T, \mathfrak{D}).$$

(i_1) holds as soon as the integral on the left exists, while (i_2) holds as soon as one of the two integrals involved exists. $H(z)$ is any finite-valued, measurable function in the z-plane. The formulas can be also used if $H(z)$ fails to be finite on a set of measure zero, as explained in IV.4.5.

$$(ii_1) \qquad \iint_{\mathfrak{D}} J(w) = \iint \nu(z, T, \mathfrak{D}).$$

$$(ii_2) \qquad \iint_{\mathfrak{D}} | J(w) | = \iint \kappa(z, T, \mathfrak{D}).$$

If $\mathfrak{D}$ is the interior $\mathfrak{R}^0$ of a bounded, finitely-connected Jordan region $\mathfrak{R}$, if T is defined and continuous on $\mathfrak{R}$, and if $| T(\mathfrak{R} - \mathfrak{R}^0) | = 0$, then we have the following formulas (where (iii_1) is subject to the same remarks as (i_1)):

$$(iii_1) \qquad \iint_{\mathfrak{R}^0} H[t(w)]J(w) = \iint H(z)\mu(z, T, \mathfrak{R}).$$

$$(iii_2) \qquad \iint_{\mathfrak{R}^0} J(w) = \iint \mu(z, T, \mathfrak{R}).$$

If the image, under T, of each one of the boundary curves of $\mathfrak{R}$ is rectifiable, then we have the formulas (B is the properly oriented boundary of $\mathfrak{R}$):

$$iii_3) \qquad \iint_{\mathfrak{R}^0} J(w) = \iint \mu(z, T, \mathfrak{R}) = \int_B x\, dy = -\int_B y\, dx = \frac{1}{2}\int_B x\, dy - y\, dx.$$

Let us note that the relation $J(w, T) = \mathcal{J}_e(w, T)$ a.e. in $\mathfrak{D}$ yields, in view of IV.3.14, a number of geometrical interpretations for the ordinary Jacobian $J(w, T)$. Further geometrical information about $J(w, T)$ is obtained from IV.3.45, and of course from the preceding formulas (ii_1), (ii_2), (iii_2), (iii_3).

IV.4.22. We apply the general results obtained so far to special types of transformations. Let us first consider the case of a transformation $T : z = t(w)$, $w \in \mathfrak{D}$, given as in IV.2.1, that is sAC in $\mathfrak{D}$ (see IV.4.1). Then T is also eAC in $\mathfrak{D}$, and furthermore the generalized Jacobians $\mathcal{J}_s(w, T)$, $\mathcal{J}_e(w, T)$ are equal to each other a.e. in $\mathfrak{D}$. If the ordinary Jacobian $J(w, T)$ happens to exist a.e. in $\mathfrak{D}$, then $J(w, T) = \mathcal{J}_s(w, T)$ a.e. in $\mathfrak{D}$ (see IV.3.22, IV.3.47 for the preceding statements). As a consequence of these remarks, we can replace, in the various transformation formulas derived for eAC transformations, the essential generalized Jacobian $\mathcal{J}_e(w, T)$ by $\mathcal{J}_s(w, T)$, and also by $J(w, T)$ if the ordinary Jacobian exists a.e. in $\mathfrak{D}$. Thus we obtain the following statements:

(a) If T is sAC in $\mathfrak{D}$, then the transformation formulas listed in IV.4.21 hold with $J(w)$ replaced by $\mathcal{J}_s(w)$.

(b) If T is sAC in $\mathfrak{D}$, and if the ordinary Jacobian $J(w)$ exists a.e. in $\mathfrak{D}$, then the transformation formulas listed in IV.4.21 hold.

The question arises as to what improvements can be derived from the stronger assumption that T is sAC in $\mathfrak{D}$. We shall make a few remarks concerning this point presently.

IV.4.23. Continuation. T being assumed to be sAC in $\mathfrak{D}$, the function $\nu(z, T, \mathfrak{D})$ admits of a simpler interpretation. Indeed, since $N(z, T, \mathfrak{D})$ is now summable (cf. IV.2.39, IV.2.13), we have $N(z, T, \mathfrak{D}) < +\infty$ a.e. in the z-plane. Let z be a point such that $N(z, T, \mathfrak{D}) < +\infty$. Then clearly $T^{-1}(z) \subset \mathfrak{g}$ (see IV.1.56), and hence (see IV.1.76) we have $i_e(w) = i_s(w)$ for $w \in T^{-1}(z)$. It follows now readily from the definition of $\nu(z, T, \mathfrak{D})$ (see IV.3.2) that $\nu(z, T, \mathfrak{D}) = \sum i_s(w)$, $w \in T^{-1}(z)$, a.e. in the z-plane.

IV.4.24. Continuation. Since T is now AC$\mathfrak{D}$ (see IV.4.1), we should expect that $\kappa(z, T, \mathfrak{D})$ can be replaced by $N(z, T, \mathfrak{D})$ in the formulas described in IV.4.22(a), (b). *It is not known at present whether this is true.* However, we have the following special statements:

(i) *If T is sAC in $\mathfrak{D}$ and $| T(\mathfrak{g}_0) | = 0$, then $\kappa(z, T, \mathfrak{D}) = N(z, T, \mathfrak{D})$ a.e. in the z-plane by* IV.1.78, IV.3.20. Hence in this case the formulas described in IV.4.22 (a), (b) hold with $\kappa(z, T, \mathfrak{D})$ replaced by $N(z, T, \mathfrak{D})$.

(ii) The same is true if T is sAC in $\mathfrak{D}$ and is also totally differentiable a.e. in $\mathfrak{D}$ (cf. IV.3.49). Indeed, by IV.3.51, we have then $| J(w) | = D(w, \mathfrak{D})$ a.e. in $\mathfrak{D}$, and hence, by IV.3.47,

(1) $$| \mathfrak{J}_s(w) | = | J(w) | = D(w, \mathfrak{D}) \qquad \text{a.e. in } \mathfrak{D}.$$

By IV.2.53, applied with $\mathfrak{B} = \mathfrak{D}$, $H(z) \equiv 1$,

(2) $$\iint_{\mathfrak{D}} D(w, \mathfrak{D}) = \iint N(z, T, \mathfrak{D}).$$

By IV.4.22(b)

(3) $$\iint_{\mathfrak{D}} | J(w) | = \iint \kappa(z, T, \mathfrak{D}).$$

(1), (2), (3) imply that

(4) $$\iint \kappa(z, T, \mathfrak{D}) = \iint N(z, T, \mathfrak{D}).$$

Since $0 \leqq \kappa(z, T, \mathfrak{D}) \leqq N(z, T, \mathfrak{D})$, (4) implies that $\kappa(z, T, \mathfrak{D}) = N(z, T, \mathfrak{D})$ a.e. in the z-plane. Thus we can replace $\kappa(z, T, \mathfrak{D})$ by $N(z, T, \mathfrak{D})$ in the formulas referred to in IV.4.22(b).

IV.4.25. For convenient reference, we state explicitly the following corollary to IV.4.24. Suppose that (α) T is sAC in $\mathfrak{D}$, (β) the ordinary Jacobian $J(w)$ exists a.e. in $\mathfrak{D}$, and (γ) either $| T(\mathfrak{g}_0) | = 0$ or T is totally differentiable a.e. in $\mathfrak{D}$. Then

(i_2^*) $$\iint_{\mathfrak{D}} H[t(w)] \, | J(w) | = \iint H(z) N(z, T, \mathfrak{D})$$

for every finite-valued, measurable function $H(z)$, as soon as one of the two integrals involved exists, and in particular

$$(\mathrm{ii}_2^*) \qquad \iint_{\mathfrak{D}} | J(w) | = \iint N(z, T, \mathfrak{D}).$$

Formula (i_2^*) applies also to functions $H(z)$ that fail to be finite on a set of measure zero, as explained in IV.4.5.

IV.4.26. Comparison reveals that the stronger property sAC yields, as against the property eAC, certain conceptual simplifications (see IV.4.23, IV.4.24), and under additional assumptions also the further transformation formulas IV.4.25 (i_2^*), (ii_2^*). On the other hand, *no closure theorems are known for sAC transformations* that would correspond to the closure theorems for eAC transformations (see IV.4.12, IV.4.15, IV.4.16, IV.4.17, IV.4.18, IV.4.19). As a consequence we have no means to determine whether various important classes of transformations occurring in the literature are or are not comprised in the class of sAC transformations. In other words, the class of sAC transformations seems to be lacking in scope, and the results obtained for sAC transformations cannot be used to account for results derived in the literature by other means for many important types of transformations. This situation may be merely a consequence of our present state of ignorance, of course. On the other hand, the class of eAC transformations appears to be quite comprehensive, due to the closure theorems just referred to. We proceed to verify that certain special transformations, needed in the study of surface area, belong to the class of eAC transformations.

IV.4.27. Given $T : z = t(w)$, $w \in \mathfrak{D}$, as in IV.2.1, we shall say that T is Lipschitzian in $\mathfrak{D}$, if there exists a positive constant λ such that $| t(w'') - t(w') | \leqq \lambda | w'' - w' |$ for any two points w', w'' in $\mathfrak{D}$, provided that the straight segment with end points w', w'' is contained in $\mathfrak{D}$ (note that this is automatically the case if $\mathfrak{D}$ happens to be convex).

IV.4.28. Continuation. *Suppose that T is Lipschitzian in $\mathfrak{D}$. Then T is sAC in $\mathfrak{D}$, and the formulas referred to in* IV.4.22(b) *hold. Furthermore, the formulas* IV.4.25(i_2^*), (ii_2^*) *also hold.*

Proof. In the first place, T is totally differentiable a.e. in $\mathfrak{D}$ by I.3.14, and hence it is sufficient to verify that T is sAC in $\mathfrak{D}$. Let s be any oriented square in $\mathfrak{D}$, with center w_0 and side length l. If $w \in s$, then $| w - w_0 | \leqq 2^{1/2}l/2$ and hence $| t(w) - t(w_0) | \leqq \lambda 2^{1/2}l/2$ (cf. IV.4.27). Thus $T(s)$ is contained in a circular disc with center $t(w_0)$ and radius $\lambda 2^{1/2}l/2$, and hence $| T(s) | \leqq (\lambda^2/2)\pi | s |$. Clearly it follows that T is sAC in $\mathfrak{D}$ (cf. IV.2.39).

IV.4.29. *Given $T : z = t(w)$, $w \in \mathfrak{D}$ as in* IV.2.1, *suppose that the following conditions hold*: (i) *The partial derivatives x_u, x_v, y_u, y_v exist and are continuous in $\mathfrak{D}$.* (ii) *The ordinary Jacobian $J(w) = x_u y_v - x_v y_u$ is summable in $\mathfrak{D}$. Then T is sAC in $\mathfrak{D}$ and all the formulas referred to in* IV.4.22(b) *hold. Furthermore, the formulas* IV.4.25(i_2^*), (ii_2^*) *also hold.*

PROOF. Let s be any oriented square in $\mathfrak{D}$. Let S be an oriented square such that $s \subset S^0$, $S \subset \mathfrak{D}$. Then the partial derivatives x_u, x_v, y_u, y_v are bounded on S, and hence T is clearly Lipschitzian in S^0. By IV.4.28 (applied to S^0) it follows that

$$\iint_{S^0} | J(w) | = \iint N(z, T, S^0),$$

$$| T(s) | \leqq \iint N(z, T, s) \leqq \iint N(z, T, S^0) = \iint_{S^0} | J(w) |.$$

Since S was any oriented square such that $s \subset S^0$, $S \subset \mathfrak{D}$, it follows that

$$| T(s) | \leqq \iint_s | J(w) |.$$

Since $J(w)$ is summable in $\mathfrak{D}$, it follows that T is sAC in $\mathfrak{D}$ (cf. IV.2.39, I.3.13). As T is now totally differentiable everywhere in $\mathfrak{D}$ (cf. I.3.14), the formulas of IV.4.25 apply.

IV.4.30. Let $\mathfrak{R}$ denote a bounded, finitely-connected Jordan region in the w-plane, and let T be given in the form

$$T : x = x(u, v),\ y = y(u, v), \qquad (u, v) \in \mathfrak{R},$$

where $x(u, v)$, $y(u, v)$ are continuous on $\mathfrak{R}$ (and not merely on $\mathfrak{R}^0$). We shall say that T is *quasi-linear* on $\mathfrak{R}$ if the following conditions hold:

(i) $\mathfrak{R}$ is bounded by polygons.

(ii) There exists a rectilinear triangulation $\mathfrak{J}$ of $\mathfrak{R}$, such that $x(u, v)$, $y(u, v)$ are linear functions of u, v in each one of the triangles of $\mathfrak{J}$.

Clearly, if T is quasi-linear in $\mathfrak{R}$, then T is Lipschitzian in $\mathfrak{R}^0$, in the sense of IV.4.27, and hence IV.4.28 applies. The following remark will be useful in the sequel. Let $t_1, \cdots, t_m$ be the triangles of $\mathfrak{J}$ (cf. (ii)). Then for a given j, the set $T(t_j) = \Delta_j$ is either a single point, or a straight segment, or else a nondegenerate (rectilinear) triangle in the z-plane. In the first two cases $| \Delta_j | = 0$, while in the third case $| \Delta_j |$ is the area of Δ_j in the elementary sense. Let w_j be any interior point of t_j. We assert that

$$(1) \qquad | \Delta_j | = \iint_{\Delta_j} N(z, T, t_j) = | J(w_j) | \, | t_j |.$$

Indeed, if Δ_j is degenerate, then (1) is obvious. If Δ_j is nondegenerate, then T is a biunique affine transformation if considered on t_j. Thus $N(z, T, t_j)$ is the characteristic function of Δ_j while $| J(w_j) | = | \Delta_j | / | t_j |$ by elementary geometry. Thus (1) is obvious in this case also. Summation yields now the formula

$$\iint_{\mathfrak{R}} | J(w) | = \sum_{j=1}^{m} | \Delta_j |.$$

IV.4.31. Given $T : z = t(w) = x(u, v) + iy(u, v)$, $u + iv = w \in \mathfrak{D}$, as in IV.2.1, let us consider the *integral means* (cf. III.2.65)

$$x^h(u, v) = \frac{1}{4h^2} \int_{-h}^{h} \int_{-h}^{h} x(u + \xi, v + \eta) \, d\xi \, d\eta,$$

$$y^h(u, v) = \frac{1}{4h^2} \int_{-h}^{h} \int_{-h}^{h} y(u + \xi, v + \eta) \, d\xi \, d\eta.$$

If R is any oriented rectangle in $\mathfrak{D}$, then for h small enough the functions $x^h(u, v)$, $y^h(u, v)$ will be defined and continuous, together with their partial derivatives of the first order, on R (see III.2.66). Let us put

$$J_h(w) = \frac{\partial x^h}{\partial u} \frac{\partial y^h}{\partial v} - \frac{\partial x^h}{\partial v} \frac{\partial y^h}{\partial u}.$$

It should be noted that $J_h(w)$ is not defined as an integral mean. *Let us now make the following assumptions*:

(i) *The ordinary Jacobian* $J(w, T)$ *exists a.e. in* $\mathfrak{D}$ *and is summable in* $\mathfrak{D}$.

(ii) *For every oriented rectangle* $R \subset \mathfrak{D}$ *we have*

$$\iint_R | J_h(w) | \underset{h \to 0}{\longrightarrow} \iint_R | J(w, T) |.$$

Then T *is eAC in* $\mathfrak{D}$ (*and hence the formulas listed in* IV.4.21 *hold*).

PROOF. Let $\mathfrak{R}_n$ be a sequence of finitely-connected Jordan regions that fill up $\mathfrak{D}$ from the interior. For every n we have then an $h_n > 0$ such that the transformation

$$T^h : x = x^h(u, v), \; y = y^h(u, v), \qquad (u, v) \in \mathfrak{R}_n,$$

is defined on $\mathfrak{R}_n$ for $h < h_n$. For each n we choose an $h > 0$ such that $h < h_n$, $h < 1/n$, and we denote the corresponding transformation T^h by T_n. In view of III.2.66, T_n satisfies in $\mathfrak{R}_n^0$ the assumption made in IV.4.29 and hence T_n is sAC and hence *a fortiori* eAC in $\mathfrak{R}_n^0$. If F is any closed set in $\mathfrak{D}$, then $\rho(T_n, T, F) \to 0$ for $n \to \infty$ by III.2.66. In view of condition (ii) above, it follows that the special closure theorem of IV.4.17 applies (with $\mathfrak{D}_n = \mathfrak{R}_n^0$), and hence T is eAC in $\mathfrak{D}$ (observe that $J_h(w) = J(w, T^h)$).

REMARK. In this application of the closure theorem, the approximating transformations T_n are known to be sAC and still we can only assert that T itself is eAC, because no closure theorem is available for sAC transformations.

IV.4.32. Given $T : z = t(w) = x(u, v) + iy(u, v)$, $u + iv = w \in \mathfrak{D}$, as in IV.2.1 let us denote by H_0 (hypothesis H_0) the following set of conditions:

(i) $x(u, v)$ and $y(u, v)$ are ACT (see III.2.64) in every oriented rectangle $R \subset \mathfrak{D}$. As a consequence, the partial derivatives x_u, x_v, y_u, y_v exist a.e. in $\mathfrak{D}$, and are summable on every oriented rectangle R and hence on every closed set in $\mathfrak{D}$ (cf. III.2.50, III.2.49).

(ii) The ordinary Jacobian $J(w) = x_u y_v - x_v y_u$ is summable in $\mathfrak{D}$.

It has been surmised that if T satisfies the hypothesis H_0 in $\mathfrak{D}$, then T is eAC in $\mathfrak{D}$. *Since practically all the special transformations studied previously in the literature satisfy the hypothesis H_0, the verification of this surmise would be a matter of considerable interest.* In the absence of a general verification of the surmise, we have to discuss some special cases explicitly. If T satisfies the hypothesis H_0 in $\mathfrak{D}$, then the following remark may be used to simplify the application of IV.4.31. Let R be any oriented rectangle in $\mathfrak{D}$. Since $x(u, v)$, $y(u, v)$ are ACT on R, we have, by III.2.67, $J_h(w) \rightarrow_{h\rightarrow 0} J(w, T)$ a.e. in R. Hence, to make sure that condition (ii) in IV.4.31 holds, it is sufficient to show (see I.3.11) that the family of functions $J_h(w)$ possesses the property (V) in R with respect to h, and for this purpose it is sufficient to show that this is true for both of the families

$$(1) \qquad \frac{\partial x^h}{\partial u}\frac{\partial y^h}{\partial v} \quad \text{and} \quad \frac{\partial x^h}{\partial v}\frac{\partial y^h}{\partial u}.$$

IV.4.33. Given T as in IV.4.32, let us make the following assumptions:

(a) T satisfies the hypothesis H_0 in $\mathfrak{D}$ (see IV.4.32).

(b) For every oriented rectangle $R \subset \mathfrak{D}$, the two factors in each one of the products $x_u y_v$, $x_v y_u$ belong to associated Lebesgue classes L^p, L^q in R (see I.3.10).

Then for every oriented rectangle $R \subset \mathfrak{D}$ we have (see IV.4.31 for notations)

$$(1) \qquad \iint_R | J_h(w) - J(w, T) | \underset{h\rightarrow 0}{\rightarrow} 0,$$

$$(2) \qquad \iint_R | J_h(w) | \underset{h\rightarrow 0}{\rightarrow} \iint_R | J(w, T) |,$$

and hence, by IV.4.31, T is eAC in $\mathfrak{D}$.

Proof. Since $J_h(w) \rightarrow J(w, T)$ a.e. in R, as noted in IV.4.32, the relations (1) and (2) are equivalent by I.3.11, and hence it is sufficient to establish (2). Proceeding according to the plan outlined in IV.4.32, we propose to show that the families IV.4.32(1) possess the property (V) with respect to h in every oriented rectangle $R \subset \mathfrak{D}$. Let us consider the first family $(\partial x^h/\partial u)(\partial y^h/\partial v)$. Let R be given by

$$R : a \leqq u \leqq b, c \leqq v \leqq d.$$

Let R_δ denote the rectangle

$$R_\delta : a - \delta \leqq u \leqq b + \delta, c - \delta \leqq v \leqq d + \delta.$$

Let $\delta > 0$ be so small that $R_\delta \subset \mathfrak{D}$ (note that $R \subset \mathfrak{D}$ by assumption). Such a δ being fixed, let us restrict h by the inequality $0 < h < \delta$. By assumption, the functions x_u, y_v belong to associated Lebesgue classes L^p, L^q where by definition (cf. I.3.10)

$$p > 1, q > 1, 1/p + 1/q = 1.$$

Since $|x_u|^p$, $|y_v|^q$ are summable in R_δ we have (cf. I.3.13) for every $\epsilon > 0$ a $\lambda = \lambda(\epsilon) > 0$ such that

(3) $$\iint_e |x_u|^p\,du\,dv, \qquad \iint_e |y_v|^q\,du\,dv < \epsilon$$

for every measurable set e such that

$$|e| < \lambda,\ e \subset R_\delta\,.$$

Now let e be a measurable set such that $|e| < \lambda$, $e \subset R$. In view of III.2.67, I.3.10 we obtain

$$\left|\frac{\partial x^h(u, v)}{\partial u}\right| \leqq \frac{1}{4h^2}\int_{-h}^{h}\int_{-h}^{h} |x_u(u + \xi, v + \eta)|\,d\xi\,d\eta$$

$$\leqq \frac{1}{4h^2}\left[\int_{-h}^{h}\int_{-h}^{h} |x_u(u + \xi, v + \eta)|^p\,d\xi\,d\eta\right]^{1/p}\left[\int_{-h}^{h}\int_{-h}^{h} 1^q\,d\xi\,d\eta\right]^{1/q},$$

and hence

(4) $$\left|\frac{\partial x^h}{\partial u}\right|^p \leqq \frac{1}{4h^2}\int_{-h}^{h}\int_{-h}^{h} |x_u(u + \xi, v + \eta)|^p\,d\xi\,d\eta.$$

Let us denote by $e(\xi, \eta)$ the set obtained from e by the translation with components ξ, η. Since $e \subset R$ and $0 < h < \delta$, it follows that for all values ξ, η relevant for (4), $e(\xi, \eta) \subset R_\delta$ and of course $|e(\xi, \eta)| = |e| < \lambda$. From (3), (4) it follows then that

(5) $$\begin{aligned}\iint_e \left|\frac{\partial x^h}{\partial u}\right|^p du\,dv &\leqq \frac{1}{4h^2}\int_{-h}^{h}\int_{-h}^{h}\left[\iint_e |x_u(u + \xi, v + \eta)|^p\,du\,dv\right]d\xi\,d\eta \\ &= \frac{1}{4h^2}\int_{-h}^{h}\int_{-h}^{h}\left[\iint_{e(\xi,\eta)} |x_u(u, v)|^p\,du\,dv\right]d\xi\,d\eta < \epsilon.\end{aligned}$$

Similarly it follows that

(6) $$\iint_e \left|\frac{\partial y^h}{\partial v}\right|^q du\,dv \leqq \epsilon.$$

The Hölder inequality (see I.3.10) yields now, in view of (5), (6),

(7) $$\begin{aligned}\iint_e \left|\frac{\partial x^h}{\partial u}\frac{\partial y^h}{\partial v}\right| du\,dv &\leqq \left[\iint_e \left|\frac{\partial x^h}{\partial u}\right|^p du\,dv\right]^{1/p}\left[\iint_e \left|\frac{\partial y^h}{\partial v}\right|^q du\,dv\right]^{1/q} \\ &< \epsilon^{1/p+1/q} = \epsilon.\end{aligned}$$

This inequality holds, for given $\epsilon > 0$, for every measurable set $e \subset R$ such that

$|e| < \lambda = \lambda(\epsilon)$. Thus the family $(\partial x^h/\partial u)(\partial y^h/\partial v)$ possesses the property (V) in R with respect to h. The family $(\partial x^h/\partial v)(\partial y^h/\partial u)$ is treated in the same manner.

IV.4.34. Let us replace, in IV.4.33, condition (b) by the following condition:

(b*) For every oriented rectangle $R \subset \mathfrak{D}$, one of the functions $x(u, v)$, $y(u, v)$ is Lipschitzian in R (see I.3.14).

Then the same conclusions hold as in IV.4.33.

PROOF. Proceeding as in IV.4.33, we introduce R_δ as before. By assumption, one of $x(u, v)$, $y(u, v)$ is Lipschitzian on R_δ, say $x(u, v)$. Then we have a constant M such that (cf. I.3.14)

$$(1) \qquad \left|\frac{\partial x}{\partial u}\right|, \left|\frac{\partial x}{\partial v}\right| < M \qquad \text{a.e. in } R_\delta .$$

On the other hand, $y(u, v)$ is ACT on R_δ, and hence for every $\epsilon > 0$ we have a $\mu = \mu(\epsilon) > 0$ such that

$$(2) \qquad \iint_e \left|\frac{\partial y}{\partial u}\right| du\, dv, \qquad \iint_e \left|\frac{\partial y}{\partial v}\right| du\, dv < \epsilon$$

for every measurable set e such that $|e| < \mu$, $e \subset R_\delta$. Now let e be any measurable set such that $|e| < \mu$, $e \subset R$. By (1), (2) and III.2.67 we have (using again the notation $e(\xi, \eta)$ as in IV.4.33)

$$\left|\frac{\partial x^h}{\partial u}\right| \leqq \frac{1}{4h^2}\int_{-h}^{h}\int_{-h}^{h} |x_u(u + \xi, v + \eta)|\, d\xi\, d\eta < M \qquad \text{in } R,$$

$$\iint_e \left|\frac{\partial x^h}{\partial u}\frac{\partial y^h}{\partial v}\right| du\, dv \leqq M \iint_e \left|\frac{\partial y^h}{\partial v}\right| du\, dv$$

$$\leqq M \frac{1}{4h^2}\int_{-h}^{h}\int_{-h}^{h}\left[\iint_e |y_v(u + \xi, v + \eta)|\, du\, dv\right] d\xi\, d\eta$$

$$= M \frac{1}{4h^2}\int_{-h}^{h}\int_{-h}^{h}\left[\iint_{e(\xi,\eta)} |y_v(u, v)|\, du\, dv\right] d\xi\, d\eta < M\epsilon.$$

Since M is fixed and ϵ arbitrary, it follows that the family $(\partial x^h/\partial u)(\partial y^h/\partial v)$ possesses the property (V) in R with respect to h. The family $(\partial x^h/\partial v)(\partial y^h/\partial u)$ is treated in the same manner.

IV.4.35. Given T as in IV.4.32, let us suppose that T satisfies the hypothesis H_0 in $\mathfrak{D}$. We mentioned there the surmise that T should then be eAC in $\mathfrak{D}$. We shall discuss now a curious theorem that may be interpreted as supporting evidence in favor of the plausibility of this surmise. In preparation, we derive a simple lemma. Let T satisfy the hypothesis H_0 in $\mathfrak{D}$, and let $\mathfrak{R}$ be a finitely-connected Jordan region in $\mathfrak{D}$. Let us put again

$$x^h(u, v) = \frac{1}{4h^2}\int_{-h}^{h}\int_{-h}^{h} x(u + \xi, y + \eta)\, d\xi\, d\eta.$$

If h is small enough, then by III.2.66, x^h will be defined and continuous, together with its first partial derivatives, in $\mathfrak{R}$. Assuming that h is so chosen, we assert that the transformation

$$T_h : x = x^h(u, v); \; y = y(u, v), \qquad (u, v) \in \mathfrak{R}^0,$$

is eAC in $\mathfrak{R}^0$ (let us note that T_h is different from the transformation T^h considered in IV.4.31).

PROOF. Since the partial derivatives of the first order of $x^h(u, v)$ exist and are continuous on $\mathfrak{R}$, clearly $x^h(u, v)$ is Lipschitzian in $\mathfrak{R}^0$ and hence in every oriented rectangle $R \subset \mathfrak{R}^0$. On the other hand, on every such rectangle $y(u, v)$ is ACT by assumption. Finally, the ordinary Jacobian $J(w, T_h)$ is summable in $\mathfrak{R}^0$, since x_u^h, x_v^h are bounded, in absolute value, by a fixed constant in $\mathfrak{R}$, and y_u, y_v are summable in $\mathfrak{R}$ (cf. IV.4.32(i)). Thus T_h satisfies the assumptions of IV.4.34 (with $\mathfrak{D}$ replaced by $\mathfrak{R}^0$), and hence T_h is eAC in $\mathfrak{R}^0$.

IV.4.36. Given $T : z = t(w) = x(u, v) + iy(u, v)$, $u + iv = w \in \mathfrak{D}$, as in IV.2.1, let us suppose that T satisfies the hypothesis H_0 (see IV.4.32). Let $\mathfrak{R}$ be a finitely-connected Jordan region in $\mathfrak{D}$. We take an auxiliary Jordan region $\mathfrak{R}_*$ such that $\mathfrak{R}_* \subset \mathfrak{D}$ and $\mathfrak{R} \subset \mathfrak{R}_*^0$. Clearly we have a $\delta > 0$ such that if (u, v) is any point in $\mathfrak{R}$, then the point $(u + \alpha, v + \beta)$ is in $\mathfrak{R}_*$ and the point $(u + \alpha + \xi, v + \beta + \eta)$ is in $\mathfrak{D}$ if $|\alpha|, |\beta|, |\xi|, |\eta| < \delta$. Let us introduce the square

$$Q_\delta : -\delta \leqq \alpha \leqq \delta, \; -\delta \leqq \beta \leqq \delta$$

in an auxiliary $\alpha\beta$-plane. Then the transformation

$$T_{\alpha\beta} : \begin{cases} x = x_{\alpha\beta}(u, v) = x(u + \alpha, v + \beta), \\ y = y(u, v), \end{cases} \qquad (u, v) \in \mathfrak{R},$$

is defined for every choice of the constants α, β such that $(\alpha, \beta) \in Q_\delta$. We shall establish presently the following curious fact.

THEOREM. *Under the conditions just described the transformation $T_{\alpha\beta}$ is eAC in $\mathfrak{R}^0$ for a.e. choice of the point $(\alpha, \beta) \in Q_\delta$.*

We subdivide the proof into several steps.

IV.4.37. CONTINUATION. We first show that $T_{\alpha\beta}$ satisfies the hypothesis H_0 (see IV.4.32) in $\mathfrak{R}^0$ for a.e. $(\alpha, \beta) \in Q_\delta$. Let $s(u, v, \delta)$ denote the oriented square with center (u, v) and side length 2δ. Clearly $s(u, v, \delta) \subset \mathfrak{R}_*$ if $(u, v) \in \mathfrak{R}$, and hence for any point $(u, v) \in \mathfrak{R}$

$$(1) \qquad \begin{aligned} \iint_{Q_\delta} |x_u(u + \alpha, v + \beta)| \, d\alpha \, d\beta &= \iint_{s(u,v,\delta)} |x_u(U, V)| \, dU \, dV \\ &\leqq \iint_{\mathfrak{R}_*} |x_u(U, V)| \, dU \, dV, \end{aligned}$$

where the last integral exists by IV.4.32(i). Using (1) and the theorem of Tonelli (see I.3.10) we obtain

$$\iint_{\mathfrak{R}} \iint_{Q_\delta} |\, x_u(u + \alpha, v + \beta) \,||\, y_v(u, v) \,|\, d\alpha\, d\beta\, du\, dv$$

$$= \iint_{\mathfrak{R}} \left[\iint_{Q_\delta} |\, x_u(u + \alpha, v + \beta) \,|\, d\alpha\, d\beta \right] |\, y_v(u, v) \,|\, du\, dv$$

$$\leqq \left[\iint_{\mathfrak{R}^*} |\, x_u(U, V) \,|\, dU\, dV \right] \left[\iint_{\mathfrak{R}} |\, y_v(u, v) \,|\, du\, dv \right].$$

The last term in this relation is finite by IV.4.32(i), and hence the existence of the quadruple integral follows, and furthermore (see I.3.10) the quadruple integral is equal to

$$\text{(2)} \qquad \iint_{Q_\delta} \left[\iint_{\mathfrak{R}} |\, x_u(u + \alpha, v + \beta) \,||\, y_v(u, v) \,|\, du\, dv \right] d\alpha\, d\beta = \iint_{Q_\delta} G(\alpha, \beta)\, d\alpha\, d\beta,$$

where we have put

$$\text{(3)} \qquad G(\alpha, \beta) = \iint_{\mathfrak{R}} |\, x_u(u + \alpha, v + \beta) \,||\, y_v(u, v) \,|\, du\, dv.$$

In view of the existence of the iterated integral (2), it follows by I.3.10 that $G(\alpha, \beta)$ is finite a.e. in Q_δ and is summable in Q_δ. In other words, for a.e. point $(\alpha, \beta) \in Q_\delta$, the integral (3) exists. A similar argument shows that for a.e. point $(\alpha, \beta) \in Q_\delta$ the integral of the product $|\, x_v(u + \alpha, v + \beta)| \,|\, y_u(u, v) \,|$ is summable in $\mathfrak{R}$ (as a function of u, v). It follows that the Jacobian

$$J(w, T_{\alpha\beta}) = x_u(u + \alpha, v + \beta) y_v(u, v) - x_v(u + \alpha, v + \beta) y_u(u, v)$$

is summable in $\mathfrak{R}$ for a.e. choice of the point $(\alpha, \beta) \in Q_\delta$. Thus for a.e. $(\alpha, \beta) \in Q_\delta$, $T_{\alpha\beta}$ satisfies condition IV.4.32(ii) in $\mathfrak{R}^0$. Since $x(u, v)$, $y(u, v)$ are ACT on every oriented rectangle in $\mathfrak{D}$, obviously $x(u + \alpha, v + \beta)$, $y(u, v)$ are ACT on every oriented rectangle in $\mathfrak{R}^0$. Thus $T_{\alpha\beta}$ satisfies the hypothesis H_0 in $\mathfrak{R}^0$ for a.e. choice of the point $(\alpha, \beta) \in Q_\delta$.

IV.4.38. Continuation. We assert that for a.e. choice of the point $(\alpha, \beta) \in Q_\delta$, the family of functions

$$\text{(1)} \quad f_h(u, v; \alpha, \beta) = \left[\frac{1}{4h^2} \int_{-h}^{h} \int_{-h}^{h} |\, x_u(u + \alpha + \xi, v + \beta + \eta) \,|\, d\xi\, d\eta \right] |\, y_v(u, v) \,|$$

possesses the property (V) in $\mathfrak{R}$ with respect to h (cf. I.3.11). The same statement holds if x_u, y_v are replaced by x_v, y_u respectively.

Proof. Consider the function $G(\alpha, \beta)$ in IV.4.37(3). As noted in IV.4.37, $G(\alpha, \beta)$ is summable in Q_δ, and hence (see III.2.65)

(2) $$\frac{1}{4h^2}\int_{-h}^{h}\int_{-h}^{h} G(\alpha + \xi, \beta + \eta)\, d\xi\, d\eta \underset{h\to 0}{\longrightarrow} G(\alpha, \beta)$$

for a.e. point $(\alpha, \beta) \in Q_\delta$. Let (α, β) be a point such that (2) holds. Keeping (α, β) fixed, and using the theorem of Tonelli (see I.3.10), (2) can be written in the form

(3) $$\iint_{\mathfrak{R}} f_h(u, v; \alpha, \beta)\, du\, dv \underset{h\to 0}{\longrightarrow} \iint_{\mathfrak{R}} |\, x_u(u + \alpha, v + \beta)\, ||\, y_v(u, v)\, du\, dv.$$

By III.2.65 we have (note that α, β are fixed)

(4) $$f_h(u, v; \alpha, \beta) \underset{h\to 0}{\longrightarrow} |\, x_u(u + \alpha, v + \beta)\, ||\, y_v(u, v)\, | \qquad \text{a.e. in } \mathfrak{R}.$$

(3) and (4) imply, by I.3.11, that the family (1) possesses the property (V) in $\mathfrak{R}$ with respect to h. Of course, the same reasoning applies if x_u, y_v are replaced by x_v, y_u.

IV.4.39. CONTINUATION. We consider now the auxiliary transformations

$$T_{\alpha\beta h}: \begin{cases} x = x^h_{\alpha\beta}(u, v), \\ y = y(u, v), \end{cases} \qquad (u, v) \in \mathfrak{R}^0,$$

where $x^h_{\alpha\beta}(u, v)$ is the integral mean

$$x^h_{\alpha\beta}(u, v) = \frac{1}{4h^2}\int_{-h}^{h}\int_{-h}^{h} x_{\alpha\beta}(u + \xi, v + \eta)\, d\xi\, d\eta$$

$$= \frac{1}{4h^2}\int_{-h}^{h}\int_{-h}^{h} x(u + \alpha + \xi, v + \beta + \eta)\, d\xi\, d\eta,$$

and h is restricted by the inequality $0 < h < \delta$. Clearly, $T_{\alpha\beta h}$ is defined in $\mathfrak{R}^0$. Now by IV.4.37, $T_{\alpha\beta}$ satisfies the hypothesis H_0 in $\mathfrak{R}^0$ for a.e. point $(\alpha, \beta) \in Q_\delta$. Let (α, β) be a point in Q_δ for which this is true. This point (α, β) will be fixed in the rest of the reasoning. Let now $\mathfrak{R}_n$ be a sequence of finitely-connected Jordan regions that fill up $\mathfrak{R}^0$ from the interior (see IV.1.41). For each n the transformation $T_{\alpha\beta h}$ is eAC in $\mathfrak{R}^0_n$ for h sufficiently small, by IV.4.35 (applied to $T_{\alpha\beta}$ considered in $\mathfrak{R}^0$). For each n let us choose an h_n such that $0 < h_n < 1/n$ and such that $T_{\alpha\beta h}$ is eAC in $\mathfrak{R}^0_n$ for $h = h_n$. Let us denote the transformation $T_{\alpha\beta h}$, $h = h_n$, by $T^n_{\alpha\beta}$. Then we have the following facts at our disposal:

(a) $T^n_{\alpha\beta}$ is eAC in $\mathfrak{R}^0_n$, $n = 1, 2, \cdots$.

(b) Since $h_n \to 0$, clearly $\rho(T^n_{\alpha\beta}, T_{\alpha\beta}, F) \to 0$ for $n \to \infty$, for every closed set $F \subset \mathfrak{R}^0$.

(c) Since $x_{\alpha\beta}(u, v)$ is ACT in every oriented rectangle in $\mathfrak{R}^0$ (cf. IV.4.37), we have the relations (cf. III.2.67)

$$\frac{\partial x^h_{\alpha\beta}}{\partial u} \underset{h\to 0}{\longrightarrow} \frac{\partial x_{\alpha\beta}}{\partial u}, \qquad \frac{\partial x^h_{\alpha\beta}}{\partial v} \underset{h\to 0}{\longrightarrow} \frac{\partial x_{\alpha\beta}}{\partial v}$$

a.e. in $\mathfrak{R}^0$, and hence (note that $0 < h_n < 1/n$)

$$J(w, T^n_{\alpha\beta}) \underset{n\to\infty}{\longrightarrow} \frac{\partial x_{\alpha\beta}}{\partial u}\frac{\partial y}{\partial v} - \frac{\partial x_{\alpha\beta}}{\partial v}\frac{\partial y}{\partial u} = J(w, T_{\alpha\beta}) \qquad \text{a.e. in } \mathfrak{R}^0. \tag{1}$$

(d) Since $x_{\alpha\beta}(u, v)$ is ACT in every oriented rectangle in $\mathfrak{R}^0$, we have by III.2.67 the formulas

$$\frac{\partial x^h_{\alpha\beta}(u, v)}{\partial u} = \frac{1}{4h^2}\int_{-h}^{h}\int_{-h}^{h} x_u(u + \alpha + \xi, v + \beta + \eta)\, d\xi\, d\eta,$$

$$\frac{\partial x^h_{\alpha\beta}(u, v)}{\partial v} = \frac{1}{4h^2}\int_{-h}^{h}\int_{-h}^{h} x_v(u + \alpha + \xi, v + \beta + \eta)\, d\xi\, d\eta,$$

a.e. in $\mathfrak{R}^0$. From IV.4.38 it follows therefore immediately that the family of the Jacobians $J(w, T^n_{\alpha\beta})$, $n = 1, 2, \cdots$, possesses the property (V) in $\mathfrak{R}^0$. In view of (1) it follows that (cf. I.3.11)

$$\iint_R | J(w, T^n_{\alpha\beta}) | \underset{n\to\infty}{\longrightarrow} \iint_R | J(w, T_{\alpha\beta}) |$$

for every oriented rectangle $R \subset \mathfrak{R}^0$.

(e) Let us note finally that $J(w, T_{\alpha\beta})$ is summable in $\mathfrak{R}^0$, since (α, β) has been so chosen that $T_{\alpha\beta}$ satisfies the hypothesis H_0 in $\mathfrak{R}^0$.

The information contained in (a) to (e) enables us to state that the transformations $T_{\alpha\beta}$, $T^n_{\alpha\beta}$ considered in the domains $\mathfrak{R}^0$, $\mathfrak{R}^0_n$, respectively, satisfy the assumptions of the closure theorem in IV.4.17. Hence $T_{\alpha\beta}$ is eAC in $\mathfrak{R}^0$, and the proof is complete.

IV.4.40. The theorem of IV.4.36 suggests itself, in view of the closure theorem of IV.4.17, as a promising tool in attempting to verify the surmise stated in IV.4.32. Only results of a special type were obtained so far in this direction. We restrict ourselves to deriving one such special result that is of independent importance. In preparation, we first establish a lemma. Let $H(\alpha, \beta)$ be a non-negative summable function of α, β in the square $Q_\delta : -\delta \leqq \alpha \leqq \delta,\ -\delta \leqq \beta \leqq \delta$, such that

$$\frac{1}{4h^2}\int_{-h}^{h}\int_{-h}^{h} H(\alpha, \beta)\, d\alpha\, d\beta \underset{h\to 0}{\longrightarrow} 0. \tag{1}$$

Let E be a subset of measure zero in Q_δ, and let $\epsilon > 0$ be given arbitrarily. Let G_ϵ be the subset of $Q_\delta - E$ on which $H(\alpha, \beta) < \epsilon$. Then the point $(0, 0)$ is a limit point of G_ϵ.

Indeed, if the assertion were false, we should have an $\eta > 0$ such that

$$H(\alpha, \beta) \geqq \epsilon \text{ for } (\alpha, \beta) \in (Q_\delta - E)Q_\eta,$$

Q_η is the square $-\eta \leqq \alpha \leqq \eta$, $-\eta \leqq \beta \leqq \eta$. Clearly we can assume $\eta < \delta$. Since $| E | = 0$, we should have then $H(\alpha, \beta) \geqq \epsilon$ a.e. in Q_η, and hence obviously

$$\frac{1}{4h^2}\int_{-h}^{h}\int_{-h}^{h} H(\alpha, \beta)\, d\alpha\, d\beta \geqq \epsilon \qquad \text{for } 0 < h < \eta$$

in contradiction to (1).

Now that (0, 0) is shown to be a limit point of G_ϵ there follows the existence of a point (α, β) such that $|\alpha|, |\beta| < \epsilon$, $0 \leqq H(\alpha, \beta) < \epsilon$, $(\alpha, \beta) \in Q_\delta - E$. Applying this result with $\epsilon = 1/n$, $n = 1, 2, \cdots$, we see that there exists a sequence of points (α_n, β_n) such that

$$(\alpha_n, \beta_n) \to (0, 0),\ 0 \leqq H(\alpha_n, \beta_n) < 1/n,\ (\alpha_n, \beta_n) \in Q_\delta - E.$$

IV.4.41. Given $T : z = t(w) = x(u, v) + iy(u, v)$, $u + iv = w \in \mathfrak{D}$, as in IV.2.1, let us suppose that T satisfies the hypothesis H_0 (cf. IV.4.32). Let us put

$$Z^*(w, \alpha, \beta) = x_u(u + \alpha, v + \beta)y_v(u, v) - x_v(u + \alpha, v + \beta)y_u(u, v).$$

Let us further assume that

$$(1) \qquad \lim_{h\to 0}\frac{1}{4h^2}\int_{-h}^{h}\int_{-h}^{h}\left[\iint_R |Z^*(w, \alpha, \beta) - J(w, T)|\, du\, dv\right] d\alpha\, d\beta = 0$$

for every oriented rectangle $R \subset \mathfrak{D}$. Then T is eAC in $\mathfrak{D}$.

Proof. Let $\mathfrak{R}$ be a finitely-connected Jordan region in $\mathfrak{D}$. Using the notations of IV.4.36, we restrict the point (α, β) by the condition $(\alpha, \beta) \in Q_\delta$, and we consider in $\mathfrak{R}^0$ the transformation $T_{\alpha\beta}$ of IV.4.36. Clearly

$$Z^*(w, \alpha, \beta) = J(w, T_{\alpha\beta}), \qquad w \in \mathfrak{R}^0.$$

Thus the assumption (1) may be written in the form

$$(2) \qquad \lim_{h\to 0}\frac{1}{4h^2}\int_{-h}^{h}\int_{-h}^{h}\left[\iint_R |J(w, T_{\alpha\beta}) - J(w, T)|\, du\, dv\right] d\alpha\, d\beta = 0$$

Let us observe that the existence of the quadruple integral follows from the argument used in IV.4.37. Since $\mathfrak{R} \subset \mathfrak{D}$, we can choose a finite number of oriented rectangles $R_1, \cdots, R_m$ such that $\mathfrak{R}^0 \subset R_1 + \cdots + R_m \subset \mathfrak{D}$. From (2) it follows then immediately that

$$(3) \qquad \lim_{h\to 0}\frac{1}{4h^2}\int_{-h}^{h}\int_{-h}^{h} H(\alpha, \beta)\, d\alpha\, d\beta = 0,$$

where we have put $H(\alpha, \beta) = \iint_{\mathfrak{R}^0} |J(w, T_{\alpha\beta}) - J(w, T)|\, du\, dv$. By the argument used in IV.4.37 we ascertain that $H(\alpha, \beta)$ exists for a.e. point $(\alpha, \beta) \in Q_\delta$ and is summable in Q_δ. By IV.4.36 we know that $T_{\alpha\beta}$ is eAC in $\mathfrak{R}^0$ for a.e. choice of $(\alpha, \beta) \in Q_\delta$. Let E be the set of those points $(\alpha, \beta) \in Q_\delta$ where either $H(\alpha, \beta)$ fails to exist or $T_{\alpha\beta}$ fails to be eAC in $\mathfrak{R}^0$. Then $|E| = 0$, and hence by IV.4.40 we have, as a consequence of (3), a sequence of points (α_n, β_n) such that

$$(4) \qquad (\alpha_n, \beta_n) \to (0, 0),\ H(\alpha_n, \beta_n) \to 0,\ (\alpha_n, \beta_n) \in Q_\delta - E.$$

Let us denote by T_n the transformation $T_{\alpha\beta}$ for $\alpha = \alpha_n$, $\beta = \beta_n$. As an immediate consequence of (4), we have then the following relations (cf. IV.4.36).

(a) $\rho(T_n, T, F) \to 0$ for $n \to \infty$, for every closed set $F \subset \mathfrak{R}^0$.

(b) T_n is eAC in $\mathfrak{R}^0$, $n = 1, 2, \cdots$.

(c) $$\iint_{\mathfrak{R}^0} | J(w, T_n) - J(w, T) | \, du \, dv \to 0 \qquad \text{for } n \to \infty.$$

A fortiori, this holds for every oriented rectangle $R \subset \mathfrak{R}^0$.

In view of (a), (b), (c), it follows by the closure theorem of IV.4.18 that T is eAC in $\mathfrak{R}^0$. Since $\mathfrak{R}$ was an arbitrary Jordan region in $\mathfrak{D}$, it follows by IV.4.20 that T is eAC in $\mathfrak{D}$.

IV.4.42. We shall discuss presently certain convergence theorems involving Jacobians. Let there be given a transformation $T : z = t(w)$, $w \in \mathfrak{D}$, as in IV.2.1, and also a sequence of transformations $T_n : z = t_n(w)$, $w \in \mathfrak{D}_n$, such that the following conditions hold:

(i) If F is any closed set in $\mathfrak{D}$, then $F \subset \mathfrak{D}_n$ for n sufficiently large, and $\rho(T_n, T, F) \to 0$ for $n \to \infty$ (cf. IV.1.1).

(ii) The ordinary Jacobian $J(w, T)$ exists a.e. in $\mathfrak{D}$ and is summable in $\mathfrak{D}$.

(iii) T_n is eAC in $\mathfrak{D}_n$, $n = 1, 2, \cdots$.

THEOREM. *Under the conditions just stated, there exists a sequence of measurable sets V_n, such that $V_n \subset \mathfrak{D}_n$, $n = 1, 2, \cdots$, and*

$$\iint_{V_n} \mathcal{J}_e(w, T_n) \underset{n \to \infty}{\to} \iint_{\mathfrak{D}} J(w, T).$$

The proof will be made in several steps. Let us note that we work with the ordinary Jacobian for T and with the generalized essential Jacobian for T_n. It would be very desirable to replace in this theorem, if possible, $J(w, T)$ by $\mathcal{J}_e(w, T)$. The above statement is merely the best that is available at present.[1]

IV.4.43. CONTINUATION. Let $\mathfrak{D}_*$ be any domain in $\mathfrak{D}$. We define

$$\lambda_n(\mathfrak{D}_*) = \underset{E}{\text{gr.l.b.}} \left| \iint_{\mathfrak{D}_*} J(w, T) - \iint_E \mathcal{J}_e(w, T_n) \right|,$$

where the greatest lower bound is taken with respect to all sets E such that (α) E is measurable (possibly empty), and (β) $E \subset \mathfrak{D}_* \mathfrak{D}_n$. It may happen, of course, that $\mathfrak{D}_* \mathfrak{D}_n = 0$, and in this case the set $E = 0$ is the only set that satisfies the conditions (α), (β). To take care of this case, we agree that an integral taken over the empty set is equal to zero. Then

$$\lambda_n(\mathfrak{D}_*) = \left| \iint_{\mathfrak{D}_*} J(w, T) \right| \qquad \text{if } \mathfrak{D}_* \mathfrak{D}_n = 0.$$

[1] The problem just referred to has been solved since by P. V. Reichelderfer.

If $\mathfrak{D}_*\mathfrak{D}_n \neq 0$, then we can choose E as a set of measure zero, and we obtain

$$\lambda_n(\mathfrak{D}_*) \leqq \left| \iint_{\mathfrak{D}_*} J(w, T) \right|.$$

Thus in either case we have the inequalities

$$(1) \qquad 0 \leqq \lambda_n(\mathfrak{D}_*) \leqq \left| \iint_{\mathfrak{D}_*} J(w, T) \right| \leqq \iint_{\mathfrak{D}_*} | J(w, T) |.$$

We define further

$$(2) \qquad \lambda(\mathfrak{D}_*) = \limsup_{n \to \infty} \lambda_n(\mathfrak{D}_*).$$

By (1) we have then the inequalities

$$(3) \qquad 0 \leqq \lambda(\mathfrak{D}_*) \leqq \left| \iint_{\mathfrak{D}_*} J(w, T) \right| \leqq \iint_{\mathfrak{D}_*} | J(w, T) |.$$

IV.4.44. CONTINUATION. Let now $s_1, \cdots, s_j, \cdots$ be a sequence of oriented squares with the following properties: (a) $s_j \subset \mathfrak{D}$, $j = 1, 2, \cdots$. (b) $s_j^0 s_k^0 = 0$ for $j \neq k$. (c) $| \mathfrak{D} - \sum s_j | = 0$. We assert that

$$(1) \qquad \lambda(\mathfrak{D}) \leqq \sum_{j=1}^{\infty} \lambda(s_j^0).$$

PROOF. Give any $\epsilon > 0$. By the definition of $\lambda_n(s_j^0)$ we have in $s_j^0\mathfrak{D}_n$ a measurable (possibly empty) set E_{nj} such that

$$\left| \iint_{s_j} J(w, T) - \iint_{E_{nj}} \mathfrak{J}_e(w, T_n) \right| < \lambda_n(s_j^0) + \epsilon/2^j.$$

Let us put $E_n = \sum E_{nj}$, the summation being taken for fixed n with respect to j. Then $E_n \subset \mathfrak{D}\mathfrak{D}_n$, and hence, in view of conditions (a), (b), (c),

$$\begin{aligned} \lambda_n(\mathfrak{D}) &\leqq \left| \iint_{\mathfrak{D}} J(w, T) - \iint_{E_n} \mathfrak{J}_e(w, T_n) \right| \\ (2) \qquad &= \left| \sum_{j=1}^{\infty} \iint_{s_j} J(w, T) - \sum_{j=1}^{\infty} \iint_{E_{nj}} \mathfrak{J}_e(w, T_n) \right| \\ &\leqq \sum_{j=1}^{\infty} \left| \iint_{s_j} J(w, T) - \iint_{E_{nj}} \mathfrak{J}_e(w, T_n) \right| < \sum_{j=1}^{\infty} \lambda_n(s_j^0) + \epsilon. \end{aligned}$$

Now take any positive integer i. We have then (see IV.4.43(1))

$$\sum_{j=1}^{\infty} \lambda_n(s_j^0) \leqq \sum_{j=1}^{i} \lambda_n(s_j^0) + \sum_{j=i+1}^{\infty} \iint_{s_j} | J(w, T) |.$$

For $n \to \infty$ there follows the inequality (cf. IV.4.43(2))

(3) $$\limsup_{n\to\infty} \sum_{j=1}^{\infty} \lambda_n(s_j^0) \leqq \sum_{j=1}^{i} \lambda(s_j^0) + \sum_{j=i+1}^{\infty} \iint_{s_j} | J(w, T) |.$$

By conditions (a), (b), (c) we have

$$\iint_{\mathfrak{D}} | J(w, T) | = \sum_{j=1}^{\infty} \iint_{s_j} | J(w, T) |.$$

Thus the infinite summation on the right in (3) is the remainder of a convergent infinite series, and hence converges to zero for $i \to \infty$. Thus (3) yields, for $i \to \infty$,

(4) $$\limsup_{n\to\infty} \sum_{j=1}^{\infty} \lambda_n(s_j^0) \leqq \sum_{j=1}^{\infty} \lambda(s_j^0).$$

(2) and (4) yield, for $n \to \infty$, the inequality

$$\lambda(\mathfrak{D}) \leqq \sum_{j=1}^{\infty} \lambda(s_j^0) + \epsilon.$$

Since ϵ was arbitrary, the inequality (1) follows. Let us note that the infinite series in (1) is convergent; indeed, by IV.4.43(3),

$$\sum_{j=1}^{\infty} \lambda(s_j^0) \leqq \sum_{j=1}^{\infty} \iint_{s_j} | J(w, T) | = \iint_{\mathfrak{D}} | J(w, T) |.$$

IV.4.45. Continuation. Now let us define a subset H of $\mathfrak{D}$ as follows. A point $w \in \mathfrak{D}$ belongs to H if and only if there exists a sequence of oriented squares $s_1(w), \cdots, s_i(w), \cdots$ with the following properties:

(1) The center of $s_i(w)$ is w, $i = 1, 2, \cdots$.

(2) $| s_i(u) | \to 0$ for $i \to \infty$.

(3) $\lambda[s_i^0(w)]/| s_i(w) | \to 0$ for $i \to \infty$,

where $s_i^0(w)$ denotes the interior of $s_i(w)$. We shall prove that

(1) $$| \mathfrak{D} - H | = 0,$$

but we want to verify first that if (1) is established, then the theorem stated in IV.4.42 follows easily. Suppose, indeed, that (1) is known to hold. Give any $\epsilon > 0$. Let w be a point in H, and let $s_i(w)$ be a sequence of squares with the properties (1), (2), (3). By discarding, if necessary, a finite number of these squares, we shall have

$$s_i(w) \subset \mathfrak{D}, \qquad \frac{\lambda[s_i^0(w)]}{|s_i(w)|} < \epsilon$$

for all the remaining (infinitely many) squares. Let now $\mathfrak{F}$ be the family of those oriented squares $s \subset \mathfrak{D}$ for which $\lambda(s^0)/| s | < \epsilon$. By the preceding remarks, the squares of the family $\mathfrak{F}$ cover the set H in the manner required by the Vitali covering theorem (see I.3.3), and hence $\mathfrak{F}$ contains a sequence $s_1, \cdots, s_j, \cdots$ of oriented squares such that $s_j \subset \mathfrak{D}$, $j = 1, 2, \cdots$, $s_j^0 s_k^0 = 0$ for $j \neq k$, and

$$\left| H - \sum_{j=1}^{\infty} s_j \right| = 0, \qquad \frac{\lambda(s_j^0)}{| s_j |} < \epsilon, \qquad j = 1, 2, \cdots .$$

In view of (1) it follows that

$$\left| \mathfrak{D} - \sum_{j=1}^{\infty} s_j \right| = 0.$$

By IV.4.44 we have therefore

$$\lambda(\mathfrak{D}) \leqq \sum_{j=1}^{\infty} \lambda(s_j^0) < \epsilon \sum_{j=1}^{\infty} | s_j | = \epsilon | \mathfrak{D} |.$$

Since ϵ was arbitrary, it follows that $\lambda(\mathfrak{D}) = 0$. Now by definition $\lambda(\mathfrak{D}) = \lim\sup \lambda_n(\mathfrak{D})$, and since $\lambda_n(\mathfrak{D}) \geqq 0$, it follows that

(2) $$\lambda_n(\mathfrak{D}) \to 0 \qquad \text{for } n \to \infty.$$

By the definition of $\lambda_n(\mathfrak{D})$ (see IV.4.43), we have a measurable (possibly empty) set V_n such that

(3) $$V_n \subset \mathfrak{D}\mathfrak{D}_n, \quad \left| \iint_{\mathfrak{D}} J(w, T) - \iint_{V_n} \mathfrak{J}_e(w, T_n) \right| < \lambda_n(\mathfrak{D}) + \frac{1}{n}.$$

(2) and (3) imply that these sets V_n possess the properties stated in the theorem of IV.4.42. As far as the preceding argument is concerned, several or even all of the sets V_n may be empty. If this is undesirable for some reason, we can replace each empty V_n by any subset of measure zero of $\mathfrak{D}_n$, and the modified sequence is then adequate for the theorem and consists of nonempty sets.

IV.4.46. CONTINUATION. It remains to prove that $| \mathfrak{D} - H | = 0$ (see IV.4.45). Let G denote the subset of $\mathfrak{D}$ defined as follows. A point $w_0 \in \mathfrak{D}$ belongs to G if and only if the following conditions hold:

(i) $J(w_0, T)$ exists.

(ii) There exists a sequence of oriented squares s_j, $j = 1, 2, \cdots$, with center at w_0, such that $| s_j | \to 0$ and $m(s_j) \to 0$ for $j \to \infty$. The symbol $m(s_j)$ is used here in the sense of IV.3.31 and is relative to the auxiliary affine transformation

$$T^* : \begin{cases} x = x_0 + a(u - u_0) + b(v - v_0), \\ y = y_0 + c(u - u_0) + d(v - v_0), \end{cases}$$

where $x_0 + iy_0 = t(w_0)$, $u_0 + iv_0 = w_0$, and $a = x_u(u_0, v_0)$, $b = x_v(u_0, v_0)$, $c = y_u(u_0, v_0)$, $d = y_v(u_0, v_0)$. Note that these partial derivatives exist by (i).

(iii) $\iint_s J(w, T)/| s | \to J(w_0, T)$ for every sequence of oriented squares s such that $| s | \to 0$ and $w_0 \in s$.

Since by assumption $J(w, T)$ exists a.e. in $\mathfrak{D}$, (i) and (ii) hold for a.e. point

$w_0 \in \mathfrak{D}$ (see I.3.13, I.3.14). Since by assumption $J(w, T)$ is summable in $\mathfrak{D}$, (iii) also holds for a.e. point $w_0 \in \mathfrak{D}$ (see I.3.13). Hence $|\mathfrak{D} - G| = 0$, and thus the relation $|\mathfrak{D} - H| = 0$ will be proved if we can show that

(1) $$G \subset H.$$

Let us first verify that the conditions

(2) $$w_0 \in G \text{ and } J(w_0, T) = 0$$

imply that $w_0 \in H$. Indeed, let us assume that (2) holds. Let s_j be a sequence of oriented squares satisfying condition (ii). By IV.4.43 we have

$$0 \leqq \lambda(s_j^0) \leqq \left| \iint_{s_j} J(w, T) \right|.$$

By (2) and condition (iii) it follows that

$$0 \leqq \limsup_{j \to \infty} \frac{\lambda(s_j^0)}{|s_j|} \leqq |J(w_0, T)| = 0.$$

Thus $\lambda(s_j^0)/|s_j| \to 0$ for $j \to \infty$, and hence $w_0 \in H$ (cf. IV.4.45). Hence, in proving (1), it is sufficient to consider points $w_0 \in G$ that satisfy the further condition

(iv) $J(w_0, T) \neq 0$.

Thus we have to prove that if a point $w_0 \in \mathfrak{D}$ satisfies the conditions (i), (ii), (iii), (iv), then $w_0 \in H$.

IV.4.47. Continuation. Let $w_0 \in \mathfrak{D}$ be a point that satisfies these conditions. We put (cf. IV.4.46(ii))

(1) $$\Delta = ad - bc = J(w_0, T) \neq 0.$$

We shall use the auxiliary affine transformation T^* described in IV.4.46(ii). In view of (1), the constant $k > 0$ of IV.3.28 is available. Let S be an oriented square in $\mathfrak{D}$ with center at w_0, and let s_j be a sequence of squares that satisfy the condition IV.4.46(ii). Since $|s_j| \to 0$ and $m(s_j) \to 0$, we can assume without loss of generality that

(2) $$s_j \subset S^0, \qquad m(s_j) < k/2^{1/2}, \qquad j = 1, 2, \cdots.$$

Furthermore, in view of IV.4.42(i), we have an n_0 such that

(3) $$S \subset \mathfrak{D}_n \qquad \text{for } n > n_0.$$

Since T_n is eAC in $\mathfrak{D}_n$ (see IV.4.42(iii)), T_n is also eAC in S^0 (see IV.4.1). Now let us consider one of the squares s_j. In view of (2), we can introduce an auxiliary smaller square s_j' as in IV.3.32. Let us put (cf. IV.4.46(ii))

$$\bar{p}_j' = T^*(s_j').$$

By IV.4.46(ii) and IV.3.30 we have then

(4) $$|\bar{p}_j'| = |J(w_0, T)|\,|s_j'|.$$

In view of IV.4.46(iv), we have by IV.3.35 $\mu(z, T, s_j) = \text{sgn } J(w_0, T) \neq 0$ for $z \in \bar{p}_j'$. By IV.4.42(i) and IV.3.36 we have an $n(j)$ such that

(5) $$\mu(z, T_n, s_j) = \text{sgn } J(w_0, T) \neq 0 \qquad \text{for } z \in \bar{p}_j',\ n > n(j).$$

By IV.3.15 (applied to T_n considered in s_j), it follows from (2), (3), (5) that

$$\mu(z, T_n, s_j) = \nu(z, T_n, s_j^0) \qquad \text{a.e. in } \bar{p}_j',\ n > n(j).$$

Integration yields, in view of (5), the formula

(6) $$\iint_{\bar{p}_j'} \nu(z, T_n, s_j^0) = |\bar{p}_j'| \text{ sgn } J(w_0, T), \qquad n > n(j).$$

(4) and (6) yield

(7) $$\iint_{\bar{p}_j'} \nu(z, T_n, s_j^0) = J(w_0, T) \mid s_j' \mid, \qquad n > n(j).$$

Now let $H(z)$ denote the characteristic function of the set $\bar{p}_j'$. Since T_n is eAC in S^0 and hence also in s_j^0 by (2), we have by IV.4.4 the formula

(8) $$\iint_{s_j} H[t_n(w)]\, \mathcal{J}_e(w, T_n) = \iint H(z)\nu(z, T_n, s_j^0), \qquad n > n_0.$$

Now $H(z) = 1$ in $\bar{p}_j'$ and $H(z) = 0$ elsewhere. Hence, by (7),

(9) $$\iint H(z)\nu(z, T_n, s_j^0) = J(w_0, T) \mid s_j' \mid, \qquad n > n(j).$$

On the other hand, in s_j^0 the function $H[t_n(w)]$ clearly coincides with the characteristic function of the set $E_n = s_j^0 T_n^{-1}(\bar{p}_j')$. Hence

(10) $$\iint_{s_j} H[t_n(w)]\, \mathcal{J}_e(w, T_n) = \iint_{E_n} \mathcal{J}_e(w, T_n), \qquad n > n_0.$$

(8), (9), (10) yield

$$\iint_{E_n} \mathcal{J}_e(w, T_n) = J(w_0, T) \mid s_j' \mid, \qquad n > n(j).$$

Since $E_n \subset s_j^0$, it follows that (cf. IV.4.43)

$$\lambda_n(s_j^0) \leqq \left| \iint_{s_j} J(w, T) - J(w_0, T) \mid s_j' \mid \right|.$$

This inequality holds, according to its derivation, if $n > n(j)$ and $n > n_0$ (cf. (3)). Keeping j fixed, we can therefore make $n \to \infty$, and we obtain (cf. IV.4.43)

$$(11) \qquad \lambda(s_j^0) \leqq \left| \iint_{s_j} J(w, T) - J(w_0, T) \mid s_j' \mid \right|.$$

Now we make $j \to \infty$. By IV.3.32 we have

$$| s_j' | = \left(1 - \frac{2^{1/2} m(s_j)}{k}\right)^2 | s_j |.$$

Since $m(s_j) \to 0$ for $j \to \infty$ (see IV.4.46(ii)), it follows that

$$\frac{| s_j' |}{| s_j |} \to 1 \qquad \text{for } j \to \infty.$$

On the other hand (see IV.4.46(iii))

$$\iint_{s_j} J(w, T)/| s_j | \to J(w_0, T) \qquad \text{for } j \to \infty.$$

From (11) there follows now the relation

$$\lambda(s_j^0)/| s_j | \to 0 \qquad \text{for } j \to \infty.$$

Thus $w_0 \in H$ (cf. IV.4.45), and the proof is complete.

IV.4.48. On account of its importance in applications, we state a special case of the theorem in IV.4.42 explicitly. *Let there be given a transformation $T : z = t(w)$, $w \in \mathfrak{D}$, as in* IV.2.1. *Let $T_n : z = t_n(w)$, $w \in \mathfrak{R}_n$, be a sequence of quasi-linear transformations* (see IV.4.30). *Suppose that the following conditions hold:*

(i) *If F is a closed set in $\mathfrak{D}$, then $F \subset \mathfrak{R}_n^0$ for n sufficiently large, and $\rho(T_n, T, F) \to 0$ for $n \to \infty$* (cf. IV.1.1).

(ii) *The ordinary Jacobian $J(w, T)$ exists a.e. in $\mathfrak{D}$ and is summable in $\mathfrak{D}$. Then there exists a sequence V_n of measurable sets, such that $V_n \subset \mathfrak{R}_n$, $n = 1, 2, \cdots$, and*

$$\iint_{V_n} J(w, T_n) \to \iint_{\mathfrak{D}} J(w, T) \qquad \textit{for } n \to \infty.$$

Proof. By IV.4.30, IV.4.28, T_n is sAC and hence eAC in $\mathfrak{R}_n^0$. Since T_n is quasi-linear, $J(w, T_n)$ exists a.e. in $\mathfrak{R}_n^0$ and, by IV.3.42, $J(w, T_n) = \mathcal{J}_e(w, T_n)$ a.e. in $\mathfrak{R}_n^0$. Thus the above statement is a direct corollary of IV.4.42. Let us note a simplification in the proof if one wants to prove this special theorem independently. Since T_n is quasi-linear, the transformation formula used in IV.4.47 (8) is practically trivial, and thus IV.4.47(8) can be used without reference to the general theory of eAC transformations.

IV.4.49. Let us point out a significant relationship between the theorem in IV.4.42 and the lemma in IV.4.11. Let the transformations T, T_n satisfy the assumptions of IV.4.42. Let R be an oriented rectangle in $\mathfrak{D}$. For n large enough, we shall have $R \subset \mathfrak{D}_n$. Hence, for n greater than a certain n_0, we can consider the transformations T, T_n merely in R^0, and the assumptions of IV.4.42 are clearly still satisfied (with $\mathfrak{D}$, $\mathfrak{D}_n$ all replaced by R^0). Hence, by IV.4.42, we have a sequence of measurable sets $V_n \subset R^0$, such that

$$\iint_{R^0} J(w\ T) = \lim_{n\to\infty} \iint_{V_n} \mathcal{J}_e(w, T_n),$$

and hence

$$(1) \qquad \left| \iint_{R^0} J(w, T) \right| \leqq \liminf_{n\to\infty} \iint_{R^0} | \mathcal{J}_e(w, T_n) |.$$

Now since $J(w, T)$ is summable in $\mathfrak{D}$, we can, for given $\epsilon > 0$, select a finite number of nonoverlapping oriented rectangles $R_1, \cdots, R_m$ in $\mathfrak{D}$, such that (see I.3.10)

$$(2) \qquad \iint_{\mathfrak{D}} | J(w, T) | < \sum_{i=1}^{m} \left| \iint_{R_i} J(w, T) \right| + \epsilon.$$

From (1) and (2) we obtain

$$(3) \qquad \iint_{\mathfrak{D}} | J(w, T) | \leqq \epsilon + \liminf_{n\to\infty} \sum_{i=1}^{m} \iint_{R_i} | \mathcal{J}_e(w, T_n) |.$$

On the other hand, $R_1 + \cdots + R_m \subset \mathfrak{D}_n$ for n sufficiently large. Since ϵ was arbitrary, there follows therefore from (3) the inequality

$$(4) \qquad \iint_{\mathfrak{D}} | J(w, T) | \leqq \liminf_{n\to\infty} \iint_{\mathfrak{D}_n} | \mathcal{J}_e(w, T_n) |.$$

Now suppose that we replace, in IV.4.42, condition (ii) by the condition:

(ii*) $\mathcal{J}_e(w, T)$ exists a.e. in $\mathfrak{D}$ and is summable in $\mathfrak{D}$. As noted previously, it is not known at present whether the theorem of IV.4.42 remains valid (with $J(w, T)$ replaced by $\mathcal{J}_e(w, T)$ in the conclusion, of course). The significant fact is that we can still prove that (4) remains valid if $J(w, T)$ is replaced by $\mathcal{J}_e(w, T)$. Indeed, let w_0 be a fixed point in $\mathfrak{D}$. Let us denote by $\mathfrak{D}_n^*$ the component of $\mathfrak{D}\mathfrak{D}_n$ that contains w_0. In view of IV.4.42(i), $\mathfrak{D}_n^* \neq 0$ for n sufficiently large, say $n > n_0$. Thus

$$(5) \qquad \iint_{\mathfrak{D}_n^*} | \mathcal{J}_e(w, T_n) | \leqq \iint_{\mathfrak{D}_n} | \mathcal{J}_e(w, T_n) | \qquad \text{for } n > n_0 .$$

We assert that

$$(6) \qquad \iint_{\mathfrak{D}} | \mathcal{J}_e(w, T) | \leqq \liminf_{n\to\infty} \iint_{\mathfrak{D}_n} | \mathcal{J}_e(w, T_n) |.$$

Indeed, if the lim inf is infinite, then (5) is obvious. If the lim inf is finite, then (6) follows directly from IV.4.11 in view of (5). This argument also shows that (6) remains valid if condition IV.4.42(ii) is replaced by the condition:

(ii′) $$\liminf_{n\to\infty} \iint_{\mathfrak{D}_n} | \mathcal{J}_e(w, T_n) | < +\infty.$$

Indeed, condition (ii′) implies, by IV.4.11, condition (ii*). Thus it appears that the inequality (6) holds under assumptions that are not known at present to be sufficient for the validity of the theorem of IV.4.42.

IV.4.50. Let $\mathfrak{R}$ be a bounded, finitely-connected Jordan region in the w-plane, and let $T : z = t(w)$, $w \in \mathfrak{R}$, $T_n : z = t_n(w)$, $w \in \mathfrak{R}$, $n = 1, 2, \cdots$, be continuous transformations in $\mathfrak{R}$ (and not merely in $\mathfrak{R}^0$). Let us assume that the following conditions hold:

(i) T and T_n are eAC in $\mathfrak{R}^0$, $n = 1, 2, \cdots$.
(ii) $| T(\mathfrak{R} - \mathfrak{R}^0) | = 0$.
(iii) $\rho(T_n, T, \mathfrak{R}) \to 0$ for $n \to \infty$ (cf. IV.1.1).

(iv) $$\iint_{\mathfrak{R}^0} | \mathcal{J}_e(w, T_n) | \to \iint_{\mathfrak{R}^0} | \mathcal{J}_e(w, T) | \text{ for } n \to \infty.$$

Under these circumstances we assert that (cf. IV.1.24)

(1) $$\iint | \mu(z, T, \mathfrak{R}) - \mu(z, T_n, \mathfrak{R}) | \to 0 \qquad \text{for } n \to \infty.$$

PROOF. By conditions (ii), (iii) and by IV.1.25

(2) $$\mu(z, T_n, \mathfrak{R}) \underset{n\to\infty}{\to} \mu(z, T, \mathfrak{R}) \qquad \text{a.e. in the } z\text{-plane}.$$

By I.3.11, (1) will follow from (2) if we can show that the sequence $\mu(z, T_n, \mathfrak{R})$ possesses the property (V), and we proceed to establish this property. By IV.3.17 it is sufficient to verify that the sequence $\kappa(z, T_n, \mathfrak{R}^0)$ possesses the property (V). Now by IV.1.52 (cf. condition (iii)),

(3) $$\liminf_{n\to\infty} \kappa(z, T_n, \mathfrak{R}^0) \geqq \kappa(z, T, \mathfrak{R}^0).$$

On the other hand, condition (iv) yields, in view of IV.4.7,

(4) $$\iint \kappa(z, T_n, \mathfrak{R}^0) \to \iint \kappa(z, T, \mathfrak{R}^0) \qquad \text{for } n \to \infty.$$

Now let us observe that, as a consequence of condition (iii), all the sets $T(\mathfrak{R})$, $T_n(\mathfrak{R})$ are comprised in some fixed circular disc in the z-plane, and thus all the functions $\kappa(z, T, \mathfrak{R}^0)$, $\kappa(z, T_n, \mathfrak{R}^0)$ vanish outside of such a disc (see IV.1.47). Thus I.3.11 applies, and (3) and (4) imply that the sequence $\kappa(z, T_n, \mathfrak{R}^0)$ possesses the property (V).

IV.4.51. The theorem of IV.4.50 suggests the following approach to the study of the transformation of double integrals. For quasi-linear transformations the formulas listed in IV.4.21 are practically trivial. The theorem of IV.4.50 makes it seem plausible that for more general transformations approximations by quasi-linear functions will yield the desired transformation formulas. It is not clear at present what the scope of such an approach would be.

IV.4.52. We shall now take up the study of transformations T that possess special features of a topological rather than metrical character. We shall restrict ourselves mostly to deriving results needed in the sequel. Let us suppose that the transformation $T : z = t(w)$, $w \in \mathfrak{D}$, given as in IV.2.1, is *biunique* in $\mathfrak{D}$. That is, $w_1 \neq w_2$ implies that $t(w_1) \neq t(w_2)$. The following facts are then obvious consequences of the definitions of the terms involved, in view of well known facts in the topology of the plane (see I.2.47, I.2.49).

(i) Since T is bounded by assumption, $T(\mathfrak{D}) = \mathfrak{D}^*$ is a bounded domain in the z-plane, and T is a homeomorphism between $\mathfrak{D}$ and $\mathfrak{D}^*$. The transformation T^{-1} is again continuous, bounded, and biunique.

(ii) If $w_0 \in \mathfrak{D}$ and $z_0 = t(w_0)$, and if $\mathfrak{R}$ is any simply-connected Jordan region in $\mathfrak{D}$ such that $w_0 \in \mathfrak{R}^0$, then $\mu(z, T, \mathfrak{R})$ (cf. IV.1.24) is $+1$ if T is sense-preserving, and -1 if T is sense-reversing.

(iii) From (ii) it follows that the sets $\mathfrak{g}$, $\mathcal{N}$, $\mathcal{E}$, $\mathcal{E}^*$ defined in IV.1.56 all coincide with $\mathfrak{D}$, and the set $\mathfrak{g}_0$ defined in IV.1.78 is empty. By IV.1.47 it follows that $\kappa(z, T, \mathfrak{D}) \equiv N(z, T, \mathfrak{D})$. Finally, the local index $i_s(w)$ (cf. IV.1.75) is equal to $+1$ if T is sense-preserving and equal to -1 if T is sense-reversing, and $i_s(w) = i_e(w)$ everywhere in $\mathfrak{D}$ (cf. IV.1.76). It follows that the concepts eBV and sBV coincide as well as the concepts eAC and sAC (cf. IV.4.1, IV.2.11, IV.2.13). In fact, eBV and sBV coincide with BV$\mathfrak{D}$ while eAC and sAC coincide with AC$\mathfrak{D}$ (cf. IV.2.11, IV.2.39).

(iv) In view of the remarks made under (iii), it is clear that the Jacobians $\mathfrak{J}_e(w)$, $\mathfrak{J}_s(w)$ coincide (see IV.3.21). Their common value will then be denoted by $\mathfrak{J}(w)$ and will be termed simply the generalized Jacobian. The common value of the local indices $i_e(w)$, $i_s(w)$ (see (iii)), will be denoted by $i(w)$.

IV.4.53. Continuation. A biunique transformation T, as described in IV.4.52, is topological (see I.2.29). In view of IV.4.52(iii), we use only $\mathfrak{D}$ itself as a base-set if T is topological. It follows immediately that a topological T is always BV$\mathfrak{D}$ in $\mathfrak{D}$. Indeed, if $R_1, \cdots, R_m$ is any system of oriented rectangles in $\mathfrak{D}$ such that $R_j^0 R_k^0 = 0$ for $j \neq k$ then $T(R_1^0), \cdots, T(R_m^0)$ are disjoint open sets in the bounded domain $\mathfrak{D}^* = T(\mathfrak{D})$, and hence (cf. IV.2.10)

$$\sum_{i=1}^{m} G(R_i) = \sum_{i=1}^{m} | T(R_i^0) | \leqq | \mathfrak{D}^* |.$$

As a consequence, the derivative $D(w, \mathfrak{D})$ exists a.e. in $\mathfrak{D}$ and is summable in $\mathfrak{D}$ (see IV.2.32). Furthermore, a.e. in $\mathfrak{D}$, $\mathfrak{J}(w) = D(w, \mathfrak{D})$ if T is sense-preserving and $\mathfrak{J}(w) = -D(w, \mathfrak{D})$ if T is sense-reversing (see IV.3.21, IV.4.52).

IV.4.54. A topological transformation T (cf. IV.4.53) will be termed *measurable* in $\mathfrak{D}$ if $T(E)$ is a measurable set as soon as E is measurable. We assert that T is measurable in $\mathfrak{D}$ if and only if $|T(e)| = 0$ for every set $e \subset \mathfrak{D}$ of measure zero.

PROOF. (i) Suppose that $|T(e)| = 0$ for every set $e \subset \mathfrak{D}$ of measure zero. Let E be any measurable set in $\mathfrak{D}$. Then we have a decomposition $E = G + e$, where G is a Borel set and $|e| = 0$ (see I.3.7). It follows that $T(E) = T(G) + T(e)$. Now $T(G)$ is measurable (see I.2.46), and $|T(e)| = 0$ by assumption. Hence $T(E)$ is measurable.

(ii) Suppose that T is measurable. Let e be any set of measure zero in $\mathfrak{D}$. Suppose that $T(e)$ is not of measure zero. Since T is measurable, $T(e)$ is measurable. Hence (see I.3.7) $T(e)$ contains some nonmeasurable set $\bar{E}$. Since T is topological, the set $E = T^{-1}(\bar{E})$ is a subset of e. Hence $|E| = 0$, and thus E is a measurable set whose image $T(E) = \bar{E}$ is nonmeasurable. This contradicts the assumption that T is measurable.

IV.4.55. *A topological transformation* T (see IV.4.53) *is* $AC\mathfrak{D}$ *in* $\mathfrak{D}$ *if and only if it is measurable in* $\mathfrak{D}$ (cf. IV.4.54).

PROOF. (i) Suppose that T is AC$\mathfrak{D}$ in $\mathfrak{D}$. Let E be any measurable set in $\mathfrak{D}$. By IV.2.48, $N(z, T, E)$ is then measurable. Now $T(E)$ is precisely the set where $N(z, T, E) = 1$. Thus $T(E)$ is measurable.

(ii) Suppose T is measurable in $\mathfrak{D}$. Then $|T(e)| = 0$ whenever $e \subset \mathfrak{D}$ and $|e| = 0$, by IV.4.54. Since T is BV$\mathfrak{D}$ in $\mathfrak{D}$ by IV.4.53, it follows by IV.2.42 that T is AC$\mathfrak{D}$ in $\mathfrak{D}$.

IV.4.56. Given a topological transformation T (see IV.4.53), let $\mathfrak{D}_\infty$ be the subset of $\mathfrak{D}$ where $\bar{D}(w, \mathfrak{D}) = +\infty$ (see IV.2.58). Then T is AC with respect to the base-set $\mathfrak{D} - \mathfrak{D}_\infty$ by IV.4.53, IV.2.61, and by IV.2.65, IV.4.53 it follows that T is AC$\mathfrak{D}$ in $\mathfrak{D}$ if and only if $|T(\mathfrak{D}_\infty)| = 0$. Equivalently (cf. IV.4.55), T is measurable in $\mathfrak{D}$ if and only if $|T(\mathfrak{D}_\infty)| = 0$.

IV.4.57. If $T: z = t(w)$, $w \in \mathfrak{D}$, is a topological transformation, then clearly T^{-1} is topological in $\mathfrak{D}^* = T(\mathfrak{D})$ (cf. IV.4.53). However, if T is measurable (see IV.4.54), then it does not generally follow that T^{-1} is also measurable. Let now $\mathfrak{D}_0$ be the subset of $\mathfrak{D}$ on which $D(w, \mathfrak{D}) = 0$. Suppose T itself is measurable. *Then we assert that* T^{-1} *is measurable if and only if* $|\mathfrak{D}_0| = 0$.

PROOF. (i) Suppose $|\mathfrak{D}_0| = 0$. Let e^* be any set of measure zero in $\mathfrak{D}^* = T(\mathfrak{D})$, and put $e = T^{-1}(e^*)$. Let $H(z)$ be the characteristic function of e^*. Then clearly $H[t(w)]$ is the characteristic function of e, and since T is topological, we have $H(z) = N(z, T, e)$. Since T is measurable, there follows by IV.4.55, IV.2.53 the formula

$$\iint_e D(w, \mathfrak{D}) = |e^*| = 0.$$

Hence $D(w, \mathfrak{D}) = 0$ on $e - e_0$, where e_0 is a certain subset of measure zero of e. Then $e - e_0 \subset \mathfrak{D}_0$, and since e_0 and $\mathfrak{D}_0$ are both of measure zero, it follows that e itself is of measure zero. Thus $|e^*| = 0$ implies that $|e| = |T^{-1}(e^*)| = 0$. Hence T^{-1} is measurable by IV.4.54.

(ii) Suppose conversely that T^{-1} is measurable. Since $D(w, \mathfrak{D})$ is Borel measurable (see III.1.24), the set $\mathfrak{D}_0$ is a Borel set. Since T itself is measurable and hence AC$\mathfrak{D}$ in $\mathfrak{D}$ (see IV.4.55), we have by IV.2.44 the formula

$$(1) \qquad 0 = \iint_{\mathfrak{D}_0} D(w, \mathfrak{D}) = \mu(\mathfrak{D}_0).$$

Now since T is topological, we have (cf. IV.2.21)

$$(2) \qquad \mu(\mathfrak{D}_0) = | T(\mathfrak{D}_0) |.$$

(1) and (2) imply that $| T(\mathfrak{D}_0) | = 0$. Since T^{-1} is measurable by assumption, it follows by IV.4.54 (applied to T^{-1}) that $\mathfrak{D}_0 = T^{-1}[T(\mathfrak{D}_0)]$ is also of measure zero.

IV.4.58. Let $T : z = t(w)$, $w \in \mathfrak{D}$, be topological and measurable in $\mathfrak{D}$ (see IV.4.53, IV.4.54). By IV.4.55, T is then AC$\mathfrak{D}$ in $\mathfrak{D}$, and thus sAC in $\mathfrak{D}$ (see IV.4.1). Hence all the results derived previously for sAC transformations apply to T. In writing the various transformation formulas for T, the following obvious simplifications take place due to the fact that T is topological (cf. IV.4.52).

(i) If T is sense-preserving, then $\nu(z, T, \mathfrak{D}) = +1$ for $z \in \mathfrak{D}^* = T(\mathfrak{D})$, and $\nu(z, T, \mathfrak{D}) = 0$ elsewhere. If T is sense-reversing, then $\nu(z, T, \mathfrak{D}) = -1$ for $z \in \mathfrak{D}^* = T(\mathfrak{D})$, and $\nu(z, T, \mathfrak{D}) = 0$ elsewhere.

(ii) $\kappa(z, T, \mathfrak{D}) = N(z, T, \mathfrak{D}) = 1$ for $z \in \mathfrak{D}^* = T(\mathfrak{D})$, and $\kappa(z, T, \mathfrak{D}) = N(z, T, \mathfrak{D}) = 0$ elsewhere.

(iii) If $H(z)$ is a finite-valued, measurable function in $\mathfrak{D}^* = T(\mathfrak{D})$, not defined outside of $\mathfrak{D}^*$, then our transformation formulas still apply, even though we assumed previously that $H(z)$ was defined and measurable in the whole z-plane. Indeed, in view of (ii), the values of $H(z)$ outside of $\mathfrak{D}^*$ are not used at all in the transformation formulas. We may also note that we can extend the definition of $H(z)$ to the whole plane by setting $H(z) = 0$ outside of $\mathfrak{D}^*$, the extended function being then clearly measurable in the whole plane.

In view of these simplifications, IV.4.22 yields the following results. We have (cf. IV.4.52(iv))

$$\iint_{\mathfrak{D}} H[t(w)] \mid \mathfrak{J}(w) \mid = \iint_{T(\mathfrak{D})} H(z)$$

for every finite-valued, measurable function $H(z)$ in $\mathfrak{D}^* = T(\mathfrak{D})$ as soon as one of the two integrals involved exists. The formula may be used also for functions $H(z)$ that fail to be finite on some set of measure zero, as explained in IV.4.5. We have further (case $H \equiv 1$)

$$\iint_{\mathfrak{D}} \mid \mathfrak{J}(w) \mid = \mid T(\mathfrak{D}) \mid.$$

If the ordinary Jacobian $J(w) = x_u y_v - x_v y_u$ happens to exist a.e. in $\mathfrak{D}$, then $\mathfrak{J}(w)$ can be replaced by $J(w)$ in these formulas (see IV.4.22). If T is sense-preserving, then $|\mathcal{J}(w)|$ can be replaced by $\mathcal{J}(w)$, and if T is sense-reversing, then $|\mathcal{J}(w)|$ can be replaced by $-\mathcal{J}(w)$ in the preceding formulas (cf. IV.4.53, IV.4.58).

IV.4.59. Given T as in IV.2.1, let w be a point in $\mathfrak{D}$. Let $\lambda > 0$ be so small that $|w - w^*| < \lambda$ implies $w^* \in \mathfrak{D}$. Let us define, for $0 < \rho < \lambda$ and given w,

$$L(w, \rho) = \text{l.u.b.} \frac{|t(w^*) - t(w)|}{|w^* - w|} \qquad \text{for } 0 < |w^* - w| \leqq \rho.$$

Clearly, $L(w, \rho)$ decreases as ρ decreases, and hence we can define

$$L(w) = \lim L(w, \rho) \qquad \text{for } \rho \to 0.$$

$L(w)$ may be of course infinite.

IV.4.60. Suppose that $T : z = t(w)$, $w \in \mathfrak{D}$, is topological (see IV.4.53) and $L(w) < +\infty$ everywhere in $\mathfrak{D}$ (see IV.4.59). Then T is sAC in $\mathfrak{D}$, and the ordinary Jacobian $J(w)$ exists a.e. in $\mathfrak{D}$.

Proof. Let w_0 be any point of $\mathfrak{D}$, and let s_n be any sequence of oriented squares such that $w_0 \in \overset{0}{s}_n$, $s_n \subset \mathfrak{D}$, $|s_n| \to 0$. If l_n is the side length of s_n then (cf. IV.4.59)

$$|t(w^*) - t(w_0)| \leqq L(w_0, 2^{1/2}l_n)\,|w^* - w_0| \leqq L(w_0, 2^{1/2}l_n)\cdot 2^{1/2}l_n$$

for $w^* \in s_n$. Thus $T(s_n)$ is contained in a circular disc of center $t(w_0)$ and radius $r_n = L(w_0, 2^{1/2}l_n)\cdot 2^{1/2}l_n$. Hence

$$|T(\overset{0}{s}_n)| \leqq 2\pi\,|s_n|\,[L(w_0, 2^{1/2}l_n)]^2.$$

For $n \to +\infty$ there follows the inequality

$$\limsup_{n\to\infty} \frac{|T(\overset{0}{s}_n)|}{|s_n|} \leqq 2\pi L(w_0)^2 < +\infty.$$

Since this holds for every sequence s_n such that $w_0 \in \overset{0}{s}_n \subset \mathfrak{D}$, $|s_n| \to 0$, we have $\overline{D}(w_0, \mathfrak{D}) < +\infty$. Since w_0 was any point in $\mathfrak{D}$, it follows that the set $\mathfrak{D}_\infty$ of IV.4.56 is empty, and hence T is sAC in $\mathfrak{D}$ (see IV.4.56). The existence of $J(w)$ a.e. in $\mathfrak{D}$ follows from I.3.14.

IV.4.61. Suppose that T, given in the form $T : z = t(w) = x(u, v) + iy(u, v)$, $w = u + iv \in \mathfrak{D}$, satisfies the following conditions:

(i) T is bounded and topological in $\mathfrak{D}$. As a consequence, $\mathfrak{D}^* = T(\mathfrak{D})$ is a bounded domain in the z-plane.

(ii) The partial derivatives x_u, x_v, y_u, y_v exist and are continuous everywhere in $\mathfrak{D}$.

(iii) The ordinary Jacobian $J(w) = x_u y_v - x_v y_u$ is different from zero in $\mathfrak{D}$. As a consequence, $J(w)$ has constant sign in $\mathfrak{D}$.

Then T^{-1} satisfies the same conditions in $\mathfrak{D}^* = T(\mathfrak{D})$ and T and T^{-1} are both sAC.

Proof. T is sAC as a direct consequence of IV.4.60. Indeed, the inequality $L(w) < +\infty$ follows readily from the Lagrange mean value theorem in view of condition (ii). T^{-1} is sAC by IV.4.57, IV.4.55, and condition (iii). If T^{-1} is written in the form

$$T^{-1} : u = u(x, y),\ v = v(x, y), \qquad (x, y) \in \mathfrak{D}^* = T(\mathfrak{D}),$$

then the existence and continuity of u_x, u_y, v_x, v_y follows by well known theorems on implicit functions by condition (iii). Finally, if $\bar{J}(z)$ denotes the ordinary Jacobian of T^{-1}, then direct computation yields the formula $J[T^{-1}(z)]\bar{J}(z) = 1$. Hence $\bar{J}(z) \neq 0$ everywhere in $\mathfrak{D}^*$.

IV.4.62. Given a topological transformation $T : z = t(w)$, $w \in \mathfrak{D}$ (see IV.4.53), the inverse transformation T^{-1} is bounded and topological in $\mathfrak{D}^* = T(\mathfrak{D})$. We shall say that T is *bimeasurable* in $\mathfrak{D}$ if T, T^{-1} are both measurable in $\mathfrak{D}$ and $\mathfrak{D}^*$ respectively (cf. IV.4.54).

Suppose that T is bimeasurable in $\mathfrak{D}$. Let us denote by $\mathcal{J}(w)$, $\mathcal{J}^*(z)$ the generalized Jacobians of T, T^{-1} respectively (cf. IV.3.21, IV.4.55, IV.4.52). *We assert that*

$$(1) \qquad \begin{aligned} \mathcal{J}(w)\mathcal{J}^*[T(w)] &= 1 \qquad \textit{a.e. in } \mathfrak{D}, \\ \mathcal{J}^*(z)\mathcal{J}[T^{-1}(z)] &= 1 \qquad \textit{a.e. in } \mathfrak{D}^*. \end{aligned}$$

Proof. By reasons of symmetry it is clearly sufficient to prove (1). By IV.4.55, T and T^{-1} are both sAC. Let now s be any oriented square in $\mathfrak{D}$. Then T^{-1} is sAC in $T(s^0)$ (cf. IV.4.1), and hence we have the formula (see IV.4.58)

$$(2) \qquad |\, s^0 \,| = \iint_{T(s^0)} |\, \mathcal{J}^*(z) \,|.$$

Furthermore, T is sAC in s^0. Hence, by IV.4.58,

$$(3) \qquad \iint_{s^0} |\, \mathcal{J}^*[t(w)] \,|\, |\, \mathcal{J}(w) \,| = \iint_{T(s^0)} |\, \mathcal{J}^*(z) \,|.$$

(2) and (3) yield

$$\iint_{s^0} |\, \mathcal{J}^*[t(w)] \,|\, |\, \mathcal{J}(w) \,| = |\, s^0 \,|.$$

Since this holds for every oriented square $s \subset \mathfrak{D}$, it follows that (cf. I.3.13)

$$(4) \qquad |\, \mathcal{J}(w) \,|\, |\, \mathcal{J}^*[t(w)] \,| = 1 \qquad \text{a.e. in } \mathfrak{D}.$$

Now T and T^{-1} are either both sense-preserving or both sense-reversing. Hence (4) yields, in view of IV.4.53, the formula (1). A remark is in order concerning (3). To write this formula, we have to know (cf. IV.4.58) that $\mathcal{J}^*(z)$ is finite a.e. in $\mathfrak{D}^*$. This is of course the case, since T^{-1} is sAC in $\mathfrak{D}^*$. However, we can even assert that $\mathcal{J}^*[t(w)]$ is finite a.e. in $\mathfrak{D}$, as a consequence of the fact that sets of measure zero are carried into sets of measure zero by both T and T^{-1} (see IV.4.55, IV.4.54).

IV.4.63. Let there be given two transformations $T : z = t(w)$, $w \in \mathfrak{D}$, $T_* : z = t_*(w_*)$, $w_* \in \mathfrak{D}_*$, as in IV.2.1. These transformations are not assumed to be topological. On the other hand, we assume that they are *topologically similar*; that is, we assume that there exists a homeomorphism $\tau(\mathfrak{D}) = \mathfrak{D}_*$ such that

$$t(w) = t_*(\tau(w)) \qquad \text{for } w \in \mathfrak{D},$$

and hence also

$$t_*(w_*) = t(\tau^{-1}(w_*)) \qquad \text{for } w_* \in \mathfrak{D}_* .$$

The following statements are then obvious consequences of the terms involved.

(i) If $E \subset \mathfrak{D}$, and $E_* = \tau(E)$, then $N(z, T, E) = N(z, T_*, E_*)$ (cf. IV.1.1).

(ii) If γ is an essential maximal model continuum for a point z under T in $\mathfrak{D}$, then $\gamma_* = \tau(\gamma)$ is an essential maximal model continuum for the same point z under T_* in $\mathfrak{D}_*$ (cf. IV.1.46).

(iii) Let $i_e(w)$, $i_{e*}(w_*)$ denote the essential local indices relative to T, T_* (see IV.1.64). Then $i_e(w) = i_{e*}(\tau(w))$ if τ is sense-preserving, and $i_e(w) = -i_{e*}(\tau(w))$ if τ is sense-reversing.

(iv) $\kappa(z, T, \mathfrak{D}^*) = \kappa(z, T_*, \tau(\mathfrak{D}^*))$, for every z and for every subdomain $\mathfrak{D}^*$ of $\mathfrak{D}$ (cf. IV.1.47).

(v) As a consequence of (iv), T is eBV in $\mathfrak{D}$ if and only if T_* is eBV in $\mathfrak{D}_*$ (cf. IV.4.1).

(vi) Let $\mathcal{N}$, $\mathcal{E}$, $\mathcal{E}^*$ be the sets associated with T in the sense of IV.1.56. Then $\tau(\mathcal{N})$, $\tau(\mathcal{E})$, $\tau(\mathcal{E}^*)$ coincide with $\mathcal{N}_*$, $\mathcal{E}_*$, $\mathcal{E}_*^*$ associated in the same sense with T_*.

(vii) Let $\mathfrak{B}$ be a Borel set in $\mathfrak{D}$. Since τ is topological, $\mathfrak{B}_* = \tau(\mathfrak{B})$ is then a Borel set in $\mathfrak{D}_*$ (see I.2.46). As a consequence of (1) and IV.2.13, T is BV$\mathfrak{B}$ in $\mathfrak{D}$ if and only if T_* is BV$\mathfrak{B}_*$ in $\mathfrak{D}_*$.

IV.4.64. Continuation. Let us now add the assumption that the homeomorphism τ is *bimeasurable* (see IV.4.62). Let us denote by $j(w)$, $j_*(w_*)$ the generalized Jacobians of τ, τ^{-1} respectively. By IV.4.62 we have then

$$(1) \qquad \begin{aligned} j(w)j_*(\tau(w)) &= 1 \qquad \text{a.e. in } \mathfrak{D}, \\ j_*(w_*)j(\tau^{-1}(w_*)) &= 1 \qquad \text{a.e. in } \mathfrak{D}_* . \end{aligned}$$

Let us now assume that T is eBV in $\mathfrak{D}$. By IV.4.63(v), T_* is then eBV in $\mathfrak{D}_*$. Let $\mathfrak{J}_e(w)$, $\mathfrak{J}_{e*}(w_*)$ be the essential generalized Jacobians relative to T, T_*. *We assert the formulas*

$$(2) \qquad \mathfrak{J}_e(w) = \mathfrak{J}_{e*}(\tau(w))j(w) \qquad \text{a.e. in } \mathfrak{D},$$

$$(3) \qquad \mathfrak{J}_{e*}(w_*) = \mathfrak{J}_e(\tau^{-1}(w_*))j_*(w_*) \qquad \text{a.e. in } \mathfrak{D}_* .$$

Proof. Let s be any oriented square in $\mathfrak{D}$. By IV.4.63(iv) we have

$$(4) \qquad \iint \kappa(z, T, s^0) = \iint \kappa(z, T_*, \tau(s^0)).$$

By IV.4.8 (applied to T_* in $\tau(s^0)$) we have the inequality

(5) $$\iint_{\tau(s^0)} | \mathcal{J}_{e*}(w_*) | \leqq \iint \kappa(z, T_*, \tau(s^0)).$$

Since τ is measurable and hence sAC, we have by IV.4.58

(6) $$\iint_{\tau(s^0)} | \mathcal{J}_{e*}(w_*) | = \iint_{s^0} | \mathcal{J}_{e*}(\tau(w)) || j(w) |.$$

(4), (5), (6) yield

$$\iint \kappa(z, T, s^0) \geqq \iint_{s^0} | \mathcal{J}_{e*}(\tau(w)) || j(w) |.$$

Since this holds for every oriented square s in $\mathfrak{D}$, we obtain by IV.3.13, IV.3.21, IV.3.25, I.3.13 the inequality

(7) $$| \mathcal{J}_e(w) | \geqq | \mathcal{J}_{e*}(\tau(w)) | | j(w) | \qquad \text{a.e. in } \mathfrak{D}.$$

A similar argument yields the inequality

(8) $$| \mathcal{J}_{e*}(w_*) | \geqq | \mathcal{J}_e(\tau^{-1}(w_*)) | | j_*(w_*) | \qquad \text{a.e. in } \mathfrak{D}_*.$$

Now let e_* be the set in $\mathfrak{D}_*$ where (8) fails to hold, and let us put $e = \tau^{-1}(e_*)$. Since τ^{-1} is measurable, e is also of measure zero by IV.4.54. For $w \in \mathfrak{D} - e$, we can use (8) with $w_* = \tau(w)$, and we obtain

(9) $$| \mathcal{J}_{e*}(\tau(w)) | \geqq | \mathcal{J}_e(w) | | j_*(\tau(w)) | \qquad \text{a.e. in } \mathfrak{D}.$$

(7), (9) and (1) yield

(10) $$\begin{aligned} | \mathcal{J}_e(w) | \geqq | \mathcal{J}_{e*}(\tau(w)) | | j(w) | &\geqq | \mathcal{J}_e(w) | | j_*(\tau(w)) | | j(w) | \\ &= | \mathcal{J}_e(w) | \qquad \text{a.e. in } \mathfrak{D}. \end{aligned}$$

Clearly, (10) implies that

(11) $$| \mathcal{J}_e(w) | = | \mathcal{J}_{e*}(\tau(w)) | | j(w) | \qquad \text{a.e. in } \mathfrak{D}.$$

Let us note that the quantities involved in (11) exist a.e. in $\mathfrak{D}$; in the rest of the proof it is understood that w is a point where all these quantities exist and (11) holds.

Case (a). $\mathcal{J}_{e*}(\tau(w)) = 0$. Then $\mathcal{J}_e(w) = 0$ by (11), and hence (2) holds.

Case (b). $j(w) = 0$. Then $\mathcal{J}_e(w) = 0$ by (11), and hence (2) holds.

Case (c). $\mathcal{J}_{e*}(\tau(w)) \neq 0$ and $j(w) \neq 0$. Then $\mathcal{J}_e(w) \neq 0$ by (11). We have then (see IV.3.21, IV.4.63(iii))

$$\operatorname{sgn} \mathcal{J}_e(w) = \operatorname{sgn} \imath_e(w) = \operatorname{sgn} j(w) \cdot \operatorname{sgn} \imath_{e*}(\tau(w)),$$

$$\operatorname{sgn} \mathcal{J}_{e*}(\tau(w)) = \operatorname{sgn} \imath_{e*}(\tau(w)).$$

There follows the formula

$$(12) \qquad \operatorname{sgn} \mathcal{J}_e(w) = \operatorname{sgn}[\mathcal{J}_{e_*}(\tau(w))j(w)].$$

(11) and (12) imply (2). The formula (3) is proved in a similar manner.

IV.4.65. Continuation. Let us now assume that τ is bimeasurable. Then T is eAC in $\mathfrak{D}$ if and only if T_* is eAC in $\mathfrak{D}_*$.

Proof. Since the relation between T and T_* is symmetrical, it is sufficient to show that if T is eAC in $\mathfrak{D}$, then T_* is eAC in $\mathfrak{D}_*$. The assumption that T is eAC means that T is AC$\mathcal{E}^*$ in $\mathfrak{D}$ (see IV.4.1). Now let e_* be any set of measure zero in $\mathfrak{D}_*$. Since τ^{-1} is topological and measurable, we have then (cf. IV.4.54, IV.4.63(vi))

$$| \tau^{-1}(e_* \mathcal{E}_*^*) | = | \tau^{-1}(e_*)\tau^{-1}(\mathcal{E}_*^*) | \leqq | \tau^{-1}(e_*) | = 0,$$

$$\tau^{-1}(e_* \mathcal{E}_*^*) = \tau^{-1}(e_*)\tau^{-1}(\mathcal{E}_*^*) = \tau^{-1}(e_*)\mathcal{E}^*.$$

Thus $\tau^{-1}(e_* \mathcal{E}_*^*)$ is a subset of measure zero of $\mathcal{E}^*$. Since T is AC$\mathcal{E}^*$, we have by IV.2.42

$$(1) \qquad | T\tau^{-1}(e_* \mathcal{E}_*^*) | = 0.$$

On the other hand, we have for any set $E_* \subset \mathfrak{D}_*$ (cf. IV.4.63)

$$(2) \qquad T\tau^{-1}(E_*) = T_*\tau\tau^{-1}(E_*) = T_*(E_*).$$

(1), (2) yield, for $E_* = e_* \mathcal{E}_*^*$, the equation $| T_*(e_* \mathcal{E}_*^*) | = 0$. Since e_* was any set of measure zero in $\mathfrak{D}_*$ it follows that T_* is eAC in $\mathfrak{D}_*$ (cf. IV.4.63(vi), IV.4.1, IV.2.42, IV.4.63(vii)).

IV.4.66. Continuation. The following very special cases are important.

(i) τ represents a change of the coordinate system in the w-plane, either to new Cartesian coordinates, or to any other coordinates, such that the assumptions of IV.4.61 are satisfied by τ.

(ii) τ is a conformal mapping from $\mathfrak{D}$ onto $\mathfrak{D}_*$. The assumptions of IV.4.61 are then clearly satisfied.

In either case, τ is bimeasurable by IV.4.61 (cf. IV.4.62), and the results derived in IV.4.63, IV.4.64, IV.4.65 apply.

IV.4.67. We shall now discuss, to the extent needed in the sequel, transformations of the special form (cf. IV.1.80, IV.1.84)

$$T : x = f(u, v),\ y = v, \qquad (u, v) \in R_0,$$

where R_0 is an oriented rectangle

$$R_0 : a_0 \leqq u \leqq b_0,\ c_0 \leqq v \leqq d_0,$$

and $f(u, v)$ is continuous in R_0 (and not merely in R_0^0). *We first assert that T is sBV in R_0^0 if and only if $f(u, v)$ is BVTu in R_0* (cf. IV.4.1, III.2.49).

Proof. Since $f(u, v)$ is continuous and hence bounded in R_0, we have a finite constant M such that

$$(1) \qquad \tilde{T}(R_0) \subset \bar{R}_0 ,$$

where

$$\bar{R}_0 : -M \leqq x \leqq M, \ c_0 \leqq y \leqq d_0 .$$

For convenience, let us write $N(x, y, R_0^0)$ instead of $N(z, R_0^0)$, $z = x + iy$, with similar conventions for $\kappa(x, y, T, R_0^0)$, and so on. Clearly, as a consequence of (1), $N(x, y, T, R_0^0)$ vanishes outside of $\bar{R}_0$, and hence in the course of integrations we can restrict ourselves to $\bar{R}_0$.

(i) Let us suppose that T is sBV in R_0^0. Then by IV.2.13 the integral

$$(2) \qquad \int_{-M}^{M} \int_{c_0}^{d_0} N(x, y, R_0^0)\, dx\, dy = \int_{c_0}^{d_0} \left[\int_{-M}^{M} N(x, y, R_0^0)\, dx \right] dy$$

exists (cf. I.3.10). In view of IV.1.80, $N(x, y, R_0^0) = N(x, y, R_0)$ a.e. in the xy-plane. Indeed, the set $T(R_0 - R_0^0)$ is a closed set which has at most two points on any line $y =$ constant, except for $y = c_0$ and $y = d_0$, and hence $| T(R_0 - R_0^0) | = 0$. Thus R_0^0 can be replaced by R_0 in (2). Thus (2) implies that the integral

$$(3) \qquad \int_{-M}^{M} N(x, y, R_0)\, dx$$

exists for a.e. y in the interval $c_0 \leqq y \leqq d_0$. If $y = \lambda$ is a value in this interval for which the integral (3) exists, then by III.2.12 the function $f(u, \lambda)$ is of bounded variation with respect to u in the interval $a_0 \leqq u \leqq b_0$, and (see III.2.23, III.2.45)

$$(4) \qquad V_u(a_0 , b_0 , \lambda, f) = \int_{-M}^{M} N(x, \lambda, R_0)\, dx.$$

In view of the existence of the double integral in (2), it follows that $V_u(a_0 , b_0 , \lambda, f)$ is summable in the interval $c_0 \leqq \lambda \leqq d_0$, and hence $f(u, v)$ is BVTu in R_0 (see III.2.49). Integrating (4) over the interval $c_0 \leqq \lambda \leqq d_0$, we obtain further the formula (see III.2.51)

$$(5) \qquad W_u(R_0 , f) = \iint N(x, y, R_0^0)\, dx\, dy,$$

where the integral on the right is now thought of as being extended over the whole xy-plane (cf. (1)).

(ii) Suppose conversely that $f(u, v)$ is BVTu in R_0. We have then, for a.e. λ in the interval $c_0 \leqq \lambda \leqq d_0$, $V_u(a_0 , b_0 , \lambda, f) < \infty$. Whenever this holds, we have by III.2.23

$$\int_{-M}^{M} N(x, \lambda, R_0)\, dx = V_u(a_0 , b_0 , \lambda, f).$$

Now the expression on the right is a summable function of λ for $c_0 \leqq \lambda \leqq d_0$, since $f(u, v)$ is BVTu in R_0. Hence $N(x, y, R_0)$ is a summable function too (cf. I.3.10), and hence T is sBV in R_0^0 (cf. IV.2.13, IV.4.1).

IV.4.68. Suppose now that T, given as in IV.4.67, is sBV in R_0^0. Then the derivative $D(w, R_0^0)$ exists a.e. in R_0 (see IV.4.1, IV.2.31, IV.2.32, IV.2.34). On the other hand, by IV.4.67, $f(u, v)$ is BVTu in R_0, and hence by III.2.50 the partial derivative $f_u(u, v)$ exists a.e. in R_0. *We assert that*

$$(1) \qquad D(w, R_0^0) = |f_u(u, v)| \qquad \textit{a.e. in } R_0,$$

where $w = u + iv$.

PROOF. Since $f(u, v)$ is BVTu in R_0, it is also BVTu in every oriented rectangle $R \subset R_0$. Hence we have (cf. the argument used in IV.4.67)

$$(2) \qquad W_u(R, f) = \iint N(x, y, R^0)\, dx\, dy$$

for every oriented rectangle $R \subset R_0$. (2) expresses the fact that two rectangle functions agree on every oriented rectangle $R \subset R_0$. It follows that the derivatives of these rectangle functions are identical. By III.2.52, IV.2.31, IV.2.34 the formula (1) follows.

IV.4.69. *Suppose that T, given as in* IV.4.67, *is* sBV *in R_0^0. Then* (cf. IV.3.21)

$$(1) \qquad \mathcal{J}_s(w) = f_u(u, v) \qquad \text{a.e. in } R_0,$$

where $w = u + iv$.

PROOF. We have, by IV.3.21 and IV.4.68,

$$(2) \qquad \mathcal{J}_s(w) = i_s(w)\, |f_u(u, v)| \qquad \text{a.e. in } R_0.$$

Case (i) $f_u(u, v) = 0$. Then $\mathcal{J}_s(u, v) = 0$ by (2), and hence (1) holds.

Case (ii) $f_u(u, v) \neq 0$. Then $i_s(w) = \operatorname{sgn} f_u(u, v)$ by IV.1.83, and again (1) follows from (2).

IV.4.70. Suppose that T, given as in IV.4.67, is sBV in R_0^0. Then $W_u(R, f) < +\infty$ for every oriented rectangle $R \subset R_0$ (cf. IV.4.67). *We assert that* (cf. IV.1.43, IV.4.67)

$$(1) \qquad W_u(R, f) = \iint \kappa(x, y, T, R^0).$$

PROOF. Let R be given by the inequalities $a \leqq u \leqq b$, $c \leqq v \leqq d$. By IV.1.82, we have

$$(2) \qquad \kappa(x, y, T, R^0) = N(x, y, R^0)$$

with the exception of a point-set $\bar{e}$ in the xy-plane that is intersected in a countable (possibly empty) set by every line $y = \text{constant} \neq c, d$. Since the functions involved in (2) are measurable (cf. IV.1.51, IV.2.6), and vanish outside of a sufficiently large circular disc, it follows that (2) holds a.e. in the xy-plane. Integration of (2) yields, in view of IV.4.68(2), the equation (1).

IV.4.71. Suppose that T, given as in IV.4.67, is sBV in R_0^0. Let $R : a \leqq u \leqq b$, $c \leqq v \leqq d$ be any oriented rectangle in R_0. We assert the inequality

$$(1) \qquad \int_c^d | f(b, v) - f(a, v) | \, dv \leqq W_u(R, f) \leqq \iint \kappa(x, y, T, R^0).$$

PROOF. Clearly (cf. III.2.45)

$$| f(b, v) - f(a, v) | \leqq V_u(a, b, v, f).$$

Integration yields, in view of IV.4.70, the inequalities (1).

IV.4.72. *Given T as in* IV.4.67, *T is sAC in R_0^0 if and only if $f(u, v)$ is ACTu in R_0* (cf. IV.4.1, III.2.49, IV.2.39).

PROOF. (i) Suppose that T is sAC in R_0^0. Then (cf. IV.2.45)

$$(1) \qquad \iint_{R_0} D(w, R_0^0) = \iint N(x, y, R_0^0).$$

Now T is also sBV in R_0^0 (cf. IV.4.1, IV.2.39). Hence, by IV.4.68(1), IV.4.68(2), we can rewrite (1) in the form

$$(2) \qquad \iint_{R_0} | f_u(u, v) | \, du \, dv = W_u(R_0, f).$$

In writing (2), we used the relation $N(x, y, R_0^0) = N(x, y, R_0)$ which holds a.e. in the xy-plane, since obviously $| T(R_0 - R_0^0) | = 0$. By III.2.57, (2) implies that $f(u, v)$ is ACTu in R_0.

(ii) Suppose conversely that $f(u, v)$ is ACTu in R_0. By III.2.56 we have then the formula (2). Now $f(u, v)$ is also BVTu in R_0 (see III.2.49), and hence T is sBV in R_0^0 by IV.4.67. In view of IV.4.68(1), IV.4.68(2) it follows that (2) implies (1). Finally, by IV.2.45, IV.4.1, (1) implies that T is sAC in R_0^0.

IV.4.73. Suppose that T, given as in IV.4.67, is eBV in R_0^0. Then T is also sBV in R_0^0. The converse is obvious (see IV.4.1).

PROOF. By IV.4.1 the assumption implies that $\kappa(x, y, T, R_0^0)$ is summable. By IV.1.82, we have $\kappa(x, y, T, R_0^0) = N(x, y, R_0^0)$ except on a set $\bar{e}$ that is intersected in a countable (possibly empty) set by every line $y = \text{constant} \neq c_0, d_0$. Since $\kappa(x, y, T, R_0^0)$ and $N(x, y, R_0^0)$ are measurable (see IV.1.51, IV.2.6), it follows that $\kappa(x, y, T, R_0^0) = N(x, y, R_0^0)$ a.e. in the xy-plane. Thus $N(x, y, R_0^0)$ is also summable, and hence T is sBV in R_0^0 (see IV.2.13, IV.4.1).

IV.4.74. Suppose that T, given as in IV.4.67, is eAC in R_0^0. Then T is also sAC in R_0^0. The converse is obvious (see IV.4.1).

PROOF. By IV.4.1 and IV.2.45 we have

$$(1) \qquad \iint_{R_0} D(w, \mathcal{E}^*) = \iint N(x, y, \mathcal{E}^*).$$

On the other hand, by IV.1.82, $N(x, y, \mathcal{E}^*) = N(x, y, R_0^0)$ a.e. in the xy-plane (cf. the remarks in IV.4.73 concerning the exceptional set). Hence

$$(2) \qquad \iint N(x, y, \mathcal{E}^*) = \iint N(x, y, R_0^0).$$

(1), (2), IV.2.31, IV.3.9 yield

$$\iint_{R_0} D(w, \mathcal{E}^*) \leqq \iint_{R_0} D(w, R_0^0) \leqq \iint N(x, y, R_0^0) = \iint_{R_0} D(w, \mathcal{E}^*).$$

There follows the equation

$$\iint_{R_0} D(w, R_0^0) = \iint N(x, y, R_0^0).$$

By IV.4.1, IV.2.45 it follows that T is sAC in R_0^0.

IV.4.75. For transformations T given as in IV.4.67, it follows from IV.4.73, IV.4.74 that the concepts eBV, eAC coincide with sBV, sAC respectively. As a consequence, we have for transformations given as in IV.4.67, the following theorems.

(i) T is eBV in R_0^0 if and only if $f(u, v)$ is BVTu in R_0 (cf. IV.4.67).

(ii) T is eAC in R_0^0 if and only if $f(u, v)$ is ACTu in R_0 (cf. IV.4.72).

(iii) If T is eBV in R_0^0 then $\mathcal{J}_\bullet(w) = f_u(u, v)$ a.e. in R_0 (cf. IV.4.69, IV.4.73, IV.3.22).

(iv) $\kappa(z, T, R_0^0) = N(z, T, R_0)$ a.e. in the z-plane (cf. IV.1.82 and the remarks in IV.4.72).

IV.4.76. The discussion in IV.4.67 to IV.4.75 applies, with obvious modifications, to transformations of the form $T : x = u,\ y = f(u, v),\ (u, v) \in R_0$.

CHAPTER IV.5. GENERAL COMMENTS ON PLANE TRANSFORMATIONS

IV.5.1. Even a superficial survey of the literature reveals that practically every worker in the theory of surface area devoted a considerable part of his efforts to the study of plane transformations. There are several reasons for this phenomenon. In the first place, comparison of the familiar formulas for length and area (see I.1.2) shows that in the formula for arc length the coordinate functions $x(u)$, $y(u)$, $z(u)$ of the curve appear *individually*, while in the formula for surface area the coordinate functions $x(u, v)$, $y(u, v)$, $z(u, v)$ are *combined in pairs* by means of the Jacobians $\partial(y, z)/\partial(u, v)$, $\partial(z, x)/\partial(u, v)$, $\partial(x, y)/\partial(u, v)$. Thus it is to be expected that the conditions for the validity of the area formula will involve *pairs of coordinate functions*, rather than individual coordinate functions as in the case of the length formula (see part III). In the second place, if a surface is given by equations $x = x(u, v)$, $y = y(u, v)$, $z = z(u, v)$, then its orthogonal projections upon the coordinate planes are bound to play a significant role. The orthogonal projection upon the xy-plane, for instance, being given by the formulas $x = x(u, v)$, $y = y(u, v)$, it is clear that the pairing off of the coordinate functions corresponds to the orthogonal projections upon the three coordinate planes. Finally, the pair $x(u, v)$, $y(u, v)$, for example, may be thought of as determining a *flat surface* $x = x(u, v)$, $y = y(u, v)$, $z = 0$, and thus the study of the *plane transformation* determined by the formulas $x = x(u, v)$, $y = y(u, v)$, may be thought of as *a preliminary study of surface area in the special case of flat surfaces.*

As a matter of fact, the study of plane transformations, given by formulas of the type $x = x(u, v)$, $y = y(u, v)$, seems to be an indispensable foundation in the theory of surface area. In the present part IV, we shall go beyond the indispensable minimum for two important reasons. (i) We propose to base surface area theory upon appropriate *two-dimensional concepts of bounded variation and absolute continuity*, in analogy with the theory of arc length where one-dimensional concepts of bounded variation and absolute continuity play a fundamental role. Before applying the two-dimensional concepts in the study of general surfaces, we propose to test the fitness of these concepts in the relatively simple case of flat surfaces, that is, plane transformations. (ii) In view of the number and diversity of results in the literature on plane transformations, an effort to construct a comprehensive theory seemed to be justified. Accordingly, the present part IV has been planned as a systematic study, rather than as a discussion of scattered results.

IV.5.2. We begin with a brief survey of the subject from *the elementary point of view.* Accordingly, the meaning of the terms used will be vague, and the conditions for the validity of the formulas will not be stated. In a sense, the purpose of part IV is to fill such gaps.

Let $\mathfrak{R}$ be a bounded, simply-connected region in the uv-plane. The boundary curve of $\mathfrak{R}$, oriented in the counterclockwise sense, will be denoted by C. Let T be a transformation, from $\mathfrak{R}$ into a Cartesian xy-plane, given by formulas

$$T : x = x(u, v),\ y = y(u, v), \qquad (u, v) \in \mathfrak{R}. \tag{1}$$

Case 1. The mapping T is biunique in $\mathfrak{R}$, and the Jacobian

$$J(u, v) = x_u y_v - x_v y_u \tag{2}$$

is positive in $\mathfrak{R}$. If we denote by $| T(\mathfrak{R}) |$ the measure of the image of $\mathfrak{R}$, then we have the familiar formula

$$\iint_{\mathfrak{R}} J \, du \, dv = | T(\mathfrak{R}) |. \tag{3}$$

Partial integration yields the equally familiar formula

$$\iint_{\mathfrak{R}} J \, du \, dv = \frac{1}{2} \int_C (x \, dy - y \, dx). \tag{4}$$

Combining (3) and (4), we obtain

$$| T(\mathfrak{R}) | = \frac{1}{2} \int_C (x \, dy - y \, dx). \tag{5}$$

Now let $H(x, y)$ be a function defined on $T(\mathfrak{R})$. We have then the transformation formula

$$\iint_{T(\mathfrak{R})} H(x, y) \, dx \, dy = \iint_{\mathfrak{R}} H[x(u, v), y(u, v)] J \, du \, dv. \tag{6}$$

Analogous formulas hold if $J < 0$ in $\mathfrak{R}$.

Case 2. Let us now drop the assumption that T is biunique in $\mathfrak{R}$, and let us also permit the Jacobian J to change its sign in $\mathfrak{R}$. Every student in Calculus is warned that he should *not* use the transformation formula (6) under these conditions. And yet, the degree of generality just indicated is indispensable for many purposes. In particular, if the mapping T represents the orthogonal projection, upon the xy-plane, associated with a surface (see IV.5.1), then T cannot be expected to be biunique. As a matter of fact, the formulas (3)-(6) do admit of plausible and important extensions to the present general case. Let $N(x, y)$ denote the number of distinct points in $\mathfrak{R}$ that are mapped by T into an assigned point (x, y). We have then the plausible formula

$$\iint_{\mathfrak{R}} | J | \, du \, dv = \iint_{T(\mathfrak{R})} N(x, y) \, dx \, dy, \tag{3*}$$

in analogy with formula (3). As regards formula (4), its derivation is a purely formal matter (involving formal partial integrations) and does not presuppose in any way that T is biunique. So we have again

$$(4^*) \qquad \iint_{\mathfrak{R}} J\, du\, dv = \frac{1}{2} \int_{C} (x\, dy - y\, dx).$$

Now let (x, y) be a point in the xy-plane that does not lie on $T(C)$. While a point (u, v) describes the boundary curve C of $\mathfrak{R}$ in the counterclockwise sense, its image under T describes a closed (possibly self-intersecting) curve which will encircle the point xy a certain number of times, say k times, where k is an integer (positive, negative, or zero). In fact, k is merely the topological index of the point (x, y) relative to the image of C (see chapter II.4). This integer k, which is a function of x and y, will be denoted by $\mu(x, y)$. In analogy with (5), we have the plausible formula

$$(5^*) \qquad \iint_{T(\mathfrak{R})} \mu(x, y)\, dx\, dy = \frac{1}{2} \int_{C} (x\, dy - y\, dx).$$

Combining (4*) and (5*), we obtain the formula

$$(5_*) \qquad \iint_{\mathfrak{R}} J\, du\, dv = \iint_{T(\mathfrak{R})} \mu(x, y)\, dx\, dy,$$

which is analogous to (3) too. We may think of (3*) as giving the *absolute area* and of (5_*) as giving the *signed area* of the image, in terms of the Jacobian J. For (6) we have the plausible generalizations

$$(6^*) \qquad \iint_{T(\mathfrak{R})} H(x, y)\mu(x, y)\, dx\, dy = \iint_{\mathfrak{R}} H[x(u, v), y(u, v)]J\, du\, dv,$$

$$(6_*) \qquad \iint_{T(\mathfrak{R})} H(x, y)N(x, y)\, dx\, dy = \iint_{\mathfrak{R}} H[x(u, v), y(u, v)] \mid J \mid du\, dv.$$

For convenient reference, let us put

$$(7^*) \qquad \alpha_T(\mathfrak{R}) = \iint_{T(\mathfrak{R})} N(x, y)\, dx\, dy,$$

$$(8^*) \qquad \sigma_T(\mathfrak{R}) = \iint_{T(\mathfrak{R})} \mu(x, y)\, dx\, dy.$$

Then $\alpha_T(\mathfrak{R})$, $\sigma_T(\mathfrak{R})$ may be thought of as the *absolute area* and *the signed area*, respectively, of the image of $\mathfrak{R}$ under T.

The utility and convenience of the preceding generalized formulas is obvious even in mathematical fields of a classical type. For example, if $f(w)$ is an analytic function of the complex variable $w = u + iv$, then on setting $z = x + iy$ the formula $z = f(w)$ defines a (generally not biunique) mapping whose geometrical properties are of great value in the study of the function $f(w)$, and the generalized formulas (3*)-(8*) prove most useful in this connection. And yet, rigorous proofs of these generalized formulas are seldom given, even in situations where the context would seem to require absolute precision. In part IV we give a comprehensive discussion of these formulas and of further generalizations. Let us add a few remarks intended to motivate some of the general concepts of the theory. Let (u, v) be an interior point of $\mathfrak{R}$, and let s be a small square that contains (u, v). The formulas (3*), (7*) and (5_*), (8*), applied to s, yield

$$(9^*) \qquad |J(u, v)| = \lim_{|s| \to 0} \frac{\alpha_T(s)}{|s|}, \qquad J(u, v) = \lim_{|s| \to 0} \frac{\sigma_T(s)}{|s|},$$

providing interesting geometrical interpretations for $|J|$ and J respectively in terms of *two-dimensional derivatives.* Let us now take a point (x, y) in $T(\mathfrak{R})$, and let E be the set of those points $(u, v) \in \mathfrak{R}$ that are mapped into the point (x, y) by T. If (u_0, v_0) is a non-isolated point of E (that is, if every vicinity of (u_0, v_0) contains further points of E), then $J(u_0, v_0) = 0$. Indeed, if $J(u_0, v_0)$ were different from zero, then by a well known theorem on implicit functions the mapping T should be biunique in a sufficiently small vicinity of (u_0, v_0), and thus (u_0, v_0) should be an isolated point of the set E. These remarks reveal the significant fact that inspection of *topological* properties of T yields information concerning points where J must vanish. Inspection of the formulas (9*) reveals that J and $|J|$ may be determined by *two-dimensional limit processes,* in contradistinction to the *linear limit processes* indicated by the formula $J = x_u y_v - x_v y_u$. These observations explain a great deal of the involved procedure followed in part IV.

IV.5.3. The formulas discussed in IV.5.2 have been subjected to a searching analysis by many mathematicians. We shall review presently one of the main lines of development. To achieve adequate generality, and also to defer a discussion of boundary phenomena, we consider a bounded continuous mapping

$$(1) \qquad T : x = x(u, v),\ y = y(u, v), \qquad (u, v) \in \mathfrak{D},$$

where $\mathfrak{D}$ is a bounded domain (connected open set) in the uv-plane, and $x(u, v)$, $y(u, v)$ are continuous and bounded in $\mathfrak{D}$ (only the real case will be considered). Then the image $T(\mathfrak{D})$ of $\mathfrak{D}$ is a bounded set in the Cartesian xy-plane. However, no assumptions will be made a priori concerning the existence of the partial derivatives x_u, x_v, y_u, y_v, and unless the contrary is explicitly stated, T will not be assumed to be biunique. Under such conditions, the ordinary Jacobian $J = x_u y_v - x_v y_u$ may not exist anywhere in $\mathfrak{D}$ (see I.1.13), and thus the derivation of formulas of the type of those discussed in IV.5.2 seems to be a hopeless undertaking. On the other hand, the formulas IV.5.2(9*) suggest a promising approach: the *two-dimensional limit processes* appearing in those formulas may converge even though the partial derivatives x_u, x_v, y_u, y_v fail to exist. As a matter of fact, this actually happens in the example discussed in I.1.13. The first systematic development of this idea is due to Rademacher [1], [2]. Rademacher assumes that T is biunique. Let then (u, v) be a point of $\mathfrak{D}$, and let s be a small square, with sides parallel to the u, v axes respectively, that contains (u, v). Rademacher shows that the limit

$$(2) \qquad D(u, v) = \lim_{|s| \to 0} \frac{|T(s)|}{|s|}$$

exists in $\mathfrak{D}$ with the possible exception of a set of measure zero. To simplify matters, let us assume that T is sense-preserving. Comparison with the formulas IV.5.2(9*) indicates that $D(u, v)$ may be adopted as a *generalized Jacobian* for the mapping T, and the results of Rademacher justify this view fully. He finds that $D(u, v)$ is summable in $\mathfrak{D}$, and in fact

$$\iint_{\mathfrak{D}} D(u, v)\, du\, dv \leqq |T(\mathfrak{D})|,$$

where the sign of equality holds if and only if T is *measurable*. The term *measurable* means that every measurable set $E \subset \mathfrak{D}$ has a measurable image $T(E)$. Equivalently, T is measurable if and only if for every set $e \subset \mathfrak{D}$ of measure zero the image set $T(e)$ is also of measure zero. Under the assumption that T is measurable, Rademacher establishes a generalization of the transformation formula IV.5.2(6), where of course the ordinary Jacobian J is now replaced by the generalized Jacobian $D(u, v)$. In case the ordinary Jacobian J happens to exist, there arises the question of the relation between J and $D(u, v)$. For the important special case when T satisfies a Lipschitz condition, Rademacher shows that $J = D$ with the possible exception of a set of measure zero. These results constitute a theory of biunique plane transformations that is surprisingly general and accounts for all elementary results by a wide margin. On the other hand, the scope of the applications (in particular in surface area theory) is severely restricted by the requirement of *biuniqueness* (cf. IV.5.2, case 2). Still, the results and methods of Rademacher served as a model in the subsequent development of the theory.

IV.5.4. The next important advance is due to Banach [1], [2] (see also G. Vitali [1]). Let

$$(1) \qquad T : x = x(u, v),\ y = y(u, v), \qquad (u, v) \in \mathfrak{D},$$

be given as in IV.5.3, except that T is now *not assumed to be biunique*. Let R be a generic notation for an oriented rectangle (a rectangle with sides parallel to the u, v axes respectively). R^0 will denote the interior of R, while $|R|$ will denote the area of R. Let $R_1, R_2, \cdots$ be any finite system of oriented rectangles in $\mathfrak{D}$ such that $R_i^0 R_j^0 = 0$ for $i \neq j$. If there exists a finite constant M such that $|T(R_1^0)| + |T(R_2^0)| + \cdots < M$ for all such systems $R_1, R_2, \cdots$, then T is termed of *bounded variation* in the Banach sense. If for every $\epsilon > 0$ there exists an $\eta(\epsilon) > 0$ such that for every such system $R_1, R_2, \cdots$ we have $|T(R_1^0)| + |T(R_2^0)| + \cdots < \epsilon$ whenever $|R_1| + |R_2| + \cdots < \eta(\epsilon)$, then T is termed *absolutely continuous* in the Banach sense. These definitions differ slightly from those given by Banach, but the difference is irrelevant. The preceding wording has been chosen for more convenient comparison with subsequent modifications. With the same end in view, let us say that T is sBV (*strongly of bounded variation*) if T is of bounded variation in the Banach sense. Similarly, let us say that T is sAC (*strongly absolutely continuous*) if T is absolutely continuous in the Banach sense. Comparison with III.2.10 shows the complete analogy of these concepts of bounded variation and absolute continuity with those employed in the theory of functions of a single real variable. The work of Banach and others revealed the surprising scope of this analogy. Let $D(u, v)$ be defined as in IV.5.3(2), except that only oriented squares are used. Then the results of Banach include the following statements. If T is sBV in $\mathfrak{D}$, then $D(u, v)$ exists a.e. (almost everywhere) in $\mathfrak{D}$, is summable in $\mathfrak{D}$, and

$$\iint_{\mathfrak{D}} D(u, v)\, du\, dv \leqq \iint_{T(\mathfrak{D})} N(x, y)\, dx\, dy < \infty, \tag{2}$$

where $N(x, y)$ denotes the number of distinct points in $\mathfrak{D}$ that are mapped by T into an assigned point (x, y). Further, T is sBV in $\mathfrak{D}$ if and only if $N(x, y)$ is summable on the set $T(\mathfrak{D})$. The sign of equality in (2) holds if and only if T is sAC in $\mathfrak{D}$. The preceding sample statements suffice to indicate the beauty and the scope of the results of Banach. A further important step was taken by Schauder [3]. Let us note that in the present case $D(u, v)$ cannot be considered as an adequate generalized Jacobian, since T is not assumed to be biunique, and hence $D(u, v)$ corresponds merely to the *absolute value* of the ordinary Jacobian (cf. IV.5.2, case 2). Now let (u_0, v_0) be a point of $\mathfrak{D}$, and let (x_0, y_0) be the image of (u_0, v_0). Let us first assume that there exists some small vicinity V of (u_0, v_0) such that no point in V (except for (u_0, v_0)) is mapped into (x_0, y_0). Let then C be a simple closed curve in V that contains (u_0, v_0) in its interior. If a point (u, v) describes C in the counterclockwise sense, then the image point describes a closed curve that encircles (x_0, y_0) a certain number of times, say k times, where k is some integer (positive, negative, or zero). This number k is defined as the *local index* $i_s(u_0, v_0)$, where we use the subscript s to distinguish this local index from one to be introduced later on. Note that a precise formal definition of the local index may be given in terms of a topological index (see chapter II.4). It follows readily that $i_s(u_0, v_0)$ is independent of the particular choice of the curve C. If (u_0, v_0) does not possess a vicinity V as required above, then we put $i_s(u_0, v_0) = 0$ (cf. the concluding remarks in IV.5.2). In terms of $D(u, v)$ and $i_s(u, v)$, Schauder defines a *generalized Jacobian* $\mathfrak{J}_s(u, v) = i_s(u, v)D(u, v)$, and derives theorems which yield the formulas of IV.5.2 for sAC mappings T, where the ordinary Jacobian J is now replaced by the generalized Jacobian $\mathfrak{J}_s$. An important observation is in order at this time. Since T is not assumed to be biunique, the local index $i_s(u, v)$ may take on any integral values. As a consequence, Schauder restricted T by the requirement that $|\, i_s(u, v)\, |$ be *bounded* in $\mathfrak{D}$. This restriction was removed by T. Radó [9], [12], who showed that $i_s(u, v) =$

± 1 or 0, except possibly on a *countable* set in $\mathfrak{D}$. Also, T. Radó [12] clarified the relation between the generalized Jacobian $\mathcal{J}_s$ and the ordinary Jacobian $J = x_u y_v - x_v y_u$ by showing that if T is sAC in $\mathfrak{D}$ and J exists a.e. in $\mathfrak{D}$, then $\mathcal{J}_s = J$ a.e. in $\mathfrak{D}$. These brief comments suffice to indicate that the theory of plane transformations initiated by Banach is indeed a beautiful and comprehensive theory, exhibiting far-reaching analogies with the corresponding classical theory for functions of a single variable. And yet, the need for further generalizations soon became evident. Indeed, far-reaching results concerning special classes of plane transformations, derived from entirely different sets of assumptions by various authors, could not be accounted for in terms of the Banach theory. Apparently, the concepts sBV, sAC introduced by Banach were too exacting. This becomes strikingly evident if one attempts applications in the theory of surface area: *the fundamental conflict between the projection principle and the lower semi-continuity principle* (cf. I.1.14) *arises immediately, the Banach concepts being found biased in favor of the projection principle.*

The theory of plane transformations, presented in part IV, originated with the desire to eliminate the conflict just referred to. We noted in I.1.15 that Geöcze [3] discovered a *modified projection principle* that did not conflict with the lower semi-continuity principle. It is then natural to attempt to apply the fundamental idea of Geöcze in the study of a plane transformation. This program, initiated by T. Radó [3], [4], [5], [6], is based on the observation (see above) that the mapping T is sBV (sAC) in the Banach sense if the rectangle function $| T(R^0) |$ is BV (AC) (cf. III.1.2), and hence less exacting concepts may be obtained by replacing $| T(R^0) |$ by a *smaller* rectangle function. Now the Geöcze concept of a modified projection principle suggests the substitution, for $T(R^0)$, of the *kernel* of the image of R under T (see IV.1.3), and the use of the measure of the kernel as the fundamental rectangle function in terms of which bounded variation and absolute continuity are to be defined. However, as the following rapid survey will show, many adjustments must be made and many problems must be solved before an adequately balanced set of concepts is obtained.

IV.5.5. Given a bounded continuous transformation T as in IV.5.3, it will be convenient to introduce the complex variables $w = u + iv$, $z = x + iy$ to condense statements and formulas. The mapping T may then be described by a formula

$$(1) \qquad T : z = f(w), \qquad w \in \mathfrak{D},$$

where $f(w)$ is a single-valued, complex-valued, bounded and continuous function of w in the bounded domain $\mathfrak{D}$ (where the term domain is used in the sense of connected open set). Let $N(z, T, \mathfrak{D})$ denote the number of distinct points $w \in \mathfrak{D}$ that are mapped by T into an assigned point z (thus $N(z, T, \mathfrak{D})$ may be infinite for certain points z). This *multiplicity function* $N(z, T, \mathfrak{D})$, which plays a fundamental role in the Banach theory (see IV.5.4), is replaced in the theory to be reviewed presently by an *essential multiplicity function* $\kappa(z, T, \mathfrak{D})$ (see IV.1.43). This essential multiplicity function was introduced (in a somewhat restricted setting) by T. Radó [6], as a substitute for certain very involved concepts defined by Geöcze [3] in connection with his modified projection principle (cf. I.1.15). Chapter IV.1 is devoted to a study of the essential multiplicity function and of closely related topics from the topological point of view. Always $\kappa(z, T, \mathfrak{D}) \leqq N(z, T, \mathfrak{D})$, and $\kappa(z, T, \mathfrak{D})$ possesses a remarkable set of continuity properties (see IV.1.51-IV.1.53), while $N(z, T, \mathfrak{D})$ is generally lacking in such properties. *Great caution must be exercised in dealing with* $\kappa(z, T, \mathfrak{D})$. For example, the image $T(\mathfrak{D})$ of $\mathfrak{D}$ may contain a square and yet $\kappa(z, T, \mathfrak{D})$ may vanish identically (cf. IV.1.43, IV.1.13). A first systematic topological study of the essential multiplicity function was undertaken by P. V. Reichelderfer [1], who established, in particular, the fundamental theorem in IV.1.47 which yields an *intrinsic characterization of* $\kappa(z, T, \mathfrak{D})$, in contradistinction to the original definition (see IV.1.43, IV.1.4) which involves all continuous mappings that differ little from the given mapping T. Given a point z, the multiplicity function $N(z, T, \mathfrak{D})$ counts the number of distinct points of the inverse set $T^{-1}(z)$, while *the essential multiplicity function* $\kappa(z, T, \mathfrak{D})$ *is found to count the number of those components of the inverse set* $T^{-1}(z)$ *that are essential in an appropriately defined sense* (see IV.1.46). The substitution of $\kappa(z, T; \mathfrak{D})$ for $N(z, T, \mathfrak{D})$ necessitates, of course, a series of further adjustments. Thus the *strong local index* $i_s(w)$ of Schauder [3] (cf. IV.5.4, IV.1.75) is replaced by an *essential local index* $i_e(w)$ (see IV.1.64), which was introduced and studied by T. Radó and P. V. Reichelderfer [17]. With the possible exception of countable subsets of the domain $\mathfrak{D}$, $i_s(w)$ and $i_e(w)$ take on only the values ± 1, 0 (see IV.1.67, IV.1.79). The replacement of $N(z, T, \mathfrak{D})$ by $\kappa(z, T, \mathfrak{D})$ may be thought of as a *downward revision in the z-plane* which must be matched by *a downward revision in the w-plane,* consisting of a replacement of the domain $\mathfrak{D}$ by an *essential subset.* The guiding idea in this connection is based on the remark, made in IV.5.2, that in elementary cases the ordinary Jacobian $J = x_u y_v - x_v y_u$ is bound to vanish at certain points w for topological reasons, the set of such points constituting an *unessential set* for the mapping. For a general continuous mapping T, several plausible candidates for the role of the *essential set* are available (see IV.1.56). An important special class of mappings is obtained by formulas of the type $x = g(u, v)$, $y = v$, where $g(u, v)$ is a (real-valued) continuous function. For such mappings, the general theory may be simplified considerably (see IV.1.80-IV.1.84). For instance, the multiplicity functions $N(z, T, \mathfrak{D})$ and $\kappa(z, T, \mathfrak{D})$ are found to agree with each other except for a negligible set of points (see IV.1.82). Except for details of exposition, the contents of chapter IV.1 are taken from T. Radó [9], [12], T. Radó and P. V. Reichelderfer [17], and P. V. Reichelderfer [8].

It is interesting to see the simplifications that take place in the case of a mapping $x = g(u)$, where $g(u)$ is real-valued, bounded and continuous in some open interval, as compared with the case of a plane mapping as studied here. The result in III.2.9 is readily found to imply that in the case of a mapping $x = g(u)$ the multiplicity functions N and κ (the latter being defined in analogy with $\kappa(z, T, \mathfrak{D})$) agree, except possibly on a *countable* set. *As a consequence, the issues* *that necessitated the modification of the Banach theory of plane transformations do not arise at all in the case of mappings $x = g(u)$.*

Several problems arise in connection with the essential multiplicity function $\kappa(z, T, \mathfrak{D})$ that seem to be relevant and interesting. Some of these problems will be mentioned presently.

(i) Let S, M^* be metric spaces (see I.2.10), and let $T(S) = S^*$ be a continuous mapping from S into M^* (that is, $S^* \subset M^*$). Let $N(p^*, T, S)$ denote the number of distinct points $p \in S$ that are mapped by T into an assigned point $p^* \in M^*$. In analogy with the concept of the essential multiplicity function of a plane transformation, the following definitions suggest themselves. For each non-negative integer k, let us define the *kernel* $\mathfrak{K}(k, T, S)$ as the set of those points $p^* \in M^*$ for which the following holds: there exists an $\epsilon = \epsilon(p^*, k, T, S) > 0$ such that $N(p^*, T_*, S) \geqq k$ for every continuous mapping $T_*(S) \subset M^*$ that is within ϵ of T on S (that is, $\rho[T(p), T_*(p)] < \epsilon$ for every point $p \in S$). The kernel $\mathfrak{K}(\infty, T, S)$ is defined as the product of all the kernels $\mathfrak{K}(k, T, S)$, $k = 0, 1, 2, \cdots$. Finally, an *essential multiplicity function* $N_e(p^*, T, S)$ is defined as follows: $N_e(p^*, T, S) = k$ if $p^* \in \mathfrak{K}(k, T, S) - \mathfrak{K}(k + 1, T, S)$, and $N_e(p^*, T, S) = \infty$ if $p^* \in \mathfrak{K}(\infty, T, S)$. Comparison with IV.1.4 reveals that the essential multiplicity function $\kappa(z, T, \mathfrak{R})$ of IV.1.4 is a special instance of the preceding general definition. On the other hand, the essential multiplicity function $\kappa(z, T, \mathfrak{D})$ (see IV.1.43) is defined in a somewhat different manner, and thus there arises the question whether $\kappa(z, T, \mathfrak{D}) = N_e(z, T, \mathfrak{D})$. Unpublished results of P. V. Reichelderfer concerning this question indicate that in a detailed study of $N_e(p^*, T, S)$ the compactness of S may be a relevant factor.

(ii) Extension of the theory, presented in part IV, to transformations in Euclidean n-space would necessitate a thorough study of $N_e(p^*, T, S)$ for the case when S and M^* are Euclidean n-spaces. The resulting issues seem to be related to certain important topological investigations of H. Hopf [1]. An intrinsic characterization of $N_e(p^*, T, S)$, corresponding to the theorem in IV.1.47, would perhaps constitute the first major objective.

(iii) The definition of the kernel $\mathfrak{K}(k, T, \mathfrak{D})$ (see IV.1.43) has been suggested by certain complicated concepts due to Geöcze, as we noted above. However, the precise relation of our concepts to those of Geöcze [3] has not been fully clarified as yet, and interesting and difficult questions may arise in this connection.

IV.5.6. The *downward revisions*, referred to in IV.5.5, lead to the topological problems studied in chapter IV.1, and also lead to metric problems that are studied in a general form in chapter IV.2. Given a continuous mapping $T : z = f(w)$, $w \in \mathfrak{D}$, as in IV.5.5, the concepts of bounded variation and absolute continuity are defined in the Banach theory in terms of the rectangle function $| T(R^0) |$ (see IV.5.4). Let now $\mathfrak{B}$ be an arbitrarily selected but fixed Borel set in the w-plane. On replacing, in the definitions of Banach, the rectangle function $| T(R^0) |$ by the rectangle function $| T(R^0\mathfrak{B}) |$, we obtain the concepts BV$\mathfrak{B}$ (*bounded variation with respect to the base set* $\mathfrak{B}$) and AC$\mathfrak{B}$ (*absolute continuity with* *respect to the base set* $\mathfrak{B}$), which were introduced and studied by T. Radó and P. V. Reichelderfer [17]. If $\mathfrak{B}$ is chosen as the whole w-plane, then the concepts BV$\mathfrak{B}$, AC$\mathfrak{B}$ reduce to the concepts sBV, sAC due to Banach. If $\mathfrak{B}$ is a proper subset of the domain $\mathfrak{D}$ in which T is considered, then the concepts BV$\mathfrak{B}$, AC$\mathfrak{B}$ are less exacting than the concepts sBV, sAC, and thus an extension of the scope of the Banach theory may be expected to result in this manner. If $\mathfrak{B}$ is chosen as the empty set, then of course every continuous T is AC$\mathfrak{B}$, and thus *the relevancy of the generalization depends upon the choice of the base set* $\mathfrak{B}$. In fact, *the intention is to choose* $\mathfrak{B}$ *to fit the individual mapping* T, and the various essential sets referred to in IV.5.5 will play an important role in this respect. However, in chapter IV.2 the concepts BV$\mathfrak{B}$, AC$\mathfrak{B}$ are studied in full generality. It appears that a complete generalization of the Banach theory is possible, and in fact several improvements are obtained. As compared with the presentation in T. Radó and P. Reichelderfer [17], a number of simplifications and improvements could be achieved, due to a large extent to suggestions of E. J. Mickle. In particular, the transformation formula in IV.2.27 seems to represent, in a concise form, the metric foundation of the transformation formulas for simple and multiple integrals. Investigations of E. J. Mickle, as yet unpublished, indicate that the metric theory presented in chapter IV.2 can be extended to abstract metric spaces.

IV.5.7. In chapter IV.3, the topological and metric information obtained in chapters IV.1, IV.2 is combined in a study of *two-dimensional derivatives*, suggested by the remarks in IV.5.2 concerning the geometrical interpretations of the ordinary Jacobian $J = x_u y_v - x_v y_u$ in terms of *area-derivatives*. As we observed in IV.5.2, one may use either the *absolute area* or the *signed area* to obtain such interpretations. In view of the downward revisions, in both the w-plane and the z-plane, described in IV.5.5, IV.5.6, a whole series of *two-dimensional derivatives* and *generalized Jacobians* must be studied and compared with each other, as well as with the ordinary Jacobian $J = x_u y_v - x_v y_u$ if the latter happens to exist. Except for various improvements of details, the contents of chapter IV.3 are taken from T. Radó [12], T. Radó and P. V. Reichelderfer [17], and P. V. Reichelderfer [8]. The papers just quoted contain further references to relevant literature. As regards methods, the work of Rademacher [2] on total differentiability and the generalizations due to Stepanoff [1], [2] and Haslam-Jones and Burkill [1], [2] served as a model in the study of the relations between the ordinary Jacobian and the generalized Jacobians. *The interested reader will find a number of relevant unsolved problems in Chapter* IV.3.

IV.5.8. Chapter IV.4 brings the theory of plane transformations according to the program stated at the end of IV.5.4. Thus *this theory represents a generalization, suggested by fundamental ideas of Geöcze, of the theory initiated by Banach.* The fundamental class of mappings T in this theory is the class K_2 defined and studied by T. Radó and P. V. Reichelderfer [17]. The definition of this class K_2, based originally upon heterogeneous conditions, has been put into a strikingly elegant form by P. V. Reichelderfer [8], who found that this class may be defined as the class of those mappings T that are absolutely continuous with respect to a properly defined base-set $\mathcal{E}^*$, where $\mathcal{E}^*$ depends upon the individual mapping T (see IV.4.1). Such a mapping is termed eAC (*essentially absolutely continuous*). The theory of eAC transformations is found to yield a complete generalization of the Banach theory, where of course all concepts involved undergo modifications in accordance with the downward revisions described in IV.5.5, IV.5.6. But while the Banach theory seemed to lack adequate scope, the present theory is found to account for all previous major results in this field. The reason is that the class of eAC transformations has certain important *closure properties*, stated in the theorems of IV.4.12, IV.4.15-IV.4.20. Furthermore, if the ordinary Jacobian $J = x_u y_v - x_v y_u$ is available, then the generalized essential Jacobian $\mathcal{J}_e$, which occurs in the present theory, agrees with J almost everywhere, according to results derived in part IV.3. The list of theorems in IV.4.2-IV.4.11 and in IV.4.21, concerned with transformation formulas corresponding to those described in IV.5.2, shows that the scope of the theory was not achieved at the price of a loss of content. As a matter of fact, application to special situations, previously studied in the literature, generally yields improvements and not merely verifications of known results. A brief review of some of the relevant applications follows, with remarks on unsolved problems.

(i) Given the bounded continuous transformation $T : z = f(w)$, $w \in \mathfrak{D}$, let us assume that T is sAC in $\mathfrak{D}$ (that is, T is absolutely continuous in the Banach sense). Then obviously T is also eAC. The generalized Jacobian $\mathcal{J}_s$, introduced by Schauder, agrees a.e. with our generalized essential Jacobian $\mathcal{J}_e$, and our results are found to include all the results of the Banach theory. Let us call attention to one significant improvement. As a consequence of IV.4.6 we have the formula

$$\text{(1)} \qquad \iint_{\mathfrak{D}} | J_s(w) | \, du \, dv = \iint \kappa(z, T, \mathfrak{D}) \, dx \, dy,$$

while in view of the general setting of the Banach theory we should expect the formula

$$\text{(2)} \qquad \iint_{\mathfrak{D}} | J_s(w) | \, du \, dv = \iint N(z, T, \mathfrak{D}) \, dx \, dy.$$

As a matter of fact, formula (2) is still not established for all sAC mappings, a fact which represents a serious gap in the Banach theory. Of course, (2) would follow if we could show that $\kappa(z, T, \mathfrak{D}) = N(z, T, \mathfrak{D})$ a.e. for sAC mappings. This problem, which has been solved so far only in special cases (see IV.4.24), is of interest for the additional reason that its solution seems to depend upon improved topological insight, even though the problem itself is primarily of metric character. As matters stand at present, the essential multiplicity function seems to be relevant within the Banach theory itself, by virtue of the formula (1).

(ii) The literature contains a large number of studies concerned with mappings T given in the form

(3) $$T : x = x(u, v),\ y = y(u, v), \quad a \leqq u \leqq b,\ c \leqq v \leqq d,$$

where the functions $x(u, v)$, $y(u, v)$ are assumed to satisfy, *at least*, the following conditions in the closed rectangle $R : a \leqq u \leqq b,\ c \leqq v \leqq d$.

(a) $x(u, v)$, $y(u, v)$ are ACT in R (see III.2.64). As a consequence, the ordinary Jacobian $J = x_u y_v - x_v y_u$ exists a.e. in R.

(b) The ordinary Jacobian J is summable in R.

Let us term the set of conditions (a), (b) *the hypothesis* H_0.

In the type of literature just referred to, various further restrictions are used by the various authors, and far-reaching results are obtained concerning transformation formulas of the nature of those described in IV.5.2, case 2. As compared with the Banach theory and our own theory, the important feature is that the concept of absolute continuity used (absolute continuity in the Tonelli sense) is, essentially, *iterated linear absolute continuity*, rather than *two-dimensional absolute continuity* as in the case of the concepts sAC, eAC. Comparison of results based upon such diverse concepts is therefore a matter of interest. The case studied by Rademacher [1], where $x(u, v)$, $y(u, v)$ are assumed to be Lipschitzian, is of course readily recognized as a special case of sAC mappings. The situation is different with respect to a large number of cases studied by W. H. Young [1]-[6], McShane [3], C. B. Morrey [1], T. Radó [8]. By virtue of the *closure theorems* on eAC mappings, referred to above, *all these cases can be identified as special eAC mappings, but some of these cases were found to lie definitely beyond the scope of the Banach theory.* The systematic verification of the fact that all such cases represent special eAC mappings was carried out by R. G. Helsel [2] and R. G. Helsel and T. Radó [22]. This work, which is based essentially upon approximations by integral means, is discussed in IV.4.32-IV.4.41. Now consider the following case: the functions $x(u, v)$, $y(u, v)$ in formula (3) satisfy the hypothesis H_0, and furthermore the partial derivatives x_u, x_v, y_u, y_v are summable with their squares in R. Then T is eAC in R, as a special case of IV.4.33. On the other hand, Cesari [1] found examples to show that T need not be sAC. This special case is however very important in the theory of surface area, and thus the lack of scope of the Banach theory, in addition to its conflict with the lower semi-continuity principle, seems to justify the efforts devoted to the construction of a more comprehensive theory.

In each case in the literature where the above hypothesis H_0 occurs, the verification of the fact that we have a special instance of an eAC mapping requires, as a rule, only part of the further assumptions made by the respective authors. *Thus there arises the problem to determine whether the hypothesis* H_0 *alone implies that the mapping is eAC.* The evidence available at present seems to be inconclusive, but an affirmative solution would represent a great step in the integration of the literature of this field.

(iii) Given the continuous mapping $T : z = f(w)$, $w \in \mathfrak{D}$, let us assume that

T is *biunique* in $\mathfrak{D}$. This is the classical case studied by Rademacher. In sections IV.4.52-IV.4.66 we find that in this case the concepts sAC, eAC reduce to the concept of a measurable T, due to Rademacher. Introducing the concept of a *bimeasurable* T (see IV.4.62), we establish results which show that bimeasurable mappings obey all the essential laws that obtain in the elementary theory of the change of the independent variables in double integrals.

(iv) The comparison of the concept eAC (essential absolute continuity of two-dimensional character) with the concept ACT (absolute continuity in the Tonelli sense, of the character of iterated linear absolute continuity) is of course a matter of interest, especially as regards applications in surface area theory. This comparison was carried out by P. V. Reichelderfer [8], whose work is closely followed in sections IV.4.67-IV.4.76. As a sample, we quote the following result. Let $g(u, v)$ be real-valued and continuous in a closed rectangle $R : a \leqq u \leqq b$, $c \leqq v \leqq d$. Let us define mappings T^u, T^v by the formulas $T^u : x = g(u, v), y = v$, and $T^v : x = u, y = g(u, v)$ respectively. Then $g(u, v)$ is absolutely continuous in the Tonelli sense if and only if T^u, T^v are both essentially absolutely continuous. Further details may be found in IV.4.67-IV.4.76, but it is evident that the concepts of Tonelli lie well within the scope of our theory.

IV.5.9. The preceding comments on the relationships between various aspects of the theory of plane transformations are not meant to give a complete account of the very extensive literature. A few comments should be added, however, concerning a series of papers by Caccioppoli (see Bibliography). As noted above, the program of applying and further developing the fundamental ideas of Geöcze on the modification of the projection principle has been initiated by T. Radó [4], [5], [6] in 1928. In 1930, Caccioppoli [8] published a paper in which he discussed a theory of plane transformations based essentially on the same approach, and gave extensive applications to surface area theory. This paper is one of several papers of Caccioppoli in which a number of important topics were considered, in particular applications to Calculus of Variations. A general vagueness of concepts and proofs made the study of these papers quite difficult, a circumstance which may account for the fact that a first fundamental error in Caccioppoli's work was noticed only in 1941. One of the main tools of Caccioppoli is the following *extension theorem.* Let $\Phi(O)$ be a (real-valued, finite-valued) function of open sets O comprised, say, in a fixed rectangle R and satisfying the following conditions: (i) $\Phi(O) \geqq 0$ for every $O \subset R$. (ii) If $O_1, O_2, \cdots$ are disjoint open sets in R, then $\Phi(\sum O_n) = \sum \Phi(O_n)$. (iii) If $O_1 \subset O_2$, then $\Phi(O_1) \leqq \Phi(O_2)$. (iv) If O_ϵ denotes the set of those points of the open set O whose distance from the boundary of O exceeds ϵ, then $\lim \Phi(O_\epsilon) = \Phi(O)$ as $\epsilon \rightarrow 0$, for every $O \subset R$. According to Caccioppoli [1], $\Phi(O)$ *should then possess a completely additive extension to all Borel sets in R.* Unfortunately, this theorem is false (see T. Radó and P. V. Reichelderfer [17]). Since fundamental concepts in the work of Caccioppoli were defined in terms of nonexisting additive extensions, the consequences of this error were extremely serious. Further errors, equally relevant, occur in the topological statements of Caccioppoli (see P. V. Reichelderfer [1]). In retrospect, it would seem that these cumulative errors prevented Caccioppoli from recognizing the fundamental difficulties that were responsible for the relatively slow progress of other workers in this field.

IV.5.10. In IV.5.5-IV.5.8, we reported on the contents of part IV, emphasizing also various unsolved problems. Some further open problems will be mentioned presently.

(i) Let $T : x = x(u, v)$, $y = y(u, v)$, $0 \leqq u \leqq 1$, $0 \leqq v \leqq 1$, be a continuous mapping, such that the partial derivatives x_u, x_v, y_u, y_v exist a.e. in the unit square and the ordinary Jacobian $J = x_u y_v - x_v y_u$ is summable there. Let further $T_n : x = x_n(u, v)$, $y = y_n(u, v)$ be a sequence of quasi-linear mappings (see IV.4.30) such that $x_n(u, v) \to x(u, v)$, $y_n(u, v) \to y(u, v)$ uniformly in the unit square, and

$$(1) \qquad \int_0^1\int_0^1 | J - J_n | \, du \, dv \to 0 \qquad \text{for } n \to \infty,$$

where J_n is the Jacobian of the mapping T_n. As a special case of IV.4.18, it follows that T is eAC in the unit square. Suppose, conversely, that T is given as above and is known to be eAC in the unit square. *Does there exist then a sequence of quasi-linear mappings T_n, converging uniformly to T, such that* (1) *holds?* The affirmative answer would yield, in the special case where the ordinary Jacobian J exists a.e. and is summable, a characterization of eAC mappings which would be independent of the involved topological concepts that occur in the original definition of such mappings (see IV.4.1). An analogous characterization of general eAC mappings would be of course an even more desirable objective.

While the problems just suggested seem to be quite difficult, it may seem quite easy to avoid them by basing the theory of plane transformations upon some conveniently chosen class of mappings, defined in terms of approximations by quasi-linear mappings, instead of insisting upon eAC mappings as a basis of the theory. However, the theorem of V.2.61 shows that eAC mappings are inescapable in surface area theory, and thus the problems suggested above would be merely deferred by changing the form of definitions in the theory of plane transformations.

(ii) An outstanding instance of a mathematical field where Jacobians are important is the theory of double integral problems in Calculus of Variations, and there arises the question of the fitness of our generalized essential Jacobians as regards applications in Calculus of Variations. A first plausible objective would be the derivation of analogues of certain theorems on the *lower semi-continuity of double integrals* (see, also for literature, McShane [2], T. Radó [19], W. Scott [1]), where the *ordinary Jacobians* considered in the literature would be replaced by *essential generalized Jacobians*. A fundamental tool in this type of work is a lemma of McShane [2], [3], which has been generalized by T. Radó [15]. While the papers just quoted are concerned with ordinary Jacobians, the theorem in IV.4.42 gives a partial generalization involving essential generalized Jacobians. To achieve adequate generality, it seems however that the following question must first be settled. Let $x(u, v)$, $y(u, v)$, $z(u, v)$ be continuous functions in a domain $\mathfrak{D}$, such that the three mappings $T^x : y = y(u, v), z = z(u, v)$, $T^y : z = z(u, v), x = x(u, v)$, $T^z : x = x(u, v), y = y(u,v)$ are eAC in $\mathfrak{D}$. Let $\mathcal{J}_e^1$, $\mathcal{J}_e^2$, $\mathcal{J}_e^3$ be the corresponding essential generalized Jacobians. Suppose now that the Cartesian coordinate system xyz is replaced by a new Cartesian coordinate system $\xi\eta\zeta$. Then the functions $x(u, v)$, $y(u, v)$, $z(u, v)$ undergo a certain linear transformation with constant coefficients, and the ordinary Jacobians, if they happen to exist, undergo the same transformation. On the other hand, *no information is available at present concerning the effect of a change of the coordinate system upon the generalized essential Jacobians.* The question thus raised is related to some apparently very difficult questions in surface area theory, and hence a thorough study of this question may lead to important methods and results.

(iii) Let us insist again upon *the discrepancy between the concepts* eBV, eAC, already noted in IV.4.1. The concept eAC is defined there as absolute continuity relative to a properly chosen base set $\mathcal{E}^*$, but the concept eBV is *not* defined as bounded variation relative to a base set. The clarification of this situation seems desirable, especially in view of the fact that the main unsolved problems in surface area theory involve the concept of bounded variation (cf. V.2.65-V.2.67).

(iv) Extension of the theory, from plane transformations to continuous transformations in Euclidean n-space, is a natural objective. From the geometrical point of view, the discussion of the isoperimetric problem (estimate for the volume enclosed by a closed surface) seems to require such an extension, for instance. The reader will see readily that many new topological phenomena arise even in Euclidean three-space.

PART V. SURFACE AREA

CHAPTER V.1. THE LOWER AREA $a(S)$

V.1.1. Given a surface S by a representation (see II.3.44)

$$S : \mathfrak{x} = \mathfrak{x}(w) = (x_1(w), x_2(w), x_3(w)), \qquad w \in \mathfrak{R},$$

let us introduce the following permanent notations. We put

$$\mathfrak{x}^1 = (0, x_2, x_3), \mathfrak{x}^2 = (x_1, 0, x_3), \mathfrak{x}^3 = (x_1, x_2, 0).$$

We denote by z_1, z_2, z_3 the complex variables

$$z_1 = x_2 + ix_3, z_2 = x_3 + ix_1, z_3 = x_1 + ix_2.$$

We define the complex-valued functions $f^1(w)$, $f^2(w)$, $f^3(w)$ by the formulas

$$f^1(w) = x_2(w) + ix_3(w), \qquad w \in \mathfrak{R},$$

$$f^2(w) = x_3(w) + ix_1(w), \qquad w \in \mathfrak{R},$$

$$f^3(w) = x_1(w) + ix_2(w), \qquad w \in \mathfrak{R}.$$

We define three transformations T^1, T^2, T^3 by the formulas

$$T^1 : z_1 = f^1(w), \qquad w \in \mathfrak{R},$$

$$T^2 : z_2 = f^2(w), \qquad w \in \mathfrak{R},$$

$$T^3 : z_3 = f^3(w), \qquad w \in \mathfrak{R}.$$

If $\mathfrak{D}$ is any domain in $\mathfrak{R}^0$, then each one of the transformations T^j, $j = 1, 2, 3$, gives rise to an *essential multiplicity function* $\kappa(z_j, T^j, \mathfrak{D})$ in the sense of IV.1.43. For fixed j and $\mathfrak{D}$ the function $\kappa(z_j, T^j, \mathfrak{D})$ is a l.s.c. (lower semi-continuous) function of z_j (see IV.1.51). Hence, for fixed j and $\mathfrak{D}$, $\kappa(z_j, T^j, \mathfrak{D})$ is a measurable function of z_j. Since T^j is continuous on $\mathfrak{R}$, T^j is bounded on $\mathfrak{R}$ and hence also on every $\mathfrak{D} \subset \mathfrak{R}^0$. Thus the theory developed in part IV applies to each T^j.

Let us note that the transformations T^j depend upon the particular representation chosen for S.

V.1.2. Given S as in V.1.1 and a domain $\mathfrak{D} \subset \mathfrak{R}^0$, we define the quantity $g_j(\mathfrak{x}, \mathfrak{D})$, $j = 1, 2, 3$, as follows. If $\kappa(z_j, T^j, \mathfrak{D})$ fails to be summable, then we put $g_j(\mathfrak{x}, \mathfrak{D}) = +\infty$. If $\kappa(z_j, T^j, \mathfrak{D})$ is summable, then we put (cf. IV.2.1)

$$(1) \qquad g_j(\mathfrak{x}, \mathfrak{D}) = \iint \kappa(z_j, T^j, \mathfrak{D}).$$

Let us note that, for given j, z_j is of the form $x_\alpha + ix_\beta$. The integral occurring in (1) is the Lebesgue integral of $\kappa(z_j, T^j, \mathfrak{D})$ considered as a function of the two real variables x_α, x_β. Since $\kappa(z_j, T^j, \mathfrak{D})$ vanishes outside of a sufficiently large circular disc in the z_j-plane (cf. IV.3.8), the integration is performed on such a disc (cf. IV.2.1).

We define now

$$g(\mathfrak{x}, \mathfrak{D}) = \{[g_1(\mathfrak{x}, \mathfrak{D})]^2 + [g_2(\mathfrak{x}, \mathfrak{D})]^2 + [g_3(\mathfrak{x}, \mathfrak{D})]^2\}^{1/2},$$

it being understood that $g(\mathfrak{x}, \mathfrak{D}) = +\infty$ if at least one of the quantities $g_1(\mathfrak{x}, \mathfrak{D})$, $g_2(\mathfrak{x}, \mathfrak{D})$, $g_3(\mathfrak{x}, \mathfrak{D})$ fails to be finite. We have the obvious inequalities

$$g_j(\mathfrak{x}, \mathfrak{D}) \leqq g(\mathfrak{x}, \mathfrak{D}), \qquad j = 1, 2, 3,$$

$$g(\mathfrak{x}, \mathfrak{D}) \leqq g_1(\mathfrak{x}, \mathfrak{D}) + g_2(\mathfrak{x}, \mathfrak{D}) + g_3(\mathfrak{x}, \mathfrak{D}).$$

V.1.3. Given S as in V.1.1, let σ denote a finite system of disjoint domains $\mathfrak{D}_1, \mathfrak{D}_2, \cdots, \mathfrak{D}_m$ in $\mathfrak{R}^0$. We define

$$a(\mathfrak{x}, \mathfrak{R}) = \underset{\sigma}{\text{l.u.b.}} \sum_{i=1}^{m} g(\mathfrak{x}, \mathfrak{D}_i),$$

where the least upper bound is taken with respect to all possible systems σ. The quantity $a(\mathfrak{x}, \mathfrak{R})$ may be infinite. The notation $a(\mathfrak{x}, \mathfrak{R})$ indicates that the definition of this quantity involves a particular representation of S. We proceed to show that $a(\mathfrak{x}, \mathfrak{R})$ is independent of the choice of the representation for S.

V.1.4. Let

$$S : \mathfrak{x} = \mathfrak{x}_1(w_1), \qquad w_1 \in \mathfrak{R}_1, \tag{1}$$

$$S : \mathfrak{x} = \mathfrak{x}_2(w_2), \qquad w_2 \in \mathfrak{R}_2, \tag{2}$$

be two representations of a surface S, and suppose that these representations are *topologically similar* (see II.1.26). We assert that $a(\mathfrak{x}_1, \mathfrak{R}_1) = a(\mathfrak{x}_2, \mathfrak{R}_2)$.

PROOF. By assumption we have a topological transformation $\mathfrak{R}_2 = \tau(\mathfrak{R}_1)$ such that

$$\mathfrak{x}_1(w_1) = \mathfrak{x}_2(\tau(w_1)) \qquad \text{for all } w_1 \in \mathfrak{R}_1. \tag{3}$$

Let us use $f_1^j(w_1)$, $f_2^j(w_2)$, T_1^j, T_2^j to denote the functions $f^j(w)$ and the transformations T^j associated with the representations (1) and (2) respectively, in the sense of V.1.1. Then (3) implies that $f_1^j(w_1) = f_2^j(\tau(w_1))$, $j = 1, 2, 3$. By IV.1.48 it follows that $\kappa(z_j, T_1^j, \mathfrak{D}^1) = \kappa(z_j, T_2^j, \mathfrak{D}^2)$, $j = 1, 2, 3$, where $\mathfrak{D}^1$ is any domain in $\mathfrak{R}_1^0$ and $\mathfrak{D}^2 = \tau(\mathfrak{D}^1)$. Hence (cf. V.1.2) $g_j(\mathfrak{x}_1, \mathfrak{D}^1) = g_j(\mathfrak{x}_2, \mathfrak{D}^2)$, $j = 1, 2, 3$, and consequently $g(\mathfrak{x}_1, \mathfrak{D}^1) = g(\mathfrak{x}_2, \mathfrak{D}^2)$. In view of V.1.3 the assertion follows.

V.1.5. Let there be given surfaces $S : \mathfrak{x} = \mathfrak{x}(w)$, $w \in \mathfrak{R}$, $S_n : \mathfrak{x} = \mathfrak{x}_n(w)$, $w \in \mathfrak{R}$, in terms of representations such that

$$\mathfrak{x}_n(w) \to \mathfrak{x}(w) \tag{1}$$

uniformly in $\mathfrak{R}$. Then

$$a(\mathfrak{x}, \mathfrak{R}) \leqq \liminf a(\mathfrak{x}_n, \mathfrak{R}). \tag{2}$$

PROOF. Let $\mathfrak{D}$ be any domain in $\mathfrak{R}^0$. We first show that

(3) $$g_j(\mathfrak{x}, \mathfrak{D}) \leqq \liminf_{n\to\infty} g_j(\mathfrak{x}_n, \mathfrak{D}), \qquad j = 1, 2, 3.$$

Let us put

(4) $$\lambda_j = \liminf_{n\to\infty} g_j(\mathfrak{x}_n, \mathfrak{D}), \qquad j = 1, 2, 3.$$

Let us fix j. If $\lambda_j = +\infty$, then (3) is obvious. So we can assume that $\lambda_j < +\infty$. We have then a sequence of positive integers $n_1 < n_2 < \cdots < n_k < \cdots$, such that

(5) $$\lim_{k\to\infty} g_j(\mathfrak{x}_{n_k}, \mathfrak{D}) = \lambda_j < +\infty.$$

Clearly, we can assume that

(6) $$g_j(\mathfrak{x}_{n_k}, \mathfrak{D}) < +\infty, \qquad k = 1, 2, \cdots.$$

Let us denote by T'_{n_k} the transformation corresponding to the representation $S_{n_k} : \mathfrak{x} = \mathfrak{x}_{n_k}(w)$, $w \in \mathfrak{R}$, in the sense of V.1.1. In view of (6) we have

(7) $$g_j(\mathfrak{x}_{n_k}, \mathfrak{D}) = \iint \kappa(z_j, T'_{n_k}, \mathfrak{D}).$$

In view of IV.1.52, the relation (1) implies that

(8) $$\kappa(z_j, T', \mathfrak{D}) \leqq \liminf_{k\to\infty} \kappa(z_j, T'_{n_k}, \mathfrak{D}).$$

(5), (7), (8) imply, by the lemma of Fatou (see I.3.10), that $\kappa(z_j, T', \mathfrak{D})$ is summable and

(9) $$\iint \kappa(z_j, T', \mathfrak{D}) \leqq \liminf_{k\to\infty} \iint \kappa(z_j, T'_{n_k}, \mathfrak{D}).$$

Since $\kappa(z_j, T', \mathfrak{D})$ is summable, we have by definition

(10) $$g_j(\mathfrak{x}, \mathfrak{D}) = \iint \kappa(z_j, T', \mathfrak{D}).$$

(9), (10), (7), (5), (4) imply (3). Clearly, (3) implies that

(11) $$g(\mathfrak{x}, \mathfrak{D}) \leqq \liminf g(\mathfrak{x}_n, \mathfrak{D}).$$

Now let $\mathfrak{D}_1, \cdots, \mathfrak{D}_m$ be any finite system of disjoint domains in $\mathfrak{R}^0$. By (11) we have then

$$\sum_{i=1}^{m} g(\mathfrak{x}, \mathfrak{D}_i) \leqq \sum_{i=1}^{m} \liminf_{n\to\infty} g(\mathfrak{x}_n, \mathfrak{D}_i) \leqq \liminf_{n\to\infty} \sum_{i=1}^{m} g(\mathfrak{x}_n, \mathfrak{D}_i)$$

$$\leqq \liminf_{n\to\infty} a(\mathfrak{x}_n, \mathfrak{D}_i).$$

Since this holds for every finite system of disjoint domains, (2) follows.

V.1.6. Let now

$$S : \mathfrak{x} = \mathfrak{x}(w), \qquad w \in \mathfrak{R},$$
$$(1) \qquad S : \mathfrak{x} = \mathfrak{x}^*(w^*), \qquad w^* \in \mathfrak{R}^*,$$

be any two representations of a surface S. We assert that $a(\mathfrak{x}, \mathfrak{R}) = a(\mathfrak{x}^*, \mathfrak{R}^*)$.

Proof. By assumption (cf. II.3.12) we have for S a sequence of representations

$$(2) \qquad S : \mathfrak{x} = \mathfrak{x}_n(w), \qquad w \in \mathfrak{R},$$

such that for each n the representations (1) and (2) are topologically similar, and $\mathfrak{x}_n(w) \to \mathfrak{x}(w)$ uniformly in $\mathfrak{R}$. By V.1.4 and V.1.5 we obtain the relations

$$(3) \qquad a(\mathfrak{x}_n, \mathfrak{R}) = a(\mathfrak{x}^*, \mathfrak{R}^*),$$

$$(4) \qquad a(\mathfrak{x}, \mathfrak{R}) \leqq \liminf_{n \to \infty} a(\mathfrak{x}_n, \mathfrak{R}).$$

(3) and (4) imply that $a(\mathfrak{x}, \mathfrak{R}) \leqq a(\mathfrak{x}^*, \mathfrak{R}^*)$. The complementary inequality $a(\mathfrak{x}^*, \mathfrak{R}^*) \leqq a(\mathfrak{x}, \mathfrak{R})$ is derived in a similar manner. Thus it follows that $a(\mathfrak{x}, \mathfrak{R}) = a(\mathfrak{x}^*, \mathfrak{R}^*)$.

V.1.7. In view of V.1.6, the quantity $a(\mathfrak{x}, \mathfrak{R})$ has the same value for all representations of a surface S. We denote this value by $a(S)$. Thus $a(S) = a(\mathfrak{x}, \mathfrak{R})$ for every representation of S. The quantity $a(S)$, which depends only upon S, will be termed *the lower area* of S (however, see the remarks in V.1.8).

From II.3.16 and V.1.5 we infer immediately that the relation $S_n \to S$ implies that $a(S) \leqq \liminf a(S_n)$. That is, $a(S)$ is a lower semi-continuous function of S, in the class of Fréchet surfaces of the type of the 2-cell (see II.3.44).

V.1.8. In defining the lower area $a(S)$, we made explicit use of the Cartesian coordinate system x_1, x_2, x_3. There arises the question as to whether $a(S)$ is or is not independent of the choice of the coordinate system, or, equivalently (cf. II.3.42), whether $a(S)$ has the same value for congruent surfaces. The answer to this question is unknown as yet.[1] On the other hand, we shall find that $a(S)$ is a very useful tool in the study of surface area.

V.1.9. In the definition of $a(\mathfrak{x}, \mathfrak{R})$ we used *finite* systems of disjoint domains in $\mathfrak{R}^0$ (cf. V.1.3). Let us denote by $\bar{a}(\mathfrak{x}, \mathfrak{R})$ the quantity obtained by using finite or infinite systems of disjoint domains. We assert that $a(\mathfrak{x}, \mathfrak{R}) = \bar{a}(\mathfrak{x}, \mathfrak{R})$.

Proof. Let $\mathfrak{D}_1, \mathfrak{D}_2, \cdots, \mathfrak{D}_m, \cdots$ be any infinite system of disjoint domains in $\mathfrak{R}^0$. For ea ch m we have then

$$(1) \qquad \sum_{i=1}^{m} g(\mathfrak{x}, \mathfrak{D}_i) \leqq a(\mathfrak{x}, \mathfrak{R}).$$

Suppose first that $a(\mathfrak{x}, \mathfrak{R}) < +\infty$. Since (1) holds for every m, it follows that

[1]In view of results of Cesari [6] which came to the attention of the writer after the manuscript of this book had been completed, it can be shown that $a(S)$ is indeed independent of the choice of the coordinate system. See also V.2.65, V.4.8. The relationships between the fundamental concepts used in this book and those used in the work of Cesari are studied in Radó [30].

$$(2) \qquad \sum_{i=1}^{\infty} g(\mathfrak{x}, \mathfrak{D}_i) \leqq a(\mathfrak{x}, \mathfrak{R}).$$

Since (2) holds for every system $\mathfrak{D}_1, \mathfrak{D}_2, \cdots, \mathfrak{D}_m, \cdots$, it follows that

$$(3) \qquad \bar{a}(\mathfrak{x}, \mathfrak{R}) \leqq a(\mathfrak{x}, \mathfrak{R}).$$

In deriving (3), we assumed that $a(\mathfrak{x}, \mathfrak{R}) < +\infty$. If $a(\mathfrak{x}, \mathfrak{R}) = +\infty$, then (3) is obvious. Thus always $\bar{a}(\mathfrak{x}, \mathfrak{R}) \leqq a(\mathfrak{x}, \mathfrak{R})$. Since the complementary inequality $a(\mathfrak{x}, \mathfrak{R}) \leqq \bar{a}(\mathfrak{x}, \mathfrak{R})$ is obvious, the assertion $a(\mathfrak{x}, \mathfrak{R}) = \bar{a}(\mathfrak{x}, \mathfrak{R})$ follows.

V.1.10. Given S as in V.1.1, let $\mathfrak{R}_0, \mathfrak{R}_1, \cdots, \mathfrak{R}_m$ be any finite system of simply connected Jordan regions, such that

$$\mathfrak{R}_i \subset \mathfrak{R}_0 \subset \mathfrak{R}, \qquad i = 1, 2, \cdots, m,$$

$$\mathfrak{R}_i^0 \mathfrak{R}_j^0 = 0 \qquad \text{for } i \neq j,\ i \neq 0,\ j \neq 0.$$

Let us denote by S_i the surface determined by the representation

$$S_i : \mathfrak{x} = \mathfrak{x}(w),\ w \in \mathfrak{R}_i, \qquad i = 0, 1, \cdots, m.$$

Then we assert that

$$\sum_{i=1}^{m} a(S_i) \leqq a(S_0) \leqq a(S).$$

PROOF. By V.1.7, we have $a(S_i) = a(\mathfrak{x}, \mathfrak{R}_i)$, $a(S) = a(\mathfrak{x}, \mathfrak{R})$. The asserted inequality appears now as a direct consequence of the definition given in V.1.3.

V.1.11. Given S as in V.1.1, let $\mathfrak{R}_n$ be a sequence of simply-connected Jordan regions that fill up $\mathfrak{R}$ from the interior. That is, $\mathfrak{R}_n \subset \mathfrak{R}$ for every n, and if F is any closed set in $\mathfrak{R}^0$, then $F \subset \mathfrak{R}_n^0$ for n sufficiently large. Let S_n be the surface determined by the representation $S_n : \mathfrak{x} = \mathfrak{x}(w)$, $w \in \mathfrak{R}_n$. Then $a(S_n) \to a(S)$.

PROOF. Let $\mathfrak{D}$ be any domain in $\mathfrak{R}^0$. Let $\mathfrak{R}_n(\mathfrak{D})$ be a sequence of Jordan regions that fill up $\mathfrak{D}$ from the interior, in the sense of IV.1.41. We have then by definition (cf. IV.1.43) the relation

$$(1) \qquad \kappa(z_j, T^j, \mathfrak{D}) = \lim_{n\to\infty} \kappa(z_j, T^j, \mathfrak{R}_n(\mathfrak{D})), \qquad j = 1, 2, 3.$$

Let us fix j. By IV.1.44 we have

$$(2) \qquad \kappa(z_j, T^j, \mathfrak{R}_n(\mathfrak{D})) \leqq \kappa(z_j, T^j, \mathfrak{D}),$$

while IV.1.50 yields

$$(3) \qquad \kappa(z_j, T^j, \mathfrak{R}_n(\mathfrak{D})) = \kappa(z_j, T^j, \mathfrak{R}_n^0(\mathfrak{D})),$$

where $\mathfrak{R}_n^0(\mathfrak{D})$ denotes the interior of $\mathfrak{R}_n(\mathfrak{D})$. From these relations we propose to derive the formula

$$(4) \qquad g_j(\mathfrak{x}, \mathfrak{D}) = \lim_{n\to\infty} g_j(\mathfrak{x}, \mathfrak{R}_n^0(\mathfrak{D})).$$

Let us note first that $g_j(\mathfrak{x}, \mathfrak{R}_n^0(\mathfrak{D})) \leqq g_j(\mathfrak{x}, \mathfrak{D})$, $n = 1, 2, \cdots$, as a consequence of (2) and (3). Hence, if (4) is denied, then we should have

(5) $$\liminf_{n\to\infty} g_i(\mathfrak{x}, \mathfrak{R}_n^0(\mathfrak{D})) < g_i(\mathfrak{x}, \mathfrak{D}).$$

(5) implies the existence of a finite constant M and of a sequence of positive integers $n_1 < n_2 < \cdots < n_k < \cdots$ such that

(6) $$g_i(\mathfrak{x}, \mathfrak{R}_{n_k}^0(\mathfrak{D})) < M < g_i(\mathfrak{x}, \mathfrak{D}), \qquad k = 1, 2, \cdots.$$

From (6) there follows that

(7) $$\iint \kappa(z_i, T', \mathfrak{R}_{n_k}^0(\mathfrak{D})) < M < g_i(\mathfrak{x}, \mathfrak{D}), \qquad k = 1, 2, \cdots.$$

(7), (3), (1) imply, by the lemma of Fatou (I.3.10), that $\kappa(z_i, T', \mathfrak{D})$ is summable, and

(8) $$\iint \kappa(z_i, T', \mathfrak{D}) \leqq M < g_i(\mathfrak{x}, \mathfrak{D}).$$

Since $\kappa(z_i, T^i, \mathfrak{D})$ is summable, we have by definition the formula

(9) $$\iint \kappa(z_i, T', \mathfrak{D}) = g_i(\mathfrak{x}, \mathfrak{D}).$$

(8) and (9) contradict each other, and thus (4) is proved. From (4) we infer that

(10) $$g(\mathfrak{x}, \mathfrak{D}) = \lim_{n\to\infty} g(\mathfrak{x}, \mathfrak{R}_n^0(\mathfrak{D})).$$

Now let $\mathfrak{D}_1, \cdots, \mathfrak{D}_m$ be any finite system of disjoint domains in $\mathfrak{R}^0$. For each $i = 1, 2, \cdots, m$, let us take a sequence of Jordan regions $\mathfrak{R}_n(\mathfrak{D}_i)$ that fill up $\mathfrak{D}_i$ from the interior. For fixed n, the set $\mathfrak{R}_n(\mathfrak{D}_1) + \cdots + \mathfrak{R}_n(\mathfrak{D}_m)$ is a closed set in $\mathfrak{R}^0$. Hence, for large enough N, we shall have $\mathfrak{R}_n(\mathfrak{D}_1) + \cdots + \mathfrak{R}_n(\mathfrak{D}_m) \subset \mathfrak{R}_N$. Hence

$$\sum_{i=1}^{m} g(\mathfrak{x}, \mathfrak{R}_n^0(\mathfrak{D}_i)) \leqq a(\mathfrak{x}, \mathfrak{R}_N).$$

For fixed n, it follows that

$$\sum_{i=1}^{m} g(\mathfrak{x}, \mathfrak{R}_n^0(\mathfrak{D}_i)) \leqq \liminf_{N\to\infty} a(\mathfrak{x}, \mathfrak{R}_N).$$

For $n \to \infty$ we obtain, in view of (10), the inequality

$$\sum_{i=1}^{m} g(\mathfrak{x}, \mathfrak{D}_i) \leqq \liminf_{N\to\infty} a(\mathfrak{x}, \mathfrak{R}_N).$$

Since this holds for every finite disjoint system $\mathfrak{D}_1, \cdots, \mathfrak{D}_m$, it follows that

(11) $$a(\mathfrak{x}, \mathfrak{R}) \leqq \liminf_{N\to\infty} a(\mathfrak{x}, \mathfrak{R}_N).$$

On the other hand, by V.1.10 we have

(12) $$a(\mathfrak{x}, \mathfrak{R}_N) \leqq a(\mathfrak{x}, \mathfrak{R}).$$

Clearly, (11) and (12) imply that $a(\mathfrak{x}, \mathfrak{R}_N) \to a(\mathfrak{x}, \mathfrak{R})$ for $N \to \infty$. In view of V.1.7, the proof is complete.

V.1.12. Given S as in V.1.1, let E be a closed set in $\mathfrak{R}$. Let σ^* denote any (finite or infinite) system of disjoint domains $\mathfrak{D}_1^*, \mathfrak{D}_2^*, \cdots$ in $\mathfrak{R} - E$. Let us define (cf. V.1.2)

$$a(\mathfrak{x}, \mathfrak{R}, E) = \underset{\sigma^*}{\text{l.u.b.}} \sum_k g(\mathfrak{x}, \mathfrak{D}_k^*), \tag{1}$$

where the least upper bound is taken with respect to all possible systems σ^* satisfying the conditions just stated. Let us note that the summation occurring in (1) may be equal to $+\infty$ for two different reasons, namely (a) one or more terms of the summation fail to be finite, or (b) every term is finite, but the series fails to converge. In view of V.1.9, it is obvious that

$$a(\mathfrak{x}, \mathfrak{R}, E) \leqq a(\mathfrak{x}, \mathfrak{R}). \tag{2}$$

Furthermore, if E_1, E_2 are closed sets in $\mathfrak{R}$, such that $E_1 \subset E_2$, then obviously $a(\mathfrak{x}, \mathfrak{R}, E_1) \geqq a(\mathfrak{x}, \mathfrak{R}, E_2)$.

V.1.13. Using the terminology of V.1.12, let us assume that, for $\jmath = 1, 2, 3$, the set $T^\jmath(E)$ is of measure zero (cf. V.1.1). Then $a(\mathfrak{x}, \mathfrak{R}, E) = a(\mathfrak{x}, \mathfrak{R})$.

PROOF. In view of (2) in V.1.12, it is sufficient to show that

$$a(\mathfrak{x}, \mathfrak{R}) \leqq a(\mathfrak{x}, \mathfrak{R}, E). \tag{1}$$

(1) is obvious if $a(\mathfrak{x}, \mathfrak{R}, E) = +\infty$. So we can assume that

$$a(\mathfrak{x}, \mathfrak{R}, E) < +\infty. \tag{2}$$

Since E is closed, the set $\mathfrak{R}^0 - E$ is open. Let us denote by $\mathfrak{D}_1, \mathfrak{D}_2, \cdots$ the components of $\mathfrak{R}^0 - E$. Then each $\mathfrak{D}_\imath$ is a domain. Let us now fix $\jmath$. By IV.1.55 we have

$$\kappa(z_\jmath, T^\jmath, \mathfrak{R}^0) = \sum_\imath \kappa(z_\jmath, T^\jmath, \mathfrak{D}_\imath) \qquad \text{for } z_\jmath \notin T^\jmath(E).$$

Since $T^\jmath(E)$ is of measure zero by assumption, it follows that

$$\kappa(z_\jmath, T^\jmath, \mathfrak{R}^0) = \sum_\imath \kappa(z^\jmath, T_\jmath, \mathfrak{D}_\imath) \tag{3}$$

a.e. in the $z_\jmath$-plane. (2) implies that each term of the series in (3) is summable and

$$\sum_\imath \iint \kappa(z_\jmath, T^\jmath, \mathfrak{D}_\imath) = \sum_\imath g_\jmath(\mathfrak{x}, \mathfrak{D}_\imath) \leqq a(\mathfrak{x}, \mathfrak{R}, E) < +\infty. \tag{4}$$

(3) and (4) imply (see I.3.11) that $\kappa(z_\jmath, T^\jmath, \mathfrak{R}^0)$ is summable. Hence, if $\mathfrak{D}$ is any domain in $\mathfrak{R}^0$, then $\kappa(z_\jmath, T^\jmath, \mathfrak{D})$ is also summable (cf. IV.1.48, IV.1.54). As a consequence, we have, for every $\mathfrak{D} \subset \mathfrak{R}^0$,

$$g_\jmath(\mathfrak{x}, \mathfrak{D}) = \iint \kappa(z_\jmath, T^\jmath, \mathfrak{D}) < +\infty. \tag{5}$$

Let us now observe that (1) will be established if we can show that

$$(6) \qquad \sum_{s=1}^{m} g(\mathfrak{x}, \mathfrak{D}_s') \leqq a(\mathfrak{x}, \mathfrak{R}, E)$$

for every finite system of disjoint domains $\mathfrak{D}_1', \cdots, \mathfrak{D}_m'$ in $\mathfrak{R}^0$. To prove (6), consider the sets $O_s = \mathfrak{D}_s' - E$, $s = 1, 2, \cdots, m$. Each O_s is an open subset of $\mathfrak{R}^0$ (since E is closed). Let $\mathfrak{D}_{s1}', \mathfrak{D}_{s2}', \cdots$ be the components of O_s. Then each $\mathfrak{D}_{sk}'$ is a domain. The domains $\mathfrak{D}_{sk}'$ are disjoint, and are comprised in $\mathfrak{R}^0 - E$. Hence, by the definition of $a(\mathfrak{x}, \mathfrak{R}, E)$,

$$(7) \qquad \sum_s \sum_k g(\mathfrak{x}, \mathfrak{D}_{sk}') \leqq a(\mathfrak{x}, \mathfrak{R}, E) < +\infty .$$

On the other hand, in view of IV.1.55,

$$(8) \qquad \kappa(z_j, T^j, \mathfrak{D}_s') = \sum_k \kappa(z_j, T^j, \mathfrak{D}_{sk}')$$

for z_j not in $T^j(E)$. Since $T^j(E)$ is of measure zero by assumption, it follows that (8) holds a.e. in the z_j-plane. From (5), (2) it follows that

$$(9) \qquad \sum_k \iint \kappa(z_j, T^j, \mathfrak{D}_{sk}') = \sum_k g_j(\mathfrak{x}, \mathfrak{D}_{sk}') \leqq a(\mathfrak{x}, \mathfrak{R}, E) < +\infty .$$

(9) implies (see I.3.11) that termwise integration is permitted in (8). In view of (5) we obtain, by integrating (8),

$$g_j(\mathfrak{x}, \mathfrak{D}_s') = \sum_k g_j(\mathfrak{x}, \mathfrak{D}_{sk}').$$

Hence, by the Minkowski inequality (see I.3.10)

$$g(\mathfrak{x}, \mathfrak{D}_s') \leqq \sum_k g(\mathfrak{x}, \mathfrak{D}_{sk}'),$$

and finally, in view of (7),

$$\sum_{s=1}^{m} g(\mathfrak{x}, \mathfrak{D}_s') \leqq a(\mathfrak{x}, \mathfrak{R}, E).$$

Thus (6) is established, and the proof is complete.

V.1.14. Given S as in V.1.1, let $\mathfrak{R}_1, \cdots, \mathfrak{R}_m$ be a finite system of simply-connected Jordan regions in $\mathfrak{R}$, such that $\mathfrak{R}_1 + \cdots + \mathfrak{R}_m = \mathfrak{R}$ and $\mathfrak{R}_i^0\mathfrak{R}_k^0 = 0$ for $i \neq k$. Suppose that $T^j(\mathfrak{R}_s - \mathfrak{R}_s^0)$ is of measure zero, $j = 1, 2, 3$, $s = 1, 2, \cdots, m$. Then

$$(1) \qquad a(\mathfrak{x}, \mathfrak{R}) = \sum_{s=1}^{m} a(\mathfrak{x}, \mathfrak{R}_s).$$

PROOF. Put

$$E = \sum_{s=1}^{m} (\mathfrak{R}_s - \mathfrak{R}_s^0).$$

Then $| T^j(E) | = 0, j = 1, 2, 3$. Hence, by V.1.13,

$$a(\mathfrak{x}, \mathfrak{R}) = a(\mathfrak{x}, \mathfrak{R}, E). \tag{2}$$

Obviously, in view of the conditions satisfied by $\mathfrak{R}_1, \cdots, \mathfrak{R}_m$,

$$a(\mathfrak{x}, \mathfrak{R}, E) = \sum_{s=1}^{m} a(\mathfrak{x}, \mathfrak{R}_s). \tag{3}$$

(2) and (3) imply (1).

V.1.15. A representation $S : \mathfrak{x} = \mathfrak{x}(w)$, $w \in \mathfrak{R}$, will be termed eBV if the transformations T^j, $j = 1, 2, 3$, are eBV in $\mathfrak{R}^0$, and eAC if the transformations T^j are eAC in $\mathfrak{R}^0$ (cf. V.1.1 and IV.4.1).

V.1.16. *Given S as in* V.1.1, *the representation used for S is eBV if and only if* $a(S) < +\infty$.

PROOF. (i) Suppose $a(S) < +\infty$. By V.1.7 we have then $a(\mathfrak{x}, \mathfrak{R}) < +\infty$, and hence *a fortiori*

$$g(\mathfrak{x}, \mathfrak{R}^0) < +\infty. \tag{1}$$

(1) implies, by V.1.1, that $g_j(\mathfrak{x}, \mathfrak{R}^0) < +\infty$; hence $\kappa(z_j, T^j, \mathfrak{R}^0)$ is summable, and consequently (cf. IV.4.1) T^j is eBV in $\mathfrak{R}^0$, $j = 1, 2, 3$.

(ii) Suppose that $\kappa(z_j, T^j, \mathfrak{R}^0)$ is summable, $j = 1, 2, 3$. Then $\kappa(z_j, T^j, \mathfrak{D})$ is also summable for every domain $\mathfrak{D} \subset \mathfrak{R}^0$, by IV.1.54. There follows, for every system $\mathfrak{D}_1, \cdots, \mathfrak{D}_m$ of disjoint domains in $\mathfrak{R}$ (cf. IV.1.54),

$$\sum_{i=1}^{m} g_j(\mathfrak{x}, \mathfrak{D}_i) = \sum_{i=1}^{m} \iint \kappa(z_j, T^j, \mathfrak{D}_i) \leqq \iint \kappa(z_j, T^j, \mathfrak{R}^0).$$

Hence (cf. V.1.2)

$$\sum_{i=1}^{m} g(\mathfrak{x}, \mathfrak{D}_i) \leqq \sum_{j=1}^{3} \iint \kappa(z_j, T^j, \mathfrak{R}^0).$$

Since this holds for every finite system of disjoint domains in $\mathfrak{R}^0$, it follows that (cf. V.1.7)

$$a(S) = a(\mathfrak{x}, \mathfrak{R}) \leqq \sum_{j=1}^{3} \iint \kappa(z_j, T^j, \mathfrak{R}^0) < +\infty.$$

V.1.17. As an immediate corollary of V.1.16 we obtain the statement: *if one representation of a surface S is eBV, then every representation of S is eBV.*

V.1.18. Given S as in V.1.1, we shall denote by $\mathfrak{J}^j_e(w, \mathfrak{x})$ the essential generalized Jacobian corresponding to the transformation T^j (cf. IV.3.21). If the ordinary Jacobian (cf. IV.3.21) corresponding to T^j happens to exist at a point w, then it will be denoted by $J^j(w, \mathfrak{x})$. The notation is meant to stress the fact that these Jacobians are relative to a given representation of S.

V.1.19. *Given S as in* V.1.1, *suppose that* $a(S) < +\infty$. *Then the Jacobians* $\mathfrak{J}^j_e(w, \mathfrak{x})$, $j = 1, 2, 3$, *exist a.e. in* $\mathfrak{R}^0$, *are summable in* $\mathfrak{R}^0$, *and*

$$(1) \qquad \iint_{\mathfrak{R}^0} \{[\mathcal{J}^1_c(w, \mathfrak{x})]^2 + [\mathcal{J}^2_c(w, \mathfrak{x})]^2 + [\mathcal{J}^3_c(w, \mathfrak{x})]^2\}^{1/2} \leqq a(S).$$

PROOF. In view of V.1.7, we have

$$(2) \qquad a(\mathfrak{x}, \mathfrak{R}) = a(S) < +\infty.$$

Let now $R, R_1, \cdots, R_m$ be any finite system of oriented rectangles in $\mathfrak{R}^0$, such that $R_i^0 R_k^0 = 0$ for $i \neq k$, and $R_i \subset R$, $i = 1, \cdots, m$. By V.1.10 we have then (cf. V.1.7)

$$(3) \qquad \sum_{i=1}^{m} a(\mathfrak{x}, R_i) \leqq a(\mathfrak{x}, R) \leqq a(\mathfrak{x}, \mathfrak{R}) < +\infty.$$

Thus the rectangle-function $a(\mathfrak{x}, \mathfrak{R})$ is of type A in $\mathfrak{R}^0$ (see III.1.52). As a consequence, this rectangle-function has a derivative a.e. in $\mathfrak{R}^0$. Let us denote this derivative by $\alpha(w)$. Then $\alpha(w)$ is summable in R and (see III.1.28)

$$(4) \qquad \iint_R \alpha(w) \leqq a(\mathfrak{x}, R).$$

Thus (4) holds for every oriented rectangle $R \subset \mathfrak{R}^0$. Now we can choose (see I.3.2) a sequence of oriented rectangles $r_1, \cdots, r_s, \cdots$ such that

$$\sum_{s=1}^{\infty} r_s = \mathfrak{R}^0, \qquad r_i^0 r_k^0 = 0 \text{ for } i \neq k.$$

If n is any positive integer, it follows that (cf. (4), (3))

$$\sum_{s=1}^{n} \iint_{r_s} \alpha(w) \leqq \sum_{s=1}^{n} a(\mathfrak{x}, r_s) \leqq a(\mathfrak{x}, \mathfrak{R}) < +\infty.$$

In view of I.3.11, it follows that $\alpha(w)$ is summable in $\mathfrak{R}^0$ and

$$(5) \qquad \iint_{\mathfrak{R}^0} \alpha(w) \leqq a(\mathfrak{x}, \mathfrak{R}).$$

Now let R be any oriented rectangle in $\mathfrak{R}^0$. By V.1.2, V.1.3 we have then the inequality

$$(6) \qquad \left\{\left[\frac{g_1(\mathfrak{x}, R^0)}{|R|}\right]^2 + \left[\frac{g_2(\mathfrak{x}, R^0)}{|R|}\right]^2 + \left[\frac{g_3(\mathfrak{x}, R^0)}{|R|}\right]^2\right\}^{1/2} = \frac{g(\mathfrak{x}, R^0)}{|R|} \leqq \frac{a(\mathfrak{x}, R)}{|R|}$$

Now let us consider the rectangle-function $g_i(\mathfrak{x}, R^0)$. Since $a(\mathfrak{x}, \mathfrak{R}) < +\infty$, we have the formula (cf. V.1.16) $g_i(\mathfrak{x}, R^0) = \iint \kappa(z_i, T^i, R^0)$. Now T^i is eBV in $\mathfrak{R}^0$ (see V.1.16). By IV.4.1, IV.3.25, IV.3.12 it follows that the derivative of $g_i(\mathfrak{x}, R^0)$ exists and is equal to $|\mathcal{J}^i_c(w, \mathfrak{x})|$ a.e. in $\mathfrak{R}^0$. From (6) there follows now that

$$(7) \qquad \{[\mathcal{J}^1_c(w, \mathfrak{x})]^2 + [\mathcal{J}^2_c(w, \mathfrak{x})]^2 + [\mathcal{J}^3_c(w, \mathfrak{x})]^2\}^{1/2} \leqq \alpha(w)$$

a.e. in $\mathfrak{R}^0$. Integration of (7) yields, in view of (5) and (2), the inequality (1).

V.1.20. *Given S as in* V.1.1, *suppose that the given representation of S is eAC* (*see* V.1.15). *Then* $a(S) < +\infty$, *the essential generalized Jacobians* $\mathfrak{J}_e^r(w, \mathfrak{x})$ *exist a.e. in* $\mathfrak{R}^0$ *and are summable in* $\mathfrak{R}^0$, *and*

$$(1) \qquad a(S) = \iint_{\mathfrak{R}^0} W_e(w, \mathfrak{x}),$$

where we have put

$$\{[\mathfrak{J}_e^1(w, \mathfrak{x})]^2 + [\mathfrak{J}_e^2(w, \mathfrak{x})]^2 + [\mathfrak{J}_e^3(w, \mathfrak{x})]^2\}^{1/2} = W_e(w, \mathfrak{x}).$$

Proof. By assumption, the transformations T^r are eAC in $\mathfrak{R}^0$. By IV.4.1, each T^r is eBV in $\mathfrak{R}^0$. Hence the given representation is eBV. By V.1.16, V.1.19 it follows that $a(S) < +\infty$, the Jacobians $\mathfrak{J}_e^r(w, \mathfrak{x})$ exist a.e. in $\mathfrak{R}^0$, and

$$(2) \qquad \iint_{\mathfrak{R}^0} W_e(w, \mathfrak{x}) \leqq a(S).$$

Now let $\mathfrak{D}_1, \cdots, \mathfrak{D}_m$ be any finite system of disjoint domains in $\mathfrak{R}^0$. Since T^r is eAC in $\mathfrak{R}^0$ and hence in each $\mathfrak{D}_i$, we have, by IV.4.7, V.1.2,

$$g_r(\mathfrak{x}, \mathfrak{D}_i) = \iint \kappa(z_r, T^r, \mathfrak{D}_i) = \iint_{\mathfrak{D}_i} | \mathfrak{J}_e^r(w, \mathfrak{x}) |.$$

By the Minkowski inequality it follows that

$$g(\mathfrak{x}, \mathfrak{D}_i) \leqq \iint_{\mathfrak{D}_i} W_e(w, \mathfrak{x}),$$

and hence, since the domains $\mathfrak{D}_i$ are disjoint,

$$\sum_{i=1}^{m} g(\mathfrak{x}, \mathfrak{D}_i) \leqq \iint_{\mathfrak{R}^0} W_e(w, \mathfrak{x}).$$

Since this holds for every disjoint system $\mathfrak{D}_1, \cdots, \mathfrak{D}_m$ in $\mathfrak{R}^0$, there follows the inequality

$$(3) \qquad a(\mathfrak{x}, \mathfrak{R}) \leqq \iint_{\mathfrak{R}^0} W_e(w, \mathfrak{x}).$$

In view of V.1.7, the formula (1) follows from (2) and (3).

V.1.21. We shall see later on (cf. V.2.61, V.2.58) that the converse of the preceding theorem also holds under very general conditions, but it is not known whether the converse holds for every surface S such that $a(S) < +\infty$. The result of V.1.20 suggests a further important, as yet unsolved, problem. Suppose that S, given as in V.1.1, has finite lower area $a(S)$. As observed in the course of the proof in V.1.19, the rectangle-function $a(\mathfrak{x}, R)$ has then a derivative

$\alpha(w)$ a.e. in $\mathfrak{R}^0$. Let us suppose now that the given representation of S is eAC. If R is any oriented rectangle in $\mathfrak{R}^0$, let us define a surface S_R by the representation

$$S_R : \mathfrak{x} = \mathfrak{x}(w), \qquad w \in R.$$

Clearly, this representation is also eAC, and hence, by V.1.20 (applied to S_R), we obtain the formula

$$a(\mathfrak{x}, R) = \iint_{R^0} W_e(w, \mathfrak{x}).$$

By III.1.26 it follows that $\alpha(w) = W_e(w, \mathfrak{x})$ a.e. in $\mathfrak{R}^0$. The problem, referred to above, consists of deciding whether this formula holds if we only assume that the given representation of S is eBV (cf. III.3.13 concerning the analogous situation in the theory of arc-length).

V.1.22. Given S as in V.1.1, suppose that $a(S) < +\infty$ and that the ordinary Jacobians $J^j(w, \mathfrak{x})$, $j = 1, 2, 3$, exist a.e. in $\mathfrak{R}^0$ (cf. V.1.18). Let us use the notation $W_e(w, \mathfrak{x})$ in the sense of V.1.20 and let us put

$$W(w, \mathfrak{x}) = \{[J^1(w, \mathfrak{x})]^2 + [J^2(w, \mathfrak{x})]^2 + [J^3(w, \mathfrak{x})]^2\}^{1/2}.$$

By V.1.19, the essential generalized Jacobians $\mathfrak{J}^j_e(w, \mathfrak{x})$ exist a.e. in $\mathfrak{R}^0$ and

(1) $$\iint_{\mathfrak{R}^0} W_e(w, \mathfrak{x}) \leqq a(S).$$

We assert that

(2) $$W(w, \mathfrak{x}) \leqq W_e(w, \mathfrak{x}) \qquad \text{a.e. in } \mathfrak{R}^0,$$

and consequently, in view of (1),

$$\iint_{\mathfrak{R}^0} W(w, \mathfrak{x}) \leqq a(S).$$

PROOF. Since $a(S) < +\infty$, the functions $\kappa(z, , T^j, \mathfrak{R}^0)$ are summable by V.1.16. Hence, by IV.3.46, $| J^j(w, \mathfrak{x}) | \leqq | \mathfrak{J}^j_e(w, \mathfrak{x}) |$ a.e. in $\mathfrak{R}^0$, and (2) follows.

V.1.23. *Given S as in* V.1.1, *suppose that the given representation of S is eAC and that the ordinary Jacobians $J^j(w, \mathfrak{x})$ exist a.e. in $\mathfrak{R}^0$. By* V.1.20 *the lower area $a(S)$ is then finite. Using the notation $W(w, \mathfrak{x})$ introduced in* V.1.22, *we assert that*

(1) $$a(S) = \iint_{\mathfrak{R}^0} W(w, \mathfrak{x}).$$

PROOF. By assumption, the transformations T^j (cf. V.1.1) are eAC in $\mathfrak{R}^0$. Hence, by IV.4.1, IV.3.42, we have $J^j(w, \mathfrak{x}) = \mathfrak{J}^j_e(w, \mathfrak{x})$ a.e. in $\mathfrak{R}^0$, for $j = 1, 2, 3$. Thus $W(w, \mathfrak{x}) = W_e(w, \mathfrak{x})$ a.e. in $\mathfrak{R}^0$, and (1) appears as a direct consequence of V.1.20.

V.1.24. In part IV, we derived various criteria for a transformation to be eAC (cf. IV.4.28, IV.4.29, IV.4.30, IV.4.31, IV.4.33, IV.4.34, IV.4.41). These special criteria, if combined with the theorem of V.1.23, yield a number of results concerning cases where the lower area $a(S)$ is given by the usual integral formula (1) in V.1.23. We shall list presently a few simple cases of this type that will be needed in the sequel.

V.1.25. A representation

$$(1) \qquad S : \mathfrak{x} = \mathfrak{x}(w), \qquad w \in \mathfrak{R},$$

will be said to be of class C' if (i) the partial derivatives of the first order of the components of $\mathfrak{x}(w)$ exist and are continuous everywhere in $\mathfrak{R}^0$, and (ii) the function $W(w, \mathfrak{x})$ (cf. V.1.22) is summable in $\mathfrak{R}^0$. A surface S will be said to be of class C' if it admits of a representation of class C'. Such a representation will be termed a *typical representation of* S (cf. II.3.47).

Let (1) *be a typical representation of a surface* S *of class* C'. *We assert that*

$$(2) \qquad a(S) = \iint_{\mathfrak{R}^0} W(w, \mathfrak{x}).$$

PROOF. If (1) is a representation of class C', then this representation is eAC as an immediate consequence of IV.4.29. Thus (2) follows directly from V.1.23.

V.1.26. A representation

$$(1) \qquad S : \mathfrak{x} = \mathfrak{x}(w), \qquad w \in \mathfrak{R},$$

will be termed quasi-linear if (i) $\mathfrak{R}$ is bounded by a simple closed polygon, and (ii) $\mathfrak{R}$ can be subdivided into a finite number of (rectilinear) triangles $\Delta_1, \cdots, \Delta_m$, such that the components of $\mathfrak{x}(w)$ are linear functions of u and v in each one of the triangles $\Delta_1, \cdots, \Delta_m$, where $u + iv = w$. A surface S will be termed quasi-linear if it admits of a quasi-linear representation. Such a representation will be called a typical representation of S (cf. II.3.47).

Let (1) be a quasi-linear representation. By IV.4.30, the representation is then eAC, and obviously the first partial derivatives of the components of $\mathfrak{x}(w)$ exist a.e. in $\mathfrak{R}^0$. Hence by V.1.23

$$(2) \qquad a(S) = \iint_{\mathfrak{R}^0} W(w, \mathfrak{x}).$$

This result admits of the following interpretation. By assumption, $\mathfrak{R}$ can be subdivided into a finite number of rectilinear triangles $\Delta_1, \cdots, \Delta_m$, such that the components of $\mathfrak{x}(w)$ are linear functions on each one of these triangles. By means of the representation (1), each Δ_i is transformed into a point-set δ_i in $x_1x_2x_3$-space. Since the components of $\mathfrak{x}(w)$ are linear in Δ_i, the point-set δ_i is either a nondegenerate, plane, rectilinear triangle, or else a straight segment or a single point. In the first case, let $|\delta_i|$ be the area of the triangle δ_i in the elementary sense, while in the second and third cases let $|\delta_i|$ be equal to zero. Well known elementary formulas yield then

$$| \delta_i | = \iint_{\Delta_i} W(w, \mathfrak{x}).$$

Hence, by (2),

(3) $$a(S) = \sum_{i=1}^{m} | \delta_i |.$$

In other words, the lower area $a(S)$ coincides in this case with what may be considered the expected value of the area of S.

V.1.27. A representation

(1) $$S : \mathfrak{x} = \mathfrak{x}(w), \qquad w \in \mathfrak{R},$$

will be termed *Lipschitzian* if the components of $\mathfrak{x}(w)$ satisfy a Lipschitz condition in $\mathfrak{R}$ (cf. I.3.14). A surface S will be termed Lipschitzian if it admits of a Lipschitzian representation. Such a representation will be termed a typical representation of S (cf. II.3.47). In the literature, a Lipschitzian surface, in the sense just explained, is usually called rectifiable. To avoid misunderstandings that may arise from various other connotations of the term rectifiable, we shall use the term Lipschitzian.

Let (1) be a Lipschitzian representation. By IV.4.28, the first partial derivatives of the components of $\mathfrak{x}(w)$ exist a.e. in $\mathfrak{R}^0$ and the representation is eAC. Hence, by V.1.23,

(2) $$a(S) = \iint_{\mathfrak{R}^0} W(w, \mathfrak{x}).$$

From the point of view of elementary mathematics, we should be hardly justified in speaking of an *expected value* of the area of a Lipschitzian surface. On the other hand, from the point of view of advanced mathematics, a Lipschitzian representation is an exceptionally favorable case, and no definition of the area would be considered as acceptable unless it agrees with the value furnished by the usual integral formula in case of a Lipschitzian representation. Thus (2) states that the lower area $a(S)$ coincides, in the Lipschitzian case, with the expected value of the area, from the point of view of the expert.

V.1.28. Given a simply-connected, bounded Jordan region $\mathfrak{R}$ in the w-plane, we define a curvilinear triangulation of $\mathfrak{R}$ as follows. A curvilinear triangle t in $\mathfrak{R}$ is a simply-connected Jordan region, with three distinct points on the boundary of t being fixed. These three points are the vertices of the curvilinear triangle t. The three nonoverlapping arcs determined by the vertices on the boundary of t are the sides of t. A curvilinear triangulation $\mathfrak{T}$ of $\mathfrak{R}$ consists of a finite number of curvilinear triangles $t_1, \cdots, t_m$, such that the following conditions are satisfied.

(i) $t_i^0 t_j^0 = 0$ for $i \neq j$.

(ii) $t_1 + \cdots + t_m = \mathfrak{R}$.

(iii) For $i \neq j$, the set $t_i t_j$ is either empty, or is a common vertex of t_i and t_j, or else a common side of t_i and t_j.

V.1.29. A representation

$$(1) \qquad S : \mathfrak{x} = \mathfrak{x}(w), \qquad w \in \mathfrak{R},$$

is termed *polyhedral* if there exists a curvilinear triangulation $\mathfrak{J}$ of $\mathfrak{R}$, consisting of triangles $t_1, \cdots, t_m$, such that the following conditions are satisfied.

(i) On each t_i, the transformation (1) from t_i into $x_1x_2x_3$-space is topological.

(ii) The image of t_i, $i = 1, 2, \cdots, m$, under the transformation (1), is a nondegenerate, plane, rectilinear triangle Δ_i, the vertices of Δ_i being the images of the vertices of t_i.

For clarity, we shall use the symbol $(\mathfrak{x}, \mathfrak{R}, \mathfrak{J})$ to refer to a polyhedral representation and a triangulation $\mathfrak{J}$ satisfying the conditions (i), (ii). If (1) is a polyhedral representation and

$$(2) \qquad S : \mathfrak{x} = \mathfrak{x}^*(w^*), \qquad w^* \in \mathfrak{R}^*,$$

is any other representation of S, then (2) is not necessarily polyhedral, but if (2) is topologically similar to (1) (cf. II.3.19), then obviously (2) is also polyhedral.

V.1.30. *A polyhedron* (more explicitly, *a Fréchet polyhedron of the type of the 2-cell*, cf. II.3.7) is a surface that admits of a polyhedral representation, in the sense of V.1.29. Such a representation is termed a typical representation of the polyhedron (cf. II.3.47). We shall use $\mathfrak{P}$ to refer to a polyhedron.

V.1.31. Given a polyhedron

$$(1) \qquad \mathfrak{P} : \mathfrak{x} = \mathfrak{x}(w), \qquad w \in \mathfrak{R},$$

in a typical representation $(\mathfrak{x}, \mathfrak{R}, \mathfrak{J})$, let $\Delta_1, \cdots, \Delta_m$ be the rectilinear triangles that correspond to the curvilinear triangles $t_1, \cdots, t_m$ of $\mathfrak{J}$ in the sense of V.1.29. We define $E(\mathfrak{x}, \mathfrak{R}, \mathfrak{J})$, the elementary area of $\mathfrak{P}$ in terms of a given typical representation $(\mathfrak{x}, \mathfrak{R}, \mathfrak{J})$, as the sum of the areas (in the elementary sense) of the triangles $\Delta_1, \cdots, \Delta_m$. In symbols: $E(\mathfrak{x}, \mathfrak{R}, \mathfrak{J}) = |\Delta_1| + \cdots + |\Delta_m|$. The notation $E(\mathfrak{x}, \mathfrak{R}, \mathfrak{J})$ indicates that it is not clear *a priori* that this elementary area depends only upon $\mathfrak{P}$. However, we shall prove now that this is the case.

V.1.32. Under the conditions stated in V.1.31, we have $E(\mathfrak{x}, \mathfrak{R}, \mathfrak{J}) = a(\mathfrak{P})$.

PROOF. (i) Let us first suppose that $m = 1$. Then $\mathfrak{R}$ itself, with three distinct points w_1, w_2, w_3 on its boundary fixed as vertices, is the only triangle of $\mathfrak{J}$. The transformation from $\mathfrak{R}$ into $x_1x_2x_3$-space, defined by (1) in V.1.31, is then topological, and carries $\mathfrak{R}$ into a nondegenerate, plane, rectilinear triangle Δ, the points w_1, w_2, w_3 being carried into the vertices P_1, P_2, P_3 of Δ. We have to show that $a(\mathfrak{P}) = |\Delta|$. Now let Δ^* be a (nondegenerate) rectilinear triangle, with vertices w_1^*, w_2^*, w_3^* in an auxiliary w^*-plane. We have then an affine transformation $T(\Delta) = \Delta^*$ that carries the vertices P_1, P_2, P_3 of Δ into the vertices w_1^*, w_2^*, w_3^* of Δ^* respectively. Let $\tau(\mathfrak{R}) = \Delta$ be the topological transformation, defined by (1) in V.1.31, from $\mathfrak{R}$ onto Δ. Then the representation $\mathfrak{x} = \mathfrak{x}\tau^{-1}T^{-1}(w^*)$, $w^* \in \Delta^*$, is clearly topologically similar to (1) in V.1.31 (cf. II.1.26), and hence this is also a representation of $\mathfrak{P}$. On the other hand, the transformation from Δ^* onto Δ, defined by this representation, clearly agrees with the transformation $T^{-1}(\Delta^*) = \Delta$. Since T is an affine transformation, it follows that this representation is quasi-linear, and hence $a(\mathfrak{P}) = |\Delta|$ by V.1.26.

(ii) Now let us return to the general case (m an arbitrary positive integer). For $s = 1, 2, \cdots, m$, let us denote by $\mathfrak{J}_s$ the triangulation of t_s consisting of the single triangle t_s. Then the representation $\mathfrak{x} = \mathfrak{x}(w)$, $w \in t_s$, is clearly a polyhedral representation $(\mathfrak{x}, t_s, \mathfrak{J}_s)$, in the sense of V.1.29. Thus we obtain m polyhedra $\mathfrak{P}_s : \mathfrak{x} = \mathfrak{x}(w)$, $w \in t_s$, $s = 1, 2, \cdots, m$, and by the discussion just completed under (i) above, we have

$$(1) \qquad a(\mathfrak{P}_s) = |\Delta_s|, \qquad s = 1, 2, \cdots, m.$$

Now for $j = 1, 2, 3$, $s = 1, 2, \cdots, m$, the set $T^j(t_s - t_s^0)$ is merely the orthogonal projection of the perimeter of Δ_s upon the z_j-plane (cf. V.1.1), and thus surely $|T^j(t_s - t_s^0)| = 0$. Hence, by V.1.14 and V.1.7,

$$(2) \qquad a(\mathfrak{P}) = a(\mathfrak{P}_1) + \cdots + a(\mathfrak{P}_m).$$

(1) and (2) imply that $a(\mathfrak{P}) = |\Delta_1| + \cdots + |\Delta_m| = E(\mathfrak{x}, \mathfrak{R}, \mathfrak{J})$.

V.1.33. The result established in V.1.32 shows that $E(\mathfrak{x}, \mathfrak{R}, \mathfrak{J})$ has the same value for all typical representations of a given polyhedron $\mathfrak{P}$, namely the value $a(\mathfrak{P})$. We shall term this common value *the elementary area* $E(\mathfrak{P})$ of the polyhedron $\mathfrak{P}$. Thus we have the formula $E(\mathfrak{P}) = E(\mathfrak{x}, \mathfrak{R}, \mathfrak{J}) = a(\mathfrak{P})$ for every typical representation $(\mathfrak{x}, \mathfrak{R}, \mathfrak{J})$ of a polyhedron $\mathfrak{P}$. In view of V.1.7, it follows that the elementary area $E(\mathfrak{P})$ is a lower semi-continuous function in the class of polyhedra. That is, if $\mathfrak{P}_n \to \mathfrak{P}$ (cf. II.3.15), then $E(\mathfrak{P}) \leqq \liminf E(\mathfrak{P}_n)$.

V.1.34. The special surfaces considered in V.1.25, V.1.26, V.1.27, V.1.30 may be thought of as *elementary surfaces*. These elementary surfaces will be used extensively for purposes of approximation, and we shall develop presently various facts needed in the sequel. For technical reasons we introduce first a slightly more general class of surfaces, termed surfaces of class $\mathfrak{F}_0$. A surface S belongs to $\mathfrak{F}_0$ if and only if S admits of a representation

$$(1) \qquad S : \mathfrak{x} = \mathfrak{x}(w), \qquad w \in \mathfrak{R},$$

such that (i) $\mathfrak{x}(w)$ satisfies a Lipschitz condition in every simply-connected Jordan region $\mathfrak{R}_* \subset \mathfrak{R}^0$, and (ii) $W(w, \mathfrak{x})$ is summable in $\mathfrak{R}^0$ (cf. V.1.22 for the definition of W). Such a representation will be called a typical representation (cf. II.3.47) of the surface S of class $\mathfrak{F}_0$. Note that as a consequence of condition (i), the first partial derivatives of the components of $\mathfrak{x}(w)$ exist a.e. in $\mathfrak{R}^0$ (cf. I.3.14), and hence condition (ii) is meaningful.

Let (1) be a typical representation of a surface S of class $\mathfrak{F}_0$. Let $\mathfrak{R}_n$ be a sequence of simply-connected Jordan regions that fill up $\mathfrak{R}$ from the interior; then by V.1.11 we have

$$(2) \qquad a(S_n) \to a(S),$$

where

(3) $$S_n : \mathfrak{x} = \mathfrak{x}(w), \qquad w \in \mathfrak{R}_n$$

Clearly, (3) is a Lipschitzian representation, and hence by V.1.27

(4) $$a(S_n) = \iint_{\mathfrak{R}_n^0} W(w, \mathfrak{x}).$$

Obviously, since $W(w, \mathfrak{x})$ is summable in $\mathfrak{R}^0$,

(5) $$\iint_{\mathfrak{R}_n^0} W(w, \mathfrak{x}) \to \iint_{\mathfrak{R}^0} W(w, \mathfrak{x}).$$

(2), (4), (5) yield

(6) $$a(S) = \iint_{\mathfrak{R}^0} W(w, \mathfrak{x}).$$

V.1.35. Let us use the symbols $\{C'\}$, $\{Q\}$, $\{L\}$, $\{P\}$ to denote the class of surfaces defined in V.1.25, V.1.26, V.1.27, V.1.30 respectively. Thus $S \in \{C'\}$ means that S is of class C' in the sense of V.1.25, $S \in \{Q\}$ means that S is quasi-linear in the sense of V.1.26, and so forth. We assert that the classes $\{C'\}$, $\{Q\}$, $\{L\}$, $\{P\}$ are subclasses of $\mathfrak{F}_0$. For the first three classes this is obvious (cf. V.1.34). The inclusion $\{P\} \subset \mathfrak{F}_0$ follows from the next section.

V.1.36. *If $\mathfrak{P}$ is a polyhedron* (cf. V.1.30), *then $\mathfrak{P}$ admits of a representation $\mathfrak{P} : \mathfrak{x} = \mathfrak{x}(w)$, $w \in \mathfrak{R}$, with the following properties.* (i) *The boundary of $\mathfrak{R}$ is a convex polygon.* (ii) *The representation is quasi-linear* (see V.1.26).

Proof. By definition, $\mathfrak{P}$ admits of a polyhedral representation

(1) $$\mathfrak{P} : \mathfrak{x} = \mathfrak{x}_*(w_*), \qquad w_* \in \mathfrak{R}_* .$$

Let $t_{1*}, \cdots, t_{m*}$ be the triangles of a curvilinear triangulation $\mathfrak{J}_*$ of $\mathfrak{R}_*$ associated with the polyhedral representation (1) in the sense of V.1.29. Then each t_{i*} is mapped, by means of (1), topologically onto a (nondegenerate) plane rectilinear triangle Δ_i in $x_1x_2x_3$-space, the vertices of t_{i*} being mapped into the vertices of Δ_i. Let us denote this topological transformation from t_{i*} onto Δ_i by $\alpha_i(t_{i*}) = \Delta_i$.

By I.2.53 there exists a topological transformation τ that maps $\mathfrak{R}_*$ onto a simply-connected, bounded Jordan region $\mathfrak{R}$ in the w-plane, where $\mathfrak{R}$ is bounded by a convex polygon and each triangle t_{i*} is mapped onto a rectilinear triangle t_i, the vertices of t_{i*} being carried into the vertices of t_i. By means of τ, the curvilinear triangulation $\mathfrak{J}_*$ of $\mathfrak{R}_*$ is thus carried into a rectilinear triangulation $\mathfrak{J}$ of $\mathfrak{R}$. The representation $\mathfrak{x} = \mathfrak{x}_*\tau^{-1}(w)$, $w \in \mathfrak{R}$, being topologically similar to the representation (1), we obtain thus for $\mathfrak{P}$ the representation

(2) $$\mathfrak{P} : \mathfrak{x} = \mathfrak{x}_*\tau^{-1}(w), \qquad w \in \mathfrak{R}.$$

This representation is obviously polyhedral and may be denoted by the symbol $(\mathfrak{x}, \mathfrak{R}, \mathfrak{J})$ in the sense of V.1.29, the triangles Δ_i being the same for both representations (1) and (2). For each triangle t_i of $\mathfrak{J}$, we define the transformation $\beta_i(t_i) = \Delta_i$ as the (univocally determined) affine transformation that carries t_i into Δ_i and agrees with (2) at the vertices of t_i. Then $\beta_i^{-1}\alpha_i\tau^{-1}(w)$, $w \in t_i$, is clearly a topological transformation of t_i onto itself, each vertex of t_i being carried into itself. Furthermore, as a consequence of the definition of α_i and β_i, if t_i, t_j are any two triangles of $\mathfrak{J}$, then $\beta_i^{-1}\alpha_i\tau^{-1}(w) = \beta_j^{-1}\alpha_j\tau^{-1}(w)$ for $w \in t_i t_j$. Consequently, the transformations $\beta_i^{-1}\alpha_i\tau^{-1}(w)$, $i = 1, 2, \cdots, m$, make up a topological transformation of $\mathfrak{R}$ onto itself that we denote by $T(\mathfrak{R}) = \mathfrak{R}$. By means of T we obtain from (2) the representation (cf. II.3.11)

$$(3) \qquad \mathfrak{P} : \mathfrak{x} = \mathfrak{x}_*\tau^{-1}T^{-1}(w), \qquad w \in \mathfrak{R}.$$

Now let w_0 be any point in $\mathfrak{R}$. Then $w_0 \in t_i$ for some $i = 1, 2, \cdots, m$, and we obtain the formula $\mathfrak{x}_*\tau^{-1}T^{-1}(w_0) = \mathfrak{x}_*\tau^{-1}\tau\alpha_i^{-1}\beta_i(w_0) = \mathfrak{x}_*\alpha_i^{-1}\beta_i(w_0) = \alpha_i\alpha_i^{-1}\beta_i(w_0) = \beta_i(w_0)$. Now β_i is an affine transformation from t_i onto Δ_i. Thus the representation (3) coincides, in each one of the triangles t_i of $\mathfrak{J}$, with an affine transformation from t_i onto Δ_i. Hence this representation is quasi-linear, and the proof is complete. As a matter of fact, *the representation* (3) *is both quasi-linear and polyhedral.*

V.1.37. Let Q_0 denote the unit square $Q_0 : 0 \leqq u \leqq 1, 0 \leqq v \leqq 1$, in the $w = u + iv$ plane, and let $\mathfrak{P} : \mathfrak{x} = \mathfrak{x}_*(w)$, $w \in Q_0$, be any representation (not necessarily typical) of a polyhedron $\mathfrak{P}$. Given $\epsilon > 0$, we have then a quasi-linear representation $\mathfrak{P} : \mathfrak{x} = \mathfrak{x}^*(w)$, $w \in Q_0$, such that $| \mathfrak{x}^*(w) - \mathfrak{x}_*(w) | < \epsilon$ on Q_0 (cf. II.3.11).

PROOF. By V.1.36, $\mathfrak{P}$ admits of a quasi-linear representation $\mathfrak{P} : \mathfrak{x} = \mathfrak{x}(w)$, $w \in \mathfrak{R}$, such that $\mathfrak{R}$ is bounded by a convex polygon. Given any η, such that $0 < \eta < \epsilon/2$, we have then (see II.3.12) a topological transformation $\mathfrak{R} = \tau(Q_0)$, such that

$$(1) \qquad | \mathfrak{x}_*(w) - \mathfrak{x}[\tau(w)] | < \eta, \qquad w \in Q_0.$$

By I.2.54, we have a quasi-linear topological transformation $\mathfrak{R} = \phi(Q_0)$, such that

$$(2) \qquad | \tau(w) - \phi(w) | < \eta, \qquad w \in Q_0.$$

Since $\mathfrak{x}(w)$ is continuous, we can choose η so small that (2) implies

$$(3) \qquad | \mathfrak{x}[\tau(w)] - \mathfrak{x}[\phi(w)] | < \epsilon/2, \qquad w \in Q_0.$$

Since ϕ is topological, we have for $\mathfrak{P}$ the representation

$$(4) \qquad \mathfrak{P} : \mathfrak{x} = \mathfrak{x}[\phi(w)], \qquad w \in Q_0.$$

This representation is clearly quasi-linear, and (1) and (3) yield $| \mathfrak{x}_*(w) - \mathfrak{x}[\phi(w)] | < \epsilon/2 + \eta < \epsilon$. Thus the representation satisfies our requirements.

V.1.38. *Given $S \in \mathfrak{F}_0$, S admits of a typical representation upon the unit square* Q_0 (cf. V.1.34).

PROOF. By definition, S admits of a representation $S : \mathfrak{x} = \mathfrak{x}_*(w_*)$, $w_* \in \mathfrak{R}_*$, such that $\mathfrak{x}_*(w_*)$ satisfies a Lipschitz condition on every simply-connected Jordan region contained in $\mathfrak{R}_*^0$. Now let $\mathfrak{R}_* = \tau(Q_0)$ be a topological transformation from Q_0 onto $\mathfrak{R}_*$, such that τ is conformal on Q_0^0 (cf. I.3.19). Then we have for S the representation $S : \mathfrak{x} = \mathfrak{x}(w) = \mathfrak{x}_*[\tau(w)]$, $w \in Q_0$. Now let $\mathfrak{R}$ be any simply connected Jordan region in Q_0^0. We assert that $\mathfrak{x}_*[\tau(w)]$ satisfies a Lipschitz condition in $\mathfrak{R}$. Indeed, if this were not the case, we should have in $\mathfrak{R}$ a sequence of pairs of points w_1^n, w_2^n, such that

$$(1) \qquad \frac{|\mathfrak{x}_*[\tau(w_2^n)] - \mathfrak{x}_*[\tau(w_1^n)]|}{|w_2^n - w_1^n|} \underset{n\to\infty}{\longrightarrow} +\infty .$$

Since $\mathfrak{x}_*[\tau(w)]$ is clearly bounded in Q_0, it follows that $|w_2^n - w_1^n| \to 0$. Since we can extract convergent subsequences from the sequences w_1^n, w_2^n, we can assume without loss of generality that both sequences converge, necessarily to the same point, say

$$(2) \qquad w_2^n \to w_0 , \; w_1^n \to w_0 , \qquad w_0 \in \mathfrak{R}.$$

Let us put $\tau(w_2^n) = w_{2*}^n$, $\tau(w_1^n) = w_{1*}^n$, $\tau(w_0) = w_{0*}$. Then (1) and (2) yield

$$(3) \qquad \frac{|\mathfrak{x}_*(w_{2*}^n) - \mathfrak{x}_*(w_{1*}^n)|}{|w_{2*}^n - w_{1*}^n|} \cdot \frac{|w_{2*}^n - w_{1*}^n|}{|w_2^n - w_1^n|} \to +\infty ,$$

$$w_{2*}^n \to w_{0*} , \; w_{1*}^n \to w_{0*} , \qquad w_{0*} \in \mathfrak{R}_*^0 .$$

Now let c_* be a closed circular disc in $\mathfrak{R}_*^0$ with center at w_{0*}. Since τ is continuous, we have a closed circular disc $c \subset Q_0^0$ with center at w_0, such that $w \in c$ implies $\tau(w) \in c_*$. For n large enough, the points w_1^n, w_2^n will lie in c, and hence the points w_{1*}^n, w_{2*}^n will lie in c_*. But $\mathfrak{x}_*$ satisfies a Lipschitz condition on c_* and τ satisfies a Lipschitz condition on c. Thus (3) is clearly impossible. There remains to show that $W(w, \mathfrak{x})$ is summable on Q_0 (cf. V.1.34). Now $W(w, \mathfrak{x})$ exists a.e. in Q_0 by I.3.14, and since $a(S) < +\infty$ by V.1.34, the summability of $W(w, \mathfrak{x})$ on Q_0 follows from V.1.22.

V.1.39. *Given* $S \in \mathfrak{F}_0$ (cf. V.1.34), *and an* $\epsilon > 0$, *there exists a polyhedron* $\mathfrak{P}$ *such that* $|a(S) - E(\mathfrak{P})| < \epsilon$ *and* $d(S, \mathfrak{P}) < \epsilon$ (cf. II.3.15).

The proof will be given, for convenience, in several steps.

V.1.40. Let S be a surface that admits of a representation

$$(1) \qquad S : \mathfrak{x} = \mathfrak{x}(w), \qquad w \in R,$$

such that R is an oriented rectangle and the components of $\mathfrak{x}(w)$ are defined and have continuous first partial derivatives in a domain $\mathfrak{D}$ containing R. Given $\epsilon > 0$, there exists then a quasi-linear surface S^* (cf. V.1.26), such that $|a(S) - a(S^*)| < \epsilon$ and $d(S, S^*) < \epsilon$ (cf. II.3.15).

Proof. Let r be an oriented rectangle in R, with vertices (u, v), $(u + h, v)$, $(u + h, v + k)$, $(u, v + k)$, where $h > 0$, $k > 0$. Let A, B, C, D be the points that correspond, in $x_1x_2x_3$-space, to the vertices of r in the indicated order, by means of the representation (1). Let the symbol $|ABD|$ denote the area of the triangle ABD if this triangle is nondegenerate, and let us put $|ABD| = 0$ otherwise. The symbol $|BCD|$ is defined in a similar manner. Let us put

$$\alpha(r) = |ABD| + |BCD|. \tag{2}$$

Let $x_i(u, v)$, $i = 1, 2, 3$, be the components of the vector $\mathfrak{x}(w)$ appearing in (1). Let us put

$$\Phi_1(r) = \frac{x_2(u + h, v) - x_2(u, v)}{h} \cdot \frac{x_3(u, v + k) - x_3(u, v)}{k} - \frac{x_2(u, v + k) - x_2(u, v)}{k} \cdot \frac{x_3(u + h, v) - x_3(u, v)}{h},$$

and let us define the quantities $\Phi_2(r)$, $\Phi_3(r)$ by cyclic permutations of the subscripts 1, 2, 3. An elementary computation yields the formula

$$|ABD| = [\Phi_1(r)^2 + \Phi_2(r)^2 + \Phi_3(r)^2]^{1/2} |r|/2.$$

Now the first partial derivatives are continuous, and hence uniformly continuous, on the closed rectangle R. Using this fact, and the Lagrange mean value theorem, we obtain readily the result that for every $\eta > 0$ we have a $\delta = \delta(\eta) > 0$, such that

$$\left| \frac{|ABD|}{|r|} - \frac{1}{2} W(u, v, \mathfrak{x}) \right| < \eta \tag{3}$$

for every $r \subset R$ such that the diameter of r is less than δ (cf. V.1.22 for W). Now let σ be a subdivision of R into oriented rectangles r, and let $\| \sigma \|$ denote the maximum diameter of the rectangles of σ. From (3) it follows that

$$\left| \sum_{r \in \sigma} |ABD| - \frac{1}{2} \sum_{r \in \sigma} W(u, v, \mathfrak{x}) |r| \right| < \eta |R| \qquad \text{if } \| \sigma \| < \delta.$$

Since W is continuous on R, it follows that for any sequence of subdivisions σ_n, such that $\| \sigma_n \| \to 0$, we have

$$\sum_{r \in \sigma_n} |ABD| \underset{n \to \infty}{\longrightarrow} \frac{1}{2} \iint_R W(u, v, \mathfrak{x}). \tag{4}$$

An entirely analogous reasoning yields

$$\sum_{r \in \sigma_n} |BCD| \underset{n \to \infty}{\longrightarrow} \frac{1}{2} \iint_R W(u, v, \mathfrak{x}). \tag{5}$$

(2), (4), (5) yield

$$\sum_{r \in \sigma_n} \alpha(r) \underset{n \to \infty}{\longrightarrow} \iint_R W(u, v, \mathfrak{x}). \tag{6}$$

Now let us choose σ_n as follows: we subdivide the horizontal and vertical sides of

R into n equal parts, and draw horizontals and verticals through the points of division. Let us now subdivide each rectangle r of σ_n into two triangles by drawing the diagonal from the upper left to the lower right corner. We have then on R a (univocally determined) single-valued continuous vector-function $\mathfrak{x}_n^*(w)$ that agrees with $\mathfrak{x}(w)$ at the vertices of each one of the triangles so obtained, and whose components are linear functions of u, v in each one of these triangles. The surface S_n^* defined by the representation $S_n^* : \mathfrak{x} = \mathfrak{x}_n^*(w), w \in R$, is then quasi-linear, and clearly (cf. V.1.26)

$$a(S_n^*) = \sum_{r \in \sigma_n} \alpha(r). \tag{7}$$

In view of the continuity of $\mathfrak{x}(w)$, it follows readily that

$$\mathfrak{x}_n^*(w) \to \mathfrak{x}(w)$$

uniformly on R, and thus (cf. II.3.17)

$$d(S_n^*, S) \to 0. \tag{8}$$

By V.1.25 we have

$$a(S) = \iint_R W(u, v, \mathfrak{x}). \tag{9}$$

(7), (6), (9) imply that

$$a(S_n^*) \to a(S). \tag{10}$$

Given any $\epsilon > 0$, it follows from (8) and (10) that $| a(S) - a(S_n^*) | < \epsilon$, $d(S, S_n^*) < \epsilon$ for n sufficiently large, and the proof is complete.

V.1.41. Let S be a quasi-linear surface (see V.1.26). Given $\epsilon > 0$, there exists a polyhedron $\mathfrak{P}$ (see V.1.30, V.1.33) such that $| a(S) - E(\mathfrak{P}) | < \epsilon$, $d(S, \mathfrak{P}) < \epsilon$.

PROOF. Let

$$S : \mathfrak{x} = \mathfrak{x}(w), \qquad w \in \mathfrak{R}, \tag{1}$$

be a quasi-linear representation of S. Then the boundary of $\mathfrak{R}$ is a polygon, and $\mathfrak{R}$ admits of a triangulation $\mathfrak{I}$ such that (i) each one of the triangles $t_1, \cdots, t_m$ of $\mathfrak{I}$ is rectilinear, and (ii) the components of the vector $\mathfrak{x}(w)$ are linear functions of u, v in each one of the triangles t_i. Now let $w_1, \cdots, w_N$ be the vertices of the triangulation $\mathfrak{I}$ in any order, and let $V_1, \cdots, V_N$ be the points that correspond to $w_1, \cdots, w_N$ in $x_1x_2x_3$-space, by means of the representation (1). For each t_i let us define $\alpha(t_i)$ as follows. If the three points V that correspond to the vertices of t_i determine a nondegenerate rectilinear triangle, then $\alpha(t_i)$ is equal to the area of this triangle; otherwise $\alpha(t_i) = 0$. By V.1.26 we have then

$$a(S) = \alpha(t_1) + \cdots + \alpha(t_m). \tag{2}$$

For each positive integer n, let $V_1^n, \cdots, V_N^n$ be a system of points, in $x_1x_2x_3$-space, with the following properties. (a) The points $V_1^n, \cdots, V_N^n$ are all distinct, and no three of these points are collinear. (b) The distance between the points V_i and V_i^n is less than $1/n$ for every $i = 1, 2, \cdots, N$. These conditions are clearly compatible. Let us consider now a triangle t_i, with vertices w_{i1}, w_{i2}, w_{i3}. Then the rectilinear triangle Δ_i^n with vertices V_{i1}^n, V_{i2}^n, V_{i3}^n is nondegenerate. Let $\tau_i^n(t_i) = \Delta_i^n$ denote the (univocally determined) affine transformation that carries t_i into Δ_i^n, the vertices w_{i1}, w_{i2}, w_{i3} being carried into the vertices V_{i1}^n, V_{i2}^n, V_{i3}^n, in the indicated order. Clearly

$$(3) \qquad | \Delta_i^n | \underset{n\to\infty}{\longrightarrow} a(t_i), \qquad i = 1, 2, \cdots, m.$$

Since Δ_i^n is nondegenerate, τ_i^n is a homeomorphism. Let us define in $\mathfrak{R}$ a vector function $\mathfrak{x}_n(w)$ by the condition that the transformation $\mathfrak{x} = \mathfrak{x}_n(w)$, $w \in \mathfrak{R}$, agrees with τ_i^n in t_i, $i = 1, 2, \cdots, m$. Then the surface determined by the representation $\mathfrak{P}_n : \mathfrak{x} = \mathfrak{x}_n(w)$, $w \in \mathfrak{R}$, is clearly a polyhedron, and

$$(4) \qquad E(\mathfrak{P}_n) = | \Delta_1^n | + \cdots + | \Delta_m^n |.$$

Obviously $\mathfrak{x}_n(w) \to \mathfrak{x}(w)$ uniformly in $\mathfrak{R}$, and hence (cf. II.3.17)

$$(5) \qquad d(S, \mathfrak{P}_n) \to 0.$$

(2), (3), (4) yield

$$(6) \qquad E(\mathfrak{P}_n) \to a(S).$$

(5) and (6) imply that for given $\epsilon > 0$, the polyhedron $\mathfrak{P}_n$ satisfies the requirements $| a(S) - E(\mathfrak{P}_n) | < \epsilon$, $d(S, \mathfrak{P}_n) < \epsilon$ for n sufficiently large.

V.1.42. Let

$$(1) \qquad S : \mathfrak{x} = \mathfrak{x}(w), \qquad w \in R,$$

be a typical representation of a surface S of class C' (see V.1.25), where R is an oriented rectangle. Given $\epsilon > 0$, we have then a polyhedron $\mathfrak{P}$ such that $| a(S) - E(\mathfrak{P}) | < \epsilon$, $d(S, \mathfrak{P}) < \epsilon$.

PROOF. Let w_0 be the center of R, and R_n the oriented rectangle with center w_0 that is obtained from R by the transformation

$$(2) \qquad w^* = w_0 + \frac{w - w_0}{1 + 1/n} = \psi_n(w), \qquad w \in R.$$

The transformation (2) is a transformation by similarity with center w_0. Let us denote this transformation by $T_n : w^* = \psi_n(w)$, $w \in R$. We have then

$$| \psi_n(w) - w | = \frac{| w - w_0 |}{n + 1}, \qquad w \in R.$$

Since $\mathfrak{x}(w)$ is uniformly continuous on R, there follows the relation

$$(3) \qquad \mathfrak{x}[\psi_n(w)] \to \mathfrak{x}(w)$$

uniformly in R. Let us define the surface S_n by

(4) $$S_n : \mathfrak{x} = \mathfrak{x}(w), \qquad w \in R_n .$$

Since T_n is a homeomorphism between R and R_n, we have for S_n the representation (cf. II.3.19)

(5) $$S_n : \mathfrak{x} = \mathfrak{x}[\psi_n(w)], \qquad w \in R.$$

(3), (1), (5) imply, by II.3.17, the relation

(6) $$d(S, S_n) \to 0.$$

Clearly the representation (4) is of class C', since by assumption the representation (1) is of class C'. Hence, by V.1.25, we have

$$a(S) = \iint_R W(w, \mathfrak{x}), \qquad a(S_n) = \iint_{R_n} W(w, \mathfrak{x}).$$

Since W is summable on R by assumption, it follows that

(7) $$a(S_n) \to a(S).$$

Now give $\epsilon > 0$. By (7), (6) we have then an N such that

(8) $$| a(S) - a(S_N) | < \epsilon/3, \; d(S, S_N) < \epsilon/3.$$

Now the representation (cf. (4))

$$S_N : \mathfrak{x} = \mathfrak{x}(w), \qquad w \in R_N ,$$

satisfies the assumptions in V.1.40 (R_N and R^0 replacing the R and $\mathfrak{D}$ of V.1.40). Hence we have a quasi-linear surface S^* such that

(9) $$| a(S_N) - a(S^*) | < \epsilon/3, \; d(S_N , S^*) < \epsilon/3.$$

By V.1.41 we have a polyhedron $\mathfrak{P}$ such that

(10) $$| a(S^*) - E(\mathfrak{P}) | < \epsilon/3, \; d(S^*, \mathfrak{P}) < \epsilon/3.$$

(8), (9), (10) yield (cf. II.3.15) finally

$$| a(S) - E(\mathfrak{P}) | < \epsilon, \; d(S, \mathfrak{P}) < \epsilon.$$

V.1.43. We proceed to prove the statement in V.1.39. Since $S \in \mathfrak{F}_0$, we have by V.1.38 a typical representation of S of the form

(1) $$S : \mathfrak{x} = \mathfrak{x}(w), \qquad w \in Q_0 ,$$

where Q_0 is the unit square $0 \leqq u \leqq 1, 0 \leqq v \leqq 1$. Let the transformation T_n be defined by (cf. the argument in V.1.42)

$$T_n : w^* = w_0 + \frac{w - w_0}{1 + 1/n}, \qquad w \in Q_0 ,$$

where w_0 is the center of Q_0. Let Q^n be the image of Q_0 under T_n; then Q^n is an oriented square with center w_0, and $| Q^n | \to | Q_0 |$, $Q^n \subset Q_0^0$. Let us define the surface S_n by

(2) $$S_n : \mathfrak{x} = \mathfrak{x}(w), \qquad w \in Q^n.$$

The argument used in V.1.42 yields

(3) $$d(S, S_n) \to 0.$$

Clearly, $S_n \in \mathfrak{F}_0$, and the representation (2) is typical, since by assumption the representation (1) is typical. Hence we have, by V.1.34,

(4) $$a(S) = \iint_{Q_0} W(w, \mathfrak{x}), \qquad a(S_n) = \iint_{Q^n} W(w, \mathfrak{x}).$$

Since W is summable on Q_0, it follows that

(5) $$a(S_n) \to a(S).$$

Give now $\epsilon > 0$. Then by (3), (5) we have an N such that

(6) $$| a(S) - a(S_N) | < \epsilon/3, \; d(S, S_N) < \epsilon/3.$$

From this point on, N will be kept fixed. For the surface S_N we have by (2) the representation $S_N : \mathfrak{x} = \mathfrak{x}(w)$, $w \in Q^N$. We note that $Q^N \subset Q_0^0$, and that $\mathfrak{x}(w)$ satisfies a Lipschitz condition in Q^N. Hence there exists a finite positive constant M such that

$$W(w, \mathfrak{x}) < M \qquad \text{a.e. on } Q^N.$$

Now let $x_i(u, v)$, $i = 1, 2, 3$, be the components of $\mathfrak{x}(w)$. We introduce the integral means (cf. III.2.66)

$$x_i^h(u, v) = \frac{1}{4h^2} \int_0^h \int_0^h x_i(u + \xi, v + \eta) \, d\xi \, d\eta.$$

Since $Q^N \subset Q_0^0$, for h small enough we can define a surface S_N^h by

(7) $$S_N^h : \mathfrak{x} = \mathfrak{x}^h(w) = (x_1^h(u, v), x_2^h(u, v), x_3^h(u, v)), \qquad u + iv = w \in Q^N.$$

Since $\mathfrak{x}(w)$ satisfies a Lipschitz condition on Q^N, we have by III.2.66, III.2.67 the following facts at our disposal.

(a) The representation (7) is of class C'.

(b) We have a finite positive constant M^*, independent of h, such that $W(w, \mathfrak{x}^h) < M^*$ in Q^N.

(c) If $h \to 0$ through any sequence $h_1, h_2, \cdots$, then $W(w, \mathfrak{x}^h) \to W(w, \mathfrak{x})$ a.e. in Q^N and $\mathfrak{x}^h(w) \to \mathfrak{x}(w)$ uniformly in Q^N.

From (a) we obtain, by V.1.25,

(8) $$a(S_N^h) = \iint_{Q^N} W(w, \mathfrak{x}^h).$$

The last part of (c) yields

(9) $$d(S_N, S_N^h) \to 0 \qquad \text{if } h \to 0$$

through any sequence $h_1, h_2, \cdots$. In view of I.3.11, it follows from (b) and (c) that

(10) $$\iint_{Q^N} W(w, \mathfrak{x}^h) \to \iint_{Q^N} W(w, \mathfrak{x}) \qquad \text{if } h \to 0$$

through any sequence $h_1, h_2, \cdots$. From (8), (10), (4) we obtain the relation

(11) $$a(S_N^h) \to a(S_N) \qquad \text{if } h \to 0.$$

From (9), (11) we infer that

(12) $$|\, a(S_N) - a(S_N^h) \,| < \epsilon/3, \; d(S_N, S_N^h) < \epsilon/3,$$

if h is sufficiently small. We choose an h small enough for this purpose and keep it fixed. Now by (a) above, the surface S_N^h is of class C'. Hence, by V.1.42, there exists a polyhedron $\mathfrak{P}$ such that

(13) $$|\, a(S_N^h) - E(\mathfrak{P}) \,| < \epsilon/3, \; d(S_N^h, \mathfrak{P}) < \epsilon/3.$$

(6), (12), (13) yield the desired relations $|\, a(S) - a(\mathfrak{P}) \,| < \epsilon$, $d(S, \mathfrak{P}) < \epsilon$.

V.1.44. Let us note that the surface S_N^h satisfies the assumptions in V.1.40, and thus V.1.42 is unnecessary as far as the preceding proof is concerned. In fact, V.1.42 is obviously a special case of V.1.39. On the other hand, V.1.42 is an elementary case of independent interest, and for this reason it has been stated and proved explicitly. Similar remarks could be made on several other occasions throughout this book.

V.1.45. We proceed to derive further approximation theorems needed in the sequel. We shall first study a certain modification of a given surface S that we shall term *the stretching process.* In this process, use will be made of an auxiliary function $\lambda(\rho)$ of a real variable ρ, defined as follows. Let there be given two constants r, R such that $0 < r < R$. Then

(1) $$\lambda(\rho) = \begin{cases} 0 & \text{for } 0 \leqq \rho \leqq r, \\ \dfrac{R(\rho - r)}{\rho(R - r)} & \text{for } r \leqq \rho \leqq R, \\ 1 & \text{for } \rho \geqq R. \end{cases}$$

Then $\lambda(\rho)$ is defined and continuous for $0 \leqq \rho < +\infty$. The function $\lambda(\rho)$ depends upon the two parameters r, R; as long as r, R are kept fixed, it is needless to use a more involved notation like $\lambda(\rho, r, R)$. Except for $\rho = r$ and $\rho = R$, $\lambda(\rho)$ has a continuous derivative $\lambda'(\rho)$. We find by computation

$$\lambda'(\rho) = \begin{cases} 0 & \text{for } 0 < \rho < r, \\ \dfrac{Rr}{(R - r)\rho^2} & \text{for } r < \rho < R, \\ 0 & \text{for } \rho > R. \end{cases}$$

It follows that

$$(2) \qquad 0 < \lambda'(\rho) < \frac{R}{(R - r)r} \qquad \text{for } r < \rho < R.$$

Thus $\lambda(\rho)$ is a strictly increasing function in the interval $r \leqq \rho \leqq R$, and hence

$$(3) \qquad 0 < \lambda(\rho) < 1 \qquad \text{for } r < \rho < R.$$

From (2) we infer, for any two distinct points ρ_1, ρ_2 of the interval $r \leqq \rho \leqq R$, the inequality

$$(4) \qquad | \lambda(\rho_2) - \lambda(\rho_1) | < \frac{R}{(R - r)r} \cdot | \rho_2 - \rho_1 |.$$

Since $\lambda(\rho)$ is constant for $0 \leqq \rho \leqq r$ and for $\rho \geqq R$, it follows that (4) holds for any two points ρ_1, ρ_2 such that $\rho_1 \geqq 0$, $\rho_2 \geqq 0$, $\rho_1 \neq \rho_2$. In other words, $\lambda(\rho)$ satisfies a Lipschitz condition, with the factor

$$k = \frac{R}{(R - r)r}$$

on its whole range of definition $0 \leqq \rho < +\infty$. Let us note finally that (cf. (1), (3)) $0 \leqq \lambda(\rho) \leqq 1$ for $\rho \geqq 0$.

V.1.46. Let there be given two constants r, R such that $0 < r < R$, and a point A, with coordinates (a_1, a_2, a_3), in $x_1x_2x_3$-space. For convenience, we introduce the vector $\mathfrak{a} = (a_1, a_2, a_3)$. Given a surface S by a representation

$$(1) \qquad S : \mathfrak{x} = \mathfrak{x}(w), \qquad w \in \mathfrak{R},$$

we derive from this representation a new surface

$$(2) \qquad S^* : \mathfrak{x} = \mathfrak{x}^*(w), \qquad w \in \mathfrak{R},$$

where

$$(3) \qquad \mathfrak{x}^*(w) = \mathfrak{a} + \lambda(| \mathfrak{x}(w) - \mathfrak{a} |)(\mathfrak{x}(w) - \mathfrak{a}).$$

Since $| \mathfrak{x}(w) - \mathfrak{a} | \geqq 0$ and $\lambda(\rho)$ is continuous for $\rho \geqq 0$, clearly $\mathfrak{x}^*(w)$ is continuous on $\mathfrak{R}$. The process that leads from (1) to (2) by means of (3) will be termed *the stretching process* $\Omega(r, R, \mathfrak{a})$, the notation being meant to indicate that this process depends upon three parameters r, R, $\mathfrak{a}$. However, if r, R, $\mathfrak{a}$ remain fixed in the course of a discussion (as will be the case for the time being), we shall write merely Ω instead of $\Omega(r, R, \mathfrak{a})$. The term stretching process is justified by the following remark. Let P be the point that corresponds, in $x_1x_2x_3$-space, to a point $w \in \mathfrak{R}$ by means of the representation (1), and let P^* have the same meaning relative to the representation (2). Let AP, AP^* denote the distances from A to P, P^* respectively. Inspection shows that $P^* = P$ if $AP \geqq R$, $AP^* < AP$ if $r < AP < R$, and $P^* = A$ if $AP \leqq r$.

V.1.47. Using the terminology of V.1.46, let

$$(1) \qquad S : \mathfrak{x} = \mathfrak{x}_1(w_1), \qquad w_1 \in \mathfrak{R}_1,$$

$$(2) \qquad S : \mathfrak{x} = \mathfrak{x}_2(w_2), \qquad w_2 \in \mathfrak{R}_2,$$

be two topologically similar representations of a surface S, and let

$$(3) \qquad \mathfrak{x} = \mathfrak{x}_1^*(w_1), \qquad w_1 \in \mathfrak{R}_1,$$

$$(4) \qquad \mathfrak{x} = \mathfrak{x}_2^*(w_2), \qquad w_2 \in \mathfrak{R}_2,$$

be the representations derived from (1) and (2) respectively by the stretching process Ω. Then the representations (3) and (4) are also topologically similar.

PROOF. By assumption we have a topological transformation $\mathfrak{R}_2 = \tau(\mathfrak{R}_1)$ such that

$$(5) \qquad \mathfrak{x}_1(w_1) = \mathfrak{x}_2(\tau(w_1)), \qquad w_1 \in \mathfrak{R}_1.$$

We have then (cf. (3) in V.1.46)

$$\begin{aligned} \mathfrak{x}_1^*(w_1) &= \mathfrak{a} + \lambda(|\, \mathfrak{x}_1(w_1) - \mathfrak{a}\, |)(\mathfrak{x}_1(w_1) - \mathfrak{a}) \\ &= \mathfrak{a} + \lambda(|\, \mathfrak{x}_2(\tau(w_1)) - \mathfrak{a}\, |)(\mathfrak{x}_2(\tau(w_1)) - \mathfrak{a}) = \mathfrak{x}_2^*(\tau(w_1)). \end{aligned}$$

Thus (3) and (4) are topologically similar also, and in fact it follows, more precisely, that (5) implies $\mathfrak{x}_1^*(w_1) = \mathfrak{x}_2^*(\tau(w_1))$, $w_1 \in \mathfrak{R}_1$.

V.1.48. Let there be given a sequence of surfaces in terms of representations

$$(1) \qquad S_n : \mathfrak{x} = \mathfrak{x}_n(w), \qquad w \in \mathfrak{R},\ n = 0, 1, 2, \cdots,$$

such that $\mathfrak{x}_n(w) \to \mathfrak{x}_0(w)$ uniformly in $\mathfrak{R}$. Let

$$S_n^* : \mathfrak{x} = \mathfrak{x}_n^*(w), \qquad w \in \mathfrak{R},\ n = 0, 1, 2, \cdots,$$

be derived from (1) by the stretching process Ω. Then $\mathfrak{x}_n^*(w) \to \mathfrak{x}_0^*(w)$ uniformly in $\mathfrak{R}$.

PROOF. We have by definition

$$(2) \qquad \mathfrak{x}_n^*(w) = \mathfrak{a} + \lambda(|\, \mathfrak{x}_n(w) - \mathfrak{a}\, |)(\mathfrak{x}_n(w) - \mathfrak{a}).$$

Now $\lambda(\rho)$ is uniformly continuous on its range of definition $0 \leqq \rho < +\infty$, and $\mathfrak{x}_n(w) \to \mathfrak{x}_0(w)$ uniformly on $\mathfrak{R}$. Thus (2) clearly implies that $\mathfrak{x}_n^*(w) \to \mathfrak{x}_0^*(w)$ uniformly on $\mathfrak{R}$.

V.1.49. Let

$$(1) \qquad S : \mathfrak{x} = \mathfrak{x}_1(w_1), \qquad w_1 \in \mathfrak{R}_1 ,$$

$$(2) \qquad S : \mathfrak{x} = \mathfrak{x}_2(w_2), \qquad w_2 \in \mathfrak{R}_2 ,$$

be two representations of the same surface S. Let

$$(3) \qquad S_1^* : \mathfrak{x} = \mathfrak{x}_1^*(w_1), \qquad w_1 \in \mathfrak{R}_1 ,$$

$$(4) \qquad S_2^* : \mathfrak{x} = \mathfrak{x}_2^*(w_2), \qquad w_2 \in \mathfrak{R}_2 ,$$

be the representations derived from (1) and (2) respectively by the stretching process Ω. Then $S_1^* = S_2^*$.

PROOF. By assumption, we have for every positive integer n a topological transformation $\mathfrak{R}_2 = \tau_n(\mathfrak{R}_1)$ such that

$$(5) \qquad | \mathfrak{x}_1(w_1) - \mathfrak{x}_2(\tau_n(w_1)) | < 1/n, \qquad w_1 \in \mathfrak{R}_1 .$$

Let us put

$$(6) \qquad \mathfrak{x}_{n*}(w_1) = \mathfrak{x}_2(\tau_n(w_1)), \qquad w_1 \in \mathfrak{R}_1 ,$$

and let us consider the surface

$$(7) \qquad S_{n*} : \mathfrak{x} = \mathfrak{x}_{n*}(w_1), \qquad w_1 \in \mathfrak{R}_1 .$$

Let us apply the stretching process Ω to (7), obtaining

$$(8) \qquad S_{n*}^* : \mathfrak{x} = \mathfrak{x}_{n*}^*(w_1), \qquad w_1 \in \mathfrak{R}_1 .$$

In view of (6), the representations (2) and (7) are topologically similar, and thus $S_{n*} = S$, $n = 1, 2, \cdots$. By V.1.47, applied to (2) and (7), it follows that the representations (4) and (8) are also topologically similar. Hence

$$(9) \qquad S_{n*}^* = S_2^* , \qquad n = 1, 2, \cdots .$$

By (5) and (6), $\mathfrak{x}_{n*}(w_1) \to \mathfrak{x}_1(w_1)$ uniformly on $\mathfrak{R}_1$. Hence, by V.1.48, $\mathfrak{x}_{n*}^*(w_1) \to \mathfrak{x}_1^*(w_1)$ uniformly on $\mathfrak{R}_1$. Thus

$$(10) \qquad S_{n*}^* \to S_1^* .$$

Clearly, (9) and (10) imply that $S_1^* = S_2^*$.

V.1.50. By V.1.49 the stretching process Ω, applied to any representation of a surface S, yields a representation of a surface S^* that is independent of the particular representation chosen for S. We shall say that this surface S^* is derived from S by the stretching process Ω, and we shall write $S^* = \Omega(S)$, or $S^* = \Omega(S, r, R, \mathfrak{a})$ if explicit reference to the parameters r, R, $\mathfrak{a}$ of the stretching process Ω is desirable.

V.1.51. Using the terminology of V.1.50, we assert that $d(S, S^*) \leqq R$, where $S^* = \Omega(S) = \Omega(S, r, R, \mathfrak{a})$.

PROOF. Let $S : \mathfrak{x} = \mathfrak{x}(w)$, $w \in \mathfrak{R}$, be a representation of S. Then we have for S^* the representation

$$S^* : \mathfrak{x} = \mathfrak{x}^*(w) = \mathfrak{a} + \lambda(| \mathfrak{x}(w) - \mathfrak{a} |)(\mathfrak{x}(w) - \mathfrak{a}), \qquad w \in \mathfrak{R}.$$

Hence

$$(1) \qquad |\mathfrak{x}(w) - \mathfrak{x}^*(w)| = |\mathfrak{x}(w) - \mathfrak{a}|\cdot[1 - \lambda(|\mathfrak{x}(w) - \mathfrak{a}|)], \qquad w \in \mathfrak{R}.$$

By V.1.45 we have

$$1 - \lambda(|\mathfrak{x}(w) - \mathfrak{a}|) = 0 \qquad \text{if } |\mathfrak{x}(w) - \mathfrak{a}| > R,$$

$$0 \leqq 1 - \lambda(|\mathfrak{x}(w) - \mathfrak{a}|) \leqq 1 \qquad \text{if } |\mathfrak{x}(w) - \mathfrak{a}| \leqq R.$$

Hence, by (1), $|\mathfrak{x}(w) - \mathfrak{x}^*(w)| \leqq R$ in $\mathfrak{R}$, and thus (see II.3.18) *a fortiori* $d(S, S^*) \leqq R$.

V.1.52. Let S be a Lipschitzian surface (see V.1.27). Using the terminology of V.1.50, we assert that the surface $S^* = \Omega(S)$ is also Lipschitzian.

PROOF. Let $S : \mathfrak{x} = \mathfrak{x}(w)$, $w \in \mathfrak{R}$, be a typical representation of the Lipschitzian surface S. We have then a finite positive constant Γ such that

$$(1) \qquad |\mathfrak{x}(w_2) - \mathfrak{x}(w_1)| \leqq \Gamma\, |w_2 - w_1|$$

for every pair of points w_1, w_2 in $\mathfrak{R}$ such that the straight segment with end points w_1, w_2 is contained in $\mathfrak{R}$. Since $\mathfrak{x}(w)$ is continuous on $\mathfrak{R}$, we have further a finite constant M such that (cf. V.1.46)

$$(2) \qquad |\mathfrak{x}(w) - \mathfrak{a}| \leqq M, \qquad w \in \mathfrak{R}.$$

For S^* we have the representation

$$S^* : \mathfrak{x} = \mathfrak{x}^*(w) = \mathfrak{a} + \lambda(|\mathfrak{x}(w) - \mathfrak{a}|)(\mathfrak{x}(w) - \mathfrak{a}), \qquad w \in \mathfrak{R}.$$

We obtain therefore

$$(3) \qquad \begin{aligned} |\mathfrak{x}^*(w_2) - \mathfrak{x}^*(w_1)| &= \Big| \lambda(|\mathfrak{x}(w_2) - \mathfrak{a}|)(\mathfrak{x}(w_2) - \mathfrak{a}) \\ &\qquad - \lambda(|\mathfrak{x}(w_1) - \mathfrak{a}|)(\mathfrak{x}(w_1) - \mathfrak{a}) \Big| \\ &\leqq \lambda(|\mathfrak{x}(w_2) - \mathfrak{a}|)\, |\mathfrak{x}(w_2) - \mathfrak{x}(w_1)| + |\mathfrak{x}(w_1) - \mathfrak{a}|\, \Big| \lambda(|\mathfrak{x}(w_2) - \mathfrak{a}|) \\ &\qquad - \lambda(|\mathfrak{x}(w_1) - \mathfrak{a}|) \Big|. \end{aligned}$$

By (4) in V.1.45 we have

$$(4) \qquad \begin{aligned} &\Big| \lambda(|\mathfrak{x}(w_2) - \mathfrak{a}|) - \lambda(|\mathfrak{x}(w_1) - \mathfrak{a}|) \Big| \\ &\qquad < \frac{R}{(R - r)r} \cdot \Big|\, |\mathfrak{x}(w_2) - \mathfrak{a}| - |\mathfrak{x}(w_1) - \mathfrak{a}| \,\Big| \\ &\qquad \leqq \frac{R}{(R - r)r} \cdot |\mathfrak{x}(w_2) - \mathfrak{x}(w_1)|. \end{aligned}$$

Since $0 \leqq \lambda(\rho) \leqq 1$ for all $\rho \geqq 0$, we obtain from (1), (2), (3), (4) the inequality

$$| \mathfrak{x}^*(w_2) - \mathfrak{x}^*(w_1) | \leqq \Gamma\left(1 + \frac{MR}{(R - r)r}\right) | w_2 - w_1 |.$$

Thus the representation defining S^* is Lipschitzian, and hence S^* is Lipschitzian.

V.1.53. *Let* $\mathfrak{P}$ *be a polyhedron* (see V.1.30), *and let* $\Omega(\mathfrak{P}) = S^*$ *be the surface derived from* $\mathfrak{P}$ *by the stretching process* Ω (note that S^* is not a polyhedron in general). *Then* S^* *is a Lipschitzian surface, and*

$$a(S^*) \leqq \frac{R}{R - r} E(\mathfrak{P}).$$

The proof will be given in several steps.

V.1.54. Continuation. By V.1.36, $\mathfrak{P}$ admits of a quasi-linear representation

$$\mathfrak{P} : \mathfrak{x} = \mathfrak{x}(w), \qquad w \in \mathfrak{R}, \tag{1}$$

where $\mathfrak{R}$ is bounded by a convex polygon. As noted at the end of V.1.36, we have in fact a representation (1) that is both quasi-linear and polyhedral. That is, $\mathfrak{R}$ admits of a rectilinear triangulation $\mathfrak{J}$ such that each one of the (rectilinear) triangles t_i, $i = 1, \cdots, m$, of $\mathfrak{J}$ is carried, by means of (1), topologically into a (nondegenerate) rectilinear triangle Δ_i in $x_1x_2x_3$-space. By V.1.46, V.1.50 we have for S^* the representation

$$S^* : \mathfrak{x} = \mathfrak{x}^*(w), \qquad w \in \mathfrak{R}, \tag{2}$$

where

$$\mathfrak{x}^*(w) = \mathfrak{a} + \lambda(| \mathfrak{x}(w) - \mathfrak{a} |)(\mathfrak{x}(w) - \mathfrak{a}). \tag{3}$$

Since (1) is quasi-linear and hence Lipschitzian, by V.1.52 the representation (2) is also Lipschitzian. Writing for brevity (cf. V.1.22) $W = W(w, \mathfrak{x})$, $W^* = W(w, \mathfrak{x}^*)$, we have therefore by V.1.27, V.1.34,

$$E(\mathfrak{P}) = \iint_{\mathfrak{R}^\circ} W, \qquad a(S^*) = \iint_{\mathfrak{R}^\circ} W^*. \tag{4}$$

V.1.55. Continuation. If μ is any non-negative real number, then we shall denote by e_μ the subset of $\mathfrak{R}$ on which $| \mathfrak{x}(w) - \mathfrak{a} | = \mu$. The set e_μ is closed (possibly empty), and the set $\mathfrak{R}^0 - e_\mu$ is open. Since the representation (1) in V.1.54 is both quasi-linear and polyhedral, the set e_μ is of a very simple structure. Indeed, for each one of the triangles t_i (see V.1.54), the set $e_\mu t_i$ is carried, by means of (1) in V.1.54, into the set G_i of those points of Δ_i whose distance from the fixed point A (see V.1.46) is equal to μ. Since the correspondence between t_i and Δ_i is affine, it follows that $e_\mu t_i$ is a closed set of measure zero. Hence, for each $\mu \geqq 0$, the set e_μ is a (possibly empty) closed set of measure zero.

V.1.56. Continuation. Let us put $e = e_0 + e_r + e_R + b$, where b is the sum of the sides of the triangles t_i (see V.1.54). By V.1.55, the set e is a closed set of measure zero. On the set $\mathfrak{R}^0 - e$, the components of $\mathfrak{x}(w)$ have continuous partial derivatives of all orders, and $|\mathfrak{x}(w) - \mathfrak{a}| \neq 0, r, R$ for $w \in \mathfrak{R}^0 - e$. Since $\lambda(\rho)$ has continuous derivatives of all orders for $0 < \rho \neq r, R$, it follows that the components of $\mathfrak{x}^*(w)$ have continuous partial derivatives of all orders on $\mathfrak{R}^0 - e$. Hence, on the set $\mathfrak{R}^0 - e$ the computations which we shall carry out presently are clearly legitimate.

V.1.57. Continuation. We assume until further notice that

$$w \in \mathfrak{R}^0 - e.$$

We shall use the conventional notations in dealing with vectors. Thus $\mathfrak{x}_u$ denotes the vector function whose components are the partial u-derivatives of the components of $\mathfrak{x}$; $\mathfrak{x}_u\mathfrak{x}_v$, $\mathfrak{x}_u \times \mathfrak{x}_v$ denote the scalar and vector products of $\mathfrak{x}_u$ and $\mathfrak{x}_v$, and so forth. We put $E = \mathfrak{x}_u^2$, $F = \mathfrak{x}_u\mathfrak{x}_v$, $G = \mathfrak{x}_v^2$, with similar definitions for E^*, F^*, G^* in terms of $\mathfrak{x}^*$. We have then the well known identities $W = (EG - F^2)^{1/2}$, $W^* = (E^*G^* - F^{*2})^{1/2}$. We shall write, for brevity, λ for $\lambda(|\mathfrak{x}(w) - \mathfrak{a}|)$ and λ' for $\lambda'(|\mathfrak{x}(w) - \mathfrak{a}|)$, and put $\rho = |\mathfrak{x}(w) - \mathfrak{a}|$. By (3) in V.1.54 we have then the formulas

$$\mathfrak{x}_u^* = \lambda'\rho_u(\mathfrak{x} - \mathfrak{a}) + \lambda\mathfrak{x}_u,$$

$$\mathfrak{x}_v^* = \lambda'\rho_v(\mathfrak{x} - \mathfrak{a}) + \lambda\mathfrak{x}_v.$$

A straightforward computation yields then the formulas

$$\text{(1)} \quad \begin{aligned} W^{*2} &= \lambda^4 W^2 + \lambda'\lambda^2\rho(\lambda'\rho + 2\lambda)(\rho_u^2 G - 2\rho_u\rho_v F + \rho_v^2 E), \\ \rho_u^2 G - 2\rho_u\rho_v F + \rho_v^2 E &= (\rho_u\mathfrak{x}_v - \rho_v\mathfrak{x}_u)^2. \end{aligned}$$

By a well known identity we have

$$(\mathfrak{x} - \mathfrak{a}) \times (\mathfrak{x}_v \times \mathfrak{x}_u) = [(\mathfrak{x} - \mathfrak{a})\cdot\mathfrak{x}_u]\mathfrak{x}_v - [(\mathfrak{x} - \mathfrak{a})\cdot\mathfrak{x}_v]\mathfrak{x}_u.$$

Direct computation yields

$$\rho_u\mathfrak{x}_v - \rho_v\mathfrak{x}_u = \frac{[(\mathfrak{x} - \mathfrak{a})\cdot\mathfrak{x}_u]\mathfrak{x}_v - [(\mathfrak{x} - \mathfrak{a})\cdot\mathfrak{x}_v]\mathfrak{x}_u}{\rho}.$$

If we put, for brevity, $\mathfrak{y} = \mathfrak{x}_u \times \mathfrak{x}_v$, then the preceding formulas yield the equation

$$\text{(2)} \quad \rho_u^2 G - 2\rho_u\rho_v F + \rho_v^2 E = \frac{[(\mathfrak{x} - \mathfrak{a}) \times \mathfrak{y}]^2}{\rho^2}.$$

By the identity of Lagrange we have

$$\text{(3)} \quad [(\mathfrak{x} - \mathfrak{a}) \times \mathfrak{y}]^2 = (\mathfrak{x} - \mathfrak{a})^2\mathfrak{y}^2 - [(\mathfrak{x} - \mathfrak{a})\cdot\mathfrak{y}]^2 \leqq (\mathfrak{x} - \mathfrak{a})^2\mathfrak{y}^2 = \rho^2\mathfrak{y}^2,$$

$$\text{(4)} \quad \mathfrak{y}^2 = (\mathfrak{x}_u \times \mathfrak{x}_v)^2 = \mathfrak{x}_u^2\mathfrak{x}_v^2 - (\mathfrak{x}_u\mathfrak{x}_v)^2 = EG - F^2 = W^2.$$

Hence, by (2), (3), (4),

$$\text{(5)} \quad \rho_u^2 G - 2\rho_u\rho_v F + \rho_v^2 E \leqq W^2.$$

Since $\lambda' \geqq 0$ and $\lambda \geqq 0$ (cf. V.1.45), (1) and (5) yield

$$W^* \leqq \lambda(\lambda + \rho\lambda')W. \tag{6}$$

Let us recall that $w \in \mathfrak{R}^0 - e$ and hence $\rho = | \mathfrak{x}(w) - \mathfrak{a} | \neq 0, r, R$. Using this fact and V.1.45, we obtain readily the inequality

$$\lambda(\lambda + \rho\lambda') < \frac{R}{R - r}. \tag{7}$$

(6) and (7) yield

$$W^* \leqq \frac{R}{R - r} W. \tag{8}$$

This inequality holds, according to the preceding derivation, on the set $\mathfrak{R}^0 - e$. Since e is of measure zero, it follows that (8) holds a.e. in $\mathfrak{R}^0$. Integration yields therefore, by (4) in V.1.54, the inequality

$$a(S^*) \leqq \frac{R}{R - r} E(\mathfrak{P}),$$

and the proof of V.1.53 is complete.

V.1.58. Given a polyhedron $\mathfrak{P}$, a point A in $x_1x_2x_3$-space, and an $\epsilon > 0$, we can determine the positive constants r, R of the stretching process (see V.1.46) in such a manner that

$$| E(\mathfrak{P}) - a(S^*) | < \epsilon, \; d(\mathfrak{P}, S^*) < \epsilon, \tag{1}$$

where $S^* = \Omega(\mathfrak{P})$ (cf. V.1.50).

PROOF. The polyhedron $\mathfrak{P}$, the constant $\epsilon > 0$, and the vector $\mathfrak{a} = (a_1, a_2, a_3)$ (where a_1, a_2, a_3 are the coordinates of the point A) are fixed, while r, R are to be determined to satisfy (1). We choose a sequence of positive numbers R_n such that $R_n \to 0$. For each n, we choose

$$r_n = \frac{R_n}{n + 1}.$$

Then

$$0 < r_n < R_n, \qquad \frac{R_n}{R_n - r_n} = 1 + \frac{1}{n}.$$

Let us denote by Ω_n the stretching process with parameters r_n, R_n, $\mathfrak{a}$, and let us put $S_n^* = \Omega_n(\mathfrak{P})$. By V.1.51, V.1.53 we have then the inequalities

$$d(\mathfrak{P}, S_n^*) \leqq R_n, \; a(S_n^*) \leqq (1 + 1/n)E(\mathfrak{P}). \tag{2}$$

Thus $d(\mathfrak{P}, S_n^*) \to 0$. Hence (see V.1.33, V.1.7)

$$E(\mathfrak{P}) \leqq \liminf a(S_n^*). \tag{3}$$

(2) and (3) imply $a(S_n^*) \to E(\mathfrak{P})$. Hence, for n large enough, we shall have the desired inequalities $| E(\mathfrak{P}) - a(S_n^*) | < \epsilon$, $d(\mathfrak{P}, S_n^*) < \epsilon$.

V.1.59. The result of V.1.20 indicates the importance of eAC representations. We shall now derive a sufficient condition for a given representation

$$S : \mathfrak{x} = \mathfrak{x}(w), \qquad w \in Q_0\,, \tag{1}$$

to be eAC, where Q_0 is the unit square $Q_0 : 0 \leqq u \leqq 1,\ 0 \leqq v \leqq 1$. This sufficient condition, which is analogous to the closure theorem of IV.4.16, will be proved in a more general and satisfactory form later on. Let us assume that the representation (1) satisfies the following conditions.

(i) $a(\mathfrak{x}, Q_0) < +\infty$ (cf. V.1.15). By V.1.16, V.1.19 the essential generalized Jacobians $\mathfrak{J}_e^j(w, \mathfrak{x})$, $j = 1, 2, 3$, exist then a.e. in Q_0^0 and are summable on Q_0^0. As a consequence, the function

$$W_e(w, \mathfrak{x}) = \{[\mathfrak{J}_e^1(w, \mathfrak{x})]^2 + [\mathfrak{J}_e^2(w, \mathfrak{x})]^2 + [\mathfrak{J}_e^3(w, \mathfrak{x})]^2\}^{1/2}$$

is also summable in Q_0^0 (and hence in Q_0), and by V.1.19 we have

$$\iint_{Q_0} W_e(w, \mathfrak{x}) \leqq a(\mathfrak{x}, Q_0) = a(S).$$

(ii) There exists a sequence

$$S_n : \mathfrak{x} = \mathfrak{x}_n(w), \qquad w \in Q_0\,,\ n = 1, 2, \cdots, \tag{2}$$

such that (a) for every n the representation (2) is eAC, (b) $\mathfrak{x}_n(w) \to \mathfrak{x}(w)$ uniformly on Q_0, and (c) we have the relation

$$\iint_{Q_0} W_e(w, \mathfrak{x}_n) \to \iint_{Q_0} W_e(w, \mathfrak{x}). \tag{3}$$

Then the representation (1) is eAC.

PROOF. For each oriented rectangle $R \subset Q_0$ we put

$$\phi^j(R) = \iint_R |\, \mathfrak{J}_e^j(w, \mathfrak{x}) \,|, \qquad j = 1, 2, 3, \tag{4}$$

$$\phi_n^j(R) = \iint_R |\, \mathfrak{J}_e^j(w, \mathfrak{x}_n) \,|, \qquad j = 1, 2, 3;\ n = 1, 2, \cdots. \tag{5}$$

$$\omega(R) = \{[\phi^1(R)]^2 + [\phi^2(R)]^2 + [\phi^3(R)]^2\}^{1/2},$$

$$\omega_n(R) = \{[\phi_n^1(R)]^2 + [\phi_n^2(R)]^2 + [\phi_n^3(R)]^2\}^{1/2}.$$

By I.3.10 we have the inequality

$$\phi_n^j(R) \leqq \omega_n(R) \leqq \iint_R W_e(w, \mathfrak{x}_n) \leqq \iint_{Q_0} W_e(w, \mathfrak{x}_n).$$

By (3), (4) it follows that

$$\liminf_{n\to\infty} \iint_R | \mathfrak{J}_e^j(w, \mathfrak{x}_n) | < +\infty, \qquad j = 1, 2, 3.$$

Since the representation (2) is eAC, it follows by IV.4.11 that

$$(6) \qquad \iint_R | \mathfrak{J}_e^j(w, \mathfrak{x}) | \leqq \liminf_{n\to\infty} \iint_R | \mathfrak{J}_e^j(w, \mathfrak{x}_n) |, \qquad j = 1, 2, 3.$$

(4), (5), (6) yield

$$\phi^j(R) \leqq \liminf_{n\to\infty} \phi_n^j(R), \qquad j = 1, 2, 3.$$

By III.1.5, III.1.50 we have

$$(7) \qquad U(Q_0, \omega) = \iint_{Q_0} W_e(w, \mathfrak{x}),$$

$$(8) \qquad U(Q_0, \omega_n) = \iint_{Q_0} W_e(w, \mathfrak{x}_n).$$

(7), (8), (3) yield $U(Q_0, \omega_n) \to U(Q_0, \omega)$. Clearly the rectangle functions $\phi^j(R)$, $\phi_n^j(R)$ are additive on oriented rectangles. Thus all the assumptions of III.1.10 are satisfied, and hence

$$(9) \qquad U(Q_0, \phi_n^j) \underset{n\to\infty}{\to} U(Q_0, \phi^j), \qquad j = 1, 2, 3.$$

Clearly

$$U(Q_0, \phi_n^j) = \iint_{Q_0} | \mathfrak{J}_e^j(w, \mathfrak{x}_n) |,$$

$$U(Q_0, \phi^j) = \iint_{Q_0} | \mathfrak{J}_e^j(w, \mathfrak{x}) |.$$

Hence, by (9),

$$\iint_{Q_0} | \mathfrak{J}_e^j(w, \mathfrak{x}_n) | \to \iint_{Q_0} | \mathfrak{J}_e^j(w, \mathfrak{x}) |, \qquad j = 1, 2, 3.$$

Since the representations (2) are eAC, it follows by IV.4.16 that each one of the transformations T^j, $j = 1, 2, 3$, associated with the representation (1) in the sense of V.1.1, is eAC. Thus the representation (1) is eAC (see V.1.15).

V.1.60. Let $\mathfrak{R}$ be a fixed, bounded, simply-connected Jordan region in the w-plane, and let $\mathfrak{Z}$ be the class of all continuous transformations from $\mathfrak{R}$ into $x_1x_2x_3$-space (cf. II.2.99). Using vector notation, each $T \in \mathfrak{Z}$ can be given in the form $T : \mathfrak{x} = \mathfrak{x}(w)$, $w \in \mathfrak{R}$, where the vector function $\mathfrak{x}(w) = (x_1(w), x_2(w),$

$x_3(w))$ is single-valued and continuous on $\mathfrak{R}$. Equivalently, T may be given by the formulas

$$T : \begin{cases} x_1 = x_1(w), \\ x_2 = x_2(w), \\ x_3 = x_3(w), \end{cases} \qquad w \in \mathfrak{R}.$$

The vector function $\mathfrak{x}(w)$ determines a surface $S : \mathfrak{x} = \mathfrak{x}(w)$, $w \in \mathfrak{R}$. We define (cf. V.1.3)

$$f(T) = a(\mathfrak{x}, \mathfrak{R}).$$

We proceed to verify that $f(T)$ satisfies the conditions required in the cyclic additivity theorem of II.2.113, II.2.114. Clearly $f(T)$ is non-negative. By V.1.5 it is clear that $f(T)$ is lower semi-continuous. The further conditions stated in II.2.113, II.2.114 will be discussed in the following sections.

V.1.61. Continuation. We assert that $f(T)$ is additive with respect to continua of constancy, in the sense of II.2.105.

Proof. Let $T \in \mathfrak{J}$ be given. Let Γ be a continuum in $\mathfrak{R}$ such that $T(\Gamma)$ is a single point A. Let a_1, a_2, a_3 be the coordinates of A, and let us define a vector $\mathfrak{a}$ by the formula $\mathfrak{a} = (a_1, a_2, a_3)$. Then $\mathfrak{x}(w) = \mathfrak{a}$ for $w \in \Gamma$. Let D be a component of $\mathfrak{R} - \Gamma$. Let the vector functions $\mathfrak{x}_1(w)$, $\mathfrak{x}_2(w)$ be defined as follows:

$$\mathfrak{x}_1(w) = \begin{cases} \mathfrak{x}(w) & \text{for } w \in D, \\ \mathfrak{a} & \text{for } w \in \mathfrak{R} - D, \end{cases}$$

$$\mathfrak{x}_2(w) = \begin{cases} \mathfrak{a} & \text{for } w \in D, \\ \mathfrak{x}(w) & \text{for } w \in \mathfrak{R} - D. \end{cases}$$

We have to show (see II.2.105) that

(1) $$a(\mathfrak{x}_1, \mathfrak{R}) + a(\mathfrak{x}_2, \mathfrak{R}) = a(\mathfrak{x}, \mathfrak{R}).$$

Let us define the sets E, E_1, E_2 as follows:

(2) $$E = \Gamma,\ E_1 = \mathfrak{R} - D,\ E_2 = D + \Gamma.$$

Then E, E_1, E_2 are closed, nonempty subsets of $\mathfrak{R}$, and

$$\begin{aligned} \mathfrak{x}(w) &= \mathfrak{a} && \text{for } w \in E, \\ \mathfrak{x}_1(w) &= \mathfrak{a} && \text{for } w \in E_1, \\ \mathfrak{x}_2(w) &= \mathfrak{a} && \text{for } w \in E_2. \end{aligned}$$

By V.1.13 we have therefore the relations

$$\begin{aligned} a(\mathfrak{x}, \mathfrak{R}) &= a(\mathfrak{x}, \mathfrak{R}, E), \\ a(\mathfrak{x}_1, \mathfrak{R}) &= a(\mathfrak{x}_1, \mathfrak{R}, E_1), \\ a(\mathfrak{x}_2, \mathfrak{R}) &= a(\mathfrak{x}_2, \mathfrak{R}, E_2). \end{aligned}$$

In view of the definition of $a(\mathfrak{x}_1, \mathfrak{R}, E_1)$ (see V.1.12), this quantity depends only upon the values of $\mathfrak{x}_1(w)$ on the set $\mathfrak{R} - E_1 = D$. But $\mathfrak{x}_1(w) = \mathfrak{x}(w)$ on D, and hence $a(\mathfrak{x}_1, \mathfrak{R}, E_1) = a(\mathfrak{x}, \mathfrak{R}, E_1)$. Similarly $a(\mathfrak{x}_2, \mathfrak{R}, E_2) = a(\mathfrak{x}, \mathfrak{R}, E_2)$. Thus (1) is equivalent to

(3) $$a(\mathfrak{x}, \mathfrak{R}, E_1) + a(\mathfrak{x}, \mathfrak{R}, E_2) = a(\mathfrak{x}, \mathfrak{R}, E)$$

and hence it is sufficient to prove (3). We first show that

(4) $$a(\mathfrak{x}, \mathfrak{R}, E_1) + a(\mathfrak{x}, \mathfrak{R}, E_2) \leqq a(\mathfrak{x}, \mathfrak{R}, E).$$

Clearly (cf. V.1.2, V.1.12), (4) will be established if we show that

(5) $$\sum_s g(\mathfrak{x}, \mathfrak{D}_s^1) + \sum_t g(\mathfrak{x}, \mathfrak{D}_t^2) \leqq a(\mathfrak{x}, \mathfrak{R}, E),$$

for every choice of the disjoint domains $\mathfrak{D}_1^1, \cdots, \mathfrak{D}_s^1, \cdots$ in $\mathfrak{R} - E_1 = D$ and for every choice of the disjoint domains $\mathfrak{D}_1^2, \cdots, \mathfrak{D}_t^2, \cdots$ in $\mathfrak{R} - E_2 = \mathfrak{R} - (D + \Gamma)$. Now if two such systems are chosen, then the domains $\mathfrak{D}_1^1, \cdots, \mathfrak{D}_s^1, \cdots, \mathfrak{D}_1^2, \cdots, \mathfrak{D}_t^2, \cdots$ are disjoint since $(\mathfrak{R} - E_1)\cdot(\mathfrak{R} - E_2)$ is clearly empty, and all these domains lie in $\mathfrak{R} - E$ since $\mathfrak{R} - E_1 = D \subset \mathfrak{R} - \Gamma = \mathfrak{R} - E$, $\mathfrak{R} - E_2 = \mathfrak{R} - (D + \Gamma) \subset \mathfrak{R} - \Gamma = \mathfrak{R} - E$. Thus (5) follows from the definition of $a(\mathfrak{x}, \mathfrak{R}, E)$ (see V.1.12). Next we show that

(6) $$a(\mathfrak{x}, \mathfrak{R}, E_1) + a(\mathfrak{x}, \mathfrak{R}, E_2) \geqq a(\mathfrak{x}, \mathfrak{R}, E).$$

In view of V.1.12, it is sufficient to show that

(7) $$\sum_k g(\mathfrak{x}, \mathfrak{D}_k) \leqq a(\mathfrak{x}, \mathfrak{R}, E_1) + a(\mathfrak{x}, \mathfrak{R}, E_2)$$

for every choice of the disjoint domains $\mathfrak{D}_1, \cdots, \mathfrak{D}_k, \cdots$ in $\mathfrak{R} - E = \mathfrak{R} - \Gamma$. Now since each $\mathfrak{D}_k$ is connected, each $\mathfrak{D}_k$ is a subset of a component of $\mathfrak{R} - \Gamma$. One such component is D (see above). Let D^* be the sum of the other components of $\mathfrak{R} - \Gamma$. Of course D^* may be empty; if this happens, then clearly $\Gamma = \mathfrak{R} - D$ and hence $E = E_1$ (see (2)). Thus in this special case $a(\mathfrak{x}, \mathfrak{R}, E_1) = a(\mathfrak{x}, \mathfrak{R}, E)$ and (6) is obvious. So we can assume that $D^* \neq 0$. Then each $\mathfrak{D}_k$ is either a subset of D or a subset of D^*. Let σ, σ^* denote the sum of those terms $g(\mathfrak{x}, \mathfrak{D}_k)$ for which $\mathfrak{D}_k \subset D$ and $\mathfrak{D}_k \subset D^*$ respectively. Then

(8) $$\sum_k g(\mathfrak{x}, \mathfrak{D}_k) = \sigma + \sigma^*.$$

On the other hand, $D \subset \mathfrak{R} - E_1$ and $D^* \subset \mathfrak{R} - E_2$ (cf. (2)). Hence (cf. V.1.12)

(9) $$\sigma \leqq a(\mathfrak{x}, \mathfrak{R}, E_1), \quad \sigma^* \leqq a(\mathfrak{x}, \mathfrak{R}, E_2).$$

(8) and (9) imply (7). Thus (6) is established. (4) and (6) imply (1), and the proof is complete.

V.1.62. Using the terminology of V.1.60 and V.1.1, let us suppose that

(1) $$| T^j(\mathfrak{R}) | = 0, \qquad j = 1, 2, 3.$$

We assert then that

$$(2) \qquad f(T) = a(\mathfrak{x}, \mathfrak{R}) = 0.$$

PROOF. By IV.1.4, IV.1.50 we have

$$0 \leqq \kappa(z_j, T^j, \mathfrak{R}^0) \leqq N(z_j, T^j, \mathfrak{R}) = 0$$

for z_j not in $T^j(\mathfrak{R})$. In view of (1) it follows that

$$\kappa(z_j, T^j, \mathfrak{R}^0) = 0 \qquad \text{a.e. in the } z_j\text{-plane.}$$

If $\mathfrak{D}$ is any domain in $\mathfrak{R}^0$, then it follows that (cf. IV.1.54) $\kappa(z_j, T^j, \mathfrak{D}) = 0$ a.e. in the z_j-plane. Hence (see V.1.2) $g_j(\mathfrak{x}, \mathfrak{D}) = 0$, $j = 1, 2, 3$, for every domain $\mathfrak{D} \subset \mathfrak{R}^0$. In view of V.1.3 the assertion (2) follows.

V.1.63. Using the terminology of V.1.60, let us assume that in the monotone-light factorization of T the middle-space $\mathfrak{M}$ (cf. II.1.18, II.1.20) reduces to a simple arc. Then $f(T) = 0$.

PROOF. In view of II.1.21, we can assume that $\mathfrak{M}$ is a closed segment $\mathfrak{M}$: $0 \leqq \xi \leqq 1$. The monotone factor M of T is then given by a formula $M : \xi = \xi(w)$, $w \in \mathfrak{R}$, and the light factor L of T is given by a formula $L : \mathfrak{x} = \mathfrak{y}(\xi)$, $0 \leqq \xi \leqq 1$, where $\xi(w)$ is a continuous real scalar function of w and $\mathfrak{y}(\xi)$ is a continuous vector function of ξ, and $M(\mathfrak{R}) = \mathfrak{M}$. For each positive integer n, let $\mathfrak{y}_n(\xi)$, $0 \leqq \xi \leqq 1$, be a vector function defined as follows: (i) $\mathfrak{y}_n(\xi) = \mathfrak{y}(\xi)$ for $\xi = k/n$, $k = 0, 1, \cdots, n$, and (ii) the components of $\mathfrak{y}_n(\xi)$ are linear on each subinterval $(k - 1)/n \leqq \xi \leqq k/n$, $k = 1, 2, \cdots, n$. Let us define a surface S_n by the formula $S_n : \mathfrak{x} = \mathfrak{x}_n(w) = \mathfrak{y}_n(\xi(w))$, $w \in \mathfrak{R}$, and let us denote by T_n the corresponding transformation $T_n : \mathfrak{x} = \mathfrak{y}_n(\xi(w))$, $w \in \mathfrak{R}$. Clearly, $\mathfrak{y}_n(\xi) \to \mathfrak{y}(\xi)$ uniformly for $0 \leqq \xi \leqq 1$, and thus $\mathfrak{x}_n(w) = \mathfrak{y}_n(\xi(w)) \to \mathfrak{y}(\xi(w)) = \mathfrak{x}(w)$ uniformly on $\mathfrak{R}$. Hence (cf. V.1.5)

$$(1) \qquad a(\mathfrak{x}, \mathfrak{R}) \leqq \liminf a(\mathfrak{x}_n, \mathfrak{R}).$$

Obviously, the set $T_n(\mathfrak{R})$ consists (at most) of a finite number of straight segments in $x_1x_2x_3$-space, and may even reduce to a single point. Hence, by V.1.62,

$$(2) \qquad a(\mathfrak{x}_n, \mathfrak{R}) = 0, \qquad n = 1, 2, \cdots.$$

(1) and (2) imply that $0 = a(\mathfrak{x}, \mathfrak{R}) = f(T)$.

V.1.64. Using the terminology of V.1.60, let us assume that in the monotone-light factorization of T (cf. II.1.18) the middle space $\mathfrak{M}$ reduces to a single proper cyclic element (by II.2.91, $\mathfrak{M}$ is then either a 2-cell or a 2-sphere). We assert then that $f(T) > 0$. The proof requires various preliminary lemmas.

V.1.65. Let $\mathcal{P}$ be a Peano space (see I.2.33). Let $y_1(P)$, $y_2(P)$, $y_3(P)$ be three single-valued, real-valued, continuous functions on $\mathcal{P}$. Then we obtain on $\mathcal{P}$ a continuous vector function

$$\mathfrak{y}(P) = (y_1(P), y_2(P), y_3(P)), \qquad P \in \mathcal{P}.$$

$\mathfrak{y}(P)$ will be termed nondegenerate if it is not constant on any nondegenerate continuum on $\mathcal{P}$.

V.1.66. Continuation. We associate with $\mathfrak{y}(P)$ the following three single-valued, complex-valued, continuous functions:

$$\phi_1(P) = y_2(P) + iy_3(P),$$
$$\phi_2(P) = y_3(P) + iy_1(P),$$
$$\phi_3(P) = y_1(P) + iy_2(P).$$

V.1.67. Continuation. A continuum Γ on $\mathcal{P}$ will be termed an *indicator continuum* for $\mathfrak{y}(P)$ if at least one of the associated functions $\phi_1(P)$, $\phi_2(P)$, $\phi_3(P)$ (see V.1.66) fails to satisfy the condition (Arg, Γ) (see II.4.3). If an indicator continuum Γ is a simple closed curve, then we shall call it an *indicator curve* for $\mathfrak{y}(P)$.

V.1.68. Continuation. If there exists an indicator continuum for $\mathfrak{y}(P)$, then there also exists an indicator curve for $\mathfrak{y}(P)$.

Proof. Let Γ be an indicator continuum for $\mathfrak{y}(P)$. Then one of the associated functions $\phi_j(P)$, $j = 1, 2, 3$, say $\phi_1(P)$, fails to satisfy the condition (Arg, Γ). That is (cf. II.4.3), there exists a complex constant ζ_1 such that

(i) $\phi_1(P) - \zeta_1 \neq 0$ for $P \in \Gamma$, and

(ii) $\phi_1(P) - \zeta_1$ does not possess a single-valued continuous argument on Γ.

By II.4.18, it follows that there exists a simple closed curve C such that $\phi_1(P) - \zeta_1 \neq 0$ on C and there exists no single-valued continuous argument for $\phi_1(P) - \zeta_1$ on C. By definition (see V.1.67) C is then an indicator curve for $\mathfrak{y}(P)$.

V.1.69. Lemma. *Let $\mathfrak{y}(P)$ be a nondegenerate continuous vector function on a Peano space $\mathcal{P}$ (see V.1.65). Suppose that for every totally disconnected closed set F on $\mathcal{P}$ the set $\mathcal{P} - F$ is connected. Then there exists an indicator curve for $\mathfrak{y}(P)$ on $\mathcal{P}$.*

Proof. Since $\mathfrak{y}(P)$ is nondegenerate, it is not constant on $\mathcal{P}$. Hence at least one of the components of $\mathfrak{y}(P)$, say $y_1(P)$, is not constant on $\mathcal{P}$. Since $\mathcal{P}$ is compact, $y_1(P)$ takes on a maximum value M_1 at a point A_* and a minimum value m_1 at a point A. Then

$$(1) \qquad y_1(A) = m_1\,,\; y_1(A_*) = M_1\,,\; m_1 < M_1\,.$$

Let us put

$$(2) \qquad \xi = \frac{m_1 + M_1}{2},$$

and let us consider the set E of those points P where $y_1(P) < \xi$. Then E is a nonempty open set, and clearly (cf. I.2.14)

$$A \in E,\; A_* \in \mathcal{P} - c(E).$$

Let $\mathfrak{D}$ be the component of E that contains the point A. Then $\mathfrak{D}$ is a domain (connected open set) that contains A. Clearly (cf. I.2.15)

(3) $$y_1(P) = \xi \qquad \text{for } P \in fr(\mathfrak{D}).$$

We assert that $fr(\mathfrak{D})$ separates the points A and A_* (that is, A and A_* lie in different components of $\mathcal{P} - fr(\mathfrak{D})$). In the first place, in view of (1), (2), (3), both A and A_* lie in $\mathcal{P} - fr(\mathfrak{D})$. Suppose then that A and A_* lie in the same component G of $\mathcal{P} - fr(\mathfrak{D})$. Then we have the relations $A + A_* \subset G$, $A \in \mathfrak{D}$, $A_* \in \mathcal{P} - E \subset \mathcal{P} - \mathfrak{D}$. Thus $G\mathfrak{D} \neq 0$, $G(\mathcal{P} - \mathfrak{D}) \neq 0$. Since G is connected, it follows that $Gfr(\mathfrak{D}) \neq 0$ (see I.2.40). This is a contradiction, since $G \subset \mathcal{P} - fr(\mathfrak{D})$. Thus $fr(\mathfrak{D})$ separates the points A and A_*. We assert that $fr(\mathfrak{D})$ is not totally disconnected. Indeed, if $fr(\mathfrak{D})$ were totally disconnected, then since $fr(\mathfrak{D})$ is closed it would follow, by assumption, that $\mathcal{P} - fr(\mathfrak{D})$ is connected, in contradiction to the fact that the points A and A_* lie in different components of $\mathcal{P} - fr(\mathfrak{D})$. Thus $fr(\mathfrak{D})$ has at least one component G_0 that does not reduce to a single point. Since $y_1(P) = \xi =$ constant on G_0 (cf. (3)), and since $\mathfrak{y}(P)$ is nondegenerate by assumption, it follows that at least one of the components $y_2(P)$, $y_3(P)$, say $y_2(P)$, is not constant on G_0.

V.1.70. Continuation. Summing up, we obtained a continuum G_0 with the following properties.

(i) G_0 is a nondegenerate continuum.

(ii) $y_1(P) = \xi =$ constant on G_0.

(iii) $y_2(P)$ is not constant on G_0.

(iv) $G_0 \subset fr(\mathfrak{D})$, where $\mathfrak{D}$ is a domain on which $y_1(P) < \xi$. On the basis of these properties, an indicator continuum for $\mathfrak{y}(P)$ may be constructed as follows. In view of (i) and (iii), $y_2(P)$ takes on a minimum value m_2 at a point $B \in G_0$ and a maximum value M_2 at a point $B_* \in G_0$, where the terms minimum and maximum are relative to G_0. In view of (iii) we have then

(1) $$y_2(B) = m_2,\ y_2(B_*) = M_2,\ B \in G_0,\ B_* \in G_0, \quad m_2 < M_2,\ B \neq B_*.$$

Now let us use again the symbols $\omega(y_1(P), F)$, $\omega(y_2(P), F)$ to denote the oscillations of the functions $y_1(P)$, $y_2(P)$ respectively on a set $F \subset \mathcal{P}$. Since $y_1(P)$, $y_2(P)$ are continuous and $m_2 < M_2$, we have an $r > 0$ such that (cf. I.2.14, I.2.10)

(2) $$\omega[y_i(P), c(U(B, r))] < \frac{M_2 - m_2}{4}, \qquad i = 1, 2,$$

(3) $$\omega[y_i(P), c(U(B_*, r))] < \frac{M_2 - m_2}{4}, \qquad i = 1, 2.$$

(4) $$U(B, r) \cdot U(B_*, r) = 0.$$

Since $B, B_* \in G_0 \subset fr(\mathfrak{D})$, the sets $\mathfrak{D} \cdot U(B, r)$ and $\mathfrak{D} \cdot U(B_*, r)$ are not empty. Hence we have two points Q, Q_* such that

$$Q \in \mathfrak{D} \cdot U(B, r),\ Q_* \in \mathfrak{D} \cdot U(B_*, r).$$

In view of (4) the points Q and Q_* are distinct. Since $\mathfrak{D}$ is a domain and $\mathcal{P}$ is a Peano space, we have in $\mathfrak{D}$ a simple arc γ with end points Q, Q_* (cf. I.2.41). Since $y_1(P) < \xi$ in $\mathfrak{D}$ and γ is a compact subset of $\mathfrak{D}$, $y_1(P)$ has a maximum value ξ_0 on γ which is less than ξ. Thus

(5) $$y_1(P) \leqq \xi_0 < \xi \qquad \text{for } P \in \gamma.$$

In view of I.2.33, we can assume that the spherical neighborhoods $U(B, r)$, $U(B_*, r)$, and hence also their closures $c(U(B, r))$, $c(U(B_*, r))$, are connected. By (1) the set

(6) $$\Gamma = G_0 + c(U(B, r)) + \gamma + c(U(B_*, r))$$

is then a continuum. Let us define a complex number ζ_3 by the formula

(7) $$\zeta_3 = \frac{\xi_0 + \xi}{2} + i\frac{M_2 + m_2}{2}.$$

V.1.71. We proceed to verify that the continuum Γ, defined by (6) in V.1.70, is an indicator continuum for $\mathfrak{y}(P)$. In view of V.1.67, this fact is established if we can show that the function

(1) $$\phi_3(P) = y_1(P) + iy_2(P)$$

fails to satisfy the condition (Arg, Γ), and this latter fact is established (cf. II.4.3) if we prove the following two statements (cf. (7) in V.1.70):

(i) $\phi_3(P) - \zeta_3 \neq 0$ for $P \in \Gamma$.

(ii) $\phi_3(P) - \zeta_3$ does not possess a single-valued continuous argument on Γ.

We first verify the inequalities

(2) $$\Re[\phi_3(P) - \zeta_3] > 0 \qquad \text{on } G_0,$$

(3) $$\Im[\phi_3(P) - \zeta_3] < 0 \qquad \text{on } c[U(B, r)],$$

(4) $$\Im[\phi_3(P) - \zeta_3] > 0 \qquad \text{on } c[U(B_*, r)],$$

(5) $$\Re[\phi_3(P) - \zeta_3] < 0 \qquad \text{on } \gamma,$$

where $\Re$, $\Im$ denote real and imaginary part respectively. To verify these inequalities, observe that we have on G_0 (cf. (1), V.1.70(ii), and V.1.70(7))

$$\Re[\phi_3(P) - \zeta_3] = \xi - \frac{\xi_0 + \xi}{2} = \frac{\xi - \xi_0}{2} > 0,$$

since $\xi_0 < \xi$. On $c[U(B, r)]$ we have (cf. (1), V.1.70(7), (1), (2))

$$\Im[\phi_3(P) - \zeta_3] = y_2(P) - \frac{M_2 + m_2}{2} \leqq y_2(B) + \omega[y_2(P), c(U(B, r))] - \frac{M_2 + m_2}{2}$$

$$\leqq m_2 + \frac{M_2 - m_2}{4} - \frac{M_2 + m_2}{2} = \frac{m_2 - M_2}{4} < 0.$$

On $c[U(B_*, r)]$ we have similarly

$$\mathfrak{I}[\phi_3(P) - \zeta_3] = y_2(P) - \frac{M_2 + m_2}{2}$$

$$\geqq y_2(B_*) - \omega[y_2(P), c(U(B_*, r))] - \frac{M_2 + m_2}{2}$$

$$\geqq M_2 - \frac{M_2 - m_2}{4} - \frac{M_2 + m_2}{2} = \frac{M_2 - m_2}{4} > 0.$$

On γ we have (see (1) and V.1.70(5), (7))

$$\mathfrak{R}[\phi_3(P) - \zeta_3] = y_1(P) - \frac{\xi_0 + \xi}{2} \leqq \xi_0 - \frac{\xi_0 + \xi}{2} = \frac{\xi_0 - \xi}{2} < 0.$$

Thus (2), (3), (4), (5) are verified. In view of V.1.70(6), it follows that $\phi_3(P) - \zeta_3 \neq 0$ on Γ. Let us note the following further inequalities:

(6) $$\mathfrak{R}[\phi_3(B) - \zeta_3] > 0,$$

(7) $$\mathfrak{I}[\phi_3(B) - \zeta_3] < 0,$$

(8) $$\mathfrak{R}[\phi_3(B_*) - \zeta_3] > 0,$$

(9) $$\mathfrak{I}[\phi_3(B_*) - \zeta_3] > 0.$$

(6) and (8) follow directly from (2), since $B \in G_0$, $B_* \in G_0$, while (7) and (9) follow from (3) and (4) respectively.

Let us assume now, in contradiction with (ii) above, that $\phi_3(P) - \zeta_3$ does possess a single-valued continuous argument $\alpha(P)$ on Γ. We have then (see II.4.3)

(10) $$\phi_3(P) - \zeta_3 = | \phi_3(P) - \zeta_3 | (\cos \alpha(P) + i \sin \alpha(P)) \qquad \text{on } \Gamma.$$

In view of (6) and (7) we can assume that

(11) $$-\pi/2 < \alpha(B) < 0.$$

Consider now $\alpha(P)$ on the continuum $G_0 \subset \Gamma$. We assert that

(12) $$-\pi < \alpha(P) < \pi \qquad \text{on } G_0.$$

Indeed, these inequalities hold at one point of G_0, namely at B by (11). If these inequalities should fail to hold at some other point of G_0, then $\alpha(P)$ would assume at some point P^0 of G_0 one of the values $-\pi$, π, since $\alpha(P)$ is continuous and G_0 is a continuum (see I.2.45). At such a point P^0 we should have (cf. (10))

$$\mathfrak{R}[\phi_3(P^0) - \zeta_3] < 0, \qquad P^0 \in G_0,$$

in contradiction with (2). Thus (12) holds. In particular it follows that

(13) $$-\pi < \alpha(B_*) < \pi.$$

In view of (8), (9), (10), it follows from (13) that

$$(14) \qquad 0 < \alpha(B_*) < \pi/2.$$

Consider now $\alpha(P)$ on the continuum $c[U(B, r)] + \gamma + c[U(B_*, r)]$. We assert that

$$(15) \qquad -2\pi < \alpha(P) < 0 \qquad \text{on } c[U(B, r)] + \gamma + c[U(B_*, r)].$$

Indeed, (15) holds at one point of this continuum, namely at B by (11). Therefore if (15) should fail to hold, then $\alpha(P)$ would assume on this continuum one of the values $0, -2\pi$, which is however impossible by (10), (3), (4), (5). Thus (15) holds. It follows in particular that

$$(16) \qquad -2\pi < \alpha(B_*) < 0.$$

Since (14) and (16) contradict each other, it is established that $\phi_3(P) - \zeta_3$ does not possess a single-valued continuous argument on Γ, and the proof of the lemma stated in V.1.69 is complete (see V.1.68).

V.1.72. Now let us consider a representation $S : \mathfrak{x} = \mathfrak{x}(w)$, $w \in \mathfrak{R}$, and (cf. V.1.60) the corresponding transformation $T : \mathfrak{x} = \mathfrak{x}(w)$, $w \in \mathfrak{R}$. Let us consider further a monotone-light factorization $T : \mathfrak{x} = LM(w)$, $w \in \mathfrak{R}$. Let us put $M(\mathfrak{R}) = \mathfrak{M}$. Then the middle space $\mathfrak{M}$ is a Peano space, and the light factor L is a continuous transformation from $\mathfrak{M}$ into $x_1x_2x_3$-space. Hence L can be represented in the form (cf. V.1.65)

$$L : \mathfrak{x} = \mathfrak{y}(P), \qquad P \in \mathfrak{M}.$$

Suppose that there exists an indicator continuum for $\mathfrak{y}(P)$ on $\mathfrak{M}$. Then we assert that there exists an indicator curve for $\mathfrak{x}(w)$ on $\mathfrak{R}$ (cf. V.1.67).

Proof. Let $y_1(P)$, $y_2(P)$, $y_3(P)$ be the components of $\mathfrak{y}(P)$. Then the components $x_1(w)$, $x_2(w)$, $x_3(w)$ of $\mathfrak{x}(w)$ are given by the formulas

$$x_j(w) = y_j[M(w)], \qquad j = 1, 2, 3.$$

Let us put (cf. V.1.66, V.1.1)

$$\phi_1(P) = y_2(P) + iy_3(P), \qquad f^1(w) = x_2(w) + ix_3(w),$$

$$\phi_2(P) = y_3(P) + iy_1(P), \qquad f^2(w) = x_3(w) + ix_1(w),$$

$$\phi_3(P) = y_1(P) + iy_2(P), \qquad f^3(w) = x_1(w) + ix_2(w).$$

Clearly

$$(1) \qquad f^j(w) = \phi_j[M(w)], \qquad j = 1, 2, 3.$$

By assumption there exists an indicator continuum Γ for $\mathfrak{y}(P)$ on $\mathfrak{M}$. Then one of the functions $\phi_1(P)$, $\phi_2(P)$, $\phi_3(P)$, say $\phi_1(P)$, fails to satisfy the condition (Arg, Γ). Hence there exists a complex number ζ_1 such that

(i) $\phi_1(P) - \zeta_1 \neq 0$ on Γ.

(ii) $\phi_1(P) - \zeta_1$ does not possess a single-valued continuous argument on Γ.

Now let us put $\gamma = M^{-1}(\Gamma)$. By II.1.2, γ is then a continuum. By (1) and (i)

$$f^1(w) - \zeta_1 \neq 0 \qquad \text{for } w \in \gamma. \tag{2}$$

Suppose now that $f^1(w) - \zeta_1$ possesses a single-valued continuous argument $\alpha(w)$ on γ. We assert that the formula

$$\alpha^*(P) = \alpha(M^{-1}(P)), \qquad P \in \Gamma, \tag{3}$$

defines a single-valued continuous argument $\alpha^*(P)$ of $\phi_1(P) - \zeta_1$ on Γ. Let us first show that $\alpha^*(P)$ is single-valued. Clearly this fact is established if we show that

$$\alpha(w') = \alpha(w'') \text{ if } M(w') = M(w'') = P_0 \in \Gamma. \tag{4}$$

Now $M^{-1}(P_0)$ is a continuum $\gamma_0 \subset M^{-1}(\Gamma) = \gamma$, and $M(\gamma_0) = P_0$. Hence, by (1), $f^1(w)$ is constant on γ_0. Hence (cf. II.4.7) $\alpha(w)$ is also constant on γ_0, and consequently $\alpha(w') = \alpha(w'')$, since $w', w'' \in \gamma_0 = M^{-1}(P_0)$ by (4). From (1) it follows now that $\alpha^*(P)$ is an argument of $\phi_1(P) - \zeta_1$ for every $P \in \Gamma$. The continuity of $\alpha^*(P)$ follows readily from (3) once we know that $\alpha^*(P)$ is single-valued on Γ. Thus $\alpha^*(P)$ would be a single-valued continuous argument of $\phi_1(P) - \zeta_1$ on Γ, in contradiction to (ii). Thus it follows that $f^1(w) - \zeta_1$ cannot possess a single-valued continuous argument on $\gamma = M^{-1}(\Gamma)$. In view of (2) this shows that γ is an indicator continuum for $\mathfrak{x}(w)$ on $\mathfrak{R}$. The existence of an indicator curve for $\mathfrak{x}(w)$ on $\mathfrak{R}$ follows now by V.1.68.

V.1.73. Given a representation $S : \mathfrak{x} = \mathfrak{x}(w)$, $w \in \mathfrak{R}$, suppose that there exists an indicator curve for $\mathfrak{x}(w)$ in $\mathfrak{R}$ (cf. V.1.67). Then $a(\mathfrak{x}, \mathfrak{R}) > 0$ (cf. V.1.3).

Proof. Let C be an indicator curve for $\mathfrak{x}(w)$ on $\mathfrak{R}$. Then one of the functions $f^r(w)$ (see V.1.1) fails to satisfy the condition (Arg, C). Suppose that this is the case for $f^1(w)$. Then there exists a complex number ζ_1, such that $f^1(w) - \zeta_1 \neq 0$ on C, and $f^1(w) - \zeta_1$ does not possess a single-valued continuous argument on C. Now C is a simple closed curve in $\mathfrak{R}$, and bounds a simply-connected Jordan region $\mathfrak{R}^* \subset \mathfrak{R}$. Clearly (cf. II.4.34, II.4.25, IV.1.24, V.1.1), $\mu(\zeta_1, T^1, \mathfrak{R}^*) \neq 0$. Hence (see IV.1.26), $\kappa(\zeta_1, T^1, \mathfrak{R}^*) > 0$, and thus *a fortiori* $\kappa(\zeta_1, T^1, \mathfrak{R}^0) = \kappa(\zeta_1, T^1, \mathfrak{R}) > 0$ (see IV.1.50, IV.1.14). Since $\kappa(z_1, T^1, \mathfrak{R}^0)$ is a lower semi-continuous function of z_1 (see IV.1.51), it follows that $\kappa(z_1, T^1, \mathfrak{R}^0) > 0$ on some open set containing ζ_1. Hence (cf. V.1.2), $g_1(\mathfrak{x}, \mathfrak{R}^0) > 0$. By V.1.3 it follows that $a(\mathfrak{x}, \mathfrak{R}) > 0$.

V.1.74. We are now ready to prove the statement in V.1.64. So let us suppose that in the monotone-light factorization

$$T : \mathfrak{x} = \mathfrak{x}(w) = LM(w), \qquad w \in \mathfrak{R}, \ M(\mathfrak{R}) = \mathfrak{M},$$

the middle space $\mathfrak{M}$ is a single proper cyclic element. By II.2.91, $\mathfrak{M}$ is then either a topological 2-cell or a topological 2-sphere. In either case (see I.2.32) for every totally disconnected closed set $F \subset \mathfrak{M}$ the set $\mathfrak{M} - F$ is connected. Hence, for the light factor

$$L : \mathfrak{x} = \mathfrak{y}(P), \qquad P \in \mathfrak{M},$$

there follows from V.1.69 the existence of an indicator continuum Γ. By V.1.72 there follows the existence of an indicator curve for $\mathfrak{x}(w)$ in $\mathfrak{R}$, and finally V.1.73 yields the result $f(T) = a(\mathfrak{x}, \mathfrak{R}) > 0$.

V.1.75. By II.2.92, V.1.60, V.1.61, V.1.62, V.1.63, V.1.64 the function $f(T)$ of V.1.60 satisfies all the conditions stated in II.2.113, II.2.114. In view of V.1.7, we thus obtain the following statements. Given a surface S by a representation

$$S : \mathfrak{x} = \mathfrak{x}(w), \qquad w \in \mathfrak{R},$$

let $\mathfrak{M}$ denote the middle space that arises in the monotone-light factorization of the corresponding transformation T (cf. V.1.60). By II.1.28, $\mathfrak{M}$ is determined by S up to a homeomorphism.

CHARACTERIZATION THEOREM FOR SURFACES S OF ZERO LOWER AREA $a(S)$. *The lower area $a(S)$ is equal to zero if and only if the middle space $\mathfrak{M}$ reduces to a dendrite.*

CYCLIC ADDITIVITY THEOREM FOR THE LOWER AREA $a(S)$. *Suppose that $a(S) > 0$. Then the middle space $\mathfrak{M}$ has at least one proper cyclic element, and hence the cyclic decomposition $\Delta(S)$ of S is not vacuous (cf.* II.3.20). *Let $\Delta(S) = [S_1, \cdots]$ be the cyclic decomposition of S. Then*

$$a(S) = \sum_n a(S_n).$$

If the series on the right diverges, then the formula is understood to mean that $a(S) = +\infty$.

CHAPTER V.2. THE LEBESGUE AREA $A(S)$.

V.2.1. Given a surface S (cf. II.3.44), there exists a sequence S_n of quasi-linear surfaces such that $S_n \to S$ (cf. V.1.26, II.3.15).

PROOF. By II.3.11, we have for S a representation $S : \mathfrak{x} = \mathfrak{x}(w)$, $w \in Q_0$, where Q_0 is the unit square $0 \leqq u \leqq 1$, $0 \leqq v \leqq 1$. For each positive integer n, let us choose a rectilinear triangulation $\mathfrak{T}^n$ (cf. V.1.28); that is, the triangles t_1^n, $\cdots$, $t_{m_n}^n$ of $\mathfrak{T}^n$ are rectilinear (nondegenerate) triangles. Let us denote by $\| \mathfrak{T}^n \|$ the maximum side-length of the triangles of $\mathfrak{T}^n$. The quantity $\| \mathfrak{T}^n \|$ may be termed the *norm* of $\mathfrak{T}^n$. If $\epsilon_n > 0$ is a sequence such that $\epsilon_n \to 0$, then we can choose $\mathfrak{T}^n$ subject to the condition

$$(1) \qquad \| \mathfrak{T}^n \| < \epsilon_n .$$

Let $\mathfrak{x}_n(w)$ be the (univocally determined) quasi-linear vector function in Q_0 that agrees with $\mathfrak{x}(w)$ at the vertices of $\mathfrak{T}^n$ and whose components are linear functions of u, v (where $u + iv = w$) in each one of the triangles t_i^n of $\mathfrak{T}^n$. Clearly (1) implies that $\mathfrak{x}_n(w) \to \mathfrak{x}(w)$ uniformly in Q_0. Hence, if we define a surface S_n by the representation

$$S_n : \mathfrak{x} = \mathfrak{x}_n(w), \qquad w \in Q_0 ,$$

then $S_n \to S$ (cf. II.3.17), and clearly S_n is quasi-linear.

V.2.2. Given a surface S (cf. II.3.44), there exists a sequence of polyhedra $\mathfrak{P}_n$ such that $\mathfrak{P}_n \to S$ (cf. V.1.30, II.3.15).

PROOF. By V.2.1, we have a sequence S_n of quasi-linear surfaces such that $S_n \to S$. By V.1.35, V.1.39 we have for each n a polyhedron $\mathfrak{P}_n$ such that $d(S_n, \mathfrak{P}_n) < 1/n$. Clearly $\mathfrak{P}_n \to S$.

V.2.3. Given a surface S, we have by V.2.2 a sequence of polyhedra $\mathfrak{P}_n$ such that $\mathfrak{P}_n \to S$. We define a quantity $A(S)$ by the formula (cf. V.1.33)

$$A(S) = \text{gr.l.b.} \liminf E(\mathfrak{P}_n),$$

where the greatest lower bound is taken with respect to all sequences of polyhedra $\mathfrak{P}_n$ such that $\mathfrak{P}_n \to S$. The quantity $A(S)$ will be termed *the Lebesgue area* $A(S)$ *of* S. Of course, $A(S)$ *may be infinite.* Since no particular representation of S was used in defining $A(S)$, this quantity is a function of S and is independent of the particular representation in terms of which S may be given.

V.2.4. CONTINUATION. There exists a sequence of polyhedra $\mathfrak{P}_n$ such that

$$(1) \qquad \mathfrak{P}_n \to S \text{ and } E(\mathfrak{P}_n) \to A(S).$$

PROOF. If $A(S) = +\infty$, then clearly (1) holds for every sequence $\mathfrak{P}_n$ such that $\mathfrak{P}_n \to S$. So we can assume that $A(S) < +\infty$. Given then any positive integer n, we have by definition a sequence of polyhedra $\mathfrak{P}'$ such that for $j \to \infty$

$$\mathfrak{P}' \to S \text{ and } \liminf E(\mathfrak{P}') < A(S) + 1/n.$$

Hence we have a j such that

$$d(\mathfrak{P}', S) < 1/n \text{ and } E(\mathfrak{P}') < A(S) + 1/n.$$

This j depends upon n. Let us denote the corresponding polyhedron $\mathfrak{P}'$ by $\mathfrak{P}_n$. We have then the inequalities

$$d(\mathfrak{P}_n, S) < 1/n, \ E(\mathfrak{P}_n) < A(S) + 1/n.$$

It follows that for $n \to \infty$

(2) $$\mathfrak{P}_n \to S, \ \limsup E(\mathfrak{P}_n) \leqq A(S).$$

On the other hand, by the definition of $A(S)$,

(3) $$A(S) \leqq \liminf E(\mathfrak{P}_n).$$

(2) and (3) imply (1).

V.2.5. CONTINUATION. If $A(S) < +\infty$, then for every $\epsilon > 0$ we have a polyhedron $\mathfrak{P}$ such that $d(S, \mathfrak{P}) < \epsilon$, $| A(S) - E(\mathfrak{P}) | < \epsilon$. This is an immediate consequence of V.2.4.

V.2.6. CONTINUATION. $A(S)$ is a lower semi-continuous funotion of S. That is, $S_n \to S$ implies that $A(S) \leqq \liminf A(S_n)$.

PROOF. If $\liminf A(S_n) = +\infty$, then the assertion is obvious. So we can assume that $\liminf A(S_n) < +\infty$. Then the lim inf is unchanged if we discard those terms S_n for which $A(S_n) = +\infty$. Thus we can assume, without loss of generality, that $A(S_n) < +\infty$ for every n. By V.2.5 we have then for every n a polyhedron $\mathfrak{P}_n$ such that

$$d(S_n, \mathfrak{P}_n) < 1/n, \ | A(S_n) - E(\mathfrak{P}_n) | < 1/n.$$

Clearly $\mathfrak{P}_n \to S$, and

(1) $$\liminf E(\mathfrak{P}_n) = \liminf A(S_n).$$

By the definition of $A(S)$ we have

(2) $$A(S) \leqq \liminf E(\mathfrak{P}_n).$$

(1) and (2) yield the desired relation $A(S) \leqq \liminf A(S_n)$.

V.2.7. Given a surface S, there exists a sequence of quasi-linear surfaces S_n such that $S_n \to S$ (cf. V.2.1). Now for quasi-linear surfaces the lower area coincides with what may be referred to as the expected value of the area (see V.1.26). There arises the question: What happens if we replace, in the definition of the Lebesgue area $A(S)$, polyhedra by quasi-linear surfaces? Let us define (cf. V.1.7)

$$A_*(S) = \text{gr.l.b.} \liminf a(S_n),$$

where the greatest lower bound is taken with respect to all sequences S_n of quasi-linear surfaces such that $S_n \to S$. Since the class of polyhedra is comprised in the class of quasi-linear surfaces (see V.1.36), it is clear that

(1) $$A_*(S) \leqq A(S).$$

Now a simple argument, entirely analogous to that used in V.2.4, shows that there exists a sequence of quasi-linear surfaces S_n such that

(2) $$S_n \to S \text{ and } a(S_n) \to A_*(S).$$

By V.1.41 we have for each n a polyhedron $\mathfrak{P}_n$ such that

(3) $$d(S_n, \mathfrak{P}_n) < 1/n, \ | a(S_n) - E(\mathfrak{P}_n) | < 1/n.$$

(2), (3) imply that $\mathfrak{P}_n \to S$, $E(\mathfrak{P}_n) \to A_*(S)$. Hence, by the definition of $A(S)$,

(4) $$A(S) \leqq \liminf E(\mathfrak{P}_n) = A_*(S).$$

(1) and (4) yield $A(S) = A_*(S)$. In other words, $A(S)$ remains unchanged if in its definition we replace polyhedra by quasi-linear surfaces (the elementary area of polyhedra being replaced by the lower area of quasi-linear surfaces).

V.2.8. The Lebesgue area $A(S)$ is independent of the choice of the Cartesian coordinate system $x_1x_2x_3$, or equivalently, congruent surfaces (see II.3.42) have equal Lebesgue areas.

Proof. Let a surface S be given by a representation

$$S : \mathfrak{x} = [x_1(w), x_2(w), x_3(w)], \qquad w \in \mathfrak{R}.$$

If $\overline{S}$ is a surface congruent to S, then for $\overline{S}$ we have a representation (cf. II.3.42) $\overline{S} : \mathfrak{x} = [\overline{x}_1(w), \overline{x}_2(w), \overline{x}_3(w)]$, $w \in \mathfrak{R}$, where $x_j(w)$, $\overline{x}_j(w)$, $j = 1, 2, 3$, are related by equations of the form

$$\overline{x}_j(w) = c_j + \sum_{k=1}^{3} c_{jk} x_k(w), \qquad j = 1, 2, 3,$$

the constants c_j, c_{jk} being real and the matrix $\| c_{jk} \|$ being orthogonal. Using the fact that $A(S) = A_*(S)$, $A(\overline{S}) = A_*(\overline{S})$, established in V.2.7, there follows the existence of a sequence of quasi-linear surfaces S^n such that

(1) $$S^n \to S, \ a(S^n) \to A(S).$$

For each n, let

$$S^n : \mathfrak{x} = [y_1^n(w), y_2^n(w), y_3^n(w)], \qquad w \in \mathfrak{R}^n,$$

be a quasi-linear representation of S^n (this implies of course that $\mathfrak{R}^n$ is bounded by a simple closed polygon). Let us define $\overline{S}^n$ by the representation

$$\overline{S}^n : \mathfrak{x} = [\overline{y}_1^n(w), \overline{y}_2^n(w), \overline{y}_3^n(w)], \qquad w \in \mathfrak{R}^n,$$

where

$$\overline{y}_j^n(w) = c_j + \sum_{k=1}^{3} c_{jk} y_k^n(w), \qquad j = 1, 2, 3.$$

In view of the elementary interpretation of the lower area of a quasi-linear surface (see V.1.26), obviously

(2) $$a(S^n) = a(\overline{S}^n).$$

By II.3.42 we have the relation

(3) $$d(S, S^n) = d(\overline{S}, \overline{S}^n).$$

Hence, by (1), (2), (3), and in view of V.2.7,

$$\overline{S}^n \to \overline{S},\ a(\overline{S}^n) \to A(S) = A_*(S).$$

Hence, by the definition of A_*, $A(S) = A_*(S) = \lim a(\overline{S}^n) \geqq A_*(\overline{S}) = A(\overline{S})$. Thus $A(S) \geqq A(\overline{S})$. An analogous argument yields the complementary inequality $A(\overline{S}) \geqq A(S)$. Hence $A(S) = A(\overline{S})$.

V.2.9. If $\mathfrak{P}$ is a polyhedron, then $E(\mathfrak{P}) = A(\mathfrak{P})$ (cf. V.1.33).

Proof. By V.2.4 we have a sequence of polyhedra $\mathfrak{P}_n$ such that $\mathfrak{P}_n \to \mathfrak{P}$, $E(\mathfrak{P}_n) \to A(\mathfrak{P})$. By V.1.33 it follows that

(1) $$E(\mathfrak{P}) \leqq \lim\inf E(\mathfrak{P}_n) = A(\mathfrak{P}).$$

Let us put $\mathfrak{P}_n^* = \mathfrak{P}$, $n = 1, 2, \cdots$. Then $\mathfrak{P}_n^* \to \mathfrak{P}$, and hence by the definition of $A(\mathfrak{P})$

(2) $$A(\mathfrak{P}) \leqq \lim\inf E(\mathfrak{P}_n^*).$$

Since $E(\mathfrak{P}_n^*) = E(\mathfrak{P})$, (1) and (2) imply that $E(\mathfrak{P}) = A(\mathfrak{P})$.

V.2.10. *For every surface S, we have the inequality $a(S) \leqq A(S)$.*

Proof. By V.2.4, we have a sequence of polyhedra $\mathfrak{P}_n$ such that

(1) $$\mathfrak{P}_n \to S,\ E(\mathfrak{P}_n) \to A(S).$$

By V.1.33

(2) $$E(\mathfrak{P}_n) = a(\mathfrak{P}_n).$$

By V.1.7

(3) $$a(S) \leqq \lim\inf a(\mathfrak{P}_n).$$

(1), (2), (3) yield $a(S) \leqq A(S)$.

V.2.11. If $S \in \mathfrak{F}_0$ (cf. V.1.34), then $a(S) = A(S)$.

Proof. By V.1.39 there exists, for every positive integer n, a polyhedron $\mathfrak{P}_n$ such that

$$d(S, \mathfrak{P}_n) < 1/n,\ | a(S) - E(\mathfrak{P}_n) | < 1/n,$$

and hence we have the relations $\mathfrak{P}_n \to S$, $E(\mathfrak{P}_n) \to a(S)$. By the definition of $A(S)$ we have therefore

(1) $$A(S) \leqq \lim\inf E(\mathfrak{P}_n) = a(S).$$

By V.2.10

(2) $$a(S) \leqq A(S).$$

(1) and (2) yield $a(S) = A(S)$.

V.2.12. If $S \in \mathfrak{F}_0$ (cf. V.1.34), then by V.2.11 and V.1.34 we have

$$A(S) = a(S) = \iint_{\mathfrak{R}^\circ} W(w, \mathfrak{x}),$$

for every typical representation $S : \mathfrak{x} = \mathfrak{x}(w)$, $w \in \mathfrak{R}$. In view of V.1.35, *this result may be interpreted to mean that in all elementary (or familiar) cases the areas* $a(S)$, $A(S)$ *coincide with the expected value of the area.*

V.2.13. *Given a surface* S *by a representation* $S : \mathfrak{x} = \mathfrak{x}(w)$, $w \in \mathfrak{R}$, *suppose that* $A(S) < +\infty$. *Then* (cf. V.1.20 for the definition of W_e)

(1) $$\iint_{\mathfrak{R}^\circ} W_e(w, \mathfrak{x}) \leqq a(S) \leqq A(S).$$

Proof. By V.2.10 we have $a(S) < +\infty$, and thus (1) is a direct consequence of V.1.19 and V.2.10.

V.2.14. Continuation. Suppose further that the ordinary Jacobians $J^j(w, \mathfrak{x})$, $j = 1, 2, 3$, exist a.e. in $\mathfrak{R}^0$ (cf. V.1.18). Then

$$\iint_{\mathfrak{R}^\circ} W(w, \mathfrak{x}) \leqq \iint_{\mathfrak{R}^\circ} W_e(w, \mathfrak{x}) \leqq a(S) \leqq A(S).$$

This is a direct consequence of V.2.13 and V.1.22.

V.2.15. Given a surface S by a representation

(1) $$S : \mathfrak{x} = \mathfrak{x}(w), \qquad w \in \mathfrak{R},$$

let $\mathfrak{R}_1, \cdots, \mathfrak{R}_m$ be any finite system of simply-connected Jordan regions in $\mathfrak{R}$, such that

(2) $$\mathfrak{R}_i^0 \mathfrak{R}_j^0 = 0 \qquad \text{for } i \neq j.$$

Let us define S_i by the representation

(3) $$S_i : \mathfrak{x} = \mathfrak{x}(w), \qquad w \in \mathfrak{R}_i, \ i = 1, 2, \cdots, m.$$

Then

(4) $$A(S_1) + \cdots + A(S_m) \leqq A(S).$$

Proof. *Case* (i). $S \in \mathfrak{F}_0$ and (1) is a typical representation of S (cf. V.1.34). Clearly, $S_i \in \mathfrak{F}_0$ and (3) is a typical representation of S_i. Hence, by V.2.12,

(5) $$A(S) = \iint_{\mathfrak{R}^\circ} W(w, \mathfrak{x}), \qquad A(S_i) = \iint_{\mathfrak{R}_i^\circ} W(w, \mathfrak{x}).$$

Thus (4) follows directly from (2) and (5).

Case (ii). $S \in \mathfrak{F}_0$, but the representation (1), instead of being typical, is merely topologically similar to a typical representation

$$(6) \qquad S : \mathfrak{x} = \mathfrak{x}_*(w_*), \qquad w_* \in \mathfrak{R}_* .$$

Then (cf. II.1.26) there exists a topological transformation $\tau(\mathfrak{R}) = \mathfrak{R}_*$, such that $\mathfrak{x}(w) = \mathfrak{x}_*[\tau(w)]$, $w \in \mathfrak{R}$. Let us put $\mathfrak{R}_{i*} = \tau(\mathfrak{R}_i)$, $i = 1, 2, \cdots, m$. Then we have for S_i the representation (cf. II.3.11) $S_i : \mathfrak{x} = \mathfrak{x}_*(w_*)$, $w_* \in \mathfrak{R}_{i*}$, $i = 1, 2, \cdots, m$. Since clearly $\mathfrak{R}^0_{i*}\mathfrak{R}^0_{j*} = 0$ for $i \neq j$ by (2), and since (6) is a typical representation, (4) follows by case (i).

Case (iii). *General Case.* By V.2.4 we have a sequence of polyhedra $\mathfrak{P}_n$, such that $\mathfrak{P}_n \to S$, $E(\mathfrak{P}_n) \to A(S)$. By V.1.36 we have for $\mathfrak{P}_n$ a representation

$$(7) \qquad \mathfrak{P}_n : \mathfrak{x} = \mathfrak{x}_{n*}(w_{n*}), \qquad w_{n*} \in \mathfrak{R}_{n*},$$

such that $\mathfrak{R}_{n*}$ is a convex polygon and the representation is quasi-linear. Since $\mathfrak{P}_n \to S$, we have by II.3.16 for each n a representation

$$(8) \qquad \mathfrak{P}_n : \mathfrak{x} = \mathfrak{x}_n(w), \qquad w \in \mathfrak{R},$$

such that (7) and (8) are topologically similar and $\mathfrak{x}_n(w) \to \mathfrak{x}(w)$ uniformly on $\mathfrak{R}$. Let us define

$$\bar{S}^n_i : \mathfrak{x} = \mathfrak{x}_n(w), \qquad w \in \mathfrak{R}_i, \; i = 1, 2, \cdots, m.$$

Then

$$(9) \qquad \bar{S}^n_i \to S_i \qquad \text{for } n \to \infty.$$

Since the representation (7) is quasi-linear, it is clearly a typical representation of $\mathfrak{P}_n$ considered as a surface of class $\mathfrak{F}_0$ (cf. V.1.34, V.1.26). Hence, by case (ii),

$$(10) \qquad A(\bar{S}^n_1) + \cdots + A(\bar{S}^n_m) \leqq A(\mathfrak{P}_n) = E(\mathfrak{P}_n).$$

In view of (9) we have (cf. V.2.6)

$$(11) \qquad A(S_i) \leqq \liminf_{n \to \infty} A(\bar{S}^n_i), \qquad i = 1, 2, \cdots, m.$$

Since $E(\mathfrak{P}_n) \to A(S)$, clearly (10) and (11) imply (4).

V.2.16. Given S by a representation $S : \mathfrak{x} = \mathfrak{x}(w)$, $w \in \mathfrak{R}$, let $\mathfrak{R}^*$ be a simply-connected Jordan region in $\mathfrak{R}$, and let us define $S^* : \mathfrak{x} = \mathfrak{x}(w)$, $w \in \mathfrak{R}^*$. Then $A(S^*) \leqq A(S)$. This is merely a special case of V.2.15.

V.2.17. Given S by a representation $S : \mathfrak{x} = \mathfrak{x}(w)$, $w \in \mathfrak{R}$, let $\mathfrak{R}_n$ be a sequence of simply-connected Jordan regions in $\mathfrak{R}$ that fill up $\mathfrak{R}$, in the sense of IV.1.41. Let us define $S_n : \mathfrak{x} = \mathfrak{x}(w)$, $w \in \mathfrak{R}_n$. Then $A(S_n) \to A(S)$ (note that our assumptions do not imply the relation $S_n \to S$).

PROOF. Given a positive integer j, we have by II.3.44 a simply-connected Jordan region $\mathfrak{R}^*_j \subset \mathfrak{R}^0$ such that the surface

$$S^*_j : \mathfrak{x} = \mathfrak{x}(w), \qquad w \in \mathfrak{R}^*_j,$$

satisfies the condition

(1) $$d(S, S_j^*) < 1/j.$$

For fixed j, we have $\mathfrak{R}_j^* \subset \mathfrak{R}_n$ for n large, and hence (cf. V.2.16)

$$A(S_j^*) \leqq A(S_n) \qquad \text{for } n \text{ large.}$$

Hence, for $n \to \infty$,

$$A(S_j^*) \leqq \liminf A(S_n).$$

For $j \to \infty$ we obtain the inequality

(2) $$\liminf_{j\to\infty} A(S_j^*) \leqq \liminf_{n\to\infty} A(S_n).$$

In view of (1), we have by V.2.6

(3) $$A(S) \leqq \liminf_{j\to\infty} A(S_j^*).$$

Since $\mathfrak{R}_n \subset \mathfrak{R}$, we have by V.2.15, V.2.16

(4) $$A(S_n) \leqq A(S).$$

(2), (3), (4) imply that $A(S_n) \to A(S)$.

V.2.18. Given S by a representation

(1) $$S : \mathfrak{x} = \mathfrak{x}(w), \qquad w \in \mathfrak{R},$$

let R be any oriented rectangle in $\mathfrak{R}^0$. Let us define

$$S_R : \mathfrak{x} = \mathfrak{x}(w), \qquad w \in R,$$

$$A(R, \mathfrak{x}) = A(S_R).$$

The representation (1) being fixed, $A(R, \mathfrak{x})$ is a rectangle-function defined for every oriented rectangle $R \subset \mathfrak{R}^0$. From V.2.15, V.2.16 it follows that if $A(S) < +\infty$, then $A(R, \mathfrak{x})$ is of type A in $\mathfrak{R}^0$ (see III.1.52). Hence, a.e. in $\mathfrak{R}^0$, $A(R, \mathfrak{x})$ has a derivative $A'(w, \mathfrak{x})$ which is summable in $\mathfrak{R}^0$ by an argument entirely analogous to that used in V.1.19. Now let us suppose that $A(S) < +\infty$, and let R be any oriented rectangle in $\mathfrak{R}^0$. By V.2.13, V.2.16 we have then

(2) $$\iint_R W_e(w, \mathfrak{x}) \leqq a(S_R) \leqq A(S_R) = A(R, \mathfrak{x}).$$

Since (1) holds for every oriented rectangle $R \subset \mathfrak{R}^0$, it follows that

$$W_e(w, \mathfrak{x}) \leqq A'(w, \mathfrak{x}) \qquad \text{a.e. in } \mathfrak{R}^0.$$

It is not known whether the sign of equality holds a.e. in $\mathfrak{R}^0$ as soon as $A(S) < +\infty$.

V.2.19. Given a surface S in terms of a representation

(1) $$S : \mathfrak{x} = \mathfrak{x}(w), \qquad w \in \mathfrak{R},$$

let us suppose that $A(S) < +\infty$. By V.2.10, V.1.19 the generalized essential Jacobians $\mathfrak{J}_e^j(w, \mathfrak{x})$, $j = 1, 2, 3$, exist then a.e. in $\mathfrak{R}^0$, and the quantity

$$W_e(w, \mathfrak{x}) = \{[\mathfrak{J}_e^1(w, \mathfrak{x})]^2 + [\mathfrak{J}_e^2(w, \mathfrak{x})]^2 + [\mathfrak{J}_e^3(w, \mathfrak{x})]^2\}^{1/2}$$

is summable in $\mathfrak{R}^0$. On the other hand, no general information is available concerning the existence of the ordinary Jacobians $J^j(w, \mathfrak{x})$, $j = 1, 2, 3$ (cf. V.1.18). If the ordinary Jacobians happen to exist a.e. in $\mathfrak{R}^0$, then by V.1.22, V.2.10 the quantity

$$W(w, \mathfrak{x}) = \{[J^1(w, \mathfrak{x})]^2 + [J^2(w, \mathfrak{x})]^2 + [J^3(w, \mathfrak{x})]^2\}^{1/2}$$

is summable in $\mathfrak{R}^0$, and in fact $W(w, \mathfrak{x}) \leqq W_e(w, \mathfrak{x})$ a.e. in $\mathfrak{R}^0$. These remarks lead, in view of the inequalities in V.2.13, V.2.14, to the following questions.

Question (i). Given S, does there exist a representation (1) such that the ordinary Jacobians $J^j(w, \mathfrak{x})$, $j = 1, 2, 3$, exist a.e. in $\mathfrak{R}^0$?

Question (ii). Given S, let (1) be a representation such that the ordinary Jacobians $J^j(w, \mathfrak{x})$, $j = 1, 2, 3$, exist a.e. in $\mathfrak{R}^0$ and are summable in $\mathfrak{R}^0$. Under what conditions will the formula

$$\iint_{\mathfrak{R}^0} W(w, \mathfrak{x}) = A(S) \tag{2}$$

hold?

Question (iii). Given S, does there exist a representation (1) such that the ordinary Jacobians $J^j(w, \mathfrak{x})$, $j = 1, 2, 3$, exist a.e. in $\mathfrak{R}^0$ and are summable in $\mathfrak{R}^0$, and the formula (2) holds?

We shall discuss presently these questions in a preliminary fashion. More satisfactory answers will be obtained later on, but it should be noted that the information available at this time is far from complete. Analogous questions arise in connection with the generalized essential Jacobians $\mathfrak{J}_e^j(w, \mathfrak{x})$, and the answers to these latter questions are a great deal more satisfactory, as will appear later on.

V.2.20. By II.3.44, II.3.11 every surface S has a representation $S : \mathfrak{x} = \mathfrak{x}_*(w_*)$, $w_* \in Q_{0*}$, where Q_{0*} is the unit square $0 \leqq u_* \leqq 1$, $0 \leqq v_* \leqq 1$ in the w_*-plane. Let $\phi(t)$, $0 \leqq t \leqq 1$, denote the so-called Cantor ternary function. The function $\phi(t)$ has the following properties:

(i) $\phi(t)$ is continuous for $0 \leqq t \leqq 1$.

(ii) $\phi(0) = 0$, $\phi(1) = 1$.

(iii) If $t_1 < t_2$, then $\phi(t_1) \leqq \phi(t_2)$.

(iv) There exists a sequence of disjoint closed intervals $I_1, I_2, \cdots$ on the interval $0 \leqq t \leqq 1$, such that $|I_1| + |I_2| + \cdots = 1$, and $\phi(t)$ is constant on each one of the intervals $I_1, I_2, \cdots$.

(v) As a consequence of (iv), $\phi'(t)$ exists and is equal to zero a.e. in the interval $0 \leqq t \leqq 1$.

For each positive integer n, let us denote by τ_n the transformation

$$\tau_n : \begin{cases} u_* = \dfrac{\phi(u) + u/n}{1 + 1/n}, \\ \\ v_* = \dfrac{\phi(v) + v/n}{1 + 1/n}, \end{cases}$$

where the point (u, v) ranges over the unit square

$$Q_0 : 0 \leqq u \leqq 1, 0 \leqq v \leqq 1.$$

For conciseness, let us write τ_n in the form $\tau_n : w_* = g_n(w)$, $w \in Q_0$. Clearly, τ_n is a topological transformation from Q_0 onto Q_{0*}. Furthermore, if we put $g(w) = \phi(u) + i\phi(v)$, then clearly

$$(1) \qquad | g(w) - g_n(w) | \leqq \frac{2^{1/2}}{n+1}, \qquad w \in Q_0 .$$

Let us define the transformation μ by the formula

$$\mu : w_* = g(w), \qquad w \in Q_0 .$$

Clearly, μ is a continuous transformation from Q_0 onto Q_{0*} (of course μ is not a homeomorphism). Let us now define a surface S^* by the representation

$$(2) \qquad S^* : \mathfrak{x} = \mathfrak{x}_*[g(w)], \qquad w \in Q_0 .$$

Now since τ_n is a homeomophism, we have by II.3.11 for S the representation

$$(3) \qquad S : \mathfrak{x} = \mathfrak{x}_*[g_n(w)], \qquad w \in Q_0 .$$

From (1) it follows that $\mathfrak{x}_*[g_n(w)] \to \mathfrak{x}_*[g(w)]$ uniformly on Q_0. By II.3.13 it follows therefore from (2) and (3) that $S^* = S$. In other words, S has the representation $S : \mathfrak{x} = \mathfrak{x}(w) = \mathfrak{x}_*[g(w)]$, $w \in Q_0$. Now let $x_{j*}(u_*, v_*)$, $j = 1, 2, 3$, be the components of $\mathfrak{x}_*(w_*)$. Then the components $x_j(u, v)$ of $\mathfrak{x}_*[g(w)] = \mathfrak{x}(w)$ are given by the formula

$$x_j(u, v) = x_{j*}[\phi(u), \phi(v)].$$

Now let (u_0, v_0) be a point in Q_0^0, such that u_0 is an interior point of an interval $u_1 \leqq u \leqq u_2$ on which $\phi(u)$ is constant (see (iv) above). Then $x_j(u, v_0) = x_j(u_0, v_0)$ for $u_1 < u < u_2$, and hence $\partial x_j/\partial u$ exists and is equal to zero at the point (u_0, v_0). In view of (iv) above it follows that $\partial x_j/\partial u$ exists and is equal to zero a.e. in Q_0. A similar reasoning shows that $\partial x_j/\partial v$ exists and is equal to zero a.e. in Q_0. As a consequence, the ordinary Jacobians $J^i(w, \mathfrak{x})$ exist and are equal to zero a.e. in Q_0.

V.2.21. Thus the answer to question (i) in V.2.19 is as follows. The construction in V.2.20 yields for every surface S a representation upon the unit square Q_0 such that the ordinary Jacobians $J^i(w, \mathfrak{x})$ and hence the quantity $W(w, \mathfrak{x})$ exist and are equal to zero a.e. in Q_0.

Hence, every surface S such that $A(S) > 0$ has what may be referred to as a deceptive representation, namely a representation upon Q_0 for which $W(w, \mathfrak{x})$ exists a.e. in Q_0, but the integral

$$\iint_{Q_0} W(w, \mathfrak{x})$$

fails to yield the value of the area $A(S)$. This remark shows that the questions (ii) and (iii) in V.2.19 are relevant.

V.2.22. Taking up question (ii) in V.2.19, let S be a surface, and let $S : \mathfrak{x} = \mathfrak{x}(w)$, $w \in \mathfrak{R}$, be a representation such that the ordinary Jacobians $J'(w, \mathfrak{x})$ exist a.e. in $\mathfrak{R}^0$ and are summable on $\mathfrak{R}^0$. $W(w, \mathfrak{x})$ is then summable in $\mathfrak{R}^0$ and

$$\iint_{\mathfrak{R}^0} W(w, \mathfrak{x}) \leqq A(S). \tag{1}$$

Indeed, if $A(S) = +\infty$, then (1) is obvious; if $A(S) < +\infty$, then (1) holds by V.2.14.

THEOREM. *Under the conditions just stated, the sign of equality holds in* (1) *if and only if there exists a sequence of surfaces* S_n *such that*

$$S_n \to S, \qquad \limsup_{n\to\infty} A(S_n) \leqq \iint_{\mathfrak{R}^0} W(w, \mathfrak{x}). \tag{2}$$

PROOF. *Necessity.* Suppose we have

$$\iint_{\mathfrak{R}^0} W(w, \mathfrak{x}) = A(S). \tag{3}$$

Then $A(S) < +\infty$. By V.2.4 we have a sequence of polyhedra $\mathfrak{P}_n$ such that $\mathfrak{P}_n \to S$, $E(\mathfrak{P}_n) \to A(S)$. Since $E(\mathfrak{P}_n) = A(\mathfrak{P}_n)$ by V.2.9, the sequence $S_n = \mathfrak{P}_n$ satisfies the conditions (2).

Sufficiency. Suppose we have a sequence S_n such that (2) holds. By (1) and V.2.6 we have then

$$\iint_{\mathfrak{R}^0} W(w, \mathfrak{x}) \leqq A(S) \leqq \liminf A(S_n) \leqq \limsup A(S_n) \leqq \iint_{\mathfrak{R}^0} W(w, \mathfrak{x}),$$

and the formula (3) follows.

V.2.23. The preceding result, combined with the method of approximation by integral means, leads to a variety of useful statements. Let there be given a surface S in terms of a representation $S : \mathfrak{x} = \mathfrak{x}(w)$, $w \in \mathfrak{R}$. Let $x_j(w) = x_j(u, v)$, $u + iv = w$, be the components of $\mathfrak{x}(w)$. For $h > 0$, let us put (see III.2.65)

$$x_j^h(u, v) = \frac{1}{4h^2} \int_{-h}^{h} \int_{-h}^{h} x_j(u + \xi, v + \eta)\, d\xi\, d\eta, \qquad j = 1, 2, 3.$$

Let $\mathfrak{R}_*$ be a simply-connected Jordan region such that $\mathfrak{R}_* \subset \mathfrak{R}^0$. Let $\mathfrak{D}_*$ be a domain such that $\mathfrak{R}_* \subset \mathfrak{D}_*$, and such that the closure of $\mathfrak{D}_*$ is contained in $\mathfrak{R}^0$. Then for h small enough, the functions $\overset{h}{x}_j(u, v)$ will be defined in $\mathfrak{D}_*$ and will have continuous partial derivatives of the first order in $\mathfrak{D}_*$ (see III.2.66). Assuming h to be small enough for this purpose, let us put

$$\mathfrak{x}_h(w) = [\overset{h}{x}_1(u, v), \overset{h}{x}_2(u, v), \overset{h}{x}_3(u, v)],$$

(1) $$S_{h*} : \mathfrak{x} = \mathfrak{x}_h(w), \qquad w \in \mathfrak{R}_* .$$

Clearly the representation (1) is of class C' in the sense of V.1.25, and hence (see V.1.35, V.2.12)

$$A(S_{h*}) = \iint_{\mathfrak{R}_*^0} W(w, \mathfrak{x}_h).$$

After these preliminary remarks, we proceed to state an application of V.2.22.

V.2.24. THEOREM. *Let $S : \mathfrak{x} = \mathfrak{x}(w)$, $w \in \mathfrak{R}$, be a representation with the following properties.*

(i) *The ordinary Jacobians $J^j(w, \mathfrak{x})$, $j = 1, 2, 3$, exist a.e. in $\mathfrak{R}^0$ and are summable in $\mathfrak{R}^0$. As a consequence, $W(w, \mathfrak{x})$ exists a.e. in $\mathfrak{R}^0$ and is summable in $\mathfrak{R}^0$.*

(ii) *$\mathfrak{x}_h(w)$ being defined as in* V.2.23, *we have (cf.* V.1.18)

(1) $$\iint_R | J^j(w, \mathfrak{x}) - J^j(w, \mathfrak{x}_h) | \underset{h \to 0}{\longrightarrow} 0, \qquad j = 1, 2, 3,$$

for every oriented rectangle $R \subset \mathfrak{R}^0$.

Then $A(S) < +\infty$ and

(2) $$A(S) = \iint_{\mathfrak{R}^0} W(w, \mathfrak{x}).$$

PROOF. Let us first note that

(3) $$\iint_F | J^j(w, \mathfrak{x}) - J^j(w, \mathfrak{x}_h) | \underset{h \to 0}{\longrightarrow} 0, \qquad j = 1, 2, 3,$$

for every closed set $F \subset \mathfrak{R}^0$, as an immediate consequence of (1). Let $\mathfrak{R}_*$, S_{h*} have the same meaning as in V.2.23. As noted there we have, for h small enough,

(4) $$A(S_{h*}) = \iint_{\mathfrak{R}_*^0} W(w, \mathfrak{x}_h).$$

By (3) the integral in (3) converges to zero if taken over $\mathfrak{R}_*$ and hence, *a fortiori*, if taken over $\mathfrak{R}_*^0$. That is,

(5) $$\iint_{\mathfrak{R}_*^0} | J^j(w, \mathfrak{x}) - J^j(w, \mathfrak{x}_h) | \underset{h \to 0}{\longrightarrow} 0, \qquad j = 1, 2, 3.$$

Now by the triangle inequality

$$(6)\qquad |W(w, \mathfrak{x}) - W(w, \mathfrak{x}_h)| \leqq \sum_{j} |J^j(w, \mathfrak{x}) - J^j(w, \mathfrak{x}_h)|.$$

(5) and (6) yield, if h_n is a sequence of positive numbers converging to zero,

$$(7)\qquad \iint_{\mathfrak{R}^0_*} |W(w, \mathfrak{x}) - W(w, \mathfrak{x}_{h_n})| \underset{n\to\infty}{\longrightarrow} 0.$$

Now let us denote by S_* the surface $S_* : \mathfrak{x} = \mathfrak{x}(w)$, $w \in \mathfrak{R}_*$. We have $\mathfrak{x}_{h_n}(w) \to \mathfrak{x}(w)$ uniformly in $\mathfrak{R}_*$ (see III.2.66), and hence

$$(8)\qquad S_{h_n*} \to S_* \qquad \text{for } n \to \infty.$$

(4) and (7) yield

$$A(S_{h_n*}) = \iint_{\mathfrak{R}^0_*} W(w, \mathfrak{x}_{h_n}) \leqq \iint_{\mathfrak{R}^0_*} W(w, \mathfrak{x}) + \iint_{\mathfrak{R}^0_*} |W(w, \mathfrak{x}_{h_n}) - W(w, \mathfrak{x})|,$$

$$(9)\qquad \limsup_{n\to\infty} A(S_{h_n*}) \leqq \iint_{\mathfrak{R}^0_*} W(w, \mathfrak{x}).$$

(8) and (9) imply, by V.2.22, that

$$(10)\qquad A(S_*) = \iint_{\mathfrak{R}^0_*} W(w, \mathfrak{x}).$$

Now let $\mathfrak{R}_{k*}$ be a sequence of Jordan regions that fill up $\mathfrak{R}$ from the interior. By V.2.17 we have then

$$(11)\qquad A(S_{k*}) \to A(S),$$

where $S_{k*} : \mathfrak{x} = \mathfrak{x}(w)$. $w \in \mathfrak{R}_{k*}$. Clearly

$$(12)\qquad \iint_{\mathfrak{R}^0_{k*}} W(w, \mathfrak{x}) \to \iint_{\mathfrak{R}^0} W(w, \mathfrak{x}).$$

By (10) we have

$$(13)\qquad A(S_{k*}) = \iint_{\mathfrak{R}^0_{k*}} W(w, \mathfrak{x}).$$

(11), (12), (13) imply the formula (2).

V.2.25. We shall state now some applications of the preceding theorem. Let $S : \mathfrak{x} = \mathfrak{x}(w)$, $w \in \mathfrak{R}$, be a representation with the following properties.

(i) The ordinary Jacobians $J^j(w, \mathfrak{x})$, $j = 1, 2, 3$, exist a.e. in $\mathfrak{R}^0$ and are summable in $\mathfrak{R}^0$.

(ii) The components of $\mathfrak{x}(w)$ are ACT on every oriented rectangle $R \subset \mathfrak{R}^0$ (cf. III.2.64). We shall term the set of conditions (i) and (ii) the hypothesis H_0.

V.2.26. Suppose that the representation $S : \mathfrak{x} = \mathfrak{x}(w)$, $w \in \mathfrak{R}$, satisfies the hypothesis H_0 of V.2.25. Suppose that on each oriented rectangle $R \subset \mathfrak{R}^0$ it is true that in each one of the six products that appear in the three Jacobians $J^j(w, \mathfrak{x})$, $j = 1, 2, 3$, the two partial derivatives involved belong to associated Lebesgue classes L^p, L^q (cf. I.3.10). Then $A(S) < +\infty$, and

$$(1) \qquad A(S) = \iint_{\mathfrak{R}^0} W(w, \mathfrak{x}).$$

Proof. Each one of the transformations T^j, $j = 1, 2, 3$, associated with the given representation of S (cf. V.1.1), satisfies the assumptions of IV.4.33; hence by IV.4.33 the given representation of S satisfies the assumptions of V.2.24, and the formula (1) follows.

V.2.27. In view of its importance, we state the special case $p = q = 2$ of V.2.26 explicitly. If a representation $S : \mathfrak{x} = \mathfrak{x}(w)$, $w \in \mathfrak{R}$, satisfies the hypothesis H_0 of V.2.25, and if the first partial derivatives of the components of $\mathfrak{x}(w)$ belong to the Lebesgue class L^2 on every oriented rectangle $R \subset \mathfrak{R}^0$, then

$$A(S) = \iint_{\mathfrak{R}^0} W(w, \mathfrak{x}).$$

V.2.28. The remarks in IV.4.34, combined with V.2.24, yield immediately the following very special but quite useful result. Given a representation $S : \mathfrak{x} = \mathfrak{x}(w)$, $w \in \mathfrak{R}$, suppose that the following conditions hold.

(i) The ordinary Jacobians $J^j(w, \mathfrak{x})$ exist a.e. on $\mathfrak{R}^0$ and are summable on $\mathfrak{R}^0$.

(ii) Two of the components of $\mathfrak{x}(w)$ satisfy a Lipschitz condition (cf. I.3.14), and the remaining component is ACT on every oriented rectangle $R \subset \mathfrak{R}^0$.

Under these conditions, $A(S) < +\infty$ and we have the formula

$$A(S) = \iint_{\mathfrak{R}^0} W(w, \mathfrak{x}).$$

Proof. By IV.4.34, the given representation satisfies the assumptions of V.2.24.

V.2.29. We proceed to discuss a number of topics in preparation for a study of question (iii) in V.2.19. A representation $S : \mathfrak{x} = \mathfrak{x}(w)$, $w \in \mathfrak{R}$, will be termed *generalized conformal* if the following conditions are satisfied.

(i) The first partial derivatives of the components $x_j(u, v)$, $j = 1, 2, 3$, of $\mathfrak{x}(w)$ exist a.e. in $\mathfrak{R}^0$. The vector functions

$$\mathfrak{x}_u = \left(\frac{\partial x_1}{\partial u}, \frac{\partial x_2}{\partial u}, \frac{\partial x_3}{\partial u}\right), \qquad \mathfrak{x}_v = \left(\frac{\partial x_1}{\partial v}, \frac{\partial x_2}{\partial v}, \frac{\partial x_3}{\partial v}\right),$$

are then defined a.e. in $\mathfrak{R}^0$. Furthermore the ordinary Jacobians $J^j(w, \mathfrak{x})$, $j = 1, 2, 3$, the function $W(w, \mathfrak{x})$, and the scalar products

$$\mathfrak{x}_u^2 = \left(\frac{\partial x_1}{\partial u}\right)^2 + \left(\frac{\partial x_2}{\partial u}\right)^2 + \left(\frac{\partial x_3}{\partial u}\right)^2,$$

$$\mathfrak{x}_u \mathfrak{x}_v = \frac{\partial x_1}{\partial u}\frac{\partial x_1}{\partial v} + \frac{\partial x_2}{\partial u}\frac{\partial x_2}{\partial v} + \frac{\partial x_3}{\partial u}\frac{\partial x_3}{\partial v},$$

$$\mathfrak{x}_v^2 = \left(\frac{\partial x_1}{\partial v}\right)^2 + \left(\frac{\partial x_2}{\partial v}\right)^2 + \left(\frac{\partial x_3}{\partial v}\right)^2$$

are also defined a.e. in $\mathfrak{R}^0$. Following the practice in Differential Geometry, we put $E = \mathfrak{x}_u^2$, $F = \mathfrak{x}_u \mathfrak{x}_v$, $G = \mathfrak{x}_v^2$. We have then the well known identity $W^2 = EG - F^2$, and the inequality $0 \leqq W \leqq (EG)^{1/2} \leqq (E + G)/2$.

(ii) The ordinary Jacobians $J^j(w, \mathfrak{x})$, $j = 1, 2, 3$, are summable in $\mathfrak{R}^0$. As a consequence, $W(w, \mathfrak{x})$ is also summable in $\mathfrak{R}^0$.

(iii) On every oriented rectangle $R \subset \mathfrak{R}^0$, the components of $\mathfrak{x}(w)$ are ACT and the first partial derivatives of these components belong to the Lebesgue class L^2.

(iv) $E = G$, $F = 0$ a.e. in $\mathfrak{R}^0$.

V.2.30. CONTINUATION. *Suppose that* $S : \mathfrak{x} = \mathfrak{x}(w)$, $w \in \mathfrak{R}$, *is a generalized conformal representation. Then* $A(S) < +\infty$, *and*

$$A(S) = \frac{1}{2} \iint_{\mathfrak{R}^0} (\mathfrak{x}_u^2 + \mathfrak{x}_v^2).$$

PROOF. By V.2.27, conditions (i), (ii), (iii) imply that $A(S) < +\infty$ and

$$A(S) = \iint_{\mathfrak{R}^0} W(w, \mathfrak{x}).$$

By condition (iv), we have $W = E = G$ and hence $W = (E + G)/2 = (\mathfrak{x}_u^2 + \mathfrak{x}_v^2)/2$ a.e. in $\mathfrak{R}^0$.

V.2.31. We shall need the following lemma. Let $g(u, v)$, $g_n(u, v)$ be (real-valued) continuous functions on an oriented rectangle

$$R : a \leqq u \leqq b,\ c \leqq v \leqq d,$$

such that the following conditions hold.

(i) $g_n(u, v) \to g(u, v)$ uniformly on R.

(ii) $g_n(u, v)$ is ACT on R (see III.2.64), $n = 1, 2, \cdots$.

(iii) The partial derivatives $g_{nu} = \partial g_n/\partial u$, $g_{nv} = \partial g_n/\partial v$ belong to the Lebesgue class L^2 on R (note that these partial derivatives exist a.e. on R by (ii)).

(iv) $$\iint_R (g_{nu}^2 + g_{nv}^2) < H \text{ for all } n, \text{ where } H \text{ is a finite constant.}$$

Then $g(u, v)$ is ACT on R, the partial derivatives g_u, g_v belong to the Lebesgue class L^2 on R, and

(1) $$\iint_R (g_u)^2 \leqq \liminf \iint_R (g_{nu})^2, \qquad \iint_R (g_v)^2 \leqq \liminf \iint_R (g_{nv})^2.$$

PROOF. By condition (ii) we have (cf. III.2.58)

$$W_u(R, g_n) = \iint_R |g_{nu}|.$$

Hence, by the Hölder inequality (see I.3.10), and by condition (iv),

(2) $$W_u(R, g_n) \leqq |R|^{1/2}H^{1/2}.$$

By III.2.53 we have the inequality

(3) $$W_u(R, g) \leqq \liminf_{n\to\infty} W_u(R, g_n).$$

(2) and (3) show that $g(u, v)$ is BVTu in R (cf. III.2.49). Similarly it follows that $g(u, v)$ is BVTv on R. Hence $g(u, v)$ is BVT on R. As a consequence (see III.2.50), the partial derivatives g_u, g_v exist a.e. on R, are summable on R, and we have

(4) $$\iint_r |g_u| \leqq W_u(r, g), \qquad \iint_r |g_v| \leqq W_v(r, g),$$

for every oriented rectangle $r \subset R$. Now (see III.2.53)

(5) $$W_u(r, g) \leqq \liminf_{n\to\infty} W_u(r, g_n).$$

Since g_n is ACT on R, we obtain, by the Hölder inequality,

(6) $$W_u(r, g) \leqq \liminf \iint_r |g_{nu}| \leqq |r|^{1/2} \liminf \left(\iint_r (g_{nu})^2\right)^{1/2}.$$

Let D be any subdivision of R into oriented rectangles $r_1, \cdots, r_m$. By (4) and (6) we have then

$$\iint_{r_i} |g_u| \leqq |r_i|^{1/2} \liminf_{n\to\infty} \left(\iint_{r_i} g_{nu}^2\right)^{1/2},$$

and hence

(7) $$\sum_{i=1}^{m} \frac{\left(\iint_{r_i} |g_u|\right)^2}{|r_i|} \leqq \sum_{i=1}^{m} \liminf_{n\to\infty} \iint_{r_i} g_{nu}^2 \leqq \liminf_{n\to\infty} \iint_R g_{nu}^2.$$

Since (7) holds for every subdivision D of R, the first inequality (1) follows by I.3.10. The second inequality (1) is proved in a similar way.

Now let $r_1, \cdots, r_m$ be any finite system of oriented rectangles in R, such that $r_i^0 r_j^0 \doteq 0$ for $i \neq j$. Using the inequality of Schwarz, we obtain from (6)

$$\sum_{i=1}^{m} W_u(r_i, g) \leqq \liminf_{n\to\infty} \sum_{i=1}^{m} |r_i|^{1/2} \left(\iint_{r_i} |g_{nu}|^2 \right)^{1/2}$$

$$\leqq \liminf_{n\to\infty} \left(\sum_{i=1}^{m} |r_i| \right)^{1/2} \left(\sum_{i=1}^{m} \iint_{r_i} |g_{nu}|^2 \right)^{1/2}.$$

Hence, by condition (iv),

$$\sum_{i=1}^{m} W_u(r, g) \leqq H^{1/2} \left(\sum_{i=1}^{m} |r_i| \right)^{1/2}.$$

Hence the rectangle function $W_u(r, g)$ is absolutely continuous in R; similarly it follows that $W_v(r, g)$ is absolutely continuous in R. Hence, by III.2.55, $g(u, v)$ is ACT in R, and the proof is complete.

V.2.32. Let S and S_n, $n = 1, 2, \cdots$, be surfaces given by representations

(1) $$S : \mathfrak{x} = \mathfrak{x}(w), \qquad w \in \mathfrak{R},$$

(2) $$S_n : \mathfrak{x} = \mathfrak{x}_n(w), \qquad w \in \mathfrak{R},$$

such that the following conditions hold.

(i) $\mathfrak{x}_n(w) \to \mathfrak{x}(w)$ uniformly on $\mathfrak{R}$.

(ii) For every n, the representation (2) is generalized conformal (see V.2.29).

(iii) $A(S) < +\infty$.

(iv) $A(S_n) \to A(S)$.

Then the representation (1) is also generalized conformal.

PROOF. By V.2.30, $A(S_n) < +\infty$ and

(3) $$A(S_n) = \frac{1}{2} \iint_{\mathfrak{R}^0} (\mathfrak{x}_{nu}^2 + \mathfrak{x}_{nv}^2).$$

As a consequence of (iii) and (iv), the sequence $A(S_n)$ is bounded. Hence, in view of (3), there exists a finite constant H such that

(4) $$\iint_{\mathfrak{R}^0} (\mathfrak{x}_{nu}^2 + \mathfrak{x}_{nv}^2) < H, \qquad n = 1, 2, \cdots.$$

Now let R be any oriented rectangle in $\mathfrak{R}^0$. By (i), (ii) and (4), the components of the vectors $\mathfrak{x}(w)$, $\mathfrak{x}_n(w)$ satisfy the assumptions of V.2.31 on R, and hence we obtain the following statements.

(a) The components of $\mathfrak{x}(w)$ are ACT on R.

(b) The first partial derivatives of the components of $\mathfrak{x}(w)$ belong to the Lebesgue class L^2 on R.

(c) $\iint_R (\mathfrak{x}_u^2 + \mathfrak{x}_v^2) \leqq \liminf_{n\to\infty} \iint_R (\mathfrak{x}_{nu}^2 + \mathfrak{x}_{nv}^2)$.

Since R was any oriented rectangle in $\mathfrak{R}^0$, it follows that the first partial derivatives of the components of $\mathfrak{x}(w)$ exist a.e. in $\mathfrak{R}^0$. Now let F be any closed set in $\mathfrak{R}^0$. We have then a finite system of nonoverlapping oriented rectangles $R_1, \cdots, R_m$ in $\mathfrak{R}^0$ that cover F. By (c) we obtain therefore

$$\begin{aligned}\iint_F (\mathfrak{x}_u^2 + \mathfrak{x}_v^2) &\leqq \sum_{i=1}^m \iint_{R_i} (\mathfrak{x}_u^2 + \mathfrak{x}_v^2) \leqq \liminf_{n\to\infty} \sum_{i=1}^m \iint_{R_i} (\mathfrak{x}_{nu}^2 + \mathfrak{x}_{nv}^2) \\ &\leqq \liminf_{n\to\infty} \iint_{\mathfrak{R}^0} (\mathfrak{x}_{nu}^2 + \mathfrak{x}_{nv}^2).\end{aligned} \tag{5}$$

By V.2.30

$$\iint_{\mathfrak{R}^0} (\mathfrak{x}_{nu}^2 + \mathfrak{x}_{nv}^2) = 2A(S_n). \tag{6}$$

(5) and (6) yield, in view of condition (iv), the inequality

$$\iint_F (\mathfrak{x}_u^2 + \mathfrak{x}_v^2) \leqq 2A(S). \tag{7}$$

By (iii) and V.2.14, the ordinary Jacobians $J'(w, \mathfrak{x})$ are summable on $\mathfrak{R}^0$. By (a), (b) and V.2.27, it follows therefore that

$$\iint_{\mathfrak{R}^0} W(w, \mathfrak{x}) = A(S). \tag{8}$$

Since (7) holds for every closed set in $\mathfrak{R}^0$, we have

$$\iint_{\mathfrak{R}^0} (\mathfrak{x}_u^2 + \mathfrak{x}_v^2) \leqq 2A(S). \tag{9}$$

(8), (9) yield

$$A(S) = \iint_{\mathfrak{R}^0} W(w, \mathfrak{x}) \leqq \frac{1}{2} \iint_{\mathfrak{R}^0} (\mathfrak{x}_u^2 + \mathfrak{x}_v^2) \leqq A(S).$$

Hence

$$\iint_{\mathfrak{R}^0} W(w, \mathfrak{x}) = \frac{1}{2} \iint_{\mathfrak{R}^0} (\mathfrak{x}_u^2 + \mathfrak{x}_v^2). \tag{10}$$

Since

$$W(w, \mathfrak{x}) \leqq (\mathfrak{x}_u^2 + \mathfrak{x}_v^2)/2 \qquad \text{a.e. in } \mathfrak{R}^0,$$

it follows from (10) that

$$W(w, \mathfrak{x}) = (EG - F^2)^{1/2} = (E + G)/2 \qquad \text{a.e. in } \mathfrak{R}^0. \tag{11}$$

Simple computation leads from (11) to the formula

$$(E - G)^2 + 4F^2 = 0 \qquad \text{a.e. in } \mathfrak{R}^0. \tag{12}$$

Clearly, (12) implies that $E = G$, $F = 0$ a.e. in $\mathfrak{R}^0$, and the proof is complete.

V.2.33. We proceed to discuss the problem of the existence of generalized conformal representations for a given surface. We begin with a study of polyhedra. Given a polyhedron $\mathfrak{P}$, we have by V.1.36 a representation

$$\mathfrak{P} : \mathfrak{x} = \mathfrak{x}(w), \qquad w \in \mathfrak{R}, \tag{1}$$

with the following properties.

(i) $\mathfrak{R}$ is bounded by a convex polygon.

(ii) There is given a rectilinear triangulation $\mathfrak{Z}$ of $\mathfrak{R}$, comprised of rectilinear triangles $t_1, \cdots, t_m$ (cf. V.1.28, V.1.36), such that the following holds. Considered on any one of the triangles t_i, the representation (1) determines a biunique affine transformation from t_i onto a (nondegenerate) rectilinear triangle Δ_i in $x_1x_2x_3$-space.

(iii) $E(\mathfrak{P}) = a(\mathfrak{P}) = A(\mathfrak{P}) = |\Delta_1| + \cdots + |\Delta_m|$ (see V.1.33, V.2.9).

While the correspondence between t_i and Δ_i is biunique, of course the representation (1) need not define a biunique transformation if considered over the whole of $\mathfrak{R}$. In other words, two triangles Δ_i, Δ_j, $i \neq j$, may intersect each other and may in fact have interior points in common.

As a consequence of the general theory of conformal maps of abstract Riemann surfaces (I.3.19), the following existence theorem holds. Let w_1, w_2, w_3 be three distinct points on the boundary of $\mathfrak{R}$, such that the points w_1, w_2, w_3 follow upon each other in the counterclockwise sense around the boundary of $\mathfrak{R}$. Let K_* denote the unit circular disc $|w_*| \leqq 1$ in an auxiliary $w_* = u_* + iv_*$ plane, and let w_{1*}, w_{2*}, w_{3*} be three distinct points on the perimeter of K_* that follow upon each other in the counterclockwise sense. Then there exists a topological transformation $\tau(K_*) = \mathfrak{R}$, such that $w_j = \tau(w_{j*})$, $j = 1, 2, 3$, and such that the representation (cf. II.3.11)

$$\mathfrak{P} : \mathfrak{x} = \mathfrak{x}_*(w_*) = \mathfrak{x}[\tau(w_*)], \qquad w_* \in K_*, \tag{2}$$

has the following properties.

(a) Let us put $t_{i*} = \tau^{-1}(t_i)$, $i = 1, 2, \cdots, m$. Then the triangles t_{i*} make up a curvilinear triangulation $\mathfrak{Z}_*$ of K_*. The sides of each t_{i*} are simple analytic arcs (including their vertices), and are not tangent to each other at the vertices of t_{i*}.

(b) For each $i = 1, \cdots, m$, the representation (2), if considered on t_{i*} only, defines a biunique and continuous transformation from t_{i*} onto Δ_i which is conformal at interior points of t_{i*}

(c) For each $i = 1, \cdots, m$, the components of $\mathfrak{x}_*(w_*)$ are harmonic functions, in t_{i*}, of u_* and v_* ($u_* + iv_* = w_*$), and we have, on putting

$$E_* = \left(\frac{\partial \mathfrak{x}_*}{\partial u_*}\right)^2, \qquad F_* = \frac{\partial \mathfrak{x}_*}{\partial u_*}\frac{\partial \mathfrak{x}_*}{\partial v_*}, \qquad G_* = \left(\frac{\partial \mathfrak{x}_*}{\partial v_*}\right)^2,$$

the equations $E_* = G_*$, $F_* = 0$ for $w_* \in t_{i*}^0$, $i = 1, 2, \cdots, m$.

(d) If w_{0*} is a point on the boundary of a t_{i*}, but w_{0*} is not a vertex of t_{i*}, then the first partial derivatives of the components of $\mathfrak{x}_*(w_*)$ approach finite limits if w_{0*} is approached from the interior of t_{i*}.

V.2.34. A representation with the properties (a), (b), (c), (d) will be termed a *quasi-conformal representation of the polyhedron* $\mathfrak{P}$. We assert that such a representation is generalized conformal in the sense of V.2.29. Indeed, let e_* be the subset of K_* consisting of all the sides of the triangles t_{i*}, $i = 1, \cdots, m$. By (a) in V.2.33, $| e_* | = 0$, and hence by (c) in V.2.33 the first partial derivatives of the components of $\mathfrak{x}_*(w_*)$ exist a.e. in K_*^0. By (iii) in V.2.33 and by V.2.14 it follows that the ordinary Jacobians $J^j(w_*, \mathfrak{x}_*)$, $j = 1, 2, 3$, and the function $W(w_*, \mathfrak{x}_*)$ are summable in K_*^0. By (c) in V.2.33 we have, since $| e_* | = 0$, the equations $E_* = G_*$, $F_* = 0$ a.e. in K_*^0. As a consequence we have $E_* = G_* = W(w_*, \mathfrak{x}_*)$ a.e. in K_*^0. Since $W(w_*, \mathfrak{x}_*)$ is summable on K_*^0, it follows that the first partial derivatives of the components of $\mathfrak{x}_*(w_*)$ belong to the Lebesgue class L^2 on K_*^0, and hence *a fortiori* on every rectangle $R_* \subset K_*^0$. There remains to show that each component of $\mathfrak{x}_*(w_*)$ is ACT on every oriented rectangle R_* : $a \leqq u_* \leqq b$, $c \leqq v_* \leqq d$, in K_*^0. Now let u_{0*} be any value between a and b, such that the line $u_* = u_{0*}$ does not contain any vertex of the triangulation $\mathfrak{T}_*$. Since the sides of the triangles of $\mathfrak{T}_*$ are simple analytic arcs, it follows that the segment $I_0 : u_* = u_{0*}$, $c \leqq v_* \leqq d$, contains at most a finite number of points that lie on the perimeter of some triangle of $\mathfrak{T}_*$. Since I_0 does not contain any vertex of $\mathfrak{T}_*$, it follows by (d) in V.2.33 that I_0 is the sum of a finite number of segments on each of which the components of $\mathfrak{x}_*(w_*)$ are absolutely continuous as functions of v_*. Thus these components are absolutely continuous on I_0 as functions of v_*. Since this holds for all but a finite number of points u_{0*}, and since the first partial derivatives of the components are summable (and in fact of the Lebesgue class L^2) on R_*, it follows (see III.2.59) that the components are ACTv_* on R_*. A similar argument shows that these components are ACTu_* on R_*.

V.2.35. CONTINUATION. In view of V.2.30 it follows that for a quasi-conformal representation $\mathfrak{P} : \mathfrak{x} = \mathfrak{x}_*(w_*)$, $w_* \in K_*$, we have the formula

$$E(\mathfrak{P}) = A(\mathfrak{P}) = \frac{1}{2} \iint_{K_*^0} (E_* + G_*).$$

V.2.36. Let $\mathfrak{P}$ be a polyhedron and $\mathfrak{P} : \mathfrak{x} = \mathfrak{x}_*(w_*)$, $w_* \in \mathfrak{R}_*$, be any polyhedral representation of $\mathfrak{P}$ (cf. V.1.29). Let w_{1*}, w_{2*}, w_{3*} be three distinct points on the boundary of $\mathfrak{R}_*$ that follow upon each other in the counterclockwise sense. Let K be the unit disc $| w | \leqq 1$ in the w-plane, and let w_1, w_2, w_3 be three distinct points on the perimeter of K that follow upon each other in the counterclockwise sense. Then there exists a topological transformation $\tau(K) = \mathfrak{R}_*$ such that the representation $\mathfrak{P} : \mathfrak{x} = \mathfrak{x}_*[\tau(w)]$, $w \in K$, is quasi-conformal, and $\tau(w_j) = w_{j*}$, $j = 1, 2, 3$.

This is an immediate consequence of V.1.36 and V.2.33. Let us note that the use of an intermediate quasi-linear representation is really unnecessary, and serves only the purpose of simplifying the geometrical picture.

V.2.37. We propose to derive further existence theorems by approximating to more general surfaces by means of polyhedra. In preparation, we establish several simple lemmas. Let $\mathfrak{P}$ be a polyhedron, and let $\mathfrak{P} : \mathfrak{x} = \mathfrak{x}(w)$, $w \in K$, be a quasi-conformal representation of $\mathfrak{P}$, where K is the unit disc $| w | \leqq 1$. Let w_0 be any point in K. For $0 < \rho < 3^{1/2}/2$, we define a simple closed curve $C(\rho, w_0)$ as follows. We draw the circle $k(\rho, w_0)$ with center w_0 and radius ρ. If $k(\rho, w_0) \subset K$, then $C(\rho, w_0) = k(\rho, w_0)$. If $k(\rho, w_0)$ is not contained in K, then (due to the restriction $0 < \rho < 3^{1/2}/2$) $k(\rho, w_0)$ intersects the perimeter of K in two points and thus divides the perimeter of K into two arcs, one of which is interior to $k(\rho, w_0)$ except for its end points. This arc will be denoted by $C_1(\rho, w_0)$. Similarly, the perimeter of K divides $k(\rho, w_0)$ into two arcs, one of which lies in K. This arc will be denoted by $C_2(\rho, w_0)$. Finally, $C(\rho, w_0)$ is the sum of the two arcs $C_1(\rho, w_0)$, $C_2(\rho, w_0)$.

V.2.38. CONTINUATION. Let us use again the symbol $\omega(\mathfrak{x}(w), F)$ to denote the oscillation of $\mathfrak{x}(w)$ on a set $F \subset K$. That is,

$$\omega(\mathfrak{x}(w), F) = \text{l.u.b.} \mid \mathfrak{x}(w_1) - \mathfrak{x}(w_2) \mid,$$

where the least upper bound is taken with respect to all pairs of points w_1, w_2 in F. Suppose now that we find a point $w_0 \in K$ and two positive constants ρ_0, δ, where $\rho_0 < 3^{1/2}/2$, such that

$$\omega(\mathfrak{x}(w), C_2(\rho, w_0)) > \delta \qquad \text{for } \rho_0 < \rho < 3^{1/2}/2. \tag{1}$$

Then we assert the inequality

$$E(\mathfrak{P}) > \frac{\delta^2}{4\pi} \log \frac{3^{1/2}}{2\rho_0}. \tag{2}$$

PROOF. For clarity, let us first suppose that the point w_0 lies on the perimeter of K. Let us denote by $\mathfrak{D}$ the domain bounded by the arcs $C_2(\rho_0, w_0)$, $C_2(3^{1/2}/2, w_0)$ and the perimeter of K. Let us introduce polar coordinates ρ, θ according to the formulas

$$u = u_0 + \rho \cos \theta, \; v = v_0 + \rho \sin \theta, \tag{3}$$

where $u_0 + iv_0 = w_0$. By proper choice of θ we can arrange for the following situation to hold. By means of (3) the domain $\mathfrak{D}$ appears as the topological image of a domain $\mathfrak{D}^*$ in the ρ, θ plane, where ρ and θ are considered as Cartesian coordinates. $\mathfrak{D}^*$ is bounded, in the Cartesian ρ, θ plane, by the lines $\rho = \rho_0$, $\rho = 3^{1/2}/2$, and by two arcs given by formulas $\theta = \theta_1(\rho)$, $\theta = \theta_2(\rho)$, where $\theta_1(\rho)$ and $\theta_2(\rho)$ are single-valued continuous functions of ρ that satisfy the inequalities

$$0 < \theta_2(\rho) - \theta_1(\rho) < \pi, \; \rho_0 \leqq \rho \leqq 3^{1/2}/2. \tag{4}$$

The transformation (3) being topological, and the first partial derivatives of u, v with respect to ρ, θ being clearly continuous in $\mathfrak{D}^*$, this transformation satisfies, by an ample margin, the conditions needed to apply the transformation formula for double integrals (cf. IV.4.58, IV.4.60, IV.4.61, IV.4.24) and thus we have the formula

$$(5)\qquad \iint_{\mathfrak{D}} g(u, v)\, du\, dv = \iint_{\mathfrak{D}^*} g(u_0 + \rho \cos \theta, v_0 + \rho \sin \theta)\rho\, d\rho\, d\theta,$$

for every function $g(u, v)$ that is summable in $\mathfrak{D}$. In writing (5), we used the formula

$$\frac{\partial(u, v)}{\partial(\rho, \theta)} = \rho.$$

In view of the special form of the domain $\mathfrak{D}^*$, (5) may be rewritten (cf. I.3.10) in the form

$$(6)\qquad \iint_{\mathfrak{D}} g(u, v)\, du\, dv = \int_{\rho_0}^{3^{1/2}/2} \left[\int_{\theta_1(\rho)}^{\theta_2(\rho)} g(u_0 + \rho \cos \theta, v_0 + \rho \sin \theta)\, d\theta \right] \rho\, d\rho.$$

We consider now the function $g(u, v) = \mathfrak{x}_u^2 + \mathfrak{x}_v^2$, which is summable in K by V.2.35, and hence *a fortiori* summable in $\mathfrak{D}$. Now let us take any point w in $\mathfrak{D}$ that does not lie on a side of any triangle of the triangulation associated with the given quasi-conformal representation of $\mathfrak{P}$ (cf. V.2.34, V.2.35). Then the components of $\mathfrak{x}(w)$ are harmonic functions on a sufficiently small open disc with center w, and thus we can compute $\partial\mathfrak{x}/\partial\theta$ in terms of θ, ρ by using the usual formulas in Calculus. We find in this manner the formula

$$(7)\qquad \frac{\partial \mathfrak{x}}{\partial \theta} = -\rho(\mathfrak{x}_u \sin \theta - \mathfrak{x}_v \cos \theta).$$

This formula holds a.e. in $\mathfrak{D}^*$. There follows the inequality

$$\left| \frac{\partial \mathfrak{x}}{\partial \theta} \right| \leqq (\mathfrak{x}_u^2 + \mathfrak{x}_v^2)^{1/2} \rho \qquad \text{a.e. in } \mathfrak{D}^*.$$

By (6), (7) we obtain (IV.4.5)

$$(8)\qquad \iint_{\mathfrak{D}} (\mathfrak{x}_u^2 + \mathfrak{x}_v^2)\, du\, dv \geqq \int_{\rho_0}^{3^{1/2}/2} \left[\int_{\theta_1(\rho)}^{\theta_2(\rho)} \left| \frac{\partial \mathfrak{x}}{\partial \theta} \right|^2 d\theta \right] \frac{d\rho}{\rho}.$$

Now let us consider a ρ between ρ_0 and $3^{1/2}/2$, such that the arc $C_2(\rho, w_0)$ does not pass through any vertex of the triangulation $\mathfrak{J}$ associated with the given quasi-conformal representation of $\mathfrak{P}$. Then $C_2(\rho, w_0)$ intersects the sides of the triangles of $\mathfrak{J}$ in at most a finite number of points, and consequently (cf. (d) in V.2.33) it follows that the components of $\mathfrak{x}(w)$, considered as functions of θ on $C_2(\rho, w_0)$, are absolutely continuous on $C_2(\rho, w_0)$. For the oscillation $\omega(\mathfrak{x}(w), C_2(\rho, w_0))$ there follows the inequality

$$(9)\qquad \omega(\mathfrak{x}(w), C_2(\rho, w_0)) \leqq \int_{\theta_1(\rho)}^{\theta_2(\rho)} \left| \frac{\partial \mathfrak{x}}{\partial \theta} \right| d\theta.$$

By the Hölder inequality we obtain

$$\left[\int_{\theta_1(\rho)}^{\theta_2(\rho)} \left| \frac{\partial \mathfrak{x}}{\partial \theta} \right| d\theta \right]^2 \leqq (\theta_2(\rho) - \theta_1(\rho)) \int_{\theta_1(\rho)}^{\theta_2(\rho)} \left| \frac{\partial \mathfrak{x}}{\partial \theta} \right|^2 d\theta.$$

In view of (1), (4), (9) we get the inequality

$$(10)\qquad \int_{\theta_1(\rho)}^{\theta_2(\rho)} \left| \frac{\partial \mathfrak{x}}{\partial \theta} \right|^2 d\theta \geqq \frac{\delta^2}{\pi}.$$

According to its derivation, (10) holds for all but a finite number of values of ρ between ρ_0 and $3^{1/2}/2$. Hence (8), (10) yield

$$(11)\qquad \iint_{\mathfrak{D}} (\mathfrak{x}_u^2 + \mathfrak{x}_v^2)\, du\, dv \geqq \frac{\delta^2}{\pi} \int_{\rho_0}^{3^{1/2}/2} \frac{d\rho}{\rho} = \frac{\delta^2}{\pi} \log \frac{3^{1/2}}{2\rho_0}.$$

On the other hand, we have by V.2.35

$$(12)\qquad \iint_{\mathfrak{D}} (\mathfrak{x}_u^2 + \mathfrak{x}_v^2)\, du\, dv \leqq \iint_{K^0} (\mathfrak{x}_u^2 + \mathfrak{x}_v^2)\, du\, dv = 2E(\mathfrak{P}).$$

(11) and (12) imply the inequality (2) (in fact with the denominator 2π).

Now let us assume that w_0 is interior to K. Inspection reveals that the preceding argument remains valid, except that (4) is to be replaced by the inequality $0 \leqq \theta_2(\rho) - \theta_1(\rho) \leqq 2\pi$, $\rho_0 \leqq \rho \leqq 3^{1/2}/2$. As a consequence, we obtain by the preceding argument the inequality (2), this time with the denominator 4π.

V.2.39. Let K_* denote the unit disc $| w_* | \leqq 1$ in the w_*-plane. Let w_{1*}, w_{2*}, w_{3*} be three points on the perimeter of K_* that are the vertices of an equilateral triangle. Let us note that the sides of this triangle have the length $3^{1/2}$. Now let w_*', w_*'' be any two distinct points on the perimeter of K_*, such that

$$(1)\qquad 0 < | w_*' - w_*'' | < 3^{1/2}/2.$$

These points w_*', w_*'' divide the perimeter of K_* into two arcs γ_{s*}, γ_{l*}, one of which will contain at most one of the three points w_{1*}, w_{2*}, w_{3*}. Indeed, note that in view of (1), at least one of the points w_*', w_*'', say w_*'', is different from the points w_{1*}, w_{2*}, w_{3*}. Hence, if γ_{s*}, γ_{l*} both contain more than one of the points w_{1*}, w_{2*}, w_{3*}, then necessarily w_*' must coincide with one of the points w_{1*}, w_{2*}, w_{3*}, say $w_*' = w_{1*}$, and w_*'' must lie on that arc $w_{2*}w_{3*}$ on the perimeter of K_* that does not contain $w_{1*} = w_*'$. Clearly this is impossible in view of (1). Let the notation be so chosen that γ_{s*} contains at most one of the points w_{1*}, w_{2*}, w_{3*}. If ζ_{1*}, ζ_{2*} are any two points on γ_{s*}, then we assert that

(2) $$|\zeta_{2*} - \zeta_{1*}| \leqq |w''_* - w'_*|.$$

PROOF. The inequality would be obvious if we should know that γ_{s*} is the shorter one of the two arcs γ_{s*}, γ_{l*}. Let us assume that γ_{l*} is the shorter one of these two arcs (note that by (1) the arcs γ_{s*}, γ_{l*} have unequal lengths). Now if γ_{l*} is the shorter arc, then for any two points $\zeta'_{1*}, \zeta''_{2*}$ on γ_{l*} we have the inequality

(3) $$|\zeta''_{2*} - \zeta'_{1*}| \leqq |w''_* - w'_*|.$$

But γ_{l*} contains at least two of the points w_{1*}, w_{2*}, w_{3*} (since by assumption γ_{s*} contains at most one of them). Thus we can choose two of the points w_{1*}, w_{2*}, w_{3*} as our points $\zeta''_{2*}, \zeta'_{1*}$, and then (3) yields $3^{1/2} \leqq |w''_* - w'_*|$, in contradiction to (1). Thus γ_{s*} is the shorter one of the two arcs γ_{s*}, γ_{l*}, and the inequality (2) follows.

V.2.40. *Let a surface S be given by a representation*

(1) $$S : \mathfrak{x} = \mathfrak{x}_*(w_*), \qquad w_* \in K_*,$$

such that the following conditions hold.

(i) *K_* is the unit disc $|w_*| \leqq 1$.*

(ii) *The representation* (1) *is nondegenerate* (see II.3.21).

(iii) *$A(S) < +\infty$.*

Then S admits of a generalized conformal representation (cf. V.2.29) *upon the unit disc $K : |w| \leqq 1$.*

PROOF. Let w_{1*}, w_{2*}, w_{3*} be three points on the perimeter of K_* that follow upon each other in the counterclockwise sense and are the vertices of an equilateral triangle. Similarly, let w_1, w_2, w_3 be three points on the perimeter of K that follow upon each other in the counterclockwise order and are the vertices of an equilateral triangle. By V.2.4 we have a sequence of polyhedra $\mathfrak{P}_n$ such that

(2) $$\mathfrak{P}_n \to S, \; E(\mathfrak{P}_n) \to A(S).$$

By II.3.16, we have for each n a representation

(3) $$\mathfrak{P}_n : \mathfrak{x} = \mathfrak{x}_{n*}(w_*), \qquad w_* \in K_*,$$

such that the representation (3) is polyhedral and

(4) $$\mathfrak{x}_{n*}(w_*) \to \mathfrak{x}_*(w_*) \qquad \text{uniformly on } K_*.$$

By V.2.36 we have for each n a topological transformation $\tau_n(K) = K_*$ such that

(5) $$\tau_n(w_j) = w_{j*}, \qquad j = 1, 2, 3,$$

and such that the representation (cf. II.3.11)

(6) $$\mathfrak{P}_n : \mathfrak{x} = \mathfrak{x}_n(w) = \mathfrak{x}_{n*}[\tau_n(w)], \qquad w \in K,$$

is quasi-conformal. Let us put

$$\mu_n = \max | \mathfrak{x}_*(w_*) - \mathfrak{x}_{n*}(w_*) |, \qquad w_* \in K_* . \tag{7}$$

By (4) we have then

$$\mu_n \to 0 \qquad \text{for } n \to \infty. \tag{8}$$

Since $\mathfrak{x}_*(w_*)$ is continuous on K_*, $\mathfrak{x}_*(w_*)$ is also bounded on K_*, and from (4) it follows thus immediately that the sequence $\mathfrak{x}_{n*}(w_*)$ is uniformly bounded on K_*. Consequently, the sequence $\mathfrak{x}_n(w)$ is uniformly bounded on K. As a further consequence of (4), the sequence $\mathfrak{x}_{n*}(w_*)$ is equicontinuous on K_* (cf. I.2.44). Thus for every $\epsilon > 0$ we have a $\delta(\epsilon) > 0$, such that for all n and w_*'', $w_*' \in K_*$,

$$| \mathfrak{x}_{n*}(w_*'') - \mathfrak{x}_{n*}(w_*') | < \epsilon \text{ if } | w_*'' - w_*' | < \delta(\epsilon).$$

Since the representation (1) is nondegenerate, by II.1.8 we have, for every $\epsilon > 0$, an $\eta(\epsilon) > 0$ such that if Γ is any continuum in K_*, then (cf. II.3.21, II.1.8)

$$d(\Gamma) > \epsilon \text{ implies } \omega[\mathfrak{x}_*(w_*), \Gamma] > \eta(\epsilon). \tag{9}$$

Now by (7) we have $\omega[\mathfrak{x}_{n*}(w_*), \Gamma] \geqq \omega[\mathfrak{x}_*(w_*), \Gamma] - 2\mu_n$. Thus, by (9), for every n, $d(\Gamma) > \epsilon$ implies $\omega[\mathfrak{x}_{n*}(w_*), \Gamma] > \eta(\epsilon) - 2\mu_n$. For given $\epsilon > 0$, we have by (8) an $n_0(\epsilon)$ such that $\mu_n < \eta(\epsilon)/4$ for $n > n_0(\epsilon)$. Hence

$$\omega[\mathfrak{x}_{n*}(w_*), \Gamma] > \eta(\epsilon)/2 \text{ if } d(\Gamma) > \epsilon,\ n > n_0(\epsilon). \tag{10}$$

After these preliminary remarks, the proof will be given in two steps.

V.2.41. Continuation. We assert that the sequence $\tau_n(w)$ is equicontinuous on K (cf. I.2.44).

Proof. If the assertion is denied, then there should exist a positive constant λ, a sequence of subscripts $n_1 < n_2 < \cdots < n_k < \cdots$, and a sequence of pairs of points w_{n_k}'', w_{n_k}' in K, such that the following conditions hold.

(i) $| w_{n_k}'' - w_{n_k}' | < 1/k$, $k = 1, 2, \cdots$.

(ii) $| \tau_{n_k}(w_{n_k}'') - \tau_{n_k}(w_{n_k}') | > \lambda > 0$, $k = 1, 2, \cdots$.

If λ is a constant for which this holds, then any positive constant less than λ satisfies (ii) also. Thus we can assume that $\lambda < 3^{1/2}$. Since we can extract convergent subsequences from the sequences w_{n_k}'', w_{n_k}', we can assume without loss of generality that the sequences w_{n_k}'', w_{n_k}' are convergent. Let w_0 be their common limit (cf. (i)). The point w_0 may be an interior point of K or it may lie on the perimeter of K. Now let us take any ρ_0 such that $0 < \rho_0 < 3^{1/2}/2$, and next any ρ such that

$$\rho_0 < \rho < 3^{1/2}/2. \tag{1}$$

Using the notations of V.2.37, let us denote by $\mathfrak{R}(\rho, w_0)$ the (bounded) Jordan region bounded by the simple closed curve $C(\rho, w_0)$. Since w_{n_k}'', $w_{n_k}' \to w_0$ for $k \to \infty$, we have a $k(\rho_0)$ such that

$$w_{n_k}'', w_{n_k}' \in \mathfrak{R}(\rho_0, w_0) \qquad \text{for } k > k(\rho_0),$$

and hence also

$$w''_{n_k},\ w'_{n_k} \in \mathfrak{R}(\rho, w_0) \qquad \text{for } k > k(\rho_0).$$

By (ii), the Jordan region $\tau_{n_k}[\mathfrak{R}(\rho, w_0)]$ contains then two points whose distance exceeds λ, and consequently

$$(2) \qquad d[\tau_{n_k}(C(\rho, w_0))] > \lambda \qquad \text{for } k > k(\rho_0).$$

We assert now that (cf. V.2.37)

$$(3) \qquad d[\tau_{n_k}(C_2(\rho, w_0))] > \lambda/2 \qquad \text{for } k > k(\rho_0).$$

Indeed, if $C_2(\rho, w_0) = C(\rho, w_0)$, then the assertion is clear. Otherwise let w''_*, w'_* be the end points of the arc $\tau_{n_k}(C_2(\rho, w_0))$. Then w''_*, w'_* lie on the perimeter of K_*, and if (3) is denied, then we should have

$$(4) \qquad | w''_* - w'_* | \leqq \lambda/2.$$

Now w''_*, w'_* are also the end points of the arc $\tau_{n_k}(C_1(\rho, w_0))$ which is a subarc of the perimeter of K_*. Now $C_1(\rho, w_0)$ contains at most one of the three points w_1, w_2, w_3, on account of the condition (1) (note that w_1, w_2, w_3 are vertices of an equilateral triangle with side length $3^{1/2}$). Hence the arc $\tau_{n_k}(C_1(\rho, w_0))$ contains at most one of the points w_{1*}, w_{2*}, w_{3*}. Since $\lambda < 3^{1/2}$, it follows by (4) and V.2.39 that

$$(5) \qquad d[\tau_{n_k}(C_1(\rho, w_0))] \leqq \lambda/2.$$

Since we denied (3), we should also have

$$(6) \qquad d[\tau_{n_k}(C_2(\rho, w_0))] \leqq \lambda/2.$$

But (5) and (6) contradict (2), and thus (3) is proved. Now (3) implies, by V.2.40 (10), the inequality

$$\omega[\mathfrak{r}_{n_k*}(w_*),\ \tau_{n_k}(C_2(\rho, w_0))] > \eta(\lambda/2)/2, \qquad \text{for } n_k > n_0(\lambda/2),\ k > k(\rho_0).$$

By V.2.40(6) it follows that

$$\omega[\mathfrak{r}_{n_k}(w),\ C_2(\rho, w_0)] > \eta(\lambda/2)/2, \qquad \text{for } n_k > n_0(\lambda/2),\ k > k(\rho_0).$$

By V.2.38 there follows the inequality

$$E(\mathfrak{P}_{n_k}) > \frac{[\eta(\lambda/2)]^2}{16\pi} \log \frac{3^{1/2}}{2\rho_0}, \qquad \text{for } n_k > n_0\left(\frac{\lambda}{2}\right),\ k > k(\rho_0).$$

For $k \to \infty$ we obtain, in view of V.2.40(2),

$$A(S) \geqq \frac{[\eta(\lambda/2)]^2}{16\pi} \log \frac{3^{1/2}}{2\rho_0}.$$

This inequality should hold for all positive values of $\rho_0 < 3^{1/2}/2$. For $\rho_0 \to 0$ it would follow that $A(S) = +\infty$, in contradiction to the assumption $A(S) < +\infty$.

V.2.42. CONTINUATION. Since the sequence $\tau_n(w)$ is equicontinuous in K by V.2.41 and the sequence $\mathfrak{x}_{n*}(w_*)$ is equicontinuous in K_* (see V.2.40), it follows that the sequence $\mathfrak{x}_n(w) = \mathfrak{x}_{n*}[\tau_n(w)]$ is equicontinuous in K. As observed in V.2.40, the sequence $\mathfrak{x}_n(w)$ is uniformly bounded in K. In view of these facts, it follows (see I.2.44) that we have a subsequence $\mathfrak{x}_{m_j}(w)$, $m_1 < m_2 < \cdots$, that converges uniformly in K. Let $\mathfrak{x}(w)$ be the limit vector function, and let us consider the surface

$$S^* : \mathfrak{x} = \mathfrak{x}(w), \qquad w \in K. \tag{1}$$

Since $\mathfrak{x}_{m_j}(w) \to \mathfrak{x}(w)$, for $j \to \infty$, uniformly in K, we have the relation $\mathfrak{P}_{m_j} \to S^*$. But $\mathfrak{P}_n \to S$ by V.2.40(2). Hence $S^* = S$ (see II.3.15). In other words, we have for S the representation $S : \mathfrak{x} = \mathfrak{x}(w)$, $w \in K$, and concerning this representation we can make the following statements.

(i) $\mathfrak{x}_{m_j}(w) \to \mathfrak{x}(w)$, for $j \to \infty$, uniformly on K.

(ii) For each j, we have the representation $\mathfrak{P}_{m_j} : \mathfrak{x} = \mathfrak{x}_{m_j}(w)$, $w \in K$, and this representation is quasi-conformal.

(iii) $\mathfrak{P}_{m_j} \to S$ and $E(\mathfrak{P}_{m_j}) \to A(S)$ for $j \to \infty$ (see V.2.40(2)).

(iv) $A(S) < +\infty$ by assumption.

By V.2.32, V.2.34, V.2.35 it follows that the representation (1) is generalized conformal, and the proof of V.2.40 is complete.

V.2.43. THEOREM. *Given a surface $S : \mathfrak{x} = \mathfrak{x}(w)$, $w \in \mathfrak{R}$, suppose that the following conditions hold.*

(i) $A(S) < +\infty$.

(ii) *The middle space $\mathfrak{M}$ associated with the corresponding transformation $T : \mathfrak{x} = \mathfrak{x}(w)$, $w \in \mathfrak{R}$, is a topological 2-cell (see* II.1.20).

Then S admits of a generalized conformal representation, in the sense of V.2.29.

PROOF. By II.3.27, II.3.11, S admits of a nondegenerate representation upon the unit disc, and hence the theorem follows from V.2.40.

V.2.44. THEOREM. *Given a surface S in terms of a representation $S : \mathfrak{x} = \mathfrak{x}(w)$, $w \in \mathfrak{R}$, suppose that the following conditions hold.*

(i) $A(S) < +\infty$.

(ii) *S is bounded by a simple closed curve (cf.* II.3.39).

(iii) *The components of $\mathfrak{x}(w)$ are Lebesgue monotone on $\mathfrak{R}$ (cf.* II.3.45).

Then S admits of a generalized conformal map upon the unit disc, in the sense of V.2.29.

PROOF. By II.3.45, II.3.29, the middle space $\mathfrak{M}$ is a topological 2-cell, and hence the theorem follows from V.2.43.

V.2.45. Let $Q_0 : 0 \leqq u \leqq 1$, $0 \leqq v \leqq 1$, be the unit square in the w-plane. With every continuous vector function $\mathfrak{x}(w)$, defined in Q_0, we associate the transformation $T : \mathfrak{x} = \mathfrak{x}(w)$, $w \in Q_0$, and the surface $S : \mathfrak{x} = \mathfrak{x}(w)$, $w \in Q_0$. We define the function $F(T)$ by the formula (cf. V.2.3)

$$F(T) = A(S).$$

It will be convenient to write $A(\mathfrak{x})$ instead of $A(S)$, in order to display the vector function $\mathfrak{x}(w)$. Thus

$$F(T) = A(\mathfrak{x}) = A(S).$$

V.2.46. Continuation. Clearly, $F(T) \geqq 0$. By V.2.6, $F(T)$ is lower semi-continuous in the sense of II.2.106. We proceed to verify that $F(T)$ satisfies the various assumptions stated in connection with the cyclic additivity theorem of II.2.113, II.2.114.

V.2.47. Continuation. Suppose that the point-set $T(Q_0)$, in $x_1x_2x_3$-space, is a subset of a straight segment σ. Then $F(T) = 0$. In particular, if $T(Q_0)$ reduces to a single point, then $F(T) = 0$.

Proof. For every positive integer n, we construct in Q_0 a quasi-linear vector function $\mathfrak{x}_n(w)$ as follows. We subdivide Q_0, by horizontals and verticals, into n^2 congruent oriented squares, in each of which we draw the diagonal from the upper left to the lower right corner. There results a rectilinear triangulation $\mathfrak{J}_n$ of Q_0, and $\mathfrak{x}_n(w)$ is defined as the (univocally determined) vector function with the following properties. (i) $\mathfrak{x}_n(w) = \mathfrak{x}(w)$ at the vertices of $\mathfrak{J}_n$. (ii) The components of $\mathfrak{x}_n(w)$ are linear functions of u, v on each one of the triangles of $\mathfrak{J}_n$. For each n we have then the quasi-linear surface

$$(1) \qquad S_n : \mathfrak{x} = \mathfrak{x}_n(w), \qquad w \in Q_0 .$$

Clearly $\mathfrak{x}_n(w) \to \mathfrak{x}(w)$ uniformly on Q_0, and hence (cf. II.3.17, V.2.6)

$$(2) \qquad A(S) \leqq \liminf A(S_n).$$

On the other hand, for every triangle t_i of $\mathfrak{J}_n$, the vertices of t_i are carried, by means of the representation (1), into points of the straight segment σ. Hence, by V.2.12, V.1.26, $A(S_n) = 0$, $n = 1, 2, \cdots$. In view of (2), it follows that $F(T) = A(S) = 0$.

V.2.48. Given a surface S in terms of a representation $S : \mathfrak{x} = \mathfrak{x}(w)$, $w \in Q_0$, where Q_0 is the unit square $0 \leqq u \leqq 1$, $0 \leqq v \leqq 1$, let us suppose that $A(S) < +\infty$. Let E be any subset of Q_0 such that

$$(1) \qquad \mathfrak{x}(w) = \mathfrak{a} \qquad \text{for } w \in E,$$

where $\mathfrak{a}$ is a constant vector. Given then any $\sigma > 0$, there exists a continuous vector function $\mathfrak{y}(w)$, $w \in Q_0$, with the following properties.

(i) $\mathfrak{y}(w)$ is Lipschitzian on Q_0 (cf. V.1.27).

(ii) $| \mathfrak{y}(w) - \mathfrak{x}(w) | < \sigma$ for $w \in Q_0$.

(iii) $A(\mathfrak{y}) < A(\mathfrak{x}) + \sigma$ (cf. V.2.45).

(iv) $\mathfrak{y}(w) = \mathfrak{a}$ for $w \in E$.

Proof. Let ϵ, r, R be positive numbers such that

$$(2) \qquad 0 < \epsilon < r < R.$$

By V.2.5 we have a polyhedron $\mathfrak{P}$ such that

$$(3) \qquad d(S, \mathfrak{P}) < \epsilon/2, \ | A(S) - E(\mathfrak{P}) | < \epsilon.$$

By II.3.18 we have a representation $\mathfrak{P} : \mathfrak{x} = \mathfrak{z}_*(w)$, $w \in Q_0$, such that

$$(4) \qquad |\mathfrak{z}_*(w) - \mathfrak{x}(w)| < \epsilon/2, \qquad w \in Q_0 .$$

By V.1.37 we have for $\mathfrak{P}$ a quasi-linear representation

$$(5) \qquad \mathfrak{P} : \mathfrak{x} = \mathfrak{z}(w), \qquad w \in Q_0 ,$$

such that $|\mathfrak{z}_*(w) - \mathfrak{z}(w)| < \epsilon/2$ on Q_0. In view of (4) it follows that

$$(6) \qquad |\mathfrak{z}(w) - \mathfrak{x}(w)| < \epsilon, \qquad w \in Q_0 .$$

To the representation (5) we apply the stretching process $\Omega = \Omega(r, R, \mathfrak{a})$, where r, R, $\mathfrak{a}$ are taken from (1) and (2) (cf. V.1.46), and we obtain a vector function $\mathfrak{z}^*(w)$, $w \in Q_0$. Since $\mathfrak{z}(w)$ is quasi-linear (cf. V.1.26) and therefore Lipschitzian on Q_0, $\mathfrak{z}^*(w)$ is also Lipschitzian on Q_0, by V.1.52. By V.1.51, we have $|\mathfrak{z}(w) - \mathfrak{z}^*(w)| \leqq R$ in Q_0, and hence by (6)

$$(7) \qquad |\mathfrak{z}^*(w) - \mathfrak{x}(w)| < R + \epsilon, \qquad w \in Q_0 .$$

By V.1.53 we have the inequality

$$(8) \qquad a(S^*) \leqq \frac{R}{R - r} E(\mathfrak{P}),$$

where $S^* : \mathfrak{x} = \mathfrak{z}^*(w)$, $w \in Q_0$. Since $\mathfrak{z}^*(w)$ is Lipschitzian in Q_0, we have $a(S^*) = A(S^*)$ by V.2.12. Hence (cf. V.2.45), we obtain from (3) and (8) the inequality

$$(9) \qquad A(\mathfrak{z}^*) < \frac{R}{R - r} (A(\mathfrak{x}) + \epsilon).$$

(1) and (6) yield, for $w \in E$, $|\mathfrak{z}(w) - \mathfrak{a}| = |\mathfrak{z}(w) - \mathfrak{x}(w)| < \epsilon$. In view of (2) we have therefore $|\mathfrak{z}(w) - \mathfrak{a}| < r$ for $w \in E$. Hence, by the definition of the stretching process $\Omega = \Omega(r, R, \mathfrak{a})$ (cf. V.1.46)

$$\mathfrak{z}^*(w) = \mathfrak{a} \qquad \text{for } w \in E.$$

Thus the vector function $\mathfrak{y}(w) = \mathfrak{z}^*(w)$ satisfies all of our requirements, provided that ϵ, r, R can be chosen in such a way that the following inequalities hold (cf. (2), (7), (9)).

$$(10) \qquad 0 < \epsilon < r < R.$$

$$(11) \qquad R + \epsilon < \sigma.$$

$$(12) \qquad \frac{R}{R - r} (A(\mathfrak{x}) + \epsilon) < A(\mathfrak{x}) + \sigma.$$

Since $A(\mathfrak{x}) < +\infty$ by assumption, these conditions are easily seen to be compatible. For example, we can first choose R to satisfy

$$(13) \qquad 0 < R < \sigma/2.$$

Next we can choose r to satisfy

(14) $$0 < r < \frac{R\sigma}{2[A(\mathfrak{x}) + \sigma]}.$$

Finally, we can choose ϵ to satisfy

(15) $$0 < \epsilon < \min (r, \sigma/2).$$

It is easy to verify that the inequalities (13), (14), (15) imply the inequalities (10), (11), (12).

V.2.49. Let S be given in terms of a representation $S : \mathfrak{x} = \mathfrak{x}(w)$, $w \in Q_0$, and let Γ be a continuum in Q_0 such that $\mathfrak{x}(w) = \mathfrak{a}$ for $w \in \Gamma$, where $\mathfrak{a}$ is a constant vector. Let D be a component of $Q_0 - \Gamma$, and let us define, in Q_0, the continuous vector functions $\mathfrak{x}_1(w)$, $\mathfrak{x}_2(w)$ as follows (cf. II.2.105):

(1) $$\mathfrak{x}_1(w) = \begin{cases} \mathfrak{x}(w) & \text{for } w \in D, \\ \mathfrak{a} & \text{for } w \in Q_0 - D. \end{cases}$$

(2) $$\mathfrak{x}_2(w) = \begin{cases} \mathfrak{a} & \text{for } w \in D, \\ \mathfrak{x}(w) & \text{for } w \in Q_0 - D. \end{cases}$$

We assert that (cf. V.2.45) $A(\mathfrak{x}_1) + A(\mathfrak{x}_2) = A(\mathfrak{x})$. In other words, the function $F(T)$ of V.2.45 is *additive with respect to continua of constancy, in the sense of* II.2.105. The proof will be given in two steps.

V.2.50. CONTINUATION. We first establish the inequality

(1) $$A(\mathfrak{x}_1) + A(\mathfrak{x}_2) \leqq A(\mathfrak{x}).$$

This inequality is obvious if $A(\mathfrak{x}) = +\infty$. Hence we can assume that $A(\mathfrak{x}) < +\infty$. By V.2.48 we have then for every positive integer n a continuous vector function $\mathfrak{y}_n(w)$, $w \in Q_0$, with the following properties.

(i) $\mathfrak{y}_n(w)$ is Lipschitzian on Q_0.
(ii) $| \mathfrak{y}_n(w) - \mathfrak{x}(w) | < 1/n$, $w \in Q_0$.
(iii) $A(\mathfrak{y}_n) < A(\mathfrak{x}) + 1/n$.
(iv) $\mathfrak{y}_n(w) = \mathfrak{a}$ for $w \in \Gamma$.

Let us define the vector functions $\mathfrak{y}_{n1}(w)$, $\mathfrak{y}_{n2}(w)$ as follows:

$$\mathfrak{y}_{n1}(w) = \begin{cases} \mathfrak{y}_n(w) & \text{for } w \in D, \\ \mathfrak{a} & \text{for } w \in Q_0 - D. \end{cases}$$

$$\mathfrak{y}_{n2}(w) = \begin{cases} \mathfrak{a} & \text{for } w \in D, \\ \mathfrak{y}_n(w) & \text{for } w \in Q_0 - D. \end{cases}$$

As an immediate consequence of (i) above, $\mathfrak{y}_{n1}(w)$ and $\mathfrak{y}_{n2}(w)$ are both Lipschitzian in Q_0. Furthermore, we have

$$| \mathfrak{y}_{n1}(w) - \mathfrak{x}_1(w) | = \begin{cases} 0 & \text{for } w \in Q_0 - D. \\ | \mathfrak{y}_n(w) - \mathfrak{x}(w) | & \text{for } w \in D. \end{cases}$$

Hence, by (ii),

(2) $$| \mathfrak{y}_{n1}(w) - \mathfrak{x}_1(w) | < 1/n, \qquad w \in Q_0,$$

and similarly

(3) $$| \mathfrak{y}_{n2}(w) - \mathfrak{x}_2(w) | < 1/n, \qquad w \in Q_0 .$$

(2) and (3) imply (cf. V.2.45, V.2.46)

$$A(\mathfrak{x}_1) \leqq \liminf_{n\to\infty} A(\mathfrak{y}_{n1}),$$

$$A(\mathfrak{x}_2) \leqq \liminf_{n\to\infty} A(\mathfrak{y}_{n2}),$$

and hence

(4) $$A(\mathfrak{x}_1) + A(\mathfrak{x}_2) \leqq \liminf_{n\to\infty} (A(\mathfrak{y}_{n1}) + A(\mathfrak{y}_{n2})).$$

Since $\mathfrak{y}_n(w)$, $\mathfrak{y}_{n1}(w)$ and $\mathfrak{y}_{n2}(w)$ are Lipschitzian in Q_0, we have by V.2.11, V.1.3, V.1.7,

(5) $$A(\mathfrak{y}_n) = a(\mathfrak{y}_n , Q_0),\ A(\mathfrak{y}_{n1}) = a(\mathfrak{y}_{n1} , Q_0),\ A(\mathfrak{y}_{n2}) = a(\mathfrak{y}_{n2} , Q_0).$$

By V.1.60, V.1.61,

(6) $$a(\mathfrak{y}_{n1} , Q_0) + a(\mathfrak{y}_{n2} , Q_0) = a(\mathfrak{y}_n , Q_0).$$

(4), (5), (6) yield

(7) $$A(\mathfrak{x}_1) + A(\mathfrak{x}_2) \leqq \liminf_{n\to\infty} A(\mathfrak{y}_n).$$

By condition (iii)

(8) $$\liminf_{n\to\infty} A(\mathfrak{y}_n) \leqq A(\mathfrak{x}).$$

(7) and (8) yield (1).

V.2.51. Continuation. We shall now complete the proof of V.2.49 by establishing the complementary inequality

(1) $$A(\mathfrak{x}) \leqq A(\mathfrak{x}_1) + A(\mathfrak{x}_2).$$

This inequality is obvious if one of $A(\mathfrak{x}_1)$, $A(\mathfrak{x}_2)$ fails to be finite. So we can assume that $A(\mathfrak{x}_1) < +\infty$, $A(\mathfrak{x}_2) < +\infty$. By V.2.48 we have then, for every positive integer n, continuous vector functions $\mathfrak{z}_{n1}(w)$, $\mathfrak{z}_{n2}(w)$ with the following properties (note that $\mathfrak{x}_1(w) = \mathfrak{a}$ on $Q_0 - D$ and $\mathfrak{x}_2(w) = \mathfrak{a}$ on $D + \Gamma$):

(α) $\mathfrak{z}_{n1}(w)$, $\mathfrak{z}_{n2}(w)$ are Lipschitzian on Q_0.

(β) $| \mathfrak{z}_{ni}(w) - \mathfrak{x}_i(w) | < 1/n$, $i = 1, 2$, $w \in Q_0$.

(γ) $A(\mathfrak{z}_{n\iota}) < A(\mathfrak{x}_\iota) + 1/n, \iota = 1, 2.$
(δ) $\mathfrak{z}_{n1}(w) = \mathfrak{a}$ on $Q_0 - D$, $\mathfrak{z}_{n2}(w) = \mathfrak{a}$ on $D + \Gamma$.
We define now

$$(2) \qquad \mathfrak{z}_n(w) = \begin{cases} \mathfrak{z}_{n1}(w) & \text{for } w \in D, \\ \mathfrak{z}_{n2}(w) & \text{for } w \in Q_0 - D. \end{cases}$$

Clearly $\mathfrak{z}_n(w)$ is single-valued and Lipschitzian on Q_0. The relation between $\mathfrak{z}_n(w)$, $\mathfrak{z}_{n1}(w)$, $\mathfrak{z}_{n2}(w)$ may be restated in the following equivalent form by using condition (δ).

$$\mathfrak{z}_{n1}(w) = \begin{cases} \mathfrak{z}_n(w) & \text{for } w \in D, \\ \mathfrak{a} & \text{for } w \in Q_0 - D. \end{cases}$$

$$\mathfrak{z}_{n2}(w) = \begin{cases} \mathfrak{z}_n(w) & \text{for } w \in Q_0 - D, \\ \mathfrak{a} & \text{for } w \in D. \end{cases}$$

Clearly, in view of (δ), we have $\mathfrak{z}_n(w) = \mathfrak{a}$ on Γ, since $\Gamma \subset Q_0 - D$. Hence, by V.1.61, V.1.7,

$$(3) \qquad a(\mathfrak{z}_{n1}, Q_0) + a(\mathfrak{z}_{n2}, Q_0) = a(\mathfrak{z}_n, Q_0).$$

On the other hand, since $\mathfrak{z}_{n1}(w)$, $\mathfrak{z}_{n2}(w)$, $\mathfrak{z}_n(w)$ are Lipschitzian on Q_0, we have by V.2.11, V.1.3, V.1.7,

$$(4) \qquad a(\mathfrak{z}_{n1}, Q_0) = A(\mathfrak{z}_{n1}),\ a(\mathfrak{z}_{n2}, Q_0) = A(\mathfrak{z}_{n2}),\ a(\mathfrak{z}_n, Q_0) = A(\mathfrak{z}_n).$$

(3), (4) and (γ) yield

$$(5) \qquad A(\mathfrak{z}_n) < A(\mathfrak{x}_1) + A(\mathfrak{x}_2) + 2/n.$$

By (2) and V.2.49(1), V.2.49(2) we have

$$\mathfrak{z}_n(w) - \mathfrak{x}(w) = \begin{cases} \mathfrak{z}_{n1}(w) - \mathfrak{x}_1(w) & \text{for } w \in D, \\ \mathfrak{z}_{n2}(w) - \mathfrak{x}_2(w) & \text{for } w \in Q_0 - D. \end{cases}$$

Hence, by (β), $|\mathfrak{z}_n(w) - \mathfrak{x}(w)| < 2/n$, $w \in Q_0$. Thus $\mathfrak{z}_n(w) \to \mathfrak{x}(w)$ uniformly on Q_0, and hence by V.2.45, V.2.6,

$$(6) \qquad A(\mathfrak{x}) \leqq \liminf_{n\to\infty} A(\mathfrak{z}_n).$$

(5) and (6) imply the inequality (1).

V.2.52. By V.2.46, V.2.47, V.2.49 the function $F(T)$ of V.2.45 is additive with respect to cyclic chains in the sense of II.2.106(f). We shall consider a very special application of this fact. Let

$$\mathfrak{x} = \mathfrak{y}(\xi), \qquad \xi \in I,$$

be a nondegenerate continuous vector function on the interval

$$I : 0 \leqq \xi \leqq 1;$$

that is, $\mathfrak{y}(\xi)$ is not constant on any nondegenerate subinterval of I. Let $\mu(Q_0) = I$ be a continuous and monotone transformation from the unit square Q_0 onto I (see II.1.1). Consider the surface

$$S : \mathfrak{x} = \mathfrak{x}(w) = \mathfrak{y}(\mu\,(w)), \qquad w \in Q_0\,,$$

and the associated transformation (cf. V.2.45) $T : \mathfrak{x} = \mathfrak{y}(\mu(w))$, $w \in Q_0$. Clearly the middle space $\mathfrak{M}$ now coincides with the interval I. Let I^* be a nondegenerate subinterval of I. Then I^* is a cyclic chain of I (cf. II.2.60). Let T^* be the transformation $T \mid I^*$ (cf. II.2.100). Then T^* is given by a formula $T^* : \mathfrak{x} = \mathfrak{y}[\mu^*(w)]$, $w \in Q_0$, where μ^* is a continuous monotone transformation from Q_0 onto I^*. Thus $\mu^*(Q_0) = I^*$. Let us now consider a subdivision of I into a finite number of nonoverlapping intervals $I_1^*, I_2^*, \cdots, I_m^*$, and let us denote by μ_1^*, $\mu_2^*, \cdots, \mu_m^*$ the corresponding transformations from Q_0 onto $I_1^*, I_2^*, \cdots, I_m^*$ respectively. Let us define $S_j^* : \mathfrak{x} = \mathfrak{x}_j^*(w) = \mathfrak{y}[\mu_j^*(w)]$, $w \in Q_0$, $j = 1, 2, \cdots, m$. Since the function $F(T)$ of V.2.45 is additive with respect to cyclic chains as noted above, we have

$$A(\mathfrak{x}) = A(\mathfrak{x}_1^*) + \cdots + A(\mathfrak{x}_m^*).$$

Now let us make the further assumption that the point set $\mathfrak{y}(I_j^*)$ is a straight segment σ_j^* in $x_1x_2x_3$-space, $j = 1, 2, \cdots, m$. By V.2.47 we have then $A(\mathfrak{x}_j^*) = 0$, $j = 1, 2, \cdots, m$, and hence also $A(\mathfrak{x}) = 0$.

V.2.53. Using the terminology of V.2.45, *let us assume that the middle space $\mathfrak{M}$, corresponding to T by means of its monotone-light factorization, reduces to a simple arc. Then $F(T) = 0$.*

PROOF. In view of II.3.11, we can assume that $\mathfrak{M}$ is the unit segment I . $0 \leqq \xi \leqq 1$. The transformation T is then factored in the form

$$T : \mathfrak{x} = \mathfrak{x}(w) = \mathfrak{y}[\mu(w)], \qquad w \in Q_0\,,$$

where $\mu(w)$ is monotone and $\mu(Q_0) = I$, while $\mathfrak{x} = \mathfrak{y}(\xi)$, $\xi \in I$, is light on I, and hence $\mathfrak{y}(\xi)$ is nondegenerate on I. For every positive integer n, let us construct a continuous vector-function $\mathfrak{y}_n(\xi)$, $\xi \in I$, as follows. With every one of the points $\xi_{nk} = k/n$, $k = 0, 1, \cdots, n$, we associate a vector $\mathfrak{a}_{nk}$ in $x_1x_2x_3$-space. Let P_{nk} be the second end point of $\mathfrak{a}_{nk}$, its first end point being at the origin (0, 0, 0). These vectors $\mathfrak{a}_{nk}$ are chosen subject to the following restrictions.

(i) No three of the points $P_{n0}, P_{n1}, \cdots, P_{nn}$ are collinear.

(ii) $\mid \mathfrak{a}_{nk} - \mathfrak{y}(\xi_{nk}) \mid < 1/n$, $k = 0, 1, \cdots, n$.

These conditions are obviously compatible. We define $\mathfrak{y}_n(\xi)$ as the (univocally determined) vector function that satisfies the following conditions. (a) $\mathfrak{y}_n(\xi_{nk}) = \mathfrak{a}_{nk}$, $k = 0, 1, \cdots, n$. (b) The components of $\mathfrak{y}_n(\xi)$ are linear functions of ξ on each one of the intervals $k/n \leqq \xi \leqq (k + 1)/n$, $k = 0, 1, \cdots, n - 1$. The transformation $\mathfrak{x} = \mathfrak{y}_n(\xi)$, $\xi \in I$, is then clearly quasi-linear, and $\mathfrak{y}_n(\xi) \to \mathfrak{y}(\xi)$ uniformly on I. Hence, if we define $S_n : \mathfrak{x} = \mathfrak{x}_n(w) = \mathfrak{y}_n[\mu(w)]$, $w \in Q_0$, $S : \mathfrak{x} = \mathfrak{x}(w) = \mathfrak{y}[\mu(w)]$, $w \in Q_0$, then we have also $\mathfrak{x}_n(w) \to \mathfrak{x}(w)$ uniformly on Q_0. Hence (cf. V.2.45, V.2.6)

$$(1) \qquad A(\mathfrak{x}) \leqq \liminf_{n\to\infty} A(\mathfrak{x}_n).$$

Let us denote by I_{nk} the interval $k/n \leqq \xi \leqq (k+1)/n$, $k = 0, 1, \cdots, n-1$. Then the point set $\mathfrak{y}_n(I_{nk})$ is a straight segment in $x_1x_2x_3$-space, by the definition of $\mathfrak{y}_n$. Hence, by V.2.52, $A(\mathfrak{x}_n) = 0$ for every n, and (1) yields the desired result $A(\mathfrak{x}) = F(T) = 0$.

V.2.54. Using the terminology of V.2.45, *let us assume that the middle space $\mathfrak{M}$, associated with the monotone-light factorization of T, reduces to a single proper cyclic element. Then $F(T) > 0$.*

PROOF. By V.2.10, V.2.45, V.1.7 we have $F(T) \geqq a(\mathfrak{x}, Q_0)$. By V.1.64, V.1.7, $a(\mathfrak{x}, Q_0) > 0$. Hence $F(T) > 0$.

V.2.55. In view of II.2.92, V.2.46, V.2.47, V.2.49, V.2.53, V.2.54, the function $F(T)$ of V.2.45 satisfies all the conditions required in II.2.113, II.2.114. Hence we have theorems for the Lebesgue area $A(S)$ which are entirely analogous to those stated for the lower area $a(S)$ in V.1.75. In V.2.45 to V.2.54 we worked with the unit square Q_0 for convenience, but the final theorems that correspond to those in V.1.75 involve only $A(S)$, the middle space $\mathfrak{M}$, and the cyclic decomposition of S. All these entities are independent of the particular representation chosen for S, except that $\mathfrak{M}$ is determined only up to a homeomorphism (cf. II.3.20). In view of these remarks, we obtain by V.2.45 and II.2.113, II.2.114, the following theorems.

Given a surface S in terms of a representation $S : \mathfrak{x} = \mathfrak{x}(w)$, $w \in \mathfrak{R}$, let T be the associated transformation $T : \mathfrak{x} = \mathfrak{x}(w)$, $w \in \mathfrak{R}$. Let $\mathfrak{M}$ be the middle space that arises in the monotone-light factorization of T.

CHARACTERIZATION THEOREM FOR SURFACES S OF ZERO LEBESGUE AREA $A(S)$. *The Lebesgue area $A(S)$ is equal to zero if and only if the middle space $\mathfrak{M}$ is a dendrite.*

CYCLIC ADDITIVITY THEOREM FOR THE LEBESGUE AREA $A(S)$. *Suppose that $A(S) > 0$. Then the middle-space $\mathfrak{M}$ has at least one proper cyclic element, and hence the cyclic decomposition of S (cf. II.3.20) is not vacuous. If $S_1, \cdots$ are the surfaces in the cyclic decomposition $\Delta(S)$ of S, then*

$$A(S) = \sum_n A(S_n).$$

In case the series diverges, this formula is understood to mean that $A(S) = +\infty$.

V.2.56. According to V.2.10, always $a(S) \leqq A(S)$. We propose to discuss conditions for the sign of equality to hold. Let there be given a surface $S : \mathfrak{x} = \mathfrak{x}(w)$, $w \in \mathfrak{R}$, and let $T : \mathfrak{x} = \mathfrak{x}(w)$, $w \in \mathfrak{R}$, be the associated transformation. Suppose that the following conditions hold.

(i) $A(S) < +\infty$.

(ii) The middle space $\mathfrak{M}$, involved in the monotone-light factorization of T, is a topological 2-cell.

Then $a(S) = A(S)$.

PROOF. In view of II.3.29, condition (ii) implies that S admits of a nondegenerate representation upon the unit disc. In view of condition (i), there follows by V.2.40 the existence of a representation

$$S : \mathfrak{x} = \mathfrak{x}^*(w^*), \qquad w^* \in K^*,$$

where K^* is the unit disc $| w^* | \leqq 1$, and the representation is generalized conformal. By V.2.30 we have then the formula

$$(1) \qquad \iint_{K^*} W(w^*, \mathfrak{x}^*) = \frac{1}{2} \iint_{K^*} \left[\left(\frac{\partial \mathfrak{x}^*}{\partial u^*} \right)^2 + \left(\frac{\partial \mathfrak{x}^*}{\partial v^*} \right)^2 \right] = A(S).$$

On the other hand, by V.2.14 we have the inequalities

$$(2) \qquad \iint_{K^*} W(w^*, \mathfrak{x}^*) \leqq a(S) \leqq A(S).$$

(1) and (2) imply that $a(S) = A(S)$.

V.2.57. Let us replace, in V.2.56, condition (ii) by the condition (ii*). The middle space $\mathfrak{M}$ is a topological 2-sphere.

Then the formula $a(S) = A(S)$ still holds.

PROOF. In view of II.3.28, condition (ii*) implies that S admits of a representation $S : \mathfrak{x} = \mathfrak{x}^*(w^*)$, $w^* \in \mathfrak{R}^*$, with the following properties.

(α) $\mathfrak{x}^*(w^*) = \mathfrak{a}$ on the boundary of $\mathfrak{R}^*$, where $\mathfrak{a}$ is a constant vector.

(β) $\mathfrak{x}^*(w^*)$ is not constant on any nondegenerate continuum in the interior of $\mathfrak{R}^*$.

Let now $\mathfrak{R}_n^* \subset \mathfrak{R}^{*0}$ be a sequence of simply-connected Jordan regions that fill up $\mathfrak{R}^*$ from the interior (cf. IV.1.41), and let us define

$$(1) \qquad S_n : \mathfrak{x} = \mathfrak{x}^*(w^*), \qquad w^* \in \mathfrak{R}_n^* .$$

By V.1.11, V.2.17, V.2.16

$$(2) \qquad a(S_n) \to a(S), \; A(S_n) \to A(S),$$

$$(3) \qquad A(S_n) \leqq A(S) < +\infty.$$

As a consequence of (β), the representation (1) is nondegenerate (cf. II.3.21). Hence (cf. II.1.18) we can take $\mathfrak{R}_n^*$ as the middle space $\mathfrak{M}_n^*$ corresponding to the representation (1). In view of (3), the representation (1) satisfies all the assumptions of V.2.56, and hence $a(S_n) = A(S_n)$. By (2) there follows the formula $a(S) = A(S)$.

V.2.58. THEOREM. *Let S be a surface such that $A(S) < +\infty$. Then $A(S) = a(S)$ (cf. V.2.3, V.I.7).*

PROOF. *Case* (i): $A(S) = 0$. Then $a(S) = 0$ too by V.2.10. *Case* (ii): $A(S) > 0$. By V.2.55, the cyclic decomposition $\Delta(S)$ of S is then not vacuous. By V.2.55, V.1.75 we have the formulas

$$(1) \qquad A(S) = \sum_n A(S_n), \qquad a(S) = \sum_n a(S_n), \qquad S_n \in \Delta(S).$$

Since $A(S) < +\infty$ by assumption, it follows that $A(S_n) < +\infty$ for every n. By II.3.36, II.3.20, the middle space $\mathfrak{M}_n$ associated with S_n is either a topological 2-cell or a topological 2-sphere. By V.2.56, V.2.57 it follows that

$$(2) \qquad A(S_n) = a(S_n) \text{ for every } n.$$

(1) and (2) imply the formula $A(S) = a(S)$.

V.2.59. THEOREM. *Let S be a surface such that $a(S) = 0$. Then $a(S) = A(S)$.*

PROOF. By V.1.75, $a(S) = 0$ implies that the middle space $\mathfrak{M}$ associated with S is a dendrite (or a single point). In either case, $A(S) = 0$ by V.2.55.

V.2.60. Inspection of the proof in V.2.58 shows that the result obtained there can be improved as follows. Let $A(S) > 0$, and let $\Delta(S) = [S_1, \cdots]$ be again the cyclic decomposition of S. If $A(S_n) < +\infty$ for every n, then $A(S) = a(S)$. Indeed, the formulas (1) and (2) are valid under these assumptions, and the assertion $A(S) = a(S)$ follows. Note that the assumption that $A(S_n) < +\infty$ for every n does not imply that $A(S) < +\infty$. We shall later on learn about another important special case where the formula $A(S) = a(S)$ can be shown to hold without the assumption $A(S) < +\infty$. In the light of the information available at present, it seems reasonable to surmise that the formula $a(S) = A(S)$ is always valid. In view of V.2.58, V.2.59, this surmise would be proved if the truth of the following statement could be established: $A(S) = +\infty$ implies that $a(S) = +\infty$. Since $a(S) = +\infty$ implies that $A(S) = +\infty$ by V.2.10, it follows that this hypothetical statement is equivalent to the following hypothetical theorem.

HYPOTHETICAL CHARACTERIZATION OF SURFACES OF INFINITE LEBESGUE AREA. The Lebesgue area $A(S)$ is infinite if and only if the lower area $a(S)$ is infinite.

V.2.61. THEOREM. *Let S be a surface such that $A(S) < +\infty$, and let*

$$(1) \qquad S : \mathfrak{x} = \mathfrak{x}(w), \qquad w \in \mathfrak{R},$$

be any representation of S. Then the following statements hold.

(i) *The essential generalized Jacobians $\mathfrak{J}^r_e(w, \mathfrak{x})$ (see V.1.18) exist a.e. in $\mathfrak{R}^0$ and are summable in $\mathfrak{R}^0$. As a consequence, the function*

$$W_e(w, \mathfrak{x}) = \{[\mathfrak{J}^1_e(w, \mathfrak{x})]^2 + [\mathfrak{J}^2_e(w, \mathfrak{x})]^2 + [\mathfrak{J}^e_3(w, \mathfrak{x})]^2\}^{1/2}$$

exists a.e. in $\mathfrak{R}^0$ and is summable in $\mathfrak{R}^0$.

$$(\text{ii}) \qquad \iint_{\mathfrak{R}^0} W_e(w, \mathfrak{x}) \leqq A(S).$$

(iii) *The formula*

$$\iint_{\mathfrak{R}^\circ} W_e(w, \mathfrak{x}) = A(S) \tag{2}$$

holds if and only if the representation (1) *is eAC* (cf. V.1.15).

Proof. By V.2.10, $a(S) \leqq A(S) < +\infty$, and thus (i) and (ii) follow directly from V.1.19. As regards (iii), suppose first that the representation (1) is eAC. Then by V.1.20

$$\iint_{\mathfrak{R}^\circ} W_e(w, \mathfrak{x}) = a(S).$$

Since $A(S) < +\infty$ by assumption, $a(S) = A(S)$ by V.2.58, and (2) follows. Let us next suppose that (2) holds; we have to show that the representation (1) is eAC. We shall make the proof in two steps.

V.2.62. Continuation. Let us first assume that $\mathfrak{R}$ coincides with the unit square $Q_0 : 0 \leqq u \leqq 1, 0 \leqq v \leqq 1$. Then (1) in V.2.61 has the form

$$S : \mathfrak{x} = \mathfrak{x}(w), \qquad w \in Q_0, \tag{1}$$

and by assumption we have the formula

$$\iint_{Q_0} W_e(w, \mathfrak{x}) = A(S). \tag{2}$$

By V.2.4 we have a sequence of polyhedra $\mathfrak{P}_n$ such that

$$\mathfrak{P}_n \to S, \ E(\mathfrak{P}_n) \to A(S). \tag{3}$$

By II.3.16 we have for the polyhedra $\mathfrak{P}_n$ representations of the form $\mathfrak{P}_n : \mathfrak{x} = \mathfrak{x}_n^*(w)$, $w \in Q_0$, such that $\mathfrak{x}_n^*(w) \to \mathfrak{x}(w)$ uniformly on Q_0. By V.1.37 we have then for $\mathfrak{P}_n$ a quasi-linear representation

$$\mathfrak{P}_n : \mathfrak{x} = \mathfrak{x}_n(w), \qquad w \in Q_0, \tag{4}$$

such that $| \mathfrak{x}_n^*(w) - \mathfrak{x}_n(w) | < 1/n$ in Q_0. Hence

$$\mathfrak{x}_n(w) \to \mathfrak{x}(w) \qquad \text{uniformly on } Q_0. \tag{5}$$

The representation (4) being quasi-linear, it is also Lipschitzian and hence surely eAC (cf. IV.4.28, IV.4.1). Hence by V.1.20, V.2.9, V.1.33,

$$\iint_{Q_0} W_e(w, \mathfrak{x}_n) = E(\mathfrak{P}_n) = A(\mathfrak{P}_n). \tag{6}$$

(2), (3) and (6) imply that

$$\iint_{Q_0} W_e(w, \mathfrak{x}_n) \underset{n\to\infty}{\to} \iint_{Q_0} W_e(w, \mathfrak{x}). \tag{7}$$

Since the representation (4) is eAC, (5) and (7) imply, by V.1.59, that the representation (1) is also eAC.

V.2.63. Continuation. Returning to the general case, we have the following situation. S is given by a representation

$$S : \mathfrak{x} = \mathfrak{x}(w), \qquad w \in \mathfrak{R}, \tag{1}$$

and by assumption we have the formula

$$\iint_{\mathfrak{R}^\circ} W_e(w, \mathfrak{x}) = A(S) < +\infty. \tag{2}$$

Now let Q_0^* be the unit square in an auxiliary w^*-plane, and let

$$\tau : w^* = \tau(w), \qquad w \in \mathfrak{R},\ \tau(\mathfrak{R}) = Q_0^*, \tag{3}$$

be a topological transformation from $\mathfrak{R}$ onto Q_0^* that is conformal on $\mathfrak{R}^0$ (cf. I.3.19). By II.3.11 we have then for S the representation

$$S : \mathfrak{x} = \mathfrak{x}^*(w^*) = \mathfrak{x}[\tau^{-1}(w^*)], \qquad w^* \in Q_0^*. \tag{4}$$

Then τ, considered as a transformation from $\mathfrak{R}^0$ onto Q_0^{*0}, is bimeasurable (see IV.4.66).

Now let us note that the representations (1) and (4) are both eBV, as a consequence of the assumption $A(S) < +\infty$ (cf. V.1.15, V.1.16, V.2.10). Hence, if we denote by $J(w)$ the Jacobian of the transformation τ, then we have by IV.4.64, IV.3.47 the formulas

$$\mathfrak{J}_e^i(w, \mathfrak{x}) = \mathfrak{J}_e^i[\tau(w), \mathfrak{x}^*]J(w) \qquad \text{a.e. in } \mathfrak{R}^0,\ i = 1, 2, 3.$$

Hence

$$W_e(w, \mathfrak{x}) = W_e[\tau(w), \mathfrak{x}^*] \mid J(w) \mid \qquad \text{a.e. in } \mathfrak{R}^0. \tag{5}$$

Since τ is bimeasurable, we have the transformation formula (cf. IV.4.58)

$$\iint_{Q_0^*} W_e(w^*, \mathfrak{x}^*) = \iint_{\mathfrak{R}^\circ} W_e[\tau(w), \mathfrak{x}^*] \mid J(w) \mid. \tag{6}$$

(5), (6), (2) yield (note that the boundary of Q_0^* is of planar measure zero)

$$\iint_{Q_0^*} W_e(w^*, \mathfrak{x}^*) = A(S).$$

Hence by V.2.62, the representation (4) is eAC. Since τ is bimeasurable, it follows by IV.4.65 that the representation (1) is also eAC.

V.2.64. Theorem. *Let S be a surface such that $A(S) < +\infty$. Let*

$$S : \mathfrak{x} = \mathfrak{x}(w), \qquad w \in \mathfrak{R}, \tag{1}$$

be a representation of S such that the ordinary Jacobians $J^i(w, \mathfrak{x})$ (see V.1.18) exist a.e. in $\mathfrak{R}^0$. Then the following statements hold.

(i) *The function*

$$W(w, \mathfrak{x}) = \{[J^1(w, \mathfrak{x})]^2 + [J^2(w, \mathfrak{x})]^2 + [J^3(w, \mathfrak{x})]^2\}^{1/2}$$

is summable in $\mathfrak{R}^0$.

(ii)
$$\iint_{\mathfrak{R}^0} W(w, \mathfrak{x}) \leqq A(S).$$

(iii) *The formula*

(2)
$$\iint_{\mathfrak{R}^0} W(w, \mathfrak{x}) = A(S)$$

holds if and only if the representation (1) *is eAC* (*cf.* V.1.15).

Proof. The statements (i) and (ii) follow directly from V.2.14. As regards (iii), let us note the inequalities (see V.2.14)

(3)
$$\iint_{\mathfrak{R}^0} W(w, \mathfrak{x}) \leqq \iint_{\mathfrak{R}^0} W_e(w, \mathfrak{x}) \leqq A(S)$$

and the formula (see V.2.58)

(4)
$$a(S) = A(S).$$

Suppose first that (2) holds. By (3) it follows that

$$\iint_{\mathfrak{R}^0} W_e(w, \mathfrak{x}) = A(S).$$

Hence the representation (1) is eAC by V.2.61. Suppose conversely that the representation (1) is eAC. By V.1.23 we have then

(5)
$$\iint_{\mathfrak{R}^0} W(w, \mathfrak{x}) = a(S).$$

(4) and (5) imply (2).

V.2.65. (Added in proof). The validity of the formula $a(S) = A(S)$ in the case when $A(S) = +\infty$ follows from results of L. Cesari which came to the attention of the writer after the manuscript of this book had been completed (see V.4.8). The general validity of the formula $a(S) = A(S)$ is thus established (see V.2.60). As a consequence, several results proved in this chapter can be improved in a significant manner. In the way of illustration, we consider the following example.

Theorem. *Let* $S : \mathfrak{x} = \mathfrak{x}(w), w \in \mathfrak{R}$, *be a representation of a surface* S. *Suppose that this representation is eBV. Then the following statements hold.*

(i) $A(S) < +\infty$.

(ii) *The essential generalized Jacobians exist a. e. in* $\mathfrak{R}^0$ *and the function* $W_e(w, \mathfrak{x})$ *(see* V.2.61) *is summable in* $\mathfrak{R}^0$.

(iii) *We have the inequality*

$$\iint_{\mathfrak{R}^0} W_e(w, \mathfrak{x}) \leqq A(S),$$

where the sign of equality holds if and only if the representation is eAC.

Indeed, by V.1.16 we have $a(S) < +\infty$, and since we now know that always $a(S) = A(S)$, it follows that $A(S) < +\infty$. Thus the theorem appears as a direct consequence of V.2.61.

V.2.66. Given a surface $S : \mathfrak{x} = \mathfrak{x}(w)$, $w \in \mathfrak{R}$, several previously established results yield sufficient conditions of a special character for $A(S) < +\infty$ (see V.2.22, V.2.24, V.2.26, V.2.27, V.2.28, V.2.30). Let us add the following statement.

If S is a nondegenerate surface (see II.3.21), then $A(S) < +\infty$ if and only if S admits of a generalized conformal representation (cf. V.2.29).

PROOF. If S admits of a generalized conformal representation, then $A(S) < +\infty$ by V.2.30. Conversely, if $A(S) < +\infty$, then S admits of a generalized conformal representation by V.2.40.

V.2.67. Let there be given a surface $S : \mathfrak{x} = \mathfrak{x}(w)$, $w \in \mathfrak{R}$. Let $\mathfrak{R}_1, \cdots, \mathfrak{R}_m$ be simply connected Jordan regions with the following properties:

(i) $\mathfrak{R}_i^0\mathfrak{R}_j^0 = 0$ for $i \neq j$.

(ii) $\mathfrak{R}_1 + \cdots + \mathfrak{R}_m = \mathfrak{R}$.

Let us define $S_i : \mathfrak{x} = \mathfrak{x}(w)$, $w \in \mathfrak{R}_i$. Let us denote by C_i the boundary of $\mathfrak{R}_i$ and by C_i^* the point set $T(C_i)$ in $x_1x_2x_3$-space, where T is the transformation $T : \mathfrak{x} = \mathfrak{x}(w)$, $w \in \mathfrak{R}$. Let us make the following further assumptions:

(iii) For every i, the orthogonal projections of C_i^* upon the three coordinate planes are point sets of (planar) measure zero.

(iv) $A(S) < +\infty$.

ADDITIVITY THEOREM. *Under the conditions just stated, we have the formula*

$$A(S) = A(S_1) + \cdots + A(S_m). \tag{1}$$

PROOF. In the first place, as a consequence of (iii), we have the formula

$$a(S) = a(S_1) + \cdots + a(S_m) \tag{2}$$

by V.1.14 and V.1.7. By V.2.16 and (iv) we have

$$A(S_i) \leqq A(S) < +\infty, \qquad i = 1, 2, \cdots, m.$$

Hence, by V.2.58,

$$a(S) = A(S),\ a(S_i) = A(S_i), \qquad i = 1, 2, \cdots, m. \tag{3}$$

(2) and (3) imply (1).

V.2.68. Let there be given a surface

$$(1) \qquad S : \mathfrak{x} = \mathfrak{x}(w), \qquad w \in \mathfrak{R}.$$

Let us denote again by T the transformation

$$(2) \qquad T : \mathfrak{x} = \mathfrak{x}(w), \qquad w \in \mathfrak{R}.$$

Suppose that the following conditions hold:

(i) The transformation T is biunique in $\mathfrak{R}$; that is, $w_1 \neq w_2$ implies that $\mathfrak{x}(w_1) \neq \mathfrak{x}(w_2)$.

(ii) The point set $T(\mathfrak{R})$ lies in a plane p in $x_1x_2x_3$-space. As a consequence of (i) and (ii), the point set $T(\mathfrak{R})$ is a bounded, simply-connected Jordan region $\mathfrak{R}_*$ in the plane p. Let $| \mathfrak{R}_*^0 |$ denote the (planar) measure of the interior of $\mathfrak{R}_*$. We assert that $A(S) = | \mathfrak{R}_*^0 |$.

PROOF. By V.2.8 we can assume, without loss of generality, that the plane p coincides with the x_1x_2-plane. Then the representation (1) and the transformation (2) may be written in the form

$$(1) \qquad S : x_1 = x_1(u, v),\ x_2 = x_2(u, v),\ x_3 = 0, \qquad (u, v) \in \mathfrak{R},$$

$$(2) \qquad T : x_1 = x_1(u, v),\ x_2 = x_2(u, v),\ x_3 = 0, \qquad (u, v) \in \mathfrak{R}.$$

Now since the transformation $T(\mathfrak{R}) = \mathfrak{R}_*$ is topological, the inverse transformation $T^{-1}(\mathfrak{R}_*) = \mathfrak{R}$ is also topological, and may be represented by formulas of the type

$$(3) \qquad T^{-1} : u = u(x_1, x_2),\ v = v(x_1, x_2), \qquad (x_1, x_2) \in \mathfrak{R}_*.$$

On replacing in (1) u and v by $u(x_1, x_2)$, $v(x_1, x_2)$, we obtain (cf. II.3.11) a representation of S upon $\mathfrak{R}_*$ which reduces, in view of (2) and (3), obviously to the form $S : x_1 = x_1,\ x_2 = x_2,\ x_3 = 0,\ (x_1, x_2) \in \mathfrak{R}_*$. As an extremely special case of V.2.12, we obtain the formula

$$A(S) = \iint_{\mathfrak{R}_*^0} 1 = | \mathfrak{R}_*^0 |.$$

V.2.69. Let there be given a surface $S : \mathfrak{x} = \mathfrak{x}(w)$, $w \in \mathfrak{R}$, and let us denote again by T the transformation $T : \mathfrak{x} = \mathfrak{x}(w)$, $w \in \mathfrak{R}$. Then there exists a surface $S^* : \mathfrak{x} = \mathfrak{x}^*(w)$, $w \in \mathfrak{R}$, such that if we denote by T^* the corresponding transformation $T^* : \mathfrak{x} = \mathfrak{x}^*(w)$, $w \in \mathfrak{R}$, the following statements hold:

(i) $T^*(\mathfrak{R}) = T(\mathfrak{R})$.

(ii) $A(S^*) = 0$.

PROOF. Let us denote by I_* the interval $I_* : 0 \leqq u_* \leqq 1$, and by Q_{0*} the unit square $Q_{0*} : 0 \leqq u_* \leqq 1,\ 0 \leqq v_* \leqq 1,\ u_* + iv_* = w_*$. The set $T(\mathfrak{R})$ is a Peano subspace of $x_1x_2x_3$-space (see I.2.43). Hence there exists a continuous transformation from I_* onto $T(\mathfrak{R})$ (see I.2.33). Using vector notation, this transformation may be written in the form

$$(1) \qquad T_1 : \mathfrak{x} = \mathfrak{y}(u_*),\ u_* \in I_*,\ T_1(I_*) = T(\mathfrak{R}).$$

Let us define a surface S^* by

(2) $$S^* : \mathfrak{x} = \mathfrak{z}(w_*) = \mathfrak{z}(u_*, v_*) = \mathfrak{y}(u_*), \qquad w_* \in Q_{0*}.$$

Let τ be a topological transformation from $\mathfrak{R}$ onto Q_{0*} (cf. I.2.50). By II.3.11 we have then for S^* the representation $S^* : \mathfrak{x} = \mathfrak{x}^*(w) = \mathfrak{z}[\tau(w)]$, $w \in \mathfrak{R}$. We assert that S^* satisfies our requirements. Indeed, we obtain for the corresponding transformation T^*, in view of (1) and (2),

$$T^*(\mathfrak{R}) = \mathfrak{z}[\tau(\mathfrak{R})] = T_1(I_*) = T(\mathfrak{R}).$$

Now let us denote, for each positive integer n, by $\mathfrak{T}_{n*}$ the rectilinear triangulation of Q_{0*} obtained by subdividing Q_{0*} into n^2 congruent oriented squares and then drawing the diagonal from the upper left to the lower right vertex in each one of these squares. Let us denote by $\mathfrak{z}_n(w_*)$ the (univocally determined) vector function in Q_{0*} that satisfies the following conditions: (α) $\mathfrak{z}_n(w_*) = \mathfrak{z}(w_*)$ at every vertex of $\mathfrak{T}_{n*}$. (β) The components of $\mathfrak{z}_n(w_*)$ are linear functions of u_*, v_* in each one of the triangles of $\mathfrak{T}_{n*}$. Let us define

$$S_n^* : \mathfrak{x} = \mathfrak{z}_n(w_*), \qquad w_* \in Q_{0*}.$$

Clearly $\mathfrak{z}_n(w_*) \to \mathfrak{z}(w_*)$ uniformly in Q_{0*}, and hence (cf. V.2.6)

$$A(S^*) \leqq \liminf_{n\to\infty} A(S_n^*).$$

On the other hand, since $\mathfrak{z}(u_*, v_*)$ is independent of v_*, clearly $A(S_n^*) = 0$ by V.2.11, V.1.26, V.1.35. Hence $A(S^*) = 0$.

V.2.70. We emphasized on several occasions the fact that the Lebesgue area $A(S)$ agrees, in simple cases, with what may be termed the *expected value* of the area of S (see V.2.12, V.1.34, V.1.35, V.2.68, V.2.67). The cases just referred to probably cover all cases in which one may speak of an expected value of the area. On the other hand, in more general cases there arise situations that justify the extreme caution that we practiced in proving theorems concerning the Lebesgue area. We proceed to illustrate this remark by several simple examples.

V.2.71. Let us consider a surface $S : \mathfrak{x} = \mathfrak{x}(w)$, $w \in \mathfrak{R}$, and the corresponding transformation $T : \mathfrak{x} = \mathfrak{x}(w)$, $w \in \mathfrak{R}$. Let $\mathcal{P}$ be any Peano subspace of $x_1x_2x_3$-space. Then we can choose $\mathfrak{x}(w)$ in such a way that $T(\mathfrak{R}) = \mathcal{P}$ (see I.2.33). By V.2.69 we have then a surface $S^* : \mathfrak{x} = \mathfrak{x}^*(w)$, $w \in \mathfrak{R}$, such that $A(S^*) = 0$ and $T^*(\mathfrak{R}) = \mathcal{P}$, where T^* is the transformation $\mathfrak{x} = \mathfrak{x}^*(w)$, $w \in \mathfrak{R}$. Let us consider the following special cases.

(i) Let us choose $\mathcal{P}$ as the unit cube $0 \leqq x_i \leqq 1$, $i = 1, 2, 3$. Then, stated picturesquely, S^* *is a cube-filling surface of zero area.*

(ii) Let us choose $\mathcal{P}$ as the unit square $0 \leqq x_1 \leqq 1$, $0 \leqq x_2 \leqq 1$, $x_3 = 0$, and let us choose

$$S : x_1 = u, x_2 = v, x_3 = 0, \qquad 0 \leqq u \leqq 1, 0 \leqq v \leqq 1.$$

By V.2.68, $A(S) = 1$, while for the corresponding surface S^* we have $A(S^*) = 0$.

From the construction of S^* (see V.2.69) it follows that under the transformation T^* some points of the unit square in the x_1x_2-plane must be covered several times (since the unit square is not the topological image of a linear interval). Thus we have, described in picturesque terms, the following situation. *The surface* S *covers the unit square just once, while* S^* *covers the unit square in such a manner that some points are covered several times, and yet* $A(S^*) < A(S)$; in fact, $A(S^*) = 0$ and $A(S) = 1$.

(iii) Let us choose $\mathcal{P}$ as a simply-connected, bounded Jordan region $\mathfrak{R}_*$ in the x_1x_2-plane, such that the boundary curve C_* of $\mathfrak{R}_*$ has positive (planar) measure (see I.3.6). Let us choose $\mathfrak{x}(w)$ in such a way that the corresponding transformation $T(\mathfrak{R}) = \mathfrak{R}_*$ is topological. Then, again in picturesque terms, the projection of the surface S upon the x_1x_2-plane has its measure equal to $|\mathfrak{R}_*|$, while the area $A(S)$ of S is less than $|\mathfrak{R}_*|$. Indeed, in view of V.2.68, $A(S) = |\mathfrak{R}_*^0| < |\mathfrak{R}_*|$.

V.2.72. Concerning surfaces of zero area $A(S)$, previous theorems yield two interesting corollaries. *Given a surface* $S : \mathfrak{x} = \mathfrak{x}(w)$, $w \in \mathfrak{R}$, *let* T *be the associated transformation* $\mathfrak{x} = \mathfrak{x}(w)$, $w \in \mathfrak{R}$.

(i) *Suppose that the orthogonal projections of the point set* $T(\mathfrak{R})$ *upon the three coordinate planes are sets of (planar) measure zero. Then* $A(S) = 0$.

PROOF. By V.1.62, V.1.7, $a(S) = 0$. Hence $A(S) = 0$ by V.2.59.

(ii) *Suppose that the transformation* T *is biunique in* $\mathfrak{R}$ (that is, $\mathfrak{x}(w_1) \neq \mathfrak{x}(w_2)$ if $w_1 \neq w_2$). *Then* $A(S)$ *is not equal to zero.*

PROOF. The middle space $\mathfrak{M}$, arising in the monotone-light factorization of T, may be taken now as $\mathfrak{R}$ itself. Thus $A(S) > 0$ by V.2.54.

CHAPTER V.3. SURFACES OF THE FORM $z = f(x, y)$

V.3.1. We take up the study of surfaces S that admit, for some choice of the Cartesian coordinate system $x_1x_2x_3$, of a representation of the form

$$S : x_3 = f(x_1, x_2), \qquad (x_1, x_2) \in \mathfrak{R}, \tag{1}$$

where $f(x_1, x_2)$ is a single-valued continuous function in the simply-connected, bounded Jordan region $\mathfrak{R}$ in the x_1x_2-plane. In conformity with the notations agreed upon in II.3.43, the representation (1) should be written explicitly in the form $S : \mathfrak{x} = \mathfrak{x}(x_1, x_2)$, $(x_1, x_2) \in \mathfrak{R}$, where $\mathfrak{x}(x_1, x_2) = (x_1, x_2, f(x_1, x_2))$.

V.3.2. Continuation. To conform to general usage, and also for reasons of technical convenience, we shall change our notations as follows. We shall replace x_1, x_2, x_3 by x, y, z; the representation of S appears then in the form

$$S : z = f(x, y), \qquad (x, y) \in \mathfrak{R},$$

and the vector $\mathfrak{x}$ of the general theory becomes

$$\mathfrak{x}(x, y) = [x, y, f(x, y)].$$

The transformations T^1, T^2, T^3 of V.1.1 will be denoted by T^x, T^y, T^z. We have then explicitly

$$T^x : y = y,\ z = f(x, y), \qquad (x, y) \in \mathfrak{R},$$

$$T^y : z = f(x, y),\ x = x, \qquad (x, y) \in \mathfrak{R},$$

$$T^z : x = x,\ y = y, \qquad (x, y) \in \mathfrak{R}.$$

If $\mathfrak{D}^*$ is a domain contained in $\mathfrak{R}^0$, then the essential multiplicity functions relative to $\mathfrak{D}^*$ and the transformations T^x, T^y, T^z, respectively, will be denoted by $\kappa(y, z, T^x, \mathfrak{D}^*)$, $\kappa(z, x, T^y, \mathfrak{D}^*)$, $\kappa(x, y, T^z, \mathfrak{D}^*)$ (cf. V.1.1). For the case when $\mathfrak{D}^* = R^0$, where $R \subset \mathfrak{R}$ is an oriented rectangle, we shall denote the quantities $g_1(\mathfrak{x}, R^0)$, $g_2(\mathfrak{x}, R^0)$, $g_3(\mathfrak{x}, R^0)$, $g(\mathfrak{x}, R^0)$ of V.1.2 by $g_1(f, R)$, $g_2(f, R)$, $g_3(f, R)$, $g(f, R)$ respectively. Clearly $g_3(f, R) = |R|$. Thus we have, by V.1.2, IV.4.70, the formulas

$$g_1(f, R) = W_x(R, f),$$

$$g_2(f, R) = W_y(R, f),$$

$$g_3(f, R) = |R|,$$

$$g(f, R) = \{[g_1(f, R)]^2 + [g_2(f, R)]^2 + [g_3(f, R)]^2\}^{1/2}.$$

If $R_1, \cdots, R_m$ is any system of nonoverlapping oriented rectangles in $\mathfrak{R}$, then we have by V.1.3, V.1.7 the inequality

$$\sum_{i=1}^{m} g(f, R_i) \leq a(S).$$

V.3.3. CONTINUATION. Let $R : a \leqq x \leqq b, c \leqq y \leqq d$, be an oriented rectangle in $\mathfrak{R}$. We define

$$\gamma_1(f, R) = \int_c^d | f(b, y) - f(a, y) | \, dy,$$

$$\gamma_2(f, R) = \int_a^b | f(x, d) - f(x, c) | \, dx,$$

$$\gamma_3(f, R) = | R |,$$

$$\gamma(f, R) = \{[\gamma_1(f, R)]^2 + [\gamma_2(f, R)]^2 + [\gamma_3(f, R)]^2\}^{1/2}.$$

Clearly (see IV.4.71)

$$\gamma_j(f, R) \leqq g_j(f, R), \qquad j = 1, 2, 3,$$

$$\gamma(f, R) \leqq g(f, R).$$

V.3.4. CONTINUATION. Let $\mathfrak{R}^*$ be a simply-connected Jordan region in $\mathfrak{R}$. Then $\mathfrak{R}^*$ gives rise to a surface $S^* : z = f(x, y), (x, y) \in \mathfrak{R}^*$. We shall use the symbols $a(f, \mathfrak{R}^*)$, $A(f, \mathfrak{R}^*)$ to denote the quantities $a(S^*)$, $A(S^*)$ respectively (cf. V.1.7, V.2.3). Let now $R_1^*, \cdots, R_m^*$ be any finite system of oriented rectangles in $\mathfrak{R}^*$ without common interior points. We define

$$\Gamma(f, \mathfrak{R}^*) = \text{l.u.b.} \sum_{i=1}^{m} \gamma(f, R_i^*),$$

$$G(f, \mathfrak{R}^*) = \text{l.u.b.} \sum_{i=1}^{m} g(f, R_i^*),$$

where the least upper bound is taken with respect to all systems of rectangles with the properties specified above. Of course, $\Gamma(f, \mathfrak{R}^*)$, $G(f, \mathfrak{R}^*)$ are not necessarily finite. In view of V.3.2, V.3.3, V.2.10 we have the inequalities

$$\Gamma(f, \mathfrak{R}^*) \leqq G(f, \mathfrak{R}^*) \leqq a(f, \mathfrak{R}^*) \leqq A(f, \mathfrak{R}^*).$$

Let $R_1, R_2, \cdots, R_n$ be any system of oriented rectangles in $\mathfrak{R}$ without common interior points. Obviously

$$\sum_{k=1}^{n} \Gamma(f, R_k) \leqq \Gamma(f, \mathfrak{R}),$$

$$\sum_{k=1}^{n} G(f, R_k) \leqq G(f, \mathfrak{R}).$$

V.3.5. CONTINUATION. Let $\mathfrak{R}^*$ be a simply-connected Jordan region such that

$$(1) \qquad \mathfrak{R}^* \subset \mathfrak{R}^0.$$

For $h > 0$ let us consider the function (cf. III.2.65)

$$f_h(x, y) = \frac{1}{4h^2} \int_{-h}^{h} \int_{-h}^{h} f(x + \xi, y + \eta)\, d\xi\, d\eta. \tag{2}$$

The region $\mathfrak{R}^*$ being fixed according to (1), we have an $h_0 > 0$ such that $f_h(x, y)$ is defined on $\mathfrak{R}^*$ and has continuous partial derivatives of the first order in $\mathfrak{R}^{*0}$ for $0 < h < h_0$ (cf. III.2.66). We assume in the sequel that h satisfies the inequalities $0 < h < h_0$. *We assert that for h sufficiently small we have the inequality*

$$\Gamma(f_h, \mathfrak{R}^*) \leqq \Gamma(f, \mathfrak{R}). \tag{3}$$

PROOF. Let (x, y) be a point in $\mathfrak{R}^*$, and let ξ and η satisfy the inequalities

$$|\xi| < h_1, |\eta| < h_1, \tag{4}$$

where $0 < h_1 < h_0$. In view of (1), we can choose h_1 so small that

$$(x, y) \in \mathfrak{R}^* \text{ implies } (x + \xi, y + \eta) \in \mathfrak{R}^0$$

if ξ and η satisfy (4). Let now R be any oriented rectangle in $\mathfrak{R}^*$, and let $R(\xi, \eta)$ be the rectangle obtained by subjecting R to the translation with components ξ, η. Clearly, if ξ, η satisfy (4), then

$$R \subset \mathfrak{R}^* \text{ implies } R(\xi, \eta) \subset \mathfrak{R}^0. \tag{5}$$

In the sequel, h is restricted by the inequality

$$0 < h < h_1 < h_0. \tag{6}$$

Let now $R_1, \cdots, R_m$ be any system of oriented rectangles in $\mathfrak{R}^*$ without common interior points. Let R_i be given by $R_i : a_i \leqq x \leqq b_i, c_i \leqq y \leqq d_i$. By (2) and V.3.3 we have then

$$\gamma_1(f_h, R_i) = \int_{c_i}^{d_i} | f_h(b_i, y) - f_h(a_i, y) |\, dy$$

$$\leqq \frac{1}{4h^2} \int_{-h}^{h} \int_{-h}^{h} \left[\int_{c_i}^{d_i} | f(b_i + \xi, y + \eta) - f(a_i + \xi, y + \eta) |\, dy \right] d\xi\, d\eta$$

$$\leqq \frac{1}{4h^2} \int_{-h}^{h} \int_{-h}^{h} \gamma_1(f, R_i(\xi, \eta))\, d\xi\, d\eta.$$

Similarly

$$\gamma_2(f_h, R_i) \leqq \frac{1}{4h^2} \int_{-h}^{h} \int_{-h}^{h} \gamma_2(f, R_i(\xi, \eta))\, d\xi\, d\eta,$$

and finally

$$\gamma_3(f_h, R_i) = |R_i| = \gamma_3(f, R_i(\xi, \eta)).$$

By the Minkowski inequality (see I.3.10) we obtain

$$\gamma(f_h, R_i) \leqq \frac{1}{4h^2} \int_{-h}^{h} \int_{-h}^{h} \gamma(f, R_i(\xi, \eta)) \, d\xi \, d\eta,$$

and hence

$$\sum_{i=1}^{m} \gamma(f_h, R_i) \leqq \frac{1}{4h^2} \int_{-h}^{h} \int_{-h}^{h} \left[\sum_{i=1}^{m} \gamma(f, R_i(\xi, \eta)) \right] d\xi \, d\eta. \tag{7}$$

Now, in view of (5) and (6), $R_1(\xi, \eta), \cdots, R_m(\xi, \eta)$ is a system of oriented rectangles in $\mathfrak{R}^0$ without common interior points. Hence (see V.3.4)

$$\sum_{i=1}^{m} \gamma(f, R_i(\xi, \eta)) \leqq \Gamma(f, \mathfrak{R}). \tag{8}$$

(7) and (8) yield

$$\sum_{i=1}^{m} \gamma(f_h, R_i) \leqq \Gamma(f, \mathfrak{R}).$$

Since the system $R_1, \cdots, R_m$ was arbitrary, the inequality (3) follows.

V.3.6. *Given a surface $S : z = f(x, y)$, $(x, y) \in \mathfrak{R}$, suppose that the first partial derivatives f_x, f_y exist and are continuous and summable in $\mathfrak{R}^0$. Then*

$$\Gamma(f, \mathfrak{R}) = A(f, \mathfrak{R}). \tag{1}$$

Proof. Let R_0 be an oriented rectangle such that $R_0 \subset \mathfrak{R}^0$. The first partial derivatives f_x, f_y are then uniformly continuous on R_0. Hence if we put

$$\eta_1(\delta, R_0) = \max |f_x(x_2, y_2) - f_x(x_1, y_1)|,$$
$$\eta_2(\delta, R_0) = \max |f_y(x_2, y_2) - f_y(x_1, y_1)|,$$
$$\eta(\delta, R_0) = \eta_1(\delta, R_0) + \eta_2(\delta, R_0),$$

where the maximum is taken with respect to all pairs of points (x_2, y_2), (x_1, y_1) such that

$$(x_1, y_1), (x_2, y_2) \in R_0, \; [(x_2 - x_1)^2 + (y_2 - y_1)^2]^{1/2} \leqq \delta,$$

then

$$\eta(\delta, R_0) \to 0 \qquad \text{for } \delta \to 0. \tag{2}$$

Now let $R : a \leqq x \leqq b$, $c \leqq y \leqq d$, be any oriented rectangle in R_0. By the mean-value theorem we obtain (cf. V.3.3)

$$\gamma_1(f, R) = |f(b, \eta) - f(a, \eta)| (d - c)$$
$$= |f_x(\xi, \eta)| (b - a)(d - c) = |f_x(\xi, \eta)| \, |R|,$$

where (ξ, η) is some point in R. If $d(R)$ denotes the diameter of R, it follows that

$$|\gamma_1(f, R) - |f_x(a, c)||R|| \leqq \eta(d(R), R_0)|R|.$$

Similarly

$$|\gamma_2(f, R) - |f_y(a, c)||R|| \leqq \eta(d(R), R_0)|R|.$$

Hence (cf. V.3.3)

(3) $$|\gamma(f, R) - (1 + f_x(a, c)^2 + f_y(a, c)^2)^{1/2}|R|| \leqq 2\eta(d(R), R_0)|R|.$$

Now let R_0 be subdivided into oriented rectangles $R_1, \cdots, R_m$ whose diameters are less than a fixed $\epsilon > 0$. From (3) it follows that

(4) $$\sum_{i=1}^{m} \gamma(f, R_i) \geqq \left\{ \sum_{i=1}^{m} (1 + f_x(a_i, c_i)^2 + f_y(a_i, c_i)^2)^{1/2} |R_i| \right\} - 2\eta(\epsilon, R_0)|R_0|.$$

By V.3.4 we have

(5) $$\sum_{i=1}^{m} \gamma(f, R_i) \leqq \Gamma(f, R_0).$$

Since f_x, f_y are continuous in R_0, for given $\sigma > 0$ the summation on the right in (4) will exceed

(6) $$\iint_{R_0} (1 + f_x^2 + f_y^2)^{1/2}\, dx\, dy - \frac{\sigma}{2}$$

if ϵ is small enough. In view of (5) and (6) it follows therefore from (4) that

$$\Gamma(f, R_0) > \iint_{R_0} (1 + f_x^2 + f_y^2)^{1/2}\, dx\, dy - \sigma$$

if ϵ is small enough. Since σ is arbitrary, it follows that

(7) $$\Gamma(f, R_0) \geqq \iint_{R_0} (1 + f_x^2 + f_y^2)^{1/2}\, dx\, dy$$

for every oriented rectangle $R_0 \subset \mathfrak{R}^0$. Now since f_x, f_y are summable in $\mathfrak{R}^0$, the function $(1 + f_x^2 + f_y^2)^{1/2}$ is also summable in $\mathfrak{R}^0$. Hence for given $\delta > 0$ we have a finite system of oriented rectangles $R_0^1, \cdots, R_0^n$ in $\mathfrak{R}^0$, without common interior points, such that

$$\iint_{\mathfrak{R}^0} (1 + f_x^2 + f_y^2)^{1/2}\, dx\, dy < \delta + \sum_{k=1}^{n} \iint_{R^k{}_0} (1 + f_x^2 + f_y^2)^{1/2}\, dx\, dy.$$

In view of (7) and V.3.4, it follows that

$$\Gamma(f, \Re) \geqq \sum_{k=1}^{n} \Gamma(f, R_0^k) \geqq \sum_{k=1}^{n} \iint_{R_0^k} (1 + f_x^2 + f_y^2)^{1/2}\, dx\, dy$$

$$> \iint_{\Re^\circ} (1 + f_x^2 + f_y^2)^{1/2}\, dx\, dy - \delta.$$

Since δ is arbitrary, we obtain

(8) $$\Gamma(f, \Re) \geqq \iint_{\Re^\circ} (1 + f_x^2 + f_y^2)^{1/2}\, dx\, dy.$$

By V.2.12 (cf. V.1.35)

(9) $$\iint_{\Re^\circ} (1 + f_x^2 + f_y^2)^{1/2}\, dx\, dy = A(f, \Re).$$

By V.3.4

(10) $$\Gamma(f, \Re) \leqq A(f, \Re).$$

(8), (9), (10) imply (1).

V.3.7. Theorem. *Given a surface*

(1) $$S : z = f(x, y), \qquad (x, y) \in \Re,$$

we have the formulas (*cf.* V.3.4)

(2) $$\Gamma(f, \Re) = G(f, \Re) = a(f, \Re) = A(f, \Re).$$

In particular, for every surface S of the form (1), *the lower area $a(S)$ and the Lebesgue area $A(S)$ are equal to each other.*

Proof. Let $\Re^*$ be any simply-connected Jordan region in $\Re^0$. Let $f_h(x, y)$ be defined as in V.3.5. For h sufficiently small we have by V.3.5

(3) $$\Gamma(f_h, \Re^*) \leqq \Gamma(f, \Re).$$

By III.2.66, for h sufficiently small, f_h has continuous first partial derivatives in some domain containing $\Re^*$. Hence, by V.3.6,

(4) $$\Gamma(f_h, \Re^*) = A(f_h, \Re^*).$$

If h approaches zero through any sequence h_n, then $f_{h_n} \to f$ uniformly in $\Re^*$ (see III.2.66). Hence, by V.2.6,

(5) $$A(f, \Re^*) \leqq \liminf_{n \to \infty} A(f_{h_n}, \Re^*).$$

(3), (4), (5) imply the inequality

(6) $$A(f, \Re^*) \leqq \Gamma(f, \Re).$$

Now let $\mathfrak{R}_n^*$ be a sequence of simply-connected Jordan regions that fill up $\mathfrak{R}$ from the interior. By (6) we have

$$A(f, \mathfrak{R}_n^*) \leqq \Gamma(f, \mathfrak{R}). \tag{7}$$

By V.2.17

$$A(f, \mathfrak{R}_n^*) \underset{n\to\infty}{\longrightarrow} A(f, \mathfrak{R}). \tag{8}$$

(7) and (8) yield

$$A(f, \mathfrak{R}) \leqq \Gamma(f, \mathfrak{R}). \tag{9}$$

By V.3.4

$$\Gamma(f, \mathfrak{R}) \leqq G(f, \mathfrak{R}) \leqq a(f, \mathfrak{R}) \leqq A(f, \mathfrak{R}). \tag{10}$$

(9) and (10) yield (2).

V.3.8. Let $f(x, y)$ be continuous in a simply-connected, bounded, Jordan region $\mathfrak{R}$. We shall say that $f(x, y)$ is BVT in $\mathfrak{R}$ (of bounded variation in $\mathfrak{R}$ in the Tonelli sense) if the following conditions hold (cf. III.2.64):

(i) $f(x, y)$ is BVT in every oriented rectangle $R \subset \mathfrak{R}$.

(ii) There exists a finite constant M such that (cf. III.2.51)

$$\sum_{i=1}^{m} W_x(R_i, f) < M, \qquad \sum_{i=1}^{m} W_y(R_i, f) < M,$$

for every system of oriented rectangles $R_1, \cdots, R_m$ in $\mathfrak{R}$ without common interior points. Clearly if $\mathfrak{R}$ itself is an oriented rectangle, then this definition is consistent with the definition already given in III.2.64.

V.3.9. THEOREM. *Given a surface $S : z = f(x, y)$, $(x, y) \in \mathfrak{R}$, the Lebesgue area $A(f, \mathfrak{R})$ of S is finite if and only if $f(x, y)$ is BVT in $\mathfrak{R}$, in the sense of* V.3.8.

PROOF. *Necessity.* Suppose that $A(f, \mathfrak{R}) < +\infty$. If R is any oriented rectangle in $\mathfrak{R}$, then we have also $A(f, R) < +\infty$ by V.2.16. By V.3.7 it follows that $G(f, R) < +\infty$ and hence *a fortiori* $g(f, R) < +\infty$. Hence, by V.3.2, $W_x(R, f) < +\infty$, $W_y(R, f) < +\infty$. Thus $f(x, y)$ is BVT in R (see III.2.64). Now let $R_1, \cdots, R_m$ be any system of oriented rectangles in $\mathfrak{R}$ without common interior points. We have then, by V.3.7, V.3.4,

$$\sum_{i=1}^{m} g(f, R_i) \leqq A(f, \mathfrak{R}) < +\infty.$$

Hence, *a fortiori*

$$\sum_{i=1}^{m} W_x(R_i, f) \leqq A(f, \mathfrak{R}), \qquad \sum_{i=1}^{m} W_y(R_i, f) \leqq A(f, \mathfrak{R}).$$

Sufficiency. If R is any oriented rectangle in $\mathfrak{R}$, then (cf. V.3.2, V.3.4)

$$g(f, R) \leqq W_x(R, f) + W_y(R, f) + |R|.$$

Hence, if $f(x, y)$ is BVT in $\mathfrak{R}$, it follows immediately that $G(f, \mathfrak{R}) < +\infty$. Hence, by V.3.7, $A(f, \mathfrak{R}) = G(f, \mathfrak{R}) < +\infty$.

V.3.10. Let $f(x, y)$ be continuous in a simply-connected, bounded, Jordan region $\mathfrak{R}$. We shall say that $f(x, y)$ is ACT in $\mathfrak{R}$ (absolutely continuous in $\mathfrak{R}$ in the Tonelli sense) if the following conditions hold.

(i) $f(x, y)$ is BVT in $\mathfrak{R}$ (see V.3.8).

(ii) $f(x, y)$ is ACT in every oriented rectangle $R \subset \mathfrak{R}$ (see III.2.64).

Clearly if $\mathfrak{R}$ itself is an oriented rectangle, then this definition is consistent with the definition already given in III.2.64.

V.3.11. *Let there be given a surface $S : z = f(x, y)$, $(x, y) \in \mathfrak{R}$. If $f(x, y)$ is ACT in $\mathfrak{R}$, then*

$$(1) \qquad A(f, \mathfrak{R}) = \iint_{\mathfrak{R}^0} (1 + f_x^2 + f_y^2)^{1/2} \, dx \, dy.$$

Proof. Note that $f(x, y)$ is also BVT in $\mathfrak{R}$ (cf. V.3.10). By III.2.50 it follows that the partial derivatives f_x, f_y exist a.e. in $\mathfrak{R}^0$. Since $A(f, \mathfrak{R}) < +\infty$ by V.3.9, f_x, f_y are summable in $\mathfrak{R}^0$ by V.2.14. Now the components of the vector (cf. V.3.2) $\mathfrak{x} = (x, y, f(x, y))$ clearly satisfy the assumptions of V.2.28; indeed, x and y are Lipschitzian in $\mathfrak{R}$, while $f(x, y)$ is ACT on every oriented rectangle $R \subset \mathfrak{R}$ by assumption. Hence (1) holds by V.2.28.

V.3.12. Given a surface $S : z = f(x, y)$, $(x, y) \in \mathfrak{R}$, suppose that the following conditions hold.

(i) The partial derivatives f_x, f_y exist a.e. in $\mathfrak{R}^0$ and are summable in $\mathfrak{R}^0$. As a consequence, $(1 + f_x^2 + f_y^2)^{1/2}$ is then also summable in $\mathfrak{R}^0$.

$$(ii) \qquad A(f, \mathfrak{R}) = \iint_{\mathfrak{R}^0} (1 + f_x^2 + f_y^2)^{1/2} \, dx \, dy.$$

Then $f(x, y)$ is ACT in $\mathfrak{R}$, in the sense of V.3.10.

Proof. (ii) implies that $A(f, \mathfrak{R}) < +\infty$. Hence $f(x, y)$ is BVT in $\mathfrak{R}$ by V.3.9. We still have to prove that $f(x, y)$ is ACT in every oriented rectangle $R \subset \mathfrak{R}$. We first show that

$$(1) \qquad A(f, R) = \iint_R (1 + f_x^2 + f_y^2)^{1/2} \, dx \, dy$$

for every oriented rectangle $R \subset \mathfrak{R}$. Let us first note that surely

$$A(f, R) \geqq \iint_R (1 + f_x^2 + f_y^2)^{1/2} \, dx \, dy$$

by V.2.14. Suppose now that we have

$$(2) \qquad A(f, R) = \iint_R (1 + f_x^2 + f_y^2)^{1/2} \, dx \, dy + \epsilon, \qquad \epsilon > 0,$$

for some rectangle $R \subset \mathfrak{R}$. We have then a sequence of oriented rectangles $R_1, \cdots, R_m, \cdots$ without common interior points, such that $R_1 + \cdots + R_m + \cdots = \mathfrak{R}^0 - R$. We have then (cf. V.2.14)

$$\tag{3} \begin{aligned} \iint\limits_{\mathfrak{R}^0 - R} (1 + f_x^2 + f_y^2)^{1/2}\, dx\, dy &= \sum_{m=1}^{\infty} \iint\limits_{R_m} (1 + f_x^2 + f_y^2)^{1/2}\, dx\, dy \\ &\leqq \sum_{m=1}^{\infty} A(f, R_m). \end{aligned}$$

(2) and (3) yield

$$\tag{4} \iint\limits_{\mathfrak{R}^0} (1 + f_x^2 + f_y^2)^{1/2}\, dx\, dy \leqq A(f, R) + \sum_{m=1}^{\infty} A(f, R_m) - \epsilon.$$

By V.2.15 we have, for every positive integer N,

$$\tag{5} A(f, R) + \sum_{m=1}^{N} A(f, R_m) \leqq A(f, \mathfrak{R}).$$

Since N is arbitrary, (4) and (5) imply that

$$\iint\limits_{\mathfrak{R}^0} (1 + f_x^2 + f_y^2)^{1/2}\, dx\, dy \leqq A(f, \mathfrak{R}) - \epsilon,$$

in contradiction to the assumption (ii). Hence (1) holds for every oriented rectangle $R \subset \mathfrak{R}$. From (1) we infer (cf. V.3.2, V.3.4)

$$\begin{aligned} W_x(R, f) = g_1(f, R) \leqq g(f, R) \leqq G(f, R) \leqq A(f, R) \\ = \iint\limits_{R} (1 + f_x^2 + f_y^2)^{1/2}\, dx\, dy. \end{aligned}$$

Thus

$$\tag{6} W_x(R, f) \leqq \iint\limits_{R} (1 + f_x^2 + f_y^2)^{1/2}\, dx\, dy,$$

and similarly

$$\tag{7} W_y(R, f) \leqq \iint\limits_{R} (1 + f_x^2 + f_y^2)^{1/2}\, dx\, dy.$$

Now let R_0 be a fixed oriented rectangle in $\mathfrak{R}$. From (6) and (7) it follows that the rectangle functions $W_x(R, f)$, $W_y(R, f)$ are absolutely continuous in R_0 (cf. I.3.13). Hence, by III.2.55, $f(x, y)$ is ACT in R_0, and the proof is complete.

V.3.13. The results obtained in V.3.9, V.3.11, V.3.12 may be summarized as follows (cf. V.3.8, V.3.10, V.2.14).

THEOREM. *Let there be given a surface* $S : z = f(x, y),\ (x, y) \in \mathfrak{R}$. *Then* $A(S)$ *is finite if and only if* $f(x, y)$ *is BVT in* $\mathfrak{R}$. *If* $A(S)$ *is finite, then the first partial derivatives* f_x, f_y *exist a.e. in* $\mathfrak{R}^0$ *and are summable in* $\mathfrak{R}^0$. *As a consequence,* $(1 + f_x^2 + f_y^2)^{1/2}$ *is also summable in* $\mathfrak{R}^0$. *Furthermore*

$$A(S) \geqq \iint_{\mathfrak{R}^0} (1 + f_x^2 + f_y^2)^{1/2}\, dx\, dy,$$

and the sign of equality holds if and only if $f(x, y)$ *is ACT in* $\mathfrak{R}$.

V.3.14. THEOREM (*cf.* V.1.15, V.3.8, V.3.10). *Let there be given a surface*

$$(1) \qquad S : z = f(x, y), \qquad (x, y) \in \mathfrak{R}.$$

Then the following statements hold:

(i) *The representation* (1) *is eBV if and only if* $f(x, y)$ *is BVT in* $\mathfrak{R}$.

(ii) *The representation* (1) *is eAC if and only if* $f(x, y)$ *is ACT in* $\mathfrak{R}$.

PROOF. (i) *Sufficiency.* Suppose $f(x, y)$ is BVT in $\mathfrak{R}$. Then $a(S) = A(S) < +\infty$ by V.3.7, V.3.9. Hence the representation is eBV by V.1.16.

Necessity. Suppose the representation is eBV. Then $A(S) = a(S) < +\infty$ by V.3.7, V.1.16. Hence $f(x, y)$ is BVT in $\mathfrak{R}$ by V.3.9.

(ii) *Sufficiency.* Suppose the representation is ACT in $\mathfrak{R}$. Then

$$\iint_{\mathfrak{R}^0} (1 + f_x^2 + f_y^2)^{1/2}\, dx\, dy = A(S) < +\infty$$

by V.3.13. Hence the representation is eAC by V.2.64.

Necessity. Suppose the representation is eAC. Then it is also eBV, and hence $f(x, y)$ is BVT in $\mathfrak{R}$ by (i) which we just proved. By V.3.13, it follows that f_x, f_y exist a.e. in $\mathfrak{R}^0$. Hence, by V.3.7, V.1.23,

$$A(S) = a(S) = \iint_{\mathfrak{R}^0} (1 + f_x^2 + f_y^2)^{1/2}\, dx\, dy.$$

By V.3.13 it follows that $f(x, y)$ is ACT in $\mathfrak{R}$.

V.3.15. Given a surface $S : z = f(x, y),\ (x, y) \in \mathfrak{R}$, we introduced previously a variety of rectangle functions associated with the representation in (1) (cf. V.3.2, V.3.3, V.3.4). We shall study presently relationships amongst these rectangle functions, the purpose being to establish contact with the theory of the Burkill integral (see III.1.12). Let

$$(1) \qquad R : a \leqq x \leqq b,\ c \leqq y \leqq d,$$

be an oriented rectangle in $\mathfrak{R}$. Let us recall some definitions used in the general theory of the Burkill integral (see III.1.12). We consider subdivisions of R of the following general character. A subdivision $D(R)$ of R consists of a finite number of oriented rectangles $r_1, \cdots, r_m$, such that $r_i^0 r_j^0 = 0$ for $i \neq j$, and $r_1 + \cdots + r_m = R$. The norm $\| D(R) \|$ of $D(R)$ is defined as max $d(r_i)$, $i = 1, 2, \cdots, m$, where $d(r_i)$ is the diameter of r_i. If $\psi(r)$ is a rectangle function defined for all oriented rectangles $r \subset R$, then we shall use the symbol

$$\sum \psi(r), \qquad r \in D(R),$$

to refer to the summation

$$\sum_{i=1}^{m} \psi(r_i).$$

Now in the theory of the Burkill integral we did not admit rectangle functions that assumed the values $\pm \infty$. On the other hand, some of the rectangle functions introduced in connection with surface area may not be finite. We shall first take care of the complications that arise in this manner.

V.3.16. CONTINUATION. We have the relations (cf. V.3.2, V.3.3, V.3.4)

$$(1) \qquad \sum_{r \in D_n(R)} \gamma_j(f, r) \to g_j(f, R), \qquad j = 1, 2, 3,$$

for every sequence of subdivisions $D_n(R)$ of R such that $\| D_n(R) \| \to 0$.

PROOF. For $j = 3$, clearly

$$\sum_{r \in D_n(R)} \gamma_3(f, r) = | R | = g_3(f, R),$$

and the assertion is obvious. Since the cases $j = 1$ and $j = 2$ are entirely analogous, we shall consider in detail only the case $j = 1$. Let $D(R)$ be any subdivision of R, and let us consider the summation

$$(2) \qquad \sum_{r \in D(R)} \gamma_1(f, r).$$

Now the rectangle function $\gamma_1(f, r)$, while not additive, has a property that may be termed vertical additivity. If r is subdivided, by a horizontal line, into two rectangles r', r'', then clearly (cf. V.3.3)

$$\gamma_1(f, r) = \gamma_1(f, r') + \gamma_1(f, r'').$$

It follows that the summation (2) is unchanged if $D(R)$ is replaced by the subdivision obtained by producing the horizontal sides of the rectangles $r \in D(R)$ to the right and left until they meet the vertical sides of R. In other words, we can replace $D(R)$ by a subdivision that is obtained by first subdividing R into a finite number of horizontal bands, and then subdividing each band into a finite number of rectangles. Thus we can assume, without loss of generality, that $D(R)$ itself is of this special character. The summation (2) admits then of the following interpretation. Let η be a number between c and d (cf. V.3.15(1)). Let us consider the interval $I_\eta : a \leqq x \leqq b$, $y = \eta$. Let us assume, for clarity, that I_η does not pass through any vertex of the subdivision $D(R)$. Then I_η is subdivided, by the vertical sides of the rectangles r that are crossed by I_η, into a finite number of segments. If x' and $x'' > x'$ are the x-coordinates of the end points of such a segment, then we put $v[\eta, D(R)] = \sum | f(x'', \eta) - f(x', \eta) |$, where the summation is extended over all the segments into which I_η is subdivided. The function $v[y, D(R)]$ is thus defined for $c \leqq y \leqq d$, with the exception of a finite number of y-values. For these exceptional values, we agree to put $v[y, D(R)] = 0$. Then $v[y, D(R)]$ is defined for $c \leqq y \leqq d$, and is clearly bounded and measurable (in fact continuous except for a finite number of y-values). Obviously (cf. V.3.3)

$$(3) \qquad \sum_{r \in D(R)} \gamma_1(f, r) = \int_c^d v[y, D(R)]\, dy.$$

Furthermore (cf. III.2.17, III.2.45), if $D_n(R)$ is a sequence of subdivisions of R, such that $\| D_n(R) \| \to 0$, then

$$(4) \qquad v[y, D_n(R)] \to V_x(a, b, y, f),\ v[y, D_n(R)] \leqq V_x(a, b, y, f),$$

with the possible exception of a countable set of y-values in $c \leqq y \leqq d$. Now let us take any sequence $D_n(R)$ such that $\| D_n(R) \| \to 0$, and let us put

$$(5) \qquad \lambda = \liminf_{n \to \infty} \sum_{r \in D_n(R)} \gamma_1(f, r).$$

Case (i). $\lambda = +\infty$. By (3), (4), (5) it follows that $V_x(a, b, y, f)$ is not summable in the interval $c \leqq y \leqq d$, and hence (see III.2.51)

$$g_1(f, R) = W_x(R, f) = +\infty.$$

Clearly $\lambda = +\infty$ implies that

$$\sum_{r \in D_n(R)} \gamma_1(f, r) \underset{n \to \infty}{\longrightarrow} +\infty.$$

Thus (1) holds for $j = 1$. For $j = 2$ the proof is similar.

Case (ii). $\lambda < +\infty$. Then (3), (4) imply (see I.3.10, I.3.11) that $V_x(a, b, y, f)$ is summable in the interval $c \leqq y \leqq d$ and (cf. V.3.2)

$$\sum_{r \in D_n(R)} \gamma_1(f, r) = \int_c^d v[y, D_n(R)]\, dy \underset{n \to \infty}{\longrightarrow} \int_c^d V_x(a, b, y, f)\, dy$$

$$= W_x(R, f) = g_1(f, R).$$

Thus (1) again holds for $j = 1$.

V.3.17. Continuation. We have the relation (cf. V.3.2, V.3.3, V.3.4)

$$(1) \qquad \sum_{r \in D_n(R)} \gamma(f, r) \to \Gamma(f, R),$$

$$(2) \qquad \sum_{r \in D_n(R)} g(f, r) \to G(f, R),$$

for every sequence of subdivisions $D_n(R)$ of R such that $\| D_n(R) \| \to 0$.

PROOF. Let us note the obvious inequalities

$$\gamma(f, r) \leqq \gamma_1(f, r) + \gamma_2(f, r) + \gamma_3(f, r), \tag{3}$$

$$\gamma_j(f, r) \leqq \gamma(f, r), \qquad j = 1, 2, 3. \tag{4}$$

We shall now distinguish between two cases to establish first (1).

Case (i). $\Gamma(f, R) = +\infty$. Then, by V.3.7, V.3.9, $f(x, y)$ is not BVT in R, and hence, by V.3.2, one at least of $g_1(f, R)$, $g_2(f, R)$ fails to be finite. Suppose, for example, that $g_1(f, R) = +\infty$. By V.3.16 it follows that

$$\sum_{r \in D_n(R)} \gamma_1(f, r) \to +\infty,$$

and hence *a fortiori* (cf. (4))

$$\sum_{r \in D_n(R)} \gamma(f, r) \to +\infty = \Gamma(f, R).$$

Case (ii). $\Gamma(f, R) < +\infty$. Then, by V.3.7, V.3.9, $f(x, y)$ is BVT in R. We proceed to verify that the rectangle function $\gamma(f, r)$ is Burkill integrable in R.

(a) The rectangle function $\gamma(f, r)$ is continuous in R, in the sense of III.1.2.

PROOF. Give $\epsilon > 0$. Since $f(x, y)$ is uniformly continuous in R, we have an $\eta_1 > 0$ such that

$$| f(x_2, y) - f(x_1, y) | < \frac{\epsilon}{d(R)} \qquad \text{if } | x_2 - x_1 | < \eta_1, \tag{5}$$

for any two points (x_1, y), (x_2, y) in R ($d(R)$ denotes the diameter of R). Since $f(x, y)$ is bounded in R, we have an inequality of the form

$$| f(x, y) | < M \qquad \text{for } (x, y) \in R,$$

where M is a finite constant. Let us put

$$\eta = \min (\eta_1, \epsilon/2M), \tag{6}$$

and let $r : \alpha_1 \leqq x \leqq \alpha_2$, $\beta_1 \leqq y \leqq \beta_2$, be any oriented rectangle in R such that

$$| r | < \eta^2. \tag{7}$$

Then either $\alpha_2 - \alpha_1 < \eta$ or $\beta_2 - \beta_1 < \eta$. In the first case, by (5), (6),

$$\gamma_1(f, r) \leqq \frac{\epsilon}{d(R)} (\beta_2 - \beta_1) < \epsilon.$$

In the second case, by (6),

$$\gamma_1(f, r) < 2M(\beta_2 - \beta_1) < 2M\eta \leqq \epsilon.$$

Thus (7) implies that $\gamma_1(f, r) < \epsilon$. The continuity of $\gamma_2(f, r)$ is shown in the same way. Since $\gamma_3(f, r) = | r |$, the continuity of $\gamma(f, r)$ follows by (3).

(b) Let $D(R)$ be any subdivision of R. Then (cf. V.3.15), in view of (3),

$$\sum_{r\in D(R)} \gamma(f,r) \leqq \sum_{r\in D(R)} \gamma_1(f,r) + \sum_{r\in D(R)} \gamma_2(f,r) + |R|$$
$$\leqq \sum_{r\in D(R)} g_1(f,r) + \sum_{r\in D(R)} g_2(f,r) + |R|$$
$$= g_1(f,R) + g_2(f,R) + |R|.$$

Since $f(x, y)$ is now BVT in R, $g_1(f, R)$ and $g_2(f, R)$ are both finite, and thus condition (iv) in III.1.19 is satisfied.

(c) Let R_0 be any oriented rectangle in R, and let $D(R_0)$ be any subdivision of R_0. We assert that

$$\gamma_1(f, R_0) \leqq \sum_{r\in D(R_0)} \gamma_1(f,r). \tag{8}$$

Indeed, let $D_1(R_0)$ be the subdivision that is obtained from $D(R_0)$ by producing the horizontal sides of all the rectangles $r \in D(R_0)$ until they meet the vertical sides of R_0. Clearly,

$$\sum_{r\in D_1(R_0)} \gamma_1(f,r) = \sum_{r\in D(R_0)} \gamma_1(f,r). \tag{9}$$

Now let $D_2(R_0)$ be the subdivision that is obtained from $D_1(R_0)$ by deleting all vertical lines of division in $D_1(R_0)$ except the vertical sides of R_0 itself. Clearly

$$\sum_{r\in D_1(R_0)} \gamma_1(f,r) \geqq \sum_{r\in D_2(R_0)} \gamma_1(f,r) = \gamma_1(f, R_0). \tag{10}$$

(9) and (10) imply (8). The inequality

$$\gamma_2(f, R_0) \leqq \sum_{r\in D(R_0)} \gamma_2(f,r) \tag{11}$$

is derived in a similar manner, and finally the relation

$$\gamma_3(f, R_0) = \sum_{r\in D(R_0)} \gamma_3(f,r) \tag{12}$$

is obvious. From (8), (11), (12) we obtain (cf. I.3.10)

$$\gamma(f, R_0) \leqq \left[\left(\sum_{r\in D(R_0)} \gamma_1(f,r)\right)^2 + \left(\sum_{r\in D(R_0)} \gamma_2(f,r)\right)^2 + \left(\sum_{r\in D(R_0)} \gamma_3(f,r)\right)^2\right]^{1/2}$$
$$\leqq \sum_{r\in D(R_0)} \gamma(f,r).$$

Thus the rectangle function $\gamma(f, r)$ increases by subdivision (cf. III.1.5).

As a consequence of (a), (b), (c), $\gamma(f, r)$ is Burkill integrable in R (see III.1.23). Since $\gamma(f, r)$ increases by subdivision (see (c) above), it follows by III.1.23 that $\Gamma(f, R)$ is equal to the Burkill integral of $\gamma(f, r)$ over R, and thus (1) is established. As regards (2), let us note the relations (cf. V.3.2, V.3.3, V.3.4, V.3.7)

$$\sum_{r\in D_n(R)} \gamma(f,r) \leqq \sum_{r\in D_n(R)} g(f,r) \leqq G(f, R) = \Gamma(f, R). \tag{13}$$

Clearly (1) and (13) imply (2).

V.3.18. THEOREM. *Given a surface $S : z = f(x, y)$, $(x, y) \in \mathfrak{R}$, let R be an oriented rectangle in $\mathfrak{R}$. Let $R_1, \cdots, R_m$ be oriented rectangles such that $R_i^0 R_j^0 = 0$ for $i \neq j$, and $R_1 + \cdots + R_m = R$. Then* (*cf.* V.3.4)

$$(1) \qquad A(f, R) = A(f, R_1) + \cdots + A(f, R_m).$$

PROOF. By V.3.7, (1) is equivalent to

$$(2) \qquad \Gamma(f, R) = \Gamma(f, R_1) + \cdots + \Gamma(f, R_m).$$

On the other hand, (2) is an obvious consequence of V.3.17(1).

V.3.19. THEOREM. *Given a surface $S : z = f(x, y)$, $(x, y) \in \mathfrak{R}$, such that $A(S) < +\infty$, let $R : a \leqq x \leqq b$, $c \leqq y \leqq d$, be an oriented rectangle in $\mathfrak{R}$. Then the rectangle function $A(f, r)$ is continuous in R, in the sense of* III.1.2.

PROOF. Since $A(f, r) = \Gamma(f, r)$ by V.3.7, it follows from V.3.17 that $A(f, r)$ is the Burkill integral of the rectangle function $\gamma(f, r)$. Since $\gamma(f, r)$ was shown to be continuous in the course of the proof in V.3.17, the continuity of $A(f, r)$ follows by III.1.17.

V.3.20. *Given a surface $S : z = f(x, y)$, $(x, y) \in \mathfrak{R}$, suppose that $A(S) < +\infty$. Then the derivative* (cf. III.1.24) *of the rectangle function $A(f, r)$ exists and is equal to $(1 + f_x^2 + f_y^2)^{1/2}$ a.e. in $\mathfrak{R}^0$.*

PROOF. Clearly it is sufficient to show that the derivative $A'(x, y, f)$ of $A(f, r)$ exists and is equal to $(1 + f_x^2 + f_y^2)^{1/2}$ a.e. in any assigned oriented rectangle $R \subset \mathfrak{R}$. Now $A(f, R) \leqq A(f, \mathfrak{R}) < +\infty$ by V.2.16. By V.3.17(2), $A(f, r)$ is the Burkill integral of the rectangle function $g(f, r)$ in R. Hence, by III.1.27, it is sufficient to show that the derivative of $g(f, r)$ exists and is equal to $(1 + f_x^2 + f_y^2)^{1/2}$ a.e. in R. Now since $A(f, R) < +\infty$, by V.3.13 the function $f(x, y)$ is BVT in R. Hence, by III.2.52, V.3.2, the rectangle functions $g_1(f, r)$, $g_2(f, r)$ have $|f_x|$, $|f_y|$ for their derivatives a.e. in R, and hence $g(f, r)$ has the derivative $(1 + f_x^2 + f_y^2)^{1/2}$ a.e. in R.

V.3.21. *Given a surface $S : z = f(x, y)$, $(x, y) \in \mathfrak{R}$, such that $A(S) < +\infty$, let R be an oriented rectangle in $\mathfrak{R}^0$. Let us put*

$$f_h(x, y) = \frac{1}{4h^2} \int_{-h}^{h} \int_{-h}^{h} f(x + \xi, y + \eta)\, d\xi\, d\eta.$$

Then $f_h(x, y)$ is defined and continuous, together with its partial derivatives of the first order, in R for h sufficiently small, and we have the formula

$$(1) \qquad A(f, R) = \lim_{h \to 0} \iint_R \left[1 + \left(\frac{\partial f_h}{\partial x}\right)^2 + \left(\frac{\partial f_h}{\partial y}\right)^2\right]^{1/2} dx\, dy,$$

where $\partial f_h/\partial x$, $\partial f_h/\partial y$ are given explicitly by the formulas

$$\frac{\partial f_h}{\partial x} = \frac{1}{4h^2} \int_{-h}^{h} [f(x + h, y + \eta) - f(x - h, y + \eta)]\, d\eta,$$

$$\frac{\partial f_h}{\partial y} = \frac{1}{4h^2} \int_{-h}^{h} [f(x + \xi, y + h) - f(x + \xi, y - h)]\, d\xi.$$

PROOF. The formulas for $\partial f_h/\partial x$, $\partial f_h/\partial y$ were already discussed in III.2.66. Since $R \subset \mathfrak{R}^0$, $f_h(x, y)$ will have continuous first partial derivatives in some domain containing R in its interior if h is small enough. Hence (cf. V.2.12)

$$A(f_h, R) = \iint_R \left[1 + \left(\frac{\partial f_h}{\partial x}\right)^2 + \left(\frac{\partial f_h}{\partial y}\right)^2\right]^{1/2} dx\, dy. \tag{2}$$

Since $f_h(x, y) \to f(x, y)$ uniformly in R for $h \to 0$ (see III.2.66), it follows by V.2.6 that, if $h \to 0$ through any sequence h_n,

$$A(f, R) \leqq \liminf A(f_h, R). \tag{3}$$

Now let δ be a positive number, and let R_δ be the oriented rectangle

$$R_\delta : a - \delta \leqq x \leqq b + \delta,\ c - \delta \leqq y \leqq d + \delta,$$

where R itself is given by $R : a \leqq x \leqq b,\ c \leqq y \leqq d$. We choose δ so small that the inclusion $R_\delta \subset \mathfrak{R}^0$ holds. By V.3.5 we have then, for h small enough, the inequality $\Gamma(f_h, R) \leqq \Gamma(f, R_\delta)$. Hence, by V.3.7,

$$\limsup_{h \to 0} A(f_h, R) \leqq A(f, R_\delta). \tag{4}$$

Now since $A(S) < +\infty$, we have $A(f, R_\delta) \to A(f, R)$ for $\delta \to 0$, as an immediate consequence of the additivity and continuity of the rectangle function $A(f, r)$ (cf. V.3.18, V.3.19). Thus (4) holds with R_δ replaced by R, and (1) follows in view of (2), (3).

V.3.22. To simplify the presentation, *we assume from here on that $\mathfrak{R}$ itself is an oriented rectangle that we shall denote by R_0.* Then S is given in the form $S : z = f(x, y)$, $(x, y) \in R_0$, where

$$R_0 : a_0 \leqq x \leqq b_0,\ c_0 \leqq y \leqq d_0.$$

We further assume that $A(S) < +\infty$. As a consequence, $f(x, y)$ is BVT in R_0 (cf. V.3.9). In the sequel, R_0 will be kept fixed. It will be a matter of great convenience to have $f(x, y)$ defined in the whole xy-plane and not merely in R_0. Therefore we extend the definition of $f(x, y)$ in the following manner. We first reflect R_0 on its right-hand vertical side, taking along the values of $f(x, y)$ in the process. Thus $f(x, y)$ is defined, by symmetry, in a larger rectangle R_0^1. Thereupon we repeat the process of extension by symmetry, this time by reflecting R_0^1 on its upper horizontal side. After this step, $f(x, y)$ is defined on a still larger rectangle $R_0^2 : a_0 \leqq x \leqq 2b_0 - a_0,\ c_0 \leqq y \leqq 2d_0 - c_0$. Furthermore, the extended function $f(x, y)$ satisfies the relations

$$f(a_0, y) = f(2b_0 - a_0, y) \quad \text{for } c_0 \leqq y \leqq 2d_0 - c_0,$$

$$f(x, c_0) = f(x, 2d_0 - c_0) \quad \text{for } a_0 \leqq x \leqq 2b_0 - a_0.$$

Hence we can now continue the extension of $f(x, y)$ to the whole plane univocally by requiring that the extended function $f(x, y)$ be periodic with period $2(b_0 - a_0)$ in x and $2(d_0 - c_0)$ in y. Clearly, if R is any oriented rectangle in the xy-plane, then the extended function $f(x, y)$ is BVT on R. Furthermore, if $f(x, y)$ happens to be ACT in R_0, then the extended function $f(x, y)$ is ACT on every oriented rectangle R in the xy-plane. In the sequel, $f(x, y)$ will refer to the extended function obtained in this precise manner. The notation $A(f, R)$, introduced in V.3.4, applies now to every oriented rectangle R in the xy-plane.

V.3.23. Continuation. By its definition, $A(f, R_0)$ is the smallest number that can be obtained as the limit of the areas of polyhedra converging to S. Clearly this definition, as it stands, does not give any information as to the choice of a sequence of polyhedra whose areas converge to this smallest possible limit $A(f, R_0)$. Furthermore, it is desirable to obtain for $A(f, R_0)$ analytic expressions in terms of standard operations in Analysis (differentiation, integration, and so on). We proceed presently to discuss results obtained in these directions. Of course, many of the results presented in the preceding theory are direct contributions along these lines, and we shall first summarize the most relevant of these results.

V.3.24. Given $f(x, y)$ as described in V.3.22, let R be any oriented rectangle in the plane. Then $f(x, y)$ is BVT on R, and we have at this time the following processes at our disposal to compute $A(f, R)$:

(i) By V.3.17 we have the relations

(1) $$\sum_{r \in D_n(R)} \gamma(f, r) \underset{n\to\infty}{\to} A(f, R),$$

(2) $$\sum_{r \in D_n(R)} g(f, r) \underset{n\to\infty}{\to} A(f, R),$$

where $D_n(R)$ is any sequence of subdivisions of R such that $\| D_n(R) \| \to 0$. For example, to make the process quite definite, we may construct $D_n(R)$ by subdividing the horizontal and vertical sides of R into n equal parts and then drawing verticals and horizontals through the points of division.

(ii) If $f_h(x, y)$ is the integral mean used in V.3.21, then we have, as shown there,

(3) $$A(f_h, R) \underset{h\to 0}{\to} A(f, R),$$

where $A(f_h, R)$ is given by the explicit formulas listed in V.3.21.

(iii) If $f(x, y)$ happens to be ACT in R, and only then (see V.3.13), we can use the standard integral formula

(4) $$A(f, R) = \iint_R (1 + f_x^2 + f_y^2)^{1/2}\, dx\, dy.$$

Of the formulas (1), (2), (3), (4), the last one is thus not generally available. Of the formulas (1), (2), (3), perhaps (1) may be considered as the simplest, and certainly only (1) and (3) are elementary formulas in the sense that they involve only integrations bearing upon continuous functions.

V.3.25. We are led to interesting questions and important results by experimenting with the following approach. Let h and k be two numbers different from zero. Then the partial difference quotients

$$\frac{f(x+h,y)-f(x,y)}{h}, \qquad \frac{f(x,y+k)-f(x,y)}{k}$$

are approximations to the partial derivatives f_x and f_y, and on replacing, in the familiar integral formula for the area, the derivatives f_x, f_y by the corresponding difference quotients, we may obtain a useful approximation to the area. We propose to investigate this situation. Let us put

$$\Phi_x(x,y,h) = \left|\frac{f(x+h,y)-f(x,y)}{h}\right|,$$

$$\Phi_y(x,y,k) = \left|\frac{f(x,y+k)-f(x,y)}{k}\right|,$$

$$\Phi(x,y,h,k) = \{1+[\Phi_x(x,y,h)]^2+[\Phi_y(x,y,k)]^2\}^{1/2},$$

$$I(h,k,R) = \iint_R \Phi(x,y,h,k)\,dx\,dy,$$

$$I^*(R) = \limsup I(h,k,R), \qquad h\to 0,\ k\to 0,$$

$$I_*(R) = \liminf I(h,k,R), \qquad h\to 0,\ k\to 0.$$

These various quantities depend also upon the function $f(x, y)$; but $f(x, y)$ will remain fixed in this investigation, and the notations used will be satisfactory. We shall study the following two questions.

(i) Under what conditions shall we have the formula

$$I(h,k,R) \to \iint_R (1+f_x^2+f_y^2)^{1/2}\,dx\,dy, \tag{1}$$

for $h\to 0$, $k\to 0$?

(ii) Under what conditions shall we have the formula

$$I(h,k,R) \to A(f,R), \tag{2}$$

for $h\to 0$, $k\to 0$?

Let us note that $f(x, y)$ is only assumed to be BVT on every oriented rectangle in the xy-plane, and thus these two questions are distinct. Let us also note that we want the formulas (1) and (2) respectively to hold if h and k converge to zero in any manner, independently of each other. The answers to these questions require considerable preparation.

V.3.26. CONTINUATION. Let us note that $f(x, y)$ is BVT on every oriented rectangle. Let

$$r : \alpha_1 \leqq x \leqq \alpha_2, \beta_1 \leqq y \leqq \beta_2, \tag{1}$$

be any such rectangle. We assert that (see V.3.2)

$$\lim_{h\to 0} \iint_r \left| \frac{f(x + h, y) - f(x, y)}{h} \right| dx\, dy = g_1(f, r), \tag{2}$$

$$\lim_{k\to 0} \iint_r \left| \frac{f(x, y + k) - f(x, y)}{k} \right| dx\, dy = g_2(f, r). \tag{3}$$

PROOF. In discussing (2), there is a slight notational difference depending upon the sign of h. We shall consider explicitly the case when $h > 0$; the case $h < 0$ is entirely analogous. Let δ be a positive number. Since $f(x, y)$ is BVT, the quantity $V_x(\alpha_1, \alpha_2, y, f)$, where $f(x, y)$ is considered as a function of x for given y (see III.2.45), is finite for a.e. y in the interval $\beta_1 \leqq y \leqq \beta_2$, and hence, by III.2.42,

$$\int_{\alpha_1}^{\alpha_2} \left| \frac{f(x + h, y) - f(x, y)}{h} \right| dx \begin{cases} \xrightarrow[h\to 0]{} V_x(\alpha_1, \alpha_2, y, f), \\ \leqq V_x(\alpha_1, \alpha_2 + \delta, y, f) \quad \text{if } 0 < h < \delta, \end{cases} \tag{4}$$

for a.e. y in the interval $\beta_1 \leqq y \leqq \beta_2$. As $f(x, y)$ is BVT in every oriented rectangle, the function $V_x(\alpha_1, \alpha_2 + \delta, y, f)$ is summable in the interval $\beta_1 \leqq y \leqq \beta_2$ (cf. III.2.49). Hence, by a termwise integration that is legitimate in view of (4) (see I.3.11), we obtain from (4) the formula (2) (cf. V.3.2). The formula (3) is proved in a similar way.

V.3.27. CONTINUATION. We assert the inequality (cf. V.3.25)

$$I_*(R) \geqq A(f, R).$$

PROOF. Let $D(R)$ be any subdivision of R (cf. V.3.15). By I.3.10 we obtain the inequalities

$$I(h, k, R) = \sum_{r\in D(R)} I(h, k, r) \geqq \sum_{r\in D(R)} \left\{ |r|^2 + \left[\iint_r \Phi_x(x, y, h)\, dx\, dy \right]^2 + \left[\iint_r \Phi_y(x, y, k)\, dx\, dy \right]^2 \right\}^{1/2}.$$

By V.3.25, V.3.26 there follows, for fixed $D(R)$ and $h \to 0$, $k \to 0$, the inequality

$$I_*(R) \geqq \sum_{r\in D(R)} g(f, r).$$

Since this holds for every subdivision $D(R)$ of R, it follows that (cf. V.3.25, V.3.7) $I_*(R) \geqq G(f, R) = A(f, R)$.

V.3.28. CONTINUATION. We proceed to derive an estimate for $I^*(R)$ (see V.3.25); a little later we shall find the precise value of $I^*(R)$ (see V.3.36). Let R be given by

$$R : a \leqq x \leqq b,\ c \leqq y \leqq d. \tag{1}$$

Let us introduce an auxiliary oriented rectangle $\overline{R}$ such that $R \subset \overline{R}^0$. By V.3.18, V.3.19, V.3.2, the rectangle functions $g_1(f, r)$, $g_2(f, r)$, $A(f, r)$ are finite, non-negative, additive and continuous for $r \subset \overline{R}$. By III.1.37, III.1.43, these rectangle functions give rise therefore to completely additive functions of Borel sets in $\overline{R}$, which will be denoted by $g_1(f, B)$, $g_2(f, B)$, $A(f, B)$ respectively, where B is a generic notation for a Borel set in $\overline{R}$. We have then, for each one of these set-functions, a Lebesgue decomposition on $\overline{R}$, and we shall use the symbols $g_1^0(f, B)$, $g_2^0(f, B)$, $A^0(f, B)$ to denote the singular set functions that are involved in these decompositions (see I.3.16). By V.3.20, III.1.29, the derivatives of $g_1(f, B)$, $g_2(f, B)$, $A(f, B)$ exist and are equal to $|f_x|$, $|f_y|$, $(1 + f_x^2 + f_y^2)^{1/2}$ respectively a.e. in $\overline{R}$. Hence, by III.1.32,

$$g_1(f, B) = \iint_B |f_x(x, y)|\, dx\, dy + g_1^0(f, B),$$

$$g_2(f, B) = \iint_B |f_y(x, y)|\, dx\, dy + g_2^0(f, B),$$

$$A(f, B) = \iint_B (1 + f_x^2 + f_y^2)^{1/2}\, dx\, dy + A^0(f, B),$$

for every Borel set $B \subset \overline{R}$. Furthermore, we have in $\overline{R}$ Borel sets e_1, e_2, e of measure zero, such that (cf. I.3.16)

$$g_1^0(f, B) = g_1(f, Be_1),\ g_2^0(f, B) = g_2(f, Be_2),\ A^0(f, B) = A(f, Be)$$

for every Borel set $B \subset \overline{R}$. Finally, we have in $\overline{R}$ a set e' of measure zero such that (see above) the derivatives of $g_1(f, B)$, $g_2(f, B)$, $A(f, B)$ exist and are equal to $|f_x|$, $|f_y|$, $(1 + f_x^2 + f_y^2)^{1/2}$ respectively on the set $\overline{R} - e'$. Let then e'' be a Borel set of measure zero such that (cf. I.3.7) $e'' \supset e_1 + e_2 + e + e'$. Let us put $e_0 = e''R$. Then e_0 is a Borel subset of measure zero of R, and the following statements hold in view of the preceding remarks.

(i) $e_0 \subset R$, $|e_0| = 0$, and e_0 is a Borel set.

(ii) The derivatives of $g_1(f, B)$, $g_2(f, B)$, $A(f, B)$ exist and are equal to $|f_x|$, $|f_y|$, $(1 + f_x^2 + f_y^2)^{1/2}$ respectively on the set $R - e_0$.

(iii) For every Borel set $B \subset R$, we have the formulas

$$g_1^0(f, B) = g_1(f, Be_0),\ g_2^0(f, B) = g_2(f, Be_0),\ A^0(f, B) = A(f, Be_0).$$

In particular, for $B = R$,

$$g_1^0(f, R) = g_1(f, e_0),\ g_2^0(f, R) = g_2(f, e_0),\ A^0(f, R) = A(f, e_0).$$

V.3.29. CONTINUATION. Let $\delta > 0$ be so small that the rectangle (cf. V.3.28(1)) $R^\delta : a - \delta \leqq x \leqq b + \delta, c - \delta \leqq y \leqq d + \delta$, is comprised in $\bar{R}$. Then $f(x, y)$ is BVT on R^δ, and hence, for a.e. y in the interval $c - \delta \leqq y \leqq d + \delta$, $f(x, y)$ is of bounded variation, as a function of x, on the interval $a - \delta \leqq x \leqq b + \delta$. Let $V_x^0(a - \delta, b + \delta, y, f)$ be the singular variation of $f(x, y)$, considered as a function of x for given y, on the interval $a - \delta \leqq x \leqq b + \delta$ (see III.2.26). We assert the formula

$$\int_{c-\delta}^{d+\delta} V_x^0(a - \delta, b + \delta, y, f)\, dy = g_1^0(f, R^\delta).$$

A similar statement holds for $g_2^0(f, R^\delta)$.

PROOF. By III.2.26 we have

$$V_x^0(a - \delta, b + \delta, y, f) = V_x(a - \delta, b + \delta, y, f) - \int_{a-\delta}^{b+\delta} | f_x(x, y) |\, dx.$$

Integration yields (cf. V.3.2, V.3.28)

$$\int_{c-\delta}^{d+\delta} V_x^0(a - \delta, b + \delta, y, f)\, dy = g_1(f, R^\delta) - \iint_{R^\delta} | f_x(x, y) |\, dx\, dy = g_1^0(f, R^\delta).$$

V.3.30. CONTINUATION. For $\delta \to 0$, we have

$$g_1^0(f, R^\delta) \to g_1^0(f, R),\ g_2^0(f, R^\delta) \to g_2^0(f, R).$$

PROOF. If we note that $g_j^0(f, B) \leqq g_j(f, B)$, $j = 1, 2$, for every Borel set B (cf. V.3.28), then we see that the assertions follow immediately from the continuity of the rectangle functions $g_j(f, r)$, $j = 1, 2$ (see V.3.2, V.3.19).

V.3.31. CONTINUATION. Since $f(x, y)$ is BVT in $\bar{R}$, the partial derivatives f_x, f_y are summable in $\bar{R}$ (see III.2.50). For $\lambda > 0$, let us put

$$\Omega_1(\lambda) = \text{l.u.b.} \iint_E | f_x |\, dx\, dy,$$

$$\Omega_2(\lambda) = \text{l.u.b.} \iint | f_y |\, dx\, dy,$$

$$\Omega(\lambda) = \Omega_1(\lambda) + \Omega_2(\lambda),$$

where the l.u.b. is taken with respect to all measurable sets E such that $E \subset \bar{R}$, $| E | < \lambda$. By I.3.13 we have then the relation $\Omega(\lambda) \to 0$ for $\lambda \to 0$.

V.3.32. CONTINUATION. Let e be a Borel set in R. Let $\delta > 0$ be determined as in V.3.29, and let $h \neq 0$, $k \neq 0$ satisfy the inequalities $| h | < \delta$, $| k | < \delta$. We assert then the inequalities (cf. V.3.28)

$$(1) \qquad \iint_e \left| \frac{f(x + h, y) - f(x, y)}{h} \right| dx\, dy \leqq \Omega(| e |) + g_1^0(f, R^\delta),$$

$$(2) \qquad \iint\limits_{e} \left| \frac{f(x, y+k) - f(x, y)}{k} \right| dx\, dy \leqq \Omega(|e|) + g_2^0(f, R^\delta).$$

PROOF. Let us consider (1), and let us assume that $h > 0$ (the case $h < 0$ is disposed of in a similar manner). For a.e. y in the interval $c \leqq y \leqq d$, $f(x, y)$ is of bounded variation, as a function of x, on the interval $a \leqq x \leqq b$, and hence

$$(3) \qquad \begin{aligned} | f(x+h, y) - f(x, y) | &\leqq V_x(x, x+h, y, f) \\ &= \int_0^h | f_x(x+\xi, y) | \, d\xi + V_x^0(x, x+h, y, f) \end{aligned}$$

(see III.2.26). Let e_y denote the intersection of e with the horizontal line at the given altitude y. Integration of (3) yields (cf. I.3.10)

$$(4) \qquad \begin{aligned} \int_{e_y} \left| \frac{f(x+h, y) - f(x, y)}{h} \right| dx &\leqq \frac{1}{h} \int_0^h \left[\int_{e_y} | f_x(x+\xi, y) | \, dy \right] d\xi \\ &\quad + \frac{1}{h} \int_a^b V_x^0(x, x+h, y, f) \, dx. \end{aligned}$$

Using the inequality (cf. III.2.43)

$$\frac{1}{h} \int_a^b V_x^0(x, x+h, y, f) \, dx \leqq V_x^0(a-\delta, b+\delta, y, f),$$

we obtain, by integrating (4) with respect to y from c to d, and using V.3.29,

$$(5) \qquad \begin{aligned} &\iint\limits_{e} \left| \frac{f(x+h, y) - f(x, y)}{h} \right| dx\, dy \\ &\qquad \leqq \frac{1}{h} \int_0^h \left[\iint\limits_{e} | f_x(x+\xi, y) | \, dx\, dy \right] d\xi + g_1^0(f, R^\delta). \end{aligned}$$

Now if $e(\xi)$ is the set obtained from e by a horizontal translation of magnitude ξ, then we obtain by V.3.31 the inequality (note that $0 \leqq \xi \leqq h < \delta$, $|e(\xi)| = |e|$)

$$(6) \qquad \iint\limits_{e} | f_x(x+\xi, y) | \, dx\, dy = \iint\limits_{e(\xi)} | f_x(x, y) | \, dx\, dy \leqq \Omega(|e(\xi)|) = \Omega(|e|).$$

(5) and (6) imply (1). (2) is proved in a similar way.

V.3.33. CONTINUATION. If e is any Borel set in R, then

$$\limsup_{h \to 0} \iint\limits_{e} \left| \frac{f(x+h, y) - f(x, y)}{h} \right| dx\, dy \leqq \Omega(|e|) + g_1^0(f, R),$$

$$\limsup_{k \to 0} \iint\limits_{e} \left| \frac{f(x, y+k) - f(x, y)}{k} \right| dx\, dy \leqq \Omega(|e|) + g_2^0(f, R).$$

This is an immediate consequence of V.3.32 and V.3.30.

V.3.34. CONTINUATION. We assert the inequality (cf. V.3.25)

$$(1) \qquad I^*(R) \leqq \iint_R (1 + f_x^2 + f_y^2)^{1/2}\, dx\, dy + g_1^0(f, R) + g_2^0(f, R).$$

PROOF. Let e_0 be the set defined in V.3.28. On $R - e_0$, the partial derivatives f_x and f_y exist, and hence, for any sequence $h_n \to 0$, $k_n \to 0$, $h_n \neq 0$, $k_n \neq 0$,

$$(2) \qquad \frac{f(x + h_n, y) - f(x, y)}{h_n} \underset{n\to\infty}{\to} f_x(x, y), \qquad (x, y) \in R - e_0,$$

$$(3) \qquad \frac{f(x, y + k_n) - f(x, y)}{k_n} \underset{n\to\infty}{\to} f_y(x, y), \qquad (x, y) \in R - e_0.$$

By the theorem of Egoroff (see I.3.17), we can therefore split the set $R - e_0$, for given $\epsilon > 0$, as follows:

(i*) $R - e_0 = F_\epsilon^* + e_\epsilon^*$, where F_ϵ^* is closed, $| e_\epsilon^* | < \epsilon$, and $F_\epsilon^* e_\epsilon^* = 0$.

(ii*) The sequences (2), (3) converge uniformly on F_ϵ^*. Let us put $e_\epsilon = R - F_\epsilon^*$. Then the following statements hold:

(i) $R = F_\epsilon^* + e_\epsilon$, where F_ϵ^* is closed, $| e_\epsilon | < \epsilon$, and $F_\epsilon^* e_\epsilon = 0$.

(ii) The sequences (2), (3) converge uniformly on F_ϵ^*.

(iii) $e_0 \subset e_\epsilon$.

Now let us write

$$(4) \quad I(h_n, k_n, R) = \iint_{F_\epsilon^*} \Phi(x, y, h_n, k_n)\, dx\, dy + \iint_{F_\epsilon^*} \Phi(x, y, h_n, k_n)\, dx\, dy.$$

By (ii) above, there follows the relation (cf. V.3.25)

$$(5) \quad \begin{aligned} \iint_{F_\epsilon^*} \Phi(x, y, h_n, k_n)\, dx\, dy &\underset{n\to\infty}{\to} \iint_{F_\epsilon^*} (1 + f_x^2 + f_y^2)^{1/2}\, dx\, dy \\ &\leqq \iint_R (1 + f_x^2 + f_y^2)^{1/2}\, dx\, dy. \end{aligned}$$

Clearly (cf. V.3.25)

$$\begin{aligned} \iint_{e_\epsilon} \Phi(x, y, h_n, k_n)\, dx\, dy \leqq | e_\epsilon | &+ \iint_{e_\epsilon} \left| \frac{f(x + h_n, y) - f(x, y)}{h_n} \right| dx\, dy \\ &+ \iint_{e_\epsilon} \left| \frac{f(x, y + k_n) - f(x, y)}{k_n} \right| dx\, dy. \end{aligned}$$

Since $| e_\epsilon | < \epsilon$, we obtain by V.3.33 the inequality

$$(6) \quad \limsup_{n\to\infty} \iint_{e_\epsilon} \Phi(x, y, h_n, k_n)\, dx\, dy \leqq \epsilon + 2\Omega(\epsilon) + g_1^0(f, R) + g_2^0(f, R).$$

(4), (5), (6) yield (since the sequences $h_n \to 0$, $k_n \to 0$ were arbitrary)

$$I^*(R) \leqq \iint_R (1 + f_x^2 + f_y^2)^{1/2}\, dx\, dy + g_1^0(f, R) + g_2^0(f, R) + \epsilon + 2\Omega(\epsilon).$$

Since $\epsilon > 0$ was arbitrary, (1) follows (cf. V.3.31).

V.3.35. THEOREM. *Given $f(x, y)$ as in* V.3.22, *the formula* (*cf.* V.3.25)

$$(1) \qquad \iint_R (1 + f_x^2 + f_y^2)^{1/2}\, dx\, dy = \lim I(h, k, R) \qquad \text{for } h \to 0,\ k \to 0$$

holds if and only if $f(x, y)$ is ACT on R.

PROOF. By V.3.13, V.3.27, V.3.34 we have the inequalities

$$(2) \qquad \begin{aligned} \iint_R (1 + f_x^2 + f_y^2)^{1/2}\, dx\, dy &\leqq A(f, R) \leqq I_*(R) \leqq I^*(R) \\ &\leqq \iint_R (1 + f_x^2 + f_y^2)^{1/2}\, dx\, dy \\ &\qquad + g_1^0(f, R) + g_2^0(f, R). \end{aligned}$$

(i) Suppose first that $f(x, y)$ is ACT on R. Then (see V.3.28, V.3.2) $g_1^0(f, R) = 0$, $g_2^0(f, R) = 0$, and hence (2) yields (1).

(ii) Suppose next that (1) holds. Then

$$(3) \qquad I^*(R) = I_*(R) = \iint_R (1 + f_x^2 + f_y^2)^{1/2}\, dx\, dy.$$

(2) and (3) imply that

$$(4) \qquad A(f, R) = \iint_R (1 + f_x^2 + f_y^2)^{1/2}\, dx\, dy.$$

(4) implies, by V.3.13, that $f(x, y)$ is ACT on R.

V.3.36. The preceding result gives the complete answer to question (i) in V.3.25. The answer to question (ii) in V.3.25 is more involved. *We first improve upon* V.3.34 *by establishing the formula*

$$I^*(R) = \iint_R (1 + f_x^2 + f_y^2)^{1/2}\, dx\, dy + g_1^0(f, R) + g_2^0(f, R).$$

PROOF. Let O be a relatively open subset of R (that is, $R - O$ is closed).

By I.3.2, we have then a sequence of oriented rectangles $r_1, \cdots, r_m, \cdots$ such that $r_1 + \cdots + r_m + \cdots = O$, $r_i^0 r_j^0 = 0$ for $i \neq j$. By V.3.28, III.1.48 we have

(1) $$g_1(f, O) = \sum_{m=1}^{\infty} g_1(f, r_m),$$

$$g_2(f, O) = \sum_{m=1}^{\infty} g_2(f, r_m).$$

By V.3.26 it follows that for every positive integer N

(2) $$\liminf_{h \to 0} \iint_O \left| \frac{f(x+h, y) - f(x, y)}{h} \right| dx\, dy \geqq \liminf_{h \to 0} \sum_{m=1}^{N} \iint_{r_m} \left| \frac{f(x+h, y) - f(x, y)}{h} \right| dx\, dy = \sum_{m=1}^{N} g_1(f, r_m).$$

Since N is arbitrary, (1) and (2) imply that

(3) $$\liminf_{h \to 0} \iint_O \left| \frac{f(x+h, y) - f(x, y)}{h} \right| dx\, dy \geqq g_1(f, O).$$

A similar reasoning yields

(4) $$\liminf_{k \to 0} \iint_O \left| \frac{f(x, y+k) - f(x, y)}{k} \right| dx\, dy \geqq g_2(f, O).$$

Now give $\epsilon > 0$. Since the set e_0 of V.3.28 is of measure zero, we can split R into a sum of two sets F, e as follows:

(5) $$R = F + e, \ Fe = 0,$$

(6) $$F \text{ closed}, \ |e| < \epsilon,$$

(7) $$e \supset e_0 .$$

We choose such sets F, e and keep them fixed. Since $R - e = F$ is closed, we have by (3), (7) and V.3.28 the inequalities

$$\liminf_{h \to 0} \iint_e \left| \frac{f(x+h, y) - f(x, y)}{h} \right| dx\, dy \geqq g_1(f, e) \geqq g_1(f, e_0) = g_1^0(f, R).$$

Hence, for a given positive integer n we can determine h_n to satisfy the conditions

(8) $$\iint_e \left| \frac{f(x+h_n, y) - f(x, y)}{h_n} \right| dx\, dy > g_1^0(f, R) - \frac{1}{n},$$

$$0 < |h_n| < \frac{1}{n}.$$

After h_n has been so chosen, we proceed as follows. Since $e \supset e_0$ and $|e_0| = 0$, we have a set e_n such that

$$e \supset e_n \supset e_0 ,$$

$$\iint_{e-e_n} \left| \frac{f(x+h_n , y) - f(x, y)}{h_n} \right| dx\, dy > g_1^0(f, R) - \frac{1}{n}, \tag{9}$$

$$R - e_n \text{ is closed.} \tag{10}$$

Note that (9) can be achieved, in view of (8), by making $|e_n|$ sufficiently small, and (10) can be achieved since $R - e$ is closed (see (6)). After e_n has been so chosen, we have by (4), (7), and V.3.28,

$$\liminf_{k\to 0} \iint_{e_n} \left| \frac{f(x, y+k) - f(x, y)}{k} \right| dx\, dy \geqq g_2(f, e_n) \geqq g_2(f, e_0) = g_2^0(f, R).$$

Hence, for the same n, we can determine k_n to satisfy the conditions

$$(11)\qquad \begin{gathered} \iint_{e_n} \left| \frac{f(x, y+k_n) - f(x, y)}{k_n} \right| dx\, dy > g_2^0(f, R) - \frac{1}{n}, \\ 0 < |k_n| < \frac{1}{n}. \end{gathered}$$

Now let us write

$$(12)\qquad \begin{aligned} I(h_n , k_n , R) = \iint_{F} \Phi(x, y, h_n , k_n)\, dx\, dy &+ \iint_{e-e_n} \Phi(x, y, h_n , k_n)\, dx\, dy \\ &+ \iint_{e_n} \Phi(x, y, h_n , k_n)\, dx\, dy. \end{aligned}$$

By (9), (11) and V.3.25 we obtain

$$(13)\qquad \begin{aligned} \iint_{e-e_n} \Phi(x, y, h_n , k_n)\, dx\, dy &\geqq \iint_{e-e_n} \left| \frac{f(x+h_n , y) - f(x, y)}{h_n} \right| dx\, dy \\ &> g_1^0(f, R) - \frac{1}{n}, \end{aligned}$$

$$(14)\qquad \begin{aligned} \iint_{e_n} \Phi(x, y, h_n , k_n)\, dx\, dy &\geqq \iint_{e_n} \left| \frac{f(x, y+k_n) - f(x, y)}{k_n} \right| dx\, dy \\ &> g_2^0 (f, R) - \frac{1}{n}. \end{aligned}$$

Observe that $\Phi(x, y, h_n, k_n) \to (1 + f_x^2 + f_y^2)^{1/2}$ a.e. on R and hence on F. By I.3.10 it follows that

(15) $$\liminf_{n\to\infty} \iint_F \Phi(x, y, h_n, k_n)\, dx\, dy \geqq \iint_F (1 + f_x^2 + f_y^2)^{1/2}\, dx\, dy.$$

(12), (13), (14), (15) yield (cf. V.3.25)

(16) $$I^*(R) \geqq \iint_F (1 + f_x^2 + f_y^2)^{1/2}\, dx\, dy + g_1^0(f, R) + g_2^0(f, R).$$

Now since $(1 + f_x^2 + f_y^2)^{1/2}$ is summable on R (note that $f(x, y)$ is BVT on R), the difference

$$\iint_R (1 + f_x^2 + f_y^2)^{1/2}\, dx\, dy - \iint_F (1 + f_x^2 + f_y^2)^{1/2}\, dx\, dy$$

$$= \iint (1 + f_x^2 + f_y^2)^{1/2}\, dx\, dy$$

will be arbitrarily small if ϵ was chosen small enough (cf. (6)). Since (16) holds for an arbitrary choice of ϵ, it follows that

$$I^*(R) \geqq \iint_R (1 + f_x^2 + f_y^2)^{1/2}\, dx\, dy + g_1^0(f, R) + g_2^0(f, R).$$

In view of V.3.34, it follows that the sign of equality holds, and the proof is complete.

V.3.37. *Given $f(x, y)$ as in* V.3.22, *the formula*

(1) $$A(f, R) = \lim I(h, k, R) \qquad \text{for } h \to 0,\ k \to 0,$$

holds if and only if (cf. V.3.28)

(2) $$g_1^0(f, R) + g_2^0(f, R) = A^0(f, R).$$

Proof. By V.3.27, V.3.36 we have the inequalities

(3) $$A(f, R) \leqq I_*(R) \leqq I^*(R) = \iint_R (1 + f_x^2 + f_y^2)^{1/2}\, dx\, dy + g_1^0(f, R) + g_2^0(f, R).$$

By V.3.28 we have

(4) $$A(f, R) = \iint_R (1 + f_x^2 + f_y^2)^{1/2}\, dx\, dy + A^0(f, R).$$

(i) Suppose that (1) holds. Then we have, by (3),

$$(5)\qquad A(f, R) = \iint_R (1 + f_x^2 + f_y^2)^{1/2}\, dx\, dy + g_1^0(f, R) + g_2^0(f, R).$$

(4) and (5) imply (2).

(ii) Suppose next that (2) holds. Then (2) and (4) imply that

$$(6)\qquad A(f, R) = \iint_R (1 + f_x^2 + f_y^2)^{1/2}\, dx\, dy + g_1^0(f, R) + g_2^0(f, R).$$

(3) and (6) imply that $A(f, R) = I_*(R) = I^*(R)$.

V.3.38. We propose to transform the condition V.3.37(2). We shall need a lemma that may be thought of as a generalization of the trivial fact that if the sum of two sides is equal to the third side in a (rectilinear) triangle, then the triangle must be degenerate.

Let there be given, in an oriented rectangle R, three functions $\phi_1(B)$, $\phi_2(B)$, $\phi_3(B)$ of Borel sets B in R, with the following properties:

(α) $\phi_j(B) \geqq 0$, $j = 1, 2, 3$.

(β) $\phi_j(B)$ is continuous on oriented rectangles in R (cf. III.1.30, III.1.2), $j = 1, 2, 3$.

(γ) $\phi_j(B)$ is completely additive on Borel sets in R (cf. I.3.16). Under these conditions, we consider the rectangle function

$$\psi(r) = [\phi_1(r)^2 + \phi_2(r)^2 + \phi_3(r)^2]^{1/2},$$

for oriented rectangles $r \subset R$. From III.1.51 it follows that $\psi(r)$ is Burkill integrable in R. Let $\alpha(r)$ be the indefinite Burkill integral of $\psi(r)$. Then $\alpha(r)$ is non-negative, and $\alpha(r)$ is additive on oriented rectangles in R (see III.1.14). We have the obvious inequalities

$$(1)\qquad \alpha(r) \leqq \phi_1(r) + \phi_2(r) + \phi_3(r),$$

$$(2)\qquad \phi_j(r) \leqq \alpha(r), \qquad j = 1, 2, 3.$$

As a consequence of (β) and (1), $\alpha(r)$ is continuous on oriented rectangles in R. By III.1.37, III.1.43 it follows that $\alpha(r)$ admits of a non-negative, completely additive extension to Borel sets in R that we shall denote by $\alpha(B)$.

LEMMA. *Let e_0 be a Borel set in R. Then the relation*

$$(3)\qquad \phi_1(e_0) + \phi_2(e_0) + \phi_3(e_0) = \alpha(e_0)$$

holds if and only if there exists a decomposition of e_0 into a sum of three Borel sets e_1, e_2, e_3 (some of which may be empty), such that

$$(4)\qquad e_0 = e_1 + e_2 + e_3,$$

$$(5)\qquad e_1e_2 = e_2e_3 = e_3e_1 = 0,$$

$$(6)\qquad \phi_j(e_i) = 0 \qquad \text{for } i \neq j,\ i > 0,\ j > 0.$$

Proof. *Necessity.* Suppose that (3) holds. As a consequence of (2), we have (cf. III.1.45) $\phi_j(B) \leqq \alpha(B)$, $j = 1, 2, 3$, for every Borel set $B \subset R$. Thus $\phi_j(B)$ is absolutely continuous with respect to $\alpha(B)$, and hence (see I.3.17), we have three non-negative, Borel measurable point functions f_1, f_2, f_3 in R, such that

$$\phi_j(B) = \int_B f_j d\alpha, \qquad j = 1, 2, 3, \tag{7}$$

for every Borel set $B \subset R$. Now let R^* be any oriented rectangle in R, and let $D_n(R^*)$ be the subdivision of R^* obtained by subdividing the sides of R^* into n equal parts and then drawing horizontals and verticals through the points of division. By III.1.50 we have then

$$\sum_{r \in D_n(R^*)} \left[\left(\int_r f_1 d\alpha\right)^2 + \left(\int_r f_2 d\alpha\right)^2 + \left(\int_r f_3 d\alpha\right)^2\right]^{1/2} \underset{n \to \infty}{\longrightarrow} \int_{R^*} (f_1^2 + f_2^2 + f_3^2)^{1/2} \, d\alpha. \tag{8}$$

Since α is the Burkill integral of $\psi = (\phi_1^2 + \phi_2^2 + \phi_3^2)^{1/2}$, we also have

$$\sum_{r \in D_n(R^*)} [\phi_1(r)^2 + \phi_2(r)^2 + \phi_3(r)^2]^{1/2} \underset{n \to \infty}{\longrightarrow} \alpha(R^*). \tag{9}$$

(7), (8), (9) yield

$$\alpha(R^*) = \int_{R^*} (f_1^2 + f_2^2 + f_3^2)^{1/2} \, d\alpha$$

for every oriented rectangle $R^* \subset R$. By I.3.17, III.1.47 there follows the formula

$$\alpha(B) = \int_B (f_1^2 + f_2^2 + f_3^2)^{1/2} \, d\alpha$$

for every Borel set $B \subset R$. Let us apply this result to the Borel set e_0. Since (3) holds now by assumption, we obtain the equation (cf. (7))

$$\int_{e_0} [f_1 + f_2 + f_3 - (f_1^2 + f_2^2 + f_3^2)^{1/2}] \, d\alpha = 0.$$

Since the integrand is clearly non-negative, it follows that

$$f_1 + f_2 + f_3 = (f_1^2 + f_2^2 + f_3^2)^{1/2} \text{ on } e_0 - \bar{e}_0, \tag{10}$$

where $\bar{e}_0$ is a Borel subset of e_0 such that

$$\alpha(\bar{e}_0) = 0. \tag{11}$$

From (10) we infer, by squaring, that $f_2 f_3 + f_3 f_1 + f_1 f_2 = 0$ on $e_0 - \bar{e}_0$. Since $f_j \geqq 0$, it follows that at every point of the set $e_0 - \bar{e}_0$ at least two of the three functions f_1, f_2, f_3 vanish. Let now e_1^* be the subset of $e_0 - \bar{e}_0$ on which $f_1 > 0$. Then, by the preceding remark,

$$f_2 = 0, f_3 = 0 \qquad \text{on } e_1^* . \tag{12}$$

Similarly, if e_2^* is the subset of $(e_0 - \bar{e}_0) - e_1^*$ on which $f_2 > 0$, then

$$f_3 = 0, f_1 = 0 \qquad \text{on } e_2^* . \tag{13}$$

Finally, if $e_3^* = [(e_0 - \bar{e}_0) - e_1^*] - e_2^*$, then clearly

$$f_1 = 0, f_2 = 0 \qquad \text{on } e_3^* . \tag{14}$$

Clearly, e_1^*, e_2^*, e_3^* are disjoint (possibly empty) Borel sets. If we put $e_1 = e_1^* + \bar{e}_0$, $e_2 = e_2^*$, $e_3 = e_3^*$, then e_1, e_2, e_3 are Borel sets such that $e_1 + e_2 + e_3 = e_0$, $e_1e_2 = e_2e_3 = e_3e_1 = 0$. By (7), (11), (12), (13), (14) we obtain readily the formulas (6).

Sufficiency. Let us assume now that we have a decomposition of e_0 that satisfies the conditions (4), (5), (6). In view of III.1.45, there follow from (1), (2) the inequalities

$$\alpha(B) \leqq \phi_1(B) + \phi_2(B) + \phi_3(B), \tag{15}$$

$$\phi_j(B) \leqq \alpha(B), \qquad j = 1, 2, 3, \tag{16}$$

for every Borel set $B \subset R$. Using (4), (5), (16), we obtain

$$\phi_1(e_0) = \phi_1(e_1) \leqq \alpha(e_1),$$

and similarly $\phi_2(e_0) \leqq \alpha(e_2)$, $\phi_3(e_0) \leqq \alpha(e_3)$. Addition yields $\phi_1(e_0) + \phi_2(e_0) + \phi_3(e_0) \leqq \alpha(e_0)$, while the complementary inequality follows directly from (15). Hence (3) holds.

V.3.39. Given $f(x, y)$ as in V.3.22, the formula

$$A(f, R) = \lim I(h, k, R) \qquad \text{for } h \to 0, k \to 0, \tag{1}$$

holds if and only if there exist in R two Borel sets B_1, B_2 such that (cf. V.3.28, I.3.16)

(i) $B_1 + B_2 = R$, and

(ii) $g_1(f, B)$ is absolutely continuous on B_1 and $g_2(f, B)$ is absolutely continuous on B_2.

PROOF. *Necessity.* Suppose that (1) holds. By V.3.37 we have then

$$g_1^0(f, R) + g_2^0(f, R) = A^0(f, R), \tag{2}$$

which may be written, by V.3.28, in the form

$$g_1(f, e_0) + g_2(f, e_0) = A(f, e_0). \tag{3}$$

Now let us put

$$\phi_1(B) = g_1(f, B), \phi_2(B) = g_2(f, B), \phi_3(B) = | B |, \alpha(B) = A(f, B). \tag{4}$$

By V.3.28, V.3.19, V.3.17, the set functions ϕ_1, ϕ_2, ϕ_3, α satisfy then all the assumptions of V.3.38. Since $|e_0| = 0$, we can write (3) in the form

$$\phi_1(e_0) + \phi_2(e_0) + \phi_3(e_0) = \alpha(e_0). \tag{5}$$

(5) implies, by V.3.38, the existence of three Borel sets e_1, e_2, e_3 with the following properties.

$$e_0 = e_1 + e_2 + e_3,$$

$$e_1e_2 = e_2e_3 = e_3e_1 = 0,$$

$$\phi_1(e_2) = \phi_1(e_3) = \phi_2(e_1) = \phi_2(e_3) = 0. \tag{6}$$

Let us put $B_1 = (R - e_0) + e_2 + e_3$, $B_2 = e_1$. Now let E be any Borel set of measure zero contained in B_1. Let us write

$$E = (R - e_0)E + e_2E + e_3E.$$

By (4), (6) and V.3.28 we have, since $|E| = 0$,

$$g_1(f, (R - e_0)E) = g_1^0(f, (R - e_0)E) = g_1(f, (R - e_0)Ee_0) = 0,$$

$$g_1(f, e_2E) \leqq g_1(f, e_2) = 0,$$

$$g_1(f, e_3E) \leqq g_1(f, e_3) = 0.$$

Addition yields the result that $g_1(f, E) = 0$ if E is any Borel subset of measure zero of B_1. Hence (see I.3.16), $g_1(f, B)$ is absolutely continuous on B_1. Since $g_2(f, B_2) = 0$ by (4) and (6), clearly $g_2(f, B)$ is absolutely continuous on B_2.

Sufficiency. Suppose conversely that there exist Borel sets B_1, B_2 satisfying (i) and (ii). Using again the definition (4), let us consider the set e_0 defined in V.3.28, and let us put

$$e_1 = (B_2 - B_1)e_0, \; e_2 = B_1e_0, \; e_3 = 0.$$

We have then $\phi_1(e_2) = g_1(f, B_1e_0) = 0$, since g_1 is absolutely continuous on B_1 and $|B_1e_0| = 0$. Furthermore $\phi_1(e_3) = 0$ since $e_3 = 0$. Next we have $\phi_2(e_1) = g_2(f, (B_2 - B_1)e_0) = 0$, since g_2 is absolutely continuous on B_2 and $|(B_2 - B_1)e_0| = 0$, $(B_2 - B_1)e_0 \subset B_2$. Since $e_3 = 0$, we have also $\phi_2(e_3) = 0$. The relations $\phi_3(e_1) = \phi_3(e_2) = 0$ are obvious, since $|e_1| = |e_2| = 0$. By V.3.38 we obtain now successively (5), (3), and (2), and finally (1) by V.3.37.

V.3.40. Let us say that $f(x, y)$ is ACY on $R: a \leqq x \leqq b, c \leqq y \leqq d$ (absolutely continuous on the oriented rectangle R, in the sense of L. C. Young), if the following conditions hold.

(i) $f(x, y)$ is BVT on R.

(ii) There exist two Borel sets B_1, B_2, such that (α) $f(x, y)$ is absolutely continuous, as a function of x, on the intersection B_{1y} of B_1 with the horizontal line at altitude y (see III.2.33), for a.e. y in the range $c \leqq y \leqq d$; (β) $f(x, y)$ is absolutely continuous, as a function of y, on the intersection B_{2x} of B_2 with the vertical line corresponding to a given x, for a.e. x in the range $a \leqq x \leqq b$; (γ) $B_1 + B_2 = R$.

Given $f(x, y)$ as in V.3.22, let us assume that $f(x, y)$ is ACY in R. Let B be any Borel subset of the set B_1 (see (ii)). By V.3.2, III.2.61 we have then

$$g_1(f, B) = \int_c^d V_x(B_y, y, f)\, dy.$$

By condition (ii) we have, in view of III.2.34,

$$V_x(B_y, y, f) = \int_{B_y} | f_x(x, y) |\, dx$$

for a.e. y in the interval $c \leqq y \leqq d$. Hence

$$g_1(f, B) = \iint_B | f_x(x, y) |\, dx\, dy$$

for every Borel set $B \subset B_1$. Thus obviously g_1 is absolutely continuous on B_1, and similarly g_2 is absolutely continuous on B_2.

Conversely, suppose that

(i*) $f(x, y)$ is BVT in R, and

(ii*) there exist two Borel sets B_1, B_2, such that $g_1(f, B)$ is absolutely continuous on B_1, $g_2(f, B)$ is absolutely continuous on B_2, and $B_1 + B_2 = R$. By I.3.10, V.3.28 we have then

$$g_1(f, B_1) = \iint_{B_1} | f_x(x, y) |\, dx\, dy = \int_c^d \left[\int_{B_{1y}} | f_x(x, y) |\, dx \right] dy,$$

$$g_1(f, B_1) = \int_c^d V_x(B_{1y}, y, f)\, dx.$$

Subtraction yields

$$\int_c^d \left[V_x(B_{1y}, y, f) - \int_{B_{1y}} | f_x(x, y) |\, dx \right] dy = 0.$$

Since the integrand is non-negative for a.e. y in $c \leqq y \leqq d$ by III.2.26, it follows that for a.e. y in this range we have

$$V_x(B_{1y}, y, f) = \int_{B_{1y}} | f_x(x, y) |\, dx.$$

By III.2.34, it follows that $f(x, y)$ is absolutely continuous, as a function of x, on the set B_{1y} for a.e. y in the range $c \leqq y \leqq d$. Similarly it follows that $f(x, y)$ is absolutely continuous, as a function of y, on the set B_{2x} for a.e. x in the range $a \leqq x \leqq b$. Thus $f(x, y)$ is ACY in R.

Summing up: If $f(x, y)$ is a continuous function in R, then $f(x, y)$ is ACY in R if and only if the conditions (i*) and (ii*) hold.

V.3.41. THEOREM. *Given $f(x, y)$ as in* V.3.22, *the formula* (*cf.* V.3.25)

$$A(f, R) = \lim I(h, k, R) \qquad \text{for } h \to 0,\ k \to 0,$$

holds if and only if $f(x, y)$ is ACY *on R* (*cf.* V.3.40).

This is an immediate consequence of V.3.40 and V.3.39.

V.3.42. *Given $f(x, y)$ as in* V.3.22, *we assert the inequality* (cf. V.3.25, V.3.28)

$$I^*(R) \leqq A(f, R) + (2^{1/2} - 1)A^0(f, R). \tag{1}$$

PROOF. If λ, μ are any two positive numbers, then we have the elementary inequality $\lambda + \mu \leqq 2^{1/2}(\lambda^2 + \mu^2)^{1/2}$. Hence, for any oriented rectangle $r \subset R$ (cf. V.3.4)

$$g_1(f, r) + g_2(f, r) \leqq 2^{1/2}[g_1(f, r)^2 + g_2(f, r)^2]^{1/2} \leqq 2^{1/2}g(f, r) \leqq 2^{1/2}A(f, r).$$

By III.1.47, V.3.28 it follows that we have

$$g_1(f, B) + g_2(f, B) \leqq 2^{1/2}A(f, B)$$

for every Borel set $B \subset R$. Applying this result to the set e_0 defined in V.3.28 we obtain the inequality

$$g_1^0(f, R) + g_2^0(f, R) \leqq 2^{1/2}A^0(f, R). \tag{2}$$

By V.3.36 we have

$$I^*(R) = \iint_R (1 + f_x^2 + f_y^2)^{1/2}\, dx\, dy + g_1^0(f, R) + g_2^0(f, R). \tag{3}$$

(2) and (3) yield

$$I^*(R) \leqq \iint_R (1 + f_x^2 + f_y^2)^{1/2}\, dx\, dy + 2^{1/2}A^0(f, R). \tag{4}$$

By V.3.28

$$\iint_R (1 + f_x^2 + f_y^2)^{1/2}\, dx\, dy = A(f, R) - A^0(f, R). \tag{5}$$

(4) and (5) imply (1).

V.3.43. In V.3.36 we obtained an explicit formula for $I^*(R)$. No explicit formula is known for $I_*(R)$ at present. In V.3.27 we derived the inequality $I_*(R) \geqq A(f, R)$, and there are examples to show that if $f(x, y)$ is only known to be BVT in R, then generally $I_*(R) > A(f, R)$.

V.3.44. Given $f(x, y)$ as in V.3.22, let $R : a \leqq x \leqq b,\ c \leqq y \leqq d$, be an oriented rectangle. Then $f(x, y)$ is BVT on R, and $A(f, R) < +\infty$ (cf. V.3.9). Let $\phi(x,$

$y, R)$ serve as a generic notation for a function that is quasi-linear in R. That is, $\phi(x, y, R)$ is continuous in R, and there exists a rectilinear triangulation $\mathfrak{J}$ of R such that $\phi(x, y, R)$ is a linear function of x, y in every triangle of $\mathfrak{J}$. The equation $z = \phi(x, y, R)$, $(x, y) \in R$, defines then a polyhedron in the sense of V.1.30. The elementary area of this polyhedron will be denoted by $E(\phi, R)$. By V.1.33, V.2.9 we have then $E(\phi, R) = A(\phi, R)$ (cf. V.3.4). Let us define

$$\overline{A}(f, R) = \text{gr.l.b.} \liminf_{n\to\infty} E(\phi_n, R),$$

where the greatest lower bound is taken with respect to all sequences of quasi-linear functions $\phi_n(x, y, R)$ such that $\phi_n(x, y, R) \to f(x, y)$ uniformly on R. .

V.3.45. CONTINUATION. We shall say that the polyhedron $z = \phi(x, y, R)$, $(x, y) \in R$, is inscribed in the surface $z = f(x, y)$, $(x, y) \in R$, if there exists a rectilinear triangulation $\mathfrak{J}$ of R such that the following conditions hold.

(i) $\phi(x, y, R)$ is linear in every triangle of $\mathfrak{J}$.

(ii) $\phi(x, y, R) = f(x, y)$ at every vertex of the triangulation $\mathfrak{J}$. Note that if we have such a triangulation $\mathfrak{J}$, and $\mathfrak{J}'$ is obtained by refining $\mathfrak{J}$, then condition (i) will hold with $\mathfrak{J}$ replaced by $\mathfrak{J}'$, but generally condition (ii) will fail to hold for $\mathfrak{J}'$.

V.3.46. CONTINUATION. Let us define

$$A^*(f, R) = \text{gr.l.b.} \liminf_{n\to\infty} E(\phi_n, R),$$

where the greatest lower bound is taken with respect to all sequences $\phi_n(x, y, R)$ of quasi-linear functions such that $\phi_n(x, y, R) \to f(x, y)$ uniformly in R, and for every n the polyhedron $z = \phi_n(x, y, R)$, $(x, y) \in R$, is *inscribed* in the surface $z = f(x, y)$, $(x, y) \in R$, in the sense of V.3.45. Obviously (cf. V.3.4)

$$A(f, R) \leqq \overline{A}(f, R) \leqq A^*(f, R). \tag{1}$$

Indeed, each one of these three quantities is defined as a greatest lower bound in terms of elementary areas of polyhedra converging to the surface $z = f(x, y)$, $(x, y) \in R$, but the class of polyhedra that may be used is subjected to more and more restrictions. In the case of $A(f, R)$ we use all possible polyhedra (cf. V.2.3, V.3.4). In the case of $\overline{A}(f, R)$ we use only polyhedra admitting of a representation of the form $z = \phi(x, y, R)$, $(x, y) \in R$. In the case of $A^*(f, R)$ we add the further restriction that the polyhedra should be inscribed in the surface $z = f(x, y)$, $(x, y) \in R$. We shall investigate presently the question whether the sign of equality holds in (1). By assumption, $f(x, y)$ is BVT in R, but this is not a relevant restriction as far as the question just raised is concerned. Indeed, if $f(x, y)$ were not BVT in R, then we should have $A(f, R) = +\infty$ by V.3.9, and clearly the sign of equality would hold throughout (1). *Thus we shall assume that* $f(x, y)$ *is as described in* V.3.22.

V.3.47. Under the conditions stated in V.3.44, we assert that for every $\epsilon > 0$ there exists a quasi-linear function $\phi(x, y, R)$ such that

(i) $| \phi(x, y, R) - f(x, y) | < \epsilon$ in R,

(ii) $E(\phi, R) < \overline{A}(f, R) + \epsilon$.

PROOF. If $\overline{A}(f, R) = +\infty$, then the assertion is obvious. So we can assume that $\overline{A}(f, R) < +\infty$. By the definition of $\overline{A}(f, R)$, there exists a sequence $\phi_n(x, y, R)$ such that $\phi_n(x, y, R) \to f(x, y)$ uniformly in R, and

$$\text{(1)} \qquad \liminf_{n\to\infty} E(\phi_n , R) < \overline{A}(f, R) + \epsilon.$$

If N is a sufficiently large positive integer, then the function $\phi_n(x, y, R)$ will satisfy condition (i) for $n > N$. On the other hand, once N has been chosen, we have, in view of (1), infinitely many values of $n > N$ such that $\phi_n(x, y, R)$ satisfies condition (ii) also.

V.3.48. Given $f(x, y)$ as in V.3.22, let us suppose that $f(x, y)$ has continuous first partial derivatives in the whole xy-plane. Then $\overline{A}(f, R) = A(f, R)$.

PROOF. Let us subdivide the sides of R into n equal parts, and let us draw horizontals and verticals through the points of division. In each one of the resulting rectangles in R, let us draw the diagonal from the upper left to the lower right corner. We obtain in this manner a rectilinear triangulation $\mathfrak{T}_n$ of R. Let $\phi_n(x, y, R)$ be the (univocally determined) function that is linear in every triangle of $\mathfrak{T}_n$ and agrees with $f(x, y)$ at the vertices of $\mathfrak{T}_n$. By the argument used in V.1.40, it follows readily that

$$\phi_n(x, y, R) \underset{n\to\infty}{\to} f(x, y) \qquad \text{uniformly in } R,$$

$$E(\phi_n , R) \underset{n\to\infty}{\to} \iint_R (1 + f_x^2 + f_y^2)^{1/2}\, dx\, dy.$$

Hence, by the definition of $\overline{A}(f, R)$ (cf. V.3.44), there follows the inequality (cf. V.1.25)

$$\overline{A}(f, R) \leqq \iint_R (1 + f_x^2 + f_y^2)^{1/2}\, dx\, dy = A(f, R).$$

By V.3.46(1), the equation $\overline{A}(f, R) = A(f, R)$ follows.

V.3.49. THEOREM. *Given* $f(x, y)$ *as in* V.3.22, *we have* $\overline{A}(f, R) = A(f, R)$.

PROOF. Let h_n be a sequence of positive numbers converging to zero. Let $f_{h_n}(x, y)$ be defined as in V.3.5. Then (see V.3.21)

$$\text{(1)} \qquad f_{h_n}(x, y) \to f(x, y) \qquad \text{uniformly in } R,$$

$$\text{(2)} \qquad A(f_{h_n} , R) \to A(f, R).$$

Now the first partial derivatives of $f_{h_n}(x, y)$ are continuous in the whole xy-plane. Hence (see V.3.48)

$$\text{(3)} \qquad A(f_{h_n} , R) = \overline{A}(f_{h_n} , R).$$

By V.3.47 we have, for every positive integer n, a quasi-linear function $\phi_n(x, y, R)$ such that

(4) $$| f_{h_n}(x, y) - \phi_n(x, y, R) | < 1/n \qquad \text{in } R,$$

(5) $$E(\phi_n, R) < \overline{A}(f_{h_n}, R) + 1/n.$$

(1) and (4) imply that $\phi_n(x, y, R) \to f(x, y)$ uniformly in R. Hence, by the definition of $\overline{A}(f, R)$,

(6) $$\overline{A}(f, R) \leqq \liminf E(\phi_n, R).$$

(6), (5), (3), (2) yield $\overline{A}(f, R) \leqq A(f, R)$. By V.3.46(1) the assertion $\overline{A}(f, R) = A(f, R)$ follows.

V.3.50. The problem of showing that $A^*(f, R) = A(f, R)$ (cf. V.3.46), which has been termed *the problem of Geöcze* in the literature, is still unsolved. We shall discuss a method that yields the best result now available. Given $f(x, y)$ as in V.3.22, consider a rectangle

$$R : a \leqq x \leqq b, \; c \leqq y \leqq d.$$

Given two positive integers m, n, let us subdivide the horizontal sides of R into m equal parts, the vertical sides into n equal parts, and let us draw verticals and horizontals through the points of division. In each one of the resulting mn rectangles we draw the diagonal from the upper left to the lower right corner, obtaining a rectilinear triangulation $\mathfrak{T}_{mn}$ of R. Let $\phi_{mn}(x, y, R)$ be the function that is linear in every triangle of $\mathfrak{T}_{mn}$ and agrees with $f(x, y)$ at the vertices of $\mathfrak{T}_{mn}$. Let $A(m, n, R)$ denote the elementary area of the polyhedron

$$P(m, n, R) : z = \phi_{mn}(x, y, R), \qquad (x, y) \in R.$$

By using elementary formulas in analytic geometry, $A(m, n, R)$ can be computed readily. The result may be written in the following form if use is made of the function $\Phi(x, y, h, k)$ defined in V.3.25.

$A(m, n, R)$

(1) $$= \frac{(b-a)(d-c)}{2mn} \sum_{i=0}^{m-1} \sum_{j=0}^{n-1} \left[\Phi\left(a + \frac{i(b-a)}{m}, c + \frac{j(d-c)}{n}, \frac{b-a}{m}, \frac{d-c}{n}\right) \right.$$
$$\left. + \Phi\left(a + \frac{(i+1)(b-a)}{m}, c + \frac{(j+1)(d-c)}{n}, -\frac{b-a}{m}, -\frac{d-c}{n}\right) \right].$$

Now let there be given two real numbers ξ, η. Let us introduce the rectangle

$$R(\xi, \eta) : a + \xi \leqq x \leqq b + \xi, \; c + \eta \leqq y \leqq d + \eta.$$

Let $P(m, n, R(\xi, \eta))$, $A(m, n, R(\xi, \eta))$ have the same meaning relative to $R(\xi, \eta)$ as that of $P(m, n, R)$, $A(m, n, R)$ relative to R. From (1) we obtain then the formula

$A(m, n, R(\xi, \eta))$

$$= \frac{(b-a)(d-c)}{2mn} \sum_{i=0}^{m-1} \sum_{j=0}^{n-1} \left[\Phi\left(a+\xi+\frac{i(b-a)}{m}, c+\eta+\frac{j(d-c)}{n}, \frac{b-a}{m}, \frac{d-c}{n}\right) \right.$$

(2)

$$+ \Phi\left(a+\xi+\frac{(i+1)(b-a)}{m}, c+\eta \right.$$

$$\left.\left. + \frac{(j+1)(d-c)}{n}, -\frac{b-a}{m}, -\frac{d-c}{n}\right)\right].$$

V.3.51. CONTINUATION. Let us restrict ξ, η by the inequalities

(1) $$0 < \xi < \frac{b-a}{m}, \qquad 0 < \eta < \frac{d-c}{n}.$$

Let a polyhedron $P'(m, n, \xi, \eta)$ be defined as follows: $P(m, n, R(\xi, \eta))$ is given by

$$P(m, n, R(\xi, \eta)) : z = \phi_{mn}(x, y, R(\xi, \eta)), \qquad (x, y) \in R(\xi, \eta).$$

Let $\mathfrak{I}_{mn}(\xi, \eta)$ be the rectilinear triangulation that gave rise to $\phi_{mn}(x, y, R(\xi, \eta))$. Let us omit from $\mathfrak{I}_{mn}(\xi, \eta)$ those triangles that meet the upper horizontal or the right vertical side of R. In view of (1), the remaining triangles of $\mathfrak{I}_{mn}(\xi, \eta)$ form a rectilinear triangulation $\bar{\mathfrak{I}}_{mn}(\xi, \eta)$ of the rectangle

$$a + \xi \leqq x \leqq a + \xi + \frac{(m-1)(b-a)}{m},$$

(2)

$$c + \eta \leqq y \leqq c + \eta + \frac{(n-1)(d-c)}{n}.$$

We complete this triangulation to a triangulation of R itself, by producing the horizontal and vertical lines of division in $\bar{\mathfrak{I}}_{mn}(\xi, \eta)$ until they meet the sides of R, and by drawing the diagonal from the upper left to the lower right corner in each one of the resulting new rectangles that lie in R but are not comprised in the rectangle (2). Let $\mathfrak{I}'_{mn}(\xi, \eta)$ be the triangulation of R obtained in this manner, and let

$$P'(m, n, \xi, \eta) : z = \phi_{mn}^{\xi\eta}(x, y, R), \qquad (x, y) \in R,$$

be the polyhedron that arises from this triangulation and is inscribed in the surface $z = f(x, y)$, $(x, y) \in R$. Let $A'(m, n, \xi, \eta)$ be the elementary area of this polyhedron $P'(m, n, \xi, \eta)$. For $\lambda > 0$, let us put

$$\omega(\lambda) = \max | f(x_2, y_2) - f(x_1, y_1) |,$$

where the maximum is taken with respect to all pairs of points (x_2, y_2), (x_1, y_1) such that $(x_2, y_2), (x_1, y_1) \in R$, $[(x_2 - x_1)^2 + (y_2 - y_1)^2]^{1/2} \leqq \lambda$. We assert the inequality

$$A'(m, n, \xi, \eta) \leqq A(m, n, R(\xi, \eta)) + 2\left(\frac{1}{m} + \frac{1}{n} + \frac{2}{mn}\right) | R | \tag{3}$$

$$+ 2d(R)\omega\left(\frac{d(R)}{m}\right)\left(\frac{m}{n} + 3\right) + 2d(R)\omega\left(\frac{d(R)}{n}\right)\left(\frac{n}{m} + 3\right),$$

where $d(R)$ is the length of the diagonal of R.

PROOF. $A'(m, n, \xi, \eta)$ is a sum of areas of rectilinear triangles, some of which were used in computing $A(m, n, R(\xi, \eta))$ and some of which were not so used. Let σ' be the sum of the areas of the second type of triangles, and let Δ' be a generic notation for a triangle in this category. Let t' be the orthogonal projection of Δ' upon the xy-plane. By the construction of $P'(m, n, \xi, \eta)$, it follows that t' has a vertex on one of the sides of R. Let us first consider the case where t' has a vertex on the side $x = b$ of R. To estimate the area of the corresponding triangle Δ', let Δ_1', Δ_2', Δ_3' denote the orthogonal projections of Δ' upon the xy, yz, zx planes respectively. We have then

$$| \Delta' | = [| \Delta_1' |^2 + | \Delta_2' |^2 + | \Delta_3' |^2]^{1/2} \leqq | \Delta_1' | + | \Delta_2' | + | \Delta_3' |.$$

Clearly (cf. (1))

$$| \Delta_1' | \leqq \frac{1}{2}\frac{b - a}{m}\frac{d - c}{n} = \frac{1}{2mn} | R |,$$

$$| \Delta_2' | \leqq \frac{1}{2}\frac{d - c}{n}\omega\left(\frac{b - a}{m}\right) \leqq \frac{1}{2n} d(R)\omega\left(\frac{d(R)}{m}\right),$$

$$| \Delta_3' | \leqq \frac{1}{2}\frac{b - a}{m}\omega\left(\frac{d - c}{n}\right) \leqq \frac{1}{2m} d(R)\omega\left(\frac{d(R)}{n}\right).$$

The number of triangles Δ' for which the corresponding t' has a vertex on the side $x = b$ of R is $2(n + 1)$. For the sum of the areas of these triangles Δ' we obtain thus the estimate

$$\frac{n + 1}{mn} | R | + \frac{n + 1}{m} d(R)\omega\left(\frac{d(R)}{n}\right) + \frac{n + 1}{n} d(R)\omega\left(\frac{d(R)}{m}\right). \tag{4}$$

Working with the side $x = a$ of R, we obtain the same bound, while the sides $y = c$, $y = d$ of R lead each to the bound (obtained by exchanging m and n in (4))

$$\frac{m + 1}{mn} | R | + \frac{m + 1}{n} d(R)\omega\left(\frac{d(R)}{m}\right) + \frac{m + 1}{m} d(R)\omega\left(\frac{d(R)}{n}\right).$$

Since $m \geqq 1$, $n \geqq 1$, the inequality (3) follows.

V.3.52. CONTINUATION. Now let m_k, n_k, $k = 1, 2, \cdots$, be two sequences of positive integers, such that

$$m_k \to +\infty, \quad n_k \to +\infty, \tag{1}$$

$$(2) \qquad 0 < p' < \frac{m_k}{n_k} < p'' < +\infty, \qquad k = 1, 2, \cdots,$$

where p', p'' are two constants (independent of k). For each k, let ξ_k, η_k be two numbers such that

$$(3) \qquad 0 < \xi_k < \frac{b-a}{m_k}, \qquad 0 < \eta_k < \frac{d-c}{n_k}.$$

Let us denote by $\phi_k(x, y, R)$ the function $\phi_{mn}^{\xi\eta}(x, y, R)$, introduced in V.3.51, that corresponds to $m = m_k$, $n = n_k$, $\xi = \xi_k$, $\eta = \eta_k$. Clearly $\phi_k(x, y, R) \to f(x, y)$ uniformly in R. Hence, by the definition of $A^*(f, R)$ (cf. V.3.46, V.3.51),

$$A^*(f, R) \leqq \liminf_{k\to\infty} A'(m_k, n_k, \xi_k, \eta_k).$$

Since, on account of the uniform continuity of $f(x, y)$ in R, $\omega(\lambda) \to 0$ for $\lambda \to 0$ (cf. V.3.51), the inequality V.3.51 (3) yields, in view of (1), (2), (3),

$$\liminf_{k\to\infty} A'(m_k, n_k, \xi_k, \eta_k) \leqq \liminf_{k\to\infty} A(m_k, n_k, R(\xi_k, \eta_k)).$$

Thus we obtain the inequality

$$(4) \qquad A^*(f, R) \leqq \liminf_{k\to\infty} A(m_k, n_k, R(\xi_k, \eta_k)).$$

V.3.53. Continuation. For fixed m, n the quantity $A(m, n, R(\xi, \eta))$, defined in V.3.50, is clearly a continuous function of ξ, η. We consider the integral mean of this function over the rectangle

$$(1) \qquad \rho: 0 \leqq \xi \leqq \frac{b-a}{m}, \qquad 0 \leqq \eta \leqq \frac{d-c}{n}.$$

Using the formula V.3.50(2), we obtain after elementary manipulations the identity

$$(2) \qquad \frac{1}{|\rho|}\iint_\rho A(m, n, R(\xi, \eta))\, d\xi\, d\eta = \frac{1}{2}\left[I\left(\frac{b-a}{m}, \frac{d-c}{n}, R\right) + I\left(-\frac{b-a}{m}, -\frac{d-c}{n}, R\left(\frac{b-a}{m}, \frac{d-c}{n}\right)\right)\right].$$

Since $A(m, n, R(\xi, \eta))$ is a continuous function of ξ and η, we have in the interior of the rectangle ρ a point (ξ_{mn}, η_{mn}), such that $A(m, n, R(\xi_{mn}, \eta_{mn}))$ is exactly equal to the mean value (2). We have then, by (1) and (2), the following relations:

$$(3) \qquad 0 < \xi_{mn} < \frac{b-a}{m}, \qquad 0 < \eta_{mn} < \frac{d-c}{n},$$

$$A(m, n, R(\xi_{mn}, \eta_{mn}))$$

(4) $$= \frac{1}{2}\left[I\left(\frac{b-a}{m}, \frac{d-c}{n}, R\right) + I\left(-\frac{b-a}{m}, -\frac{d-c}{n}, R\left(\frac{b-a}{m}, \frac{d-c}{n}\right)\right)\right].$$

Now let us take any positive number δ, and let us consider the rectangle $R^\delta : a - \delta \leqq x \leqq b + \delta, c - \delta \leqq y \leqq d + \delta$, and let us choose any two sequences of positive integers m_k, n_k, such that the conditions V.3.52(1), (2) are satisfied (for example, we may choose $m_k = n_k = k$, $k = 1, 2, \cdots$). Let us denote by ξ_k, η_k the numbers ξ_{mn}, η_{mn} that correspond to $m = m_k$, $n = n_k$ (cf. (4)). We have then (cf. (3), (4))

(5) $$0 < \xi_k < \frac{b-a}{m_k}, \qquad 0 < \eta_k < \frac{d-c}{n_k},$$

$$A(m_k, n_k, R(\xi_k, \eta_k))$$

(6) $$= \frac{1}{2}\left[I\left(\frac{b-a}{m_k}, \frac{d-c}{n_k}, R\right) + I\left(-\frac{b-a}{m_k}, -\frac{d-c}{n_k}, R\left(\frac{b-a}{m_k}, \frac{d-c}{n_k}\right)\right)\right].$$

For k sufficiently large, the rectangle

$$R\left(\frac{b-a}{m_k}, \frac{d-c}{n_k}\right)$$

is contained in the rectangle R^δ. Hence (6) yields, for k sufficiently large, the inequality

$$A(m_k, n_k, R(\xi_k, \eta_k))$$

$$\leqq \frac{1}{2}\left[I\left(\frac{b-a}{m_k}, \frac{d-c}{n_k}, R^\delta\right) + I\left(-\frac{b-a}{m_k}, -\frac{d-c}{n_k}, R^\delta\right)\right].$$

For $k \to \infty$ there follows, in view of V.3.52(4) and V.3.25, the inequality $A^*(f, R) \leqq I^*(R^\delta)$. Since δ was arbitrary, we obtain for $\delta \to 0$ (cf. V.3.30, V.3.36)

(7) $$A^*(f, R) \leqq I^*(R).$$

V.3.54. Continuation. V.3.53(7), V.3.46, V.3.36 yield now the inequalities

$$A(f, R) \leqq A^*(f, R)$$

(1) $$\leqq \iint_R (1 + f_x^2 + f_y^2)^{1/2}\, dx\, dy + g_1^0(f, R) + g_2^0(f, R) = I^*(R).$$

Noting that $A^0(f, R) \leqq A(f, R)$ (cf. V.3.28), we have by V.3.42 the inequalities

(2) $$A(f, R) \leqq A^*(f, R) \leqq 2^{1/2} A(f, R).$$

These inequalities yield several relevant results concerning the problem of Geöcze (see V.3.50) that we shall state presently.

V.3.55. THEOREM. *Given $f(x, y)$ as in* V.3.22, *we have $A^*(f, R) = A(f, R)$ for every oriented rectangle R in which $f(x, y)$ is ACY, in the sense of* V.3.40.

PROOF. If $f(x, y)$ is ACY in R, then $I^*(R) = A(f, R)$ by V.3.41, and thus the assertion follows directly from V.3.54(1).

V.3.56. THEOREM. *Given $f(x, y)$ as in* V.3.22, *the quantity $A^*(f, R)$ is a function of the oriented rectangle R. The derivative of this rectangle function exists and is equal to $(1 + f_x^2 + f_y^2)^{1/2}$ almost everywhere. In other words, the rectangle functions $A(f, R)$ and $A^*(f, R)$ have the same derivative almost everywhere (cf.* V.3.20).

PROOF. Let (x_0, y_0) be a point in a fixed oriented rectangle $\bar{R}$, and let Q_n be any sequence of oriented squares, containing (x_0, y_0), with side-lengths converging to zero. By V.3.20, V.3.28, III.1.31, I.3.13 we have then, for (x_0, y_0) not in a certain exceptional set of measure zero in $\bar{R}$, the relations

$$\frac{g_1^0(f, Q_n)}{|Q_n|} \to 0, \qquad \frac{g_2^0(f, Q_n)}{|Q_n|} \to 0, \qquad \frac{A(f, Q_n)}{|Q_n|} \to (1 + f_x^2 + f_y^2)^{1/2},$$

$$\frac{1}{|Q_n|} \iint_{Q_n} (1 + f_x^2 + f_y^2)^{1/2}\, dx\, dy \to (1 + f_x^2 + f_y^2)^{1/2}.$$

Hence, by V.3.54(1), it follows that

$$\frac{A^*(f, Q_n)}{|Q_n|} \to (1 + f_x^2 + f_y^2)^{1/2}.$$

Thus the derivative of $A^*(f, R)$ is equal to $(1 + f_x^2 + f_y^2)^{1/2}$ a.e. in $\bar{R}$. Since $\bar{R}$ was arbitrary, the proof is complete.

V.3.57. If $f(x, y)$, given as in V.3.22, is subjected to no further restrictions, then it is not known at present whether $A^*(f, R) = A(f, R)$. By V.3.54(2) we merely know that $A^*(f, R)$ is finite and in fact cannot exceed $A(f, R)$ by more than 42%. The theorems of V.3.55, V.3.56 represent the best information available at this time. In view of the fact that generally $I_*(R) > A(f, R)$, it is perhaps not clear whether the method used to obtain the theorem of V.3.55 is adequate to lead to a complete solution of the problem of Geöcze.

CHAPTER V.4. GENERAL COMMENTS ON SURFACE AREA

V.4.1. In the introductory chapter I.1 we already gave a non-technical survey of the problems that are bound to arise in any comprehensive theory of surface area. At that time, fundamental concepts were used in a vague intuitive sense, and we observed that in each instance there may exist several plausible formal definitions for the concept under consideration. Thus any particular theory of surface area may be thought of as depending upon a number of *parameters*, each parameter representing the choice of a particular formal definition for some fundamental concept with an intuitive or familiar connotation.

Analogous situations arise of course in many mathematical fields. For instance, in introducing the real number system, the concept of a real number may be formally defined in terms of Cauchy sequences, or in terms of Dedekind cuts, or in terms of nested sequences of closed intervals, and in still other ways. But the particular choice of the formal definition turns out to be essentially irrelevant, since the resulting systems are readily recognized as being isomorphic. In other words, the choice of the formal definition of a real number is a matter of exposition and does not affect the theory itself. *The situation is entirely different in surface area theory.* Taking the concept of surface area itself, we already commented, in Chapter I.1, on the overwhelming and still increasing number of formal definitions for this fundamental concept. About a dozen of these definitions may be referred to as *major definitions* on account of the authority of the mathematicians who proposed them and on account of the obvious relevancy of the underlying principles. Of course, any two major definitions of surface area agree with each other in the strictly elementary range (cf. I.1.11) and hence they are equivalent as far as mathematical disciplines of a classical type are concerned. On the other hand, the comments in Chapter I.1 suggest that discrepancies are bound to arise beyond the strictly elementary range. This point has been studied in detail by Nöbeling [2] who established the distressing fact that *most of the presently known major definitions of surface area conflict with each other in relatively simple nonelementary cases.* Let us note that all such conflicts, and indeed all so-called paradoxes involving surface area, could be readily eliminated by properly restricting the concept of a surface. For example, the cube-filling surface of zero area constructed by Geöcze (see V.2.71) and the volume-filling simple closed surface of finite area constructed by Besicovitch [1] owe their existence to the generality of the concept of surface. From the elementary point of view, it would be entirely justified to refuse membership in the family of surfaces to a geometrical being that fills a cube or occupies a locus of positive three-dimensional measure. Needless to say, the purpose of a general mathematical theory is not to eliminate novel phenomena but rather to discover the general laws that govern them. From this point of view, the so-called paradoxes involving surface area are comparable to phenomena like the continuous function of Weierstrass which has no derivative anywhere, or the one-sided Moebius band.

Even though we accept this view, the fact remains that theories of surface area based upon different choices of the fundamental concepts differ essentially beyond the elementary range. In other words, *there exists at present no unified general theory of surface area.* As a matter of fact, few of the major definitions of surface area gave rise, up to the present, to a body of studies that may properly be termed a theory. From this point of view, the Lebesgue area $A(S)$, whose study is the primary objective of this book, is an outstanding exception. However, *concentration upon $A(S)$ should not be interpreted as a rejection of alternative approaches.* On the contrary, it is hoped that a systematic presentation of the theory of $A(S)$ will serve as a stimulus, by analogy and by contrast, for the development of alternative theories of comparable completeness. In particular, it is hoped that the lack of a comparable theory of surface area based upon measure-theoretical concepts will be more keenly realized. Significant and promising beginnings in this direction include the recent work of Federer [1], [2], Besicovitch [1], Busemann [1], as well as some earlier papers amongst which the work of Gross [1], [2] seems to deserve especial attention.

V.4.2. Turning to a detailed survey of the results achieved and of the major problems yet open in the theory of the Lebesgue area $A(S)$, let us first recall that the term *surface* is used in the sense of Fréchet surface of the type of the 2-cell (see II.3.44). As explained in Chapter II.5, the selection of this concept of a surface implies that we decided to use path-surfaces rather than point-surfaces (cf. also I.1.5). Furthermore, in defining convergent sequences of surfaces, we adopted the Fréchet distance in the space of our surfaces (see II.3.15). Let us point out that the restriction to F-surfaces of the type of the 2-cell is entirely unjustified from the geometrical point of view, and we adopted this restricted setting not merely to avoid excessive detail but rather because the topological facts needed in a comparable treatment of the theory for F-surfaces of the type of a general two-dimensional manifold are not yet available. Furthermore, it is desirable for many purposes, in particular for potential applications in Calculus of Variations, to consider oriented F-surfaces (cf. II.5.4). The topological issues just referred to are now being studied by J. W. T. Youngs whose results justify the assumption that a complete extension of the theory of the Lebesgue area to F-surfaces of the type of a general two-dimensional manifold will be soon possible.

V.4.3. Even though we are primarily interested in the Lebesgue area, we had to devote a whole Chapter V.1 to the study of *the auxiliary lower area $a(S)$*. Let us insist again (cf. V.1.7, V.1.8) that the notation $a(S)$ is not justified a priori, since the definition of $a(S)$ involves the coordinate system xyz, and thus $a(S)$ is actually a functional depending upon S and also upon the choice of the coordinate system xyz. This point being understood, we use the notation $a(S)$ as a matter of convenience. Of course, it is hoped that ultimately $a(S)$ will be shown to depend solely upon S. The lower area $a(S)$, as defined in V.1.7, is the result of successive modifications of the original concept of lower area proposed by

Geöcze [3], intermediate concepts being due to T. Radó [4], [5], [6], and P. V. Reichelderfer [8]. The lower area $a(S)$, as defined in V.1.7, is an upward revision of the definition used by P. V. Reichelderfer (cf. T. Radó [27]). The precise relationships between these various lower areas are not fully known at present, except that our $a(S)$ is obviously the largest of all the lower areas just referred to. While it is plausible that all these lower areas will ultimately turn out to agree with each other, our $a(S)$ seems to offer technical advantages which account for its selection as a most useful tool in our theory. The lower area $a(S)$ exhibits a number of remarkable additivity properties (see V.1.13, V.1.14, V.1.61), and as a consequence the cyclic additivity theorem of V.1.75 (cf. II.2.113) follows with ease, while the corresponding properties of the Lebesgue area $A(S)$ present serious problems (cf. the comments in II.5.10 concerning the work of C. B. Morrey and J. W. T. Youngs). Also, the lower area $a(S)$ turns out to be a most efficient tool in the study of the important and curious phenomena related to *surfaces of zero area* (cf. I.1.13, I.1.16). The study of this fundamental topic in V.1.64-V.1.74 is based upon the work of Geöcze [11] and a subsequent investigation of T. Radó [24]. Section V.1.75 summarizes the results concerning surfaces of zero area and the cyclic additivity of the lower area $a(S)$. Sections V.1.25-V.1.44 are concerned with results showing that the lower area $a(S)$ agrees with what may be termed *the expected value* of the area in simple cases. The theorems in V.1.15-V.1.23, due essentially to P. V. Reichelderfer [8], relate $a(S)$ to the fundamental concepts of *essential bounded variation* and *essential absolute continuity*. In preparation for the study of the Lebesgue area $A(S)$, a number of approximation theorems for $a(S)$ are derived in V.1.36-V.1.43. The interesting theorem on polyhedra in V.1.36 is taken from Huskey [3]. The fundamental approximation theorem of V.1.59 is due to P. V. Reichelderfer [8]. An important method of approximation is obtained by means of the *stretching process* studied in V.1.45-V.1.58 (see C. B. Morrey [3], T. Radó and P. V. Reichelderfer [13], J. W. T. Youngs [1], [3] for the interesting historical background).

V.4.4. Chapter V.2 is concerned with the Lebesgue area itself, for the case of surfaces given in terms of general parametric representations. The formal definition of $A(S)$ is given in V.2.3, and the invariance of $A(S)$ under changes of the Cartesian coordinate system xyz (or alternatively under distance-preserving transformations) is verified in V.2.8. Sections V.2.11, V.2.12 are concerned with results showing that *in the elementary range $A(S)$ agrees with generally accepted values of the area*, even though this statement should be qualified by the observation that there arise doubtful situations, even on a very elementary level, as regards the generally accepted value of surface area (see I.1.11). Generally speaking, $A(S)$ exhibits a tendency, in borderline cases, to agree with the value furnished by the familiar integral formula. On the other hand, this integral formula should be viewed with suspicion outside of the strictly elementary range. Indeed, as noted by J. W. T. Youngs [4], *every* surface S admits of a representation where the ordinary Jacobians exist and are equal to zero almost everywhere, and hence the classical integral formula yields the value zero (see V.2.20). In a general way, great caution must be exercised outside of the strictly elementary range. The curious phenomena mentioned in I.1.16 and discussed in detail in V.2.68-V.2.71 show that extreme caution and complete accuracy are indispensable in dealing with $A(S)$.

The definition of $A(S)$ involves the concept of a *polyhedron* $\mathfrak{P}$. There arises the question whether a different functional would be obtained if our formal definition of a polyhedron (see V.1.30) were replaced by some other plausible concept. This question has been studied by Huskey [3], the answer being that all plausible definitions of a polyhedron lead to the same functional $A(S)$ (cf. V.2.7). Once the concepts of surface S (see II.3.44), polyhedron $\mathfrak{P}$ (see V.1.30) and elementary area $E(\mathfrak{P})$ of a polyhedron $\mathfrak{P}$ (see V.1.33) are agreed upon, the formal definition of the Lebesgue area $A(S)$ may be motivated as follows. Let K be the class of all surfaces S (Fréchet surfaces of the type of the 2-cell). Let us ask for a functional $\alpha(S)$ with the following properties.

(P_1) $\alpha(S)$ is defined and non-negative for every surface $S \in K$. For certain surfaces S, $\alpha(S)$ may be infinite.

(P_2) $\alpha(\mathfrak{P}) = E(\mathfrak{P})$ for every polyhedron $\mathfrak{P}$ (cf. V.1.30, V.1.33).

(P_3) $\alpha(S)$ is a lower semi-continuous functional. That is, the relation $S_n \to S$ (see II.3.15) implies the relation $\alpha(S) \leqq \liminf \alpha(S_n)$.

(P_4) For every $S \in K$, there exists a sequence of polyhedra $\mathfrak{P}_n$ such that $\mathfrak{P}_n \to S$ and $E(\mathfrak{P}_n) \to \alpha(S)$.

The intention is, of course, to *find an area-functional with properties analogous to the properties of arc length* (cf. I.1.6). From this point of view, the insistence upon the properties (P_1)-(P_4) is entirely reasonable. It is not a priori obvious that such a functional $\alpha(S)$ exists. Indeed, (P_1), (P_2), (P_3) imply that the elementary area $E(\mathfrak{P})$ is lower semi-continuous in the class of polyhedra. This fact is true, but the proof is by no means trivial, even if the proof given in V.1.33 is abstracted for a direct proof. Now by V.2.2-V.2.6 and V.2.9 the Lebesgue area $A(S)$ possesses all the properties (P_1)-(P_4), and the point we wish to make is that $A(S)$ *is univocally determined by these properties.* Indeed, if $\alpha(S)$ is any functional with these properties, then we have by (P_4) two sequences of polyhedra $\mathfrak{P}_n'$, $\mathfrak{P}_n''$ such that $\mathfrak{P}_n' \to S$, $E(\mathfrak{P}_n') \to \alpha(S)$ and $\mathfrak{P}_n'' \to S$, $E(\mathfrak{P}_n'') \to A(S)$. By ($P_2$) and ($P_3$) we infer that $\alpha(S) \leqq \liminf \alpha(\mathfrak{P}_n'') = \liminf E(\mathfrak{P}_n'') = A(S)$, and hence $\alpha(S) \leqq A(S)$. The complementary inequality $A(S) \leqq \alpha(S)$ follows in a similar way. The preceding remarks furnish a *motivation of the definition of* $A(S)$ *in terms of postulates* represented by (P_1)-(P_4).

V.4.5. Let us turn now to the *representation problem,* that is, the problem of determining the conditions under which $A(S)$ is given by the classical integral formula (cf. I.1.2). We shall survey presently the extensive literature of this problem. Let a surface S be given by a representation

$$(1) \qquad S : x = x(u, v),\ y = y(u, v),\ z = z(u, v), \qquad 0 \leqq u \leqq 1,\ 0 \leqq v \leqq 1,$$

where the unit square is taken as the parameter range merely to simplify the language. Let us also assume that the first partial derivatives x_u, x_v, y_u, y_v,

z_u, z_v exist almost everywhere in the unit square. Let us denote by $W(u, v)$ the square root of the sum of the squares of the ordinary Jacobians (see V.1.18), and let us ask for conditions under which the formula

$$\int_0^1\int_0^1 W(u, v)\, du\, dv = A(S) \tag{2}$$

will hold. In view of the remarks in I.1.13, the assumption of the availability of the ordinary Jacobians represents a severe restriction upon the scope of the problem. Still, the literature of this restricted problem is most remarkable, and the results will be found very fruitful for the general theory. Curiously, some of the most relevant contributions to this problem were made by W. H. Young who thoroughly disapproved of the Lebesgue area $A(S)$ and studied in fact a surface area which he proposed himself (see W. H. Young [1], [2]). The surface area in his sense will be denoted by $A_Y(S)$. We shall not need the explicit definition, and merely note that the main objective of W. H. Young is the determination of a sequence of surfaces

$$S_n : x = x_n(u, v),\ y = y_n(u, v),\ z = z_n(u, v), \qquad 0 \leqq u \leqq 1,\ 0 \leqq v \leqq 1, \tag{3}$$

subject to the following conditions. (i) Each S_n is of an elementary character (quasi-linear, or of class C', and so forth). (ii) $x_n(u, v) \to x(u, v)$, $y_n(u, v) \to y(u, v)$, $z_n(u, v) \to z(u, v)$, uniformly in the unit square. (iii) the relation

$$\int_0^1\int_0^1 W_n(u, v)\, du\, dv \to \int_0^1\int_0^1 W(u, v)\, du\, dv \tag{4}$$

holds. Let us term, for brevity, a sequence (3) with these properties a Y-sequence. The existence of a Y-sequence is established, of course, on the basis of various additional assumptions concerning the representation (1) of the given surface S. The importance of the work of W. H. Young for the theory of the Lebesgue area $A(S)$ has been recognized by McShane [2], [3] and C. B. Morrey [1], who obtained far-reaching results by a process that may be described in outline as follows. If (3) is a Y-sequence, then due to the elementary character of the representations (3) we have the formulas

$$\int_0^1\int_0^1 W_n(u, v)\, du\, dv = A(S_n). \tag{5}$$

The lower semi-continuity of the Lebesgue area yields then, in view of (4),

$$\int_0^1\int_0^1 W(u, v)\, du\, dv \geqq A(S). \tag{6}$$

Hence (2) will follow if we can establish the complementary inequality

$$\int_0^1\int_0^1 W(u, v)\, du\, dv \leqq A(S), \tag{7}$$

and as a matter of fact some of the assumptions concerning the representation (1) are chosen, in the work of McShane and of Morrey, for the purpose of securing the inequality (7). Subsequently, T. Radó [11] found that (7) holds *automatically* as soon as the quantities involved exist. As a consequence, substantial improvements of previous results follow. Sections V.2.22-V.2.28 are concerned with results obtained by the various authors mentioned in this section.

V.4.6. We turn now to investigations aiming at the *solution of the representation problem in terms of a two-dimensional concept of absolute continuity*, in analogy with the theory of the arc length (see I.1.8). Given S in terms of a representation as in V.4.5(1), let $W_e(u, v)$ denote the square root of the sum of the squares of the generalized essential Jacobians whenever these exist at a point (u, v) (cf. V.1.15), and let $W(u, v)$ have the same meaning relative to the ordinary Jacobians. If one attempts to derive for $A(S)$ theorems analogous to those stated for arc length in I.1.8, one discovers that parts of these theorems follow rather readily for the lower area $a(S)$ and other parts for the Lebesgue area $A(S)$. In particular, fundamental results of this type for $a(S)$ are due to P. V. Reichelderfer [8]. Thus it is clear that the complete theorems for $A(S)$ will follow whenever we can show that $a(S) = A(S)$, a relation that has been first studied by Geöcze (for the lower area originally considered by him). We shall return to the relation $a(S) = A(S)$ below, and proceed presently to state the two principal results showing the relevancy of the concept of essential absolute continuity (see V.2.61-V.2.64 for the proofs). In the following statements, $W_e(u, v)$ and $W(u, v)$ have the meaning explained earlier in this section.

THEOREM A (*see T. Radó* [27]). *Given a surface*

$$S: x = x(u, v),\ y = y(u, v),\ z = z(u, v), \qquad 0 \leqq u \leqq 1,\ 0 \leqq v \leqq 1,$$

suppose that $A(S) < \infty$. *Then the representation is of essential bounded variation* (*see* V.1.15). *Furthermore, the essential generalized Jacobians exist almost everywhere in the unit square,* $W_e(u, v)$ *is summable there, and*

$$\int_0^1\int_0^1 W_e(u, v)\, du\, dv \leqq A(S).$$

The sign of equality holds if and only if the representation is essentially absolutely continuous (see V.1.15).

THEOREM B (*see T. Radó* [27]). *Given a surface*

$$S: x = x(u, v),\ y = y(u, v),\ z = z(u, v), \qquad 0 \leqq u \leqq 1,\ 0 \leqq v \leqq 1,$$

suppose that $A(S) < \infty$. *Suppose also that the first partial derivatives* x_u, x_v, y_u, y_v, z_u, z_v *exist almost everywhere in the unit square. Then* $W(u, v)$ *is summable there and*

$$\int_0^1\int_0^1 W(u, v)\, du\, dv \leqq A(S).$$

The sign of equality holds if and only if the representation is essentially absolutely continuous.

Comparison with the corresponding theorems for arc length (see I.1.8) reveals complete analogy, with the added insight that the essential generalized Jacobians are superior tools as compared with the ordinary Jacobians. In fact, Theorem A may be construed as confirmation of the appropriateness of the concepts of essential absolute continuity and essential generalized Jacobian. In view of the fundamental role, throughout Analysis, of the corresponding classical concepts for functions of a single variable, there arises the problem of testing our two-dimensional concepts in theories involving Jacobians, especially in the theory of double integral problems in Calculus of Variations. A first step in this direction should be, perhaps, a study of the lower semi-continuity of double integrals (see McShane [6], T. Radó [19], W. Scott [1] for recent literature).

V.4.7. Let us now return to the relation $a(S) = A(S)$ whose importance has been stressed in V.4.6. The special studies reviewed in V.4.5 acquire added significance in this connection, due to the following facts. Using the terminology of V.4.5, let us first note that we have the following improvement of V.4.5(7), due to P. V. Reichelderfer [8],

$$\int_0^1\int_0^1 W(u, v)\, du\, dv \leqq a(S) \leqq A(S) \tag{1}$$

assuming that $A(S) < \infty$ and that the first partial derivatives x_u, x_v, y_u, y_v, z_u, z_v exist almost everywhere in the unit square. Hence whenever

$$\int_0^1\int_0^1 W(u, v)\, du\, dv = A(S), \tag{2}$$

it follows from (1) that $a(S) = A(S)$. Thus there arises the question: under what conditions will a surface S admit of a representation such that (2) holds? The complete answer is unknown at present, but important special results (see V.2.43, V.2.44) were obtained by McShane [4] and C. B. Morrey [3]. In particular, the result discussed in V.2.43 (due to C. B. Morrey [3]) and the inequalities (1) yield the result that the relation $a(S) = A(S)$ holds if (i) $A(S) < \infty$ *and* (ii) the middle-space associated with S is a 2-cell (see II.3.20). To secure the relation $a(S) = A(S)$ without the restriction (ii), the cyclic additivity theorems for $A(S)$ and $a(S)$ (see V.2.55 and V.1.75) are combined. The cyclic additivity theorem for $A(S)$ was first stated by C. B. Morrey [3], but his proof is open to fundamental objections (see II.5.8). The first adequate proof is due to J. W. T. Youngs [3]; however, his work is concerned with Fréchet surfaces of the type of the 2-sphere, and the extension of his method to the 2-cell case seems to require further study (see II.5.8, II.5.9). The proof presented in V.2.45-V.2.55 avoids these added topological difficulties by using the general cyclic additivity theorem of II.2.113 which covers simultaneously the 2-sphere case and the 2-cell case. In fact, in view of a recent result of G. T. Whyburn [4], the theorem applies to all unicoherent Peano spaces.

As a result of these combined efforts, we have the following results concerning the relation $a(S) = A(S)$ at our disposal (see V.2.56-V.2.59, V.3.7).

THEOREM 1 (*P. V. Reichelderfer* [8]). *Always* $a(S) \leqq A(S)$.

THEOREM 2 (*T. Radó* [27]). *If* $A(S) < \infty$, *then* $a(S) = A(S)$.

THEOREM 3 (*T. Radó* [27]). *If* $a(S) = 0$, *then* $A(S) = 0$.

THEOREM 4 (*T. Radó* [2] *and P. V. Reichelderfer* [8]). *If S admits of a representation of the form* $z = f(x, y)$, $(x, y) \in R$, *where R is a simply-connected Jordan region and* $f(x, y)$ *is single-valued and continuous in R, then* $a(S) = A(S)$ *even if* $A(S) = \infty$.

V.4.8. (Added in January 1947.) In view of the preceding results, the gap to be yet filled is represented by the statement that *the relation* $A(S) = \infty$ *implies that* $a(S) = \infty$. In this connection, reference should be made to a series of papers by L. Cesari (see Bibliography) that appeared in Italian journals during the war years. At the time of this writing, fundamental portions of the work of Cesari are still not available to the writer, but from the material on hand it appears that his work is based on the same fundamental concepts and is pointed at the same general objectives as the theory presented in this book, even though an adequate evaluation is impossible at this time (as a matter of fact, the relevant papers of Cesari began to reach America only after the manuscript of this book had been completed). As far as can be determined at present, the contributions of Cesari fit readily into the framework of our theory, and hence we restrict ourselves to a brief consideration of what appears to be the most significant result discussed by Cesari, as far as completion of the theory presented in this book is concerned. Let us consider a surface

$$(1) \qquad S : x = x(u, v),\ y = y(u, v),\ z = z(u, v), \qquad 0 \leqq u \leqq 1,\ 0 \leqq v \leqq 1.$$

The representation (1) gives rise to the further surfaces

$$(2) \qquad S^1 : x = 0,\ y = y(u, v),\ z = z(u, v),$$

$$(3) \qquad S^2 : x = x(u, v),\ y = 0,\ z = z(u, v),$$

$$(4) \qquad S^3 : x = x(u, v),\ y = y(u, v),\ z = 0.$$

Clearly (cf. V.1.3), $a(S^1)$ for example is simply the integral of the essential multiplicity function associated with the mapping $y = y(u, v)$, $z = z(u, v)$ if this function is summable, and $a(S^1) = \infty$ otherwise. Thus the representation (1) is eBV (see V.1.15) if and only if $a(S^1) + a(S^2) + a(S^3) < \infty$. Now Cesari states the inequality

$$A(S) \leqq a(S^1) + a(S^2) + a(S^3). \tag{5}$$

Since (see V.1.16) we have $a(S) \geqq a(S^i)$, $i = 1, 2, 3$, it follows from (5) that if $A(S) = \infty$, then $a(S^i) = \infty$ for some i, and hence $a(S) = \infty$. Thus the inequality (5) of Cesari fills the last remaining gap in the general proof of the relation $a(S) = A(S)$. Another significant application of (5) should be mentioned. By V.1.16, V.2.10 we know that if $A(S) < \infty$, then the representation (1) is eBV. Conversely, if the representation is eBV, then $a(S^i) < \infty$, $i = 1, 2, 3$, as already noted, and hence $A(S) < \infty$ by (5). Thus (5) yields the proof of the theorem that $A(S) < \infty$ if and only if the representation (1) is eBV, thus materially strengthening the evidence in favor of the appropriateness of the concept of two-dimensional bounded variation adopted in this book. Of course, the general validity of the relation $a(S) = A(S)$ has further implications, pointed out at various points in this book.

V.4.9. Returning to Chapter V, let us call attention again to the results in V.2.43, V.2.44 concerning generalized conformal maps. The general result in V.2.43 is due to C. B. Morrey [3], while the theorem in V.2.44 (which appears as an application) has been discovered independently by McShane [4]. The student of conformal mapping will readily realize that a number of interesting problems will arise if generalized conformal maps are studied in more detail.

The Lebesgue area $A(S)$ and the lower area $a(S)$ are both F-invariant in the following sense: if $\mathfrak{x} = \mathfrak{x}_1(w_1)$, $w_1 \in \mathfrak{R}_1$, and $\mathfrak{x} = \mathfrak{x}_2(w_2)$, $w_2 \in \mathfrak{R}_2$, are F-equivalent representations (see II.3.7), then the corresponding values of a and A respectively are the same. J. W. T. Youngs established the remarkable fact, for F-surfaces of the type of the 2-sphere, that A is even K-invariant. According to results of R. G. Helsel, as yet unpublished, a and A are both K-invariant in the 2-cell case also. In view of these results, there arises the possibility that K-equivalence may serve instead of F-equivalence in defining our surfaces (cf. II.1.30, II.3.7). The interesting topological issues involved in this approach are now being studied by J. W. T. Youngs.

We already noted that F-surfaces of a general type should be fully studied. In particular, F-surfaces of the type of the 2-sphere arise in connection with the *isoperimetric inequality*. Using intuitive language for brevity, let Σ be a simple closed surface, A its area, and V the volume enclosed by Σ. The isoperimetric inequality asserts that $V^2 \leqq A^3/36\pi$. If one attempts to prove this inequality in a general form, then the definition of V is an important issue. Indeed, in analogy with the Osgood curve, Besicovitch [1] constructed a simple closed surface Σ such that $|\Sigma|$ (the three-dimensional measure of Σ) is positive. If $|D|$ is the three-dimensional measure of the (open) domain D enclosed by Σ, then we should distinguish between the interior volume $V_i = |D|$ and the exterior volume $V_e = |D| + |\Sigma|$, since now $V_e > V_i$. The isoperimetric inequality may be then considered in either one of the forms

$$(1) \qquad V_e^2 \leqq A^3/36\pi.$$

$$(2) \qquad V_i^2 \leqq A^3/36\pi.$$

Now Besicovitch shows that for given $\epsilon > 0$ and $G > 0$, his surface may be made to satisfy $V_e > G$, $A < \epsilon$, where A is the Lebesgue area of Σ. Thus V_e is not compatible with the Lebesgue area A as far as the inequality (1) is concerned. On the other hand, it is readily seen that in the Besicovitch example the inequality (2) *does* hold, a situation which suggests that *the concept of enclosed volume V must be properly adjusted to the concept of surface area.* As a matter of fact, in a paper to appear in the Transactions of the American Mathematical Society, the writer established the isoperimetric inequality in very general form, using the Lebesgue area. Subsequent results of J. W. T. Youngs, as yet unpublished, contain even stronger results. Let us recall in this connection the application of the Lebesgue area in the Plateau problem (for developments up to 1933, the reader may consult a monograph by T. Radó [7]). The role played by the Lebesgue area in these two classical variation problems may be considered as added evidence of the relevancy of this concept of surface area.

V.4.10. Chapter V.3 is concerned with surfaces of the form $z = f(x, y)$ (*the nonparametric case*). Historically, this case is of especial interest because the first substantial results obtained in this relatively simple case served as a model and as an incentive in the study of the more difficult parametric case. In the nonparametric case, the definition of the Lebesgue area may be stated in terms of polyhedra in nonparametric form (see V.3.44, V.3.48). This alternative definition has been freely used in the early stages of the theory, even though the explicit justification came much later (see P. V. Reichelderfer [8]). In any case, the availability of a definition of $A(S)$ in terms of nonparametric polyhedra, combined with the extreme simplicity of the topological situation in the nonparametric case, reduces the study of $A(S)$ to a problem in pure Analysis. From the analytic point of view, one has to deal with a single function $f(x, y)$ instead of the three pairs formed from the three coordinate functions in the parametric case. Accordingly, perfect analogues of the classical theorems on arc length (see I.1.8) were obtained by L. Tonelli [5]. Furthermore, in the nonparametric case the simple expressions of Geöcze (see V.3.3) yield sums analogous to the lengths of inscribed polygons in the case of arc length. These Geöcze sums were shown by Geöcze [1] to converge to $A(S)$ in certain important special cases. Verifying a surmise of Geöcze, T. Radó [2] proved that the Geöcze sums always converge to $A(S)$, and Saks [3] recognized the fact that according to this result $A(S)$ is merely the Burkill integral of a certain simple rectangle function. On this basis, Saks developed the theory of $A(S)$, in the nonparametric case, in a manner entirely free of Topology. The fact that $A(S)$ is the limit of Geöcze sums has been recognized by P. V. Reichelderfer [8] to mean that in the nonparametric case the fundamental relation $a(S) = A(S)$ always holds. In view of the topology-free presentation in Saks [6], it seemed desirable to give in chapter V.3 a discussion in which the nonparametric case is treated as a special application of the theory developed for the parametric case. In addition, several more recent developments were included. In particular, the sections V.3.25-V.3.45 are devoted to a discussion of the beautiful results of L. C. Young [1], certain details of exposition being improved in conformity with T. Radó [25]. The concluding sections V.3.50-V.3.57 are concerned with the so-called *Geöcze problem.* If one permits only *inscribed* polyhedra in the definition of $A(S)$ (see V.4.1(2)), then there results a functional $A^*(S)$ that may correspond more closely to the elementary idea of surface area than $A(S)$ itself. Clearly $A^*(S) \geqq A(S)$, and the problem of Geöcze is the problem of deciding whether always $A^*(S) = A(S)$. The first important results of Geöcze [1] and subsequent improvements by T. Radó [23] were based on methods of discouraging complexity. Curiously, much better results in a much simpler way were obtained later on by Huskey [2], whose ingenious method made use of the results of L. C. Young, already referred to. The presentation of the Huskey result in V.3.50-V.3.57 is based upon modifications proposed by T. Radó [26], and contains additional remarks (for instance, the result in V.3.56). Interesting results, closely related to the problem of Geöcze, were obtained by Kempisty [1], [3], Besicovitch [2], and studies on convergence in area by McShane [1] and T. Radó and P. V. Reichelderfer [20] may also have a bearing on this topic. Still, the general solution is not in sight as yet. Judging by the methods involved in the work up to the present, the Geöcze problem seems to be of great difficulty. Even if an example of a surface S_0 such that $A^*(S_0) > A(S_0)$ could be found, we would still have the problem of determining criteria for the relation $A^*(S) = A(S)$ to hold. In this connection, it is interesting to return to the example of Schwarz (see I.1.10). In view of that example, it may be assumed that at the time of its publication there existed a general belief that the areas of approximating inscribed polyhedra, *if chosen with some care,* always converge to the area of the surface. The problem of Geöcze, in fact, requires us to show that there exists *some* sequence of inscribed polyhedra whose areas converge to the area of the surface. Something intuitively obvious appears thus, after the intensive work of many mathematicians for more than fifty years, as a difficult unsolved problem.

BIBLIOGRAPHY

ADAMS, C. R.

[1] (with J. A. Clarkson) *On convergence in variation*, Bull. Amer. Math. Soc., vol. 40, 1934, pp. 413-417.

[2] (with H. Lewy) *On convergence in length*, Duke Math. Journal, vol. 1, 1935, pp. 19-26.

AYER, MIRIAM

[1] *On convergence in length*, Proc. Nat. Ac. Sci. USA, vol. 31, 1945, pp. 261-266.

[2] *On convergence in length*, Doctoral Dissertation, The Ohio State University, 1945.

BANACH, S.

[1] *Sur une classe de fonctions d'ensemble*, Fund. Math., vol. 6, 1924, pp. 170-188.

[2] *Sur les lignes rectifiables et les surfaces dont l'aire est finie*, Fund. Math., vol. 7, 1925, pp. 225-236.

BESICOVITCH, A. S.

[1] *On the definition and the value of the area of a surface*, Quart. J. Math., vol. 16, 1945, pp. 86-102.

[2] *On the definition of the area of a surface by means of inscribed polyhedra*, Proc. London Math. Soc., vol. 48, 1945, pp. 138-141.

BRAY, H. E.

[1] *Proof of a formula for an area*, Bull. Amer. Math. Soc., vol. 29, 1923, pp. 264-270.

BURKILL, J. C.

[1] *Functions of intervals*, Proc. London Math. Soc., vol. 22, 1923, pp. 275-310.

[2] *The expression of area as an integral*, Proc. London Math. Soc., vol. 22, 1923, pp. 311-336.

[3] (with U. S. Haslam-Jones) *The derivates and approximate derivates of measurable functions*, Proc. London Math. Soc., vol. 32, 1930, pp. 346-355.

[4] (with U. S. Haslam-Jones) *Note on the differentiability of functions of two variables*, J. London Math. Soc., vol. 7, 1932, pp. 297-305.

BUSEMANN, H.

[1] *Intrinsic area*, Bull. Amer. Math. Soc., vol. 52, 1946, p. 620.

[2] *Intrinsic area*, Annals of Math., vol. 48, pp. 234-267.

CACCIOPPOLI, R.

[1] *Sulle funzioni additive d'insieme*, Rendiconti Acc. Sci. Fis. Mat. Napoli, vol. 34, 1927, pp. 150-153.

[2] *Sulle quadratura delle superficie piane e curve*, Rendiconti Acc. Lincei, series 6, vol. 6, 1927, pp. 142-146.

[3] *Sul carattere infinitesimale delle superficie quadrabili*, Rendiconti Acc. Lincei, series 6, vol. 7, 1928, pp. 901-905.

[4] *Sulle definizione dell'area di una superficie*, Rendiconti Acc. Lincei, series 6, vol. 8, 1928, pp. 34-40.

[5] *Sull'espressione dell'area di una superficie mediante un integrale doppio*, Rendiconti Acc. Lincei, series 6, vol. 8, 1928, pp. 470-473.

[6] *Sulle coppie di funzioni a variazione limitata*, Rendiconti Acc. Sci. Fis. Mat. Napoli, vol. 34, 1928, pp. 83-88.

[7] *Teoria generale del cambiamento di variabili negli integrali doppi*, Math. Annalen, vol. 101, 1929, pp. 672-685.

[8] *Trasformazioni piane, superficie quadrabili, integrali di superficie*, Rendiconti Circ. Mat. Palermo, vol. 54, 1930, pp. 217-262.

[9] *Gli integrali doppi di forma parametrica nel calcolo delle variazioni*, Atti Reale Istituto Veneto di Sc. Lett. e Arti, vol. 93, 1933, pp. 705-730.

[10] *Rappresentazione conforme e superficie quadrabili*, Atti Acc. Lincei, series 6, vol. 22, 1935, pp. 379-384.

CARATHÉODORY, C.

[1] *Conformal representation*, Cambridge Tracts, vol. 28, 1931.

CAUCHY, A.

[1] *Mémoire sur la rectification des courbes et la quadrature des surfaces courbes*, Oeuvres complètes, series 1, vol. 2, pp. 167-177.

CESARI, L.

[1] *Sulle trasformazioni continue*, Ann. Mat. Pura Appl., vol. 21, 1942, pp. 157-188.

[2] *Su di un problema di Analysis Situs dello spazio ordinario*, Rend. Istituto Lombardo di Scienze e Lettere, vol. 74, 1941, pp. 267-291.

[3] *Sulle superficie de area finita secondo Lebesgue*, Rend. Reale Accademia d'Italia, series 7, vol. 3, 1941, pp. 350-365.

[4] *Sulle trasformazioni continue e sull'area delle superficie*, Memorie Reale Accademia d'Italia, vol. 12, 1941, pp. 1305-1397.

[5] *Sulle funzioni assolutamente continue di due variabili*, Annali R. Scoula Normale Sup. Pisa, series 2, vol. 10, 1941, pp. 91-101.

[6] *Carratterizzazione analitica delle superficie continue di area finita secondo Lebesgue*, Annali R. Scuola Normale Sup. Pisa, series 2, vol. 10, 1941, pp. 253-294 and vol. 11, 1942, pp. 1-42.

[7] *Sulla quadratura delle superficie in forma parametrica*, Boll. Un. Mat. Ital., series 2, vol. 4, 1942, pp. 109-117.

[8] *Una proprietà caratteristica delle trasformazioni a variazione limitata*, Boll. Un. Mat. Ital., series 2, vol. 4, 1942, pp. 224-235.

[9] *Sui punti di diramazione delle trasformazioni continue e sull'area delle superficie in forma parametrica*, Rend. di. Mat. e delle sue applicazioni, series 5, vol. 3, 1942, pp. 37-62.

[10] *Su di un teorema di T. Radó sulle trasformazioni continue*, Atti Reale Istituto Veneto di Sc. Lett. e Arti, vol. 101, 1942, pp. 377-403.

[11] *Sul concetto di trasformazione assolutamente continue*, Boll. Un. Mat. Ital., series 2, vol. 5, 1943, pp. 1-5.

[12] *Sui fondamenti geometrici dell'integrale classico per l'area delle superficie in forma parametrica*, Memorie Reale Accademia d'Italia, vol. 13, 1943, pp. 1323-1481.

[13] *Criteri di uguale continuità ed applicazioni al problema della quadratura delle superficie*, Annali R. Scuola Normale Sup. Pisa, series 2, vol. 11, 1943, pp. 61-84.

[14] *Una uguaglianza fondamentale per l'area delle superficie*, Atti della reale Accademia d'Italia, vol. 14, 1944, pp. 891-950.

EILENBERG, S.

[1] *Transformations continues en circonférences et la topologie du plan*, Fund. Math., vol. 26, 1936, pp. 61-112.

EISENHART, L. P.

[1] *Non-Riemannian Geometry*, Amer. Math. Soc. Colloquium Publications, vol. 8, 1927.

FEDERER, H.

[1] *Surface area*, part I, Trans. Amer. Math. Soc., vol. 55, 1944, pp. 420-437; part II, ibid. vol. 55, 1944, pp. 438-456.

[2] *Coincidence functions and their integrals*, Trans. Amer. Math. Soc. vol. 59, 1946, pp. 441-466.

FRANKLIN, P.

[1] (with N. Wiener) *Analytical approximations to topological transformations*, Trans. Amer. Math. Soc., vol. 28, 1926, pp. 762-785.

Fréchet, M.

[1] *Sur l'écart de deux courbes*, Trans. Amer. Math. Soc., vol. 6, 1905, pp. 435-449.

[2] *Sur la semi-continuité en géométrie élémentaire*, Nouvelles Ann. Math., vol. 3, 1924, pp. 1-9.

[3] *Sur la distance de deux surfaces*, Ann. Soc. Polonaise Math., vol. 3, 1924, pp. 4-19.

[4] *Sur l'aire des surfaces polyédrales*, Ann. Soc. Polonaise Math., vol. 3, 1925, pp. 1-3.

[5] *Sur une représentation paramétrique intrinsèque de la courbe continue la plus générale*, J. Math. Pure Appl., vol. 4, 1925, pp. 281-297.

[6] *Sur le prolongement des fonctionnelles semicontinues et sur l'aire des surfaces courbes*, Fund. Math., vol. 7, 1925, pp. 210-224.

[7] *L'expression la plus générale de la distance sur une droite*, Amer. J. Math., vol. 47, 1925, pp. 1-10.

[8] *Sur une définition intrinsèque de l'aire d'une surface courbe comme la limite d'aires polyédrales inscrites*, Ann. R. Scuola Norm. Sup. Pisa, series 2, vol. 8, 1939, pp. 285-300.

Geöcze, Z. de

[1] *Quadrature des surfaces courbes*, Math. Naturw. Berichte Ungarn, vol. 26, 1908, pp. 1-88.

[2] *Contributions to the quadrature of the surface* $z = f(x,y)$ (Hungarian), Math Term. Tud. Ért., vol. 26, 1908, pp. 475-512.

[3] *Recherches générales sur la quadrature des surfaces courbes*, part I, Math. Naturw. Berichte Ungarn, vol. 27, 1909, pp. 1-21; part II, ibid. pp. 131-163; part III, ibid. vol. 30, 1912, pp. 1-29.

[4] *Sur la fonction semi-continue*, Bull. Soc. Math. France, vol. 39, 1911, pp. 256-295.

[5] *Contributions à la quadrature des surfaces*, Comptes Rendus Ac. Sci. Paris, vol. 152, 1911, pp. 678-679.

[6] *On the quadrature of surfaces* (Hungarian), Math. Phys. Lapok, vol. 20, 1911, pp. 255-301, and vol. 21, 1912, pp. 25-57.

[7] *Contributions à la quadrature des surfaces courbes*, Comptes Rendus Ac. Sci. Paris, vol. 154, 1912, pp. 1211-1213.

[8] *On the theory of surface area* (Hungarian), Math. Term. Tud. Ért., vol. 31, 1913, pp. 306-318.

[9] *Sur l'exemple d'une surface dont l'aire est égale à zéro et qui remplit un cube*, Bull. Soc. Math. France, vol. 41, 1913, pp. 29-31.

[10] *Example of a surface of zero area which fills a cube* (Hungarian), Math. Phys. Lapok, vol. 23, 1914, pp. 115-117.

[11] *On the surface of zero area* (Hungarian), Math. Term. Tud. Ért., vol. 33, 1915, pp. 730-748.

[12] *On the rectifiable surface* (Hungarian), Math. Term. Tud. Ért., vol. 34, 1916, pp. 337-354.

[13] *A necessary and sufficient condition for the finiteness of the area of a surface* (Hungarian) Math. Phys. Lapok, vol. 25, 1916, pp. 61-81.

[14] *On the Peano definition of surface area* (Hungarian), Math. Term. Tud. Ért., vol. 35, 1917, pp. 325-358.

[15] *On the general surface* (Hungarian), Math. Term. Tud. Ért., vol. 35, 1917, pp. 359-360.

Gross, W.

[1] *Über das Flächenmass von Punktmengen*, Monatshefte f. Math. Phys., vol. 27, 1918, pp. 145-176.

[2] Die Minimaleigenschaft der Kugel, Monatshefte f. Math. Phys., vol. 27, 1917, pp. 77-97.

Hahn, H.

[1] *Über Annäherung an Lebesguesche Integrale durch Riemannsche Summen*, Sitzungsb. Ak. Wien, vol. 123, 1914, pp. 713-743.

[2] *Reelle Funktionen*, Part I, Leipzig, 1932.

HARDY, G. H., LITTLEWOOD, J. E., PÓLYA, G.
[1] *Inequalities*, Cambridge, 1934.
HASLAM-JONES, U. S.
[1] (with J. C. Burkill) *The derivates and approximate derivates of measurable functions*, Proc. London Math. Soc., vol. 32, 1930, pp. 346-355.
[2] (with J. C. Burkill) *Note on the differentiability of functions of two variables*, J. London Math. Soc., vol. 7, 1932, pp. 297-305.
HAUSDORFF, F.
[1] *Mengenlehre*, Berlin, 1935.
HELSEL, R. G.
[1] (with P. M. Young) *Characterization theorems for integral means*, Duke Math. Journal, vol. 10, 1943, pp. 259-269.
[2] *A geometrical application of integral means*, The Ohio State University Press, Abstracts of doctoral dissertations, No. 40, 1943.
[3] (with T. Radó) *The transformation of double integrals*, Trans. Amer. Math. Soc., vol. 54, 1943, pp. 83-102.
[4] *A theorem on surface area*, Trans. Amer. Math. Soc., vol. 61, 1947, pp. 443-453.
HOBSON, E. W.
[1] *The theory of functions of a real variable and the theory of Fourier series*, Cambridge, 1926.
HOPF, H.
[1] *Zur Topologie der Abbildungen von Mannigfaltigkeiten*, part I, Math. Annalen, vol. 100, 1928, pp. 579-608; part II, ibid. vol. 102, 1929, pp. 562-623.
HUREWICZ, W.
[1] (with H. Wallman) *Dimension theory*, Princeton University Press, 1941.
HUSKEY, H. D.
[1] *Contribution to the problem of Geöcze*, Duke Math. Journal, vol. 10, 1943, pp. 249-257.
[2] *Further contributions to the problem of Geöcze*, Duke Math. Journal, vol. 11, 1944, pp. 333-339.
[3] *Fréchet polyhedra*, Duke Math. Journal, vol. 11, 1944, pp. 417-425.
JEFFERY, R. L.
[1] *Sets of k-extent in n-dimensional space*, Trans. Amer. Math. Soc., vol. 35, 1933, pp. 629-647.
KEMPISTY, S.
[1] *Sur la méthode triangulaire du calcul de l'aire d'une surface courbe*, Bull. Soc. Math. France, vol. 64, 1936, pp. 119-132.
[2] *Sur les fonctions absolument continues d'intervalles*, Fund. Math., vol. 27, 1936, pp. 11-37.
[3] *Application des fonctions de triangle à la théorie de l'aire des surfaces courbes*, Travaux Soc. Sci. Lettres, Vilno, vol. 12, 1937, pp. 1-8.
[4] *Fonctions d'intervalle non additives*, Actualités scientifiques et industrielles, Paris, 1940.
KERÉKJÁRTÓ, B. DE
[1] *Involutions et surfaces continues*, Acta Szeged, vol. 3, 1927, pp. 49-67 and pp. 242-249.
KNESER, H.
[1] *Die Deformationssätze der einfach zusammenhängenden Flächen*, Math. Zeitschrift, vol, 25, 1926, pp. 362-372.
KURATOWSKI, C.
[1] *Une caractérization topologique de la surface de la sphère*, Fund. Math., vol. 13, 1929, pp. 307-318.
[2] (with G. T. Whyburn) *Sur les éléments cyclic et leurs applications*, Fund. Math., vol. 16, 1930, pp. 305-331.
[3] *Topologie* I, Warsaw, 1933.
LEBESGUE, H.
[1] *Sur la définition de l'aire d'une surface*, Comptes Rendus Ac. Sci. Paris, vol. 129, 1899, pp. 870-873.

[2] *Intégrale, longueur, aire*, Ann. Mat. Pura Appl., vol. 7, 1902, pp. 231-359.

[3] *Sur le problème de Dirichlet*, Rendiconti Circ. Mat. Palermo, vol. 24, 1907, pp. 371-402.

[4] *Observations sur la communication précédente* (de Geöcze), Bull. Soc. Math. France, vol. 41, 1913, pp. 31-32.

[5] *Quelques remarques sur la définition de l'aire des surfaces*, Fund. Math., vol. 8, 1926, pp. 160-165.

LEFSCHETZ, S.

[1] *Algebraic Topology*, Amer Math. Soc. Colloquium Publications, vol. 27, 1942.

LUSIN, N.

[1] *Leçons sur les ensembles analytiques et leurs applications*, Paris, 1930.

MCSHANE, E. J.

[1] *On a certain inequality of Steiner*, Ann. Math., vol. 33, 1932, pp. 125-138.

[2] *On the semi-continuity of integrals in the calculus of variations*, Ann. Math., vol. 33, 1932, pp. 460-484.

[3] *Integrals over surfaces in parametric form*, Ann. Math., vol. 34, 1933, pp. 815-838.

[4] *Parametrizations of saddle surfaces, with applications to the problem of Plateau*, Trans. Amer. Math. Soc., vol. 35, 1933, pp. 716-733.

[5] *On the analytic nature of surfaces of least area*, Ann. Math., vol. 35, 1934, pp. 456-473.

[6] *Existence theorems for double integral problems of the calculus of variations*, Trans. Amer. Math. Soc., vol. 38, 1935, pp. 549-563.

[7] *Integration*, Princeton University Press, 1944.

MENGER, K.

[1] *Kurventheorie*, Leipzig and Berlin, 1932.

MORREY, C. B. JR.

[1] *A class of representation of manifolds*, part I, Amer. J. Math., vol. 55, 1933, pp. 683-707; part II, ibid. vol. 56, 1934, pp. 275-293.

[2] *The topology of (path) surfaces*, Amer J. Math., vol. 57, 1935, pp. 17-50.

[3] *An analytic characterization of surfaces of finite Lebesgue area*, part I, Amer. J. Math. vol. 57, 1935, pp. 692-702; part II, ibid. vol. 58, 1936, pp. 313-322.

[4] *Review of J W. T. Youngs* [3], Math. Reviews, vol. 6, 1945, p. 272.

MORSE, A. P.

[1] *Convergence in variation and related topics*, Trans. Amer. Math. Soc., vol. 41, 1937, pp. 48-83.

MORSE, MARSTON

[1] *A special parametrization of curves*, Bull. Amer. Math. Soc., vol. 42, 1936, pp. 915-922.

NEWMAN, M. H. A.

[1] *Elements of the topology of plane sets of points*, Cambridge University Press, 1933.

NÖBELING, G.

[1] *Über den Flächeninhalt dehnungsbeschränkter Flächen*, Math. Zeitschrift, vol. 48, 1943, pp. 747-771.

[2] *Über die Flächenmasse im Euklidischen Raum*, Math. Annalen, vol. 118, 1943, pp. 687-701.

OSGOOD, W F.

[1] *A Jordon curve of positive area*, Trans. Amer. Math. Soc., vol. 4, 1903, pp. 107-112.

PEANO, G.

[1] *Sur une courbe qui remplit toute une aire plane*, Math. Annalen, vol. 36, 1890, pp. 157-160.

[2] *Sulle definizione dell'area di una superficie*, Rendiconti Acc. Lincei, series 4, vol. 6, 1890, pp. 54-57.

RADEMACHER, H.

[1] *Ein-eindeutige Abbildungen und Messbarkeit*, Monatshefte f. Math. Phys., vol. 27, 1916, pp. 183-291.

[2] *Über partielle und totale Differenzierbarkeit von Funktionen mehrerer Variabeln*, part I, Math. Annalen, vol. 79, 1919, pp. 340-359; part II, ibid. vol. 81, 1920, pp. 52-63.

RADÓ, T.

[1] *Bemerkung zur Arbeit des Herrn Bieberbach etc.*, Math. Annalen, vol. 90, 1923, pp. 30-37.

[2] *Sur le calcul de l'aire des surfaces courbes*, Fund. Math., vol. 10, 1927, pp. 197-210.

[3] *Sur l'aire des surfaces courbes*, Acta Szeged, vol. 3, 1927, pp. 131-169.

[4] *Sure la quadrature des surfaces courbes* (Hungarian with French abstract), Math. Term. Tud. Ért., vol. 45, 1928, pp. 225-244.

[5] *Sur l'aire des surfaces continues*, Atti del Congresso Internazionale dei Matematici, Bologna, 1928, vol. 6, pp. 355-360.

[6] *Über das Flächenmass rektifizierbarer Flächen*, Math. Annalen, vol. 100, 1928, pp. 445-479.

[7] *On the problem of Plateau*, Ergebnisse d. Math. u. ihrer Grenzgebiete, vol. 2, 1933.

[8] *A remark on the area of surfaces*, Amer. J. Math., vol. 58, 1936, pp. 598-606.

[9] *On continuous transformations in the plane*, Fund. Math., vol. 27, 1936, pp. 201-211.

[10] *A lemma on the topological index*, Fund. Math., vol. 27, 1936, pp. 212-225.

[11] *On the derivative of the Lebesgue area of continuous surfaces*, Fund. Math., vol. 30, 1938, pp. 34-39.

[12] *On absolutely continuous transformations in the plane*, Duke Math. Journal, vol. 4, 1938, pp. 189-221.

[13] (with P. V. Reichelderfer) *On a stretching process for surfaces*, Amer. J. Math., vol. 61, 1939, pp. 645-650.

[14] (with P. V. Reichelderfer) *Cyclic transitivity*, Duke Math. Journal, vol. 6, 1940, pp. 474-485.

[15] *On a lemma of McShane*, Ann. Math., vol. 42, 1941, pp. 73-83.

[16] (with P. V. Reichelderfer) *Note on an inequality of Steiner*, Bull. Amer. Math. Soc., vol. 47, 1941, pp. 102-108.

[17] (with P V. Reichelderfer) *A theory of absolutely continuous transformations in the plane*, Trans. Amer. Math. Soc., vol. 49, 1941, pp. 258-307

[18] *On semi-continuity*, Amer. Math. Monthly, vol. 49, 1942, pp. 446-450.

[19] *On the semi-continuity of double integrals in the parametric form*, Trans. Amer. Math. Soc., vol. 51, 1942, pp. 336-361.

[20] (with P. V. Reichelderfer) *Convergence in length and convergence in area*, Duke Math. Journal, vol. 9, 1942, pp. 527-565.

[21] *What is the area of a surface?* Amer. Math. Monthly, vol. 50, 1943, pp. 139-141.

[22] (with R. G. Helsel) *The transformation of double integrals*, Trans. Amer. Math. Soc., vol. 54, 1943, pp. 83-102.

[23] *On a problem of Geöcze*, Amer. J. Math., vol. 65, 1943, pp. 361-381.

[24] *On continuous path-surfaces of zero area*, Ann. Math., vol. 44, 1943, pp. 173-191.

[25] *Functions of rectangles*, Duke Math. Journal, vol. 11, 1944, pp. 487-496.

[26] *Some remarks on the problem of Geöcze*, Duke Math. Journal, vol. 11, 1944, pp. 497-506.

[27] *On surface area*, Proc. Nat. Ac. Sci. USA, vol. 31, 1945, pp. 102-106.

[28] *On continuous mappings of Peano spaces*, Trans. Amer. Math. Soc., vol. 58, 1945, pp. 420-454.

[29] *The isoperimetric inequality and the Lebesgue definition of surface area*, Trans. Amer. Math. Soc., vol. 61, 1947, pp. 530-555.

[30] *Two-dimensional concepts of bounded variation and absolute continuity*, Duke Math. Journal, vol. 14, 1947, pp. 587-608.

RANDOLPH, J. F

[1] *Carathéodory linear measure and a generalization of the Gauss-Green lemma*, Trans. Amer. Math. Soc., vol. 38, 1935, pp. 31-48.

[2] (with A. P. Morse) *The φ-rectifiable subsets of the plane*, Trans. Amer. Math. Soc., vol. 55, 1944, pp. 236-305.

REICHELDERFER, P. V.

[1] *Some properties of continuous transformations in the plane,* The Ohio State University Press, Abstracts of doctoral dissertations, No. 30, 1939.

[2] (with T. Radó) *On a stretching process for surfaces,* Amer. J. Math., vol. 61, 1939, pp. 645-650.

[3] (with T. Radó) *Cyclic transitivity,* Duke Math. Journal, vol. 6, 1940, pp. 474-485.

[4] (with T. Radó) *A theory of absolutely continuous transformations in the plane,* Trans. Amer. Math. Soc., vol. 49, 1941, pp. 258-307.

[5] (with T. Radó) *Note on an inequality of Steiner,* Bull. Amer. Math. Soc., vol. 47, 1941, pp. 102-108.

[6] (with L. A. Ringenberg) *The extension of rectangle functions,* Duke Math. Journal, vol. 8, 1941, pp. 231-242.

[7] (with T. Radó) *Convergence in length and convergence in area,* Duke Math. Journal, vol. 9, 1942, pp. 527-565.

[8] *On bounded variation and absolute continuity for parametric representations of continuous surfaces,* Trans. Amer. Math. Soc., vol. 53, 1943, pp. 251-291.

SAKS, S.

[1] *Sur une certaine classe de fonctions d'ensemble,* Bull. Ac. Polonaise, 1926, pp. 103-108.

[2] *Sur la différentiation de l'aire de surfaces,* Comptes Rendus Ac. Sci. Paris, vol. 183, 1926, pp. 850-852.

[3] *Sur l'aire des surfaces* $z = f(x,y)$, Acta Szeged, vol. 3, 1927, pp. 170-176.

[4] *Sur les fonctions d'intervalle,* Fund. Math., vol. 10, 1927, pp. 211-224.

[5] *On the surfaces without tangent planes,* Ann. Math., vol. 34, 1933, pp. 114-124.

[6] *Theory of the integral,* Warsaw, 1937

SCHAUDER, J.

[1] *The theory of surface measure,* Fund. Math., vol. 8, 1926, pp. 1-48.

[2] *Über die Halbstetigkeit des Flächenmasses,* Fund. Math., vol. 12, 1928, pp. 269-276.

[3] *Über stetige Abbildungen,* Fund. Math., vol. 12, 1928, pp. 47-74.

SCHWARZ, H. A.

[1] *Sur une définition erronée de l'aire d'une surface courbe,* Gesammelte Math. Abhandlungen, vol. I, pp. 309-311.

SCOTT, W.

[1] *A note on the lower semi-continuity of double integrals in the parametric form,* Bull. Amer. Math. Soc., vol. 48, 1942, pp. 763-767

SIERPINSKI, W.

[1] *Introduction to general topology,* The University of Toronto Press, 1934.

STEINHAUS, H.

[1] *Sur la portée pratique et théorique de quelques théorèmes sur la mesure des ensembles de droites,* Comptes Rendus du Premier Congrès des Mathématiciens des Pays Slaves, Warsaw, 1929, pp. 348-354.

[2] *Zur Praxis der Rektifikation und zum Längenbegriff,* Berichte Math. Phys. Klasse Ak. W. Leipzig, vol. 82, 1930, pp. 120-130.

STEPANOFF, W.

[1] *Über totale Differenzierbarkeit,* Math Annalen, vol. 90, 1923, pp. 318-320.

[2] *Sur les conditions d'existence de la différentielle totale,* Recueil Soc. Math. Moscou, vol. 32, 1924, pp. 511-526.

TONELLI, L.

[1] *Sulla rettificazione delle curve,* Atti Acc. Sci. Torino, vol. 43, 1908, pp. 399-416.

[2] *Sul differenziale dell'arco di curva,* Atti Acc. Lincei, vol. 25, 1916, pp. 207-213.

[3] *Fondamenti di calcolo delle variazioni,* Bologna, vol. I, 1921, vol. II, 1923.

[4] *Sur la quadrature des surfaces,* Comptes Rendus Ac. Sci. Paris, vol. 182, 1926, pp. 1198-1200.

[5] *Sulla quadratura delle superficie*, part I, Rendiconti Acc. Lincei, vol. 3, 6ª, 1926, pp 357-362; part II, ibid. pp. 445-450; part III, ibid. pp. 633-638; part IV, ibid. pp. 714-719.

[6] *Su un polinomio d'approssimazione e l'area di una superficie*, Rendiconti Acc. Lincei, vol. 5, 6ª, 1927, pp. 313-318.

[7] *Sulle funzioni di due variabili assolutamente continue*, Memorie Acc. Sci. Bologna, 8, vol. 6, 1928, pp. 1-10.

[8] *Sur la semi-continuité des intégrales doubles du calcul des variations*, Acta Math., vol. 53, 1929, pp. 325-346.

VITALI, G.

[1] *Sulle funzioni continue*, Fund. Math., vol. 8, 1926, pp. 175-188.

WEYL, H.

[1] *Die Idee der Riemannschen Fläche*, Leipzig and Berlin, 1913.

WHYBURN, G. T.

[1] (with C. Kuratowski) *Sur les éléments cyclic et leurs applications*, Fund. Math., vol. 16, 1930, pp. 305-331.

[2] *What is a curve?* Amer. Math. Monthly, vol. 49, 1942, pp. 492-497.

[3] *Analytic Topology*, Amer. Math. Soc. Colloquium Publications, vol. 28, 1942.

[4] *On monotone retractability into simple arcs*, Bull. Amer. Math. Soc., vol. 52, 1946, pp. 109-112.

YOUNG, L. C.

[1] *An expression connected with the area of a surface* $z = F(x,y)$, Duke Math. Journal, vol. 11, 1944, pp. 43-57.

YOUNG, P. M.

[1] *On the approximation of functions by integral means*, The Ohio State University Press, Abstracts of doctoral dissertations, No. 37, 1942.

[2] (with R. G. Helsel) *Characterization theorems for integral means*, Duke Math. Journal, vol. 10, 1943, pp. 259-269.

YOUNG, W. H.

[1] *On the triangulation method of defining the area of a surface*, Proc. London Math. Soc., vol. 19, 1919, pp. 117-152.

[2] *On the area of surfaces*, Proc. Royal Soc. London, series A, vol. 96, 1920, pp. 71-81.

[3] *On change of the variables in a multiple integral*, Proc. Royal Soc. London, series A, vol. 96, 1920, pp. 82-91.

[4] *On a formula for an area*, Proc. London Math. Soc., vol. 18, 1920, pp. 339-374.

[5] *On a new set of conditions for a formula for an area*, Proc. London Math. Soc., vol. 21, 1923, pp. 75-94.

[6] *Integration over the area of a curve and transformation of the variables in a multiple integral*, Proc. London Math. Soc., vol. 21, 1923, pp. 161-190.

YOUNGS, J. W. T.

[1] *The additivity of the Lebesgue area*, Bull. Amer. Math. Soc., vol. 49, 1943, pp. 779-784.

[2] *Curves and surfaces*, Amer. Math. Monthly, vol. 51, 1944, pp. 1-11.

[3] *The topological theory of Fréchet surfaces*, Ann. Math., vol. 45, 1944, pp. 753-785.

[4] *On surfaces of class* K_1, Bull. Amer. Math. Soc., vol. 51, 1945, pp. 669-673.

INDEX